T0396935

Sustainable Practices in Geoenvironmental Engineering

This third edition focuses on the application of geoenvironmental engineering procedures and practices to mitigate and reduce the adverse impacts on the geoenvironment from anthropogenic sources including emerging contaminants such as micro and nanoplastics, pharmaceuticals, and fire retarding chemicals. Thoroughly updated with three new chapters and extensive use of case studies to showcase examples of sustainable practices, this new edition discusses many activities that are still generating geoenvironmental impacts that are adverse to the quality and health of the geoenvironment. It includes new tools and procedures that have been developed to evaluate and minimize adverse impacts.

This new edition:

- Discusses the impacts of climate change and potential mitigation.
- Addresses emerging contaminants of concern.
- Introduces an entirely new chapter on sustainable nitrogen and carbon cycles.
- Includes new case studies like the Fukushima case study on sediments and microbial induced precipitation processes.
- Provides new practices and tools for sustainability to evaluate and to minimize adverse impacts
- Discusses the aspects of social sustainability and cultural aspects of the geoenvironment.

This book is intended for professionals, researchers, academics, senior undergraduate students, and graduate students in geotechnical engineering, geoenvironmental engineering, site remediation, sustainable development, and earth sciences.

Sustainable Practices in Geoenvironmental Engineering

Third Edition

Catherine N. Mulligan
Masaharu Fukue
Raymond N. Yong

CRC Press is an imprint of the
Taylor & Francis Group, an **informa** business

Designed cover image: © Catherine N. Mulligan, Masaharu Fukue, and Raymond N. Yong

Third edition published 2025
by CRC Press
2385 NW Executive Center Drive, Suite 320, Boca Raton FL 33431

and by CRC Press
4 Park Square, Milton Park, Abingdon, Oxon, OX14 4RN

CRC Press is an imprint of Taylor & Francis Group, LLC

First edition published by CRC Press 2006
Second edition published by CRC Press 2015

Library of Congress Cataloging-in-Publication Data
Names: Yong, R. N. (Raymond Nen), author. | Mulligan, Catherine N., author. | Fukue, Masaharu, author.
Title: Sustainable practices in geoenvironmental engineering /
Catherine N. Mulligan, Masaharu Fukue, and Raymond N. Yong.
Other titles: Geoenvironmental sustainability
Description: Third edition. | Boca Raton, FL : CRC Press, 2025. |
Revised edition of : Sustainable practices in geoenvironmental engineering /
Raymond N. Yong, Catherine N. Mulligan, Masaharu Fukue. Boca Raton :
CRC Press, Taylor & Francis Group, [2015]. |
Includes bibliographical references and index.
Identifiers: LCCN 2024039225 (print) | LCCN 2024039226 (ebook) |
ISBN 9781032525945 (hbk) | ISBN 9781032525952 (pbk) | ISBN 9781003407393 (ebk)
Subjects: LCSH: Environmental geotechnology. | Sustainable engineering.
Classification: LCC TD171.9 .Y66 2025 (print) | LCC TD171.9 (ebook) |
DDC 333.7–dc23/eng/20241218
LC record available at https://lccn.loc.gov/2024039225
LC ebook record available at https://lccn.loc.gov/2024039226

ISBN: 9781032525945 (hbk)
ISBN: 9781032525952 (pbk)
ISBN: 9781003407393 (ebk)

DOI: 10.1201/9781003407393

Typeset in Times
by Newgen Publishing UK

Contents

Preface

Much has happened in the past ten years since the publication of the second edition of this book, *Sustainable Practices in Geoenvironmental Engineering*. Since that time, the combination of population growth, increased exploitation of both renewable and non-renewable natural resources and increased impacts of climate change have added increased stresses on the quality and health of the geoenvironment. This is especially true when viewed in the context of the growing demand on food and shelter, and particularly on energy and critical mineral resources and their resultant effects on the natural capital of the geoenvironment. There is considerable need for governments, stakeholders, and geoenvironmental scientists and engineers to develop and implement measures needed to manage the resources of the geoenvironment to ensure that future generations of humankind are not compromised because of the lack of availability of geoenvironmental resources. Geoenvironmental engineers must work increasingly directly with users in the community.

At the Rio Summit +20 in 2012, 190 nations agreed to promote a green economy and to enhance global environmental protection and to come up with a set of sustainable development goals (SDGs). Subsequently since 2015, 17 goals with the 169 targets were devised to be accomplished by the year 2030. In this book the most relevant ones that are addressed include the following:

- Ensuring availability and sustainable management of water and sanitation for all.
- Ensuring access to affordable, reliable, sustainable, and modern energy for all.
- Building resilient infrastructure, promoting inclusive and sustainable industrialization and foster innovation.
- Making cities and human settlements inclusive, safe, resilient, and sustainable.
- Taking urgent action to combat climate change and its impacts.
- Conserving and sustainably using the oceans, seas, and marine resources for sustainable development.
- Protecting, restoring, and promoting sustainable use of terrestrial ecosystems, sustainably manage forests, combatting desertification, and halting and reversing land degradation and biodiversity loss.
- Strengthening the means of implementation and revitalizing the global partnership for sustainable development.

These goals are highly interlinked and therefore, there is a growing need for more integrated resource management frameworks. Some of the extensive challenges of sustainable development are related to a healthy and sufficient water supply for healthy people, resource availability, and protection of the natural environment. Climate change, industrialization, and population growth are all exerting pressures on resources. While the SDG goals are a priority, the means to address them are still lacking and thus there are substantial opportunities for contribution by the geoenvironmental community.

As we have pointed out in previous editions of this book, continued harvesting or exploitation of the non-renewable geoenvironmental natural resources means that we will never be able to achieve geoenvironment sustainability. The renewable natural resources and the natural capital of the geoenvironment need to be managed to ensure their sustainability. This means the development and implementation of technology and practices that seek to protect the quality and health of the natural resources and capital in the face of chemical, mechanical, hydraulic, thermal, and biogeochemical stressors originating from natural and anthropogenic sources. The material in this new edition focuses on the application of geoenvironmental engineering procedures and practice to mitigate and ameliorate the adverse impacts generated by stressors imposed on and in the geoenvironment from anthropogenic sources. New contaminants have gained attention since the last edition, such as micro and nanoplastics and per- and polyfluoroalkyl substances (PFAS). Whilst *Industry* and *Society* have made considerable efforts in recent years to adopt practices and procedures to "*protect the environment*" (*i.e., to go* "*green*"), there are still many activities that generate geoenvironment stressors, the impacts of which are adverse to the quality and health of the geoenvironment. What is needed is a set of tools and/or procedures that can be used to minimize and perhaps even eliminate the adverse consequences of the stressor impacts.

The first two chapters provide the basic background needed to address the assimilative capacity of soils, particularly in the light of management of pollutants in the ground, and also in the light of sustainable development, climate change, and land use. The intent of Chapter 1 is to provide an introduction to many of the basic issues that arise in respect to impacts and assaults on the geoenvironment as a result of anthropogenic activities associated with the production of goods and services and sustainability challenges and concepts. In Chapter 2, the focus is on contamination of the land environment as one of the key issues in the need to protect the natural capital and assets of the land environment. Important contaminants, soil contaminant interactions, and pollutant transport are highlighted.

In Chapter 3, the importance of water is emphasized. Adequate quantities of good quality water are also essential for health, agriculture, energy, and biodiversity. Characterization of water quality and methods of treatment for management are presented.

Chapter 4 directs its attention to the impacts on the geoenvironment in relation to industrial ecology. The interactions on the geoenvironment by activities associated with forestry, mineral mining, petroleum and chemical industries, health care and energy production are discussed. Insofar as geoenvironmental resources are concerned, and in respect to sustainability goals, the primary concerns are (a) use of natural resources both as raw materials and energy supply and (b) emissions and waste discharges.

Chapter 5 is confined to industrial activities associated with the extraction of non-renewable mineral, non-mineral, and energy mineral natural resources (uranium and tar sands). Activities associated with the mining, extraction, and on-site processing and storage of the extracted natural resource materials (mineral and non-mineral) contribute significantly to the inventory of potential impacts to the terrestrial ecosystem. Impacts and their management are elaborated on.

In Chapter 6, the land environment and sustainability of the land ecosystem in relation to food production are discussed. Instead, the focus is from a geoenvironmental perspective on the results of activities in food production and on geoenvironmental management of the agro-industry including tools and indicators for evaluating impacts.

Chapter 7 examines the built environment. Populations within cities require clean air and water, sewage, and waste management systems, housing, and transportation. They consume significant resources while polluting the air, land, and water. The increasing urban population is increasing pressures on the geoenvironment in the years to come. The discussion is on impact avoidance, risk minimization, and remediation of urban sites.

Chapter 8 discusses (a) the impacts on the health of the coastal sediments from the discharge of pollutants and other hazardous substance from anthropogenic activities, (b) the determination of the sediment quality, and (c) the necessary remediation techniques developed to restore the health of the coastal sediments. The purification of sediments contaminated by cesium is highlighted. A healthy coastal marine ecosystem ensures that aquatic plants and animals are healthy and thus will not pose a risk to human health when part of the food-chain.

Chapter 9 addresses the subject of land environment sustainability as it pertains to its interaction with the various waste discharges originating from industrial and urban activities. The focus is on developing concepts related to indicators, and contaminant fate and transport mechanisms and predictions.

Chapter 10 discusses the impact from the presence of pollutants in the ground, which need to be mitigated and managed – as a beginning step toward protection of the resources in the environment and also as a first step toward achievement of a sustainable geoenvironment. Physical, chemical, and biological properties relevant in soil contaminant mitigation are discussed. In particular, the emphasis is on using the properties and characteristics of the natural soil–water system in natural attenuation Lines of evidence and protocols are discussed.

Chapter 11 covers the remediation and management of contaminated soil, including physical/chemical, thermal and biological remediation technologies. The concepts of green and sustainable remediation, including frameworks and tools are included. In this edition, new chapters (12 and 13) are added. Chapters 12 and 13 introduce sustainable nitrogen and carbon cycles, and ureolytic microbial carbonate precipitation, respectively. The elements of N and C are appropriate for investigating the material cycles involved in the decomposition and excretion by living organisms, which contribute to CO_2 emissions. Chapter 13 provides an innovative approach for MICP application to engineering purposes, such as soil improvements, based on the inherent properties of biocement solution, which has not been well understood. Biocement technology has potential to contribute carbon neutrality. Chapter 14 has been updated and covers microbial-induced carbonate precipitation (MICP), which viewed as the next-generation ground improvement technology. The microbiological soil mechanics and various applications of MICP are now elucidated. This chapter deals with the MICP technology in soils, which is based on the fundamentals of MICP presented in Chapter 13.

Finally, in Chapter 15, the following are covered (a) discussion of the case of non-renewable nonliving renewable natural resources, (b) a look at some typical case histories and examples of sustainability actions, and (c) presentation of the geoenvironmental perspective of the present status of "where we are in the geoenvironmental sustainability frameworks and tools", with a view that points toward "where we need to go".

In the preparation of this book, the authors have benefited from the many interactions and discussions with their colleagues and research students, and most certainly with the professionals in the field who face the very daunting task of educating the public, industry, and the political bodies on the need for conservation and protection of our natural resources. We have discussed the serious impact of contaminants and consequences of such discharges and mentioned in many chapters the impact of excess consumption of renewable resources and the significant problems of depleting non-renewable resources – especially the energy and mineral resources. No detailed discussions on the kinds of alternative and/or substitute energy sources and the very pressing need for the development of such sources are included. That the need exists is eminently obvious. Instead, we have concentrated on the impacts of the various energy forms and other activities on the geoenvironment and sustainability of the land ecosystem such as in relation to food production, etc. It was not the intent to develop or present extensive basic theories in any one discipline area of this multidisciplinary problem – except as is necessary to support the discussion from the geoenvironmental sustainability viewpoint.

It is well understood that there is considerable effort directed toward alleviating many of the impacts described by industry, consumers, legislative bodies, the general public, local communities, and the professionals responsible for developing and implementing solutions. We wish to acknowledge these efforts and to remind all that much greater effort is needed and we must all work together. What is needed now is a deeper integration of the various disciplines such as soil physics and chemistry, microbiology, hydrogeology, and geochemistry, into the encompassing field of geoenvironmental engineering. As we have indicated before, in order to provide the kinds of technology and solutions needed to safeguard the quality and health of the geoenvironment, it is necessary for one to utilize all the pertinent science from the many disciplines.

Catherine N. Mulligan
Masaharu Fukue
Raymond N. Yong
June, 2024

About the Authors

Catherine N. Mulligan is a Distinguished Research Professor and full professor in the Department of Building, Civil and Environmental Engineering at Concordia University, Canada. She has authored more than 130 refereed papers in various journals, authored, coedited, or coauthored eight other books, holds three patents, and has supervised to completion more than 75 graduate students. Her research involves the treatment of soil, water, sediments, and mining wastes. She is the founder and Director of the Concordia Institute of Water, Energy and Sustainable Systems. The Institute trains students in sustainable development practices and performs research into new systems, technologies, and solutions for environmental sustainability. She is a Fellow of the Royal Society of Canada, the Canadian Academy of Engineering, the Engineering Institute of Canada and Canadian Society for Civil Engineering and recent winner of the Royal Society of Canada's Miroslaw Romanowski Medal.

Masaharu Fukue, BEng, MEng, PhD, is a professor emeritus at Tokai University, after he retired from the Marine Science and Technology, Tokai University, Japan in 2013. He was recommended as an honorary member of Japanese Geotechnical Society. After he established the Japanese Geotechnical Association for Housing Disaster Prevention, he serves as the Representative Director, and Research Development Director. Over the last 15 years, he has been immersed in MICP research and has achieved results. The achievement on his innovative MICP research using his Japanese patents is included in this book. His other patent, resuspension technique for remediation of sediments, has been applied to Cs removal in reservoirs, in Fukushima projects from 2016 to 2023 (to be continued). This achievement is also included in this book. As an expert, he serves as a committee member of national and local governments and as a director of business associations.

Raymond N. Yong is the William Scott Professor Emeritus at McGill University, Canada, He has authored and coauthored 14 other textbooks, over 500 refereed papers in various journals in the disciplines of geoenvironmental engineering and earth science, and holds 44 patents. He is a Fellow of the Royal Society (Canada) and a Chevalier de l'Ordre National du Québec. He and his students were among the early researchers in geoenvironmental engineering engaged in research on the physicochemical properties and behavior of soils, their use in buffer/barriers for HLW (high-level radioactive waste) and HSW (hazardous solid waste) containment and isolation and restoration/remediation of contaminated sites. He and his colleagues are currently engaged in research on geoenvironmental sustainability and natural habitat protection and restoration.

1 Sustainable Geoenvironment

1.1 INTRODUCTION

The *geoenvironment* has been referred to as the breadbasket for humans in that exploitation of its various constituents and functions provide for "*food, shelter, and clothing*" for the well-being of humans. Exploitation, for example, of (a) non-renewable natural resources such as rare earth elements and minerals and (b) renewable resources that cannot regenerate and replenish themselves to meet sustainability criteria serve to create distress to the geoenvironment. The stresses generated by these demands create, in most cases, adverse impacts on the geoenvironment, such as (a) loss of biodiversity, (b) increasing discharge of noxious gases and particulates that find their way back to the land surface, (c) loss of soil quality and soil functionality, (d) increasing generation and discharge of wastes and pollutants to the land and aquatic environments, and (e) most importantly, decreased geoenvironment carrying capacity. Considerable attention is being paid to many of these issues by researchers, policy makers, and other professionals well-versed in engineering, scientific, and socio-economic disciplines to alleviate the stresses upon the geosphere and to seek sustainable ways for society to live in harmony with the geoenvironment.

The primary focus of this book is on the *geoenvironment* and its importance as a resource base for life-support systems – with particular attention to issues relating to its carrying capacity and capability for regeneration of the geoenvironment natural resources and the natural habitat. We define the *carrying capacity of the geoenvironment* as the ability of the geoenvironment to provide the necessary resources to sustain the biotic population at hand. Whilst some degree of human intervention is required to capitalize on these natural resources – e.g., agriculture, forest management, breeding, hydro-power generation, etc. – it must be understood that the basic natural resource being exploited is the factor that enters into the calculation of the carrying capacity of the geoenvironment. Humans depend on the geoenvironment to provide the basic materials to support life. Because of the many threats and adverse impacts on the various life-support systems within the geoenvironment, there is a pressing need for one to (a) develop a better appreciation of the stresses imposed on the geoenvironment by humans, and (b) determine and implement the requirements for sensible and proper management of our geoenvironmental resources to meet the goals of a sustainable society.

Many of the terms used in this book will have slightly different meanings depending on one's background, perspective, and scientific-engineering discipline. It would be useful, at the outset, to establish what we mean by the term *geoenvironment*. The *geoenvironment* is a significant part of the various systems of the earth. Figure 1.1 shows the various *earth systems* and their relationship to the geoenvironment. The various systems shown in the Venn diagram in the figure consist of the (a) atmosphere, (b) geosphere, which is also known as the lithosphere, (c) hydrosphere, (d) biosphere, and (e) anthroposphere.

The *geoenvironment* is a specific domain of the environment, and as such, concerns itself with the various elements and interactions occurring in the domain defined by the dry solid land mass identified as terra firma. These include a significant portion of the geosphere and portions of the atmosphere, hydrosphere, biosphere, and anthroposphere.

- We consider the ***geosphere*** to include the inorganic mantle and crust of the earth, including the land mass and the oceanic crust. Also included in this category are the solid layers

DOI: 10.1201/9781003407393-1

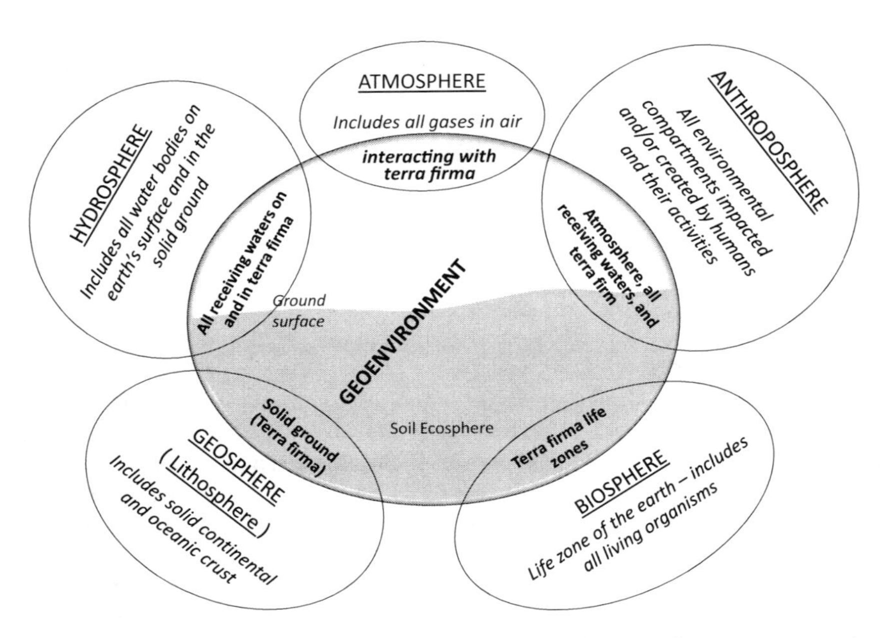

FIGURE 1.1 Venn diagram showing the various elements in the five ecosphere components (atmosphere, hydrosphere, geosphere, biosphere, and anthroposphere) that make up the geoenvironment.

(soil and rock mass) stretching downward from the mantle and crust. One could say that one part of the geoenvironment is the *terra firma* component of the geosphere – as seen in Figure 1.1.

- The ***hydrosphere*** refers to all the forms of water on Earth – i.e., oceans, rivers, lakes, ponds, wetlands, estuaries, inlets, aquifers, groundwater, coastal waters, snow, ice, etc. The geoenvironment includes all the receiving waters contained within the terra firma in the *hydrosphere*. This excludes oceans and seas, but is meant to include rivers, lakes, ponds, inlets, wetlands, estuaries, coastal marine waters, groundwater, and aquifers. The inclusion of coastal (marine) environment in the geoenvironment is predicated on the fact that these waters are impacted by the discharge of contaminants in the coastal regions via runoffs on land, and discharge of polluted waters from rivers or streams. Chapter 8 examines sustainability issues of the coastal marine environment in greater detail.
- The ***biosphere*** is the zone that includes all living organisms, and the *environment* is the biophysical system wherein all the biotic and abiotic organisms in the geosphere, hydrosphere, and atmosphere interact. The geoenvironment includes the life zones in or on terra firma from the *biosphere*.
- The ***atmosphere*** refers to all the gases and airborne substances in the air that interact with the land environment (terra firma).
- The ***anthroposphere*** refers to the various environmental domains impacted by humans and their activities.

Industrialization, urbanization, agriculture (food production), and natural resource exploitation (including energy) are basic activities associated with a living and vibrant society. We consider these basic elemental activities to be necessary to sustain life and also to be integral to *development.* In general terms, we consider *development* to (a) embody the many sets of activities associated with the production of goods and services, (b) reflect the economic growth of a nation, state, city, or society in general, and (c) serve as an indication of the output or result of activities associated with these four main elemental activities. Questions often arise as to how these activities are compatible with or are in conflict with the carrying capacity of the geoenvironment.

1.1.1 Impacts on the Geoenvironment

Almost any physical or chemical action or event that happens in the any of the earth systems shown in Figure 1.1 will impact the geoenvironment. The question of whether these impacts will add value to or detract from the functionality of the geoenvironment and its earth systems highlights some of the issues at hand. It is difficult to catalogue or list all of the impacts resulting from all the various kinds of stresses imposed on the geoenvironment. It is not only because this cannot be done, but also because we need to arrive at some sets of criteria that will tell us "*what constitutes an impact*". In many instances, we may not readily recognize or be aware of the impacts from many sets of activities or events – natural or man-made. To some extent, this is because (a) the effects of the impacts will not be immediately evident, as, for example, in the case of long-term health effects, and (b) the effects or results of the impacts cannot be recognized, i.e., we have yet to learn or recognize the results or effects of the impacts on the various biotic receptors and the environment itself. The important lesson to bear in mind is that any kind of geoenvironment impact – i.e., impact on and in the geoenvironment – may, sooner or later, produce reactions and/or conditions in the geoenvironment that may be benign, beneficial, or adverse to the geoenvironment.

1.1.1.1 Geoenvironment Impacts from Natural Events, Disasters, and Humans

The impacts on the geoenvironment from "natural" events (disasters), such as, for example, earthquakes, tsunamis, hurricanes, typhoons, forest fires, tornadoes, floods, and landslides, to name a few, are obvious – inasmuch as they are well reported in the daily newspapers. The impacts on the lives of humans and wildlife in the affected regions are often incalculable.

Some of the dramatic impacts on the geoenvironment landscape due to these natural catastrophic events include (a) collapse of man-made and natural structures and other infrastructure facilities, such as roads, pipelines, transmission towers, etc., (b) floods, (c) deposition of airborne toxic substances released from damaged structures, such as the radioactive nuclides released from the tsunami-associated damaged nuclear power reactors in the East Japan earthquake, (d) landslides, and (e) fires. Displacement of thousands and even millions of people due to loss of dwellings, and loss of life due to collapsing structures, floods and landslides, ground contamination from deposited airborne toxic substances, restriction or prohibition of agricultural and food-production activities in lands contaminated by deposition of airborne toxic substances, and ingestion of polluted waters are some of the impacts to humans. All of these events and their impacts on human life and other life forms, and local and global societal response to such events merit serious and proper consideration and attention in books and treatises devoted to the various aspects of these catastrophic events. They are not within the purview of this book. What is of direct concern in this book is the impact of anthropogenic activities on the health and carrying capacity of the geoenvironment.

The many basic issues that arise from impacts and assaults on the geoenvironment resulting primarily from anthropogenic activities, and to the kinds of geoenvironmental engineering practices oriented towards restoring and/or preserving site (or region) functionality and carrying capacity of the geoenvironment, i.e., sustainable practices. We consider the geoenvironment to consist of the terrestrial (land surface) ecosystem. As seen in Figure 1.1, this includes the aquatic ecosystems contained within and contiguous to the land mass. Many of the driving forces responsible for these impacts and assaults can be attributed to actions of *production* and *implementation* technology.

A pertinent example of this is the presence of historic and orphan toxic and non-toxic waste-polluted sites populating the land surface in many parts of the world. These are the legacy of our historic lack of appreciation of the damage done to the land environment by the many activities in support of production of *goods and services*. These kinds of goods and activities are necessary items in support of industrial development and a vibrant society. Not all of the kinds or types of natural and man-made impacts on the geoenvironment resulting from these activities can be considered. The geoenvironmental impacts that are health-threatening constitute the major focus of this book. By and large, these (impacts) result from discharges from industrial operations and urban activities. A more detailed description and discussion of the anthropogenic impacts on the geoenvironment will be considered later in the book.

1.2 GEOENVIRONMENT, ECOSYSTEMS, AND RESOURCES

1.2.1 Ecozones and Ecosystems

The term *ecosystem* refers to a system where the various individual elements and organisms interact individually or collectively to the benefit or disadvantage of the whole system. For an ecosystem to be self-sustaining, the relationships between the communities in the ecosystem need to be symbiotic, and furthermore, the interactions must be mutually beneficial. The various ecosystems that exist in the geoenvironment have functions, uses, resources, and habitats that are crucial to the production of goods and the means for ensuring life support. Habitat protection and preservation are crucial if sustainability goals are to be met.

The biological component within the ecosystems that comprise the terrestrial ecosystem do not fall within the purview of this book except insofar as they contribute to the persistence, transformation, and fate of pollutants in the ground. In particular, we will be concentrating on the land aspects of the terrestrial ecosystem, and more specifically with the land surface (landscape), the soil ecosystem, and subsurface systems. The term *pollutants* is used to denote *contaminants* that have been classified by regulatory agencies as toxic or noxious substances that are threats to the health of humans and the environment. *Contaminants* are substances that are not indigenous to the location under consideration. All pollutants are contaminants, but not all contaminants are pollutants.

Ecozones are zones that are delineated according to certain sets of established ecological characteristics. They are essentially basic units of the land or marine environment that are distinctly characterized by the living and non-living organisms within that region. *Ecozones* are geographical units that are usually several hundreds of square kilometres in spatial extent. The *ecosystem* or ecosystems bounded or resident within an ecozone deals with the mutual interactions between the living and non-living organisms in this zone. An *ecosystem* is defined herein as a discrete system that (a) contains all physical (i.e., material) entities and biological organisms, and (b) includes all the results or products of the interactions and processes of all the entities and organisms in this system. With this classification scheme, one can distinguish between the two primary ecosystems constituting the ecosphere, namely the land and aquatic ecosystems. The delineation of ecozones and ecosystems is somewhat arbitrary and can be performed or undertaken according to several guidelines. The boundaries demarking the ecozones are not fixed.

Classification and characterization of these land ecosystems can be performed according to various standards or guidelines. Classification according to the physiographic nature of the land is one of the more popular schemes available. Under such a scheme, one has therefore such ecosystems as alpine, desert, plains, coastal, arctic, boreal, prairie, etc. Another popular scheme for classification is the resource-based method of classification. This approach is based on the identification of the sets of activities or the nature of the primary or significant resources constituting the specific land environment under consideration – such as agro ecosystem and forest ecosystem. Within each ecosystem, there exist numerous elements and activities that can be examined and documented in respect to *before* and *after* ecosystem impact. One of the primary reasons for classification of any of the ecosystems is to define, bound, or document the *sphere of influence or examination wherein all the elements of the ecosystem interact, and are dependent on the welfare of each individual element for the overall state, benefit, and function of the ecosystem.*

1.2.2 Natural Resources and Biodiversity in the Geoenvironment

Natural resources in the geoenvironment are defined as commodities that have intrinsic value in their natural state. They are the natural capital of the geoenvironment. In describing the natural resources in the geoenvironment, the most obvious ones are often cited immediately. These include water, forests, minerals, coal, hydrocarbon resources, and soil. Other not so obvious resources are the developed resources, such as agricultural products and alternative energy generation (resources), such as solar, geothermal, wind, tidal, and hydro.

One natural resource that is often overlooked is biological diversity (biodiversity). This is one of the most significant natural resources in the ecosystem. We use the term biodiversity to mean the diversity of living organisms, such as plants, animals, and microbial species in a specific ecosystem. They play significant roles in the development of the many resources that we have identified above, through mediation of the flow of energy (photosynthesis) and materials, such as carbon, nitrogen, and phosphorous. According to Naeem et al. (1999), ecosystems consist of plants, animals, and microbes and their associated activities, the results of which impact on their immediate environment. They point out that a functioning ecosystem is one that exhibits biological and chemical activities characteristic for its type and give the example of a functioning forest ecosystem, which exhibits rates of plant production, carbon storage, and nutrient cycling that are characteristic of most forests. It follows that if the trees in the forest are harvested or if the forest ecosystem is converted to another type of ecosystem, the specific characteristics of a functioning forest ecosystem will no longer exist.

For the purpose of this book, we define a *functioning ecosystem* to include not only the biological and chemical activities, but also the physico-chemical activities and physical interactions characteristic of the type of ecosystem under consideration. Trevors (2003) has enumerated a noteworthy list for consideration in respect to the role of biodiversity as part of our life support system. Included in the detailed list of Trevors (2003) are considerations, such as the following:

- Maintenance of atmospheric composition and especially the production of oxygen by photosynthesis and the fixation of carbon dioxide.
- Water cycle via evaporation and plant transpiration.
- Interconnected nutrient cycles (e.g., C, N, P, S) driven by microorganisms in soils, sediments, and aquatic environments.
- Carbon sources and sinks.
- Pollination of agricultural crops and wild plants.
- Natural biocontrol agents as, for example, in microbial degradation of pollutants in soil, water, sediments, waste water, and sewage treatment facilities.

1.3 GEOENVIRONMENT SUSTAINABILITY

Since the natural resources of the geoenvironment provide the main elements and basis for production of "*food, shelter, and clothing*" for the global population, the question that begs itself is "*how sustainable is the geoenvironment*"? There is no single answer to this seemingly simple question – primarily because one needs to include in the discussions (a) the ingenuity of mankind to develop technologies, tools, and devices designed to alleviate stressful demands on the geoenvironment, and (b) the effect and influence of governmental and societal regulations on practices designed to manage the natural resources of the geoenvironment. In recognition of these two fundamental factors, the discussions in the various chapters of this book centre around: (i) The impacts on the geoenvironment due to the stresses generated by mankind (anthropogenic stressors and their impacts), and (ii) the practical ways in which one could alleviate, ameliorate, mitigate, etc., these impacts. The remedial activities required to meet the goals ensuing from item (ii) can be considered to form the basis for geoenvironment sustainable practices. The principal focus for these activities is on the land component of the geoenvironment and its included waters.

1.3.1 Geoenvironment as a Natural Resource Base

Figure 1.2 shows some of the primary natural resources in the geoenvironment in combination with various land and aquatic ecosystems. The health and accessibility of these are essential rudiments of life-support systems. The beneficial interaction between all the elements shown in Figure 1.2 is needed to produce the necessary ingredients required for production of goods and services to sustain the population in a society. The ultimate goal for all of humankind is to obtain both a sustainable society – i.e., a society that can regenerate itself without compromising the various resources for future generations.

It has been contended that since the beginning of the industrial revolution, "mankind's occupation of this planet has been markedly unsustainable" (Glasby, 2002), and that the concept of sustainable development as defined in the World Commission on Environment and Development (1987) Report is a chimera. At the present rate of exploitation of the renewable and non-renewable resources, *sustainable development* is a goal that is not readily achieved so long as depletion of

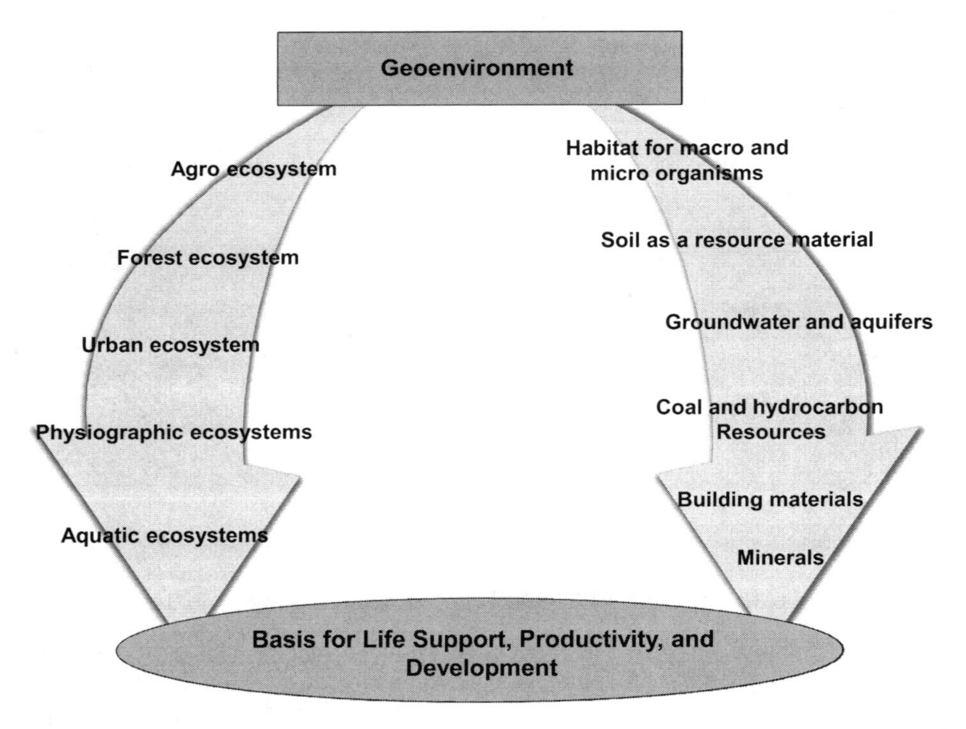

FIGURE 1.2 Some of the major ecosystems, resources, and features of the geoenvironment. Note that the physiographic ecosystems include, for example, coastal, alpine, desert, arctic, etc., ecosystems.

non-renewable resources occurs, and so long as excessive exploitation of renewable resources outstrips the replenishment rate of these resources. The term *sustainable development* is used herein to mean that *all the activities associated with development in support of human needs and aspirations, must not compromise or reduce the chances of future generations to exploit the same resource base to obtain similar or greater levels of yield.* The land environment houses the terrestrial and aquatic ecosystems. These are the fundamental components in the engine responsible for life support, one needs to determine. These issues will be addressed in the other chapters of this book from a geoenvironmental perspective.

Evidence of environmental mismanagement from past activities associated with development abound. Commoner (1971) has suggested, for example, that the wealth gained by the modern technology-based society has been obtained by short-term exploitation of the environmental system. Whilst protection of the geoenvironment may lead to a preservation of many of the features and assets of the geoenvironment, it is clear from the various resource and habitat features shown in the right-hand compartment of Figure 1.2 that protection of the geoenvironment does not necessarily lead to sustainability of the geoenvironment. We recognize that techniques, procedures, and management programmes should be structured to permit exploitation of the geoenvironment to occur with minimal adverse impact on the geoenvironment. To that end, the intent of this book is to provide one with an appreciation and understanding of (a) the geoenvironmental impacts that result from activities associated with mankind, and (b) measures, requirements, and procedures needed to avoid, minimize, and/or mitigate these impacts.

We have previously defined the *geoenvironment* to include the land environment. This includes all the geophysical (geological and geomorphological) features, together with the aquatic elements classified as receiving waters. The major ecosystems that constitute the land ecosystem, shown in the left compartments of Figure 1.2, consist of a mixture of physiographic and resource-type ecosystems. The mixed-method of presentation of ecosystems shown in the diagram has been chosen deliberately because it does not require one to detail every single physiographic unit and every single resource-type unit.

The geoenvironment contains all the elements that are vital for the sustenance and well-being of the human population. Commoner (1971) states that the ecosphere, together with the earth's mineral resources, is the source of all goods produced by human labour or wealth. It is obvious that any degradation of the ecosphere will impact negatively on the capability of the ecosphere to provide the various goods produced by human labour or wealth.

1.3.2 Impacts on the Geoenvironment

The posit that the geoenvironment is in itself a natural resource is founded on the fact that it provides the various elements necessary for life support, such as food, energy, and resources. Degradation of any of the physical and biogeochemical features that permit life-support systems to function well will be a detriment to the requirements for a sustainable society.

1.3.2.1 Impacts Due to Population Growth

The Malthusian model, for example (Malthus, 1798), links the availability of food with population growth (or reduction). The Malthusian model contends that since the rate of human population increase is geometric, there will come a time when food production will not be sufficient to meet population needs. Whilst no account was given to availability of resources and industrial output in the original model, one presumes that these were accounted for in the *ceteris paribus* condition.

Meadows et al.'s (1972) system dynamics model, which considered five specific quantities – industrialization, population, food production, pollution, and consumption of non-renewable natural resources – concluded (phase I study "The Project on the Predicament of Mankind") that, "If present growth trends in world population, industrialization, pollution, food production, and resource depletion continue unchanged, the limits to growth on this planet will be reached sometime within the next hundred years……". Whilst some have argued that the conclusions are perhaps too pessimistic, since the system dynamics model used in the Meadows et al. (1972) report, by their own admission, was, at the time of publication of the report somewhat "imperfect, oversimplified and unfinished". However, it cannot be denied that the essence of the model and the analyses were fundamentally sound. The five specific quantities examined in the model were deemed significant in view of the global concern on "accelerating industrialization, rapid population growth, widespread malnutrition, depletion of non-renewable resources, and a deteriorating environment".

1.3.2.2 Impacts from Natural Resource Exploitation

Arguments against consideration of the environment, and specifically the geoenvironment, as a limited natural resource, are generally based on a very limited appreciation of the totality of the geoenvironment as an ecosphere, and also on negligent attention to the many adverse impacts attributable to anthropogenic activities. Not all the geoenvironmental resources are non-renewable (e.g., forest resource). However, for those resources that are renewable, over-use or over-exploitation will surpass their recharge and replenishment rate, thus creating a negative imbalance. Is geoenvironmental deterioration a threat to human survival? Commoner (1971) has examined the overall environmental problem and has posed the question in terms of ecological stresses: "are present ecological stresses so strong that – if not relieved – they will sufficiently degrade the ecosystem to make the earth uninhabitable by man?" His judgment?

> "…..based on the evidence now on hand,…the present course of environmental degradation, at least in industrialized countries, represents a challenge to essential ecological systems that is so serious that, if continued, it will destroy the capability of the environment to support a reasonably civilized human society…".

The declaration issued at the beginning of U.S. National Environmental Policy Act (NEPA) of 1969 recognizes "…..the profound impact of man's activity on the interrelations of all components of the natural environment, particularly the profound influences of population growth, high-density urbanization, industrial expansion, resource exploitation, and new and expanding technological advances………" and further recognizes "…the critical importance of restoring and maintaining environmental quality to the overall welfare and development of man". The significance of this declaration cannot be overlooked. We can easily appreciate the need for environment protection and sustainability of the geoenvironment.

The geoenvironment is the resource base that serves as the engine that provides for the various elements necessary for human sustenance. Through resource exploitation and industrial activities, it is the source for everything that is necessary for the production of *food, shelter, and clothing.* It is also the habitat for various land and aquatic biota. Adverse impacts to the geoenvironment and its ecosystem need to be minimized and mitigated if one wishes to undertake the necessary steps toward *sustainable development.* Management of the geoenvironment is required if a sustainable geoenvironment is to be obtained.

The sets of forces needed to sustain a forceful economic climate, and provide for a dynamic population base or population growth, can be gathered into two main groups. These are defined by some very clear factors:

- Urban–industrial – this grouping includes those efforts and industries associated with the production of *food, shelter, clothing and economic health,* and
- Socio–economic–political – the grouping of factors that include the social, economic, and political dimensions of a society.

The subjects covered in this book will deal with stresses on the geoenvironment resulting from the various activities associated with society and its desire for development. The framework within which these will be examined will be confined to the one determined by the urban–industrial factors defined above. This by no means diminishes the significance of the socio–economic–political grouping of factors. These are important, but are not within the purview of this book. From the geoenvironmental point of view, this means dealing with the impacts to the geoenvironment from anthropogenic activities. Yong and Mulligan (2019) show that to properly address the problems and issues connected with degradation of the land environment, a knowledge of the linkages, interactions, and impacts between the human population and a healthy, robust, and sustainable land environment is required (Figure 1.3). The illustration in the figure shows the linkages and identifies some of the major issues and land environment impacts. The observations made by Yong and Mulligan (2019) regarding the major land environment or geoenvironmental issues shown in the figure are cited directly as follows:

- "Waste generation and pollution: Wastes generated from the various activities associated with resource exploitation, energy production, and industry associated with the production of goods and services will ultimately find their way into one or all of three disposal media; (a) receiving waters, (b) atmosphere, and (c) land. Land disposal of waste products and waste streams appears to be the most popular method for waste containment and management. The various impacts arising from this mode of disposal and containment include degradation of land surface environment and ground contamination by pollutants.
- Depletion of agricultural lands and loss of soil quality: This will arise because of increased urbanization and industrialization pressures, infrastructure development, exploitation of

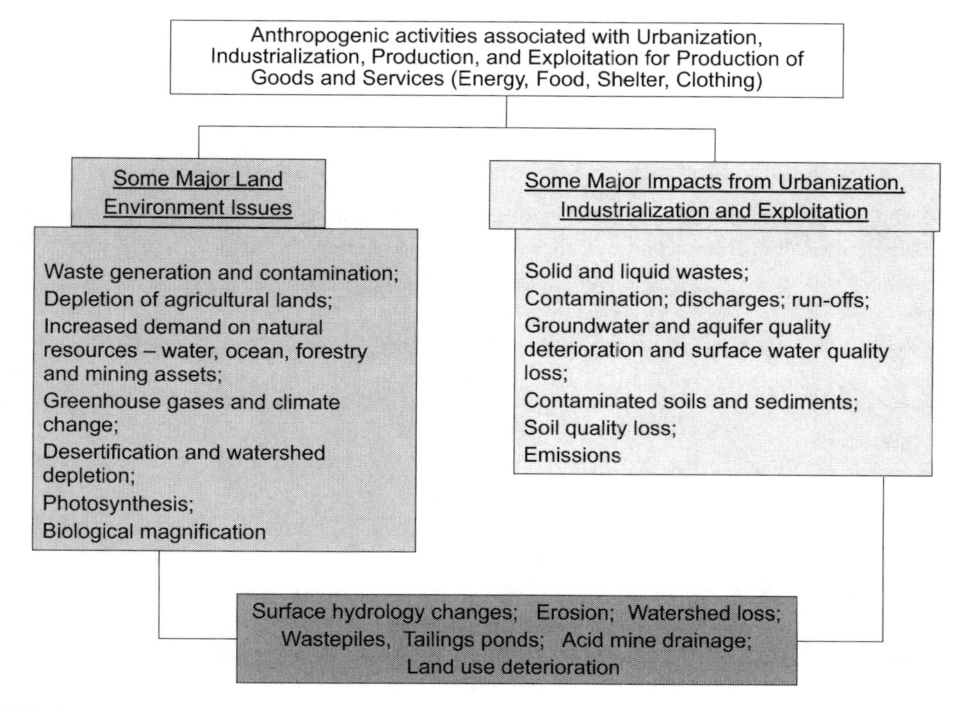

FIGURE 1.3 Some major land environment issues and impacts resulting from activities associated with urbanization, industrialization, production, and exploitation for production of goods and services. (Adapted from Yong and Mulligan, 2019.)

natural resources, and use of intensive agricultural practices. The loss of agricultural lands places greater emphasis and requirement on higher productivity per unit of agricultural land. The end result of this is the development of high-yield agricultural practices. One of the notable effects is soil quality loss. To combat this, there is an inclination to use pesticides, insecticides, fertilizers, more soil amendments and other means to enhance productivity and yield. A resultant land environment impact from such practices is contamination of the ground, groundwater and receiving waters from run-offs and transport of contaminants.

- Increased demand on natural resources and depletion of natural capital: The various issues related with all the generated exploitation activities fall into the categories of: (a) land and surface degradation associated with energy production, mining and forestry activities, and (b) water supply, delivery, and utilization. Surface hydrology changes, erosion, watershed loss, tailings and sludge ponds, acid mine drainage, etc. are some of the many land environment impacts.
- Greenhouse gases, climate change, desertification: To a very large extent, these are consequences of industrialization, urbanization, and production. Their impact on the land environment can be felt for example in acid rain (and snow) interaction with soil and undesirable changes in photosynthesis processes and erosion of coastal areas due to increasing water levels.
- Photosynthesis: The processes associated with photosynthesis are important since they in essence constitute about 20% of the available oxygen in the atmosphere. Desertification, deforestation, and many of the activities associated with mineral and other natural resource exploitations will degrade the capability of the various participants (land and aquatic plants) to engage in these processes.
- Biological magnification: This concerns itself with the concentration of various toxic elements or pollutants by plants and such biotic receptors as aquatic organisms and animals, is a problem that needs to be addressed in the containment and management of pollutants".

1.3.3 Stressors and Sources

Stresses imposed on the geoenvironment originate from various agents and sources. In respect to the land element of the geoenvironment, these stresses arise from physical, thermal, hydraulic, and mechanical actions and forces, and from physico-chemical, chemical, and biologically-mediated reactions and processes. Their actions, reactions, or processes are the result of (a) natural environmental events, such as volcanic eruption, earthquakes, tsunamis, forest fires, landslides, drought, tornados, hurricanes, etc., and (b) from human-related activities, such as deforestation and habitat destruction, containment of hazardous and radioactive wastes, soil contamination from high-intensity agro-industry practices and from waste leachates, exploitation of mineral and hydrocarbon resources, constructed facilities, etc.. The actions that result in imposition of stresses on the land compartment of the geoenvironment are called *stressors*. Since the terms *stressors* and *stresses* are used in many different disciplines and also used to describe particular situations, we should not confuse their concept when they are applied to the land compartment of the geoenvironment. Yong et al. (2012) have provided some examples of stressor sources and stressors in the soil environment – as shown in Table 1.1.

The types of stressors acting on a piece of land mass can be mechanical, hydraulic, thermal, chemical, physico-chemical, electrical, gaseous, radioactive, etc., with sources that are either natural or human-related. The Venn diagram in Figure 1.1 shows that the natural stressor sources delivering stressors can be elements from the atmosphere, anthroposphere, biosphere, and hydrosphere, and stressors generated in the individual ecospheres by actions originating in the anthroposphere.

TABLE 1.1
Examples of Stressors and Effects on Land Mass Soil Properties

Stressor Sources	Stressors for Soils	Kinetics Induced in Soils	Affected Soil Properties and Behaviour
		Natural Sources	
Volcanic eruption	Ejecta, heat	Alteration of soil, Heat transfer	Soil elements, water retention Thermal and hydraulic properties
Forest fire, deforestation	Heat	Heat and water transfer Alteration of soil elements	Soil structure, water retention Thermal and hydraulic properties
Landslides, floods, and avalanches	Soil elements, water	Mass runoff	Soil elements, soil structure
Cyclical temperature and drying	Heat, water	Heat and water transfer	Soil structure, thermal and hydraulic properties
		Anthropogenic Sources	
Industrial effluents	Leachates	Adsorption, chemical reaction	Chemical composition, soil structure and hydraulic properties
Mining and metal processing	Contaminants	Water and solute transfer	
		Constructed facilities	
Buried waste barrier	Water, solutes, heat	Water, solute and heat transfer	Hydraulic, chemical and thermal properties
Structural loads	Pressure	Shear stress, consolidation	Mechanical properties and behaviour
		Agricultural land-use	
Fertilization	Chemical compounds	Adsorption, chemical reaction	Chemical composition, soil structure
Irrigation and drainage	Water	Water and solute transfer	Salt accumulation, water retention
River embankments	Overburden pressure	Mass run-off, shear stress, consolidation	Hydraulic properties, mechanical behaviour
Underground water use	Water, solutes	Adsorption and desorption	Chemical composition, salt accumulation
		Abnormal climate	
Gas component in air	CO_2, CH_4, N_2O, etc.	Adsorption, gas transfer	Chemical composition, soil structure,
Air temperature	Heat	Heat transfer,	Water retention,
Acid rain	Acid precipitation	chemical reaction, solute transfer	Hydraulic properties

Source: From Yong et al. (2012).

1.3.3.1 Natural Stressor Sources and Stressors

The two categories of stressor sources are *natural* and *anthropogenic* as illustrated in Table 1.1. It is not often easy to distinguish between natural geoenvironment stressor sources and those associated with or related to human activities and products. Well-defined natural events that readily classify as natural geoenvironment stressor sources generally fall under the category of *natural disasters* or *geo-disasters*. These include events, such as hurricanes, earthquakes, tsunamis, tornadoes, volcanic eruptions, rainstorms, and blizzards – generating various types of stressors and stresses in a land or soil mass. For example, earthquakes generate dynamic forces, e.g., seismic force, to a soil mass through two types of body waves – primary or *P* waves and secondary or *S* waves.

Volcanic eruption and forest fires are also classified as typical natural disasters that will generate various types of stressors. They cover land surface with ejecta and ashes, apply extreme high temperature or heat, and chemical components constituting ejecta and ashes to land surfaces that will migrate into soil and underground water through percolating rainwater. Since such heat and chemical compounds are defined as stressors, volcanic eruption and forest fires can be classed as stressor sources.

Landslides and floods, which are often classified as natural disasters, can be classified as geoenvironment stressor sources. They are disasters that result from situational vulnerability to provocative events, such as earthquakes and rainstorms. Landslides and floods can also occur in instances where human activities have created vulnerable circumstances. Deforestation of slopes could make them more susceptible to landslides, and harvesting of groundwater in low-lying coastal regions can lower ground surfaces to levels that invite flooding. Since landslides and floods transport various matters, such as soil components and relevant chemicals to downstream areas, these can be classified as sources responsible for generating stressors, such as chemical elements, in the geoenvironment.

Anthropogenic activities, such as war, are very detrimental to human life and infrastructure in cities and countries. During this time, the land is subjected to many of the stressors in Table 1.1. Although not the focus of this book, geoenvironmental engineering is indispensable for the regeneration of the land when peace returns.

Regular cyclical events are events, such as winter–summer cycles, and non-regular events are wetting–drying events where repeat cycles do not adhere to any time-calendar schedule. In the case of sub-zero temperatures that accompany winter seasons, for example, under the right conditions, ground uplift due to frost heaving can occur, thus resulting in considerable damage to overlying structures. The stressor source is sub-zero freezing temperatures (winter), and the geoenvironment stressors associated with this stressor source are thermal in nature. In the case of wetting–drying events, involving extreme cases of deluge, and drought, the deluge could result in severe flooding whereas the drought event could lead to parched-earth conditions. The types of stressors generated in the affected include thermal, hydraulic, chemical, and mechanical.

1.3.3.2 Anthropogenic Stressor Sources and Stressors

Anthropogenic geoenvironment stressor sources include (a) various industries, such as mining, agro, manufacturing and processing, and industries associated with all facets of resource exploration and exploitation, and (b) human activities, such as construction of facilities, disposal and land management of waste, as well as agricultural land use, underground water use and abnormal climate, as shown in Table 1.1. How the various industries (life-supporting and manufacturing-production) and their associated activities interact with the geoenvironment can be viewed as follows:

(1) Resource extraction and processing – the various industries included in this group use the geoenvironment as a resource pool containing materials and substances that can be extracted and processed as value-added products. The common characteristic of the industries in this group is *processing of material extracted from the ground*. Included in this group are (a) the metalliferous mining industries, (b) those industries involved in extraction and processing of other resources from the ground such as non-metallic minerals (potash, refractory and clay minerals, phosphates), (c) the industries devoted to extraction of aggregates, sand and rock for production of building materials, and (d) the raw energy industries, such as those involved in the extraction of hydrocarbon-associated materials and other fossil fuels (natural gas, oil, tar sands, and coal). Included in this list is the extraction and recovery of uranium for the nuclear power industry.

(2) Utilization of land and soil as a resource material in aid of production – essentially, this group includes the agro and forest industries, and also the previously mentioned non-metallic minerals' industries.

(3) Water, groundwater, and aquifer harvesting – we include the hydroelectric facilities, and industries associated with extraction and utilization of groundwater and aquifers.

(4) Use of land as a facility – this category considers land as a facility for use, for example, in the land disposal of waste products. Broadly speaking, we can consider the land surface environment here as a resource for treatment and containment of waste products generated by all the industries populating the previous three categories.

Some of the major negative or degradative geoenvironmental impacts resulting from the various activities associated with production technology – e.g., agriculture, forestry, mining, energy, and general production – are shown in Figure 1.4. The nature of the threats to the land environment and the waste streams are shown in Figure 1.5. These affect both soil and water quality. We will discuss the nature of some of the impacts in greater detail in the next few chapters. The diagrams show the nature of the threats originating from the source-activities, and their immediate physical impact in the land environment. The bottom-most element in Figure 1.5 shows some of the required sets of action for reduction of threats, such as pollution management, and toxicity and concentration reduction. Note that problems, such as habitat protection and impacts, and air quality are not considered since the attention in this book is focused on the physical land environment itself. In that sense, for the problems and activities shown in Figure 1.5, pollution management and control will have to be exercised to minimize damage to the geoenvironment – which in most cases refer to the surface environment and the receiving surface water and groundwater. Since many of the sources of the impacts cannot be completely reduced to zero, impact management will have to be practised. Much of the discussions in succeeding chapters of this book relate to the effects of anthropogenic stressors on the geoenvironment, with particular emphasis on the soil environment, and the measures needed to counter the adverse impacts – i.e., measures that can be viewed as *sustainable geoenvironmental engineering practice*.

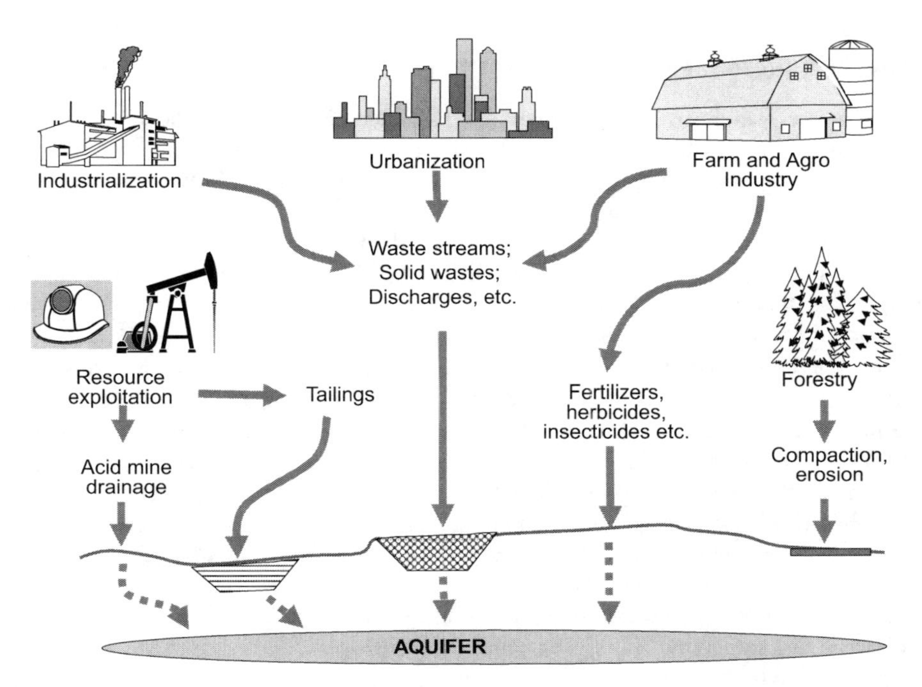

FIGURE 1.4 Nature of geoenvironmental impacts resulting from activities associated with industrialization, urbanization, resource exploitation, farm and agro industries, and timber harvesting.

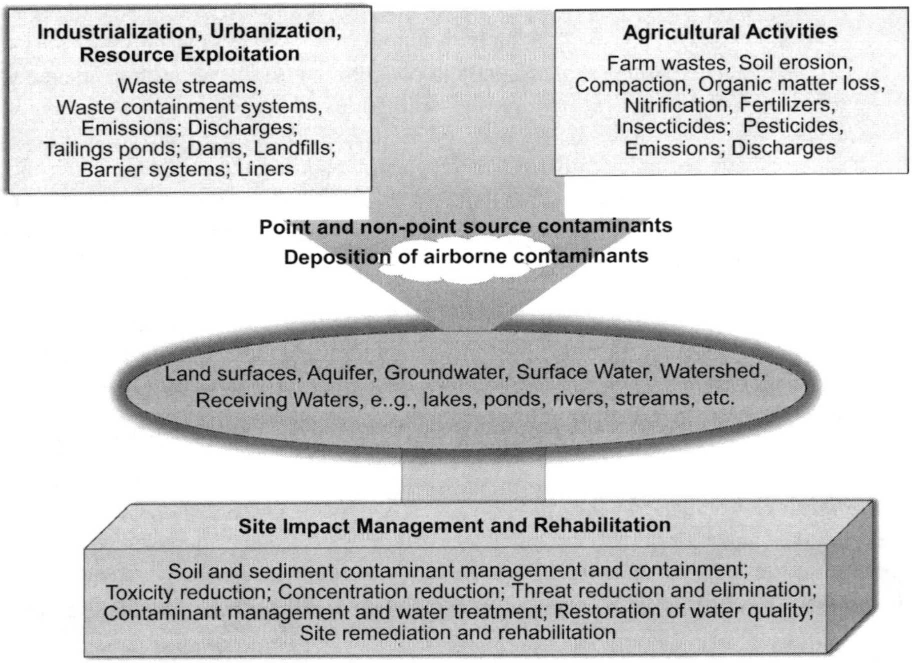

Industrialization, Urbanization, Resource Exploitation
Waste streams,
Waste containment systems,
Emissions; Discharges;
Tailings ponds; Dams, Landfills;
Barrier systems; Liners

Agricultural Activities
Farm wastes, Soil erosion,
Compaction, Organic matter loss,
Nitrification, Fertilizers,
Insecticides; Pesticides,
Emissions; Discharges

Point and non-point source contaminants
Deposition of airborne contaminants

Land surfaces, Aquifer, Groundwater, Surface Water, Watershed,
Receiving Waters, e..g., lakes, ponds, rivers, streams, etc.

Site Impact Management and Rehabilitation
Soil and sediment contaminant management and containment;
Toxicity reduction; Concentration reduction; Threat reduction and elimination;
Contaminant management and water treatment; Restoration of water quality;
Site remediation and rehabilitation

FIGURE 1.5 Threats and waste streams from anthropogenic activities in support of industrialization, urbanization, and resource exploitation, impacting on soil and water components of the physical land environment – and examples of requirements for restoration of impacted lands to pre-impact states.

1.4 GEOENVIRONMENT IMPACTS ON SOIL AND WATER RESOURCES

The sets of forces, which, by their interactions, pose potential threats to the geoenvironment, are those resulting from activities mounted in pursuit of *industrialization, urbanization, energy and resource exploitation* and *food production.* The net effect or impact of these activities and their stresses on the geoenvironment can, by their very nature, be disastrous if measures for protection of the geoenvironment are not put into practice. A reduction or degradation of geoenvironmental resources and the various ecosystems will diminish the capability of the geoenvironment to provide the elements necessary to sustain life. To avoid or to minimize the degradation, there needs to be (a) a proper audit of the geoenvironmental impacts on the prominent features that constitute the geoenvironment, and also on the ecosystem, and (b) available sensible and logical sets of tools that can be used to mitigate and ameliorate the adverse effects of the impacts.

Good land management practice: (a) Minimizes and mitigates deleterious impacts to the land environment, (b) seeks optimal land use and benefit from the land, and (c) preserves and minimizes depletion of natural capital. The obvious threat to human health linked to detrimental geoenvironment impacts comes from waste discharge and impoundment – as shown in Figures 1.4 and 1.5. Most of the activities associated with manufacturing and production of goods will generate waste in one form or another. Table 1.2 shows some typical waste streams from a representative group of industries. Impacts on the geoenvironment from these activities and discharges take the form of (a) wastewater and solid waste discharge, and spills, leaks, and other forms of discharges to the land environment (including the receiving waters in the land environment), and (b) use of chemical aids in pest control and other intensive agricultural practices, resulting in pollution of the receiving waters and excessive nitrogen and phosphate loading of the soil – leading to a consequent decrease in soil quality. Many of these issues will be discussed in detail in the various chapters to follow.

TABLE 1.2
Typical Composition of Waste Streams from Some Representative Industries

Industry	Waste Streams
Laboratories	Acids, bases, heavy metals, inorganics, ignitable wastes, solvents
Printing, etc.	Acids, bases, heavy metals, inorganic wastes, solvents, ink sludges, spent plating
Pesticide user and services	Metals, inorganics, pesticides, solvents
Construction	Acids, bases, ignitable wastes, solvents
Metal manufacture	Acids, bases, cyanide wastes, reactives, heavy metals, ignitable wastes, solvents, spent plating wastes
Formulators	Acids, bases, ignitable wastes, heavy metals, inorganics, pesticides, reactives, solvents
Chemical manufacture	Same as metal manufacture, except no plating wastes
Laundry/dry cleaning	Dry clean filtration residue, solvents

1.4.1 Impacts on Land Mass and Soil

The impacts resulting from the various activities of humans take several forms depending on the types of stressors involved. As we have noted previously, the stressors could be mechanical in nature, or hydraulic, thermal, chemical, physico-chemical, electrical, gaseous, and/or radioactive. Mechanical stressors generate physical stresses resulting in movement of land masses or pieces of a particular mass – leading to a degradation of its functionality and perhaps to an ultimate consequence of instability of the land mass, or failure of that particular unit of mass – i.e., a particular unit of soil mass. We pay particular attention to a unit of soil mass since this constitutes the basic building block of a land mass. Since the collective status of individual soil mass units defines the overall status of a particular land mass, it would be useful to pay attention to the "*health*" of a soil. In engineering terms, we can evaluate or categorize soil health in terms of its functionality.

1.4.1.1 Soil Functionality and Indicators

We define the *functionality* of a soil (i.e., *soil functionality*) as its capability to perform the various functions demanded of it, in accord with the status of the soil. For example, a major role of soil is its capability as a "plant growth medium". Loss of functionality is evident when the soil fails to yield the kinds of crops harvested in previous years. In the case of soils required to support overlying loads, loss of support capability means diminished functionality. The key to an understanding of how stressors impact on a piece of land mass is to determine the functional state of a unit of soil mass in the land mass. The use of soil functionality indices proposed by Yong et al. (2012) is a means not only to quantify the functionality of a soil, but also is a means to determine if the soil is no longer capable of meeting its planned/designed function.

The concept of using soil functionality, i.e., soil functional capability, to denote the ability of a soil or a site to function according to design or service requirements, is a novel concept, in that it addresses the performance aspects of a specific piece of soil or a soil mass. In most cases, the soil attributes used in assessment of soil functionality include (a) properties and characteristics of the particular soil under consideration, and (b) performance requirements of the soil to meet design or service specifications over the short- and long-term (periods). To determine soil functional capability – i.e., soil functionality – one needs to use *soil functionality index* (SFI) as an assessment tool. To appreciate how the SFI can be usefully utilized as a tool in assessment of soil functional capability, (a) over the long-term, or (b) in relation to stresses generated by soil environment stressors, one needs to specify soil functionality indicators. These indicators constitute the parameters for

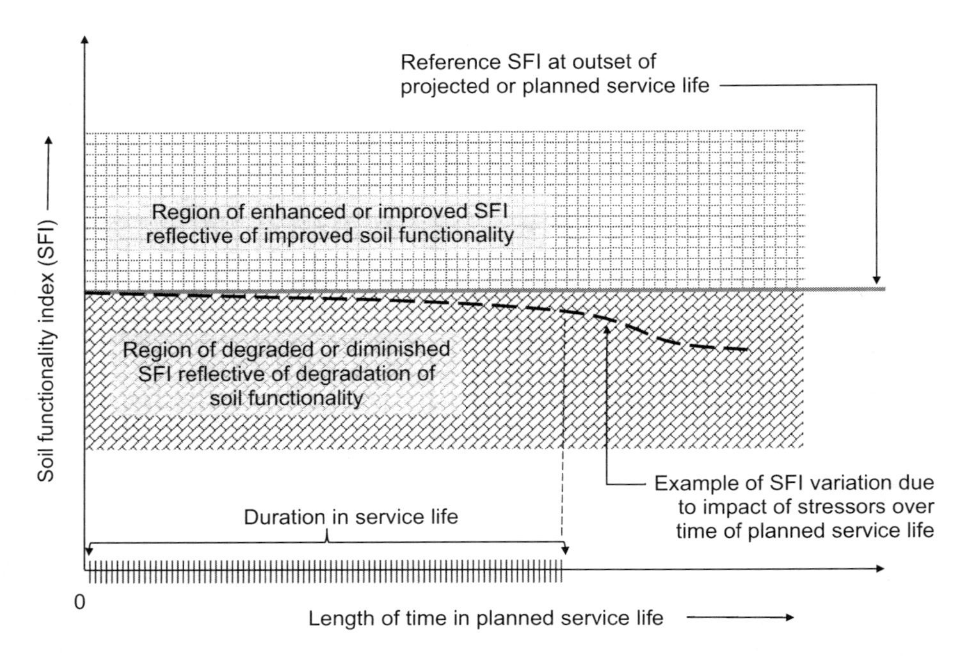

FIGURE 1.6 Example of time-SFI variation for an imaginary planned service project. Over the course of the "planned service life" shown on the abscissa, the SFI shows some level of degradation. (Adapted from Yong et al., 2012.)

computing soil functionality – using for example, single-parameter analysis, dimensional analysis, multi-variant analytical techniques, risk-based analysis, fuzzy logic, or lumped-parameter analysis. Figure 1.6 shows a graphic representation of how SFI might change with time under stressor impacts.

Indicators that are used or encountered in everyday events and situations include such common sensory ones as aural, visual, scent, and sensation. In vehicular traffic situations, for example, green lights at cross-roads indicate that one has the right of way, and red lights indicate that one should stop at the intersection. Indicators can (a) inform one of the status or nature of the situation at hand, and (b) provide guidance or insight into the performance of a system or even a particular piece of equipment. In soil science, agriculture and the earth environment, indicators are often used in soil quality and soil functionality assessments, with particular interest to soil health in relation to agricultural productivity, human health, and preservation or improvement of habitats and biodiversity. Soil quality indicators can range from complex and involved techniques using microorganisms as indicators of soil health, to more simple ones such as the use of soil colour to determine the soil constituents or components, the water content and water logging status of soil.

In the case of geoenvironmental engineering, soil contamination by pollutants and toxicants and land settlement and subsidence by overburden pressure from constructed facilities and excess withdrawal of ground water are of significant concern. Indicators are used to monitor the status of soils threatened by pollutants and toxicants, and by ground subsidence. These indicators are used as signals to inform the stakeholder that the system is functioning well, or to alert the stakeholder to potential problems. An example of the use of certain soil attributes as indicators is shown in Figure 1.7.

1.4.2 Impacts on Water and Water Resources

It has been suggested by many that in the future, conflicts amongst various groups, factions, and jurisdictions, will arise over drinking water and its availability. We need only consider the availability

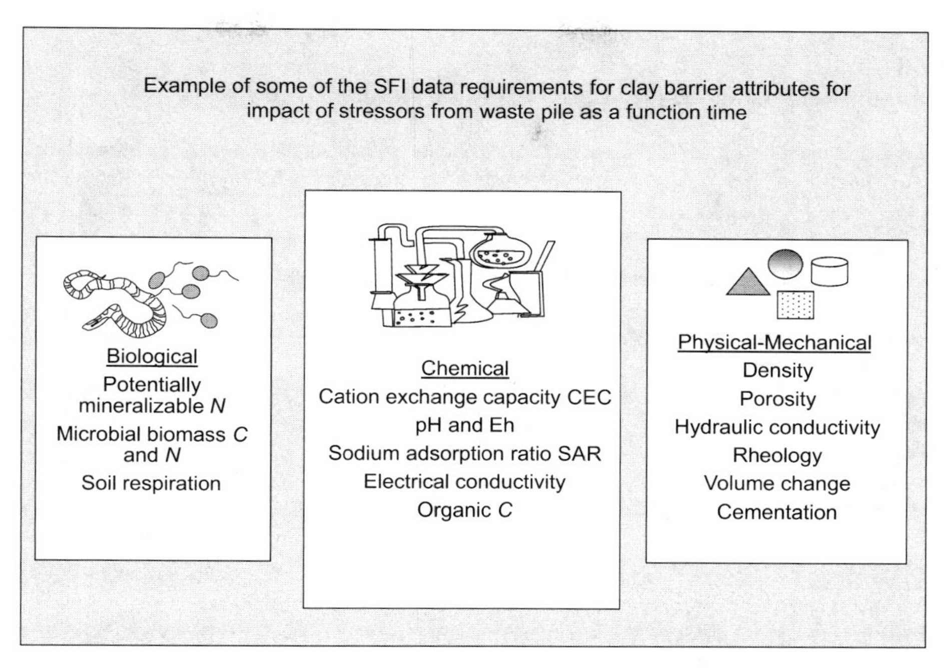

FIGURE 1.7 Some of the major soil attributes used as indicators for SFI data for a clay barrier component of multi-barrier system used for containment of waste pile in a landfill. (Adapted from Yong et al., 2012.)

and distribution of drinking water in the world to see that this suggestion has substance. Less than 5% of the global water is non-saline water. Of this less than 5% non-saline water, it is estimated that about 0.2% of the non-saline water is contained in lakes and rivers, with the remaining proportion existing as snow, ice, wetlands, and groundwater. Values reported much earlier by Leopold (1974) give numbers such as 2.7% of total volume of water (i.e., global water) as fresh water, and of that fresh water, it was estimated by Leopold that about 0.36% was "easily accessible".

Some of the more common and significant impacts to the quality of groundwater and receiving waters have been shown in Figures 1.5 and 1.6. Deterioration of the quality of these waters will not only limit their usefulness but will also cause distress to the animal and plant species that live in these waters. Considering that at least one half or more of the world's plant and animal species live in water, it is clear that any deterioration and/or decrease of water quality and water availability will have severe consequences on these species. Protection of both surface water and groundwater must be a priority. Chapter 3 discusses these and other issues in greater detail. Water usage by industry, for example, can produce liquid waste streams that are highly toxic by virtue of the chemicals contained in the waste streams, or by virtue of concentration of noxious substances. Before the liquid phase of any waste stream can be returned to the environment, it has to be treated and rendered harmless as a health threat to biotic receptors. As indicated previously, the source of these pollutants can be traced to waste streams and discharges from industrial plants, households, resource exploitation facilities, and from farms. Table 1.3 shows some chronic effects from some of these waste products on human health.

Farming and agricultural activities contribute agro additives to the receiving waters and groundwater through surface run-off and through transport in the ground (Figures 1.5 and 1.6). All the other discharges and waste streams shown in the two figures are most likely contained in storage dumps, landfills, holding ponds, tailings ponds, or other similar systems. All of these containment systems have the potential to deliver pollutants to the receiving waters (ground and surface waters)

TABLE 1.3
Chronic Effect of Some Hazardous Wastes on Human Health

Waste Type	Carcinogenic	Mutagenic	Teratogenic	Reproductive System Damage
Halogenated organic pesticide	A	A	A	H
Methyl bromide			H	
Halogenated organic phenoxy herbicide	A	A	A	A
2-4-D*				
Organophosphorus pesticide	A	A	A	
Organonitrogen herbicide	A	A	A	
Polychlorinated biphenyl	A		A	
Cyanide wastes*				
Halogenated organics	H	H		
Non-halogenated volatile organics	A	A		
Zn, Cu, Se, Cr, Ni	H			
Hg		H	H	
Cd	H			

H, A = statistically verifiable effects on humans and animals, respectively.
Adapted from Governor's Office of Appropriate Technology, Toxic Waste Assessment Group, California, 1981.
* No reportable information available.

because of eventual and inevitable leaks, discharges, and failures. Many of these phenomena will be discussed in greater detail in the later chapters of this book.

We highlight the importance of groundwater resources because it is a major water resource and as a rule, groundwater is more accessible than surface water. Furthermore, it is not uncommon for many rural communities to rely heavily on these groundwater resources as a source without proper treatment prior to use. We should note that contamination of receiving waters, such as ponds, lakes, and rivers, occurs also through leachate transport through the soil and quite obviously, from surface run-off from point and non-point pollutant sources. From the perspective of the geoenvironment, protection of both surface water and groundwater quality requires one to practice impact mitigation and management – shown, for example, in Figure 1.8 for management of liquid waste discharge into the environment to avoid impacting the receiving waters. The decision points shown in the protocol diagram include criteria, procedures, tests, etc., that need to be conducted to satisfy regulatory requirements.

To provide proper protection of the health of biotic receptors in the geoenvironment, treatment of the liquid waste streams requires detoxification and removal of all toxic and hazardous constituents, and suspended solids before discharge. Reuse of the treated waste streams is encouraged. Typical reuse schemes include irrigation (in farming and agriculture activities), process streams (such as resource extraction), and cooling towers. Waste streams that cannot be treated effectively and economically to reach acceptable discharge standards will require impoundment in secure ponds. Procedures have been developed that will reduce the liquid content of these noxious liquid waste streams. To protect the resources in the geoenvironment, the product(s) will most likely need to be incinerated or contained in secure impoundment facilities. Typical containment and impoundment facilities would be landfills. Co-disposal of these kinds of waste products with other types of waste products has been proposed as a means to accommodate these waste products. Figure 1.9 shows two typical barrier systems used to line landfill facilities. The details for these kinds of containment and waste management systems will be discussed in the next chapter.

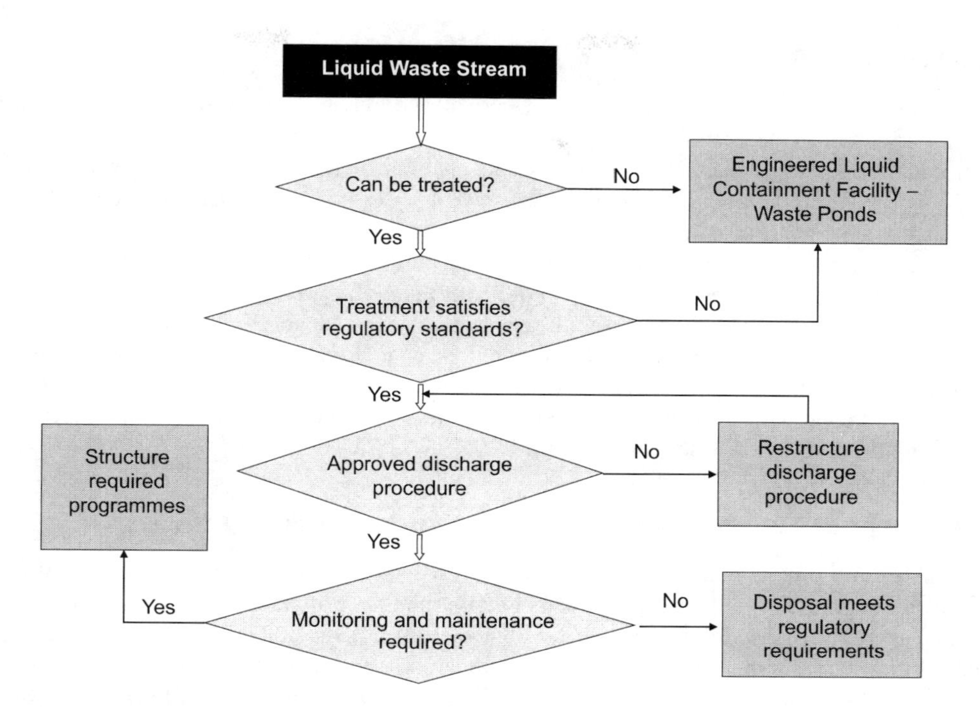

FIGURE 1.8 Programme for the management of liquid waste stream discharge into receiving waters and land surface areas.

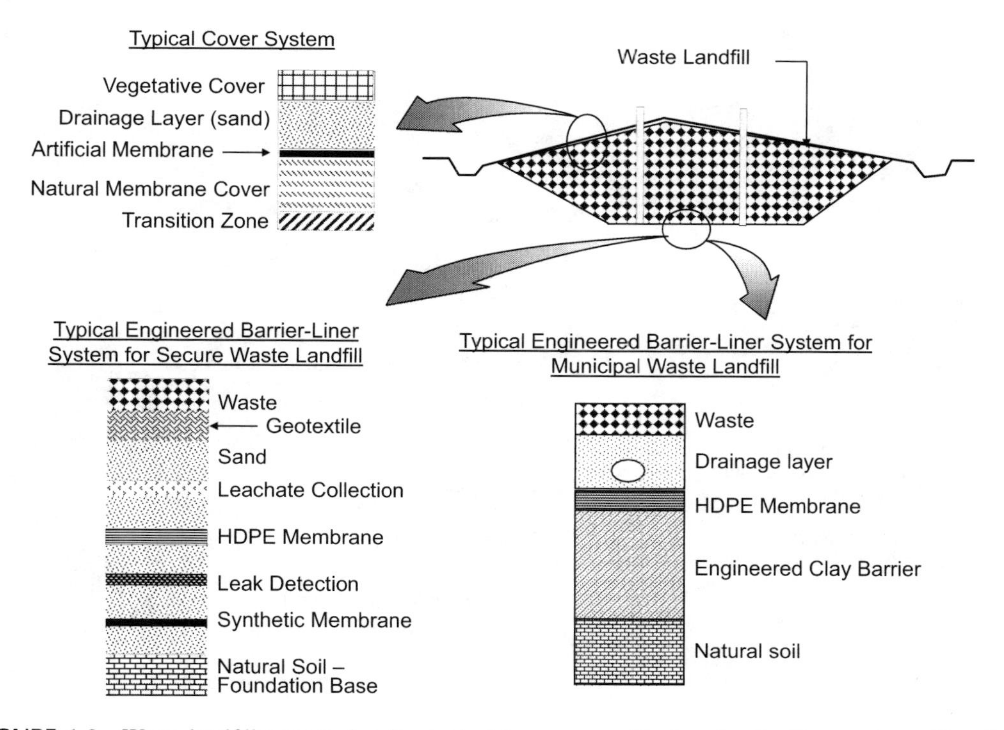

FIGURE 1.9 Waste landfill system showing typical top cover and bottom barrier-liner systems. Two liner systems are shown: the maximum security barrier system (left), and the bottom barrier system generally used for landfills containing municipal solid wastes.

1.5 SUSTAINABILITY

In a Gaean world, the earth is a living being where all the living organisms and non-living entities function and interact independently but contribute to collectively define and regulate the material conditions necessary for life. When stresses and resultant negative impacts associated with the activities of mankind in the production of *goods and services* arise, the Gaean hypothesis becomes somewhat untenable. This is because living matter in the geoenvironment will be constrained from regulating the material conditions necessary for life. Loss of species diversity is one of the major factors. The paramount terrestrial ecosystem imperatives are (a) protection and conservation of the various land environment resources, and (b) ensuring that the capability to provide life support is not degraded or diminished. From the viewpoint of the geoenvironment and ecosphere, the pressures from *development stresses* and WEHAB (*water and sanitation, energy, health, agriculture, biodiversity*) combined with the processes necessary to satisfy sustainable development objectives are summed up by the schematic shown in Figure 1.10.

1.5.1 Renewable and Non-Renewable Geoenvironment Natural Resources

Sustainable development in itself may in all probability be a chimera – a non-attainable goal and an illusion. However, this should not deny the fact that proper environmental management and conservation measures are needed if we are to strive to meet the goals and objectives of sustainability. This includes (a) resource conservation and management, and (b) natural habitat protection and preservation of diversity. Failure to do so will result in the diminution of the capability of the geoenvironment to provide the basis for life support. The case of renewable and non-renewable geoenvironmental natural resources is a good demonstration of this point.

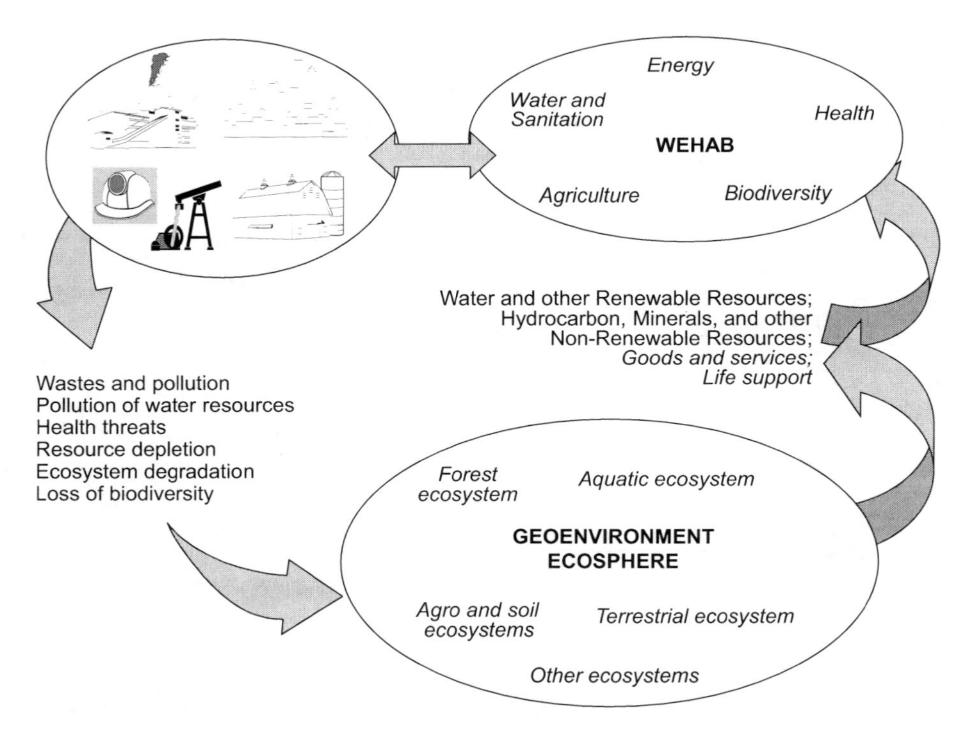

FIGURE 1.10 The continuous cycle of interaction between industry-production, WEHAB and the geoenvironment – from a geoenvironmental perspective.

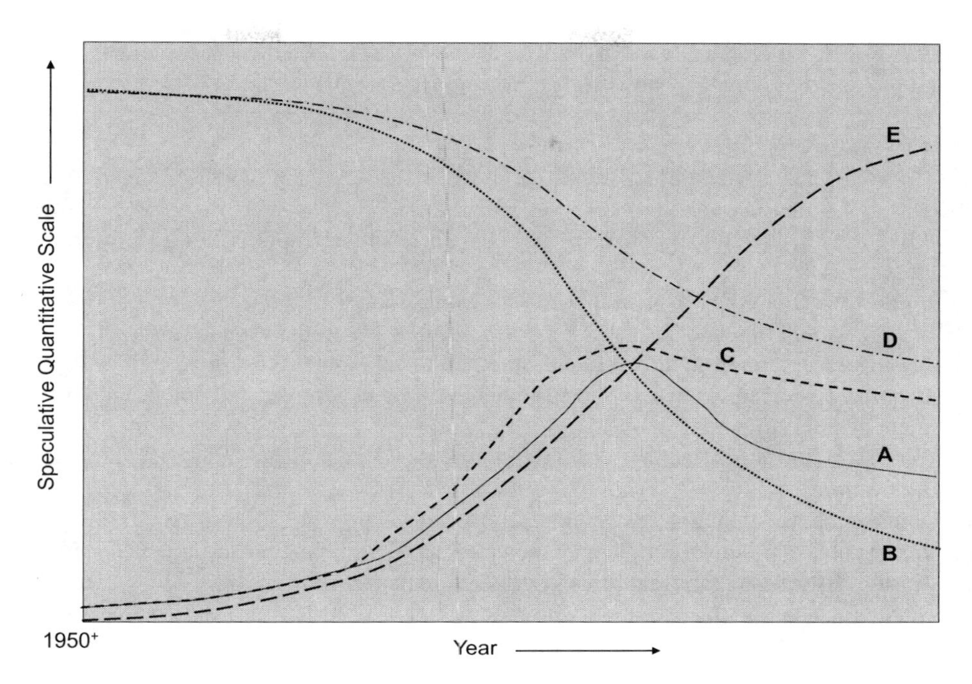

FIGURE 1.11 Speculative chart on status of global population relative to conservation of the non-renewable natural resources. **A** = status of population based on current usage of resources, **B** = speculative quantity of non-renewable resources remaining available, **C** = status of population based on conservation of resources, use of 4Rs and alternative materials and energy sources, **D** = available non-renewable natural resources using conservation management and 4Rs, **E** = assumed availability of alternative materials and "green" energy sources.

Following the spirit of the systems dynamics model predictions of Meadows et al. (1972), the chart shown in Figure 1.11 speculates on the status of the global population at some future time under conditions identified in the figure caption. The curve identified as **A** shows the status of population based on current depletion rate of non-renewable geoenvironmental natural resources in relation to some future time. The abscissa on the diagram shows "years at some future time", and the ordinate gives a qualitative appreciation of the growth or decline of the parameter under consideration. The curve identified as **B** is the speculative quantity of non-renewable resources available assuming that the depletion rate of the non-renewable geoenvironmental natural resources remains the same – i.e., constant in proportion to the population at hand. We consider the principal non-renewable geoenvironmental natural resources to consist of fossil fuels, minerals, and geologic building materials (sand, gravel, stone, soil).

Assuming that conservation measures for non-renewable geoenvironmental natural resources are in place and that these measures are bolstered by the use of the 4Rs (reduction, reuse, recycle, and recovery), we can further reduce the depletion rate of the non-renewable resources by using alternative energy sources, such as geothermal, wind, solar, etc. The curve **C** shown in Figure 1.11 indicates the status of the global population based on conservation of the non-renewable resources, use of 4Rs and use of alternative materials and energy sources, **D** = available natural geoenvironmental resources using conservation management, 4Rs, and alternative energy sources. Finally, curve **E** shown in Figure 1.11 is an assumption of the resources that would be made available through the use of alternative materials and various other alternative energy sources.

1.5.2 UN Sustainability Goals 2030

At the Rio Summit +20 in 2012, 190 nations agreed to promote a green economy and to enhance global environmental protection and to come up with a set of sustainable development goals (SDGs).

Subsequently since 2015, 17 goals with the 169 targets were devised to be accomplished by the year 2030 and include the following (https://sdgs.un.org/, accessed Jan. 2024):

1. End poverty in all its forms everywhere.
2. End hunger, achieve food security and improved nutrition, and promote sustainable agriculture.
3. Ensure healthy lives and promote well-being for all at all ages.
4. Ensure inclusive and equitable quality education and promote lifelong learning opportunities for all.
5. Achieve gender equality and empower all women and girls.
6. Ensure availability and sustainable management of water and sanitation for all.
7. Ensure access to affordable, reliable, sustainable, and modern energy for all.
8. Promote sustained, inclusive, and sustainable economic growth, full and productive employment and decent work for all.
9. Build resilient infrastructure, promote inclusive and sustainable industrialization and foster innovation.
10. Reduce inequality within and among countries.
11. Make cities and human settlements inclusive, safe, resilient, and sustainable.
12. Ensure sustainable consumption and production patterns.
13. Take urgent action to combat climate change and its impacts.
14. Conserve and sustainably use the oceans, seas, and marine resources for sustainable development.
15. Protect, restore, and promote sustainable use of terrestrial ecosystems, sustainably manage forests, combat desertification, and halt and reverse land degradation and halt biodiversity loss.
16. Promote peaceful and inclusive societies for sustainable development, provide access to justice for all and build effective, accountable, and inclusive institutions at all levels.
17. Strengthen the means of implementation and revitalize the global partnership for sustainable development.

These goals are highly interlinked and therefore there is a growing need for more integrated resource management frameworks. In particular, goals 6, 7, 9, 11, 13, 14, and 15 are the most related to the geoenvironment. Some of the extensive challenges of sustainable development are related to a healthy and sufficient water supply for healthy people, resource availability, and protection of the natural environment. Climate change, industrialization, and population growth are all exerting pressures on resources. While the SDGs are a priority, the means to address them are still lacking and thus there is substantial opportunities for contribution by the geoenvironmental community.

1.5.3 Social Sustainability Through Community Engagement

An important aspect of social sustainability is community engagement. Since Dec. 2016, in particular, there has been a substantial paradigm shift as indicated by the following statement by Prime Minister Trudeau of Canada, "Last year I committed to a renewed relationship with Indigenous Peoples, one based on the recognition of rights, respect, co-operation, and partnership. Today, we take further steps on the journey of reconciliation."

The Truth and Reconciliation Commission challenged the business sector to

- Commit to meaningful consultation, building respectful relationships, and obtaining the free, prior, and informed consent of Indigenous Peoples before proceeding with economic development projects.

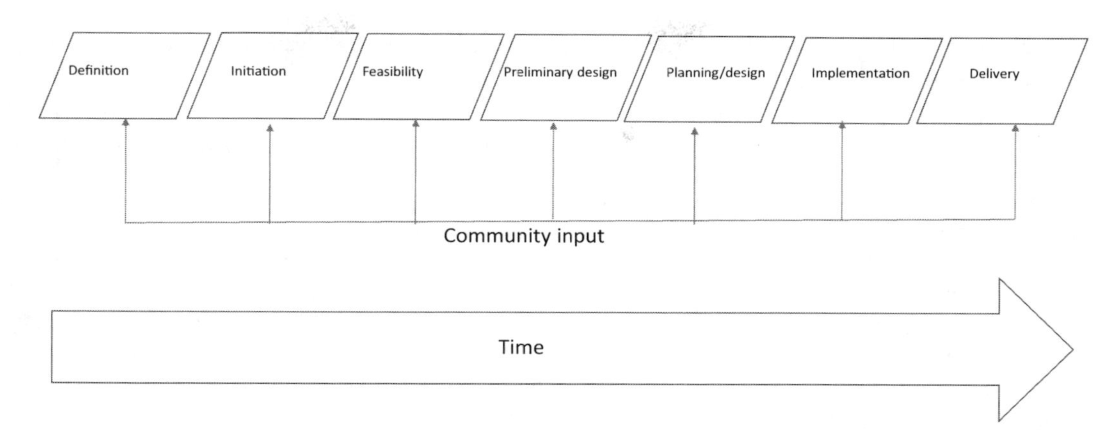

FIGURE 1.12 Community engagement in a project.

- Ensure that Indigenous Peoples have equitable access to jobs, training, and education opportunities in the corporate sector, and that Indigenous communities gain long-term sustainable benefits from economic development projects.

An important aspect is engagement. Engagement is different from consultation in form and objective. It is a process for Indigenous peoples, government, and industry to share information on issues of mutual interest. Engagement can develop effective and practical relationships even when there is no duty to consult. Early engagement can help determine if there is a duty to consult and result in a more efficient consultation process. Engagement on any major development project is now an expectation of Indigenous communities.

As a first step, the community environmental scan allows getting a snapshot of the community and getting a basic understanding of it. Indigenous people/groups must be treated as Partners and Rights Holders (not stakeholders).

Engagements should start as early as possible in the project, as shown in Figure 1.12. Information shared should be timely and accurate. There should be flexibility with regards to the realities of Indigenous communities. The lines of communication must be open and clear, following local Indigenous protocols.

There are challenges in Indigenous engagement. There may be limited knowledge, expertise, and capacity in the community. It may be difficult in understanding Indigenous political environment, Indigenous cultural norms, and determining treaty rights, interests, etc. Identifying land constraints/restrictions (traditional territory, cultural sites, archaeology, claims) may also be complicated. Some solutions to deal with these challenges are to perform an environmental scan to determine the community strengths, engage with communities before making assumptions/decisions, work with federal and provincial partners, and/or consult with and hire industry experts.

1.5.4 The United Nations Biodiversity Conference (COP 15)

In Montreal, Canada, on Dec. 19, 2022 an agreement was reached among 188 governments. It was named the Kunming-Montreal Global Biodiversity Framework (GBF) with the main objectives of reducing biodiversity and ecosystem loss, and protecting Indigenous rights (www.unep.org, accessed Jan. 2024). The GBF also features 23 targets to achieve by 2030, including those highly relevant to the geoenvironment: To accomplish this, 30% of the land, coastal areas, and oceans and restoration of 30%of degraded ecosystems are to be protected by 2030. Currently, only 17% of land and 8% of marine areas are under protection. Other objectives include reducing the loss of areas of high biodiversity importance and high ecological integrity to near zero and reducing global food waste by 50%.

1.6 CLIMATE CHANGE IMPACTS ON THE GEOENVIRONMENT AND COP 27

The temperature of the Earth has increased by 1.1 °C since the pre-industrial era (IPCC 2023, www.ipcc.ch/sr15/, accessed Feb. 2024). Population and economic growth have led to increases in CO_2 emissions from combustion of fuels and thus risks to water, energy, food, and building sectors, among others. Extreme temperatures, variability of precipitation, and rising sea levels can have serious effects on urban areas, produce greenhouse gases (GHGs) (carbon dioxide, methane, nitrogen oxide), and other harmful emissions. Droughts are occurring more frequently, which increases forest fires and further increases carbon dioxide emissions. Increasing sea levels leads to higher levels of salinization within coastal aquifers (IPCC 2007, www.ipcc.ch/report/ar4/syr/, accessed Jan. 2024). Climate change likely will lead to many changes in water quality and quantities by 2050 (OECD, 2012). Floods and droughts will influence water levels. Irrigation needs therefore could increase by 2080 up to 20% (FAO, 2011). Melting of glaciers will raise water levels initially but will subside eventually (Black and King, 2009). Contamination of fresh water by salinity intrusions will increase water supply challenges.

Changes in 25–50 year design lives of infrastructure must be considered by engineers due to increases in storms, rainfall intensity, and droughts. The U.S. Bureau of Reclamation for climate change scenarios provides guidance for reservoir design (USBR, 2016) while the NOAA (2012) provides guidance for sea level rise in coastal regions. Urban areas will be particularly impacted by heat waves and water scarcity.

According to the WEF (2020), increases of 1 °C have occurred within the last decade alone. This has significant impacts on permafrost, which is the ground with temperatures remaining below 0 °C for 2 or more consecutive years. As approximately 1/5 of the frozen soils thaws in the Arctic, releases of methane and organic matter degradation will further accelerate climate change. Large-scale impacts include erosion, subsidence and landslides, damage to infrastructure, flooding, and ecological transformations.

To address climate change, the Paris Agreement at the COP21 in Paris was adopted on Dec. 12, 2015 and went into effect on Nov. 4, 2016. All countries agreed to efforts to limit global temperature rise to below 2 °C, but to aim for an increase of only 1.5 °C.

Some aspects covered in the agreement include the following:

- A balance between anthropogenic sources, and removal by sinks of GHGs must be achieved.
- The need for developed countries to lead the efforts.
- Sinks and reservoirs as means of conservation and enhancement should be promoted.
- Measures for adaptation must be increased.

Subsequently, the Glasgow Climate Pact was reached by 197 countries on Nov. 13, 2021 that "Reaffirms the Paris Agreement temperature goal of holding the increase in the global average temperature to well below 2 °C above pre-industrial levels and pursuing efforts to limit the temperature increase to 1.5 °C above pre-industrial levels" and "Recognizes that limiting global warming to 1.5 °C requires rapid, deep and sustained reductions in global greenhouse gas emissions, including reducing global carbon dioxide emissions by 45% by 2030 relative to the 2010 level and to net zero around mid-century, as well as deep reductions in other greenhouse gases". Climate Action Tracker has estimated that with the current state, the global temperature will increase by 2.7 °C by the end of the century but could be reduced to 2.4 °C if the targets for 2030 are met.

The 2022 United Nations Climate Change Conference (known as COP 27) was held from Nov. 6 to 20, 2022 in Egypt. The impacts of climate change in 2022 were particularly highlighted such as Hurricane Ian in North America, European heat waves, and Pakistani floods. An IPCC report (2021) highlighted that Africa would be particularly threatened as a result of rainy season, drought, and precipitation pattern changes, and temperature increases on agriculture. In summary, these agreements pose challenges and opportunities for engineers to design infrastructure, processes, and projects, particularly related to the geoenvironment.

1.7 LIFE-CYCLE CONCEPT AND A CIRCULAR ECONOMY

1.7.1 Life-Cycle Concept

As a tool for determining the sustainability of a process or product, the concept of the life-cycle – the entire life from cradle to grave – must be considered. This includes raw material processing, product manufacturing, delivery to consumer, recycling, reuse, and/or final disposal (Figure 1.13). All materials must be identified at each step in the environmental impact assessment. In industrial ecology (to be discussed in Chapter 4), the life-cycle assessment (LCA) can help identify where changes can be made to reduce environmental impact. Initially, SETAC (the Society of Environmental Toxicology and Chemistry) and then International Standards Organization (ISO14040 and ISO14044), consulting organizations and academia (ISO2006 a, b) have developed the frameworks for its use and thus are employed worldwide. While it is highly advantageous for its comprehensiveness, accuracy and environmental impact detail, there can be extensive requirements for data and specialized expertise.

1.7.2 Circular Economy

A related concept to the life-cycle of a product or a process (cradle to grave), is a circular economy (considered as cradle to cradle). It involves extending the life-cycle as much as possible for by reusing, repairing, recycling, and refurbishing to tackle global challenges, such as climate change and loss in biodiversity. It has received more attention recently as a means of reducing emissions, resource requirements, and wastes has been gaining popularity because it helps to minimize emissions and

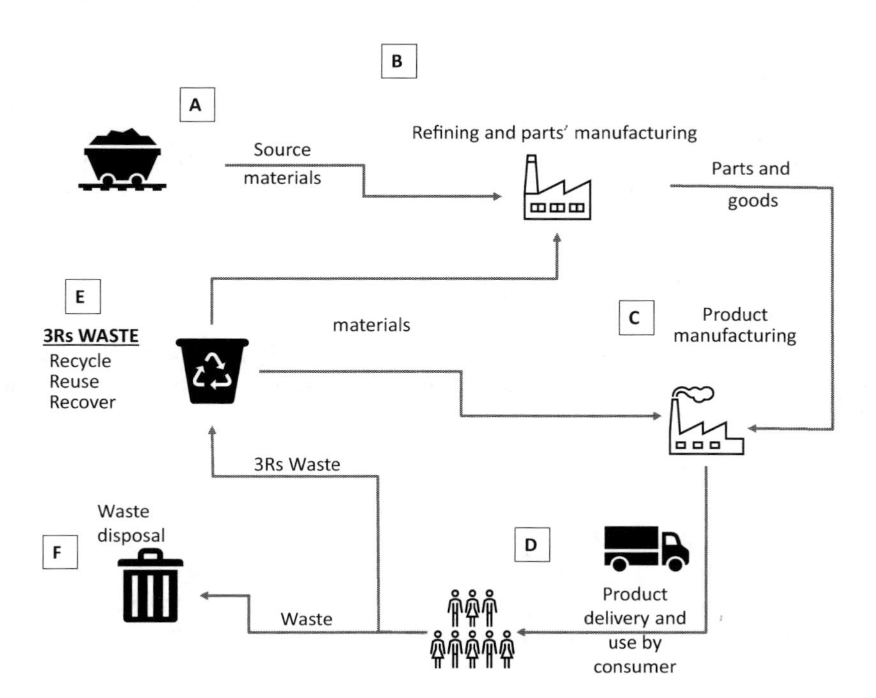

FIGURE 1.13 Concept of life-cycle assessment.

consumption of raw materials, open up new market prospects and, principally, increase the sustainability of consumption and improve resource efficiency. The business model is shifted from a linear to a circular economy. While it may be difficult if not impossible to achieve a fully circular economy, targeting certain industries such as electronics and plastics has gained momentum. It has also been argued that products should be designed from the start with reuse and recycling in mind.

1.8 CONCLUDING REMARKS

The impacts of natural and catastrophic events, such as earthquakes, tsunamis, hurricanes, and typhoons, and associated floods and landslides are not discussed in this chapter (and not in this book). This does not mean that these are minor events or impacts – in comparison to the impacts on the geoenvironment generated by human activities. It is recognized that these natural catastrophic events can and do result in considerable loss of life and physical facilities. The problems and impacts generated by these natural disasters on the geoenvironment deserve proper recognition and discussion in a textbook or treatise specifically devoted to such subjects.

Proper management of the geoenvironment is essential if the platform for almost all the life-support systems is to be protected for future generations. The principles of *sustainability* require us to recognize a fundamental fact that geoenvironmental natural and cultivated resources are *renewable* and *non-renewable*. Chapter 5 addresses this issue and the situations where renewable natural resources can become non-renewable and hence not sustainable. It is necessary to recognize that renewable geoenvironmental natural resources can be easily threatened and can become ineffective as a resource. A good case in point is *water*. Pollution of receiving waters will render such waters unacceptable for human consumption, therefore rendering this renewable geoenvironmental resource useless. The following items are some of the major issues facing us as we seek to maintain the life-support base that provides us with the various goods and services:

- Depletion of non-renewable resources or natural capital is a reality. Energy production relying on fossil fuels is an example of how non-renewable resources are continuously depleted.
- Industrial wastes and waste streams will need to be managed, and it is likely that some of the waste products will find their way into the land environment, resulting thereby in threats to the health and welfare of biotic receptors.
- Loss of soil quality due to various soil degradative forces, such as erosion and salinization and melting permafrost. In addition to reduction in capability of the soil for crop production, one faces a loss in the capability of the soil to act effectively as a carbon sink.
- Depletion of agricultural lands will occur because of urbanization pressures, thus requiring remaining agricultural lands to be more productive. Implementation of high-yield practices may exacerbate the problem of pollution of both land and water resources.
- Deforestation and inadequate replacement rates, thus contributing to the CO_2 imbalance.
- Contamination of groundwater and surface water resources can reach proportions that render such sources as health threats to biotic receptors.

In the context of the geoenvironmental perspective of environmental management, three particular points need to be stated in regard to the development environment or sustainable society problem:

- Soil is a natural resource. In combination with the other geophysical features of the land environment, they constitute at least 90% of the base for sustenance of the human population and production of energy and goods. The depletion rate of the natural capital, represented by all the natural resources, must be minimized.
- Technology and its contribution to environmental management – in addition to the various remediation and impact avoidance tools that technology can develop and contribute to

environmental management, perhaps one of the more significant contributions that technology can make would be the development of renewable resources' replacements for the non-renewable resources that are being depleted.

- Protocols and procedures for management of changes in the environment – it is becoming very evident that changes to the geoenvironment that are presently occurring may reach proportions that require one to develop technology and new social attitudes and community engagement to manage the change. A particular case in point might be, for example, global warming and the greenhouse effect.

REFERENCES

Black, M. and King, J. (2009), *The Atlas of Water: Mapping the World's Most Critical Resource*, 2nd ed, University of California Press, Berkeley, CA.

Commoner, B. (1971), *The Closing Circle, Nature, Man and Technology*, Alfred A. Knopf, New York, 326p.

FAO (2011), *The State of the World's Land and Water Resources for Food and Agriculture: Managing Systems at Risk*, FAO and Earthscan, London.

Glasby, G.P. (2002), Sustainable development: The need for a new paradigm, *Environmental Development and Sustainability*, 4:333–345.

Governor's Office of Appropriate Technology, Toxic Waste Assessment Group, California, 1981.

IPCC. (2007), Climate Change 2007: Synthesis Report. Contribution of Working Groups I, II and III to the Fourth Assessment Report of the Intergovernmental Panel on Climate Change [Core Writing Team, Pachauri, R.K and Reisinger, A. (eds.)]. IPCC, Geneva, Switzerland, 104 pp.

IPCC. (2023), Sections. In: Climate Change 2023: Synthesis Report. Contribution of Working Groups I, II and III to the Sixth Assessment Report of the Intergovernmental Panel on Climate Change [Core Writing Team, H. Lee and J. Romero (eds.)]. IPCC, Geneva, Switzerland, pp. 35–115, doi: 10.59327/IPCC/AR6-9789291691647

Leopold, L.B. (1974), *Water: A Primer*, W.H. Freeman and Co., San Francisco, 172 p.

Malthus, T. (1798), An essay on the principle of population, as it affects the future improvement of society with remarks on the speculations of Mr. Godwin, M. Condorcet, and Other Writers, London, P2rinted for J. Johnson, in St. Paul's church-yard. (HTML format by Ed Stephan, 10 Aug 1997)

Meadows, D.H., Meadows, D.L., Randers, J., and Behrens III, W.W. (1972), *The Limits to Growth*, Universe Books, New York, 205 p.

Naeem, S., Chapin III, C.F.S., Costanza, R., Ehrlich, P.R., Golley, F.B., Hooper, D.U., Lawton, J.H., Neill, R.V.O., Mooney, H.A., Sala, O.E., Symstad, A.J., and Tilman, D. (1999), Biodiversity and ecosystem functioning: Maintaining natural life support processes, *Issues in Ecology*, 4: 11.

NOAA (National Oceanic and Atmospheric Administration). (2012), Incorporating Sea Level Rise Scenario at the Local Level. www.ngs.noaa.gov/PUBS_LIB/SLCScenariosLL.pdf, accessed Aug. 1, 2023

OECD (Organization of Economic Co-operation and Development). (2012), *OECD Environmental Outlook to 2050. The Consequences of Inaction*, OECD Publishing, Paris.

Trevors, J.T. (2003), Editorial: Biodiversity and environmental pollution, *Journal of Water, Air and Soil Pollution*, 150:1–2.

USBR (U.S. Bureau of Reclamation). (2016), West-Wide Climate Risk Assessments. Hydroclimate Projections. www.usbr.gov/climate/secure/docs/2016secure/wwcra-hydroclimateprojections.pdf, accessed Aug. 1, 2023

U.S. National Environmental Policy Act (1969), 42 U.S.C. §§ 4321–4347.

World Commission on Environment and Development. (1987), *Our Common Future*, Oxford University Press, Oxford, 400 p.

World Economic Forum (WEF). (2020), Global Warming is Melting the Permafrost Layer, Creating Big Holes in the Arctic, www.weforum.org/agenda/2020/02/permafrost-ice-melt-thaw-arctic-global-warming-carbon/, accessed Nov. 28, 2023.

Yong, R.N., and Mulligan, C.N. (2019), *Natural and Enhanced Attenuation of Contaminants in Soils*, CRC Press, Boca Raton, 324 p.

Yong, R.N., Nakano, M., and Pusch, R. (2012), *Environmental Soil Properties and Behaviour*, CRC Press, Boca Raton, 435 p.

2 Stressors and Soil Contamination

2.1 INTRODUCTION

Two of the most important ecosystem components of the geoenvironment are soil and water. These two components, together with the atmospheric component provide almost all of the necessary elements needed to support life for all living species. This chapter focuses on the soil ecosystem component and the problem of stressor impacts on this component. The next chapter will pay similar attention to the *water* ecosystem component of the geoenvironment.

2.2 STRESSORS AND IMPACTS

As has been pointed out in Chapter 1, almost anything that happens, i.e., any input or any activity in the ecosphere will result in the production of some kind of stressor-related impact on the geoenvironmental landscape and its associated ecosystems, i.e., the land environment. To protect the status and manage the geoenvironment, the nature of these impacts and whether these impacts will add value to the particular ecosystem – or subtract from the functionality of that particular ecosystem in the geoenvironment – needs to be determined and better understood. A complete listing of all the impacts on the ecosystems of the geoenvironment is not possible because we are not sure if we are fully aware of all the kinds of activities and interactions that are active in a functioning ecosystem. As previously defined, a *functioning ecosystem* includes not only the biological and chemical activities in the ecosystem, but also the physico-chemical, mechanical, hydraulic, and thermal interactions that are characteristic of the type of ecosystem under consideration. With our present understanding of the geoenvironment and the associated functioning ecosystems, it is difficult (and virtually impossible) to fully catalogue all of these activities and interactions. What is possible at this stage is to examine and determine how the known activities and interactions are affected or changed because of the stresses, disturbances, alterations, etc., to the ecosystem of interest.

The major sources of impacts and the resultant nature of the impacts cannot be easily listed without specification of targets of the impacts. Some of these sources may not be immediately evident, and some of the impacts will not be readily perceived or understood – because the effects of the impacts may not be apparent, and/or because the effects or results of the impacts cannot be readily recognized. A simple case in point is the effect of buried toxic substances in the ground on human health – particularly if the impact on the health of those that come into contact with the material is mutagenic or teratogenic.

The distress and damage to the geoenvironment can be readily perceived because the energy generated in these events in the form of forces and stresses can cause substantial physical damage to the geoenvironment and considerable loss of life. Whilst many of the sources or causes of the impacts are generally obvious, there are many that are not. This is because we do not have any hard and fast rule as to what constitutes an impact, and more importantly, when the particular impact under consideration causes irreversible damage. Damage to the geoenvironment of the same scale and magnitude as earthquakes and hurricanes can be obtained as a result of man-made activities, such as landslides and pollution of ground and receiving waters. However, more often than not, the impacts to the geoenvironment resulting from anthropogenic activities associated with the

DOI: 10.1201/9781003407393-2

production of goods and services are less dramatic. That being said, they nevertheless endanger public safety and present health-threatening problems and issues.

2.2.1 Stressor Impacts on Soils

The categories of stressors include groups, such as (a) hydraulic, (b) mechanical, (c) thermal, (d) chemical, (e) geochemical, and (f) biologically mediated. Some of the main impacts resulting from the actions of these types of stressor groups that have been discussed by Yong et al. (2012) are listed below.

2.2.1.1 Hydraulic

The stressors that classify under the *hydraulic* group are directly related to water and its actions in soils – with types of actions that result in application of pressures that act directly on and within a soil mass. The impacts from these stresses could

- Initiate piping that could undermine the stability of overlying structures and facilities (adverse impact).
- Trigger erosion, landslides, and quick soil conditions from excessive porewater pressures, self-detachment of soil particles (adverse impact).
- Restructure affected soils because of the pressures resulting in changes in soil properties and behaviour (could be adverse or beneficial impact).
- Dilute or decrease contaminant concentration in contaminated soils (generally considered to be beneficial).
- Initiate or increase advective transport of contaminants in the soil (adverse or beneficial impact, depending on initial conditions).
- Detach sorbed contaminants and contribute to the environmental mobility of the contaminants (probably more adverse than beneficial – again depending on initial conditions).
- Affect biological processes through changes in natural habitat and energy sources (mostly adverse impact).
- Influence natural soil-weathering processes leading to alteration and/or transformation of susceptible minerals (could be adverse or beneficial).
- Other actions classed under the hydraulic stressor grouping could include those associated with floods, droughts (lack of water as a hydraulic stressor), excessive rainfall and water availability detrimental to agricultural productivity, etc. (adverse impact).

2.2.1.2 Mechanical

The common types of stressors included in the *mechanical* group – that act on soils – are those that produce pressures and stresses in a soil mass. Natural stressor sources are earthquakes and avalanches, whilst anthropogenic sources include those activities associated with resource, agro, primary and secondary industries, constructed facilities and activities related to urbanization. Note also that although war leads to substantial anthropogenic impacts on the environment, it is not dealt with in this book. The impacts generated by the stressors in the mechanical group include (Yong et al., 2012)

- Direct loading from a solid mass, such as an overlying structure – e.g., a bridge abutment and foundation footings for a structure, or facility. Resultant effects in the affected soil include
  (a) collapse of overlying structure due to failure of the soil to support the applied load (adverse impact),

(b) settlement of the overlying structure due to consolidation, secondary compression and/or creep of the supporting soil (adverse impact).

- Pressures on a soil mass as a result of actions related to water movement – e.g., swelling pressure in confined swelling soils, and pressures developed in the soil in unstable slopes.
  (a) For swelling soils under overlying structures, swelling pressure could undermine stable support of the structures, thereby causing collapse of the structure (adverse impact).
  (b) For unstable slopes, instability of the slope will result in slope movement or slope failure (adverse impact).
- Soil freezing and frost heaving pressures resulting from formation of ice lenses. Although these impacts can be reckoned as the results of stressors classifying under the *thermal* group, their inclusion in this portion of the discussion shows that not all stressor sources reside solely in one single category. The consequent effects of soil freezing and developed frost heaving pressures are, in a manner, similar in principle to those experienced with swelling soils. In the case of ice lens formation, the consequent effects from subsequent thawing of the ice lens can be severe if ice lenses formed in the freezing stage have created significant frost heaving in the soil.

2.2.1.3 Thermal

The *thermal* group of stressors generates heat or cooling in soils. The *natural* stressor source is the summer–winter cycle. A significant anthropogenic source is canisters containing high-level radioactive wastes embedded in underground repositories generating heat over long periods of time. The two sets of issues are (a) high temperatures, and (b) freezing temperatures, and their impacts include

- Water and vapour transfer in soils (e.g., evapotranspiration) resulting in changes in soil properties and behaviour (could be adverse or beneficial impact).
- Types and rates of chemical reactions and biological processes resulting in changes in the nature and energy status of the affected soil – consistent with soil weathering processes (could be adverse or beneficial impact).
- Soil freezing and developed frost heaving pressures – as discussed under the *mechanical* group of stressors,
- Freeze–thaw phenomena and development of ice lenses in frost susceptible soils with water availability, and thaw subsidence from disappearing ice lenses (adverse impact).

2.2.1.4 Chemical

Some of the *natural* stressors classifying under the grouping of "chemical" include acid rain, organic acids obtained from decomposing surficial organic matter, and the natural chemical constituents in soils. Waste landfills, discharge of contaminated wastewater, acidic leachates from mine heaps, acid rain, etc., are stressors attributable to human activities. Note that we include acid rain under both *natural* and *human activities* inasmuch as emissions into the atmosphere from production facilities are seen to be the principal contributors to the acid nature of rain and snow. The impact of actions and/or events in the chemical environment in soils will be felt in terms of changes in the nature and properties of the soil because of changes in (a) the nature of the soil fractions themselves, e.g., transformation of montmorillonite to illite or mixed layer clay mineral, (b) interactions between soil particles and between particles and water, and soil-water energy characteristics, (c) biological processes in the soil, leading to changes in the nature and character of the soil, and (d) chemical reactions and processes.

2.2.1.5 Geochemical

In respect to soils, the stressors in the geochemical category are leachates or contaminants from mine heaps, constructed facilities, and excess fertilization in agriculture. The impacts from these stressors generated in a soil mass are (a) decomposition of soil minerals and chemical constituents in soils, and (b) changes of soil particle surface functional groups. These decompositions and changes result in alteration of soil structure and chemical constituents in soil mass, thus affecting water, solute, and heat-transfer phenomena in soils.

2.2.1.6 Biologically Mediated

The stressors in this category are microorganisms in the soil. The impacts of their interactions with soil constituents and other energy sources are decomposition of organic matter and alteration or decomposition of clay minerals and chemical constituents of soil. In turn, the outcome would be changes in the nature of the soil – resulting in corresponding changes in not only hydraulic, mechanical, and thermal phenomena in soils, but also the relevant soil properties.

2.2.2 Soil Contamination from Chemical Stressors

Chapter 1 points out that good land management practice (a) minimizes and mitigates deleterious impacts to the land environment, (b) seeks optimal land use and benefit from the land, and (c) preserves and minimizes depletion of geoenvironmental natural capital. Probably the most significant stressor responsible for degrading the quality of the land, its ecosystems and especially the receiving waters (including groundwater), is the group of chemical stressors. The impact from these stressors is contamination of the land environment and its receiving waters by contaminants and hazardous substances, and as will be discussed in the latter part of this chapter and the next chapter, the importance of protection of the receiving waters cannot be overstated. Ground contamination – i.e., contamination of soil mass and receiving waters – by these same contaminants and hazardous substances poses threats to human health, other biotic receptors, and the environment. Our attention in this chapter will be focused on the various aspects of ground contamination and land management, together with remediation and restoration requirements needed to meet sustainability goals. The later chapters in this book will focus on actual technologies, procedures, and tools to implement sustainable geoenvironmental engineering land management practices. Once again, whilst it is recognized that depletion of non-renewable resources and threats to renewable resources render absolute sustainability an impossible goal, it is nonetheless necessary and important to undertake measures for protection of the geoenvironment and its natural capital and resources – and where necessary, to restore the geoenvironment to its pre-impact state. Failure to do so will exacerbate the conditions that have already led to a compromised geoenvironment. As with all the chapters in this book, the actions discussed and proposed in these chapters recognize the need to strive for measures and actions that will relieve the adverse pressures and stresses on the geoenvironment.

Contamination of the land environment by hazardous substances, contaminants and non-contaminants results primarily from man-made activities and events mounted to meet societal and industrial demands. Contamination from natural events can also occur. These include, for example, deposition of ash from volcanic discharge, and seepage of sulphuric acid and iron hydroxide when pyrite (FeS_2) is exposed to air and water – according to the following relationship:

$$4\,FeS_2 + 15\,O_2 + 14\,H_2O \leftrightarrow 4\,Fe\,(OH)_3 + 8\,H_2SO_4 \tag{2.1}$$

We can encounter more dramatic discharges of sulphuric acid in the phenomenon commonly referred to as *acid mine drainage* (AMD). In this instance, the contamination is considered to be associated with anthropogenic activities associated with metalliferous mining – as will be discussed in a later section and in more detail in Chapter 5.

By and large, contamination due to anthropogenic activities is without doubt the greatest contributor to overall contamination of the environment and especially the geoenvironment. Accordingly, we will pay particular attention to this problem in this chapter. We use the term *contamination* to include contamination by contaminants, toxicants, hazardous substances, and all substances

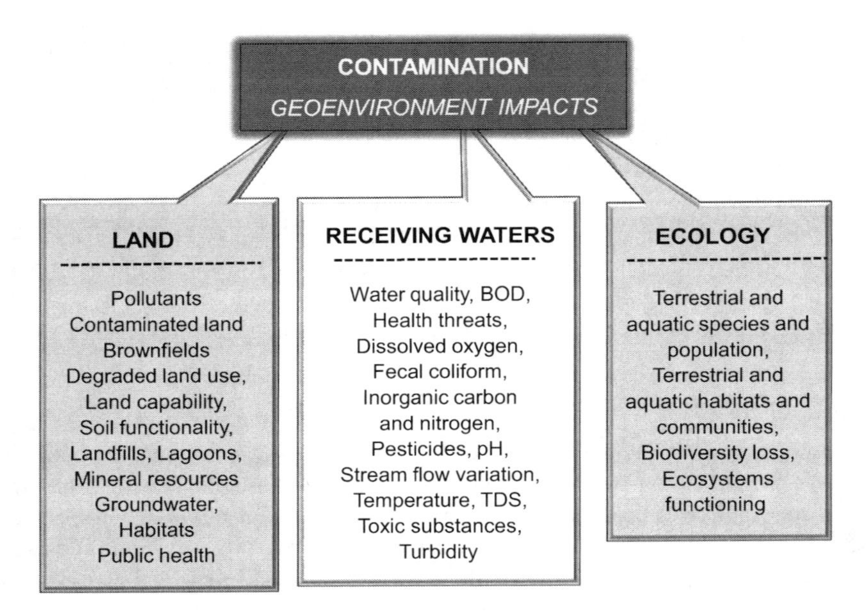

FIGURE 2.1 Simple schematic showing some of the impacts, articulated as concerns and issues, for *land*, *receiving waters*, and *ecological system* because of contamination of the land environment.

foreign to the natural state of the particular site and micro-environment. When necessary, the terms *contaminants, toxicants*, and *hazardous substances* will be used to highlight the substance under discussion or to lay emphasis to the problem. It (contamination) poses the most significant challenge in the maintenance and protection of the many land environment ecosystems and resources needed to support life on earth. Figure 2.1 shows a simple tabular sketch of some of the effects of impacts from contamination of the land environment – illustrated as concerns and issues, for *land, receiving waters*, and *ecological system.*

Because of the far-reaching primary and secondary effects of contamination of the land environment, and because these

- impact severely on our ability to implement sustainability practices for the geoenvironment,
- directly affect our means to sensibly and responsibly exploit the natural resources in the geoenvironment, and
- diminish the quality of the land environment and the very resources needed for mankind to sustain life,
- pose direct health threats to humans and other biotic receptors,
- degrade soil quality and reduce the ability of the soil to function as a resource material,

the discussions in this chapter will pay particular attention to the problem of *soil contamination due to chemical stressors.* The requirements and measures needed to mitigate contamination impacts and the remedial actions required for recovery of soil functionality will be discussed in the later chapters of this book.

In addition to the very significant problem of contamination of the land environment, there are other stressor impacts to the geoenvironment that have considerable effect on the functioning of the land environment itself. The sources of these stressors originate from a group of man-made events that have substantial physical impacts on the geoenvironment landscape features – as seen in section 2.3. The projects and events included in this group, identified as the primary group, deal directly with the landscape features of the geoenvironment. As such, the effects or impacts arising from the stressors associated with this group of events or projects most often result in threats to public safety and loss or diminution of natural geoenvironment resources and natural capital. Examples of these are the physical sets of activities associated with mineral resources and hydrocarbon resources recovery – discussed in detail in Chapters 5 through 7.

2.3 CONTAMINATION AND GEOENVIRONMENTAL IMPACTS

Included in Title 1 of the US National Environmental Policy Act (NEPA) of 1969 given in Chapter 1 are the three specific environmental requirements that we can consider as essential in the management and control of the contamination impacts to the geoenvironment. These include

- Environmental Inventory – this is essentially an environmental audit, i.e., complete description of the environment as it exists in the area where a particular proposed (or ongoing) action is being considered. The physical, biological, and cultural environments are considered to be integral to the environment under consideration.
- Environmental Assessment – the various components included in the assessment package are
  (a) prediction of anticipated change,
  (b) determination of magnitude of change, and
  (c) application of importance or significant factor to the change.
- Environmental Impact Statement (EIS) – this is a very crucial document, which needs to be written in a format specified by the specific regulatory agency responsible for oversight of the project or event. In respect to the geoenvironment, this document must contain the proper determination of the various impacts to the geoenvironment (i.e., geoenvironmental impacts) arising from implementation of the project under question or the event being investigated. In respect to the NEPA-type response, this document contains a summary of environmental inventory and findings of environmental assessment (referred to as 102 statements, i.e., the section of NEPA relating to requirements for preparation of EIS in NEPA).

To determine the nature of impacts to the geoenvironmental, it is necessary to

- Develop a frame of reference. This is essentially a series of targets or receptors that are the recipients of the impacts. The reference frame will permit one to examine the effects of the geoenvironmental impacts in relation to the members constituting the reference frame. The following members constitute the essential elements of the reference frame:
  (a) The separate compartments (terra firma and aquatic) of the land environment. By and large, one determines the impacts of projects and events on the integrity of the landscape (including receiving waters and their boundaries).
  (b) Health of the human population and other biotic receptors in the geoenvironmental compartment (land and receiving waters). This requires examination of the impacts as threats to human health and the biotic receptors. Generally, this includes a study of waste and waste-pollutant streams, and other catastrophic phenomena arising from man's activities.
  (c) Overall health of the environment. Terrestrial and aquatic habitat and community preservation are central to the health of the environment.
- Establish a general or broad impact-identification scheme. By doing so, this allows us to look for the source of the impact. Knowledge of the stressors and their source provides one with a better appreciation of the extent and details of the impact. In looking for the stressors responsible for the impacts, it is necessary to separate the *stressor source* from the *impact*, since it is not always easy to determine what constitutes the impact.

2.3.1 Reference Frame

The geoenvironmental impact reference frame provides or specifies the targets in the total land environment and the receiving waters contained therein. It provides a framework that requires questions to be raised as to whether the actions arising from projects or events will have adverse effects on the various physical and biological elements constituting the geoenvironmental compartment. The specific case of a contaminated ground shown in Figure 2.2 (from Yong and Mulligan, 2019) is a good example of the application of the reference frame for determination of geoenvironmental impacts. In this instance, one is concerned with the impacts resulting from the presence of contaminants in the ground. We define a *pollutant* to mean a contaminant that has been identified as a threat to human health and/or the environment. These are generally toxic elements, chemicals, and compounds and are most often found in priority pollutant lists of regulatory agencies. *Contaminants*, on the other hand, are defined as substances that are not natural to the site or material under consideration. These substances can include hazardous materials or elements, toxic substances, contaminants, and all other substances that are non-threatening to human health and the environment. In other words, the term *contaminants* is used when we wish to refer to non-indigenous elements, substances, etc., found at a site or in a material under investigation.

In reference to Figure 2.2, the source of the contaminant plume at the site in question is at the ground surface – e.g., dump site, landfill, etc. For this discussion, the reasons for locating the dump site, landfill, toxic dump, etc., are not addressed, even though they result in creating the contaminant source. Our interest is directed towards the impacts generated by the presence of these sources. The terms *stressor sources* and *impacts* need to be carefully differentiated. For the particular problem shown in Figure 2.2, the "event" or "action product" responsible for the contaminant plume is the landfill itself. The *stressor source* is the leachate plume containing fugitive contaminants – obtained as a result of dissolution of the waste material contained in the landfill. The *stressors,* which are the fugitive contaminants, are chemical stressors.

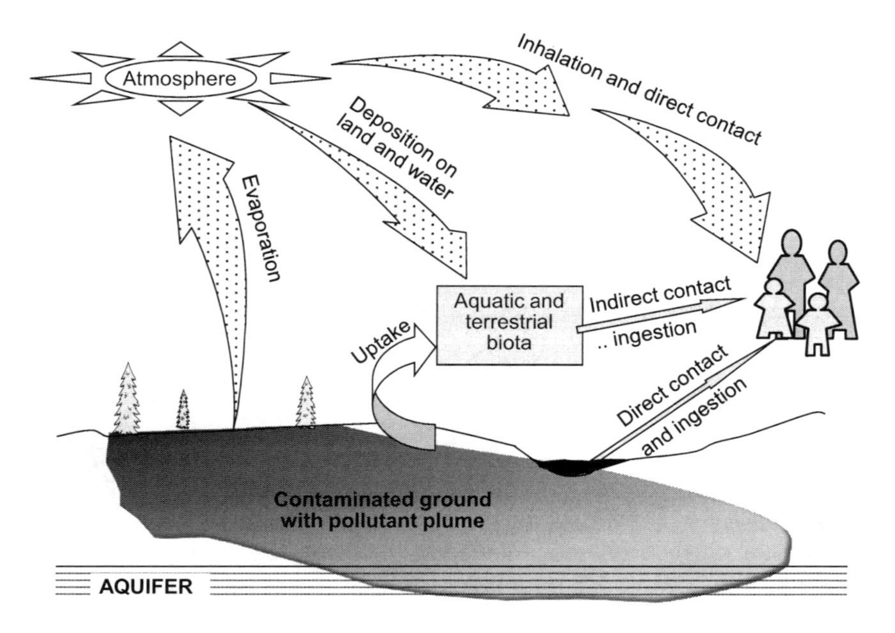

FIGURE 2.2 Schematic diagram showing contaminated ground with a pollutant plume as the source of health and environmental threats. (From Yong and Mulligan, 2019.)

Some of the impacts to the geoenvironment resulting from the contaminants in the ground shown in Figure 2.2 include

- Contaminants in the atmosphere carried into the atmosphere by evaporation and volatilization.
- Contaminants on the land surface within and outside the contaminated site, resulting from deposition of the airborne contaminants.
- Contaminants in the ground.
- Contaminated groundwater and surface water due to transport of the pollutant plume and also to deposition of contaminants in the atmosphere.
- Threats to habitats of terrestrial and aquatic biota.
- Threats to human health.

The impacts shown in the preceding list illustrate the target values of the four members (land element, receiving waters, humans, and terrestrial and aquatic biota) of the reference frame. By using these as targets, one determines not only *what is being impacted*, but also the nature of the impact. There are obviously many more impacts that can be cited. We can, for example, discuss or speculate on the impact of the contaminants and contaminants in the contaminated ground on the possible loss of biodiversity in the affected region, and the impact on the quality of the land and how this affects future land use. These impacts can be covered in the impact statements that need to be produced or developed in association with specific projects and events. These will be discussed in greater detail at various times throughout the book when specific projects and/or events are considered.

2.3.2 Characterization of Geoenvironmental Impacts

In addition to the geoenvironmental impact frame of reference, there needs to be a mechanism or means or criteria for one to determine *what constitutes an impact to the geoenvironment*. The group of questions that need to be addressed include

(1) whether reversibility (of damage) should be used as one of the decision mechanisms for determination of impacts, and
(2) how man-made improvements, amelioration, mitigation, and remediation procedures to the geoenvironment can be factored into the process of evaluation of geoenvironmental impacts and their effects.

There are many ways in which impacts to the geoenvironment can be categorized or classified. A useful and popular method is to categorize the geoenvironmental impacts in relation to natural or man-made causes, leading to events that are determined to be responsible for the impact – as shown in Figure 2.3. Described within the two major categories shown in the figure are some typical causes for events leading to geoenvironmental impacts. *Floods* and *landslides* have been singled out as typical examples of both natural and man-made causes for events, resulting in geoenvironmental impacts. For example, floods can arise naturally because of hurricanes or tsunamis generated by earthquakes, as, for example, shown by the massive floods caused in late summer of 2005 in the gulf region of Central and North America and in the US Eastern seaboard in late 2012. Floods can also occur naturally because the natural waterways (rivers, brooks, and streams) do not have the capability to carry the excessive water load produced by an undue rainfall occurring over a very short period of time. It is also possible for floods to occur because of man-made waterway constrictions and shoreline alterations that impair previously capable performance of the waterways.

Figure 2.3 shows that the anthropogenic events or stressor sources have been divided into two groups. The *primary* group refers to those anthropogenic sources for stressors are directly associated

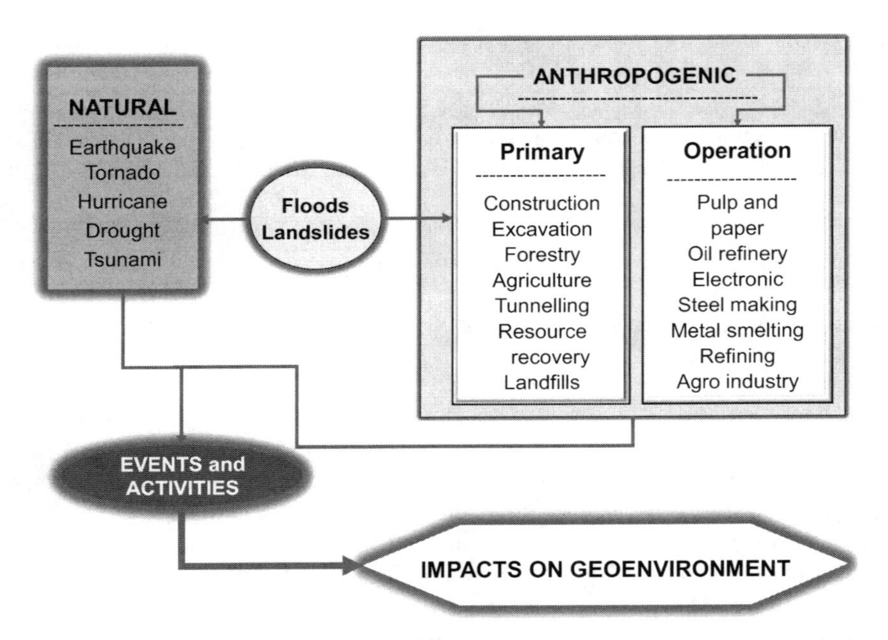

FIGURE 2.3 Categories of some typical stressor sources. Note that flood and landslides can be both natural and man-made causes for events resulting in geoenvironmental impacts.

with the physical landscape of the geoenvironment. The stressors generated are the result of various kinds of anthropogenic activities in support of physical projects, such as excavations and mining, construction of infrastructure and buildings, resource recovery, drilling, tunnelling, and waste landfills. The immediate geoenvironmental impacts are mostly physical in nature. These generally involve alteration of the surficial and subsurface landscape features. Whilst these landscape alterations are evident as physical impacts of the land surface features, they also have the ability to serve as sources for stressors – as shown previously in respect to the example of acid mine drainage (AMD) and the impact generated from physical extraction of metal ores from the ground. In this case, the AMD problem poses a very significant threat to the receiving waters and ultimately to the biotic receptors. Another less dramatic health-threat example is the undercutting of a slope to facilitate the construction of a right-of-way for a highway system. Excessive undercutting without proper analysis of the stability of the slope could produce a situation where the undercut slope could subsequently trigger a slope failure. When such occurs, one needs to be concerned with the safety of the human and animal populations in the affected region.

The other grouping for anthropogenic stressor sources shown in Figure 2.3 is identified as *operations*. The impacts to the geoenvironment arising from this stressor-source group (*operations*) are the direct result of activities associated with midstream and upstream industries, such as the agro industry (outside of physical cultivation of the land), refining, mineral dressing, manufacturing, production and process industries, etc. The effects or results of the impacts on the geoenvironment can be physical, chemical, physico-chemical, and biogeochemical in nature. One needs to factor into the analysis the resultant or potential threats to human health and other biotic receptors. Some examples of these are (a) application of pesticides and fungicides as pest controls in support of agricultural activities leading to non-point source contamination of ground and groundwater, (b) landfilling of hazardous wastes resulting in the production of contaminants in the fugitive leachate plumes, (c) discharge of waste streams from chemical and electronic industries resulting in contamination of the receiving waters, and (d) isolation-disposal of high-level nuclear wastes in underground repositories.

2.3.3 Identifying and Assessing for Impact on the Geoenvironment

A useful procedure for performing a geoenvironmental impact scoping exercise is given in Figure 2.4. The various steps shown in the diagram are guidelines and are designed to provide one with specific objectives or targets.

- Stressor Sources – knowledge of the stressor sources (i.e., cause or events) can be helpful in narrowing the field of study or investigation. Take, for example, the case of a chemical plant producing various organic chemicals where inadvertent spills and fugitive discharges during storage are suspected to have occurred. If the spills and discharges are identified as the sources, the stressors associated with the sources would be the chemicals involved in the spills or discharges. The impacts to the geoenvironment can be readily identified.
- Nature of Impacts – it is important to have a proper knowledge of the kind of impact. This allows determination or estimation of the extent of the *damage* or *improvement* of the geoenvironment resulting from the applied stressors. Using the previous chemical plant example, the damage inflicted on the geoenvironment is seen in terms of a contaminated site or ground.

Guidelines are necessary to prevent one from rendering judgements on impacts without methodical and proper analyses because the results of the impacts are dramatically visible – as opposed to impacts that do not show visible distress signs. It is not the magnitude of the distress caused by the impact that should determine what constitutes an impact. As we have pointed out in the previous chapter, the results of many impacts do not manifest themselves until many years hence. This is especially true for health-related issues. In the most general sense, the guidelines used to determine what constitutes an impact to the geoenvironment should be determined on the basis of whether the geoenvironmental impact will

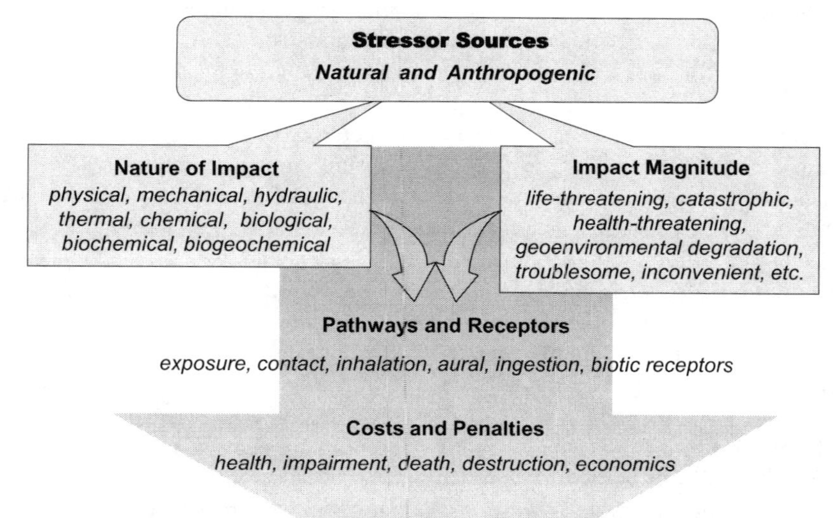

FIGURE 2.4 Useful protocol for geoenvironmental impact scoping exercise. Scoping for impact assessment or determination can omit the last step (remedial actions) if a quick assessment is needed. The last step is needed if one has the ability to influence decision-making concerning implementation of the project or event.

- Generate direct and/or indirect threats and problems relating to public health, natural habitats, and the environment. A good case in point is the contamination of receiving waters. These waters serve as habitats for aquatic species and in many instances, will serve as sources for drinking water. Not only is such an impact a direct threat to the usable water supply for the human population, it is also a direct threat to the food supply for the same population because of the likely reduction in aquatic food supply for the population.
- Diminish the functioning of the ecosystems in the geoenvironment. An example of this can be found in the degradation of soil quality due to many of the activities associated with high-yield agriculture and mineral resources' exploitation. This is particularly important since *soil quality* is a direct measure of the capability of a soil to sustain plant and animal life and their productivity within their particular natural or man-managed ecosystem. The Soil Science Society of America (SSSA) defines soil quality as "The capacity of a specific kind of soil to function, within natural or managed ecosystem boundaries, to sustain plant and animal productivity, maintain or enhance water and air quality, and support human health and habitation", (Karlen et al., 1997). Any diminution of soil quality will impact on the capability of the soil to provide the various functions, such as plant and animal life support, forestry and woodland productivity, and will also result in the loss of biodiversity and nutrients. Other prime examples of this can be found in the various activities associated with mineral extraction and energy resource development. To appreciate the impacts resulting from these activities, we can focus on the status of the biological, chemical, physico-chemical activities, and physical interactions that define the functioning of the ecosystem of interest. In addition, we can also study the changes in land capabilities or land use options.

2.3.4 Man-Made and Natural Combinations

It is not always easy or simple to distinguish between natural and anthropogenic stressor sources that impact directly or indirectly on the geoenvironment. This is because many geoenvironmental impacts are the result of a sequence combination of anthropogenic and natural sources. A very good example of this is the previously mentioned acid mine drainage (AMD) problem – a problem that is triggered by the results of mining exposure of pyrite. The presence of pyrite (FeS_2) in rock formations where coal and metalliferous mining occurs will create problems for the environment if the pyrite is exposed to both oxygen and water. Given favourable geologic and hydrologic conditions, we have the situation where oxidation of the pyrite exposed during mining operations will produce ferrous iron (Fe^{2+}) and sulphate (SO_4^{2-}). For this first chemical reaction step, we can conclude that the trigger for the first sets of reactions is a man-made event or source, i.e., mining of the rock formation. Subsequent rate determining reactions, which may or may not be catalyzed by certain bacteria (e.g., *Thiobacillus ferrooxidans*), involve oxidation of the ferrous iron (Fe^{2+}) to ferric iron (Fe^{3+}), to be followed later by hydrolysis of the ferric iron and its ensuing precipitation to ferric hydroxide [$Fe(OH)_3$] if and when the surrounding pH goes above 3.5. Throughout these processes, hydrogen ions are released into the water, thereby reducing the pH of the surrounding medium. The sum total of the reaction products and the reducing pH condition is commonly known as the acid mine drainage problem (AMD). This problem is a significantly large problem because of the many sets of mining activities conducted all over the world, and particularly because of the presence of pyrite in many of these mines. In Canada, elevated levels of arsenic in soil, sediments, surface, and groundwater have been related to mining, metal smelting, wood preservation, and power generation (Wang and Mulligan, 2006). Acid contamination of groundwater and other receiving waters creates conditions that are adverse to human health and other biotic species. Chapter 4 will discuss the problem and impact of AMD in greater detail – together with procedures for amelioration of AMD.

Another example of contamination of groundwater is the case of the arsenic contaminated aquifers in West Bengal and Bangladesh. These aquifers serve a significant portion of the population of these two countries, and ingestion of water from these contaminated aquifers has led to the development of arsenicosis in thousands of unfortunate individuals. Tube wells sunk into the aquifers constitute a major drinking water supply source for the two countries. Investigations on tube-well water supply showed concentrations of arsenic far in excess of allowable limits. Tests on these wells at time of installation showed arsenic concentrations, if any, well below threat levels. The present levels of arsenic concentration indicate arsenic poisoning developed several years after installation of the wells. By switching to groundwater resources to avoid waterborne diseases from ingestion of surface water supplies, the population now faces considerable risk of arsenic poisoning as a result of ingesting arsenic-contaminated water obtained from tube wells tapping into the groundwater.

If the aquifers were not contaminated by arsenic before extensive harvesting of the aquifer resource, how did the arsenic get there? Is the arsenic contamination due to natural causes? Or indirectly due to a man-made cause? Arsenic occurrence in the hard rock and sedimentary rock aquifers have been reported in Argentina, Nepal, Nigeria, Czech Republic, and many other countries. Whilst it is not uncommon to find small concentrations of arsenic in the groundwater in many parts of the world, the arsenic concentrations in the samples obtained from the aquifers in Bangladesh and West Bengal, for instance, are too high to be attributable to natural processes for arsenic release from arsenopyrites. In fact, the arsenic concentrations in the Bangladesh and West Bengal aquifers are also too high to be accounted for by direct human activities, such as those associated with metalliferous mining. There appears to be no doubt that the release of arsenic into the aquifers is from geologic source materials. Why is arsenic being released from the source rocks? Is this a natural process or is the release of arsenic triggered by some man-made event?

In the Bangladesh case, two factors appear prominently: (1) Presence of arsenopyrites and arseniferrous iron oxyhydroxides in the substrate material, and (2) the use of tube wells as a means for abstracting water. The two models proposed to explain the release of arsenic from the arsenic-bearing materials are shown in Figure 2.5. These include

(a) reduction mechanisms – reductive dissolution of arseniferrous iron oxyhydroxides releases the arsenic responsible for contamination of the groundwater, and

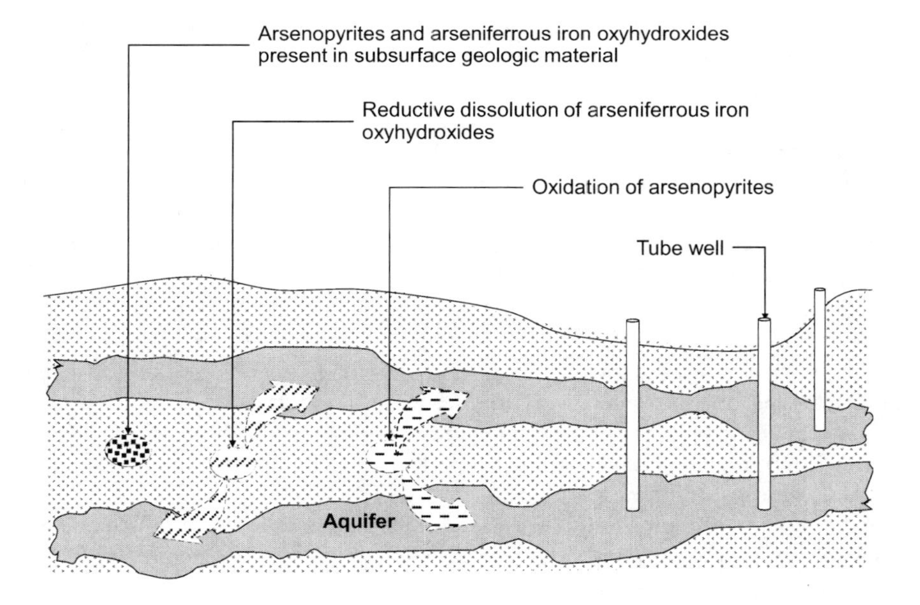

FIGURE 2.5 Schematic of speculative models for release of arsenic from arseniferrous iron oxyhydroxides and arsenopyrites into aquifers.

(b) oxidation processes – oxygen invades the groundwater because of the lowering of the groundwater from the abstracting tubewells resulting thereby in the oxidation of the arsenopyrite (FeAsS).

Based on these two possible mechanisms for arsenic release into the aquifer and based on detailed field studies to determine the presence and distribution of As(III) and As(V) and other reaction products, there are some serious questions as to whether the arsenic releases processes are totally man-induced or a case of natural processes hastened and aggravated by man-made events, i.e., abstraction of water from the tube wells. The almost equal proportions of As(III) and As(V) in the aquifer testify to almost equal sets of activity from both arsenic-release mechanisms.

2.4 WASTES, CONTAMINANTS, AND THREATS

The discharge of liquid and solid wastes from industrial and energy producing plants and facilities, together with inadvertent spills and deliberate dumping of waste materials, combine to introduce contaminants, toxicants, and other kinds of hazardous substances into the land environment – including the receiving waters contained therein. The term *contaminants* used in this section includes those contaminants that have are classified as pollutants (priority listing) by various regulatory agencies. The term *pollutant* or *pollutants* will be used when necessary, to place emphasis on the contaminant or contaminants in question. These pose significant threats not only to human health and other biotic receptors, but also to the health of the environment and the various ecosystems in the geoenvironment. In the treatment of threats to the environment and public health, there are several ways in which wastes and contaminants can be classified or categorized. One could categorize the wastes and contaminants in terms of source production, i.e., "*where they come from*". Alternatively, one could categorize them in respect to

- level of toxicity, e.g., highly toxic, carcinogenic, priority listing, etc.
- inorganic or organic substances and chemicals, e.g., heavy metals, polycyclic aromatic hydrocarbons (PAHs), perchloroethylene (PCE), etc.
- type of industry (source industry or source activity), e.g., pulp and paper, forest, electronic, pharmaceutical, etc.
- class of contaminants, e.g., pesticides, solvents, etc.
- nature of impact or threat, e.g., physical, chemical, biological, etc.
- type of receptor, e.g., land, water, human, other biota, etc.

Experience has shown that although categorization of contaminants according to any single method as described above is not practical or feasible, it is always necessary to obtain a proper identification of the types or species of contaminants. Considerable significance is placed on the potential health threat of contaminants. A popular approach that has gained the attention of many regulatory agencies is the SPR (source–pathway–receptor) method for determination of health threats and impacts created by the presence of contaminants in the geoenvironment and also by events or projects as source-contaminants. There have been questions raised as to whether the SPR approach discriminates between levels of treatment or protection from health-threat events depending on the importance of the *receptors*. A school of thought suggests very strongly that risk management should be directly linked to receptor importance and also to certainty of pathways. It is not always clear that pathways to potential receptors are well-defined. Nevertheless, one must include knowledge of the degree of certainty of pathways as an integral factor in the level of risk determination.

2.4.1 Inorganic Contaminants

Evidence of the presence of inorganic contaminants classifying as pollutants in the geoenvironment (land and water) show that these are mainly heavy metals, such as Pb, Cr, Cu, etc. Yong (2001) has indicated that whilst those elements with atomic numbers higher than Sr (atomic number 38) are classified as *heavy metals* (HMs), it is not uncommon to include elements with atomic numbers greater than 20 as heavy metals. The 38 elements commonly considered as HMs fall into three groups of atomic numbers as follows:

- from atomic numbers 22 to 34 – Ti, V, Cr, Mn, Fe, Co, Ni, Cu, Zn, Ga, Ge, As, and Se;
- from 40 to 52 – Zr, Nb, Mo, Tc, Ru, Rh, Pd, Ag, Cd, In, Sn, Sb, and Te,
- from 72 to 83 – Hf, Ta, W, Re, Os, Ir, Pt, Gu, Hg, Tl, Pb, and Bi.

The more common HMs found in the geoenvironment come as a result of anthropogenic activities, such as management and disposal of wastes in landfills, generation and storage of chemical waste leachates and sludges, extraction of metals in metalliferous industries, metal plating works, and even in municipal solid wastes. The more notable HMs include lead (Pb), cadmium (Cd), copper (Cu), chromium, (Cr), nickel (Ni), iron (Fe), mercury (Hg), and zinc (Zn).

Recently, using microorganisms, remediation technologies for metal and related contaminants in soils are investigated. Some of the studies are introduced in a later chapter. In this chapter, typical metal pollutants are explained in terms of their properties related to environmental threats.

2.4.1.1 Arsenic (As)

Strictly speaking, arsenic is a non-metal – although it is often classified as a metal. It is a metalloid (semi-metal) with atomic number 33, and is in Group 5 of the Periodic Table. Arsenic is found naturally in rocks, most often in iron ores and in sulphide form as magmatic sulphide minerals. The more common ones are arsenopyrite (FeAsS), realgar (AsS), nicolite (NiAsS), and orpiment (As_2S_3). Arsenic is also found naturally in soils in association with hydrous oxides, and sometimes in elemental form in association with gold and silver ores. Arsenic found in the geoenvironment can come directly from weathering of the arsenic-containing rocks and also from industrial sources, such as manufacturing, processing, pharmaceutical, agriculture, and mining industries. Products such as paints, dyes, preservatives, herbicides, and semi-conductors are some of the more common contributors to the arsenic found in the ground and in receiving waters. Extensive use of arsenic-containing (lead arsenate) pesticides, herbicides, and insecticides in agricultural and farm practices can contribute some considerable amounts of arsenic to the subsurface and the receiving waters.

Whilst the existent valence states for arsenic are –3, 0, +3 (arsenite), and +5 (arsenate), arsenite and arsenate are the more common forms of arsenic found in nature. Arsenic is a toxic element, and a regulatory limit of 50 μg/L in groundwater (aquifers) for drinking water has been adopted in many countries and regulatory agencies. In the United States, this is lowered to 10 μg/L for all water systems by 2006. Ingestion of arsenic for a period of time, as, for example, in the use of arsenic-contaminated waters in some regions of the world, can lead to serious health problems – e.g., mortality from hypertensive heart disease traceable to ingestion of arsenic-contaminated drinking water (Lewis et al., 1999), and arsenic-associated skin lesions of keratosis and hyperpigmentation (Mazumder et al., 1998). Similarly, inhalation of arsenic dust generated in ore refining processes can also lead to serious health problems – e.g., nasal septal perforation, and pulmonary insufficiency (U.S. EPA, 1984).

2.4.1.2 Cadmium (Cd)

Cadmium can be found in nature as greenockite (cadmium sulphide, CdS) or otavite (cadmium carbonate, $CdCO_3$) and is usually associated with zinc, lead, or copper in sulphide form. The two major groups using cadmium include (a) cadmium as a filler, alloy, or active constituent for an industrial product, e.g., nickel-cadmium batteries, enamels, fungicides, phosphatic fertilizers, motor oil, solders, paints, plastics, etc., and (b) cadmium as a coating or plating material, e.g., steel plating, metal coatings. The presence of Cd as a pollutant in the geoenvironment can be traced to

- non-point sources associated with the use of fungicides and fertilizers,
- deposition of Cd particles in the atmosphere because of mining activities and burning of coal and other Cd-containing wastes, and
- specific sources such as industrial discharges and wastes, and municipal wastes where the products manufactured and consumed include Cd as a filler, alloy, or active constituent.

From the viewpoint of human health effects and requirements, cadmium is considered to be a non-essential element. The EPA TCLP (toxicity characteristic leaching procedure) regulatory level for Cd is 1.0 mg/L (roughly equivalent to 1 ppm). The EPA specifies a threshold limit of 5 ppb for drinking water and the FDA (Food and Drug Administration) specifies a limit of 15 ppm of Cd in food colouring (ATSDR, 1999). Accumulation of Cd in the liver and kidney from oral ingestion can lead to distress to these organs.

2.4.1.3 Chromium (Cr)

Chromium is found naturally as chromite (ferrous chromic oxide, $FeCr_2O_4$) and crocoisite (lead chromate, $PbCrO_4$) minerals. It is an essential element in human nutrition. The three common valence states for chromium are 0, +3, and +6, i.e., chromium (0), chromium (III) chromium (VI). Chromium (III) is found naturally in the environment, whereas compounds of chromium are generally with chromium (VI). Trivalent chromium (chromium (III)) is stable and is considered to be relatively non-toxic Cr(III) can form various stable, inert complexes. $Cr(H_2O)_6^{3+}$, $Cr(H_2O)_5(OH)^{2+}$, $Cr(H_2O)_3(OH)_3$, and $Cr(H_2O)(OH)_4^-$. On the other hand, hexavalent chromium (chromium (VI)) is highly toxic and is considered to be a carcinogen. Oxidation of the trivalent chromium to the hexavalent chromium anions chromate (CrO_4) and dichromate (Cr_2O_7) anions will not only render the previously non-toxic trivalent chromium toxic, but will also make it more mobile. The major form of Cr(VI) is CrO_4^{2-} at pH greater than 6.5 and $HCrO_4^-$ at pH less than 6.5. Both ions are very soluble.

Other than the natural sources, chromium found in the geoenvironment can be traced to waste discharges and tailing ponds associated with chromium mining. Principal uses for chromium (Cr) and its compounds include (a) use of chromium as alloys, with iron and nickel – stainless steel and super alloys as probably the best known alloys, (b) chromium compounds used in metal plating, tanning of hides, wood preservation, glass and pottery products, and (c) production of chromic acid. Chromium in the land and aquatic compartments of the geoenvironment can be the result of production and waste discharges associated with the industries, and from the tailing ponds associated with mining activities.

2.4.1.4 Copper (Cu)

Copper is found naturally in sandstones and in other copper-bearing oxidized and sulphide ores. These include such ores as malachite [$Cu_2(CO_3)(OH)_2$], tenorite (CuO), cuprite (Cu_2O), and chalcopyrite ($CuFeS_2$) – with chalcopyrite being the most abundant. In addition to its natural occurrence in the land environment, contributions of Cu to the geoenvironment come from (a) deposition of airborne particles from mining of copper and combustion of fossil fuels and wastes, (b) discharges from industrial processes utilizing copper as a metal, and copper compounds (production of electrical products, piping, fixtures, and different alloys), and (c) industrial and domestic discharge of waste water.

Copper deposited on the surface of the land environment from the various sources discussed in the preceding will initially be attached to organic matter and clay minerals – if such are present in the landscape. Degradation of the organic matter through anaerobic or aerobic means will release copper in its monovalent or divalent form, respectively. However, if the subsurface soils contain reactive soil particles, the released copper will be bound to these particles. Environmental mobility of copper in the substratum is not generally a big factor when the soil substratum is composed of fine soil fractions consisting of clay minerals and other soil fractions with reactive particles. The presence of copper in the receiving waters is most often confined to the sediments since that copper will attach itself to the fine particles in water.

In terms of health considerations, copper is considered to be an essential trace element in both human and animal nutrition. The amounts required, however, are extremely small. Threshold limits for human ingestion of copper vary between different countries and jurisdictions, with values of about 1.3 ppm for drinking water and 0.1 mg/m^3 for airborne concentrations being reported.

2.4.1.5 Lead (Pb)

Lead is found in nature in sulphide, carbonate, and oxide forms. These are galena (lead sulphide, PbS), anglesite (lead sulphate, $PbSO_4$), cerrusite (lead carbonate, $PbCO_3$) and minium (lead oxide, Pb_3O_4). Although it has three valence states (0, +2, and +4), the most common state is +2. Compounds of Pb(II) have ionic bonds whereas the higher valence state, Pb(IV) compounds, have covalent bonds. Lead found in the land compartment of the geoenvironment will most often be bonded to reactive soil particles. It is a non-essential element.

Lead is used to a very large extent in the manufacture of lead-acid batteries and in the electronics and munitions industries. Other lesser uses for lead are in production of crystal glass, lead liner material, weights, insecticides, and in construction. Lead found in the ground and in the receiving waters can be traced to deposition of airborne lead obtained from emissions, such as burning of wastes and fuel, and from transport of lead compounds in the soil. Lead is considered to be a non-essential toxic element, and ingestion or inhalation of lead will result in consequences to the central nervous system and damages to kidneys and reproductive system.

2.4.1.6 Nickel (Ni)

Nickel is quite widely found in nature in various soil deposits – e.g., laterite deposits, and generally in mineral form in combination with oxygen or sulphur as oxides or sulphides. These include nickel sulphide (NiS), nickel arsenide (NiAs), nickel diarsenide ($NiAs_2$), and nickel thioarsenide (NiAsS). Some debate exists concerning whether nickel is, or is not, an essential element. It is maintained that small amounts of nickel are essential for maintaining the good health of animals, and to a lesser extent, humans. For example, the urease-producing bacteria need two Ni elements at the active centre of the enzyme, for urease activity, which is important in terms of microbial-induced carbonate precipitation, MICP (discussed further in Chapters 13 and 14) and microbial remediation of soil contaminants with heavy metals (in Chapter 11).

Nickel found in the geoenvironment, other than from natural sources, can be traced to fugitive atmospheric nickel and waste discharge associated with nickel mining activities, burning of waste, operation of oil-burning and coal-burning power plants, discharges from manufacturing industries using nickel alloys and compounds. Nickel does not precipitate but is sorbed onto clays, oxides of manganese and iron, and organic material occurs. Mobility increases with the formation of complexes with organic and inorganic ligands.

2.4.1.7 Zinc (Zn)

Similar to nickel, zinc is found in soil deposits and does not generally exist as a free element. Instead, it is found in mineral form in combination with oxides, sulphides, and carbonates to form zinc compounds. The sulphide form is perhaps the more common form for zinc found naturally in the environment. Some of the naturally-occurring zinc compounds are zincite (zinc oxide, ZnO), hemimorphite (zinc silicate, $2ZnO{\cdot}SiO_2H_2O$), smithsonite (zinc carbonate, $ZnCO_3$), and as sphalerite (zinc sulfide, ZnS). Natural levels of zinc in soils are 30 to 150 ppm.

Typical uses and products for zinc in element form as oxide and sulphide compounds include alloys, batteries, paints, dyes, galvanized metals, pharmaceuticals, cosmetics, plastics, electronics, and ointments. It follows that non-naturally occurring zinc found in the geoenvironment would be the zinc compounds associated with the production and use of industrial products. Deposition of airborne fugitive zinc from mining and extraction of zinc, together with discharges (spills, wastes, and waste streams) from the processing and production of products utilizing zinc compounds account for the major sources of non-naturally occurring zinc found in the geoenvironment. Some of these sources are galvanizing plant effluents, coal and waste burning, leachates from galvanized structures, natural ores, and municipal waste treatment plant discharge.

Although not as toxic as cadmium, zinc is quite often associated with this metal. Under acidic conditions below its precipitation pH, zinc is usually divalent and quite mobile. In the divalent state, sorption onto the surfaces of reactive soil particles includes ionic bonding and sequestering by organic matter. At high pH, the solubility of its organic and mineral colloids can render zinc bioavailable. Zinc hydrolyses at pH 7.0 to 7.5, forming $Zn(OH)_2$ at pH values higher than 8. Under anoxic conditions, ZnS can form upon precipitation, whereas the unprecipitated zinc can form $ZnOH^+$, $ZnCO_3$, and $ZnCl^+$.

2.4.2 Organic Chemical Contaminants

Organic chemical contaminants found in the geoenvironment classify as organic chemical pollutants. These have origins from (a) industries producing various chemicals and pharmaceuticals, e.g., refineries, production of specialty chemicals, etc. (b) waste streams and disposal of chemical products, e.g., sludges and spills, and (c) utilization of various chemical products, e.g., use of petroleum products, pesticides, organic solvents, paints, oils, creosotes, and greases, etc. There are at least a million organic chemical compounds registered in the various chemical abstracts services available, with many thousands of these in commercial use. By and large, organic chemicals found in the geoenvironment can be traced to sources and activities associated with humans. Figure 2.6 shows some of the main sources of contaminants (inorganic and organic) found in the land compartment of the geoenvironment.

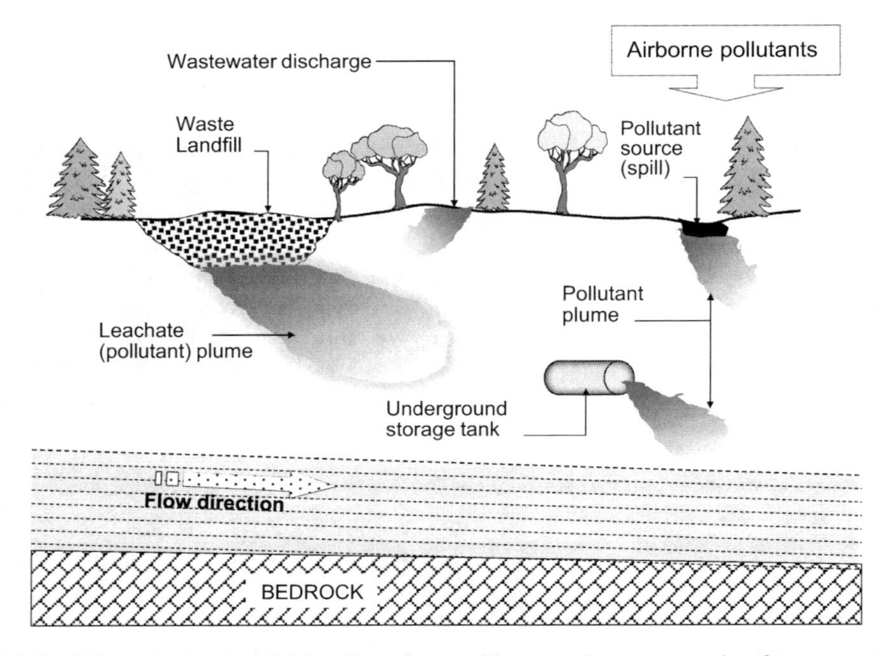

FIGURE 2.6 Schematic showing (a) leachate plume with contaminants emanating from a waste landfill, (b) pollutant plumes from a leaking underground storage tank and wastewater discharge, (c) deposition of airborne contaminants onto land surface, and (d) a surface pollutant source (spill). (Adapted from Yong and Mulligan, 2019.)

The more common organic chemicals found in the physical landscape of the land environment can be grouped as follows:

- Hydrocarbons – including the PHCs (petroleum hydrocarbons), the various alkanes and alkenes, and aromatic hydrocarbons such as benzene, MAHs (multicyclic aromatic hydrocarbons), e.g., naphthalene; and PAHs (polycyclic aromatic hydrocarbons), e.g., benzo-pyrene.
- Organohalide compounds – of which the chlorinated hydrocarbons are perhaps the best known. These include TCE (trichloroethylene), carbon tetrachloride, vinyl chloride, hexachlorobutadiene, PCBs (polychlorinated biphenyls), and PBBs (polybrominated biphenyls).
- Oxygen-containing and nitrogen-containing organic compounds, such as phenol, methanol, and TNT (trinitrotoluene).

Not all of the organic chemical contaminants are soluble in water. Those that are not, are identified as non-aqueous phase organics. Separation of the non-aqueous phase organic compounds into two classes which distinguish between whether they are lighter or denser than water is useful because it tells one about the transport characteristics of the organic compound. These non-aqueous phase organics are called non-aqueous phase liquids (NAPL), and the distinction between the lighter than water and heavier than water is given as LNAPL and DNAPL, respectively, i.e., the LNAPLs are lighter than water and the DNAPLs are heavier than water. The schematic diagram in Figure 2.7 shows that because the LNAPL is lighter than water, it stays on the surface of or above the water table. Because the DNAPL is denser than water, it will sink through the water table and will come to rest at the impermeable bottom (bedrock). Some typical LNAPLs include gasoline, heating oil, kerosene, and aviation fuel. DNAPLs include the organohalide and oxygen-containing organic compounds, such as 1,1,1-trichloroethane, creosote, carbon tetrachloride, pentachlorophenols, dichlorobenzenes, and tetrachloroethylene.

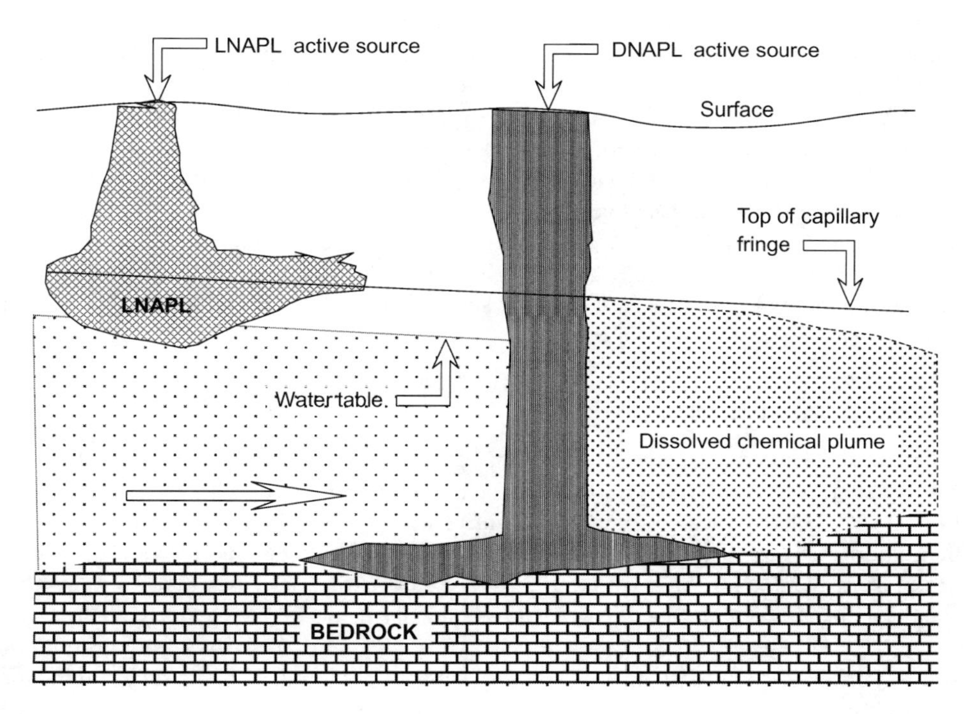

FIGURE 2.7 Schematic diagram showing LNAPL and DNAPL in the subsurface. Note that the LNAPL stays above the water table whereas the DNAPL penetrates into substratum and rests on the impermeable rock bottom. (From Yong, 2001.)

Pollutants that interfere with the endocrine system are called endocrine disruptors. Endocrine disruptors, chemicals that can replace or simulate hormones in animals or humans through water consumption, often not removed in wastewater treatment plants. They either mimic hormones in the body, affect the release of the hormones or block the hormone target site. Estrogen, taken for birth control or during menopause, is not removed during wastewater treatment and is transported to surface water (Pepper et al., 2006). Other endocrine disruptors are bisphenol A (PBA) from plastics and phthalates. Various pesticides such as atrazine are also known as endocrine disruptors (Pepper et al., 2006). These compounds can then accumulate in sediments, affecting the benthic organisms.

2.4.2.1 Persistent Organic Chemical Pollutants (POPs)

Not all the organic chemicals found in the geoenvironment will be biologically and chemically degraded. The characteristic term used to describe organic chemicals that persist in their original form or in altered forms that pose threats to human health, is *persistence*. The acronym used to describe persistent organic chemical contaminants is POPs (persistent organic pollutants). These are generally defined as organic contaminants that are toxic, persistent, and bio-accumulative. Included in the POPs are dioxins, furans, the pesticides and insecticides (aldrin, chlordane, DDT, etc.) and a whole host of industrial chemicals grouped as polycyclic aromatic hydrocarbons (PAHs), and halogenated hydrocarbons (including the chlorinated organics). The top 12 POPs that have been identified by the United Nations Environmental Programme as POPs for reduction and elimination are dioxins, furans, PCBs, hexachlorobenzene, aldrin, dieldrin, endrin, chlordane, DDT, heptachlor, mirex, toxaphene. The majority of the top 12 POPs are pesticides. Because of their heavy use in agriculture, golf courses, and even at the household level for control of insects and other pests, it is not difficult to see how these find their way into the geoenvironment. Others were added to the POP list for reduction or elimination in 2022 including perfluorohexane sulfonic acid (PFHxS), its salts, and PFHxS-related compounds.

Others include per- and polyfluoroalkyl substances (PFASs), which includes thousands of manufactured chemicals that have been used in industrial, consumer, and commercial products since the 1940s because of their useful properties in non-stick cookware, fire fighting foams, textiles, etc. Many PFASs are acids and can exist in a protonated or anionic state. Perfluorooctanoic acid (PFOA) and perfluorooctane sulfonate (PFOS), have been widely used and studied (OECD, 2013) but have been replaced due to environmental and health risks in the United States (USEPA, 2016) and other countries with other PFASs in recent years. PFASs have been found and persist over long periods of time in soil, water, air, and fish (USEPA, 2016). Soil contamination has occurred at waste disposal and manufacturing sites and where biosolids have been added to agricultural fields. Many PFASs are found in a variety of food products and have shown harmful effects and thus they are a concern for human and animal health.

2.4.3 Nano and Microplastics

According to the United Nations Environmental Program (www.unep.org/plastic-pollution), the amount of plastic waste entering aquatic ecosystems, is in the range of 19 to 23 million tonnes. The OECD (2022) has indicated that the amount of plastic waste in the environment is growing and the majority is due to inadequate disposal (82%), while 12% is due to abrasion and losses of microplastics (12%), followed by littering (5%), and marine activities (1%). In addition, the micro and nanoplastics in the water could adsorb chemicals including PAHs, PCBs, and trace metals (Engler, 2012; Teuten et al., 2007). Tire and brake wear can emit microplastic particles into the air can be transported over long distances. Therefore, humans could inhale the particles in indoor and outdoor environments (Gasperi et al., 2017; Allen et al., 2019).

2.5 SURFACE AND SUBSURFACE SOILS

Surface and subsurface soils constitute the uppermost portion of the mantle of the land environment, i.e., the unconsolidated material in the upper layer of the lithosphere. This upper layer is an integral part of the terrestrial ecosystem (Figure 2.8). When combined with the flora and fauna, this upper mantle constitutes the habitat for terrestrial living organisms. Soil, as a material, can be considered as a natural capital of the geoenvironment. There are many functions served by surface and subsurface soils. They provide the physical, chemical, and biological habitat for animals and soil microorganisms. In addition, they support growth of plants and trees, and are the vital medium for agricultural production – the virtual host for food production. Soil materials in the subsurface are very useful in the mitigation of the impacts of liquid wastes discharged on (and in) the land surface – because of their inherent chemical and physical buffering capabilities. Soil is a renewable resource that is in danger of becoming a non-renewable renewable resource (see Chapter 10).

2.5.1 Soil as a Resource Material

Soils are considered as an essential resource material. Food production, forestry, and extraction of minerals are some of the life-support activities that depend on soil. Surface and subsurface soils constitute the primary host or recipient of contaminants and contaminants. From the schematic shown in Figure 2.8, it is evident that the transport and fate of contaminants that find their way into this land compartment of the geoenvironment will be a function of (a) the properties of the soil, (b) the properties of the contaminants themselves, (c) the geological and hydrogeological settings, and (d) the microenvironment (regional controls). It is not the intent of this section or this book to deal in detail with the properties and characteristics of soils. Nor is it intended to develop the basic fundamental principles

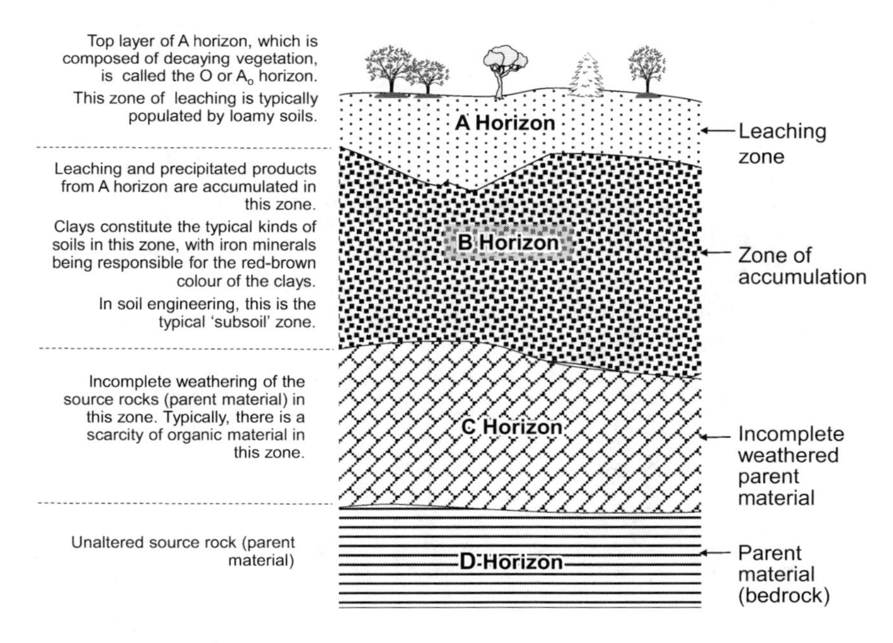

FIGURE 2.8 Soil horizons typical of a mature soil. Immature soils (soils that have not had much exposure to weathering and leaching phenomena) will not show distinct layering for zone classification. (From Yong et al., 2012.)

of contaminant–soil interactions. These can be found in Yong and Warkentin (1975), Yong (2001), and Yong and Mulligan (2004). Instead, we will focus on the aspects of soils and their interactions that help us develop a better understanding of soils as they relate to the control of contaminant fate and transport.

2.5.2 Nature of Soils

Soils are three-phased systems – solids, aqueous, and gaseous. All three phases co-exist to form a soil mass. The composition of the solids, the chemistry of the aqueous phase and the proportions of each phase at any location is a function of how the soil was formed, "*how it got there*", and the governing regional controls and climatic factors. Soils are formed by natural processes associated with the weathering of rock and the decomposition of organic matter. Weathering or disintegration of the parent rock material can be either physical or chemical. In both cases, rock disintegration will produce smaller fragments, and ultimately, soil material is formed. Unless the soil material is transported to other regions after formation, as, for example, by wind forces, by water movement or even by glacial action, the in-place and resident soil material will reflect the primary compositional features of the parent rock. Further weathering of the soil material will produce the compositional features seen in the soil in place. Weathering is at its highest intensity in the upper soil zone in temperate humid climate regions, and deeper soil zones in humid tropics. The transformations occur principally in the regolith, the region between solid rock and the topsoil.

Figure 2.9 shows the three phases (gaseous, aqueous, and solids) of soils and the various kinds of constituents that make up a soil. The various constituents in the solid phases are generally identified as *soil fractions* since each type of constituent is a fraction of the soil solids that comprise the total soil itself. The various soil fractions combine to form the natural soils that one sees at any one site. The soils may have been transported and deposited as sedimentary deposits of alluvial, fluvial, or marine action. The soils derived from these actions are appropriately called fluvial soils, alluvium, and marine soils.

Figure 2.10 shows an idealized schematic of the various soil fractions grouped into a soil unit. Not all soils have all the various fractions shown in Figure 2.9. How the proportions and distributions of

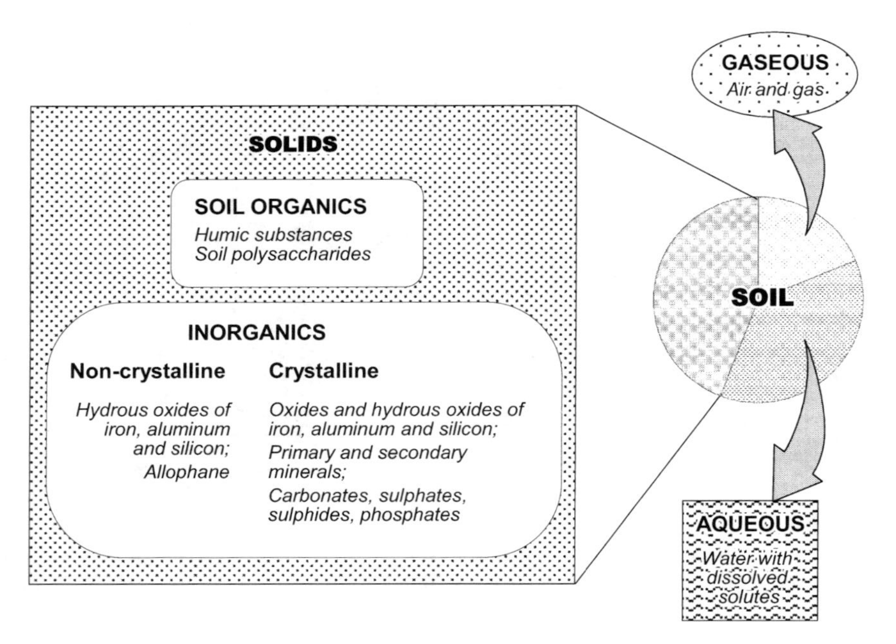

FIGURE 2.9 Soil constituents. Proportioning of the three phases (gaseous, aqueous, and solids) is approximately representative of a partly saturated soil.

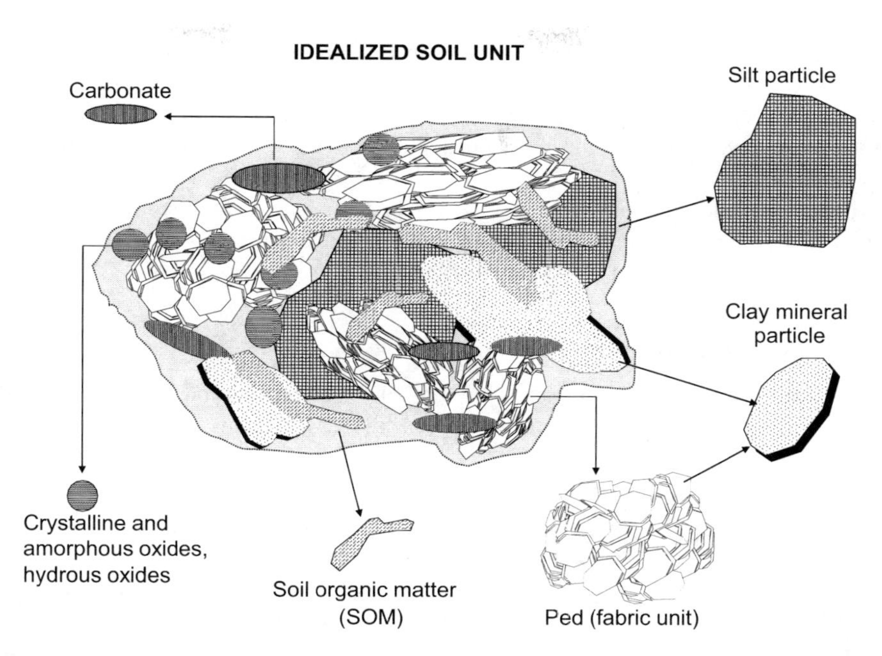

FIGURE 2.10 An idealized typical soil unit in a soil mass consisting of various soil fractions. The positions of the various fractions and the configuration define the structure of the soil. (From Yong and Mulligan, 2019.)

the different soil fractions occur will depend on not only the geologic origin of the soil, but also the regional controls and weathering processes existent at the soil location. At least five factors and four different processes are involved in the production of individual soil fractions. The five factors are

- Parent rock material – composition and texture are important. The influence of these features depends on where weathering occurs. In extreme humid conditions and temperatures, the influence of composition and texture are short lived. However, in arctic and arid regions, the influence of composition and texture of the parent rock material are long-lived, and can even remain indefinitely. Alkali and alkaline earth cations are important factors in determining the weathering products. Thus, for example, rocks containing no alkali can only produce kaolinite or lateritic soils as weathering products. On the other hand, weathering of igneous rocks, shales, slates, schists, and argillaceous carbonates will produce a large variety of weathering products because of the presence of alkalis, alkaline earth cations, alumina, silica, etc.
- Climate – temperature and rainfall are important climatic factors. Warm and humid climates encourage rapid weathering of the minerals of the parent rock material. Decaying vegetative products and organic acids contribute significantly to the weathering process.
- Topography – this affects how water infiltrates into the ground. The greater the residence time of water, as, for example, found at low-lying areas that impound water, the greater is the reactions between the solutes in water and the soil material.
- Vegetation – decaying vegetation is a significant factor since this reacts with the parent silicate minerals.
- Time – this is an important factor in situations where reaction rates are slow.

The four processes that are influential in the weathering sequences include

- Hydrolysis – this is the reaction between the H^+ and OH^- ions of water and other mineral ions, particularly for the rock-forming silicates.

- Hydration – this is important for the formation of hydrous compounds with the minerals in the rocks, such as the silicates, oxides of iron and aluminium, and the sulphates.
- Oxidation – since most rocks carry iron in the form of sulphides or oxides, oxidation of the Fe to FeS, FeS_2 could easily occur in the presence of moisture since this promotes the process of oxidation.
- Carbonation – the interaction or reaction of carbonic acid with bases will yield carbonates. The process of carbonation in silicates is accompanied by the liberation of silica. The silica may remain as quartz or may be removed as colloidal silica.

2.5.3 Soil Composition

Figures 2.9 and 2.10 show that the solid phase of soils contain both inorganic and organic constituents, and that the inorganic components can be minerals as well as other quasi-crystalline and non-crystalline materials.

2.5.3.1 Primary Minerals

We define *primary minerals* as those minerals derived in unaltered form from parent rock material – generally through physical weathering processes. The more common ones found in soils are quartz, feldspar, micas, amphiboles, and pyroxenes. By and large, primary minerals are generally found as sands and silts, with a small portion of clay-sized fractions qualifying as primary minerals. We classify particles less than 2 microns in effective diameter as clay-sized. This classification is made because it is necessary to distinguish between *clay-sized* particles and *clay minerals*.

2.5.3.2 Secondary Minerals

Secondary minerals are derived as altered products of physical, chemical, and/or biological weathering processes. These minerals are layer silicates, commonly identified as phyllosilicates, and they constitute the major portion of the clay-sized fraction of soil materials in clays. Because of the possibility for confusion in usage of terms and names, it is important to distinguish between the terms *clays*, *clay soils*, *clay-sized*, and *clay minerals*.

Clays and *clay soils* refer to soils that have particle sizes less-than-2 micron effective diameter. *Clay-sized* refers to soil particles with effective diameters less than 2 microns. No specific reference to the kind or species of particles is required, since attention is directed towards the size of the particles. *Clay minerals* refer specifically to the layer silicates. These are secondary minerals. They consist of oxides of aluminium and silicon with small amounts of metal ions substituted within the crystal structure of the minerals. Because of their size and their structure, secondary minerals have large specific surface areas and significant surface charges. The major groups of clay minerals include kaolinites, smectites (montmorillonites, beidellites, and nontronites), illites, chlorites, and vermiculites.

2.5.3.3 Soil Organic Matter

Soil organic matter (SOM) can exist in soils in proportions as low as 0.5 to 5%. Although their proportions may be small, their influence on the bonding of soil particles and aggregate groups, together with their ability to attenuate contaminants cannot be overstated. SOM originates from vegetation and animal sources, and is generally categorized in accord with its state of degradation into humic and non-humic material. Humic materials or substances are those organics that result from the chemical and biological degradation of non-humic material. Non-humic material or compounds, on the other hand, are organics that remain un-decomposed or are partly degraded. Humic substances are classified into humic acids, fulvic acids, and humins, with the distinction being made of the basis of their solubility to acid and base.

2.5.3.4 Oxides and Hydrous Oxides

The general list of oxide and hydrous oxide minerals includes the oxides, hydroxides and oxyhydroxides of iron, aluminium, manganese, titanium, and silicon. The common crystalline form of these minerals includes haematite, goethite, gibbsite, boehmite, anatase, and quartz. They differ from layer silicate minerals (secondary minerals) in that their surfaces essentially consist of broken bonds. In an aqueous environment, these broken bonds are satisfied by the OH^- groups of disassociated water molecules. The surfaces exhibit pH-dependent charges, i.e., the surfaces have variable charged properties.

2.5.3.5 Carbonates and Sulphates

The most common carbonate mineral found in soils is calcite ($CaCO_3$). Some of the other less common ones are magnesite ($MgCO_3$) and dolomite ($CaMg(CO_3)_2$). Gypsum ($CaSO_4{\cdot}2H_2O$) is the most common of sulphate minerals found soils. These and other carbonates are dealt with in later chapters.

2.5.4 Soil Properties Pertinent to Contaminant Transport and Fate

The reactions between contaminants and soil during the time when the contaminant is in contact with a soil will determine its transport through the soil, and also its fate. Figure 2.11 shows a simplified schematic of an influent contaminant leachate entering a soil unit. Interaction between the various contaminants in the leachate stream with the exposed surfaces of the soil particles of the various soil fractions will ultimately determine the transport characteristics and fate of the contaminants in the leachate stream. The controlling factors involved, other than the properties and functional groups (chemically reactive groups) of the contaminants, are listed in the diagram.

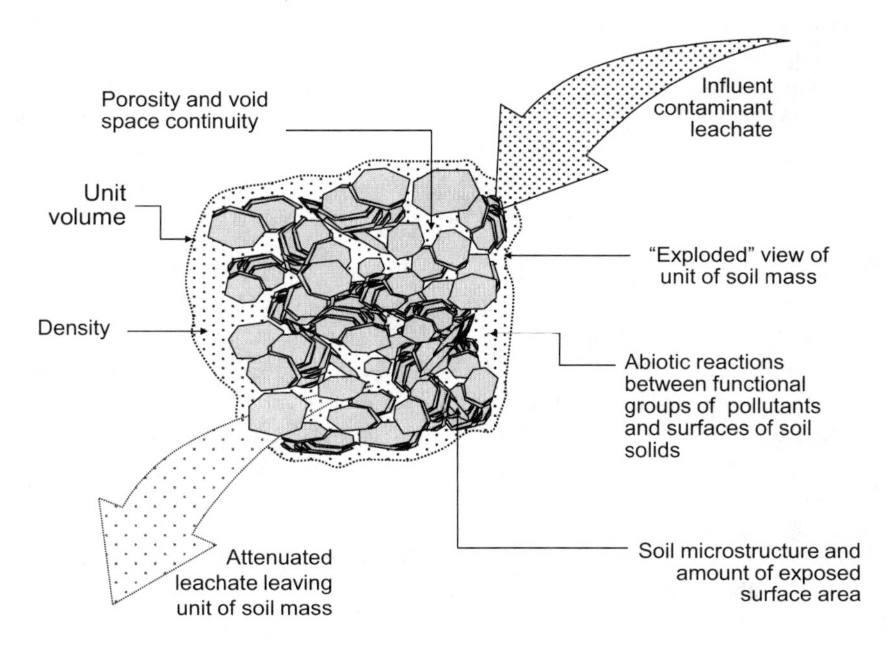

FIGURE 2.11 Schematic diagram showing the major physical soil factors involved in controlling contaminant transport through the unit soil mass.

The two types of interactions involved in contaminant transport processes are as follows:

- Physical interactions – these interactions involve fluid and contaminant movement through the soil fabric and structure. As such, the nature of the void spaces and the amount of surface area presented by the soil particles to the liquid waste and leachate stream are important factors. The principal factors include
  - Void spaces – the size distribution of voids and the continuity of voids are important factors in determining the rate of transport of the leachate stream through the soil. The nature of these voids, their continuity and their distribution are all dependent on the density and structure of the soil. These in turn are functions of soil composition and manner in which the soil was formed *in situ*.
  - Surface area and microstructure – the interactions of the various contaminants in the leachate stream with the surfaces of the soil particles will be dictated by the amount of surfaces exposed to the contaminants. Because of the existence of soil microstructural units, i.e., packets of soil particles grouped together to form aggregate groups or peds or clusters of particles, not every single particle will have its total surface area exposed to the leachate stream. The sizes and types of microstructural units that comprise the soil in question will determine the amount of surface areas exposed to the influent leachate. The detailed discussion on soil microstructure and hydraulic conductivity will be found in Section 10.5 in the discussion on mitigation of impacts from contaminant transport.
- Chemical interactions – this grouping of interactions includes all the types of chemical reactions that occur when two chemically reactive participants interact with each other. The surface properties, and especially the surface chemistry of the soil solids, are very important factors. The surface chemically reactive groups for the soil solids and for the contaminants are identified as functional groups. These will be discussed in a later section in this chapter.

From Figures 2.9 and 2.10, an appreciation of the constituents of a typical soil unit that have different soil solids known as soil fractions are obtained. Although the types of soil fractions can range from sands to clay minerals to soil organics, it is the soil fractions with reactive surfaces that are of interest in the study of transport and fate of contaminants. *Reactive surfaces*, in the present soil unit context, are defined as those surfaces which can react chemically with dissolved solutes in the pore water of the soil. We should also note that contaminants in the ground will also have reactive surfaces, and in this case, these reactive surfaces can react chemically with other dissolved solutes in the pore water and also with the soil solids.

2.5.4.1 Specific Surface Area (SSA) and Cation Exchange Capacity (CEC)

The soil fractions that have more particles with significant reactive surfaces are the clay minerals, oxides and hydrous oxides, soil organics, and carbonates. Table 2.1 gives the surface charge characteristics, SSA and CEC for some clay minerals. We define the *specific surface area* (SSA) as the total surface area of all the soil solids or particles per unit volume. Since theoretical calculations for specific surface area can become both complex and tedious, because of the irregular shapes and sizes of the soil particles, laboratory techniques are often used. A popular procedure is to determine the amount of gas or liquid (adsorbate) that forms a monolayer coating on the surface of the particles. The choice of the adsorbate and the availability of soil particles in a totally dispersed state are important factors in production of the final sets of data. Because of the dependence on techniques used, we consider a laboratory measurement of the specific surface area (SSA) of a soil sample to be an operationally defined property, i.e., dependent on technique, adsorbate used and degree to which the soil has been properly dispersed.

To explain the *reciprocal of charge density* shown in Table 2.1, we need to explain what surface charge density means. The *surface charge density* is the total number of electrostatic charges on the

TABLE 2.1
Charge Characteristics, SSA and CEC for Some Clay Minerals

Soil Fraction	Cation Exchange Capacity (CEC), meq/100 g	Surface Area, m²/g	Range of Charge meq/100 g	Reciprocal of Charge Density nm²/charge	Isomorphous Substitution	Source of Charges
Kaolinite	5–15	10–15	5–15	0.25	Dioctahedral; 2/3 of positions filled with Al.	Surface silanol and edge silanol and aluminol groups (ionization of hydroxyls and broken bonds)
Clay micas and chlorite	10–40	70–90	20–40	0.5	Dioctahedral: Al for Si Trioctahedral or mixed Al for Mg	Silanol groups, plus isomorphous substitution and some broken bonds at edges
Illite	20–30	80–120	20–40	0.5	Usually octahedral substitution Al for Si	Isomorphous substitution, silanol groups and some edge contribution
Montmorillonite[1]	80–100	800	80–100	1.0	Dioctahedral; Mg for Al	Primarily from isomorphous substitution, with very little edge contribution
Vermiculite[2]	100–150	700	100–150	1.0	Usually trioctahedral substitution Al for Si	Primarily from isomorphous substitution, with very little edge contribution

Source: Adapted from Yong (2001).

[1] Surface area includes both external and intra-layer surfaces. Ratio of external particle surface area to internal (intra-layer) surface area is approximately 5 : 80.

[2] Surface area includes both external and internal surfaces. Ratio of external to internal surface area is approximately 1 : 120. Note that ratios of external : internal surface areas are highly approximate since surface area measurements are operationally defined – i.e., they depend on the technique used to determine the measurement.

clay particles' surfaces divided by the total surface area of the particles. The common procedure is to express this surface charge density in terms of its reciprocal – as shown in Table 2.1. We have omitted the values for the hydrous oxides such as goethite [-FeOOH] and gibbsite [-Al(OH)$_3$] from the table because the range of values for these types of soil fractions are dependent upon: (a) their structure, (b) the specifically adsorbed potential-determining ions, and (c) the pH of the pore water.

The cation exchange capacity (CEC) is defined as the quantity of exchangeable ions held by a soil, and is generally equal to the amount of negative charge in the soil. This is usually expressed in terms of milliequivalents per 100 grams of soil (meq/100 g soil). Exchangeable cations are associated with clay minerals, amorphous materials, and natural soil organics. Many of the surface functional groups of these soil fractions are direct participants in cation exchange, e.g., the oxygen-containing functional groups of soil organic matter (SOM) such as the carboxyl and phenolic functional groups. Whilst not reported in Table 2.1, we see measured values for CEC ranging from 15 to 24 meq/100 g soil for Fe-oxides, from 10 to 18 meq/100 g soil for Al-oxides, and from 20 to 30 meq/100 g for allophanes. CEC values of up to 100 meq/100 g soil for goethites and hematites, and from 150 to 400 meq/100 g soil for organic matter at a pH of 8 have been reported by Appelo and Postma (1993).

Their empirical relationship for the CEC of a soil is given in terms of the percentage of clay less than 2 μm and the organic carbon as follows:

$$\text{CEC (meq/100 g soil)} = 0.7\ \text{Clay\%} + 3.5\ \text{OC\%} \quad (2.2)$$

where Clay% refers to the percentage of clay less than 2 μm and OC% refers to the percentage of organic carbon in the soil.

By combining the density of charges with the amount of surface areas available and the cation exchange capacity of the specific clay mineral, we will obtain some appreciation of the degree of reactivity of the clay mineral in question. This should not be construed as a quantitative estimate since actual field soils will not have all particles and their surfaces available for exposure to contaminants. Aggregate groups of particles, such as flocs, domains, peds, and clusters will diminish the total calculated surface area obtained from single particle theory.

2.5.5 Surface Properties

The surface properties of soils are important because it is these properties, together with those surface properties of contaminants themselves and the geometry and continuity of the pore spaces that will control the transport processes of the contaminants. We have previously defined *reactive surfaces* to mean those surfaces which by virtue of their properties are capable of reacting physically and chemically with solutes and other dissolved matter in the pore water. The chemically reactive groups, which are molecular units, are found on the surfaces of the various soil fractions, are defined as *surface functional groups*. These surface functional groups give the surfaces their reactive properties.

The soil fractions that possess significant reactive surfaces include layer silicates (clay minerals), soil organics, hydrous oxides, carbonates and sulphates. The surface hydroxyls (OH group) are the most common surface functional group in inorganic soil fractions (soil solids) such as clay minerals with disrupted layers (e.g., broken crystallites), hydrous oxides, and amorphous silicate minerals. The common functional groups for soil organic matter (SOM) include the hydroxyls, carboxyls, phenolic groups, and amines. More detailed explanations concerning the nature of these functional groups and their manner of interaction with the functional groups associated with contaminants can be found in soil science and geoenvironmental engineering textbooks (e.g., Sposito, 1984; Greenland and Hayes, 1981; Huang et al., 1995a, 1995b; Knox et al., 1993; Yong, 2001; Yong and Mulligan, 2004; Yong et al., 2012).

2.6 CONTAMINANT TRANSPORT AND LAND CONTAMINATION

The transport of contaminants (or contaminants) in soils refers to the movement of contaminants through the pore spaces in soils. Liquid contaminants, such as organic chemical compounds and inorganic/organic contaminants carried in waste streams and leachate streams that pass through the soil pore spaces will interact with the exposed surfaces of the soil fractions. The reactions arising from the interactions between contaminants and soil fractions will dictate the nature of the transport of contaminants, and indirectly or directly, the fate of the contaminants. It is useful to remember that except for liquid chemicals, water is the primary carrier or transport agent for contaminants. The liquid phase of a soil-water system, i.e., the pore water, consists of water and dissolved substances, such as free salts, solutes, colloidal material, and/or organic solutes. All dissolved ions, and probably all dissolved molecules are to some extent, surrounded by water molecules.

The questions relating to *what happens to contaminants in the ground* are perhaps the most critical concerns at hand. Will they eventually disappear? How long will they likely stay in the ground? Will they move to other locations, i.e., be transported? Will they be harmful to human health and the environment? To address the questions, it is necessary to obtain an understanding of the partitioning mechanisms, i.e., the chemical mass transfer of contaminants from the pore water to the surfaces of the soil solids, and how these are related to the soil and contaminant properties. A detailed consideration of these will be found in Chapters 9 and 10. For this section, we will highlight some of the main elements of the interactions as they relate to the mass transfer of dissolved solutes in the porewater (i.e., partitioning of dissolved solutes).

2.6.1 Mechanisms of Interaction of Heavy Metal Contaminants in Soil

The processes of transfer of metal cations from the soil pore water can be grouped as follows:

- Sorption – this includes physical adsorption (physisorption), occurring principally as a result of ion-exchange reactions and van der Waals forces, and chemical adsorption (chemisorption), which involves short-range chemical valence bonds. The general term *sorption* is used to indicate the process in which the solutes (ions, molecules, and compounds) are partitioned between the liquid phase and the soil particle interface. When it is difficult to fully distinguish between the mechanisms of physical adsorption, chemical adsorption, and precipitation, the term *sorption* is used to indicate the general transfer of material to the interfaces.

  Physical adsorption occurs when the contaminants or contaminants in the soil solution (aqueous phase, pore water) are attracted to the surfaces of the soil solids because of the unsatisfied charges of the soil particles. In the case of the heavy metals (metal cations), for example, they are attracted to the negative charges exhibited by the surfaces of the soil solids (Figure 2.12). When the cations are held primarily by electrostatic forces, this is called *non-specific* cation adsorption.

  Specific cation adsorption refers to the situation where the ions penetrate the coordination shell of the structural atom and are bonded by covalent bonds via O and OH groups to the structural cations. The valence forces are of the type which binds atoms to form chemical

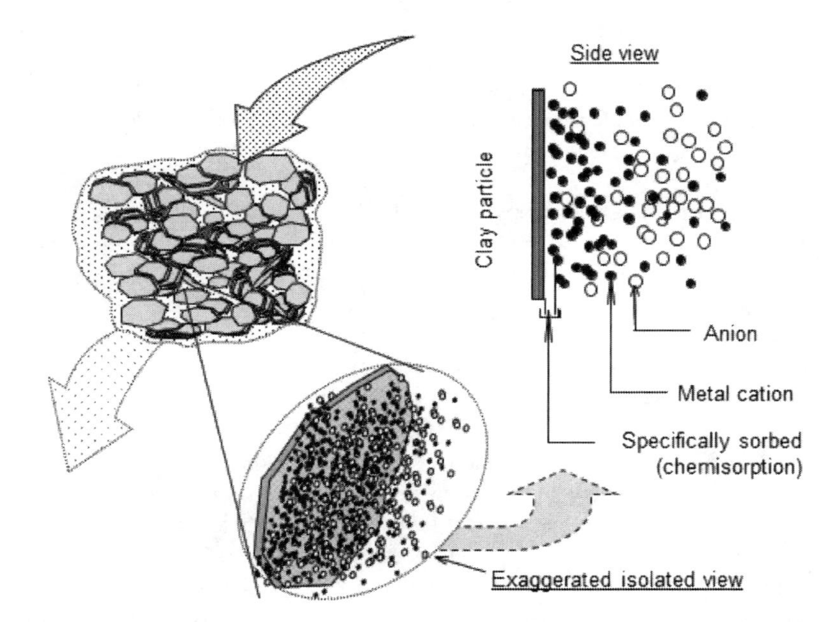

FIGURE 2.12 Exaggerated isolated view of interaction of metal cations (in the pore water) with soil particles. The soil particle shown is a clay mineral particle.

compounds of definite shapes and energies. This type of adsorption is also referred to as *chemisorption.*

- Complexation with various ligands. *Complexation* occurs when a metallic cation reacts with an anion that functions as an inorganic ligand. Metallic ions that can be complexed by inorganic ligands include the transitional metals and alkaline earth metals. The inorganic ligands which will complex with the metallic ions include most of the common anions – e.g., OH^-, Cl^-, SO_4^{2-}, CO_3^{2-}, PO_4^{3-}, etc. Complexes formed between metal ions and inorganic ligands are much weaker than those complexes formed with organic ligands.
- Precipitation – accumulation of material (solutes, substances,) on the interface of the soil solids to form new (insoluble) bulk solid phases. Precipitation occurs when the transfer of solutes from the aqueous phase to the interface results in accumulation of a new substance in the form a new soluble solid phase. The Gibbs phase rule restricts the number of solid phases that can be formed. Precipitation can occur on the surfaces of the soil solids or in the pore water.

2.6.2 Chemically Reactive Groups of Organic Chemical Contaminants

When organic chemical compounds come into contact with soil, the nature of the chemically reactive groups in the organic molecules, their shape, size, configuration, polarity, polarizability, and water solubility are important factors in determining the adsorption of these chemicals by the soil solids. These chemically reactive groups, which are also known as functional groups populate both the surfaces of contaminants and soil solids. The chemical properties of the functional groups of the organic chemicals will influence the surface acidity of the soil particles. This is important in the adsorption of ionizable organic molecules by the soil solids (clays).

The mechanisms of interaction between organic chemicals and soil fractions include (a) London–van der Waals forces, (b) hydrophobic bonding, (c) charge transfer, (d) ligand and ion exchange, and (c) chemisorption. Sorption of organic chemicals is enhanced when there is no hydration layer (of water) on the surfaces of soil particles. Further sorption of other organic chemicals occurs through van der Waals type forces and hydrogen bond formation between functional groups, such as the hydroxyl (OH) group on the soil particles and the carboxyl (COOH) group on the organic chemicals.

The *hydroxyl group* (OH) consists of a hydrogen atom and an oxygen atom bonded together. This group is by far the most common reactive surface functional group for soil fractions, such as clay minerals, amorphous silicate minerals, metal oxides, and the other oxides (oxyhydroxides and hydroxides). The hydroxyl group is also present in two groups of organic chemicals:

(a) Alcohols – methyl, ethyl, isopropyl, and n-butyl. Alcohols can be considered as hydroxyl alkyl compounds (R-OH), and are neutral in reaction since the OH group does not ionize.
(b) Phenols – monohydric (aerosols) and polyhydric (obtained by oxidation of acclimated activated sludge (pyrocatechol, trihydroxybenzene).

The two other kinds of functional groups associated with organic chemical compounds (Figure 2.13), in addition to the hydroxyl (OH) group, are as follows:

(1) Functional groups having a C-O bond. These include the carboxyl, carbonyl, methoxyl, and ester groups. Compounds possessing the *carbonyl group,* called carbonyl compounds, include aldehydes, ketones, and carboxylic acids. The carboxyl group, which combines the carbonyl and hydroxyl groups into a single unit to form a new functional group, is the characteristic functional group of carboxylic acids, e.g., benzoic and acetic acids.
(2) Nitrogen-bonding functional groups such as the amine and nitrile groups. The *amino group* NH_2 is found in primary amines. The amines may be aliphatic, aromatic, or mixed, depending on the nature of the functional groups, and are classified as

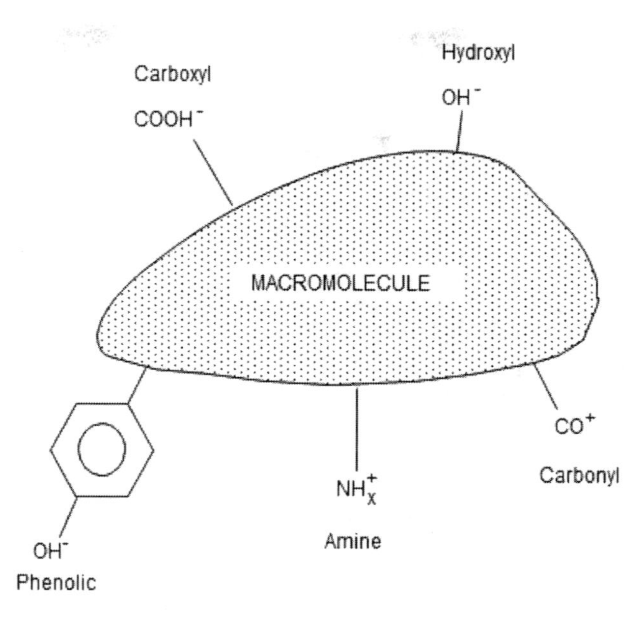

FIGURE 2.13 Some typical functional groups associated with organic chemicals. The macromolecule shown in the diagram is an organic chemical.

- primary – e.g., methylamine (primary aliphatic); aniline (primary aromatic);
- secondary – e.g., dimethylamine (secondary aliphatic); diphenylamine (secondary aromatic); and
- tertiary – e.g., trimethylamine (tertiary aliphatic).

Surface acidity is very important in the adsorption of ionizable organic molecules of clays. The chemical properties of the functional groups of the soil fractions contribute appreciably to the acidity of the soil particles. Surface acidity is an important factor in clay adsorption of amines, *s*-triazines, amides, and substituted urea. This is due to protonation on the carbonyl group – as demonstrated by the hydroxyl groups in organic chemical compounds. As shown at the beginning of this section, there are two broad classes of these compounds: (a) Alcohols – ethyl, methyl, isopropyl, etc. and (b) phenols – monohydric and polyhydric. In addition, there are two types of compound functional groups – i.e., those having a C-O bond (carboxyl, carbonyl, methoxyl, etc.) and the nitrogen-bonding group (amine and nitrile). Amine, alcohol and other organic chemicals that possess dominant carbonyl groups which are positively charged by protonation can be readily sorbed by clays. In amines, for example, the NH_2 functional group of amines can protonate in soil, thereby replacing inorganic cations from the clay complex by ion exchange. The extent of sorption of these kinds of organic molecules depends on (a) the CEC of the clay minerals, (b) the composition of the clay soil (soil organics and amorphous materials present in the soil), (c) the amount of reactive surfaces, and (d) the molecular weight of the organic cations. Because they are longer and have higher molecular weights, large organic cations are adsorbed more strongly than inorganic cations. Polymeric hydroxyl cations are adsorbed in preference to monomeric species because of the lower hydration energies and higher positive charges and stronger interactive electrostatic forces.

The unsymmetrically shared electrons in the double bond endow carbonyl compounds with dipole moments, thus allowing for hydrogen bonding between the OH group of the adsorbent (soil particles) and the carbonyl group of the ketone or through a water bridge. Sorption onto soil particles, especially clays, for the carbonyl group or organic acids (e.g., benzoic and acetic acids) occurs directly with the interlayer of cation or by formation of hydrogen bonds with the water molecules (water bridging) coordinated to the exchangeable cation of the clay complex.

2.6.3 Partitioning of Contaminants and Partition Coefficients

Section 9.5 in Chapter 9 provides a more detailed discussion of contaminant partitioning in transport through a soil. The introduction to this subject in this section provides an overall appreciation of "what happens" when waste leachate streams enter the subsurface soil that constitutes the land environment. The *partitioning* of contaminants refers to the transfer of contaminants from the pore water in the soil to the soil solids by processes that include all of those described in the previous subsections. It is important to determine the partitioning of target contaminants because this will tell us something about its distribution of the contaminants – contaminants sorbed by the soil particles and contaminants remaining in the pore water. This can be interpreted in terms of the quantity or proportion of contaminants likely to move from one location to another. Determination of partitioning of inorganic contaminants and contaminants is generally conducted using batch equilibrium tests. The conventional procedure shown in Figure 2.14 uses soil solutions. The candidate soil is used with an aqueous solution consisting of the contaminant of interest to form a soil solution. Figure 2.14 shows the various steps and analyses required. Results obtained from the tests called adsorption isotherms. Graphical representation of these results show sorbed concentration (of contaminants or contaminants) on the ordinate and equilibrium concentration of contaminants (contaminants) on the abscissa – as seen at the bottom right-hand part of Figure 2.14. The three common types of adsorption isotherms (Freundlich, Langmuir, and constant) found in reported literature on sorption characteristics of inorganic contaminants are shown in Figure 2.15. The parameter k_n in the equations shown with the various curves denotes the slope of the curves.

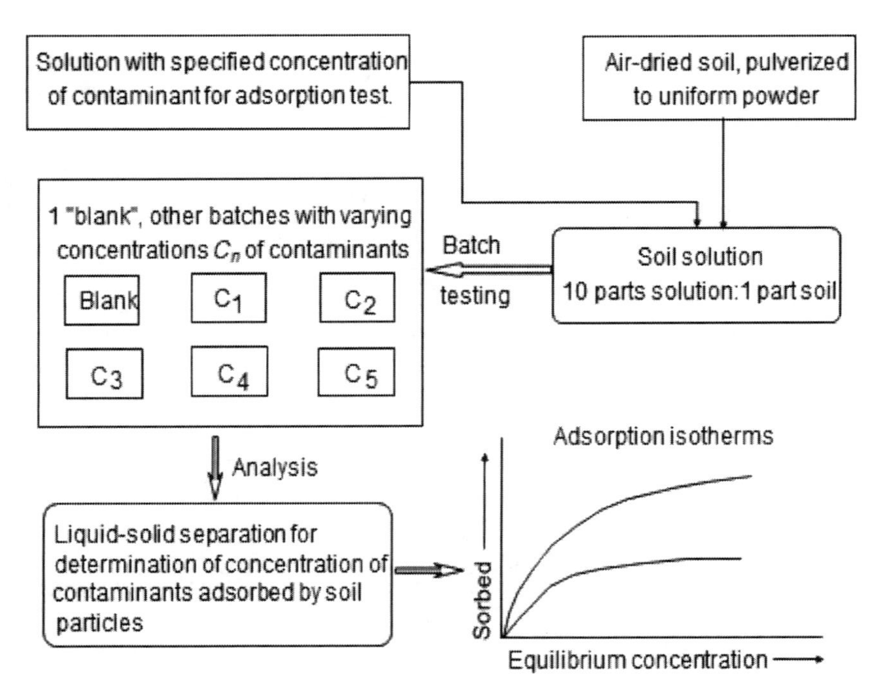

FIGURE 2.14 Batch equilibrium procedure for determination of adsorption isotherms. (From Yong and Mulligan, 2004.)

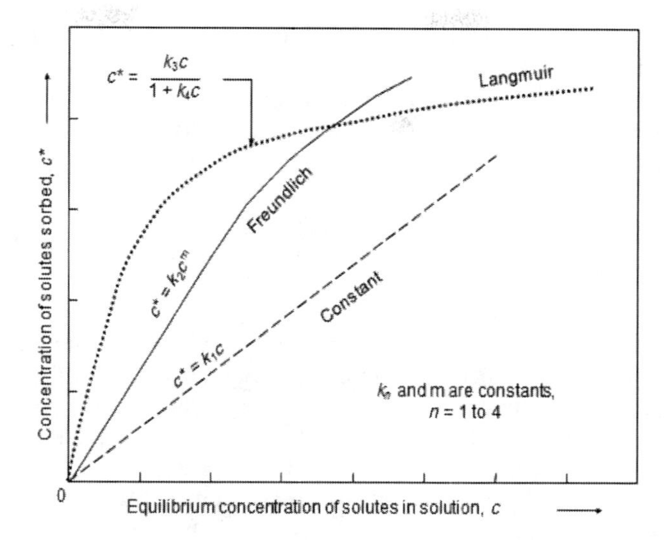

FIGURE 2.15 Freundlich, Langmuir, and Constant types of adsorption isotherms obtained from batch equilibrium tests. c = concentration of solutes or contaminants and c* = concentration of solutes or contaminants sorbed by soil fractions.

In the case of organic chemicals, partitioning is denoted by an *equilibrium partition coefficient* k_{ow}, i.e., coefficient describing the ratio of the concentration of a specific organic contaminant in other solvents to that in water. This coefficient k_{ow}, which relates the water solubility of an organic chemical with its n-octanol solubility is more correctly referred to as the n-octanol-water partition coefficient. It has also been found to be sufficiently correlated to soil sorption coefficients. The relationship for the n-octanol-water partition coefficient k_{ow} has been given in terms of the solubility S by Chiou et al. (1982) as

$$\log k_{ow} = 4.5 - 0.75 \log S \text{ (ppm)} \tag{2.3}$$

Organic chemicals with $k_{ow} < 10$ are generally considered to be relatively hydrophilic. They tend to have high water solubilities and small soil adsorption coefficients. Organic chemicals with $k_{ow} > 10^4$ have low water solubilities and are considered to be very hydrophobic. The greater the hydrophobicity, the greater is its bioaccumulation potential (Section 9.5 in Chapter 9).

2.6.4 Predicting Contaminant Transport

There are many models that purport to describe the movement of contaminants, solutes, contaminants, etc., in soil. For convenience in discussion, we will use the term *solutes* to mean contaminants and all other kinds of solutes found in the pore water. Most of the models deal with movement of solutes in saturated soils, and is best applied to inorganic contaminants and contaminants, such as heavy metals. Application of the models for determination of transport of organic chemicals has been attempted by some researchers – with varying degrees of success. So long as the movement of the contaminants is governed by Fick's law, and instantaneous equilibrium sorption processes occur, some success in prediction of transport can be obtained. The ability to properly model partitioning of the contaminants remains as one of the central issue in determination of success or failure of predictions. The problem is complicated by the fact that a reasonably complete knowledge of the initial and boundary conditions is not always available. Additionally, the presence of multi-component contaminants and their individual and collective reactions with the soil fractions will make partitioning determinations difficult. Detailed discussions of the modelling problems and the physico-chemical interactions and partitioning of contaminants in soils can be found in textbooks dedicated to the study of contaminant fate and transport in soils, e.g., Knox et al., 1993; Fetter, 1993; Huang et al., 1995a, 1995b; Yong, 2001, and Yong and Mulligan, 2019. The most common and widely used transport model has the relationship shown as

$$\frac{\partial c}{\partial t} = D_L \frac{\partial^2 c}{\partial x^2} - v\frac{\partial c}{\partial x} - \frac{\rho}{n\rho_w}\frac{\partial c^*}{\partial t} \tag{2.4}$$

where c = concentration of solutes or contaminants, t = time, D_L = longitudinal dispersion coefficient, v = advective velocity, x = spatial coordinate, ρ = bulk density of soil, ρ_w = density of water, n = porosity of soil, and c^* = concentration of solutes or contaminants adsorbed by soil fractions (see Figure 2.15). If we assume a slope constant $k_d = k_l$ as shown in Figure 2.15 for the constant adsorption isotherm, the concentration of solutes sorbed by the soil fractions c^* can be written as $c^* = k_d\, c$. The slope constant k_d is defined as the distribution coefficient and is meant to indicate the manner of distribution of the solutes being transported in the pore water of a soil-water system. Equation (2.1) can be written in a more compact form to take into account the distribution coefficient as follows:

$$R\frac{\partial c}{\partial t} = D_L \frac{\partial^2 c}{\partial x^2} - v\frac{\partial c}{\partial x} \tag{2.5}$$

where R = retardation factor = $\left[1 + \frac{\rho}{n\rho_w} k_d\right]$.

A more detailed discussion of the transport and fate of contaminants will be found in Chapter 9.

2.7 PHYSICO-CHEMICAL PROPERTIES AND PROCESSES

Water is the carrier for contaminants in the subsoil. The movement and distribution of the various contaminants in the soil depend not only on the hydrogeological setting, but also on the interactions between the contaminants carried in the liquid phase and the soil fractions. Many of these interactions have been described in the previous sections as transport processes. In this section, we will look at the physico-chemical properties and processes involved when they are introduced to the geoenvironment.

2.7.1 Solubility

The amount of solutes needed to reach a saturated state in a given quantity of solvent at a specific temperature is defined as the solubility of the given solvent. For considerations of ionic equilibrium in aqueous solutions, the solutes are those that fall into the class of sparingly or slightly soluble ionic solids. The *solubility product* k_{sp} is defined as the equilibrium constant for the equilibrium that exists between the sparingly soluble ionic solid and its ions in a saturated solution. Because of the significant electrostatic attraction between ions, crystals composed of small ions packed closely together are generally harder to pull apart than crystals made up of large ions. For example, fluorides (F^-) and hydroxides (OH^-) are less soluble than nitrates (NO_3^-) and perchlorates (ClO_4^-).

Solubility equilibria are useful in predicting whether a precipitate will form under specified conditions, and in choosing conditions under which two chemical substances in solution can be separated by selective precipitation. Substances that are more soluble are more likely to desorb from soils and less likely to volatilize from water. On the other hand, substances with no hydrogen bonding groups or little polar character, such as hydrocarbons or halogenated hydrocarbons, usually have very low solubilities compared to compounds, such as alcohols, which are capable of interaction with water. The solubility of an organic compound depends primarily upon the sorption-desorption characteristics of the sorbate (organic compound) in association with the sorbent (soil and sediment).

2.7.2 Partition Coefficients

Partition coefficients provide a measure of the distribution of a given inorganic contaminant – between sorption onto the soil solids and the porewater. For organic chemicals, the octanol-water partition coefficient is used to describe partitioning. The *octanol/water partitioning coefficient* (k_{ow}), which has been previously defined as the ratio of the amount of a solute dissolved in octanol and water in octanol-water immiscible mix. k_{ow} is well-correlated with the solubility of several organic chemicals. Log k_{ow} values normally range from –3 to 7. High water-soluble compounds, such as ethanol, have values of log $k_{ow} < 1$, and hydrophobic compounds, such as certain PCBs and chlorinated dioxin congeners, have values of 6 to 7.

2.7.3 Vapour Pressure

The vapour pressure of a liquid or solid is the pressure of the gas in equilibrium with the liquid or solid at a given temperature. Gasoline, for example, will evaporate rapidly since it has a high vapour pressure and is very volatile. Volatilization is a significant factor in disposal for compounds with vapour pressure greater than 10^{-3} mm Hg at room temperature. Chemicals with relatively low vapour pressures and high solubility in water are less likely to vaporize and become airborne. The transport of a compound from the liquid to the vapour phase is called volatilization. This could be an important pathway for chemicals with high vapour pressures or low solubilities. Evaporation depends on the equilibrium vapour pressure, diffusion, dispersion of emulsions, solubility, and temperatures.

2.8 GEOENVIRONMENTAL LAND MANAGEMENT

The major geosphere and hydrosphere features that constitute the geoenvironment components for land management attention are shown in Figure 2.16. Land management, as the term implies, is the utilization of management practices to a land environment to meet a set of *land use* objectives.

Because of the use of several common terms in the various geosciences and geoenvironmental communities, we will define the following terms as they relate to their use in *geoenvironmental land management*:

- *Land use* – utilization of a land to fit a particular set of objectives or purposes. A simple example of such can be found in such usages as eco-parks, playing fields, forests, wetlands, etc. It is important to appreciate that land use should be considered within the bounds of proper environment and natural resource protection.
- *Land capability and functionality* – the performance capability and/or functionality of a piece of land, i.e., what a land is capable of *doing* or, what it can be used for. Land capability and functionality requires one to determine the natural capital of the land and to determine how these assets can be utilized. In the context of geoenvironmental sustainability, it follows that utilization of these assets must be consistent with the principles of sustainability.

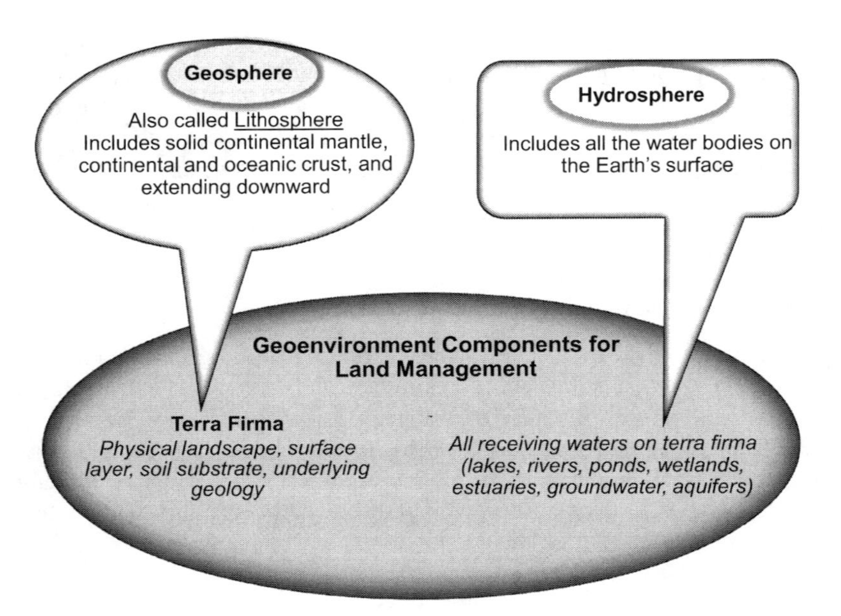

FIGURE 2.16 Geosphere and hydrosphere features included as geoenvironment components for land management.

- *Natural capital and assets* – this includes the physical attributes, such as landforms and natural resources (including biodiversity). This grouping of properties is not only the most important, but is perhaps the most difficult to fully delineate. Other than the obvious asset of a piece of land, e.g., mineral and hydrocarbon resources, what constitutes and asset or a natural capital is to some extent dependent on the sets of criteria and guidelines developed for the particular problem at hand.

The manner in which a land is exploited is dictated by several factors and forces, not the least of which is *land capability*. Figure 2.17 gives an illustration of the procedure that might be used in land use planning and implementation to satisfy geoenvironmental land management concerns. As with any project, the impacts to the geoenvironment must be evaluated, and procedures for avoidance and mitigation of these impacts need to be established. The end result of all of these must satisfy geoenvironmental land management requirements – i.e., they must ensure that there are no threats to public health and that the natural capital and assets of the site are maintained. Reassessment and re-design are necessary if initial land use plans do not satisfy geoenvironmental impact concerns. The key elements that must be satisfied are preservation of natural capital and assets.

2.9 CONCLUDING REMARKS

We have focused on contamination of the land environment as one of the key issues in the need to protect the natural capital and assets of the land environment.

- Geoenvironmental land management requires one to practice the principles of sustainability.
- Insofar as the land environment is concerned, environmental impact associated with anthropogenic activities can be in the form of changes in the quantity or quality of the various features that constitute the land environment.

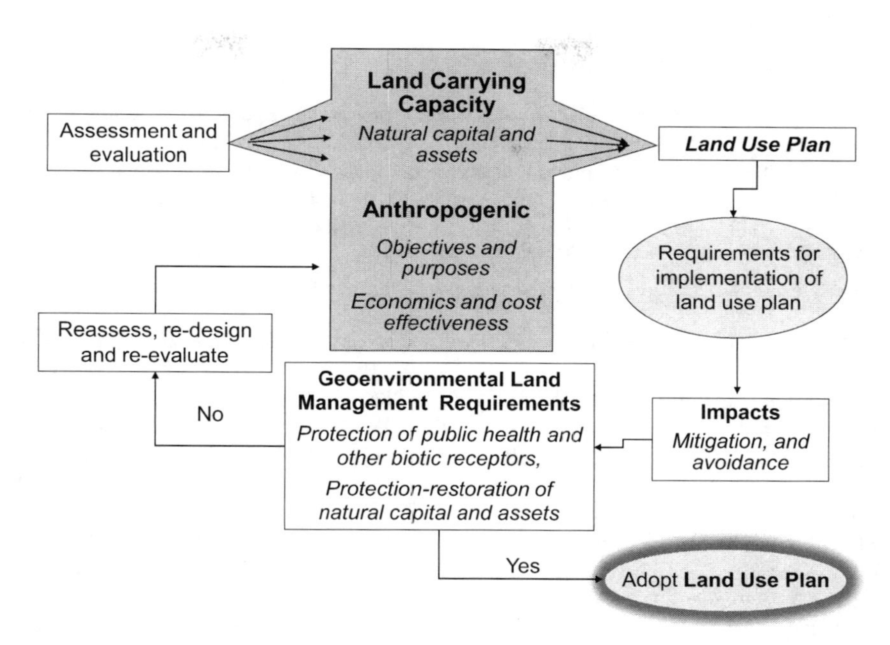

FIGURE 2.17 Example of procedure to be used to satisfy geoenvironmental land management issues in adoption and implementation of land use plans.

- Not all anthropogenic activities will result in adverse impacts on the environment.
- The degree of environmental impact due to contaminants in a contaminated ground site is dependent on (a) the nature and distribution of the contaminants, (b) the various physical, geological and environmental features of the site, and (c) existent land use.
- Each type of land use imposes different demands and requirements from the land. The ideal situation in land utilization matches land suitability with land development consistent with environmental sensitivity and sustainability requirements.
- Groundwater is an integral part of land use considerations.
- Causes and sources of groundwater contamination include wastewater discharges, injection wells, leachates from landfills and surface stockpiles, open dumps and illegal dumping, underground storage tanks, pipelines, irrigation practices, production wells, use of pesticides and herbicides, urban runoff, mining activities, etc.
- To evaluate and determine the nature of geoenvironmental impacts created by the presence of contaminants in the geoenvironment, one needs to have an appreciation of the nature of the contaminants and associated events that are responsible for the impacts. The various activities associated with the production of goods and services generate waste streams and products. In most instances, these waste streams and products find their way into the land environment – either inadvertently as in the case of waste mismanagement, run-offs and spills, or by design, i.e., constructed safe land disposal facilities.

REFERENCES

Agency for Toxic Substances and Disease Registry (ATSDR). (1999), *Toxicological Profile for Cadmium*, U.S. Department of Health and Human Services, Public Health Service, Atlanta, GA.

Allen, S. Allen, D., Phoenix, V.R ., Le Roux, G., Durántez Jiménez, P., Simonneau, A., Binet, S., and Galop, D. (2019), Atmospheric transport and deposition of microplastics in a remote mountain catchment, *Nature Geoscience*, 12(5): 339–344, https://doi.org/10.1038/s41561-019-0335-5.

Appelo, C.A.J., and Postma, D. (1993), *Geochemistry, Groundwater and Pollution*, Balkema, Rotterdam, 536 pp.

Chiou, G.T., Schmedding, D.W., and Manes, M. (1982), Partition of organic compounds on octanol-water system, *Environmental. Science & Technology*, 16: 4–10.

Engler, R. (2012), The complex interaction between marine debris and toxic chemicals in the ocean, *Environmental Science & Technology*, 46/22: 12302–12315, https://doi.org/10.1021/es3027105.

Fetter, C.W. (1993), *Contaminant hydrogeology*, Macmillan Publishing Co., New York, 458 pp.

Gasperi, J., Wright, S.L., Dris, R., Collard, F., Mandin, C., Guerrouache, M., and Langet, V. (2017), Microplastics in air: Are we breathing it in?, *Current Opinion in Environmental Science & Health*, 1: 1–5, https://doi.org/10.1016/j.coesh.2017.10.002.

Greenland, D.G., and Hayes, M.H.B. (Eds.), (1981), *The Chemistry of Soil Processes*, John Wiley and Sons, New York, 714 pp.

Huang, P.M., Berthelin, J., Bollag, J.-M., McGill, W.B., and Page, A.L. (Eds.), (1995a), *Environmental Impact of Soil Component Interactions: Natural and Anthropogenic Organics,* CRC Press, Boca Raton, 450 pp.

Huang, P.M., Berthelin, J., Bollag, J.-M., McGill, W.B., and Page, A.L. (Eds.), (1995b), *Environmental Impact of Soil Component Interactions: Metals, Other Inorganics, and Microbial Activities*, CRC Press, Boca Raton, 263 pp.

Karlen, D.L., Mausbach, M.J., Doran, J.W., Cline, R.G., Harris, R.F., and Schuman, G.E. (1997), Soil quality: A concept, definition, and framework for evaluation, *Soil Science Society of America Journal*, 61: 4–10.

Knox, R.C., Sabatini, D.A., and Canter, L.W., (1993), *Subsurface Transport and Fate Processes*, Lewis Publishers, Boca Raton, 430 pp.

Lewis, D.R., Southwick, J.W., Ouellet-Hellstrom, R., Rench, J. and Calderon, R.L. (1999), Drinking water arsenic in Utah: A cohort mortality study, *Environmental Health Perspectives*. 107: 359–365.

Mazumder, D.N.G., Haque, R., Ghosh, N., De, B.K., Santra, A., Chakraborty, D., and Smith, A.H. (1998), Arsenic levels in drinking water and the prevalence of skin lesions in West Bengal, India, *International Journal of Epidemiology*, 27:871–877.

OECD (2013), Synthesis paper on per and polyfluorinated chemicals, OECD Series on Risk Management of Chemicals, OECD Publishing, Paris, https://doi.org/10.1787/0bc75123-en

OECD (2022), *Global Plastics Outlook: Economic Drivers, Environmental Impacts and Policy Options*, OECD Publishing, Paris, https://doi.org/10.1787/de747aef-en.

Pepper, I.L., Gerba, C.P., and Brusseau, M.L. (Eds.) (2006), *Environmental and Pollution Science,* 2nd ed., Boston, MA, Academic Press.

Sposito, G. (1984), *The Surface Chemistry of Soils*, Oxford University Press, New York, 234 pp.

Teuten, E., Rowland, S.J., Galloway, T.S., and Thompson, R.C. (2007), Potential for plastics to transport hydrophobic contaminants, *Environmental Science and Technology*, 41(22): 7759–7764, https://doi.org/10.1021/es071737s.

USEPA. (1984), United States Environmental Protection Agency. Health Assessment Document for Inorganic Arsenic. EPA-600/8-83-012F. Final Report. USEPA Office of Research and Development, Office of Health and Environmental Assessment, *Environmental Criteria and Assessment*, U.S. EPA, Research Triangle Park, NC, pp. 5–20.

USEPA (U.S. Environmental Protection Agency). (2016), Risk Management for Per- and Polyfluoroalkyl Substances (PFASs) under TSCA. accessed Nov. 30, 2023. www.epa.gov/assessing-and-managing-chemicals-under-tsca/risk-management-and-polyfluoroalkyl-substances-pfas#:~:text=Risk%20Management%20for%20Per-%20and%20Polyfluoroalkyl%20Substances%20(PFAS)#:~:text=Risk%20Management%20for%20Per-%20and%20Polyfluoroalkyl%20Substances%20(PFAS), website is now updated, July 11, 2024

Wang, S., and Mulligan, C.N. (2006), Occurrence of arsenic contamination in Canada: Sources, behavior and distribution, *Science of The Total Environment*, 366(2–3): 701–721, https://doi.org/10.1016/j.scitotenv.2005.09.005.

Yong, R.N. (2001), *Geoenvironmental Engineering: Contaminated Soils, Pollutant Fate and Mitigation*, CRC Press, Boca Raton, 307 pp.

Yong, R.N., and Mulligan, C.N. (2004), *Natural Attenuation of Contaminants in Soils*, Lewis Publishers, Boca Raton, 319 pp.

Yong, R.N. and Mulligan, C.N. (2019), *Natural and Enhanced Attenuation of Contaminants in Soils*, CRC Press, Boca Raton. 324 pp.

Yong, R.N., Nakano, M., and Pusch, R., (2012), *Environmental Soil Properties and Behaviour,* CRC Press, Boca Raton, 435 pp.

Yong R.N., and Warkentin, B.P. (1975), *Soil Properties and Behaviour*, Elsevier Scientific Publishing Co., Amsterdam, 449 pp.

3 Sustainable Water Management

3.1 INTRODUCTION

As we have pointed out in the previous chapter, *water* is one of the essential ecosystem components for survival of all living species. The water component of the geoenvironment includes all rivers, lakes, ponds, inlets, wetlands, estuaries, coastal water, groundwater, and aquifers. These contribute as inputs to the oceans that make up 70% of the earth's surface water. Water is required for many needs, such as drinking, agriculture, cooking, domestic and industrial uses, transportation, recreation, electrical power production, and support for aquatic life and other wildlife. Amongst the many underlying reasons for increasing water shortages are (a) demand in excess of supply, (b) depletion of aquifers, (c) lack of rain and other forms of precipitation, (d) watershed and water resources' mismanagement, and (d) diversion of rivers. It is noted that irrigation requirements for agriculture are increasing. Water demand has tripled since the 1970s, which is faster than the rate of population growth. Agriculture use for irrigation, livestock, and aquaculture is the main user (72% in 2018 according to statista.com), followed by municipal and industrial uses.

The importance of water has been highlighted in SDG 6, which aims by 2030 to

- Ensure access to water and sanitation for all.
- Increase water use efficiency.
- Implement water resources' management and cooperation at all levels.
- Restore water-related eco-systems.
- Strengthen the participation of local communities.

Lack of water and poverty are intimately linked. The discussion in this chapter will focus on the uses (and misuse) of water in the geoenvironment. It will also examine some of the main elements required to address, contain, and manage stressor impacts to water quality – as a step towards water management for sustainability of water resources.

3.1.1 Geoenvironment Sustainable Water Management

Sustainable water management can be defined in a similar manner to the definition for *sustainable development* articulated in Section 1.3 in Chapter 1. Accordingly, we can define sustainable water management as *all the activities associated with usage of water in support of human needs and aspirations, must not compromise or reduce the chances of future generations to exploit the same resource base to obtain similar or greater levels of yield.* In the context of the geoenvironment – with particular focus on geoenvironmental engineering – we limit ourselves to the receiving waters in the geoenvironment, i.e., all water forms contained within the land surface, such as rivers, lakes, ponds, groundwater (aquifer and soil porewater), etc.

3.1.1.1 Water Availability and Quality

Water availability and water quality are central issues in survivability of living species – humans, animals, plants, etc. Lack of water and unacceptable water quality are significant threats to

DOI: 10.1201/9781003407393-3

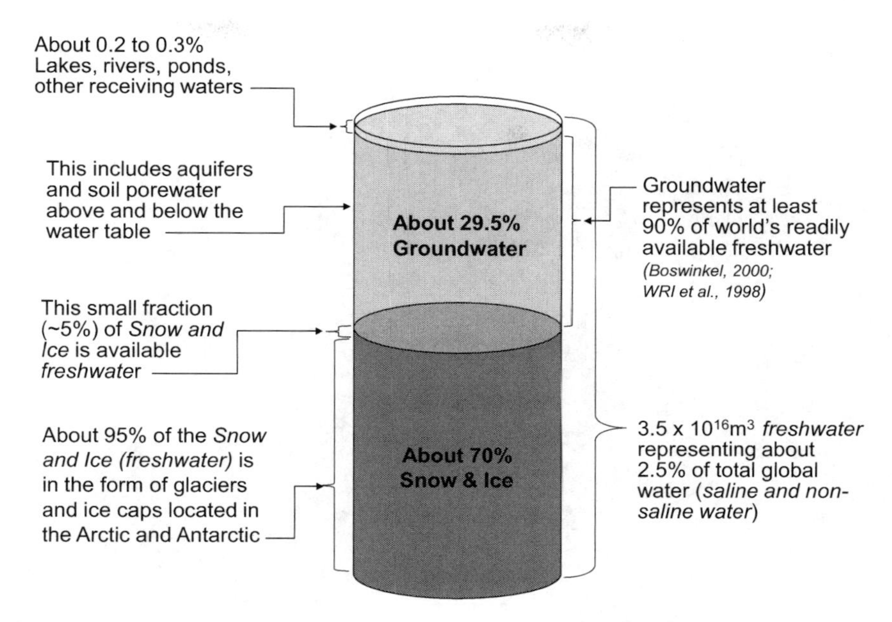

FIGURE 3.1 Sketch of distribution of global freshwater. According to Boswinkel (2000) and WRI et al. (1998) groundwater represents at least 90% of the world's readily available freshwater.

survivability. Water availability or the lack thereof is a topic that is well covered in textbooks devoted to such a subject. The problem of threats to the quality of water in the geoenvironment constitutes the central focus of discussions in this book. It has been said previously, in Chapter 2, that soil contamination does not only mean contamination of the soil solids themselves, but also contamination of the porewater, with possible extension to groundwater and the aquifers through transport of contaminants in the subsoil. Chemical stressors producing point source and non-point source contaminants (to be discussed in the succeeding chapters) will contaminate not only soils, but the receiving waters. The importance in protection of available freshwater – and especially groundwater – is demonstrated in Figure 3.1, which highlights the degree of utilization of this resource by the world's population.

3.2 USES OF WATER AND ITS IMPORTANCE

Safe and adequate amounts of water are essential. The first Dublin–Rio principle has emphasized the need for sustainable water resource practices. It states that "Fresh water is a finite and vulnerable resource, essential to sustain *life, development and the environment*". In other words, water is essential for these three categories. Note that the use of the term *environment* is meant to include the life-supporting functions of ecosystems – as discussed in Chapter 1.

3.2.1 Hydrologic Cycle

Although it is convenient to think in terms of essentially two primary sources of drinking water, i.e., surface water and groundwater (aquifer and soil porewater), it is more useful to bear in mind the total hydrologic cycle when one wishes to consider: (a) The need for water for life-support systems, and (b) the impact of mankind on the sources of water. The hydrologic cycle reflects the constant or continuous movement of water within the earth, on the earth and in the atmosphere – as shown in Figure 3.2.

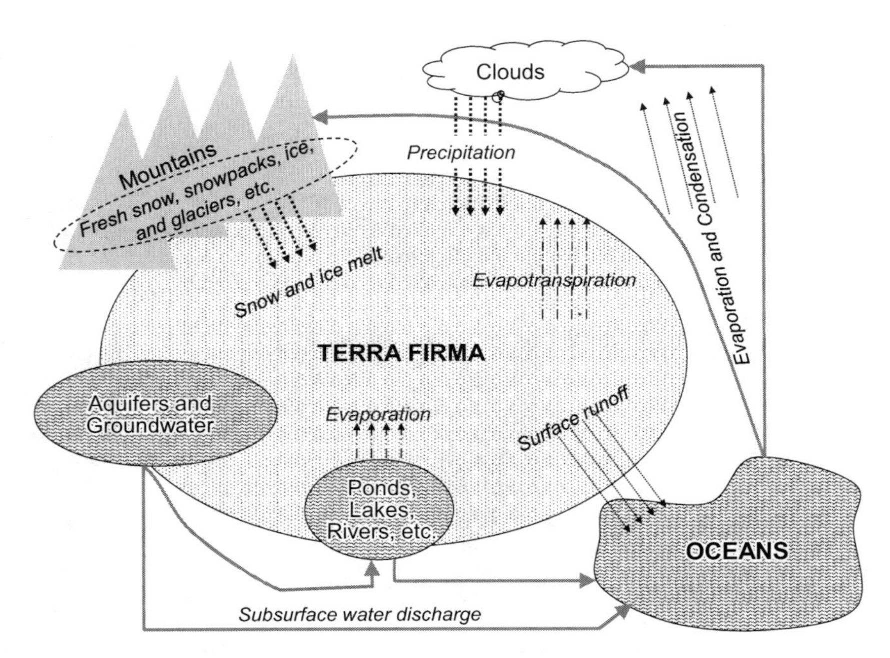

FIGURE 3.2 Sketch of basic elements and processes in the hydrologic cycle. Percolation from surface into subsurface is not shown in the sketch.

Beginning with the ocean and other surface water bodies, such as rivers, ponds, lakes, etc., together with open land surfaces, we can identify these processes as (a) evaporation and condensation, (b) evapotranspiration, (c) precipitation, (d) infiltration (percolation) from ground surface into subsurface, (e) groundwater (subsurface water) discharge into land receiving waters and oceans, (f) discharge from rivers, and (g) surface run-off. It is necessary to note that the terms *infiltration* and *percolation* are quite often used to mean the same event, i.e., entry of water into the ground surface from water on the surface, either from precipitation or from ponded water, or other similar sources. Direct anthropogenic interference in the natural hydrologic cycle occurs most often in processes such as infiltration and run-off – as for example in the quality of the water that infiltrates into the ground and also the quality of the run-off that enters the receiving waters.

3.2.1.1 Human Interference on Infiltration and Run-Off

Human interference in the hydrologic cycle is most significant in the processes of infiltration and run-off. The form of some of the major interferences on infiltration and their impacts, some of which have been shown in Figure 1.5 in Chapter 1, are as follows:

- Development, production, and construction of impermeable surface areas: These impermeable surfaces include housing and similar structures, roads and pavements, runways and aprons, parking lots and other generally paved surfaces constructed in the urban environment. The effect of these impermeable surfaces is to deny infiltration into the ground and hence denying recharge of any underlying aquifer. Run-offs obtained on the impermeable surfaces are generally fed to storm drains or other similar drainage discharge systems.
- Compacted surface layers: These are surfaces of natural soil compacted by agricultural and construction machinery and other similar devices. The effect of compacted surface layers is to reduce infiltration rate and infiltration capability in general. Surface run-off occurs when the rate of precipitation onto the surface is greater than the infiltration rate. Unlike the

paved impermeable surfaces, these run-offs are not generally fed to storm drains or other catchment facilities.

- Soil porewater and aquifer contamination by infiltration: The term soil porewater is used to mean the water in the pores of the soil matrix. Normally, in compact clay soils, this porewater is not easily or readily extractable. However, in more granular materials, such as silts and sands, this porewater can be harvested. Aquifers in general are seams or layers of primarily granular materials that are full of water. We use the term *groundwater* to mean both soil porewater (above and below the water table) and water in the aquifers, but mainly to denote water in the ground that is normally considered as a water resource. When it is necessary to talk about soil porewater, the term *porewater* is generally used. Contaminants on the land surfaces will be transported into the subsoil via infiltration and other transport mechanisms, such as dispersion and diffusion. Pesticides, herbicides, insecticides, organic and animal wastes, ground spills, other deliberate and inadvertent discharges of hazardous and noxious substances on surfaces, etc., serve as candidates for transport into the subsoil. Contamination of groundwater will occur when communication between the contaminants and these water bodies is established, i.e., when the contaminated infiltration plume reaches the groundwater – meaning that contamination of groundwater occurs the moment contaminants reach soils containing porewater. The end result is impairment of groundwater quality. Ingestion of polluted groundwater can be detrimental to human health.

As with infiltration processes, surface runoffs can occur on natural ground surfaces devoid of much human contact. This occurs when the rainfall rate exceeds the infiltration rate, and when natural surface cover (vegetation and plants, etc.) is so dense that it acts as a shield or umbrella. As we have seen from the preceding, natural infiltration properties of soils can be severely compromised by human activities resulting in run-offs. There are at least two types of run-offs: (a) Managed run-offs where the run-offs are channelled into drains and sewers, and (b) un-managed run-offs where the run-offs take directions controlled by surface topography and permeability properties of the surface cover material. In both cases, the run-offs will end up in receiving waters – lakes, rivers, oceans, etc. In the case of managed run-offs, there are at least two options for the run-offs before they meet the receiving waters: (a) Treatment (full or partial) of the run-off water, and (b) no treatment before discharge into receiving waters. The result of untreated run-offs into receiving waters is obvious – degradation of the quality of the receiving waters. When such occurs, the receiving waters will require treatment to reach drinking water quality standards. A point can be reached where the accumulated contaminants will be at a level where treatment of the degraded receiving water will not be effective – not from an economic standpoint and not from a regulatory point of view.

The combination of infiltration and run-offs as potential carriers of contaminants into the ground and receiving waters (including groundwater) is a prospect that should alert one to the need for proper management of the sources of contamination of water resources. Treatment of water to achieve levels of quality dictated by drinking water standards is only one means for water resource management. The other has to be directed towards eliminating or mitigating the sources of contamination of water resources.

3.2.1.2 Impacts Due to Climate Change

Climate change is forecast to lead to many changes in water quality, and quantities by 2050 (OECD, 2012). Water levels will rise from glacier melting (Black and King, 2009) causing floods. Droughts will also be more frequent. According to the FAO (2011), irrigation will increase by up to 20% by 2080. Fresh water contamination by sea water will increase water supply challenges.

Changes in 25–50-year design lives must be considered by engineers following the changes in precipitation patterns. The U.S. Bureau of Reclamation (USBR, 2016) and the NOAA (2012) have provided some guidance for reservoir design for various climate change scenarios and for sea level rise in coastal regions, respectively. Water and energy supply and demand will be influenced by climate change (Green, 2016; Schaeffer et al., 2012). Extreme events such as heat waves and water scarcity in urban areas will impact water supply and distribution. Reduction of the energy footprints will minimize impacts on urban and agricultural systems (De Zeeuw, 2011; Gill et al., 2007).

3.2.2 Harvesting of Groundwater

We have seen in Chapter 1 and in Figure 3.1 that only about 5% of the water in the world is fresh water – with the rest being seawater. This tells us that our drinkable fresh water resource is severely limited. The graphics in Figure 3.1 informs one that whilst groundwater represents about 30% of the available freshwater, at least 90% of the world's population rely on this as a source of freshwater (Boswinkel, 2000; WRI et al., 1998). Infiltration and run-offs carrying contaminants serve as chemical stressors whose impacts on the solid land environment are both *soil contamination* and *contamination of groundwater*. The subject of soil contamination has been discussed in Chapter 2, and the processes and geoenvironmental engineering practices for mitigation and remediation of contaminated soils will be discussed later in Chapters 9 and 10. Groundwater abstraction for drinking water purposes will require treatment aids. Since reliance on groundwater as a drinking water source is considerable, it is clear that maintaining acceptable groundwater quality is a high priority not only because of the need for treatment after abstraction, but also because once the aquifer is contaminated, clean-up of the aquifer to acceptable standards is almost impossible. When aquifers become contaminated, one key element for aquifer sustainability is lost. In effect, sustainability of that aquifer as a water resource is lost.

Depletion of aquifers is also another sustainability loss that requires attention. Depletion occurs when the rainfall is insufficient and/or aquifer recharge is exceedingly slow, i.e., input to aquifer is less than the output. Continued water use from that aquifer is thus not sustainable. This is occurring with more frequency as water use has increased by a factor of six since the beginning of the twentieth century. Excessive groundwater pumping can substantially reduce groundwater levels. For example, the USGS (Capel et al., 2018) indicated that the groundwater levels for drinking water use have decreased in Chicago from 1864 to 1980 by 274 m. From 1900 to 2008, the estimated volume of groundwater depletion in the United States is about 1000 km^3. Rates have increased substantially since 1950. Very high rates occured during the period of 2000–2008 when the depletion rate averaged almost 25 km^3 per year in comparison to an average of 9.2 km^3 per year in the period of 1900–2008.

Another study examined data from 170,000 monitoring wells and 1693 aquifers. These comprise 75% of all groundwater withdrawals (Jasechko et al., 2024). It was found that groundwater has declined substantially in about 30% of the global aquifers in the past four decades. Levels have declined by more than 0.5 m per year, particularly in the dry agricultural area in this century. However, it was also found that if groundwater can be managed appropriately, declining can be reversed.

Fresh water is depleting rapidly in countries, such as India, China, and even in the US, rivers are drying up and water table levels are decreasing. In China, there is a lack of water in more than 300 cities (WRI, 1994). It was estimated in the *Global Environment Outlook 3* report by the United Nations Environment Programme (UNEP, 2002) that about half of the world's population will not have sufficient water by the year 2032. This would be the result of the currently unsustainable practices that are currently using about 50% of the earth's fresh water supply (UN-FPA, 2001). What will happen to the ecosphere and to human life when the population reaches 9 billion by 2100 (UNDP, 2001). It has just reached 8 billion. Even now, dehydration is occurring in many areas to the extent that this will affect biodiversity and increase the requirements for agricultural irrigation.

Per hectare of land, 10 million litres of water each season are required for production of 8000 kg of corn (Pimentel et al., 1996). Water use for irrigation has decreased due to improved agricultural practices (waterfootprint.org estimates that it is currently 760 L/kg corn in the US). Treatment of groundwater as a scarce resource may be required. Water markets have been proposed (Dorf, 2001) as a means of valuing water as a resource. Conservation by farmers and urban users will increase as the price of water increases.

3.2.2.1 Excessive Groundwater Abstraction and Land Subsidence

In regions where the underlying geology consists of interlayering of soft aquitards and aquifers, excessive groundwater abstraction from the interlayered aquifers can cause subsidence of the ground surface. A good case in point is the quaternary sediments that underlie many coastal cities, such as Shanghai, Bangkok, and Jakarta. In Bangkok, the capital of Thailand, for example, the city is situated on a low lying, flat deltaic plain known as the Lower Central Plain, known also as the Lower Chao Phraya Basin – about 30 km north of the Gulf of Thailand. The basement bedrock gently inclines southwards towards the Gulf of Thailand, and the strata overlying the basement bedrock consists of a complex mix of unconsolidated and semi-consolidated sediments of the tertiary to quaternary geologic age. The thickness of the strata ranges from about 400 m in the north to more than 1800 m in the south, with a stratigraphic profile that shows five discernable separate aquifers layers overlain by a stiff surface clay layer. Excessive long-term groundwater abstraction has resulted in subsidence in the region. This causes severe flooding of the region. Yong et al. (1991) reported that with the drainage system existent at that time, extensive flooding lasting for periods of 6 to 24 hours occurs with rainfall exceeding 60 mm.

For coastal cities in similar situations, land subsidence due to prolonged excessive groundwater abstraction can reach the stage where the land surface will reach levels below sea level. When such occurs, unless containment dikes are built, seawater intrusion causing local flooding and contamination of the aquifers can occur. The impacts to human health in respect to contaminated water, in addition to physical problems associated with flooding can be severe. Buildings and other structures have been known to suffer considerable structural distress due to uneven settlement of footings and foundations. By some accounts, delivery of potable water from a central source located in a "safe" region is required when flood waters compromise the existent drinking water sources.

3.2.3 Uses of Water

3.2.3.1 Agricultural, Food, Industrial, and Domestic Uses

By all indications, if sustainable water usage is to be obtained, water treatment and reuse will need to increase, and demand will need to decrease. Decreased water quality through contamination decreases the amount of available water. In the industrialized world, although water quality is in general good, water use is increasing. Many cities lose up to 40% of their water through leaking sewer and distribution systems (United Nations Economic Commission for Europe, 1998; European Environment Agency, 2003). In developing countries, water is both scarce and unhealthy due to rapid urbanization. Water management practices must be improved to reduce scarcities and impacts on ecosystems. Factors affecting leakage include system age, main type and pressure, soil type, climate and topography, and local value of water. A programme by the water regulator in England and Wales reduced water leakage from 29 to 22% between 1992/1993 and 2000/2001. Currently, in the UK, for example, this has further been reduced to approximately 17.6% (www.utilitybidder.co.uk/our-services/business-water/environmental-damage-of-global-water-waste/#stats, accessed Feb. 2024).

Water can also be used as geothermal energy. The sites are mainly in volcanic and seismic regions. Countries such as Iceland, Spain, France, Hungary, Japan, Mexico, Russia, and the US (in the states of California and Hawaii) are exploiting this form of energy (Chamley, 2003). Iceland uses this form of energy for domestic heating, greenhouse cultivation, and electricity production. Exploitation can be from the steam of geysers or very hot liquids of 100 °C or more. Cooled water can then be reinjected for reheating and reuse. Geothermal systems can also be utilized where groundwater at a depth of 100 m is about 15 °C. This is warm enough to heat buildings in the winter using a heat pump principle. There are some hurdles that need to be overcome in exploiting geothermal energy. These include (a) transportation of warm water over long distances without major loss of heat, and (b) earthquakes may be induced. The advantages in the use of geothermal energy include (a) negligible amounts of contamination produced, (b) little or negligible production of greenhouse gases, and (c) almost no waste production.

As water serves multi-purposes, these demands can be highly competitive. Agriculture is an intensive user and thus water use must be more efficient per crop grown. Since ecosystems can be depleted for human water use, public awareness of these links is necessary if conservation and sustainability goals are to be achieved. The Sustainable Development Goals to be accomplished by the year 2030 further emphasized the need for water and food security as follows (https://sustainabledevelopment.un.org/rio20):

Goal 2: End hunger, achieve food security and improved nutrition, and promote sustainable agriculture.

Goal 6: Ensure availability and sustainable management of water and sanitation for all.

3.2.3.2 Cultural Importance of Water

As indicated by the UNESCO 2021 UN World Water Development Report, water is highly important to people of many religions and cultures world-wide. It can be seen as important for life (humans and wildlife), revitalization, renewal, and reconciliation. The connection between water and place, is also strong for many indigenous cultures. For example, in Māori culture, many tribes consider water (directly or indirectly) as the source or foundation of all life. Many cultures also connect with the moods, nature, strength, and tranquility of water.

While water can also be a source of conflict, it can also lead to peace and conflict resolutions. According to the Water as the source of life – Te Ara Encyclopedia of New Zealand (https://teara.govt.nz/en/tangaroa-the-sea/page-5), a fundamental need is the full and effective gender-sensitive participation of all stakeholders in decision-making, allowing everyone to express their own values in their own way.

In June 2023, the Second International High-Level Conference on the International Decade for Action "Water for Sustainable Development" 2018–2028 was hosted by Tajikistan in cooperation with the Netherlands. The Conference reviewed major themes on water, water science, energy, and policy issues. Its Final Declaration "From Dushanbe 2022 to New York 2023" is a milestone in the water agenda and highlights the importance of the participation of Indigenous Peoples and local communities. In the Declaration, high-level representatives of the states, organizations, major groups, and other stakeholder reiterate their determination to achieve the objectives of the Water Action Decade and declare their commitment to

> Demonstrate and scale-up solutions that lead to action underpinned by science and technology innovations, including open and citizen science, women-led, youth-led initiatives, as well as traditional and Indigenous knowledge, to achieve more effective and climate-resilient water and sanitation **management in line with national priorities and circumstances** [...]

In summary, the active participation in decision-making and management of water of all concerned is highly important.

3.3 CHARACTERIZATION OF WATER QUALITY, MANAGEMENT, AND MONITORING

3.3.1 Classes of Contaminants Characterizing Chemical Stressors

Organic contaminants in water will deplete oxygen required for fish and other water organisms. These contaminants generally originate from the discharge of domestic and industrial wastewaters into water bodies. Increases in population density near these water bodies generally result in corresponding increases in the levels of organic matter in the water. High levels of biological oxygen demand (BOD) or chemical oxygen demand (COD) deplete the oxygen in the water through microbial degradation of the organic matter. This depletion of oxygen can lead to severe effects on aquatic biota. Colour, taste, and odour of the drinking water may also be affected.

Farming and agricultural activities have added contaminant sources, such as insecticides, pesticides, fungicides, and fertilizers. For example, at least one pesticide was found in about 94% of water samples and in more than 90% of fish samples taken from streams across the Nation, and in nearly 60% of shallow wells sampled by the USGS (Capel et al., 2018). Herbicides and pesticides are (a) persistent in the environment, (b) highly mobile, and (c) can accumulate in the tissues of animals, producing a variety of ill effects. Data on their concentrations in groundwater are quite limited. So as not to appear prejudicial, all of these contaminant sources will be identified as a group under the term *agro-additives*. These agro-additives find their way into the receiving waters and groundwater through surface run-off and through transport in the ground. Nutrients such as nitrates from animal wastes from poultry or pigs are other sources of groundwater contaminants (Figure 3.3). Agricultural contaminants are typically difficult to monitor and estimate due to their dispersion and seepage through soils into the groundwater.

High levels of nutrients (nitrogen and phosphorus) from fertilizers, detergents, and other sources can reach surface waters and lead to eutrophication and excessive algal growth. Nitrogen, in particular, and, to a lesser degree, phosphorus demand in fertilizers have grown significantly in the last

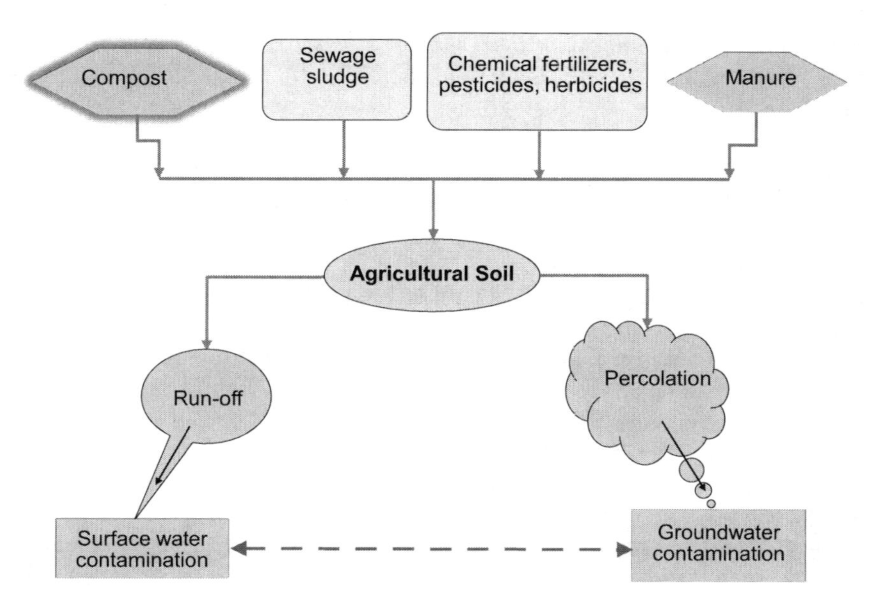

FIGURE 3.3 Contaminant sources (stressors) for surface and groundwater contamination from agricultural soil. The dashed line between groundwater and surface water contamination indicates potential contamination of groundwater from surface water or surface water contamination from groundwater.

50 years (Capel et al., 2018). Eutrophication causes oxygen depletion in the water and production of bad odours, tastes, and colours. Nitrogen, in particular, originates from fertilizer and manure addition to the soil. Although nitrates have an important role within the nitrogen cycle, over-application of manure and fertilizers impact negatively on plant and microorganism biodiversity. Although ammonium ions adsorb onto clay particles, nitrate compounds are easily transported. Movement of nitrates in groundwater depends on aquifer hydraulic conductivity, soil type and moisture, temperature, vegetation, and amount of precipitation. Shallow unconfined aquifers have been found to be highly susceptible to nitrate contamination from agricultural sources (Burkart and Stoner, 2002).

Another major class of contaminants is microorganisms from agricultural run-off, and septic and sewage systems. Microorganisms contribute turbidity, odours, and elevated levels of oxygen demand. Drinking contaminated water can lead to severe gastrointestinal illnesses and even death. The case of Walkerton, Ontario, a small town 200 km northwest of Toronto, is an example of this. In May 2000, heavy rain-washed manure into swampy land, which subsequently contaminated drinking well water. More than 2300 people became ill and seven died as a result of drinking the contaminated well water. This is discussed further in Chapter 9.

Numerous organic compounds are potentially mutagenic and carcinogenic to humans, animals, and plants. We have previously seen from Figures 1.4 and 1.5 in Chapter 1 how industrial, agricultural, and urban discharges can contaminate aquifers and surface waters. As indicated previously, the sources of these contaminants can be traced to waste streams and discharges from industrial plants, households, resource exploitation facilities, and from farms. Oil spills are a major cause of devastation to marine and land ecosystems. Some examples of contaminant groups include pesticides and herbicides, such as dichlorodiphenyltrichloroethane (DDT), aldrin, chlordane, diazonin and partinon, volatile organic compounds (VOCs), such as vinyl chloride, carbon tetrachloride and trichloroethylene (TCE) and heavy metals (e.g., chromium, cobalt, copper, iron, mercury, molybdenum, strontium, vanadium, and zinc). VOCs enter the water systems as industrial and municipal discharges. Due to their higher volatility, they are less persistent than herbicides and pesticides. Metals originate from industrial processing, run-off from mining operations and atmospheric disposition from incinerator emissions and other processes.

All other sources of contaminants shown, i.e., discharges and waste streams, are most likely contained in storage dumps, landfills, holding ponds, tailings ponds, or other similar systems. All of these containment systems have the potential to deliver contaminants to the receiving waters (ground and surface waters) because of eventual leaks, discharges, and failures. Some of these will be discussed in greater detail in a later section dealing with containment systems.

In urban regions, leakage of sewers and other wastewater sources can significantly contribute to recharge and contamination of aquifers. It was estimated that more than 985 million m^3 per year of drinking water (2.6 trillion gallons per year) or 17% of all water is lost due to broken sewers in the United States (www.watermainbreakclock.com). Aquifers under cities can be highly polluted, making them unsuitable for drinking water. This is particularly significant in regions where (a) wastewater is untreated, (b) source contaminants such as nitrates, ammonia, faecal coliforms, and dissolved organic carbon abound, and (c) urbanization is rapid and essentially uncontrolled.

Contamination of the surface waters and groundwater can occur as a result of industrial or municipal discharges or run-off from agricultural land, mining operations, or construction. Industrial contaminants in the groundwater, such as benzene, toluene, xylene, and petroleum products originate from (a) leakage of underground storage tanks, (b) chemical spills, and (c) discharges of organic chemicals and heavy metals, such as cadmium, zinc, mercury, and chromium. The numbers of affected sites in the US have been reported to be at least five or more orders of magnitude (Gleick, 1993). Run-off and seepage from mining operations can contribute significant levels of heavy metals – as, for example, in the illustration shown in Figure 3.4 of run-off of iron from a coal mine.

FIGURE 3.4 Iron run-off from a coal mine into canal area.

Natural sources of contaminants can also contribute to the contamination of groundwater. Saltwater intrusion from coastal aquifers can also degrade groundwater quality by transporting salt into the groundwater (Melloul and Goldenberg, 1997). The arsenic polluted aquifers in West Bengal and Bangladesh discussed previously in Section 2.2.4 in Chapter 2 provide good examples of a combined "*man and nature*" impact on the geoenvironment. Since these aquifers provide potable water for the majority of the population of Bangladesh and some significant proportion of population in West Bengal, it has been estimated that from 35 to 50 million people are at risk of arsenic poisoning. Often referred to as the singular most dramatic case of mass poisoning of the human race, the arsenic polluted aquifers serving the tube wells in the two countries contain arsenic concentrations far in excess of the prescribed limits of the WHO. Worldwide it was estimated that more than 94 to 220 million people in over 70 countries including the US are affected by arsenic in drinking water (WHO, 2004).

In the US, a survey showed that more than 36% of the surface water does not meet water quality objectives (USEPA, 1996). More recently in the *National Water Quality Inventory: Report to Congress* for the 2017 Reporting Cycle www.epa.gov/sites/default/files/2017-12/documents/305brtc_finalowow_08302017.pdf, US states reported that of the assessed river and stream miles, lake acres, and coastal waters wetlands about 46%, 21%, 18%, and 32%, respectively, were impacted. The common types of impairment of lakes are shown in Figure 3.5. Increasing population and economic growth are contributing to this problem. Nutrients loadings are a particular problem in the US affecting 58% of rivers and streams, 40% of lakes and 21% of coastal waters (riverstreamassessment.epa.gov/webreport/, accessed Aug. 2024). Major sources of contaminants are the agricultural industry and urban run-off from storm sewers. Although problematic in some locations, the presence of pesticides and fertilizers is currently not a huge concern. There is not a large amount of data concerning groundwater quality in the United States.

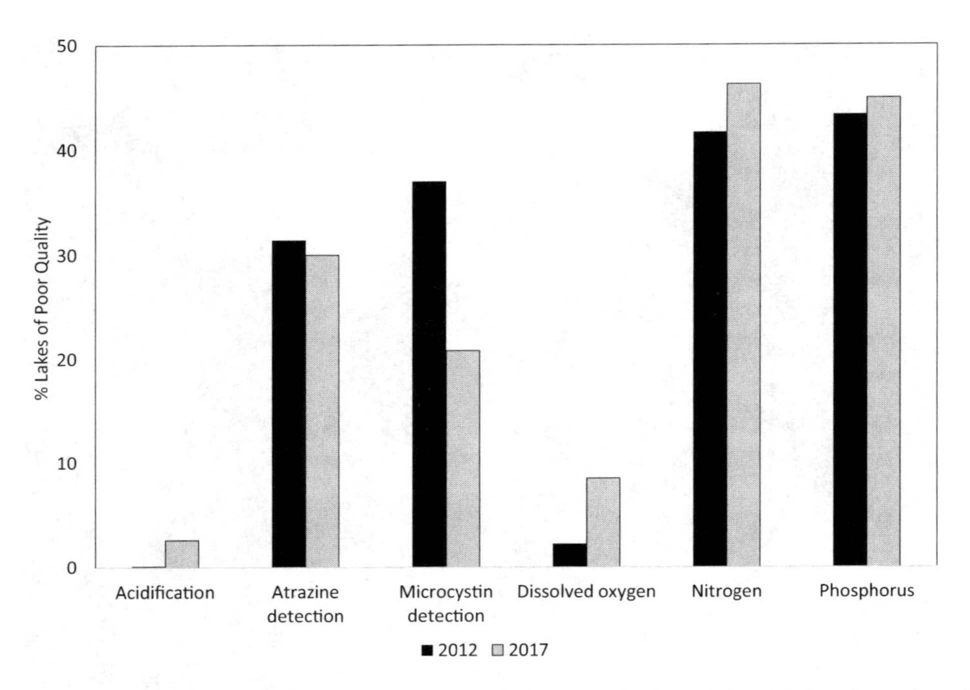

FIGURE 3.5 Percentage of lakes in poor condition according to U.S. EPA National Lakes Assessment 2017 (≥1 hectare) in 2012–2017, data retrieved Dec. 2023 from https://nationallakesassessment.epa.gov/dashboard/?&view=indicator&studypop=al&subpop=national&label=none&condition=poor&diff=2v3

3.3.2 Monitoring of Water Quality

In industrialized countries, concern over the quality of rivers has resulted in a considerable amount of public funds being invested in water quality monitoring during the last decade. Accordingly, monitoring of chemical contaminants in the environmental matrices has entered a new phase. Modifications in instrumentation, sampling, and sample preparation techniques have become essential in keeping pace with the requirements for (a) achieving low detection levels, (b) high speed analysis capability, and (c) convenience and cost-efficiency.

Environmental indicators such as water quality can be used as indicators of sustainability. The term *monitoring* is used in many different ways. In the context of monitoring of a particular site to determine whether the events expected to occur in the site have indeed transpired, *monitoring* means the gathering of all pertinent pieces of information providing evidence that those events had occurred. We interpret from the definition of monitoring in the previous chapter to mean a programme of sampling, testing, and evaluation of status of the situation being monitored. In the situation being monitored, a *management zone* needs to be established – as shown, for example, in Figure 3.6. To determine whether attenuation of contaminants in a contaminated site has been effective, it is necessary to obtain information pertaining to the nature, concentrations, toxicity, characteristics, and properties of the contaminants in the attenuation zone. The contaminants have residence in both the porewater (or groundwater) and on the surfaces of the soil solids, i.e., attached to the soil solids' surfaces. Residence (of the contaminants) associated with the soil solids can take the form of sorbates and co-precipitates. In turn, the sorbates can be complexed with the soil solids and will remain totally fixed within the structure of the soil solids. However, the sorbates can also be held by ionic forces, which can be easily disrupted, thus releasing the sorbates.

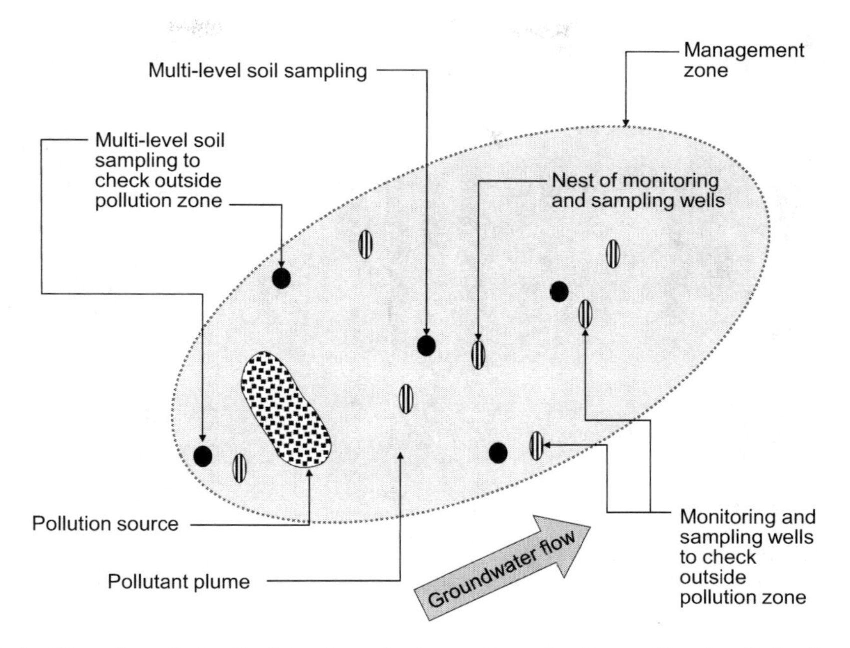

FIGURE 3.6 Plan view of distribution of monitoring wells and soil sampling boreholes for verification monitoring and long-term conformance monitoring. (Yong and Mulligan, 2019.)

What the preceding discussion of residence status of the contaminants tells us is that we need to monitor and sample not only the porewater or groundwater, but also the soil fractions in the contaminant attenuation zone. Two types of monitoring-sampling systems are needed. For porewater or groundwater, monitoring wells are generally used. These wells are necessary to provide access to groundwater at various locations (vertically and spatially) in a chosen location. The choice of type of monitoring wells and distribution or location of wells will depend on the purpose for the wells. In respect to determination of whether natural attenuation can be used as a treatment process, there are at least three separate and distinct monitoring schemes that need to be considered. These range from the initial site characterization studies to verification monitoring and long-term conformance monitoring.

The term *monitoring scheme* is used deliberately to indicate the use of monitoring and sampling devices to obtain both soil and water samples. Figure 3.7 shows some typical devices used as monitoring wells to permit monitoring groundwater at various levels. In the left-hand group, we see individual monitoring wells with sampling ports located at different depths but grouped together in a shared borehole. This is generally identified as a single bore-hole multi-level monitoring well system. The middle drawing shows a nest of single monitoring wells in their own separate boreholes, and the right-hand drawing shows a single tube system with monitoring portholes located at the desired depths. With present technological capabilities, monitoring wells and the manner of operation have reached levels of sophistication where downhole sample analysis of groundwater can be performed without the need for recovery of water samples.

Site characterization monitoring is necessary to provide information on the hydrogeology of the site. It is necessary to properly characterize sub-surface flow to fully delineate or anticipate the transport direction and extent of the contaminant plume. Determination of the direction and magnitude of groundwater flow is most important. Obviously, this means a judicious distribution of monitoring wells up-gradient and down-gradient. A proper siting of the monitoring wells and

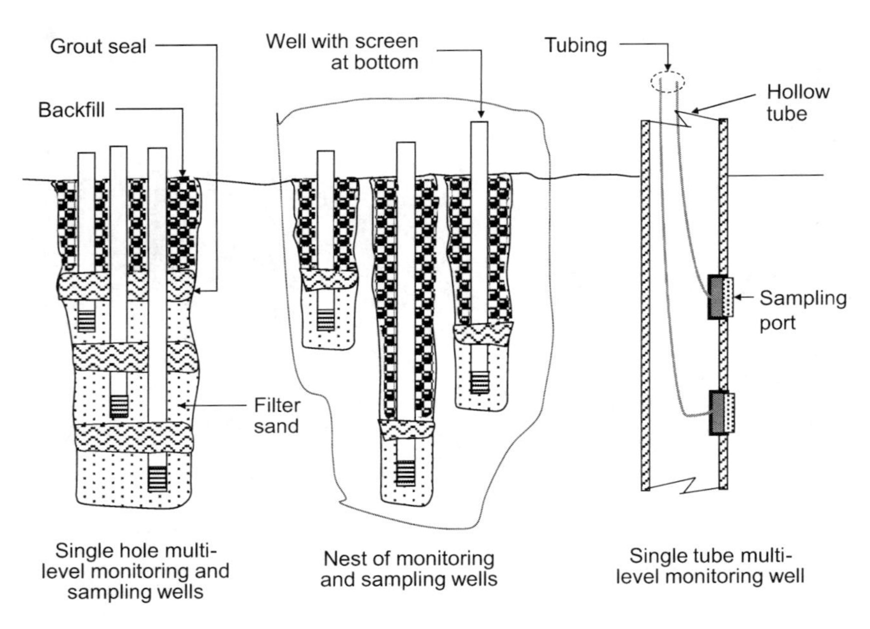

FIGURE 3.7 Some typical groundwater monitoring and sampling wells. (Yong and Mulligan, 2019.)

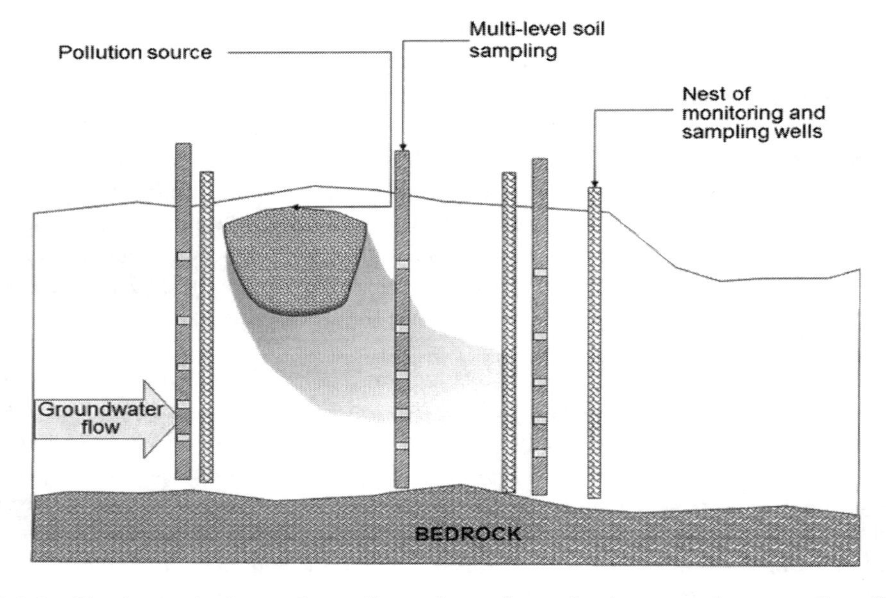

FIGURE 3.8 Simple monitoring and sampling scheme for evaluating groundwater quality. (Yong and Mulligan, 2019.)

analysis of the results should provide one with knowledge of the source of the contaminants and the characteristics of the contaminant plume.

Verification monitoring requires placement of monitoring wells and soil sampling devices within the heart of the contaminant plume and also at positions beyond the plume. Figures 3.6 and 3.8 show the vertical and plane views of how the wells and sampling stations might be distributed. It is a truism to state that the more monitoring and sampling devices there are, the better one is able to properly characterize the nature of the contaminant plume – assuming that the monitoring wells and sampling devices are properly located. The monitoring wells and sampling devices placed outside the contaminant plume, shown in Figure 3.6 will also serve as monitoring wells and sampling devices for long-term conformance assessment.

The tests required of samples retrieved from monitoring wells are designed to determine the nature of the contaminants in the pore or groundwater. These will inform one about the concentration, composition, and toxicity of the target contaminant. For prediction of further or continuous attenuation of the target contaminant, the partition coefficients and solubilities of the various contaminants are needed as input to transport and fate models. If biotransformation of the target contaminants has occurred, supporting laboratory research would be needed to determine the likely fate of the transformed or intermediate products.

3.3.2.1 Remote Sensing

Most of the existing technologies for monitoring algae and cyanobacteria rely on microscopic techniques that are laborious and highly variable. Recently, algae and algal blooms can be detected by satellite imaging systems and spectrofluorimetry (Gitelson and Yacobi, 1995). Subsequently, Millie et al. (2002) have demonstrated that absorbance and fluorescence spectra can be used to discriminate microalgae. This method can potentially be applied as part of the *in situ* monitoring programmes.

Remote sensing by Landsat satellites has been evaluated by Zhu et al. (2002) for the Pearl River in South China, Hong Kong, and Macau region and by Han and Jordan (2005) in the US. Sea color was monitored, and the images were analysed by an algorithm of "Gradient Transition". *In situ* optical, chemical, and biochemical measurements for chlorophyll-a and phytoplankton and other parameters correlated well with ocean colour. Han and Jordan (2005) found that the first derivative spectra and the best wavelengths could correlate well for determining the chlorophyll content of biofilms.

3.3.2.1.1 Chemical and Microbial Analyses

Significant advances have been made in the last decade concerning the levels of detection, information about contaminants including speciation and the speed of monitoring of contaminants in water, as regulatory requirements become more demanding. For example, the US Groundwater Rule is established to provide a risk-based regulatory strategy for groundwater systems that can be sensitive to faecal contamination.

There has been significant advancement in remote sensing of water quality retrieval due to the development of unmanned aerial vehicles (UAV), hyper-spectral sensors, and artificial intelligence (Yang et al., 2022). Landsat images can be used for small inland water areas, such as lakes, due to high resolution at 30 m. Coarser resolution images of about 1 km for MODIS, MERIS, and SeaWiFS are more useful for coastal areas. Although other programs of higher resolution exist, Landsat programs are at an advantage due to the existence of records back to the 1970s. In addition, the Landsat program is open source (Adjovu et al., 2023). Another example is the Arsenic Rule where the rule agreed that 10 μg/L of arsenic would be the standard from 2002 onward for drinking water (www.epa.gov/dwreginfo/drinking-water-arsenic-rule-history, accessed Feb. 2024), drinking water systems had to comply by 2006. Analytical methods for determining the different arsenic species include SPME and solid-phase extraction (SPE) with GS/MS, Liquid chromatography (LC)/electrospray ionization mass spectrometry (ESI-MS), LC/ICPMS and ion chromatography (IC) /inductively coupled plasma mass spectrometry (ICP-MS). Field kits have also been used, particularly in Bangladesh and West Bengal. However, there have been problems with their accuracy (Erickson, 2003). The most applicable methods for arsenic analyses are methods, such as ICP-MS, LC-MS, LC/ICP-MS, but other methods that are more novel and able to be used on site are based on biosensors, nanomaterials, and electroanalytical or electronic devices. These methods could prove to be sensitive, reliable, and cost-effective in the future (Bhat et al., 2023).

Following water contamination cases such as Walkerton in Ontario, Canada where *Escherichia coli* gastroenteritis occurred in 2000, the importance of analytical methods for microorganism detection has increased. Due to the deficiencies in the standard methods for detection of viruses, bacteria, and protozoa, particularly length of time, various new methods have been developed. These include (a) immunofluorescent antibodies techniques, (b) fluorescent *in situ* hybridization, (c) magnetic bead cell sorting, (d) electrochemiluminescence, (e) amperometric sensors, (f) various polymerase chain reactions (PCR), RT-PCR, and (g) real-time PCR methods (Straub and Chandler, 2003), and (h) immunomagnetic separation. Sandrin and Demirev (2018) have reviewed mass spectrometric techniques for characterization of microorganisms.

Analysis of herbicides and pesticides is also of interest because of their effects on human health and the environment. A field kit has been evaluated by Ballesteros et al. (2001) for analysis of triazine herbicides in water samples. Detection levels by this ELISA-based technique were 0.1 μg/L for atrazine and 0.5 μg/L for triazine. Other methods have now been developed to analyse enantiomeric pesticides. Studies were performed by enantioselective GC/MS from 1997 to 2001 on a lake that received agricultural run-off in Switzerland (Poiger et al., 2002). Pre-1998, samples showed the dominance of the racemic metolachlor, whereas samples collected in 2000 to 2001 showed a clear dominance of the S-isomer. This coincided with the commercial switch from the racemic to the S form for agricultural use. Pharmaceutically active compounds and other personal care products have also been identified in surface run-off from fields irrigated with wastewater treatment effluents (Pedersen et al., 2002). A review of analytical methods for pesticides and herbicides included extraction methods, chromatographic or mass spectrometric techniques, spectrophotometric techniques, electrochemical techniques, chemiluminescence and fluorescence methods, and biochemical assays (Wang et al., 2019).

Methods, such as GC, HPLC, GLC, FTIR, and atomic absorption spectrometry (AA), are accurate, and limits of detections are continually decreasing due to combined techniques (e.g., GC/MS, LC/MS, CE/MS, CE/ICP-MS, and ICP-MS). Polar compounds, prior to LC/MS were extremely difficult to detect. Improved methods of sample preparation have reduced solvent usage and are more environmentally friendly and rapid. Although efforts have also been made to automate and simplify many of the technologies, they must continue to enable more widespread use of the technology. Recoveries from solid-phase extraction will need to improve to enable one to obtain more reliable data from this technique. More extensive testing needs to be done with real environmental samples – not just standards – to more fully understand interferences within the samples. This will enable one to (a) monitor water quality, (b) determine the origin or source of biological and chemical contaminants, and (c) determine the transport and fate of the contaminants in the environment.

Using state-of-the-art analytical chemistry instrumentation, researchers are working to determine appropriate analytical methods to characterize and quantify total microplastics in sediment and water samples, as well as the different types of plastic polymers. This research helps inform recommendations for best practices and standardized methodologies to characterize and assess the extent of micro- and nanoplastic pollution in water.

3.3.2.2 Biomonitoring

Biomonitoring is used to indicate the effect and extent of contaminants in the water. It includes determining changes in species diversity, composition in a community, and in the mortality rates of a species. Buildup of contaminants in the tissues of individuals can also be evaluated, in addition to physiological, behavioural, and morphological changes in individuals. The effect of specific contaminants is difficult to determine. Biomonitoring involves the determination of the numbers, health, and presence of various species of algae, fish, plants, benthic macroinvertebrates, insects, or other organisms as a way of determining water quality (USEPA, 2000). Knowledge of background information is essential. Attached algae (known as periphyton) are good indicators of water quality since they grow on rocks and other plants in the water and cannot avoid any pollutants. The advantages for using these as indicators are (a) high numbers of species are available, (b) their responses to changes in the environment are well-known, (c) they respond quickly to exposures, and (d) they are easy to sample. An assessment could include (a) determination of the biomass by chlorophyll or on an ash-free dry basis, (b) species, (c) distribution of species, and (d) condition of the attached algae assemblages. Their use is becoming more widely incorporated in monitoring programmes.

Indicator species within the aquatic environment can highlight when there are sudden abrupt changes and long-term gradual shifts in water quality. Multiple indices have been developed for river and stream environments globally, mostly based on aquatic macroinvertebrates – insects and small animals that live for all or part of their lives in water. Biological sampling and observing rivers and streams can also provide opportunities for social engagement and the inclusion of local citizens in water quality monitoring in their local area.

Benthic macroinvertebrates have numerous advantages as bioindicators (protocols). They do not move very far and thus can be used for upstream-downstream studies. Their life span is about a year, enabling their use for short-term environmental changes. Sampling is easy. They are numerous and experienced biologists can easily detect changes in macroinvertebrate assemblages. In addition, different species respond differently to various contaminants. They are also food sources for fish and other commercial species. Many states in the US have more information on the relationship between invertebrates and contaminants than for fish (Southerland and Stribling, 1995).

Fish are good indicators of water quality since (a) they always live in water, (b) they live for long periods of time (2 to over 10 years) (Karr et al., 1986), (c) they are easily identifiable and easy to collect, and (d) they can quickly recover from natural disturbances. They are also consumed by humans and are of importance to sport and commercial fisherman. Fish make up almost 50% of the endangered vertebrate species in the US (Warren et al., 2000).

Aquatic plants (macrophytes) grow near or in water and many of them can serve as indicators of water quality. A lack of macrophytes can indicate quality problems caused by turbidity, excessive salinity, or the presence of herbicides (Crowder and Painter, 1991). Excessive numbers can be caused by high nutrient levels. They are good indicators since they respond to light, turbidity, contaminants such as metals and herbicides and salt. No laboratory analysis is required, and sampling can be performed through aerial photography.

Biosurveys are useful in identifying if a problem exists. Chemical and toxicity tests would then be required to determine the exact cause and source (USEPA, 1991). Routine biomonitoring can be less expensive than chemical tests over the short-term but more expensive over the long-term. Field bioassessment experts are required to obtain and interpret data. However, there are no established protocols. More knowledge is required to determine the effects of contaminants on populations of organisms and better coordination of background data before site contamination. Data on toxicity and chemicals have been combined to evaluate the sustainability of reaction pathways by Zhang et al. (2000). Risk indices were developed for aquatic life or human health as part of environmental index determination.

The USEPA (Barbour et al., 1999) developed a biological data management system linked to STORET, which can store biological data and associated analytical tools for data analysis. STORET enables storing, retrieving, and analyzing biosurvey data. The data can then be processed to evaluate the distribution, abundance, and physical condition of aquatic organisms, and their environment. This was replaced in 2018 by the Water Quality Portal (WQP), which has over 380 million water quality data records on the Water Quality Exchange (WQX) submitted by 900 federal, state, tribal, and other partners. Everyone can download data from the portal.

3.4 SUSTAINABLE WATER TREATMENT AND MANAGEMENT

To enable the adequacy of water resources for future generations, management practices must control the sources of contamination and limit water use. This requires sufficiency in recharge of aquifers and prevention of contamination of surface water and groundwater. Remediation of contaminated water is required, but as is well-known, effective and complete remediation of aquifers is not easily accomplished. As discussed previously, the quality of both surface water and groundwater need to be protected by mitigation and management procedures. Reuse of treated waste streams, in particular, needs to be practised in farming and agricultural activities by irrigation.

Real-time monitoring and remote sensing and graphical information systems (GIS) are essential for water management. Calera Belmonte et al. (1999) examined the use for GIS tools to manage water resources in an aquifer system using remote sensing from a satellite to determine the spatial distribution of irrigated crops and water pumping estimates. The information obtained enables the GIS to be used as a tool for monitoring and control of water exploitation for agricultural uses.

Machiwal et al. (2018) reviewed the integration of statistical techniques into GIS for groundwater quality. Groundwater quality indices were reviewed. Integration of statistical techniques is highly important for evaluating the water quality data spatially and temporally. Thorslund et al. (2020) have presented a global database for salinity measurements for 45,103 surface water and 208,550 groundwater locations from 1980 to 2019. This represents more than 16.3 million measurements. Salinization of freshwater is of growing importance due to climate change and other impacts. Databases like this will enable a better understanding of the impacts worldwide.

3.4.1 Techniques for Surface and Groundwater Treatment

Contamination of groundwater occurs when chemicals are introduced on or into the ground through such actions as application of pest control aids, direct and indirect discharge of liquid wastes, transport of leachates containing contaminants escaping from landfills, etc. Decontamination of the groundwater requires reduction (elimination?) of the toxicity of the contaminants by (a) sequestration, or by (b) chemical and biologically mediated transformation of the contaminants, and/or (c) removal of the noxious contaminants from the groundwater. In most instances, the groundwater of concern is porewater (water in the void spaces of compact soil). Several techniques are available to manage and control contamination of groundwater by contaminant plumes – to minimize adverse environmental and health impacts. These include (a) construction of impermeable barriers and liner systems for containment facilities, (b) remediation techniques designed to remove or reduce (attenuate) the contaminants in the ground, such as soil flushing, and (c) passive procedures relying on the properties of the ground to reduce contaminant concentrations in leachate streams and contamination plumes.

3.4.1.1 Isolation and Containment

Contaminants can be isolated and contained to (a) prevent further movement, (b) reduce the permeability to less than 1×10^{-7} m/s, and/or (c) increase the strength. (USEPA, 1994). Physical barriers made of steel, concrete, bentonite, and grout walls can be used for capping, vertical and horizontal containment. Liners and membranes are mainly used for protection of groundwater systems, particularly from landfill leachates. A variety of materials are used, including polyethylene, polyvinyl chlorides, asphalt materials, and soil-bentonite or cement mixtures. Monitoring is a key requirement to ensure that the contaminants are not mobilized.

Most *in situ* remediation techniques are potentially less expensive and disruptive than *ex situ* ones, particularly for large, contaminated areas. Natural or synthetic additives can be utilized to enhance precipitation, ion exchange, sorption and redox reactions (Mench et al., 2000). The sustainability of reducing and maintaining reduced solubility conditions is key to the long-term success of the treatment. *Ex situ* techniques are expensive and can disrupt the ecosystem and the landscape. For shallow contamination, remediation costs, worker exposure, and environmental disruption can be reduced by using *in situ* remediation techniques.

3.4.1.2 Extraction Treatment Techniques

To remove NAPLs from the groundwater, extraction of the groundwater can be performed by extraction pumping of the contaminated dissolved phase and/or free phase NAPL zone. Drinking water standards can be achieved with the method of treatment. However, substantial periods of time can be required before this occurs. To enhance the removal rates of the contaminants, extraction solutions can be infiltrated into the soil using surface flooding, sprinklers, leach fields, horizontal or vertical drains. Water with or without additives is employed to solubilize and extract the contaminants as shown in Figure 3.9. Chemical additives include organic or inorganic acids, bases, water soluble solvents, complexing agents and surfactants. Removal efficiencies are related to, and affected by, soil pH, soil type, cation exchange capacity, particle size, permeability, and the type of contaminants. High soil permeabilities (greater than 1×10^{-3} cm/sec) are considered to be beneficial for such procedures.

Volatile components can also be removed from the groundwater by air sparging. In this technique, bubbles of air are injected into the groundwater to strip NAPLs and to add oxygen for *in situ* bioremediation. It has been successfully used for dissolved hydrocarbon plumes (Bass et al., 2000). Reduction of the NAPL zone may then allow natural attenuation processes to proceed. Bioventing is a variation of this technique where lower aeration rates are used to promote aerobic biodegradation instead of volatilization.

3.4.1.3 Natural and Enhanced Natural Attenuation

The attenuation of contaminants and/or contaminants due to the assimilative processes of soils refers to the reduction of concentrations and/or toxicity of contaminants and contaminants during transport in soils. This process is discussed in detail in Section 10.7 in Chapter 10 where the use of soils as a waste management tool is addressed. For the present, we will examine some of the phenomena pertinent to the present context of water and groundwater controls. Reduction in concentrations and toxicity of contaminants in the groundwater can be accomplished by (a) dilution because of mixing with uncontaminated groundwater, (b) interactions and reactions between contaminants and soil solids that can lead to partitioning of the contaminants between the soil solids and porewater, and (c) transformations that reduce the toxicity threat posed by the original contaminants. Short of

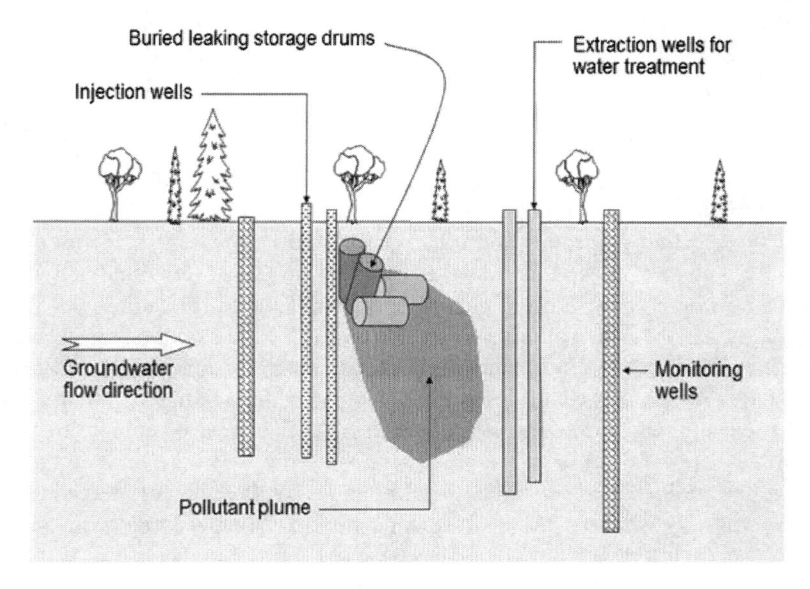

FIGURE 3.9 Schematic diagram of a soil flushing process for removal of contaminants.

overwhelming dilution with groundwater, it is generally acknowledged that partitioning is by far the more significant factor in attenuation of contaminants and/or contaminants.

Natural attenuation refers to the situation when attenuation of contaminants results because of the processes that contribute to the natural assimilative capacity of soil. This means that contaminant attenuation occurs as a result of the natural processes occurring in the soil during contaminant–soil interaction. Broadly speaking therefore, natural attenuation refers to natural processes occurring in the soil that serve to reduce the toxicity of the contaminants and/or the concentration of the contaminants. These natural processes of contaminant attenuation include dilution, partitioning of contaminants, and transformations. They involve a range of physical actions, chemical, and biologically mediated reactions, and combinations of all of these.

According to the United States National Research Council (NRC), the sustainability of natural attenuation is dependent on the sustainability of the mechanisms for immobilizing or destroying contaminants while the contaminants are being released into the groundwater. A mass balance analysis can be used to estimate the long-term destruction or immobilization rates (NRC, 2000). For hydrocarbons, the availability of electron acceptors or donors may be evaluated to determine the sustainability of remediation techniques, such as natural attenuation for hydrocarbons. However, in the case of metals and metalloids, such as arsenic, this approach is only applicable if the attenuation is biologically driven.

Monitored natural attenuation (MNA), because of its adherence to "*remedy by natural processes*" necessitates a proper understanding of the many principles involved in the natural processes that contribute to the end result. Monitoring of the contaminant plume at various positions away from the source is a key element of the use of MNA. Remembering that this (MNA) is a contaminant and soil-specific phenomenon, one generally tracks a very limited number of contaminants, and specifically the ones considered to be the most noxious. Historically, more attention has been paid to documenting the properties and characteristics of the contaminants. By and large, the contaminants tracked have primarily been the organic chemicals including, for example, chlorinated solvents (PCE, TCE, and DCE), and hydrocarbons, such as benzene, toluene, ethyl benzene, and xylene (BTEX). MNA may also be applicable for MTBE in special cases (ITRC, 2005).

When active controls or agents are introduced into the soil to render attenuation more effective, this is called *enhanced natural attenuation*. This is to be distinguished from *engineered natural attenuation* (EngNA), which is probably best illustrated by the permeable reactive barrier shown in Figure 3.10 and the barrier-liner system shown in Figure 3.11. *Enhanced natural attenuation* (ENA) refers to the situation where for example nutrient packages are added to the soil system to permit enhanced biodegradation to occur, or where catalysts are added to the soil to permit chemical reactions to occur more effectively. ENA could include biostimulation and/or bioaugmentation as follows.

Probably the simplest procedure for improving the intrinsic bioremediation capability of a soil is to provide a stimulus to the microorganisms that already exist in the site. This procedure is called *biostimulation* – i.e., adding nutrients and other growth substrates, together with electron donors and acceptors. The intent of biostimulation is to promote increased microbial activity with the set of stimuli to better degrade the organic chemical contaminants in the soil. With the addition of nitrates, Fe (III) oxides, Mn (IV) oxides, sulphates, and CO_2, for example, anaerobic degradation can proceed. This technique is used for sites contaminated with organic chemical contaminants and is perhaps one of the least intrusive of the methods of enhancement of natural attenuation. The other method of enhancement, which falls in the same class of "less-intrusive" enhancement procedure is bioaugmentation.

If the native or indigenous microbial population is not capable of degrading the organic chemicals in the soil – for whatever reason, e.g., concentrations, inappropriate consortia, etc. – other microorganisms can be introduced into the soil. These are called exogenous microorganisms. Their

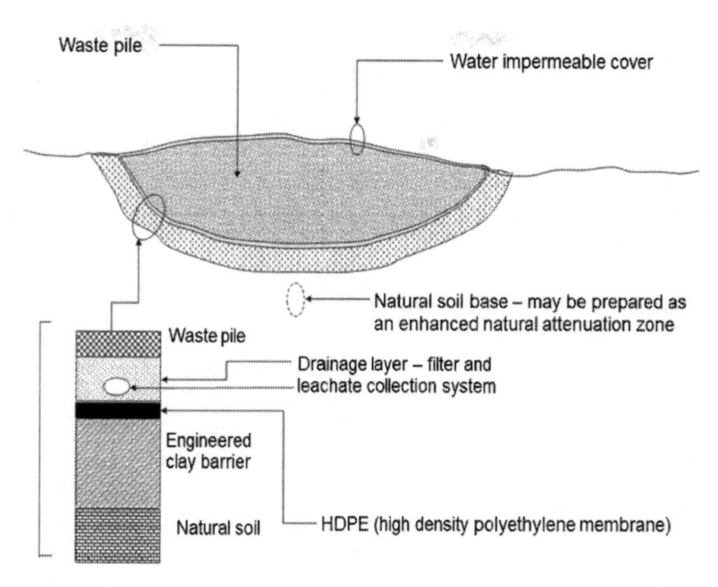

FIGURE 3.10 Enhancement of natural attenuation using treatment wells. Treatments for enhancement can be any or all of the following: geochemical intervention, biostimulation, and bioaugmentation. Treatment occurs in the contaminant plume and downgradient from the plume. (Yong and Mulligan, 2019.)

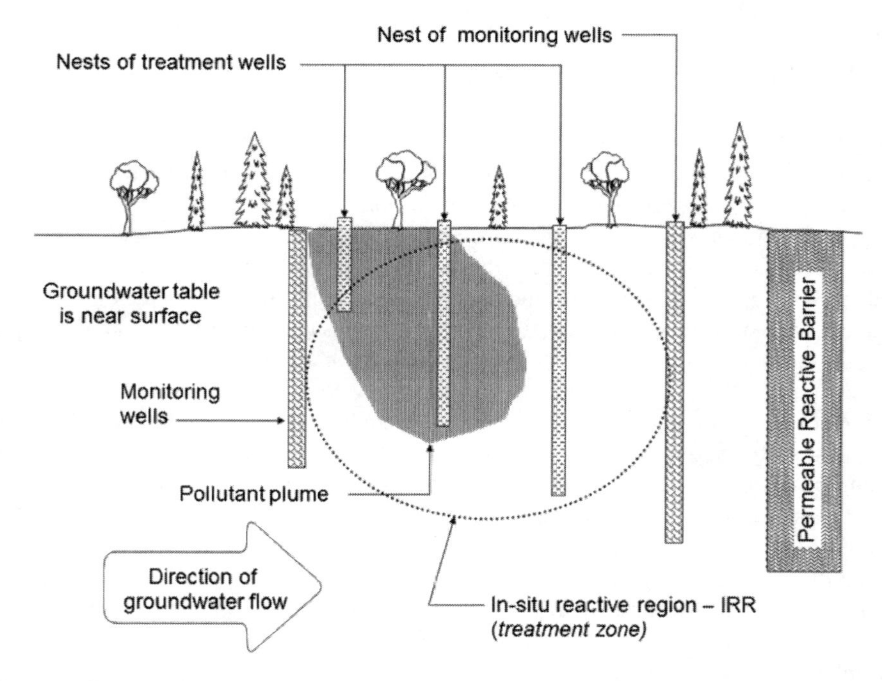

FIGURE 3.11 Contaminant attenuation layer constructed as part of an engineered barrier system. The dimensions of the attenuation layer and the specification of the various elements that constitute the "filter, membrane and leachate collection system" are generally determined by regulations or by performance criteria. (Yong and Mulligan, 2019.)

function is to augment the indigenous microbial population such that effective degradative capability can be obtained. If need be, biostimulation can also be added to the bioaugmentation to further increase the likelihood of effective degradative capability. We need to be conscious of the risks that arise when unknown results are obtained from interactions between the genetically engineered microorganisms and the various chemicals in the contaminated ground. The use of microorganisms grown in uncharacterized consortia, which include bacteria, fungi, and viruses can produce toxic metabolites (Strauss, 1991). In addition, the interaction of chemicals with microorganisms may result in mutations in the microorganisms themselves, and/or microbial adaptations.

We show in Figure 3.10 a direct application of enhanced natural attenuation (ENA) as an *in situ* remediation process. Enhanced treatment of a region (spatial and vertical) of the site down-gradient from the contaminated site permits the ENA to function as planned. The treated region is called the *in situ reactive region* (IRR) or *treatment zone* and can be used in conjunction with other treatment procedures. Figure 3.10 illustrates the use of the IRR as a treatment procedure for the contaminant plume in the region in front of the permeable reactive barrier. Treatment procedures using treatment wells or boreholes and associated technology include

- geochemical procedures, such as pH and Eh manipulation,
- soil improvement techniques, such as introduction of inorganic and organic ligands, introduction of electron acceptors and donors, and
- various other biostimulation procedures, and bioaugmentation.

The choice of any of these, or a combination of these methods of augmentation will depend on the type, distribution, and concentration of contaminants in the contaminated site, and also on the results obtained from microcosm and treatability studies.

3.4.1.4 Permeable Reactive Barriers (PRB)

We have seen from Figure 3.10 an example of the use of a treatment zone, known also as an *in situ* reactive region (IRR) – i.e., the region immediately in front of the permeable reactive barrier. The purpose of an IRR is to provide not only pre-treatment or pre-conditioning in support of another treatment procedure, but also as a post-treatment process for sites previously remediated by other technological procedures. In the drawing shown in Figure 3.10, we show the IRR used in support of the permeable reactive barrier (PRB) treatment procedure. Other treatment procedures can also be used in place of the PRB. The presence of heavy metals in combination with organic chemicals in the contaminant plume is not an uncommon occurrence. One could, for example, envisage using IRR as a treatment procedure in combination with a subsequent procedure designed to fix or remove the metals.

In application of IRR as a post-treatment process, one is looking towards the IRR as the *final cap* for some kind of design or technological process for remediation of a contaminated site. This is generally part of a multiple-treatment process – as opposed to the use of IRR in a pre-treatment or pre-conditioning process. A good example of this is the use of pump–treat procedure as the first phase of the remediation programme, followed by the IRR as a post-treatment process where the treated contaminant plume will receive its final clean-up. The efficiency of clean-up using pump-and-treat methods rapidly decreases as greater contaminant extraction from the groundwater or porewater is required. It is not unusual to remove some large proportion of the contaminants from the groundwater or porewater, and to leave the remaining proportion to be removed via natural attenuation processes in an *in situ* reactive region (IRR).

The intent of a permeable reactive barrier (PRB) is to provide treatment as a remediation procedure to a contaminant plume as it is transported through the PRB so that the plume no longer poses a threat to biotic receptors when it exits the PRB. Figure 3.12 shows a funnel and gate arrangement of a PRB application where the contaminant plume is channelled to the PRB gate by the impermeable walls. Transport of the contaminant plume through the PRB allows the various assimilative and

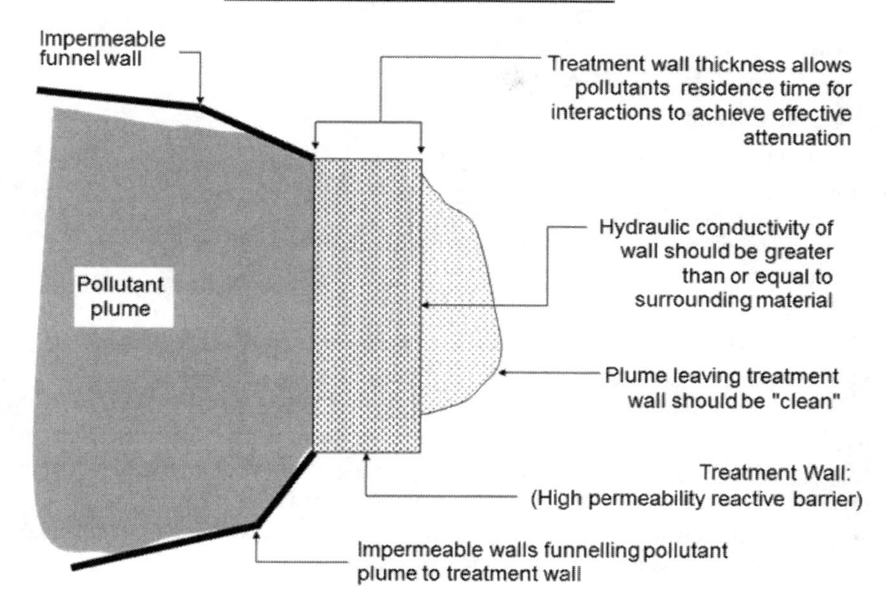

FIGURE 3.12 Funnel and gate arrangement of PRB treatment of contaminant plume. Funnel effect is provided by the impermeable walls that channel contaminant plume transport to the PRB gate. (From Yong and Mulligan, 2019.)

biodegradative mechanisms of the treatment wall to attenuate the contaminants. The PRB needs to be strategically located downgradient to intercept the contaminants.

PRBs are also known as treatment walls. The soil materials in these walls or barriers can include a range of oxidants and reductants, chelating agents, catalysts, mulch and other biological materials, microorganisms, zero-valent metals, zeolite, reactive clays, ferrous hydroxides, carbonates and sulphates, ferric oxides and oxyhydroxides, activated carbon and alumina, nutrients, phosphates, and soil organic materials. The choice of any of these treatment materials is made on the basis of site-specific knowledge of the interaction processes between the target contaminants and material in the PRB but zero-valent iron (ZVI) is the most common. Rapid clogging or deactivation of the material must be avoided. Site geology, geochemistry, and microbiology, in addition to contaminant concentrations and properties must be determined. Laboratory tests and treatability studies are essential elements of the design procedure for the treatment walls (PRBs). When designed properly, a PRB provides the capability for assimilation of the contaminants in the contaminant plume as it migrates through the barrier. In that sense, PRBs function in much the same manner as *in situ* reactive regions (IRR) – except that the region is a constructed barrier. Some of the assimilative processes in the PRB include the following:

- Inorganic contaminants: Sorption, precipitation, substitution, transformation, complexation, oxidation, and reduction.
- Organic contaminants: Sorption, abiotic transformation, biotransformation, dichlorination, abiotic degradation, biodegradation.

Use of natural attenuation for management of contaminant transport and transmission in soil: We have at least three ways in which natural attenuation can be used to manage and/or control the transport of contaminants in soil. These include monitored natural attenuation (MNA), enhanced natural attenuation (ENA), and engineered natural attenuation (EngNA). These have various benefits and are used as effective tools in the control and management of contaminant and contaminant leachate plumes, and especially in specific bioremediation schemes. More detailed discussions of these can be found in Chapter 10.

3.4.1.5 Bioremediation

Biological techniques can also be combined with the extraction techniques in biosparging (Figure 3.13) and bioslurping processes. Essentially, both techniques add another component to the bioventing technique, as, for example, shown as the solvent extraction procedure. In the bioslurping technique, another dimension to the SVE process is added by using vacuum-enhanced pumping to recover free product (NAPLs).

3.4.1.6 *Ex Situ* Processes

For extracted groundwater, treatment is required before discharge or utilization of the abstracted groundwater as drinking water. These techniques are usually quite extensive, involving extraction of substantial groundwater. Standard physical, chemical, or biological wastewater treatment techniques are utilized. Physical-chemical techniques include physical and/or chemical procedures for removal of the contaminants including precipitation, air stripping, ion exchange, reverse osmosis, electrochemical oxidation, etc.

Techniques for groundwater treatment for arsenic and waste considerations are highlighted in Table 3.1. Treatment methods need to minimize the wastes produced to ensure that these processes are sustainable. To evaluate the sustainability of these methods, several factors including materials, energy, transportation, and waste management requirements for the treatment process need to be taken into consideration. One of the principal methods is ion exchange. However, in some cases, simple ion exchange techniques are insufficient. An example of this is As(III). Oxidation of this form to As(V) must be required and performed with a preoxidation filter. Although this method is highly efficient, disposal of a toxic arsenic waste from the regeneration of these filters and the ion exchange resins as a result of the water treatment procedures impacts the sustainability of these processes. These purification activities generate significant wastes that can severely impact the

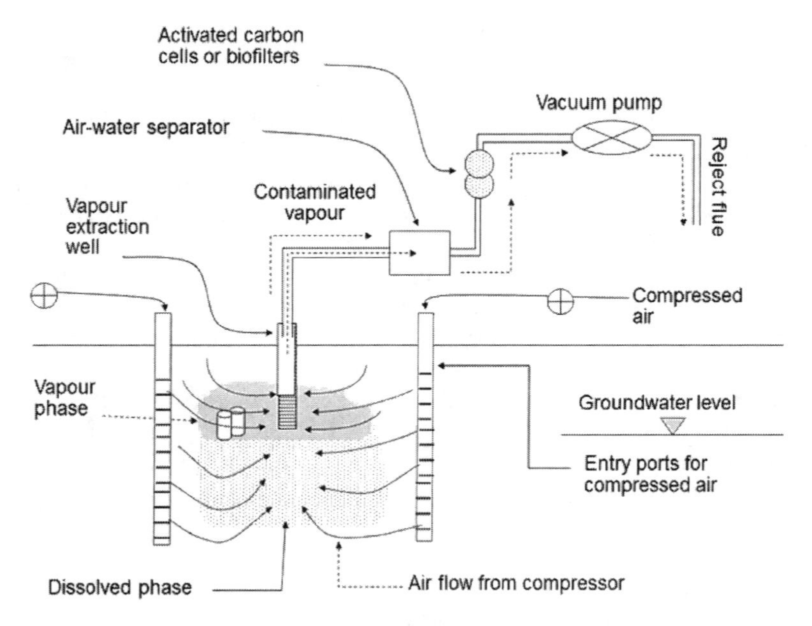

FIGURE 3.13 Schematic of a biosparging process in combination with SVE. Note that a series of compressed air wells and SVE wells can be introduced into the ground – connected in series or in parallel. The number of wells that can be introduced will depend on the capacity of the compressor and vacuum pump systems.

TABLE 3.1
Comparison of Technologies for the Remediation of Arsenic Contaminated Groundwater

Technology	Waste Stream	Treatment of Waste	Disposal Options
Coagulation/filtration	Ferric sludge, redox sensitive, 97% water content	Dewatering and drying	Landfill after dewatering, brick manufacture (Rouf and Hossain, 2003)
Activated alumina with regeneration	Alkaline and acidic liquids	Neutralization and precipitation with ferric salts	Sewer, residual into landfill
Iron oxide filters	Exhausted adsorbent, redox sensitive, < 50% solids, passes TCLP test*	No treatment	Landfill, immobilization, brick manufacture (Rouf and Hossain, 2003)
Ion exchange	Liquid saline brine	Precipitation with ferric salts	Sewer, brine discharge, landfill for residual, possible recycling of brines
Membrane techniques such as reverse osmosis or nanofiltration	Concentrated liquids can be hazardous	None performed	Sewer or brine discharge
Greensand filtration (with iron)	Particles that clog the filter	Dewatering/drying	Landfill disposal
Electrocoagulation	Sludge sometimes, electrodes need to be replaced	Sludge easily dewatered	Landfill disposal
Lime softening	Backwash water	pH adjustment	Sludge for landfill
Phytoremediation	No chemicals used	None needed	Possible disposal of plants to landfill

Source: Adapted from Driehaus (2005); Thomas et al. (2022).
*TCLP refers to the Toxicity Characteristic Leaching Procedure – see Figure 7.4 and discussion in Section 7.3.5 in Chapter 7.

environment, causing more harm than good. Due to the problems of arsenic in the groundwater, economic solutions need to be found to ensure the safety of the drinking water.

Several common treatment technologies are used for removal of inorganic contaminants, including arsenic, from drinking water supplies. Large-scale treatment facilities often use conventional coagulation with alum or iron salts followed by filtration to remove arsenic. Lime softening and iron removal also are common, conventional treatment processes that can potentially remove arsenic from source waters. Treatment options identified by EPA include ion exchange, reverse osmosis, activated alumina, nanofiltration, electrodialysis reversal, coagulation/filtration, lime softening, greensand filtration and other iron/manganese removal processes, and emerging technologies not yet identified (USEPA, 2003). New combinations of materials as sorbents such as granular activated carbon (GAC) supported nanoscale zero-valent iron (nZVI) are being developed (Chowdhury and Mulligan, 2013). Synthesis and evaluation of Fe/Cu nanoparticles with a mean diameter of 13.17 nm were performed for the removal of As(III) and As(V) at sorption capacities of 19.68 mg/g and 21.32 mg/g, arsenic from at a pH of 7.0 (Babaee et al., 2017).

Other groundwater treatment methods such as oxidation could lead to toxic by-product formation and sludge generation. Adsorption is considered as very promising techniques as it is effective, economical and socially acceptable (Liu et al., 2015; Nicomel et al., 2015). An example of a community system is the Kanchan Arsenic Filter (KAF). It is an affordable household water treatment device for rural communities for removal of arsenic and contaminants for drinking water purposes. The KAF is constructed and operated by locally trained technicians. Inexpensive, locally available, environmental-friendly materials including sand and iron nails are employed. Maintenance requirements are very low as no or very little replacement of materials is needed (CAWST, 2012; Zaman et al., 2014). The effect of water composition was evaluated by Nguyen and Mulligan (2023). Other low-cost methods were reviewed by Hassan (2023).

Methods for treatment of MTBE include bioremediation, biobarriers, *in situ* oxidation, granular activated carbon (GAC), air-stripping, ozonation, ozone/hydrogen peroxide or phytoremediation (Richardson, 2003; ITRC, 2005). These include ion exchange, reverse osmosis, activated alumina, nanofiltration, electrodialysis, coagulation/filtration, lime softening, greensand filtration, and other processes. Thermal techniques include pyrolysis and super critical water oxidation. For a detailed discussion of standard chemical and biological treatment processes, readers are referred to various textbooks dealing with wastewater treatment – e.g., Metcalf and Eddy Inc (2014).

Previously, effluent quality was the only basis for evaluating treatment capabilities of water treatment processes. Capital, energy, nutrient, energy, and other requirements need to be included to determine if the process under consideration is sustainable for future generations. Recycling of resources needs to be practised as much as possible. Mulder (2003) compared the sustainability of nitrogen removal systems that included (a) conventional activated sludge systems, (b) an activated sludge system that relies on autotrophic nitrogen removal, (c) algal or duckweed ponds, and (d) constructed wetlands (Figure 3.14). The author used six sustainability indicators: production of sludge, energy consumption, resource recovery, space requirements, and N_2O emissions. They determined that the system that combines nitrification and anaerobic ammonia oxidation (autotrophic nitrogen removal) is the most sustainable because (a) organic matter is not required, (b) sludge production is low, and (c) nitrogen removal is high.

More extensive efforts have been performed by Alimahmoodi et al. (2010). It is a decision support tool (GoldSET) developed to incorporate sustainable development principles into engineering projects. It has a new wastewater treatment module that allows the application of sustainability principles through the tool to projects within the wastewater context to provide an assessment of different project options against a number of quantitative and qualitative sustainability indicators for the four dimensions of sustainability: Environment, society, economy, and technology. Indicators provide a way of describing the situation surrounding the project, along with an associated weighting scoring scheme allowing the relative importance of each indicator to be reflected. The scoring scheme assigned to each indicator provides a mechanism to assess the performance of each option with respect to the indicator, producing a comparative graphical result of each option's sustainability performance. The developed module allows for both detailed design-engineering phase

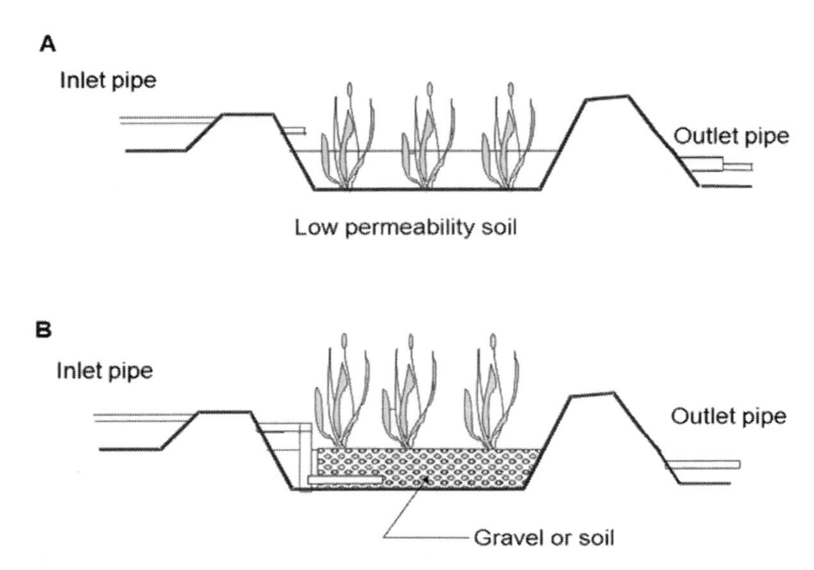

FIGURE 3.14 Overview of (A) surface and (B) subsurface flow constructed wetlands. (Adapted from Mulligan, 2002.)

option assessments and design selections, and for more general project planning decisions, or future upgrades prioritization decisions.

3.4.2 Groundwater and Water Management

An effective groundwater management policy must first involve an evaluation of present practices – beginning with a determination of (a) the basic needs for the water, and (b) laws and regulations that need to be established to ensure water quality and quantity. As shown in Figure 3.15, one must first determine if adequate quantities of groundwater are available to meet the required needs. If not, one needs to manage the system to allow for recharge before depleting this resource. If water budget analyses show that the quantities are sufficient, the quality of the groundwater will then need to be determined. It may be adequate for industrial or irrigation purposes without treatment or perhaps with *in situ* treatments – as indicated in the previous sections. Drinking water quality may require further treatment. Most often this is accomplished by extraction pumping and treatment with a suitable sustainable method – to avoid harming aspects of the environment to protect others. Evaluation of the most sustainable water treatment processes can be performed through procedures similar to that previously described.

Twelve Principles of Green Engineering have been suggested to engineers as a way to improve the sustainability of industrial processes (Anastas and Zimmerman, 2003). The second principle is particularly relevant for the prevention of water contamination. It says that "*It is better to prevent waste than to treat or clean up waste after it is formed*". In other words, processes should be designed to reduce water use and the amount of contaminants that reaches the water so that the water will not have to be treated later on. In the past, many models have been developed for prediction of the impact of certain chemicals in the environment, such as the transport of contaminants from point and non-point sources (Mihelcic et al., 2003). However, they have not focused on how to reduce or prevent the contamination. End of pipe solutions were the main waste management strategy until recently when green engineering has become more prominent.

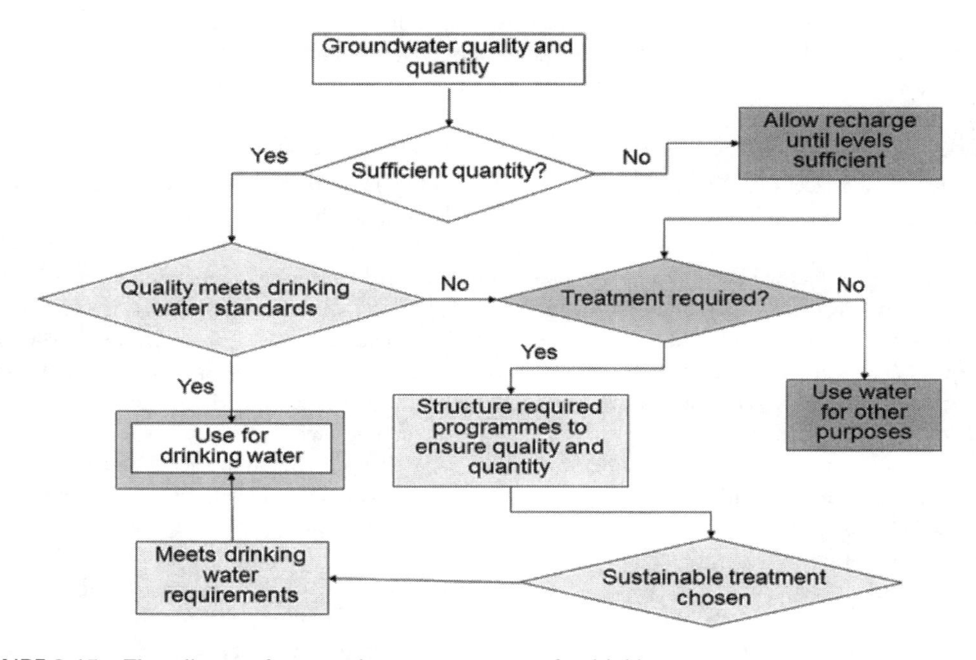

FIGURE 3.15 Flow diagram for groundwater management for drinking water purposes.

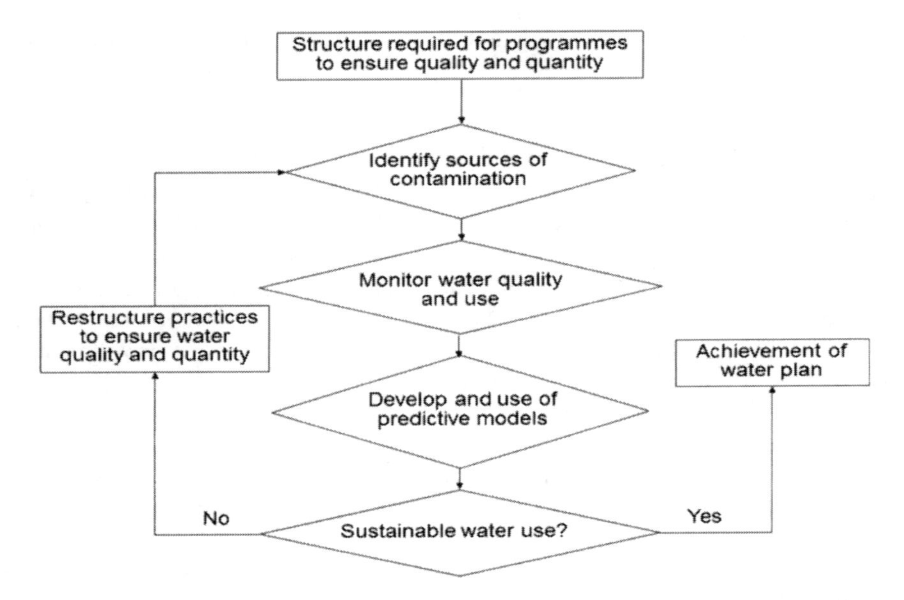

FIGURE 3.16 Flow chart demonstrating development of a programme to ensure sustainable water quantity and quality.

Upon determination of impaired water quality, strategies would need to be developed to prevent the introduction of the contaminants. Groundwater and surface water monitoring and GIS systems, as mentioned previously, will enable the development of the management strategies. The GIS would incorporate all aspects of land use, includung the types of ecosystems, landscapes, and water use. Monitoring will include determination of the quantities of water, the quality in terms of nutrient and contaminant contents and biological monitoring as described previously. Models would need to be developed to predict water discharge and recharges. All the information could then be combined to determine the water management strategy for avoidance of water contamination and optimal water use (Figure 3.16). All of these cannot be achieved however, if the society is not educated concerning water usage and its importance. Legal guidelines must also be issued and followed to protect the quantity of resource water.

3.4.2.1 Evaluation of the Sustainability of Remediation Alternatives

Attention is focused on the problem of arsenic-polluted groundwater because contamination of groundwater from arsenic is a major threat to human health, and because this is both a man-induced and a naturally occurring phenomenon. In choosing the remediation technologies to treat this problem, it is necessary to factor in the targets, exposure routes, future land use, acceptable risks, legislation, and resultant emissions. A schematic illustration of the criteria and tools for evaluating technologies and protocols for environmental management of contaminated soils and groundwater is shown in Figure 3.17. Specific comments are included in Table 3.2 for the various technologies. Other factors that need to be considered to evaluate site remediation technologies include (a) disturbance to the environment, (b) energy use and consumption, (c) solid wastes generated, (d) emissions of contaminants and greenhouse gases into the air, and (e) water and materials used and public acceptance.

The latter aspect is highly important. SDG 6 of the 17 Sustainable Development Goals specifically mentions the participation of the local community for water management. It has been estimated that water treatment often fails or are abandoned within 12 to 17 months due to lack of community engagement (Alfredo and O'Garra, 2020). Brunson et al. (2013) has indicated that there are a number of reasons for failure. Some include lack of culturally appropriate technologies, ownership, and accountability. The community is often left with the operation and maintenance of the systems.

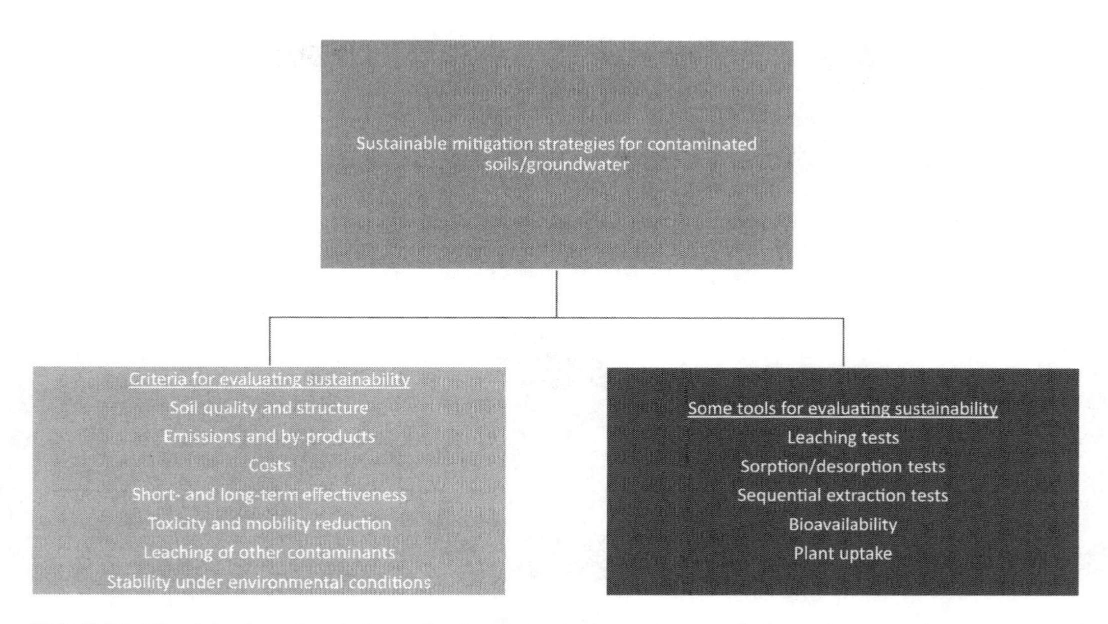

FIGURE 3.17 Criteria and tools for evaluating technologies and protocols for environmental management of contaminated soils and groundwater.

TABLE 3.2
Comparison of Remediation Technologies for Groundwater

Technology	Capital Cost*	Operating Costs	Energy Requirements	Waste Generation	Social Acceptance
Coagulation/flocculants (adsorbents)	Low	Low	Low	Medium	Medium
Biosand filtration	Medium	Low	Low	Low	Medium/high
RO/UF membranes	High	High	High	High	Medium/high
Electrocoagulation	Medium	Medium	Medium	Low	Medium

Source: Adapted from Thomas et al. (2022).

Lack of training and availability of replacement parts. Participation of the communities in the water management will enhance water sustainability.

A philosophy is to combine Indigenous knowledge and science in a kind of "two-eyed" seeing (Bartlett et al., 2012). The Indigenous idea of Two-Eyed Seeing supports a blended experience in the training that respects and weaves together the strengths of both Indigenous and Western ways of Knowing–Being–Doing. The Two-Eyed Seeing approach inherently includes a fundamental respect for differing worldviews. This approach has the capacity to simultaneously acknowledge common ground while remaining cognizant and respectful of the differences. Two-Eyed Seeing helps us recognize the idea of wholeness through each separate, whole eyes, to see things through Indigenous perspectives, in harmony with Western ways of knowing, inviting these eyes to work together in a collective and collaborative way, they see a collision of realities to gain an understanding of the whole picture. Figure 3.18 shows the relationship of social and cultural factors.

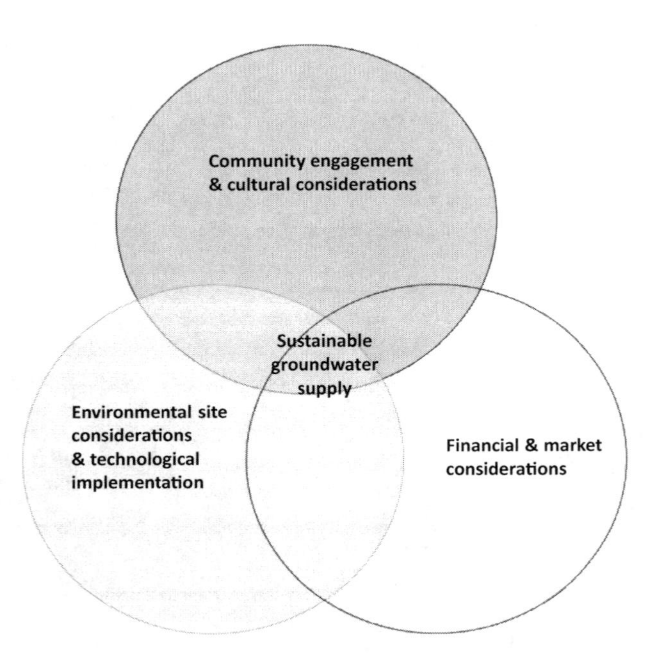

FIGURE 3.18 Sustainable water treatment and management. (Adapted from Thomas et al., 2022.)

3.5 CONCLUDING REMARKS

Water is of utmost importance because without water, living species would perish. The demand and use of water can often produce situations that result in conflicts between humankind and the environment. Degradation or impairment of water quality results from various usages associated with processes and activities associated with farming, natural resources harvesting, industrialization, and urbanization. Management and education in sustainable water usage is required, and sources of contamination must be eliminated to maintain water quality and supply for future generations. Water must be conserved and managed properly for preservation of biodiversity. Failure to do so will result in the diminution of the capability of the geoenvironment to provide the basis for life support. Various remediation tools have been developed to treat water once it has been contaminated. The choice of any of these techniques requires examination of resource depletion, energy requirements, and emissions – if the aspirations of water sustainability are to be fulfilled.

The record shows that outside of the global distribution of water resources, the two great threats to the availability and quality of water resources (surface water and groundwater) are (a) over-use of the available water resources, and (b) contamination of these same water resources. Over-use problems arise from poor management practices and lack of knowledge or ignorance of the nature of the various water budget items and how they contribute to the health of the available water. Control and/or mitigation of the impact of contamination of available water resources from the many contaminant sources are measures that must be undertaken as critical procedures in structuring water sustainability protocols and requirements.

Whilst implementation of remediation technology to improve compromised or impaired water quality is an admission that management and controls on water resource contamination have failed, it is nevertheless a remedy that needs more attention and research. Development of more capable technology to clean contaminated water resources is necessary – to meet the critical demand for clean water from the consumers. Since aquifers or groundwater in general serve as primary water resource for many developed and remote communities, it is imperative that protection of groundwater quality be mandated as the first priority by regulatory agencies. It is also essential that communities are engaged from the beginning in water management. This in essence requires (a) attention to the many forces that individually and collectively produce the contaminants that find their way into the groundwater resource, and (b) control, treatment, and remediation technology to manage these forces.

REFERENCES

Adjovu, G.E., Stephen, H., James, D., and Ahmad, S. (2023), Overview of the application of remote sensing in effective monitoring of water quality parameters, *Remote Sensor*, 15: 1938. https://doi.org/10.3390/rs15071938.

Alimahmoodi, M., Mulligan, C.N., Chalise, A., Grey, V., and Noel-de-Tilly, R. (2010), Tool for Evaluating the Sustainability of Wastewater Treatment Systems. In: *1st International Specialty Conference on Sustaining Public Infrastructure*, Edmonton, Alberta, June 6–9, 2012.

Anastas, P., and Zimmerman, J. (2003), Design through the 12 principles of green engineering, *Environmental Science and Technology*, 37: 94A–101A.

Babaee, Y., Mulligan, C.N., and Rahaman, S. (2017), Removal of arsenic (III) and arsenic (V) from aqueous solutions through adsorption by Fe/Cu nanoparticles, *Journal of Chemical Technology & Biotechnology,* 9: 63–71.

Ballesteros, B., Barcelo, D., Dankwardt, A., Schneider, P., and Marco, M.P. (2001), Evaluation of a field-test kit for triazine herbicides (SensioScreen(R) TR500) as a fast assay to detect pesticide contamination in water samples, *Analytica Chimica Acta,* 475(1–2): 105–116.

Barbour, M.T., Gerritsen, J., Snyder, B.D., and Stribling, J.B. (1999), Rapid Bioassessment Protocols for Use in Streams and Wadable Rivers: Periphyton, Benthic Macroinvertebrates and Fish, 2nd Edition. EPA 841-B-99-002. U.S. Environmental Protection Agency; Office of Water, Washington, DC.

Bartlett, C. Marshall, M., and Marshall A. (2012), Two-eyed seeing and other lessons learned within a colearning journey of bringing together indigenous and mainstream knowledges and ways of knowing, Journal of Environmental Studies and Sciences, 2: 331–340.

Bass, D.H., Hastings, N.A., and Brown, R.A. (2000), Performance of air sparging systems: A review of case studies, *Journal of Hazardous Materials,* 72(2–3): 101–120.

Bhat, S., O Hara, T., Tian, F., and Singh, B. (2023), Review of analytical techniques for arsenic detection and determination in drinking water. *Environmental Science: Advances*, 2: 171–195. https://doi.org/10.1039/D2VA00218C.

Black, M., and King, J. (2009), *The Atlas of Water: Mapping the World's Most Critical Resource*, 2nd ed., University of California Press, Berkeley, CA.

Brunson, L.R., Busenitz, L.W., Sabatini, D.A., and Spicer, P. (2013), In pursuit of sustainable water solutions in emerging regions, *Journal of Water, Sanitation and Hygiene for Development,* 3(4): 489–499. https://doi.org/10.2166/washdev.2013.136.

Boswinkel, J.A. (2000), *Information Note, International Groundwater Resources Assessment Centre (IGRAC)*, Netherlands Institute of Applied Geoscience, Netherlands.

Burkart, M.R., and Stoner, J.D. (2002), Nitrate in aquifers beneath agricultural systems, *Water Science and Technology,* 45(9): 19–29.

Calera Belmonte, A., Medrano Gonzalez, J., Vela Mayorga, A., and Castano Fernandez, S. (1999), GIS tools applied to the sustainable management of water resources: Application to the aquifer system 08-29, *Agriculture Water Management,* 40: 207–220.

Capel, P.D., McCarthy, K.A., Coupe, R.H., Grey, K.M., Amenumey, S.E., Baker, N.T., and Johnson, R.L. (2018), Agriculture — A River runs through it — The connections between agriculture and water quality: U.S., *Geological Survey Circular,* 1433: 201. https://doi.org/10.3133/cir1433.

CAWST. (2012), Biosand filter construction manual. Unpublished manuscript.

Chamley, H. (2003), *Geosciences, Environment and Management*, Elsevier, Amsterdam, 450 pp.

Chowdhury, R., and Mulligan, C.N. (2013), Removal of arsenate from contaminated water by granular activated carbon embedded with nano scale zero-valent iron, GeoMontreal Sept. 29–Oct 2, 2013, Montreal.

Crowder, A., and Painter, D.S. (1991), Submerged macrophytes in Lake Ontario: Current knowledge, importance, threats to stability and needed studies, *Canadian Journal of Fisheries and Aquatic Studies,* 48: 1539–1545.

De Zeeuw, H. (2011), Cities, climate and urban agriculture, *Urban Agriculture Magazine,* 25: 39–42.

Dorf, R.C. (2001), *Technology, Humans and Society, Toward a Sustainable World*, Academic Press, San Diego, 500 pp.

Driehaus, W. (2005), Technologies for Arsenic Removal from Potable Water. In J. Bundschuh, P. Bhattacharaya, and D. Chandrasekharam, (eds.), *Natural Arsenic in Groundwater: Occurrences, Remediation and Management*, Taylor & Francis, London, pp. 189–203.

Erickson, B.E. (2003), Field kits fail to provide accurate measure of arsenic in groundwater, *Environmental Science and Technology,* 37: 35A–38A.

European Environment Agency — Drought and water overuse in Europe (europa.eu), press release in Feb. 12, 2009, www.eea.europa.eu/media/newsreleases/drought-and-water-overuse-in-europe#:~:text=In%20Europe%20as%20a%20whol

FAO. (2011), *The State of the World's Land and Water Resources for Food and Agriculture: Managing Systems at Risk*, FAO and Earthscan, London.

Gitelson, A., and Yacobi, Y. (1995), Spectral Features of Reflectance and Algorithm Development for Remote Sensing of Chlorophyll in Lake Kinneret, *Air Toxics and Water Monitoring, Europto.* In*: SPIE,* 2503, 21 June.

Gleick, P.H. (1993), *Water in Crisis*, Oxford University Press, New York, 504 pp.

Gill, S.E., Handley, J.G., Ennos, A.R., and Pauleit, S. (2007), Adapting cities for climate change: the role of green infrastructure. *Built Environmental*, 33: 115–133.

Green, T.R. (2016), Linking Climate Change and Groundwater. In: Jakeman, A.J., Barreteau, O., Hunt, R.J., Rinaldo, J-D., and Ross, A. (eds.), *Integrated Groundwater Management: Concepts, Approaches and Challenges*, Springer International Publishing, Cham, pp. 97–114.

Han, L.H., and Jordan, K.J. (2005), Estimating and mapping chlorophyll-*a* concentration in Pensacola Bay, Florida using Landsat ETM+ data, *International Journal of Remote Sensing,* 26(23): 5245–5254.

Hassan, H.R. (2023), A review on different arsenic removal techniques used for decontamination of drinking water, *Environmental Pollutants and Bioavailability,* 35(1): 2165964. https://doi.org/10.1080/26395940.2023.2165964.

ITRC (Interstate Technology Regulatory Council). (2005), *Overview of Groundwater Remediation Technologies for MTBE and TBA, Interstate Technology and Regulatory Cooperation Work Group MTBE and Other Fuel Oxygenates Team*, Interstate Technology Regulatory Council, MTBE and Other Fuel Oxygenates Team. Available on www.itrcweb.org.

Jasechko, S., Seybold, H., Perrone, D., Fan, Y., Shamsudduha, M., Taylor, R.G., Othman Fallatah, O., and Kirchneret, J.W. (2024), Rapid groundwater decline and some cases of recovery in aquifers globally, *Nature* 625: 715–721. https://doi.org/10.1038/s41586-023-06879-8.

Karr, J.R., Fausch, K.D., Angermeier, P.L., Yant, P.R., and Schlosser, I.J. (1986), *Assessing Biological Integrity in Running Waters: A Method and its Rationale, Special Publication 5*, Illinois Natural History Survey.

lfredo, K. A., and O'Garra, T. (2020), Preferences for water treatment provision in rural India: comparing communal, pay-per-use, and labour-for-water schemes, *Water International*, 45(2): 91–11. https://doi.org/10.1080/02508060.2020.1720137.

Liu, C.-H., Chuang, Y.-H., Chen, T.-Y., Tian, Y., Li, H., Wang M.-K., and Wei Zhang, W. (2015), Mechanism of arsenic adsorption on magnetite nanoparticles from water: Thermodynamic and spectroscopic studies, *Environmental Science & Technology*, 49(13): 7726–7734. DOI: 10.1021/acs.est.5b00381

Machiwal, D., Cloutier, V., Güler, C., and Kazakis, N. (2018), A review of GIS-integrated statistical techniques for groundwater quality evaluation and protection. Environmental Earth Sciences, 77: 681. https://doi.org/10.1007/s12665-018-7872-x.

Melloul, A.J., and Goldenberg, L.C. (1997), Monitoring of seawater intrusion in coastal aquifers: Basic and local concerns. *Journal of Environmental Management,* 51: 73–86.

Mench, M., Vangronsveld, J., Clijsters, H., Lepp, N.W., and Edwards, R. (2000), In Situ Metal Immobilization and Phytostabilization of Contaminated Soils. In: N. Terry, and G. Bauelos (eds.), *Phytoremediation of Contaminated Soil and Water*, Lewis Publishers, Boca Raton, FL, pp. 323–358.

Metcalf and Eddy Inc. (2014), *Wastewater Engineering: Treatment and Resource Recovery,* 5th ed, McGraw-Hill, Dubuque, IA, 2048 pp.

Mihelcic, J.R., Crittenden, J.C., Small, M.J., Shonnard, D.R., Hokanson, D.R., Zhang, Q., Chen, H., Sorby, S.A., James, V.U., Sutherland, J.W., and Schnoor, J.L. (2003), Sustainability science and engineering: The emergence of a new metadiscipline, *Environmental Science and Technology,* 37: 5314–5324.

Millie, D.F., Schofield, O.M., Kikpatrick, G.J., Johnson, G., and Evens, T.J. (2002), Using absorbance and fluorescence spectra to discriminate microalgae, *European Journal of Phycology,* 27: 313–322.

Mulder, A. (2003), The quest for sustainable nitrogen removal technologies, *Water Science and Technology,* 48(1): 67–75.

Mulligan C. N. (2002), *Environmental Biotreatment*, Government Institutes, Rockville, MD, 395 pp.

National Research Council (NRC). (2000), *Natural Attenuation for Groundwater Remediation*, National Academy Press, Washington, DC, 292 pp.

Nguyen, M.P., and Mulligan, C.N. (2023), Study on the influence of water composition on iron nail corrosion and arsenic removal performance of the Kanchan arsenic filter (KAF), *Journal of Hazardous Materials Advances*, 10: 100285. https://doi.org/10.1016/j.hazadv.2023.100285.

Nicomel, N.R., Leus, K., Folens, K., Van Der Voort, P., and Du Laing, G. (2015), Technologies for arsenic removal from water: Current status and future perspectives. *International Journal of Environmental Research and Public Health* 13(1), 62. doi:10.3390/ijerph13010062.

NOAA (National Oceanic and Atmospheric Administration). (2012), Incorporating Sea Level Rise Scenario at the Local Level. www.ngs.noaa.gov/PUBS_LIB/SLCScenariosLL.pdf, accessed Feb. 1, 2024.

OECD Organization of Economic Co-operation and Development. (2012), *OECD Environmental Outlook to 2050, The Consequences of Inaction*, OECD Publishing, Paris.

Pedersen, J.A., Yeager, M.A., and (Mel)Suffet, I.H. (2002), Characterization and mass load estimates of organic compounds in agricultural irrigation runoff. *Water Science and Technology,* 45(9): 103–110.

Pimentel, D., Wilson, C., McCullum, C., Huang, R., Dwen, P., Flack, J., Tran, Q., Saltman T., and Cliff, B. (1996), Economic and environmental benefits of biodiversity, *BioScience,* 47(11): 747–757.

Poiger, T., Muller, M.D., and Buser, H.R. (2002), Verifying the chiral switch of the pesticide metolachlor on the basis of the enantiomer composition of environmental residues, *Chimica,* 56(6): 300–303.

Richardson, S.D. (2003), Water analysis: Emerging contaminants and current issues. *Analytical Chemistry,* 75: 2831–2857.

Rouf, Md. A., and Hossain, Md. D. (2003), Effects of Using Arsenic-Iron Sludge in Brick Making. In: *Fate of Arsenic in the EnvironmentProceedings of the BUET-UNU International Symposium*, 5–6 February 2003, Dhaka, Bangladesh.

Sandrin, T.R., and Demirev, P.A. (2018), Characterization of microbial mixtures by mass spectrometry, *Mass Spectrometry Reviews,* 37(3): 321–349, https://doi.org/10.1002/mas.21534.

Schaeffer, R., Szkio, A.S., de Luceno, A.F.P., Borba, B.S.M.C., Noguiera, L.P.P., Fleming, F.P., Troccoli, A., Harrison, M., and Boulahya, M.S. (2012), Energy sector vulnerability to climate change; A review, *Energy,* 38: 1–12.

Southerland, M.T., and Stribling, J.B. (1995), Status of Biological Criteria Development and Implementation. In W.D. Davis and T.P. Simon (eds.), *Biological Assessment and Criteria: Tools for Water Resource Planning and Decision Making*, Lewis Publishers, Boca Raton, Florida, pp. 81–96.

Straub, Ti. M., and Chandler, D. P. (2003), Towards a unified system for detecting waterborne pathogens, *Journal of Microbiological Methods,* 53 (2): 185–198.

Strauss, H. (1991), *Final report: An Overview of Potential Health Concerns of Bioremediation*, Env. Health Directorate, Health Canada, Ottawa, 54 pp.

Thomas, B., Vinka, C., Pawan, L., and Sabatini, D. (2022), Sustainable groundwater treatment technologies for underserved rural communities in emerging economies, *Science of The Total Environment,* 813: 152633, https://doi.org/10.1016/j.scitotenv.2021.152633.

Thorslund, J., and van Vliet, M.T.H. (2020), A global dataset of surface water and groundwater salinity measurements from 1980–2019, *Science Data,* 7: 231. https://doi.org/10.1038/s41597-020-0562-z.

United Nations Economic Commission for Europe, Environments and Human Settlements Division. (1998), *Burst Water Main Floods Central Manhattan ECE/ENV/98/1*, Geneva, 1 January 1998.

UNDP- United Nations Development Program. (2001), *Making New Technologies Work for Human Development: The Human Development Report 2001*, Oxford, Oxford University Press.

UNEP – United Nations Environmental Program. (2002), *GEO: Global Environment Outlook*, London, Earthscan Publisher.

UN-FPA – United Nations Populations Fund. (2001), *The State of World Population 2001 Footprints and Milestones: Population and Environmental Change*, Denmark, Phoenix-Trykkkeriet AS.

USBR (U.S. Bureau of Reclamation). (2016). West-wide Climate Risk Assessments. Hydroclimate Projections. www.usbr.gov/climate/secure/docs/2016secure/wwcra- hydroclimateprojections.pdf, accessed Feb. 1, 2024.

USEPA (1991), Technical Support Document for Water Quality Based Toxics Control. EPA 505-2-90-001, U.S. Environmental Protection Agency, Office of Water, Washington, DC.

USEPA. (1996), *Environmental Indicators of Water Quality in the U.S. EPA-841-R-95- 005,* U.S. Environmental Protection Agency, Office of Water, Washington, DC.

USEPA. (1997), Engineering Bulletin Technology Alternatives for the Remediation of Soils Contaminated With Arsenic, Cadmium, Chromium, Mercury and Lead, Office of Emergency and Remedial Response,Washington, DC, Office of Research and Development Cincinnati, OH, Superfund, EPA/540/S-97/500, August 1997

USEPA. (2000), *Methods for the Determination of Organic and Inorganic Compounds in Drinking Water, Volume I, EPA 815-R-00-014,* U.S. Environmental Protection Agency, Office of Water, Washington, DC.

USEPA. (2003), Arsenic Treatment Technology Handbook for Small Systems, Office of Water (4606M), EPA 816-R-03-014, July.

Wang, Y., Shen, L., Gong, Z., Pan, J., Zheng, X., and Xue, J. (2019), Analytical methods to analyze pesticides and herbicides. *Water Environment Research,* 91(10): 1009–1024. https://doi.org/10.1002/wer.1167.

WRI, UNEP, UNDP, and World Bank. (1998), *World Resources 1998-99 – A Guide to the Global Environment*, Oxford University Press, New York.

Warren Jr., M.L ., Burr, B.M., Walsh, S.J., Bart Jr, H.L., Cashner, R.C., Etnier, D.A., Freeman, B.J., Kuhajda, B.R., Mayden, R.L., Robison, H.W., Ross, S.T., and Starnes, W.C. (2000), Diversity, distribution, and conservation status of the native freshwater fishes of the southern United States – A comprehensive review of the diversity, distribution, and conservation status of native freshwater fishes of the southern United States reveals formidable challenges for conservation management. *Fisheries – American Fisheries Society*, 25(10): 7–31.

WHO. (2004), *Guidelines for Drinking-Water Quality*, 3rd ed, World Health Organization, Geneva.

World Resources Institute (WRI). (1994), *World Resources 1994-95*, Oxford University Press, New York, 217 pp.

Yang, H., Kong, J., Hu, H., Du, Y., Gao, M., and Chen, F. (2022), A review of remote sensing for water quality retrieval: progress and challenges. *Remote Sensors*, 14: 1770. https://doi.org/10.3390/rs14081770

Yong, R.N., Nutalaya, P., Mohamed, A.M.O., and Xu, D.M. (1991), Land subsidence and flooding in Bangkok, Proc. Fourth Int. Symp. on Land Subsidence, IAHS Publ. 200, pp. 407–416.

Zaman, S., Yeasmin, S., Inatsu, Y., Ananchaipattana, C., and Bari, M.L. (2014), Low-cost sustainable technologies for the production of clean drinking water - a review. *Journal of Environmental Protection* 5(1): 42–53. doi:10.4236/jep.2014.51006.

Zhang, Q., Crittenden, J.C., and Mihelcic, J.R. (2000), Does simplifying transport and exposure yield reliable results?An analysis of four risk assessment methods, *Environmental Science and Technology*, 35(6): 1282–1288.

Zhu, X., He, Z., and Deng, M. (2002), Remote sensing monitoring of ocean colour in Pearl River estuary. *International Journal of Remote Sensing*, 23(20): 4487–4497.

4 Industrial Ecology and the Geoenvironment

4.1 INTRODUCTION

The discussions in Chapter 1 have pointed out that the geoenvironment is an essential resource base that directly and indirectly provides the essential elements of life support for humankind and other living species. The different industries and associated activities undertaken – e.g., agro-industry, mining, forestry, manufacturing, energy, service industries, etc. – to respond to the needs and demands of humankind and other living species are therefore dependent on the health of the geoenvironment. Considering *industry* to be the primary driver for the various life-support systems needed to sustain life, it stands to reason that one needs to ensure that activities associated with *industry* do not materially degrade the geoenvironment.

Industrialization, as commonly defined, is the process whereby *industry* becomes a dominant component in a socio-economic order. The term *industry* is used here in the broadest sense to mean *the work undertaken (in a plant or facility) to produce goods using machines and/or technological aids*. A vital component of industrialization is the group of industries called *manufacturing industries*. These industries are essential to the health of the economy and to the life-support systems of a vibrant society. However, it is contended that many industrial activities conflict with the goals of a sustainable geoenvironment – a contention that arises from the perception that (a) non-renewable natural resources (materials and energy) are required to fuel the engine of *industry*, (b) non-renewable source materials are used in the production or manufacture of goods, and (c) the smoke-stack emissions and discharge of liquid and solid wastes from these industries are harmful to both human health and the environment (Figure 4.1).

There are two particular geoenvironment milestones that need to be observed in respect to *industry* and its activities: (a) Exploitation of renewable and non-renewable natural resources housed in the geoenvironment must be conducted with sufficient prudence so as not to degrade the geoenvironment, and also so as to allow future generations to continue to benefit from these resources (see Chapter 1), and (b) the discharge of waste products into the geoenvironment should not cause harm to the geoenvironment (see Chapter 2). In short, one must be conscious of the need to protect the geoenvironment and to ensure that its natural capital continues to be maintained – i.e., the functionality and health of the various constituents, compartments, elements, and units be preserved.

The discussion in this chapter concerns the geoenvironment and its role in the industrial activities mounted to provide for the needs of humankind and other living species. We will consider the interactions on the geoenvironment by activities associated with manufacturing and service industries. Since the purview of this book addresses resource use from the geoenvironment framework and not from the industry perspective, we will concentrate our attention on the land aspects of geoenvironment problems posed by *industry*.

DOI: 10.1201/9781003407393-4

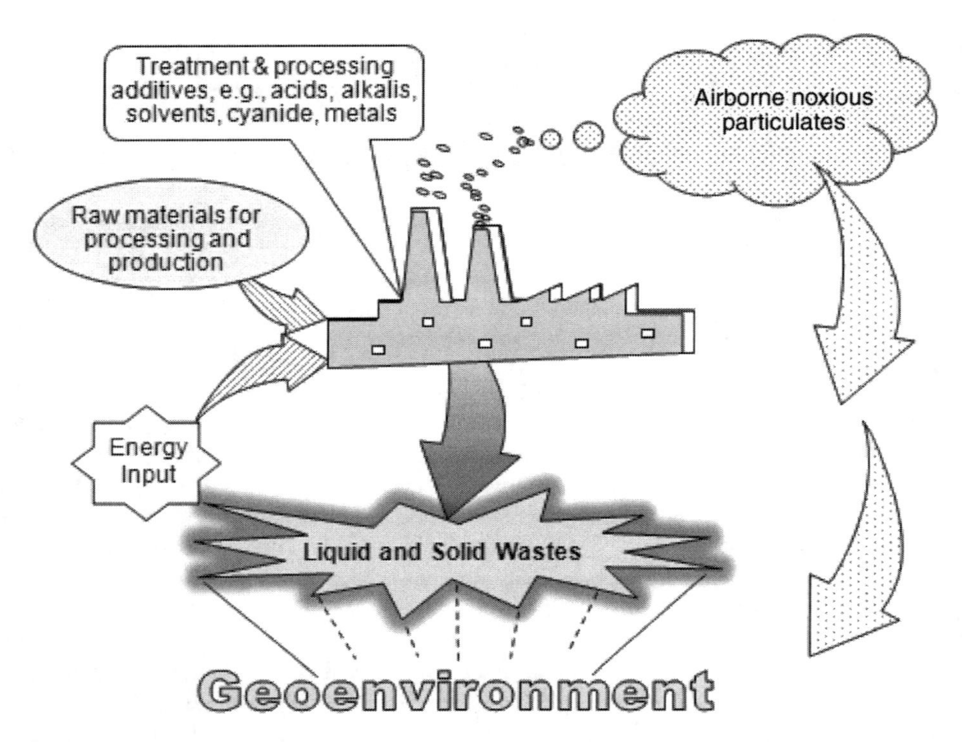

FIGURE 4.1 Interactions between manufacturing industries and geoenvironment. Primary issues are energy usage, deposition of smokestack, and other airborne noxious particulates, and discharge of liquid and solid wastes.

4.2 CONCEPT OF INDUSTRIAL ECOLOGY

The term *industrial ecology*, which stems from the initial concept of *industrial ecosystems* in a study reported by Frosch and Gallopoulous (1989), utilizes a *systems* approach to environment protection and natural resources' conservation as sustainable development objectives in processes involved in industrial production. By treating the industry-environment as an intertwined system, *industrial ecology* focuses its attention on the total picture of (a) renewable and non-renewable natural resources' exploitation and conservation at the one end (front end) of industry activities, (b) efficient industrial production through technology and resource conservation, and adherence to the 4Rs (recycle, recovery, reduction, and reuse of waste products), and (c) environmentally conscious management of emissions and disposal of waste products from industrial activities at the other end. In an ideal world, industrial ecology is a holistic approach to industrial production of goods, i.e., it takes into account the goals of environment protection and resources sustainability whilst meeting its goals of production of goods and other life-support systems to the benefit of consumers. These goals fit into the framework geoenvironment protection and preservation/conservation of the natural resources housed within the geoenvironment.

4.2.1 Geoenvironmental Life-Cycle Assessment (GLCA)

There are a number of protocols and/or methodologies that have been developed to implement the objectives contained in the concept of industrial ecology. Considering industrial ecology in respect to the geoenvironment itself, we can regard the majority of these methodologies to stem from the *life-cycle assessment* (LCA) of a particular industry or set of industries in question. The idea of using life-cycle assessment or life-cycle analysis of a particular item (industry, element, product, hardware, etc.) is not new or novel. LCAs are common tools for evaluation of items of interest, as for example economic costs over the life-cycle of a particular piece of equipment or hardware, material or mass flow over the life-cycle, and risks encountered – i.e., risk assessment of a particular set of events, items, activities, economics, etc. – over the life-cycle.

In respect to the geoenvironment, and the goals that define a sustainable geoenvironment, the principal focus of *geoenvironmental life-cycle assessment* (GLCA) methodologies is the impact of the various activities and products associated with industrial production on the health of the geoenvironment. As commonly perceived, the ultimate goal of industrial production is the transformation of raw or source materials into finished goods and products for the benefit of society. The agents or tools for industrial production include manufacturing industries, such as upstream, midstream, and downstream industries. Upstream industries produce raw or source materials that feed midstream and downstream industries – such as those discussed in Chapters 5 and 6 – to produce their goods as finished products or as inputs or source materials for further downstream industry use. The petrochemical industry is a good example of a downstream industry. The source or raw materials for the petrochemical industries include oil and gas produced or obtained from oil and gas production upstream industries.

For an assessment or analysis of the life-cycle of a particular set of products on the health of the geoenvironment, one needs to begin with the upstream phase and end with the final downstream phase. The illustrative example shown in Figure 4.2 for consumer goods involving metal products depicts the various entities that a GLCA would include. Note that insofar as the geoenvironment-related issues for the life-cycle assessment are concerned, the illustrated elements in the figure

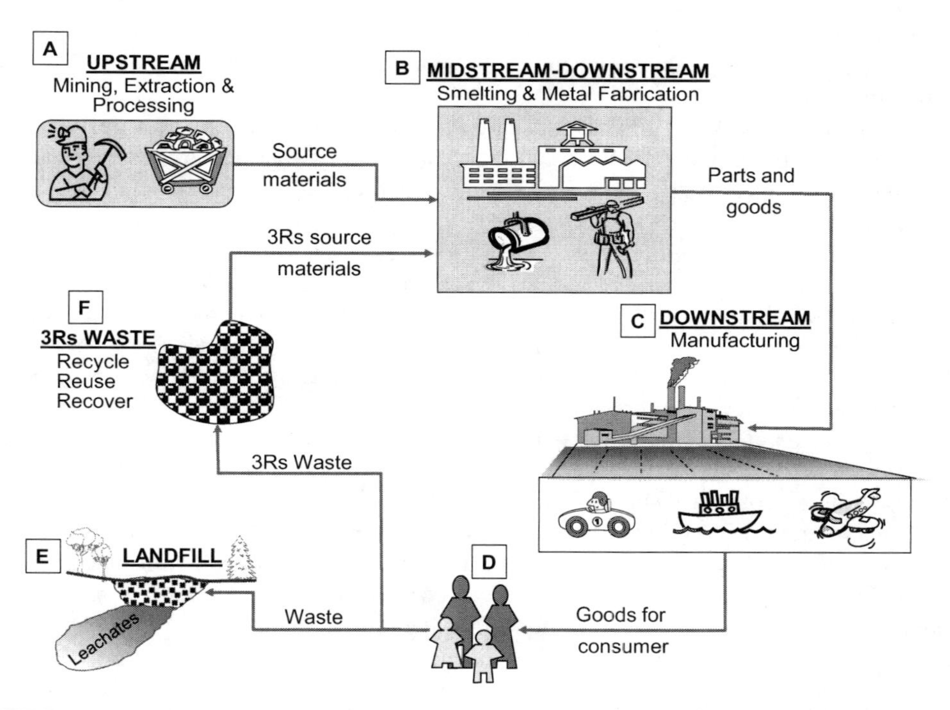

FIGURE 4.2 Illustration of life-cycle events for consumer goods involving the use of metal products. The start of the life-cycle is at the upstream end (A), and the end of the cycle is when the consumed product ends up as waste for landfilling (E) or for recycle, reuse, and recovery (F). Note that the individual life-cycle events for the various industries – upstream.

deal only with the beginning and "end-of-life" of the consumer product. The individual GLCA of each of the industries and elements involved, beginning with the upstream industry and ending with the landfill itself are not included in the illustration. The GLCA and related discussion for the upstream mining, extraction, and processing industry shown as (A) in Figure 4.2 can be found in Chapter 5, Figure 5.1. The extent of detail that a GLCA could consider can be very short or very long, depending on such factors or issues as magnitude of the project, economics, time, public, and corporate awareness of issues involved, and anticipated severity of geoenvironmental impact.

4.2.2 Geoenvironment Impacts and Sustainability

In this chapter, we will be examining the dilemma of the apparent antagonistic actions from the activities of manufacturing industries established to meet societal needs in respect to geoenvironmental sustainability indicators and/or goals. The problem reduces to a simple resolution of the required actions needed to accommodate the goals of societal and geoenvironmental sustainability – i.e., the co-existence of societal and geoenvironmental sustainability aims and requirements. A clearer understanding of the interactions between the manufacturing industries and the land environment will allow for management strategies and ameliorative and preventative actions to be implemented. These should serve to promote harmony between societal and geoenvironmental sustainability goals.

The discussion in Chapter 1 has pointed out that almost any external physical/chemical input or external sets of activity in the ecosphere will more than likely have indirect or even direct impact on the geoenvironment and its ecosystems. This is particularly true when it comes to the various activities associated with the midstream and downstream industries dealing with the manufacture of goods. Included in the group of midstream-downstream industries are (a) life-support industries, such as the agro, forest, mining, fisheries, and energy producing industries, and (b) production industries producing such varied goods as automobiles, pharmaceuticals, urban infrastructures, etc.

Although strictly speaking, service industries are not considered as downstream industries because they do not produce goods, we include them in the discussions in this chapter because these industries provide services (as goods) to the consumer. Two simple examples of sources of potential geoenvironmental problems from service industries are (a) medical services (e.g., hospitals), and (b) military services (e.g., munitions handling and storage).

The *industry* concerns addressed in this chapter centre around *land use in an industrial context* – i.e., land use in the context of activities in development of, and in support of the various kinds of life-support, manufacturing-production, and service industries. The industries discussed in the next few chapters will give examples of the interactions of various upstream and downstream industries with the geoenvironment. Since downstream industries are the consumers of the raw goods and products from upstream industries, their interactions with the geoenvironment are more in terms of "*what comes out from the processing and transformation end*". The nature and composition of the outputs (airborne noxious particulates, liquid and solid wastes) will be direct functions of (a) the nature of the product being produced, (b) the nature and composition of the source material used for the production of the product, (c) the process technology and the kinds of treatments and additives used in the process technology, and (d) the various controls on emissions, wastes and wastewater, treatments and management of the production technology and system.

Given the wide extent of the various kinds of industries, it is clear that within the context of *geoenvironmental sustainability*, a discussion of all of them would not be possible or desirable. Instead, the examination of the effects of land use and geoenvironmental management requirements will confine itself to a few representative industries in the three groups of downstream industries. It is important to stress that the terms *effects* and *impacts* are not used in a negative sense. The nature of impacts ascribed to activities and events run the gamut from beneficial through benign (neutral) to negative, depending on (a) the activity or event, and (b) the indicators, markers, and criteria used to evaluate or assess the results of the impact or impacts. Whether impacts from a specific set of activities or events will add to, or subtract value from, the particular ecosystem in the geoenvironment is the question that needs to be answered. It is important to recognize that a comprehensive listing of all the impacts on the ecosystems of the geoenvironment accruing from a particular set of activities is not possible – to a very large extent because of the lack of knowledge of all the various items, activities, and interactions that comprise a functioning ecosystem. Chapter 2 has discussed many of these concerns.

4.3 UPSTREAM, MIDSTREAM, AND DOWNSTREAM INDUSTRIES

The various kinds of industries that exist run the gamut from (a) those that produce raw goods to those that provide finished products, to (b) service industries providing a variety of services. As mentioned previously, upstream industries are primarily those devoted to production of the raw materials that need processing and transformation by midstream and downstream industries before reaching the consumer. Agricultural activities in aid of food production, for example, constitute upstream industrial activities, whereas food preparation and food processing using materials from agricultural activities can legitimately be considered as downstream industries. Downstream industries run the gamut from industries devoted to preparation of source materials as items and parts for manufacturers to transformation into final consumer products. For example, the production of automobiles requires countless numbers of parts such as tyres, engines, electronic parts, chassis, side panels, etc. Many of these parts are produced as downstream products by industries devoted to production of vital elements and parts for more intricate products. The following definitions apply:

- Upstream industries are those industries that produce the raw goods and source materials for downstream industries. Examples of these have been given in Chapters 5 and 6. Production of food, i.e., agricultural production of food (wheat, corn, barley, livestock, etc.), and mining and processing of metal ores are good examples of upstream activities and industries. The raw products can be used by the individual consumer and can also be used as the resource material for midstream or even downstream industries.
- Downstream industries include (a) those industries that use the raw goods produced by the upstream industries and prepares them as resource material for other downstream industries. Technically speaking, these can be called midstream industries. However, this term is not a popular term. A good example of this is the metal fabrication and processing industry discussed in the next section, and (b) production and assembly plants and industries that produce consumer goods and products that are directly utilized by the individual and collective consumers. These are the industries that transform source materials into consumer goods. Good examples of these are buildings, bridges, automobiles, leather goods, newsprint, electronic products, etc.

The subject of interest in the following sections relates directly to the net effect of the activities associated with these varied types of industries on the geoenvironment. The material covered in these sections will focus on some industries to highlight or demonstrate the particular geoenvironmental land use in question. At that time, we will examine some of the major consequences and impacts of these interactions on the geoenvironment, with the aim of seeking solutions that would permit us to satisfy many of the geoenvironmental sustainability requirements.

Figures 1.5 and 1.12 in Chapter 1 show that many of the activities and industries required to provide for human sustenance and needs will have direct interaction with the geoenvironment and will incur major impacts on the geoenvironment. For the purpose of examination of geoenvironmental interaction by downstream industries, we will group them into groups of industries and/or activities associated with the following:

- Mineral mining and processing industries: The upstream industries are those that deal directly with extraction of the raw materials required for the metallurgical industries. The downstream mining-processing industries are the metallurgical finishing and manufacturing-production types of industries. Also included are the non-metal mineral processing industries, such as cement production, phosphoric acid production, etc. The GLCA for the upstream industries can be found in Chapter 5.
- Agroprocessing industries: The various interactions and actions of the upstream agro industry will be discussed in Chapter 6. Associated downstream industries are generally classed as agroprocessing-type industries. These are the industries that transform the products from the agricultural, forestry, and fisheries industries. The two general categories are (a) food industries, and (b) non-food industries.
- Hydrocarbon, hydro, and other energy resources: Other energy sources include biomass, hydrogen, solar, geothermal, wind, and tidal. Electric power generation and distribution for the non-hydro power systems fall somewhere in-between upstream and downstream depending on the type of energy resource being harvested. The major downstream industries are the various kinds of petro-chemical industries. It is probably safe to say that downstream utilization of the products from this group of upstream industries is perhaps the largest of any of the categories of geoenvironmental resource usage.
- Production of goods and facilities: These are the downstream industries that transform the products issuing from upstream industries, such as hydrocarbon extraction, agricultural products, and metal ore production. The products from these downstream industries are either further transformed by other downstream industries or used directly by individual consumers.
- Public and private services: We classify these kinds of services as industries and as downstream industries even though they do not necessarily deliver hard goods to the consumer. The delivery of services, as goods, requires facilities and use of technology and goods that in one way or another will interact with the geoenvironment. The effects or results of these interactions need to be examined.

4.4 FORESTRY-RELATED INDUSTRIES

4.4.1 Lumber and Wood Processing

The forest industry is key for mitigating climate change. Tree growth captures carbon dioxide and thus forest conservation is essential for this and for ecosystem and biodiversity preservation. The implementation of sustainable forest management is Goal 15 of the Sustainable Development Goals. Forest management reduces soil erosion and enhances water retention capacity. Efforts to reduce deforestation and clearcutting practices have led to increased regulations in many countries. There are now substantial efforts for planting trees and seeding to replace harvested trees.

Climate change has numerous impacts on the forests due to higher temperatures, variable precipitation patterns, and drier conditions in the forests. Stresses on the forests make them more vulnerable to invasive species. In addition, fire seasons have lengthened as a result of the dry conditions and are more severe. This includes direct emissions of greenhouse gases (GHGs) and particulates from the fires, emissions from fire-killed tree decomposition, and reductions in carbon uptake as dead trees cannot absorb carbon. In Canada 2023, wildfires destroyed 18.4 million hectares (https://cwfis.cfs.nrcan.gc.ca/report). In 2020, managed forests and the harvested wood products in Canada were able to remove about 5.3 Mt CO_2eq from the atmosphere. According to the State of Canada Forests Annual report (NRCan, 2022), construction grade lumber contains 1 tonne of CO_{2e} for each m^3 of wood, and a Canadian single-family home has almost 30 tonnes of CO_2eq due to its wood-based construction materials.

Forestry supplies renewable raw materials for building materials, cellulose fibre, resins and other pulp chemicals, and pulp and paper products. By far, the supply of building materials and other non-paper products constitute the major portion of the overall wood industries. Some of the products include lumber, plywood, particle board, wood pellets, and wood composites. Newer products are being developed with the aim to replace plastics.

In Canada (NRCan, 2022) solid wood product production including softwood lumber and structural panels are increasing. In the pulp and paper sub-sector, there have been increases in printing and writing paper production, while wood pulp and newsprint production decreased due to the increase in digital media. Reuse of wood scraps and sawdust is important to reduce waste. Replacing solvent-based finishes with water-based finishes can help reduce VOCs and hazardous waste production. Energy-efficient equipment receiving certifications from the U.S. Environmental Protection Agency's Energy Star program will reduce GHG emissions.

4.4.2 Pulp and Paper Industry

In the US, 36% of harvested timber goes to paper and paperboard production (Bower et al., 2014). Other sources for paper making include sawmill residues and recycled papers. There is general agreement that the many thousands of pulp and paper industries in the world, when taken as a whole, constitute one of the largest contributors to water, land, and air pollution problems in the locations of those industries. Air emissions include particles, dioxins, furans, metals, hydrogen sulfide and other volatile sulfur compounds, hexachlorobenzene, and sulfur oxides and nitrogen oxides (which contribute to acid rain). Figure 4.3 shows a simplified flow diagram for typical pulp and paper processes. Debarking and chipping are the major operations in the *wood preparation* process. Recovery of the removed bark for subsequent use as fuel has reduced the waste discharge whilst decreasing energy costs.

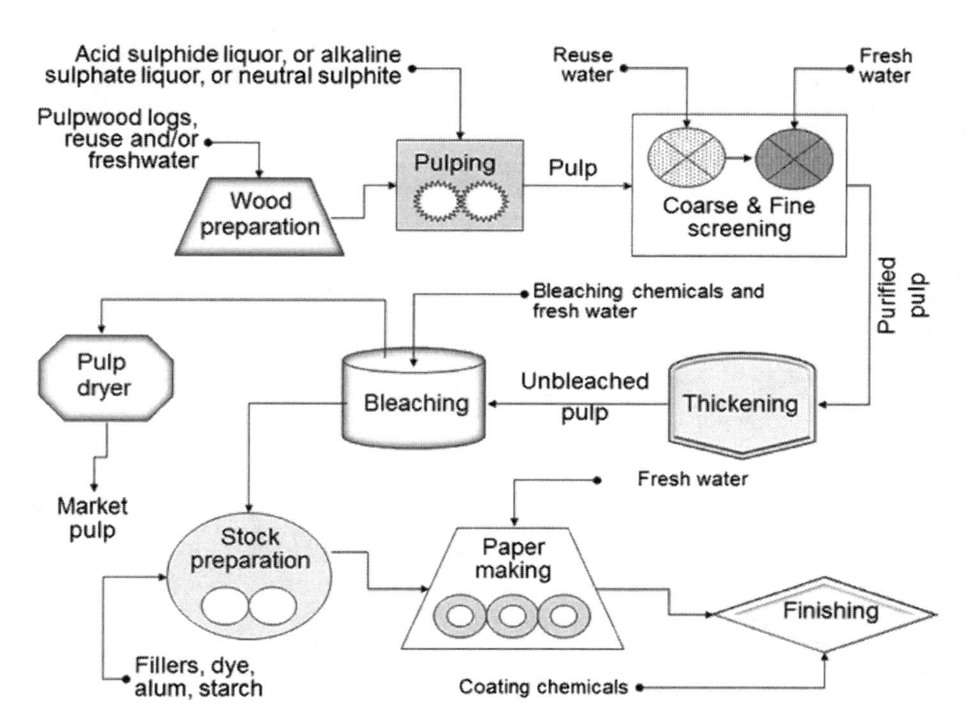

FIGURE 4.3 Flow diagram showing typical pulp and paper processes. Note inputs into the various processes and the extensive use of fresh water for many of the processes.

The pulping and bleaching processes contribute the more significant impacts to the geoenvironment. The various means for pulping range from mechanical, thermo-mechanical, and chemical. The most frequently used process is the kraft process, i.e., the sulphate process. This chemical process (sulphate process) is not to be confused with the sulphite pulping process, which is becoming less popular as a process because the source wood species (spruce, balsam, fir, and western hemlock) are not as plentiful as the other species used for the kraft process. This kraft process generates sludges high in chromium (Cr), lead (Pb), and sodium (Na).

The various processes shown in Figure 4.3 require considerable amounts of energy input and extensive use of fresh water. The industry is the fifth ranked energy consumer in the world but its use of biomass as an energy source reduces the GHG emissions. Recycled water, identified as whitewater, is used to augment the water input to the coarse screening, bleaching, and wood preparation processes. For the fine screening-washing and paper machine processes, fresh water is required. Other inputs into the different processes include nutrients, organic matter, heavy metals, acid sulphite liquor, alkaline sulphate liquor and neutral liquor, chlorine-type bleaching chemicals, such as hydrogen peroxide, ozone, peracetic acid, sodium hypochloride, chlorine dioxide, and fillers, dyes, alum, starch, and paper-coating chemicals. The wastewater discharges from the many processes include pulping liquor, mill washings and acid plant wastes, solvents and chlorine-based organic bleach compounds, and general wastewaters. The organochlorines discharged in the wastewater and solid wastes present some difficult toxic issues to the biotic receptors in the immediate geoenvironment. There have been efforts to replace chloring in the bleaching process by ozone, hydrogen peroxide, or other chemicals. Discharged mill wastewater in the rivers causes considerable breeding problems to several types of aquatic species. Solid wastes discharged include sludges from secondary treatment plants containing inks, dyes, pigment, etc., boiler ash, chemical processing, and waste fibres, including organochlorines (if chlorine has been used in the bleaching process).

Recycling of paper reduces the need for virgin paper but requires recovery of the used paper. Some paper mills now can accommodate recycled by adding a deinking process that removes the coatings on the paper. Scraps of paper are then incorporated with the deinked paper in the paper machine. Paper recycling reduces water and air pollution and virgin pulp compared to virgin paper production.

4.4.3 Land Environment Impact and Sustainability Indicators

The indicators for measurement of land environmental impacts for forestry and pulp and paper industries are similar to those shown in Figure 4.1, including emissions or discharges of (a) airborne particulates, (b) SO_2 and NO_x, (c) greenhouse gas emissions, (c) wastewater and other soluble contaminants; (d) solid wastes and other disposable solids. The composition, distribution of the various components in the wastes and particulates, and nature of the discharges are all functions of the type of process technology, regulations, technological efficiency, forestry and waste management capabilities and strategies. Many of the chemicals emitted have been mentioned in the previous section. In North America, the pulp and paper industry is tightly regulated, particularly for air and water pollution. Forest management is also heavily regulated to prevent illegal logging. In Canada, 94% of the forests are publicly owned and are managed by provinces and territories (NRCan, 2023).

Certification has grown in an effort to support the sustainability of operations in the industry. There are no international standards, however. ISO 38200 has only a standard for a wood and wood-based product chain of custody. The certification is a way of indicating that the paper and wood products come from well-managed, legally harvested forests. The Forest Stewardship Council (FSC) is one of the most well-known. Certification auditing can be a challenge for developing countries. In contrast, Canada has 158 million ha of forest certified to third-party standards of sustainable forest management. This represents 37% of the world's certified forest area (Certification Canada, 2022).

Emissions of carbon as carbon dioxide (CO_2) and as methane (CH_4) to the atmosphere are important contributors to global warming. The increase in the wood product manufacturing sub-sector is from the high demand from the home building and renovation markets. The pulp and paper manufacturing sub-sector had minimal year-over-year growth in real GDP (0.2%) in 2021. Steady demand for certain products, such as packaging and tissues and towels, offset the decline in demand for other products such as newsprint.

4.5 MINERAL MINING AND PROCESSING DOWNSTREAM INDUSTRIES

For convenience in discussion in this section, it is understood that when we use the term *industry* this will mean *downstream industry*.

4.5.1 Metallurgical Industries

The detailed treatment of resource extraction and processing associated with mineral mining and processing of the raw earth and rock materials as upstream industries can be found in the next chapter (Chapter 5). For this section, we want to outline the essential items for those downstream industries established to process the metal ores, obtained from the upstream industries, that directly or indirectly impact on the geoenvironment. From the illustrative example shown in Figure 4.2 of the industries associated with mining-extraction and processing as upstream and downstream types of industries, it is noted that the *midstream-downstream* industries in the upper right corner of the figure can be classified as either midstream or downstream industries – depending on whether the product from these industries serve as source materials for other downstream industries, or whether the products directly serve the individual consumer. A good example of this is the *parts* industry – i.e., the industries that produce products, such as parts and elements for other industries, that will assemble the parts and elements into consumer goods (assembly plants or industries). Note that midstream industries become downstream industries when their goods or products serve the individual consumer directly. An example of this from the diagram shown in Figure 4.2 is the production of metal consumer goods. Most reports on these industries are satisfied with the distinction between upstream and downstream types, and in general, these would be sufficient to encompass all the activities engaged in the exploitation and use of the products obtained from mining. However, the nature of the impacts on the geoenvironment can be vastly different depending upon whether it is a midstream industry that produces or prepares parts for use by other downstream industries or a downstream industry that concentrates on consumer production goods. For example, the impacts from contaminants that find their way into the geoenvironment from the metal fabrication and processing midstream industries can be more significant than in the production and assembly of their counterpart downstream industries.

Metallurgical industries cover a vast variety of processes depending on the final product issuing from the industry in question. The common practice of classifying these industries into three kinds of industries in relation to the material source used in the process and product produced, allows one to conveniently group the kinds of interaction of these industries with the geoenvironment. The midstream-downstream industries represented in the top right-hand corner drawing, for example, are intensive users of energy. The foundries produce smokestack emissions that are sources of (a) acid rain generation, and (b) land environment contamination when noxious airborne particulates find their way onto the solid land environment. The use of wet scrubbers and/or dry collectors can reduce harmful discharge of noxious gases and airborne particulates. Some of the discharges for foundries producing steel and aluminium include the various heavy metals, such as lead, zinc, manganese, chromium, arsenic, iron, nickel, and copper. For many other industries, in addition to the heavy metals, we will have a variety of organic chemicals included in the hazardous air pollutants (HAPs). A listing of many of these can be found at www.epa.gov (accessed Jan. 2024).

4.5.1.1 Metal Fabrication and Processing

Typical metal fabrication and processing midstream-downstream industries are those that produce value-added metals and metal goods. The types of metals include iron, aluminium, copper, lead, zinc, gold, tungsten, tin, silver, cadmium, etc., and are produced in a variety of forms. These serve as downstream products or as resource material for other downstream industries. Some of the downstream industries in this present category include (a) metal finishing industries, e.g., electroplating, anodizing, and coatings, and (b) industries and assembly plants utilizing metals for production of goods, e.g., manufacture of automobiles, planes, trains, ships, ovens, refrigerators, and tin cans – to name a few.

Interactions between downstream industries and the geoenvironment are primarily in respect to energy resources needed to satisfy the energy requirements of the industries, and the handling and disposal of the waste discharges – including the inadvertent spills and overflows during processing, manufacture, and production. The demands for energy in downstream industries, especially those in the category of material preparation and finishing, are of considerable concern in the overall strategy to reach the sustainability indicators that define the sustainability goals in the energy resource field. Chapter 5 discusses many of these concerns and strategies. Increased efficiency in manufacturing and production technology can alleviate some of the demands on energy use in these kinds of industries.

The metal finishing industries use a significantly large proportion of toxic chemicals in their various processes to produce, for example, corrosion resistance, wear resistance, electrical resistance, hardness, chemical resistance and tarnish-resistance metals. Figure 4.4 gives an example of the processes associated with metal fabrication. The sources of solid and liquid waste discharges are seen in the diagram. Most of the chemicals used in the metal finishing business end up as wastes. Considering that the various inputs to the processes include acids, solvents, alkalis, cyanide, loose metals, and complexing and emulsifying agents, it will not come as a surprise to note that the waste discharge and wastewater can contain these chemicals and various other residues – especially since many of the processes include rinsing and bathing operations. These wastes predominately result from the use of (a) organic halogenated solvents, ketones, aromatic hydrocarbons, and acids during the surface preparation stage of the overall finishing process, and (b) cyanide and metals in the form of dissolved salts in the plating baths during the surface treatment stage. These will all be found in the discharge streams shown in the bottom right of the diagram in Figure 4.3 – as sludge, solid waste, and wastewater.

In the pig iron production process, the flue dust generated is captured by wet dust cleaners. In common with most of the metal finishing processes shown typically in Figure 4.3, steel finishing involves a number of necessary processes in the production of the desired surface and mechanical characteristics for the steel. The sulphuric acid used in most pickling processes creates hazardous by-products. Pickling solutions contain free acids, ferrous sulphate, un-dissolved scale and dust, and the various inhibitors and wetting agents, as well as dissolved trace elements. The concentration of acids and their types largely depend on the type of iron being produced. Finishing operations generate effluents containing rolling oils, lubricants, and hydraulic oils that are in free and emulsified state. Other oils found in the effluents can also originate from cold reduction mills, electrolytic tin lines, and a variety of machine shop operations.

Water leaving the wet dust cleaners usually contains anywhere from 1000 mg/l to 10,000 mg/l of suspended solids, depending upon the furnace burden, furnace size, operating methods employed and type of gas washing equipment. Disposal of these onto the land environment will require treatment and containment to minimize ground, groundwater, and receiving water contamination. Some of these aspects will be discussed in Section 4.4.3 when sustainability targets are considered. The more detailed treatment will be found in Chapter 9.

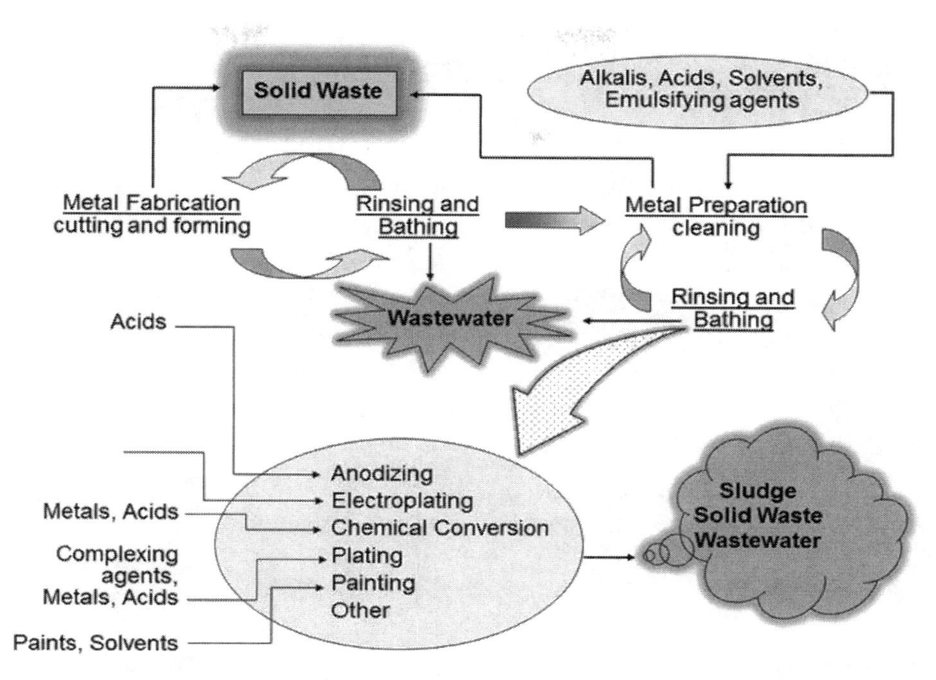

FIGURE 4.4 Processes and waste products in metal fabrication and preparation. (Adapted from EPA, *Profile of Metal Products Industry*, Washington, DC, Office of Enforcement and Compliance Assurance, 1995.)

4.5.2 Non-Metal Mineral Resources Processing

Non-metal mineral resources include (a) clays and clay minerals, (b) crushed stone, (c) sand and gravel, (d) dimensional rock slabs, such as granite and marble slabs, (e) phosphate, (f) potash, (g) gypsum, (h) peat, (i) sulphur, (j) diamond, (k) vermiculite, and natural alkali. Examples of utilization of the minerals supplied by upstream mining and processing industries include the following:

- Clays – for pottery and ceramics industries; construction industries involved in the construction of roadway fills, embankments, and clay-engineered barriers.
- Clay minerals – for paper coatings in paper industries, as catalysts for chemical industries; as expandable slurry materials (bentonites) oil exploration industries and also construction industries.
- Crushed stone, silica, sand and gravel – glass industries, smelting industries (silica used as flux material) concrete production industries, bituminous concrete industries, cement industries, construction industries.
- Dimensional slabs – construction industry.

The two minerals chosen as examples to illustrate the geoenvironmental interactions of activities of downstream industries transforming non-metal minerals obtained from upstream mining-extraction industries are the kaolin clay mineral and limestone. Figure 4.5 shows the principal steps taken to produce coatings-grade kaolin for use as kaolin-based coating pigment for the paper industry or in other industries such as latex and alkyl paints and primers. The process is not energy intensive in comparison to the metal industries and cement production industries, and the discharges from the processing technique can be controlled. These discharges are the clay impurities and the iron and

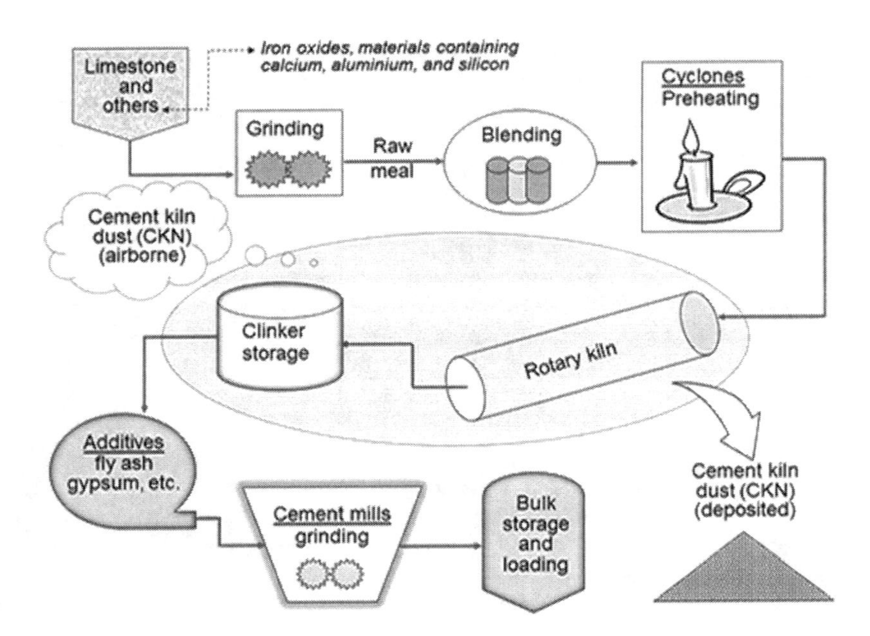

FIGURE 4.5 Production of coatings-grade kaolin using the hydrous procedure. Note that the discharge of impurities needs to be managed to reduce the impact on the geoenvironment.

titanium "impurities" that are generally associated with kaolinites. The iron and titanium can be captured and reused for other applications. Insofar as geoenvironmental interactions are concerned, this downstream industrial activity is relatively benign.

The use of limestone for production of cement is shown in a simplified schematic in Figure 4.6. In respect to geoenvironmental interactions, two significant factors are evident: (a) Intensive energy consumption is required for almost every step of the production procedure – especially in the cyclones' preheating stage and in the rotary kiln, and (b) discharge of fugitive cement kiln dust (CKN) as airborne particulates and as land discharge from electrostatic precipitators (ESPs) and other kinds of scrubbers, etc.

The significant human health issues attending the emission of CO_2, NO_x, SO_2, dioxins and furans cannot be ignored. For example, the release of metals into the atmosphere by copper-Ni refining, fossil fuel combustion and iron manufacture in the northern former USSR has contributed up to 90% of the metal loading in the European Arctic – a very sensitive environment (Pacyna, 1995). Arsenic (As), Cd, Pb, and Zn emissions in this region contribute 4.5, 2.4, 3, and 2.4%, respectively, of the global emissions. Wet and dry deposition of the metals with sulphuric acid has resulted in the accumulation of the metals in the soils, surface waters, and sediments. Acid rain enhances the environmental mobility of the metals and the bioavailability of these heavy metals. These metals may also bioaccumulate in the foods in the food chain, thus endangering the health of the consumers. Seabirds, seals, and polar bears have shown elevated levels of As, Cd, and Hg from ingesting fish and shellfish that eat contaminated algae, and plankton that accumulate heavy metals from sediments. Kansanen and Venetvaara (1991) have shown that lichens and mosses are effective bioaccumulators of heavy metals.

In 1995, more than 2 million tonnes of materials with heavy metals were emitted from a Ni-Cu smelter in Siberia. Reports indicate that the population suffers from respiratory illnesses. The precipitation of the materials in the areas has killed the lichens, which affects the grazing

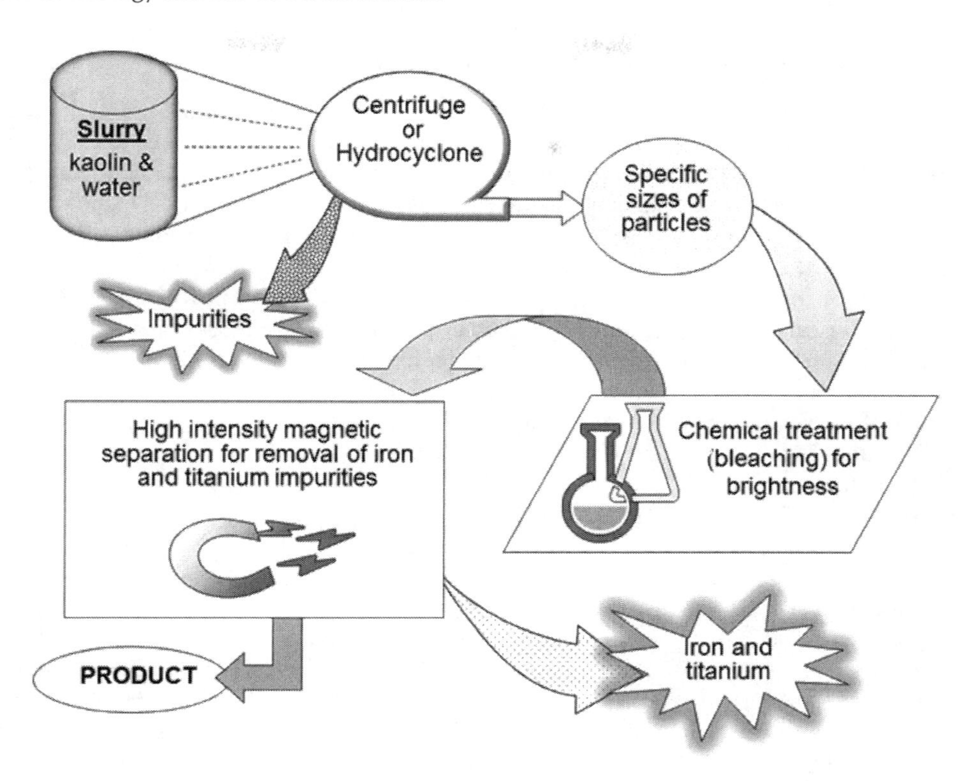

FIGURE 4.6 Basic elements in the manufacture of cement. Fugitive cement kiln dust (CKN) from the rotary kiln and clinker storage elements of the process will be airborne and captured CKN from scrubber systems will be deposited on land.

TABLE 4.1
Release of Heavy Metals Through Effluent Discharge, Emissions or Waste Disposal from Selected Industries

Industry	Heavy Metals Released
Fossil fuel combustion (electricity)	As, Cd, Hg, Pb, Sb, Se
Mining, smelting, metallurgy	As, Be, Cd, Cr, Cu, Hg, Mn, Ni, Pb, Sb, Se Ti, Tl, V, Zn
Petroleum refining	As, Co, Cr, Cu, Ni, Pb, V, Zn
Pulp and paper	Co, Cr, Hg, Ni, Pb

reindeer (Klein and Vlasova, 1992). The affected area has extended up to 70 km in the SSE direction. Heavy metals may also be discharged from industrial effluents into rivers, ponds, lakes, lagoons, wetlands, and oceans. Metals generated by various industries are shown in Table 4.1. Mercury has been particularly problematic. Fish such as swordfish are known as hyperaccumulators of Hg. Fish samples in the Smithsonian Institute contain up to 500 ppb of Hg, the limit permissible by the WHO. In Lake Ontario, Hg has been found in fish exceeding this level in the early 70s. Hg reached the lake as a result of industrial discharges.

4.5.3 Land Environment Impacts and Sustainability Indicators

The indicators that define or establish the path towards sustainability goals are specific to the activity and/or industry under consideration. One needs to define or identify the various processes or activities that impact directly or indirectly on the geoenvironment. This can be a very detailed accounting of all the activities and their outcome or could be a broad sweep of the major categories, elements, or issues. The basic elements that contribute directly as land environmental impacts for most of the types of downstream industries considered in this section, shown in Figure 4.1, include (a) deposition of airborne noxious particulates, (b) acid precipitation provoked by smokestack emissions of SO_2 and NO_x, (c) wastewater and other liquid waste discharges, and (d) solid wastes and other disposable solids. The composition, distribution of the various components in the wastes and particulates, and nature of the discharges are all functions of the type of process technology, technological efficiency, smokestack emission control, "housekeeping" efficiency, and waste management capabilities and strategies.

Acid precipitation impact on the surface of the geoenvironment (including the receiving waters contained in the land environment) is felt in several ways:

(a) Soil quality – the increased soil acidity will release metal ions and other positive ions bound to the soil particles, and also the structural ions, such as aluminium. The mechanism for such release is found in the ionic bonds formed between the charged soil particle surfaces and the positive ions (metals and salts such as Ca^{2+}, K^+, Na^+, and Mg^{2+}). The sulphate and nitrate ions from acid precipitation act as counterions and have the effect of releasing the sorbed cations. Weathering of the silicate minerals will release the structural metals, such as aluminium, manganese, and iron. The release of the salts Ca^{2+}, K^+, Na^+, and Mg^{2+} will result in nutrient depletion for plant growth and the release of aluminium especially will be harmful to aquatic life and plant growth.
(b) Biology of the forest – reduction in rates of decomposition of the forest floor, damage to roots and foliage, changes in respiration rates of soil microorganisms.
(c) Water quality and aquatic habitats – acidification of lakes and rivers, deposition of soil-released aluminium, species destruction, and alteration of food supply for higher fauna.

Other not-so-evident land environment impacts from the mining-processing downstream industries are (a) use of non-renewable mineral resources as the source material for the downstream industries, and (b) excessive use of non-renewable energy resources to drive the various processes in production of the final product. Except for the aggregate and slab production industries, metal mining, and processing downstream industries and the cement producing industries are heavy users of energy. Until alternative renewable energy resources become more available, these industries impact directly on the land environment when land energy resources are used to fuel their many process requirements. Cement kilns are perhaps the best potential users of alternative energy sources using recycled or recoverable materials. Because of the high burn temperatures required in the cement kilns, municipal biowaste and various kinds of combustible solid and liquid wastes can be used as burn energy sources.

Land environment sustainability objectives of direct relevance to the metal mining and processing downstream industries are (a) preservation or minimal use of non-renewable energy resources, (b) preservation or minimal use of metal-mineral resources, (c) elimination of smokestack emissions and airborne noxious particulates, and (d) 4Rs and non-toxic and non-hazardous discharge of liquid and solid wastes. Figure 4.7 shows some of the elements that can serve to drive downstream industries towards sustainability goals. Industry initiatives are needed for many of the elements shown, e.g., (a) better control on smokestack emissions and airborne particulate discharge to eliminate deposition of particulates and generation of acid precipitation, (b) more efficient

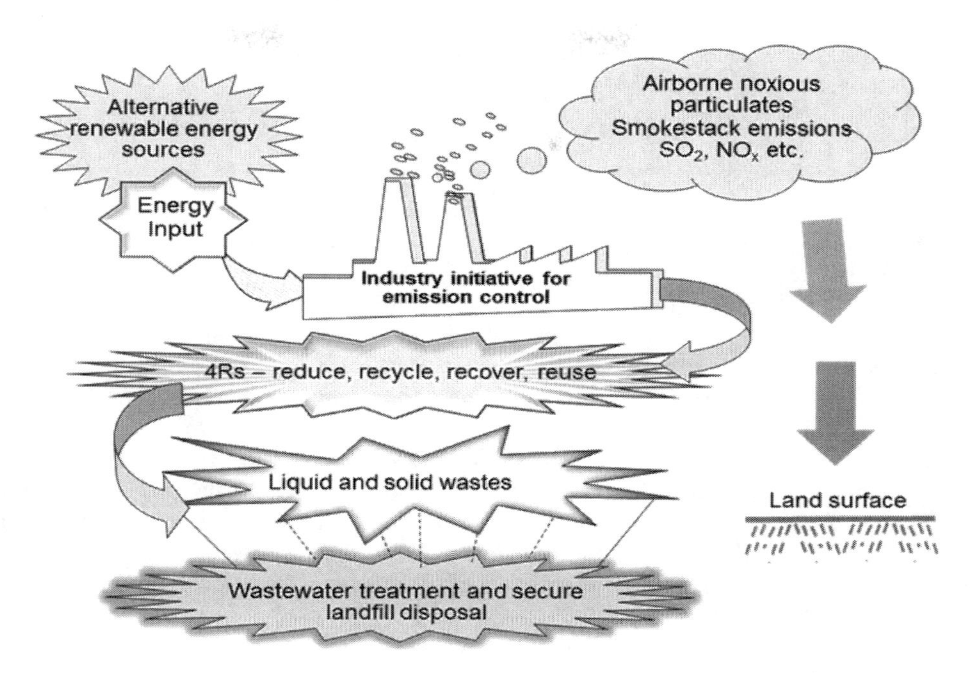

FIGURE 4.7 Industry initiatives for amelioration of geoenvironmental impacts.

use of metal-mineral resources to produce more "yield", (c) the use of the 4Rs (reduce, recover, reuse, and recycle) strategy as part of the process efficiency technology, and (d) use of alternative renewable energy sources to aid in reduction of consumption of non-renewable energy resources. Chapter 9 gives a brief discussion of these initiatives. A full treatment of these industry initiatives is not within the purview of this book. There exists much concern and interest in the development of these initiatives, and without a doubt, much is being done by industry to resolve these issues.

The direct connections between the impacts shown in Figure 4.7 and the geoenvironment are seen in terms of contamination of the land and water elements of the geoenvironment. Amelioration of the contaminant loads, and protection of the land environment are necessary requirements to ensure and maintain the quality of the land and water elements of the geoenvironment. The sustainability goals for these land environment elements are (a) maintenance of the quality of the land environment and the receiving waters, (b) protection of biodiversity and natural habitats, and (c) protection of the natural (e.g., biotic and organic) and geoenvironmental (e.g., mineral and energy) resources in the region of consideration. The procedures and protocols in respect to impact minimization, avoidance and amelioration are developed in Chapter 10, together with considerations and requirements for remediation and management.

4.6 PETROCHEMICAL AND CHEMICAL INDUSTRIES

The use of inorganic and organic chemicals as source materials for downstream industries brings with it the problems of control of production operations and management of discharges during operations and discharges as waste products. There are three groups of industries that use chemicals as source materials. These include (a) petrochemical industries that work with organic chemicals, (b) chemical industries that use inorganic chemicals as their feedstock, and (c) pharmaceutical industries that use both inorganic and organic chemicals as the source material. We should also note that pharmaceutical industries use a wide variety of source materials in addition to inorganic and organic chemicals. The interactions of pharmaceutical industries with the geoenvironment are not considered in this section.

4.6.1 Petrochemical Industries

Petrochemical industries constitute the bulk of the chemical (inorganic and organic chemicals) industries in the world. The overwhelming portion of organic chemicals is derived from feedstock obtained from crude oil, natural gas liquids, and coal. The two major processes involved in obtaining organic chemical feedstock are (a) chemical reaction and (b) purification of reaction products (USEPA, 2002). The simplest chemical reaction process is obtained in the batch reaction method where chemicals used to obtain the desired reactions and products are introduced into a reaction vessel. At completion of the reaction process, the reaction products obtained are removed from the vessel and extraneous by-products and unreacted inputs removed (USEPA, 2002). This step is identified as product separation or purification. The techniques used involve filtration, distillation, and extraction either singly or in combination. The continuous reactions' method, as the procedure implies, is a continuous reaction process technique and is suited more for greater production of reaction products – in comparison to the batch reaction method, which is used for production of smaller quantities of reaction products. As in the case of the batch method, the reaction products of the continuous reaction technique require product separation.

The feedstock obtained includes alkanes, benzene, butane, butadiene, butylene, ethane, ethylene, methane, propane, propylene, toluene, and xylene. The feedstock organic chemicals are both end-use products and also intermediate chemicals, i.e., they serve as feedstocks for production (generally by conversion processes) into other end-use organic chemicals or products. Typical end-use products include the various pesticides and fertilizers used in agriculture, the various forms and types of plastics, textiles, solvents, detergents, pharmaceuticals, appliances, synthetic lubricants, nylon, plumbing, and even chewing gum. By and large, the various kinds of plastics are perhaps the largest and most important product group emanating from petrochemical industries.

4.6.2 Chemical Industries

By definition, inorganic chemical industries manufacture chemicals that do not contain the carbon molecule. To a large extent, the feedstock or raw materials for the inorganic chemicals are mineral in origin, and the products manufactured are acids, alkalis, salts, and chemicals that are used as aids in producing other products – especially fertilizers. Ammonium nitrate (NH_4NO_3), ammonium sulphate ($[NH_4]_2SO_4$), urea ($CO[NH_2]_2$), and superphosphates are some of the fertilizers manufactured. The processes involved for production or manufacture of the fertilizers, for example, vary depending on the type of fertilizer. Neutralization of nitric acid (HNO_3) with ammonia (NH_3) will yield ammonium nitrate (NH_4NO_3) through the simple reaction:

$$NH_3 + HNO_3 \rightarrow NH_4NO_3 \tag{4.1}$$

On the other hand, there are at least three different ways or sources to obtain ammonium sulphate (USEPA, 1979). These include (a) combining anhydrous ammonia and sulphuric acid in a reactor to obtain synthetic ammonium sulphate, (b) as a by-product from production of caprolactam [$(CH_2)_5COHN$], and (c) as a coke over by-product obtained by reacting ammonia from coke oven offgas with sulphuric acid.

The acids produced are to a large extent mainly utilized as intermediates in industrial and manufacturing activities. The fertilizer industry is a big beneficiary. Nitric acid (HNO_3) and sulphuric acid (H_2SO_4) contribute significantly to the production of ammonium nitrate and phosphate respectively.

Hydrochloric acid (HCl) is used in steel pickling, etching and metal cleaning and hydrometallurgical production.

Perhaps one of the largest groupings of inorganic chemical industries is the chlorine-alkali group. This group produces chlorine, sodium hydroxide, sodium carbonate and bicarbonate, and potassium hydroxide. Their products are greatly utilized as intermediates in the organic chemical manufacturing industries (USEPA, 1995) ranging from (a) for chlorine – vinyl chloride monomer, ethylene dichloride, glycols, chlorinated solvents and methane, and (b) for caustic soda – propylene oxide, polycarbonate resin, epoxies, synthetic fibers, soaps, detergents, and rayon. The raw or source material for the industry is both natural salt deposits and seawater. Removal of the impurities, such as calcium, iron, aluminium, sulphate, magnesium, and trace metals is required before the electrolysis process used to obtain the end product (chlorine, caustic soda, and hydrogen). The three types of cells used in the electrolysis processes for the manufacture of the products are mercury, diaphragm, and membrane.

4.6.2.1 Stressors and Impacts on Geoenvironment

The major areas of concern for land environment protection are similar to the other industries described in the previous sections. These areas are (a) deposition of smokestack (point) and fugitive (from process equipment, leaks and spills) emissions of particulates and noxious substances including SO_x and NO_x, (b) discharge of wastewater and fugitive process waters, and (c) solid wastes. The nature of the chemicals, especially the organic chemicals, makes it critical to monitor both the acidity and chemistry of precipitations. For example, ammonia (NH_3) and nitric acid (HNO_3) have been recorded as emissions from processing for ammonium nitrate. NO, NO_2, and SO_2 have been detected as emissions from plants producing HNO_3 and H_2SO_4, respectively.

Wastewater from petrochemical industries manufacturing organic chemicals will contain excess chemicals, hydrocarbons, and other dissolved solids in suspended form. Also included are the wastewaters from maintenance procedures and washing of equipment. These would likely contain solvents, lubricants, and detergents. In the case of the inorganic chemical manufacturers, corresponding wastewater discharges will be obtained, except that the surplus chemicals will be inorganic chemicals. Maintenance and cleaning procedures will produce wastewater that would also contain lubricants and detergents and perhaps some solvents.

Land disposal of waste materials would generally be in a sludge form since raw materials and manufacturing processes generally do not involve solid materials. Brine muds are perhaps the greatest "solid" wastes derived from the chemical industry. These are obtained from the chlorine-alkali industry and are regularly disposed of in brine mud ponds in a manner similar to the holding ponds of the mining-metal industries.

4.6.3 Land Environment Impacts and Sustainability Indicators

The impacts to the land environment from the emissions, discharges, and land disposal of the waste items are in general similar to those for the other industries described in the previous sections. The only significant difference in the case of petrochemical and chemical industries is the added chemical nature of the various emissions and discharges. The record shows that utilization of proper scrubber systems together with stringent wastewater treatment procedures have served to reduce the levels of impact to the land environment. Indicators of proper sustainability of the land environment are obviously zero emissions and discharges.

4.7 HEALTH CARE INDUSTRIES

Service industries do not necessarily create tangible goods or products. By definition, they are industries that create or provide services to the consumer. The major service industries include (a) health – all aspects of medical, dental, and social services, (b) military, (c) educational, (d) government, (e) technical, and (f) financial. Because they do not really create goods, the interactions between these industries and the geoenvironment are limited to the operational or fugitive discard of liquid and solid items associated with the service and the "tools of the trade". Military services for example, present an added dimension to the composition of waste streams. This relates to waste products associated with munitions and the storage and use of ammunitions. A major feature of the decommissioning of military sites is the decontamination of sites contaminated by all of these. The bulk of the discards and wastes generated by the large service industries can be classified as "institutional" wastes. This category includes the paper discards and general housekeeping items. The exception to the preceding will be the health service industry. The problem arises from services associated with the care of patients in hospitals.

4.7.1 Hospital Wastes and the Geoenvironment

Outside of the regular "housekeeping-type" of waste products, such as paper consumables and kitchen waste, hospitals generate wastes that classify under the category of hazardous, toxic, and infectious wastes. The contributors to these are the biomedical, radioactive, and chemical-pharmaceutical wastes. The sources for biomedical wastes include biological, medical, and pathological. Contributors to these are services associated with surgery, pathology, biopsy, laboratories, and autopsy. For the radioactive wastes, the sources include x-ray discards, liquid scintillation vials, and all other treatment procedures and equipment utilizing radioactive materials. Sources of chemical-pharmaceutical wastes include research laboratories, pathology, and histology.

The record shows that amongst hospital wastes, infectious biomedical wastes pose the greatest threat to human health. Management and disposal of these wastes to the land environment are critical issues. Special regulations have recently been structured by most state regulatory agencies concerning hospital wastes in general and infectious wastes, in particular. Sorting, storage, transportation, treatment, disinfection, and incineration are some of the principal steps in the handling and disposal of these infectious wastes. Liquid infectious wastes are required to be treated before discharge and only non-infectious and non-anatomical wastes are permitted to be disposed of in landfills. Unhappily, technical and economic constraints may deny full and safe-secure disposal of these wastes.

4.7.2 Impact of COVID-19

The recent COVID-19 pandemic has impacted waste generation significantly. The WHO estimated that more than 87,000 tonnes of personnel protective equipment (PPE) were purchased during March 2020 and November 2021 for the health care system not including public mask use. Other wastes generated were 140 million test kits leading to 2600 tonnes of plastic wastes, and 731,000 litres of liquid waste. 8 billion vaccine doses of vaccines generated 144,000 tonnes of syringe, needle, and safety box wastes. The pandemic highlighted the substantial issues with sustainable waste management in health care. Improper management of the waste could potentially lead to spreading of COVID-19 (UNEP, 2020) and contaminant emissions from burning and incineration. The WHO report recommended the use of reusable PPE, more eco-friendly packaging, better waste treatment, recycling, and more biodegradable material use. Minimization of waste going to a landfill is key while protecting the workers who handle the waste. New policies, such as a zero-waste approach, were initiated in the EU (Zero Waste Europe, 2020). EU members are required to recycle 70 to 80% of the waste while reducing GHG emissions.

4.8 ENERGY PRODUCTION AND THE GEOENVIRONMENT

The sources of energy are of two distinct types: (a) Non-renewable, and (b) renewable. The main non-renewable source of energy is fossil fuel (hydrocarbons and coal). Another notable source of non-renewable energy is uranium – used in the production of nuclear energy. Renewable energy sources, by definition, refer to those sources that are sustainable, such as solar, wind, ocean (tidal and wave), geothermal, osmosis, and biomass. The discussion in this section deals with the impacts arising from stressors generated by the use of the non-renewable and renewable resources in energy production.

4.8.1 Fossil Fuel Energy Production

The primary sets of concerns associated with energy production industries relying on fossil fuels can be grouped into three categories: (a) Mining, drilling for extraction of the fossil fuel (source material), and delivery of the fuel, (b) conversion of the source material into energy, i.e., energy production, and (c) transmission of the energy to the consumer (infrastructure). The discussion on the stressors and their impacts, generated from extraction of the resource, can be found in the next chapter, which deals with non-renewable resources' extraction and their stressors and impacts. The sources of stressors that will generate adverse impacts on the geoenvironment from the mining and drilling operations include mine drainage, discharges, and spills from mining operations, and waste streams. There are textbooks devoted to mining, petroleum, and coal mining operations and technologies, which the reader should consult for detailed discussions on how these types of industries operate.

4.8.1.1 Geoenvironment Stressors

By and large, the sources of stressors associated with fossil fuel energy production can be traced to

- *Operations required in conversion of source material to energy:* The sources of stressors for energy production are essentially similar to most manufacturing-production industries, i.e., the sources are mostly related to processing, manufacturing, and production operations, with differences in composition, quantities, and quality of the discharges and spills attending each type of industry (discussed in Section 4.9 and in Chapters 5 and 6). Spills, discharges of liquid, and solid wastes constitute the major sources of stressors to the geoenvironment. Treatment and disposal technologies constitute the main elements of a sustainable geoenvironment strategy (see Chapter 10).
- *Transmission and delivery of fossil fuel – including land transport systems such as trucking, railways, and pipelines:* Assuming that the infrastructure for vehicular (trucks and trains) land transport systems are already in-place, outside of the various geoenvironment-related aspects associated with of construction of pipelines for transmission of liquid hydrocarbons, the main source of stressors in respect to *transmission-delivery* appears to be spills and inadvertent discharges arising from vehicular and pipeline accidents and transmission-delivery operations. These constitute major sources of chemical stressors contaminating the land compartment of the geoenvironment.

4.8.2 Nuclear Energy

The case of nuclear energy production is somewhat unique in that considerations and accounting for interactions with the geoenvironment must include the time factor as a big issue. Uranium must be mined and then extracted. Some recycling of uranium and plutonium can be done. While nuclear energy is a clean process as no GHG are emitted, the problem of disposal of high level radioactive waste is of considerable concern in that one needs to take into account: (a) The level of radioactivity of the spent fuel (high level radioactive waste from used fuel and rods), (b) the heat of the spent fuel, (c) the time required for the level of radioactivity of the spent fuel to reach acceptable limits – as shown in Figure 4.8 – i.e., below threshold levels for exposure to humans, and (d) the

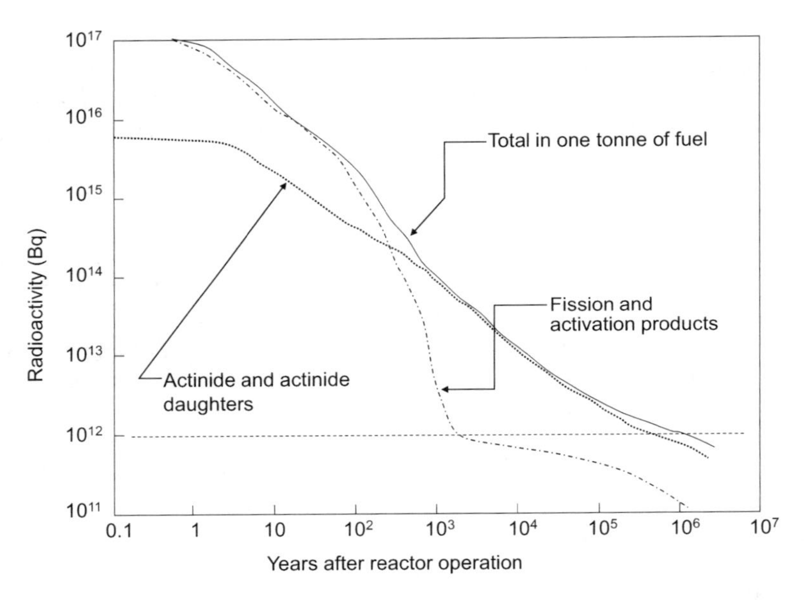

FIGURE 4.8 Level of radioactivity of spent nuclear fuel with a burn-up of 38 MWd/kg U as a function of time after reactor operation – i.e., after removal from active service. (Adapted from Hedin, 1997.)

time required for the heat to dissipate and to reach ambient temperature levels (Yong et al., 2010; Pusch et al., 2011).

Figure 4.8 shows the time-decay relationship for changes in radioactivity of the fuel rods given in units of Becquerel (Bq) with a burnup of 38 MWd/kg (megawatt days per kg). Initial radioactivity is by fission and activation products. This activity decays rapidly after 100 years. Further activity appears to be contributed primarily by actinides and actinide daughters. The horizontal line shown at a radioactivity level of 10^{12} Bq represents the radioactivity for eight tonnes of natural uranium with daughters. The half-lives of uranium 235 (^{235}U), ^{236}U, ^{238}U and iodine 129 (^{129}I), which are 7.0×10^8, 2.3×10^7, 4.5×10^9, and 1.6×10^7 years, respectively, tells one, for example, why disposal or safe containment of the burnt-up fuel rods are a special type of management problem for geoenvironmental engineers. It is useful to note that if one establishes a reduction in radioactivity level to the one matching natural uranium deposits (10^{12} Bq level) as the required containment level of security, this would require at least 100,000 years of "safe" containment isolation.

In addition to the problem of radioactivity, one needs to look at the heat generated by the decay heat, which is the residual power, or the heat generated in the fuel after cessation of operation (Figure 4.9) in relation to the time after cessation of operations. As can be expected, the fission and activation products contribute to the heat in the canister in the first 100 plus years, after which time, the main contributor to the total heat is from the actinides and actinide daughters. The discussion of containment of these radioactive wastes can be found in Chapter 10. Storage site construction is of tremendous concern from citizen groups due to potential radioactive leakage. The Chernobyl accident in 1986, led to an explosion and fallout of radioactive particles. Forests, fish, cattle, and horses died as a result and more than 100,000 people had to be relocated.

A method under development for energy is nuclear fusion currently exploited for the atomic bomb. Fusion is the bringing together of atoms. As the source of electricity is inexhaustible, it holds substantial promise. However, no reactors have been built yet (Encyclopedia Britannica www.britannica.com/science/nuclear-energy).

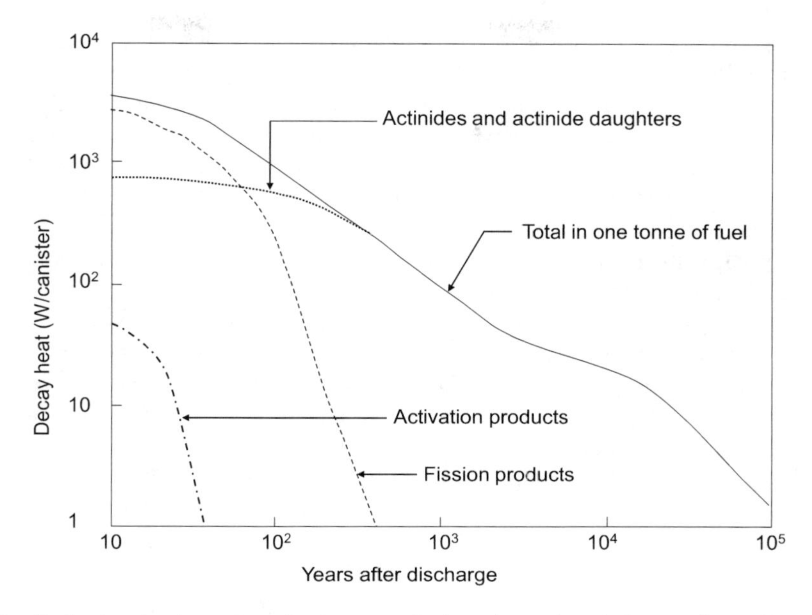

FIGURE 4.9 Reduction in decay heat in the spent-fuel canisters in relation to time after removal from operation. (Adapted from SKB, 1999.)

4.8.3 Geothermal Energy

Geothermal energy is the harnessing of heat energy from the earth. Three types of energy for direct use, geothermal heat pumps, and production of electricity. Heated water from the ground can be used directly for heating homes, swimming pools, spas, and other applications. Geothermal pumps extract heat within 300 m from the surface in the winter and cool in the summer. They are composed of a heat exchange and a pump. They are very efficient using less energy than comparable systems. Electricity production can be produced via dry steam, flash steam, or binary-cycle. More than 24 countries currently use geothermal energy for electricity generation. Geothermal energy is more reliable and available than wind and solar energies. Costs for the infrastructure can be high but costs are comparable with coal once the system is installed. The carbon footprint is low and the energy is renewable and sustainable. However, proper management is needed to the return of water into the underground. There is a very large untapped potential for this type of energy. There is a risk of earthquakes and gases can be released due to the forcing of water into the earth to better exploit the resource.

4.8.4 Methane and Methane Hydrates

Methane can be produced as a renewable source of low carbon between energy from waste. It is also a greenhouse gas (86 times more than carbon dioxide in 20 years) and thus efforts are being made to reduce emissions particularly from landfills. To reach the target of at 1.5 °C temperature increase agreed upon in Paris, levels of methane emissions need to be reduced by 30 to 60%. Agriculture, fossil fuels, followed by waste and wastewater are the major sources of methane (UNEP, 2023).

Biogas can be produced from landfills, manure management and systems for waste management (digesters). The gas contains between 50 and 70% methane, the remaining amount being mainly carbon dioxide with some other gases. The liquid digestate can be used as a soil amendment. Biogas can be used instead of coal or natural gas and could reduce water pollution. A number of mitigation methods are available. Installation of gas recovery systems is most feasible for larger landfills with significant production rates. The gas could then be used for energy for heat, heat and power, or electricity generation. Removal of carbon dioxide and the other gases from the biogas enables injection of the gas with natural gas. For example, the company Energir in the province of Quebec in Canada will be using food waste to produce renewable natural gas (RNG). It is estimated that 25,000 tonnes of GHG will be reduced by the RNG plant in Saint-Hyacinthe (Energir, 2023). It has been estimated that greenhouse gas emissions for RNG (0.30 kg CO_2 eq/GJ) are comparable to hydroelectricity GHG emissions (0.44 kg CO_2 eq/GJ).

Another form of methane, methane hydrates, are found as deposits of solid methane surrounded by water, in the arctic, Antarctic, sediments in deep lakes and the ocean floor. Despite the high potential for production of methane and water, there are no commercial installations. Reinjection of the carbon dioxide into the reservoir after methane use would enable production of a zero-carbon energy source. Japan has explored exploitation of seabed sediments for methane hydrates but there are significant risks for extraction from the ocean floor including increased risk of destabilization of the floor, tsunamis, and mudslides (Yiallourides and Partain, 2019).

Addition of gas mitigation systems such as biofilters or biocovers, diversion of organic waste from landfills and stricter regulation of close landfills. Financial resources are often lacking though for remote communities. For organic waste, alternatives are composting or anaerobic digestion.

4.8.5 Wind Turbines

The installation of wind turbines is becoming more frequent as power generation from wind does not emit any carbon dioxide emissions and thus can help in the fight against climate change. The life-cycle carbon dioxide emissions are in the order of 12 to 19 g CO_2 eq per kWh (US DOE Department of Energy Wind Technologies Office, Wind Vision Report, 2015) among the lowest of any energy source. While there can be significant spacing between turbines, the space can be used for agricultural and other purposes. Land is often cleared for other purposes than the turbine installment. Some impacts are aesthetic, potentially affecting tourism and increasing public resistance, habitat loss, noise, bird loss and effects of cultural and archeological protected areas. In addition, as the life of the fibreglass blades is about 10 to 20 years and there is no market for recycling, they will end up in landfills. In addition, mining of rare earth elements, such as neodymium for magnet manufacture, is required for certain types of turbines. Efforts, however, are being made to replace this element. Pile driving for wind turbine foundation installation can affect wildlife, particularly offshore.

4.8.6 Alternative Energy Sources and the Geoenvironment

The pursuit of alternative energy sources or means to produce energy without reliance on nonrenewable fossil fuels, such as coal and hydrocarbon resources, has gained considerable attention amongst those concerned with two principal issues: (a) Generation of GHGs from present fossil fuel energy-producing plants that have been faulted for contributing directly to "global warming", and also from the means to procure the fossil fuels themselves, and (b) the need to conserve nonrenewable hydrocarbon resources, i.e., using renewable natural resources. The types of industries or efforts mounted to generate energy for the consumer have earned the nickname of *green energy sources*, with the related name of *green energy production*. The proper designation is *alternative energy sources*, with the term *alternative* meaning *alternative to coal and hydrocarbon resources (fossil fuels)*. A listing of some of the alternative energy production-type industries includes hydroelectric, solar, geothermal, biomass, wind, and ocean (tidal and waves), some of which were discussed in previous sections. Strictly speaking, solar, geothermal, ocean, and wind are *non-depleting energy resources* – meaning that they are *always there*, and do not really undergo renewable processes over any specified time period.

The concerns in respect to the land compartment of the geoenvironment and in regard to the sources of stressors from the use of these types of energy resources are not as severe as with the non-renewable resource types of energy production. In common with the other types of energy production systems, the sources of geoenvironment stressors are from construction of the production systems and also from the construction and delivery of the harvested energy (transmission system). The various geoenvironment stressors are thought to be mostly physical in nature.

4.9 CONTAMINATING DISCHARGES AND WASTES

Table 4.2 gives a very short summary of many of the contaminating substances and chemicals found in the geoenvironment as a result of deliberate discharges, spills, leaks, emissions, and disposal of liquid waste and solid waste materials from the downstream service industries. For the processing and manufacturing industries, the discharges come during the processing and manufacturing stages. These include (a) inadvertent losses of raw and intermediate products and materials utilized during processing and manufacturing, and (b) liquid and solid waste products associated with processing and manufacturing procedures and technology. Acids, bases, heavy metals, inorganics, organic chemicals, and solvents are common to the industries shown in the table. Many of the inorganics are composed of chemical compounds that do not contain carbon as the principal element. Most of the inorganic compounds are stable and soluble in water. They tend to have rapid chemical reactions and large numbers of elements. They are generally less complex than the organic chemical compounds. Other more specialized wastes include pesticides, herbicides and the like, solvents, cyanides, reactive wastes.

Other than general wastewater, liquid waste streams emanating directly from downstream industries can be grouped into four categories: (a) Aqueous-inorganic – including brines, electroplating wastes, metal etching and caustic rinse solutions, (b) aqueous-organic – including wood preservatives, water-based dyes, rinse water from pesticide and herbicide containers, organic chemical production, etc. (c) organic – including oil-based paint wastes, production of pesticides, herbicides, fungicides etc., spent motor oil, cleaning agents, refining and reprocessing wastes, etc., and (d) high solid content and high molecular weight hydrocarbon sludges.

In respect to managed liquid and solid wastes in landfills, there are two types of liquid plumes or streams that emanate from the wastepile in the landfill. The first type, which is identified as *primary leachate*, consists of the liquid waste originally submitted to the landfill combined with dissolved constituents in the wastepile. The primary leachate may be aqueous-organic, aqueous-inorganic, or organic. Leachate generated from water entering into the wastepile is defined as *secondary leachate* and is generally composed of the percolating water and solutes from dissolution products in the

TABLE 4.2
Typical Composition of Discards, Spills, and Waste Streams from Some Representative Industrial Activities and Industries

Industry	Discards, Spills, and Waste Streams
Metal manufacturing and finishing	Acids, bases, cyanide, reactive wastes, heavy metals, ignitable wastes, solvents, spent platings, oil and grease, emulsifying agents, particulates, polishing sludges, scrubber residues, complexing agents, wastewater treatment sludges
Petrochemical and chemical industries	Acids, bases, ignitable wastes, heavy metals, inorganics, pesticides, reactive wastes, solvents, lubricants, spent catalysts, spent caustic and sweetening agents, organic waste sludges
Hospitals-medical facilities	Biomedical wastes, infectious wastes, acids, bases, radioactive materials, protective materials, solvents, heavy metals, ignitable wastes

wastepile. This leachate may be aqueous-organic or aqueous-inorganic or a combination of inorganic and inorganic solutes and compounds in the liquid phase. Since it is not really possible to distinguish between the two kinds of leachates when they exit from the bottom and/or sides of a landfill, the general term *leachate* is used – with no attempt at categorization. The predominant liquid in a leachate may be water, an organic liquid or a combination. The solutes and inorganic and organic chemicals in the leachate are the products of the dissolution of the materials in the waste pile. The relative abundance of a given dissolved component is a function of the composition of the principal liquid. Neutral non-polar organic liquids will have large carrying capacities and will easily carry other neutral non-polar organic chemicals. Aqueous liquids have very limited carrying capacity and will not be capable of carrying non-polar organics in its dissolved phase. Water, on the other hand, has a relatively large carrying capacity for polar organic chemicals (they may be miscible in each other in all proportions) and for inorganic acids, bases, and salts.

4.10 CONCLUDING REMARKS

The concept of industrial ecology is a very powerful concept – if implemented to its fullest extent. The idea that industrial activities should not only be cognizant of the need to protect the environment (and in this case, the geoenvironment) and to conserve the natural resources (both renewable and non-renewable), but also to devise and incorporate strategies and technologies that would serve these purposes, is most refreshing and forward-looking. To that end, the use of life-cycle assessment is a powerful tool. One can track all kinds of information in respect to the production and delivery of the set of goods used by the consumer. It is particularly useful in *sustainable geoenvironmental engineering practice* directed towards the application of *geoenvironment protective technology*, i.e., technology designed to protect the geoenvironment. The discussion in this chapter is designed to show the reader the importance of protection of the natural capital of the geoenvironment – from initial harvest or procurement of the resource to final manufacture-production and delivery of the consumer goods of interest. In addition, it also points out that the responsibility for protection of the geoenvironment lies with the consumer – through conservation of resources and disposal of consumed goods.

To implement the basic concepts of industrial ecology from a geoenvironment perspective, one is required to determine the connections or interactions between downstream manufacturing industries and the geoenvironment. In particular, one needs to determine the geoenvironment stressors generated from the various sources in these industries. For manufacturing and other kinds of downstream industries, the major areas requiring detailed scrutiny include (a) use of non-renewable resources as energy input and also as raw materials for the industries, (b) spills and debris, together with liquid and solid waste discharges, and (c) gaseous and noxious particulate airborne emissions. Figure 4.10 gives a schematic of what one might call *common denominator descriptors* for the industries and their interaction with the geoenvironment. The various input and output items shown in the schematic are common to most of the midstream and downstream industries. Whilst the geoenvironment perspective developed in this book does not consider industry manufacturing and processing technology, the use of non-renewable resources as raw materials and as energy sources impact directly on the mandate of industrial ecology and must therefore be identified as issues that need resolution. The issues of direct concern in this chapter in respect to the land compartment of the geoenvironment are shown in the bottom half of the diagram shown in Figure 4.10 – identified by the broad arrows leading to the bottom of the diagram. From a geoenvironment perspective, the main points that require attention in establishing geoenvironment sustainability indicators as a step towards assessment of capabilities to attain geoenvironmental sustainability objectives have been stated in the preceding paragraph. These are grouped into (a) resource utilization, and (b) discharges.

The discussions and examples given in this chapter have focused principally on the discharges since these are the agents, i.e., stressors that come directly into contact with the geoenvironment. In discussing

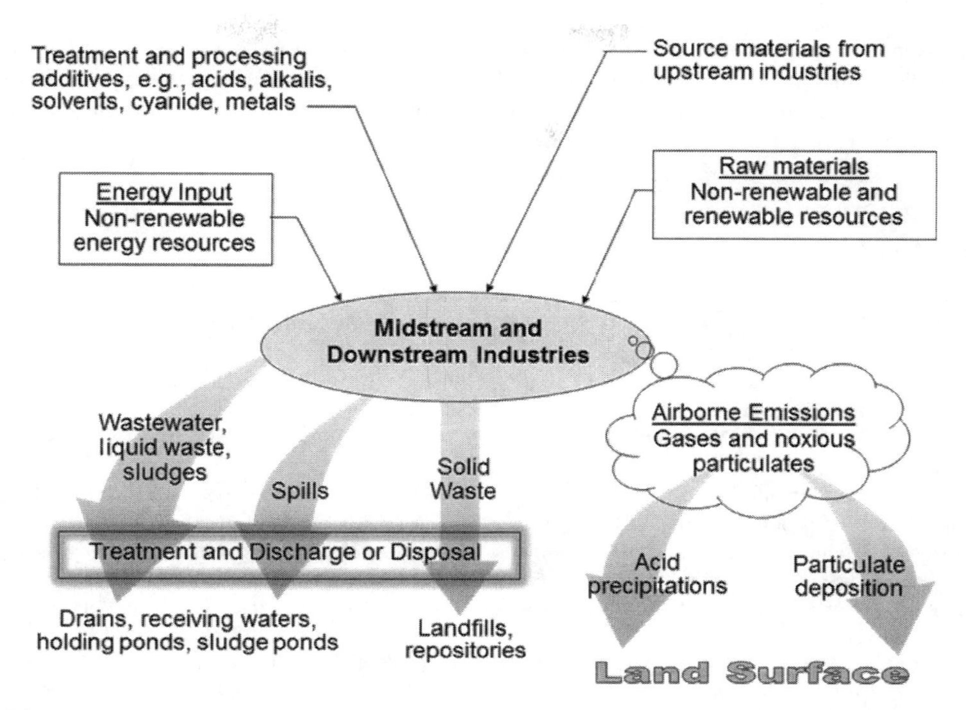

FIGURE 4.10 Common denominator descriptor identifying interactions between midstream-downstream industries and geoenvironment. Whilst all the descriptors shown are central to determination of industrial ecology and sustainability indicators, the mandate for this book does not cover industry technology for manufacturing and processing (top half of the diagram). The geoenvironmental concerns are directed to the issues identified in the bottom half of the diagram.

the various processes in this chapter, and in showing the diagrams for some of the processes, some appreciation of the many sources of interactions with the geoenvironment and the kinds of stressors generated can be gained. In the final analysis, these all fall into the *discharge* group of *points for study.* In the bottom half of Figure 4.10, the interactions with the geoenvironment are given as wastewater and liquid spills – these are treated before being discharged into receiving waters. The concern is in regard to whether treatment is capable of removing all the noxious and toxic substances.

- Liquid wastes and sludges – as noted in Table 4.2, these consist of inorganic and organic chemicals and also inorganic and organic sludges. Disposal of these in the geoenvironment is generally performed by constructing holding ponds or various kinds of secure containment ponds. Escape of these contained liquids or sludges into the subsurface environment will pose problems to the environment and to human health and other biota.
- Solid wastes – these are contained in waste landfills and in underground repositories. Leachates generated can escape into the subsurface and can pose health and environmental threats.
- Airborne emissions – gaseous and noxious particulates pose problems when they return to the land surface under gravitational forces and through rainfall and snowfall. The classic problems of acid rain are clear demonstrations of the "*return*" to land surface. In addition, GHG emissions are affecting climate change, which are having effects on geoenvironment, such as increased forest fires, melting permafrost and extreme weather events to name a few.

It is clear that the primary issues in respect to the discharges are the interactions of the inorganic and organic chemicals with the subsurface geologic material. Chapter 2 has given a brief overview of these interactions in the context of contaminant–soil interactions. These interactions are fundamental elements that govern the transport and fate of these inorganic and organic chemicals in the ground. We will examine these further in the context of the material developed in this chapter and the previous chapters. Chapters 9 and 10 consider most of these issues in terms of land environment impacts and geoenvironmental sustainability indicators.

REFERENCES

Bower, J., Howe, J., Pepke, E., Bratkovich, S., Frank, M., and Fernholz, K. (2014), Tree-Free Paper: A Path to Saving Trees and Forests? www.twosidesna.org/wp-content/uploads/sites/16/2018/05/Tree-free-paper.pdf, accessed Dec. 2023.

Certification Canada. (2022), Forest Management Certification. www.certificationcanada.org/en/certification/forest-management-certification/, accessed Dec. 2023.

Energir. (2023), Renewable Natural Gas. www.energir.com/en/about/our-energies/natural-gas/renewable-natural-gas/, accessed Dec. 2023.

Frosch, R., and Gallopoulos, N. (1989), Strategies for manufacturing, *Scientific American* 261: 144–152.

Hedin, A. (1997), Spent nuclear fuel – How dangerous is it? *SKB Technical Report*: 97–13.

Kansanen, P.H., and Venetvaara, J. (1991), Comparison of biological collectors of airborne heavy metals near ferrochrome and steel works, *Water, Air and Soil Pollution,* 60: 337–359.

Klein, D.R., and Vlasova, T.J. (1992), Lichens, a unique forage resource threatened by air pollution. *Rangifer*, 12: 21–27.

NRCan. (2022), The State of Canada's Forests (ANNUAL REPORT 2022), Canadian Forest Service. www.natural-resources.canada.ca/sites/nrcan/files/forest/sof2022/SoF_Annual2022_EN_access.pdf), accessed Dec. 2023.

NRCan. (2023), Sustainable Forest Management in Canada. www.natural-resources.canada.ca/our-natural-resources/forests/sustainable-forest-management/sustainable-forest-management-canada/24361, accessed Jan. 2024.

Pacyna, J.M. (1995), The origin of Arctic air pollutants: Lessons learned and future research, *Science of the Total Environment*, 160–161: 39–53.

Pusch, R., Yong, R.N., and Nakano, M. (2011), *High-level Radioactive Waste Disposal – A Global Challenge*, WIT Press, Southampton, UK, 299 p.

Swedish Nuclear Fuel and Waste Management Company (SKB). (1999), Deep Repository for Spent Nuclear Fuel, SR97 – Post-Closure Safety, SKB Report TR99-06, Main Report, Vol. 1.

UNEP. (2020), Waste Management an Essential Public Service in the Fight to Beat COVID-19 Pandemic. www.unep.org/news-and-stories/press-release/waste-management-essential-public-service-fight-beat-covid-19, accessed Dec. 2023.

UNEP. (2023), Facts About Methane. www.unep.org/explore-topics/energy/facts-about-methane, accessed Dec. 2023

US DOE. (2015), Wind Vision a new era for wind power in the United States, DOE/GO-102015-4557, March 2015

USEPA. (1979), Ammonium Sulfate Manufacture: Background Information For Proposed Emission Standards, North Carolina, Report EPA-450/3-79-034a.

USEPA Office of Compliance. (1995), Profile of the Inorganic Chemical Industry, Washington, DC, Report EPA/310-R-95-004.

USEPA Office of Compliance. (2002), Profile of the Organic Chemical Industry, 2nd ed, Washington, DC, Report EPA/310-R-02-001.

Yiallourides, C., and Partain, R. A. (2019, September), *Offshore Methane Hydrates in Japan: Prospects, Challenges and the Law (October 1, 2019)*, British Institute of International and Comparative Law (BIICL). www.ssrn.com/abstract=3601849 or http://dx.doi.org/10.2139/ssrn.3601849

Yong, R.N., Nakano, M., and Pusch, R., (2010), *Containment of High-Level Radioactive and Hazardous Solid Wastes with Clay Barriers*, Spon Press, Taylor and Francis, London, 468 p.

Zero Waste Europe. (2020), Zero Waste Europe Statement on Waste Management in the Context of COVID-19. www.zerowasteeurope.eu/press-release/zero-waste-europe-statement-on-waste-management-in-the-context-of-covid-19/, accessed Dec. 2023.

5 Natural Resource Extraction
Stressors and Impact Management

5.1 INTRODUCTION

The *geoenvironmental natural capital*, which refers to natural resources and processes – such as the biogeochemical cycles in the geoenvironment – that have or provide value to humans, can be considered to comprise of two major categories: (a) Renewable natural resources, and (b) non-renewable natural resources. Renewable natural resources include living resources, such as forests, plants, wildlife, marine, and other aquatic species, etc., and "non-living" resources such as soil and water. Biodiversity, as a natural resource (see Section 1.2.2) is not within the purview of this book, and is therefore not extensively included in this discussion of the natural capital in the geoenvironment. The basis for classification as a renewable natural resource is the ability of the resource to regenerate, replenish, or renew itself within a period of time from onset of initial resource harvesting to onset of repeat harvesting. It follows that if the resource is allowed to regenerate itself fully in the "between-harvest" period, it can be viewed as a sustainable resource. Failure to do so will render the renewable natural resource into the sub-category of *exhaustible (renewable) natural resource.* In reality, most of the renewable natural resources that are exploited or harvested are exhaustible. Their ability to remain inexhaustible, i.e., sustainable, is dependent on proper exploitation-management of these resources.

The prominent non-renewable natural resources include minerals and fossil fuels (coal, oil, and gas). The discussion in this chapter will be confined to industrial activities associated with the extraction or harvesting of non-renewable mineral, non-mineral, and (energy) natural resources (e.g., uranium and oil sands). The materials that constitute these resources are extracted or harvested by primary upstream industries devoted to such activities as mining, excavation, fracking (rock fracturing and extraction of hydrocarbon product), mineral and hydrocarbon extraction, processing, drilling, and pumping. The outputs from these upstream industries are raw materials for their respective midstream or downstream industries (see Chapter 7).

Resource extraction and processing industries use the geoenvironment as a resource pool containing materials and substances that can be extracted and processed as value-added products. The common characteristic of the industries in this group is *processing of material extracted from the ground.* The sources of stressors, the types of stressors, and their impacts discussed in this chapter include the following:

- Mineral-metal mining industries: Worldwide production is in the order of 1.4 billion tonnes per year for iron ore and 2700 tonnes per year of gold in 2011 (Menzie et al., 2013), which has now increased to 1.54 billion and 3300 tonnes per year, respectively in 2019 (Reichl and Schatz, 2021; Idoine et al., 2023). Some developed countries in Europe and Japan have completely depleted their underground resources.

DOI: 10.1201/9781003407393-5

- Industries involved in extraction and processing of other resources from the ground such as non-metallic minerals (potash, refractory and clay minerals, phosphates).
- Industries devoted to extraction and/or production of aggregates, sand and rock for the building-construction industry, and for production of cement.
- Raw energy industries such as the extraction and recovery of uranium for the nuclear power generation industry, and those industries involved in the extraction of hydrocarbon-associated materials, such as the oil sands, and extraction of shale gas and tight oil.

5.2 STRESSORS AND IMPACTS

To institute geoenvironmental engineering practices to limit, mitigate, or prevent deleterious impacts on the geoenvironment – i.e., sustainable geoenvironmental engineering practices – it is necessary to determine the sources and types of stressors that are likely to act on or in the geoenvironment. In the case of the geoenvironment, stressors are agents (forces, stresses, processes, etc.) that are responsible for actions that impact on a particular piece of the geoenvironment. These types of stressors can be physical, mechanical, hydraulic, or thermal, and can include processes that are chemical, physico-chemical, and/or biogeochemical in nature. Knowledge of the sources and their related stressors allows one to determine not only the types of impacts, but also the ultimate fate of the impacted body.

It is impossible to describe in detail the manner of operation of the different types of industries – not only because of their diverse natures, but also because of the different models for operational efficiency and success. The discussion presented in the following sections and chapters will only provide some of the main basic elements of operation of some of the industries involved, together with some of the main sources of stressors and their likely stressor-impacts. It is not the intent, nor is it feasible, to document a complete list of stressors and their impacts – not only because it is not realistically feasible, but also because not all of the impacts are impacts on the geoenvironment. The later sections in this chapter will provide information on the nature of the impacts on the geoenvironment together with suggestions as to how these impacts can be mitigated or managed. Activities mounted in conjunction with harvesting of energy resources, such as the development of oil-producing wells, natural gas wells, extraction of bitumen from oil sands, dams and hydroelectric facilities, also contribute their share to the potential impacts' list. Not all the potential impacts are directly due to the discharge of wastes and contaminants in leachate streams or wastewaters. The discussion in this chapter will focus primarily on the mineral and non-mineral (including oil sands) mining and processing upstream industries, and industrial activities dealing with underground *in situ* hydrocarbon extraction – i.e., industries dealing with exploitation of non-renewable natural resources.

5.2.1 Mining-Related Activities

Activities associated with the mining, extraction, and on-site processing of extracted natural resource material (mineral and non-mineral) contribute significantly to the inventory of potential impacts to the terrestrial ecosystem. In the UK, for example, more than 200 years of mining of coal and iron ore have left 500 million tonnes of residual waste from coal and iron and steel manufacturing as sterile waste tips (Barr, 1969). These tips are the result of the removal of more than 52 million tonnes of coal from the underground. Ground subsidence, water pollution, and toxic gas emissions are other impacts. Not only do these tips constitute blights on the landscape but they are also hazards when slide failures occur. There are historic reports of such failures that have resulted in human casualties. Figure 5.1 shows a simple generic illustration of operations associated with mining and extraction of metal mineral resources. Individual mineral extraction and/or beneficiation processes will differ between different types of minerals and their host ores – creating their own types of stressors and associated impacts. Specific examples will be discussed in a later section.

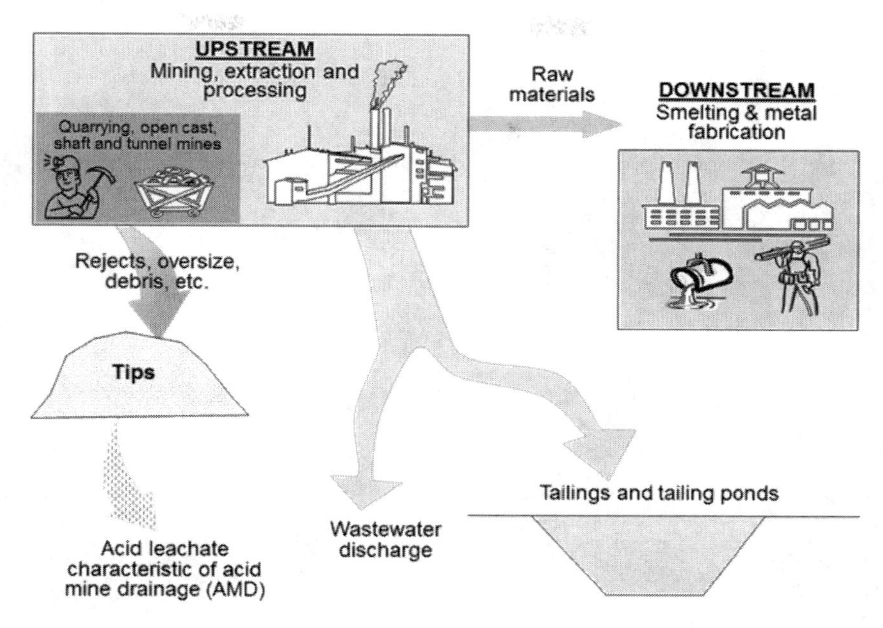

FIGURE 5.1 Illustration of some of the major features associated with mining and natural resource extraction. The example shown is typical of metalliferrous mining-extraction operations.

The two principal groups of activities of the natural resource extraction process includes (a) mining for procurement of host rocks containing the desired ores, and (b) extraction and processing the host rocks to obtain the desired minerals (generally called beneficiation). As shown in Figure 5.1, the result of these two groups of activities are generation of rejects and other debris that end up as *heaps*, and discharge of tailings' slurries and wastewater.

5.2.2 Underground and Surface Hydrocarbon Extraction

The activities of interest in deep underground *in situ* hydrocarbon extraction include (a) steam-based recovery of bitumen using the SAGD (steam-assisted gravity drainage) process, or a process known as the CSS (cyclic steam stimulated) process, and (b) hydraulic fracturing of deep underground shale and other geologic formations containing natural gas or tightly-held oil for extraction of the hydrocarbon products.

The SAGD technique is a steam-assisted heavy recovery process where parallel wells drilled into the deep underground are separated a few metres apart – with the top well providing steam to the surroundings, and the bottom well serving as a collection well capturing the gravity-assisted flow of the fluidized heavy oil as illustrated in the left-hand drawing of Figure 5.2. The steam that is fed through the upper well to soften the heavy bitumen is delivered at pressures below the fracture pressure of the host rock – thus allowing the fluidized or softened bitumen to flow under gravitational forces into the collecting well below.

In the other technique, as the name implies, the CSS process uses a somewhat similar procedure – i.e., introduction of steam into the geologic formation of interest to soften the contained bitumen. However, instead of utilizing gravity drainage of the softened bitumen into a parallel well below, the same well that provided the injected steam is used to extract the softened bitumen – meaning that only a single vertical or horizontal well per location is drilled. In short, the CSS technique uses the same well for both steam-injection and for extraction of the product bitumen. The cyclical procedure

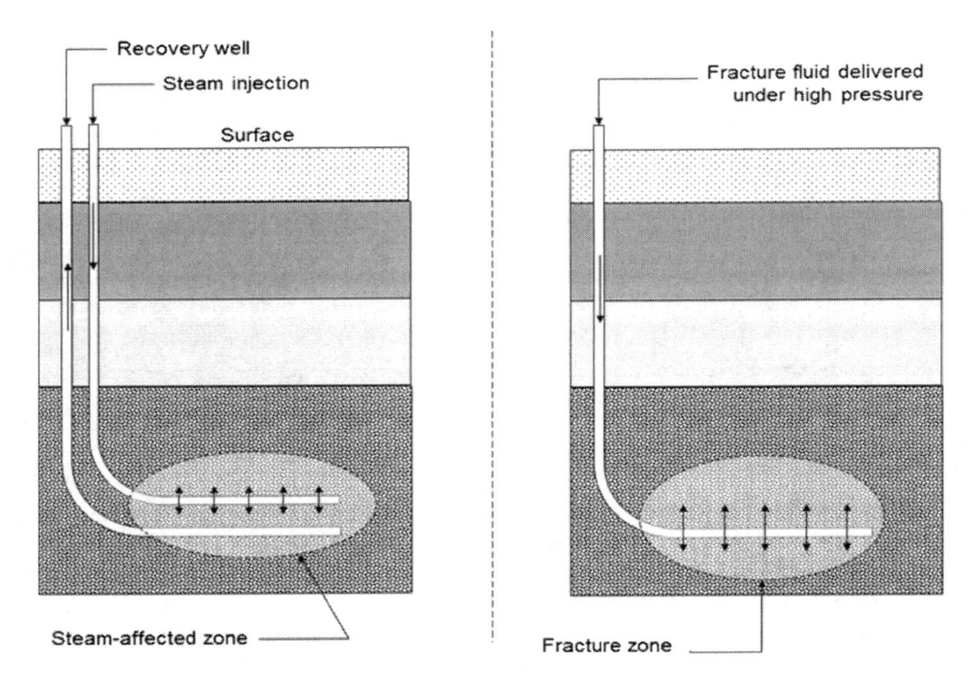

FIGURE 5.2 Elements of (a) left drawing – steam-assisted gravity drainage (SAGD) process for recovery of heavy bitumen, and (b) right drawing – hydraulic fracturing of shale or other hydrocarbon-containing geologic feature. Note that capture of the released hydrocarbons in the fracking process is affected by other capture wells.

of steam injection followed by extraction of the bitumen previously softened by the steam injection gives this procedure its name *cyclic steam stimulation process (CSS)*.

We have previously described underground in-site hydrocarbon extraction to mean the extraction of natural gas from shale and similar geologic formations, and oil (tight oil) or heavy bitumen in deep geologic deposits (using the SAGD technique or the cyclic steam stimulation process). The procedures common to all of these underground *in situ* extraction techniques is the requirement for deep drilling – in most instances, to obtain horizontal wells at the depths required for extraction of the desired product hydrocarbon. The discussion of sources of stressors and resultant impact that follows is confined to those sources associated with the extraction of the product hydrocarbons in the already-drilled horizontal wells.

5.2.2.1 Fluid Usage and Stressors

Common to the three types of underground *in situ* extraction processes is the use of high volumes or quantities of water – with or without additives. Either with or without prior addition of additives, water obtained after its accomplished purpose, is contaminated – meaning that the utilized water is a source for chemical stressors. In the case of steam-injection water used for softening the bitumen in the host material, contamination of the water by the hydrocarbons released or fluidized by the steaming process occurs – thereby rendering the used steam-water a source of chemical stressors. In the case of fracking fluids used for extraction of shale gas, for example, the additives used in the fracking fluid constitute the sources for chemical stressors. Whilst the industries involved in the implementation of these processes make every effort to recover the water used for the steam-assisted *in situ* processes and/or the fracking process for treatment at their respective surface treatment facilities, not all the contaminated water is recovered. It is the non-recovered (fugitive) water that constitutes the source of stressors to the immediate geoenvironment.

5.2.3 Hydraulic Fracturing

The right-hand sketch of Figure 5.2 shows the basic elements of the process commonly known as *fracking* – i.e., hydraulic fracturing of the host rock containing the hydrocarbons of interest (oil or natural gas). As with the CSS technique, a single horizontal well (per location) is drilled. A common technique is to use a perforating gun inserted to the end of the well to initiate small cracks in the penetrated host rock by detonating small charges. Subsequent crack generation is brought about through the use of high-pressure fracking fluids. The types of fracture-fluid commonly used consists of water and sand or other products capable of producing physical actions necessary to keep the produced fractures open to allow for recovery of the hydrocarbons of interest (natural gas or "tight oil"). This sand-water mix will generally contain various types of chemicals, such as (a) water-soluble gels, (b) chemical additives that have the capability of maintaining fractured openings open, (c) additives needed to inhibit bacterial growth, and (d) chemicals required to maintain efficiency of delivery of the fracturing fluid such as surfactants, corrosion inhibitors, and friction reducers.

Hydraulic fracturing is used to enhance difficult to extract natural gas or petroleum from shale rock, coal beds, sandstone, limestone, or dolomite rocks from 2000–6000 m below the surface. Rock fracturing induces fractures in the rock to enable extraction from highly impermeable shale reservoirs. Since 2013, hydraulic fracturing has been applied on a commercial scale to shales in the United States, Canada, and China (Wikipedia, hydraulic fracturing, https://en.wikipedia.org/wiki/Fracking, accessed Feb. 2024).

5.2.4 Sulphide Minerals and Acidic Leachates

We consider the case of production of acidic leachates from mining and extraction of sulphide minerals as a separate issue. This is because of the significant problems and stressors generated by sulphidic metalliferous rocks, and by their impacts on the geoenvironment – the most common or prominent issue being the generation of acid mine drainage (AMD) and the release of arsenic from their host rock ores.

5.2.4.1 Acid Mine Drainage

The sulphide minerals in the host rock in the mines and in the ores, such as iron sulphides, pyrites, arsenopyrites, chalcopyrites, pyrrhotites, sphalerite, and maracasites present severe environmental problems when they are exposed to water and oxygen. Figure 5.3 gives an illustrative example of what happens when pyrites (FeS_2) are exposed to oxygen and a source of water. Whilst oxygen and water are the two primary ingredients needed for the development of the phenomenon commonly described as acid mine drainage, it must be noted that microorganisms contribute significantly to the processes by way of catalyzing iron oxidation – especially at pH levels below 3.5 (Manahan, 1990). The cycle of acid contact and oxidation of the pyrite example shown in the diagram continues so long as oxidation processes can proceed.

In the series of chemical reactions reported by Manahan (1990), beginning with the oxidation of pyrite, the processes proceed as follows:

$$2Fe_2(s) + 2H_2O + 7O_2 \rightarrow 4H^+ + 4SO_4^{2-} + 2Fe^{2+} \quad (5.1)$$

It is noted that because of the low pH levels, further iron oxidation of the pyrite can be aided by various iron-oxidizing bacteria, as follows:

$$4Fe^{2+} + O_2 + 4H^+ \rightarrow 4Fe^{3+} + 2H_2O \quad (5.2)$$

$$FeS_2(s) + 14Fe^{3+} + 8H_2O \rightarrow 15Fe^{2+} + 2SO_4^{2-} + 16H^+ \quad (5.3)$$

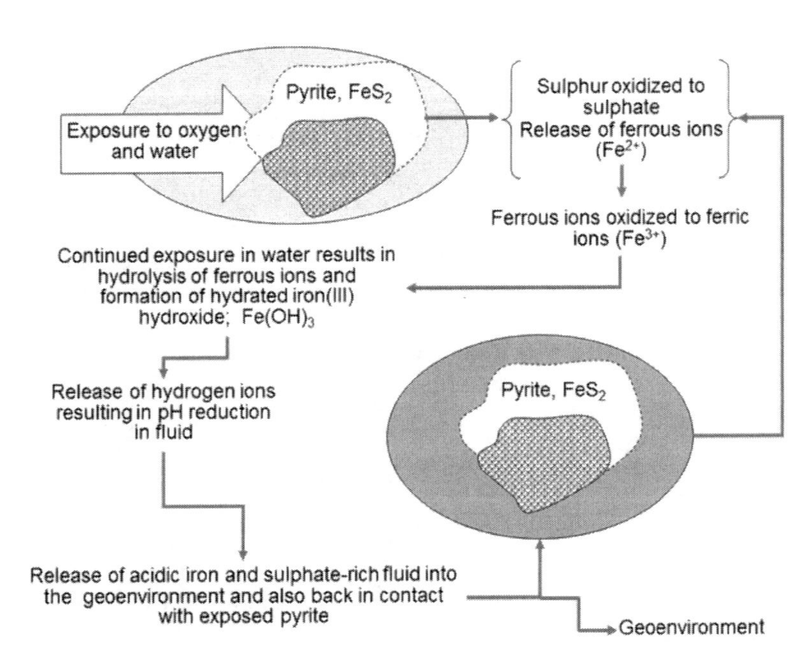

FIGURE 5.3 Effect of exposure of pyrite to oxygen and water. Continued exposure to water will result in the generation of iron hydroxide (yellowboy) and acidic solution that will be harmful to aquatic plants, animals, and will also release heavy metals previously held by the soil. Similar reactions shown in the diagram will also occur for sulphides of copper, lead, arsenic, cadmium, and zinc.

At pH values well above 3, Manahan (1990) reports that iron(III) precipitates as hydrated iron(III) oxide $Fe(OH)_3(s)$.

$$Fe^{3+} + 3H_2O \leftrightarrow Fe(OH)_3(s) + 3H^+ \tag{5.4}$$

Release of the acid fluid into the land environment within the mining site will allow the fluid to come into further contact with other exposed pyrites. The generation of acidic leachate rich in iron and sulphate (known as "yellowboy", $Fe(OH)_3(s)$) is characteristic of the outcome of the various processes that accompany oxidation of the sulphur and iron in the pyrite. So long as there is a source of these in the host rock and ores, and so long as these continue to be exposed to water, generation of "yellowboy" will continue unabated. The presence of sulphate-reducing bacteria in soil and water will exacerbate the problem. These bacteria are anaerobes that use sulphate as electron acceptors.

Whilst the generation of the acidic leachate, commonly known as acid mine drainage (AMD) or acid rock drainage (ARD) constitutes a major negative impact from mining operations, the cascading or domino effects that accrue from AMD can be severe. The domino effects arising from discharge of the leachate into the environment include (a) severe health threats to aquatic species, native habit, and plant life, (b) pollution of groundwater and drinking water, (c) deterioration of soil quality, and (d) release of trace metals and heavy metals previously retained by the soil solids in the ground. Information reported in the Interstate Mining Compact Commission study (IMCC, 1992) showed that for five Western States in the United States, there were (in 1992) about 130,000 inactive and abandoned mine sites. An example of an extreme case is a tin mine in England that was mined by the Romans more than 2000 years ago that still produces AMD from pyrite oxidation. In a later study, Skousen et al. (2019) report that approximately 20,000 km of streams and rivers in the United

States have been degraded by acid mine drainage and that 90% of AMD reaching streams originate from abandoned mines. This makes the problem of attaching ownership of the problem, and for clean-up of the sites and streams and rivers difficult.

The extent of acid generation at mine sites (underground mines, openings, leach ores, spent ores, etc.) is a function of several factors. These include (a) type and concentration of sulphide minerals in the host ore and in the spent ores and leach piles, (b) the host rock, (c) availability of oxygen, (d) site hydrogeology, (e) pH of the water in the system, and (e) presence or absence of bacteria – e.g., *Thiobacillus ferrooxidans.*

5.2.4.2 Arsenic Release

Oxidation of the mineral sulphides may result in the solubilization of trace metals and heavy metals – effectively releasing them and allowing them to be mobile in the liquid phase. It is not uncommon to find evidence of arsenic, cadmium, cobalt, copper, lead, manganese, nickel, and zinc as released metals. Arsenic poisoning of groundwater and aquifers has been reported in many parts of the world. This problem has gained considerable publicity and has been reported as "the greatest mass poisoning of mankind" in relation the poisoning of the tubewells in Bangladesh and West Bengal – as previously mentioned in Chapters 1 and 3. In this particular case, the information available points to the presence of naturally-occurring arsenopyrites (FeAsS) and arseniferrous iron oxyhydroxides in the substrate material as being the immediate source materials for the arsenic. If oxygen is available in the groundwater, oxidation of the arsenopyrites will release the arsenic. Some reports have speculated on the use of tubewells as a means for introduction of oxygen into the sub-soil strata. In the absence of oxygen, the processes associated with reductive dissolution of the arseniferrous iron oxyhydroxides will release arsenic whilst increasing the bicarbonate concentrations. This will result in arsenic pollution of the groundwater. The reports given in Appelo (2008) provide a comprehensive recounting of the world-wide occurrence of arsenic together with the geochemistry, problems, and impacts.

Mine tailings and effluents often have high concentrations of arsenic, which can then be potential sources of environmental contamination. A wide range of minerals in soils contain arsenic naturally in various inorganic forms. Common arsenic-containing minerals include arsenopyrite, mispickel (FeAsS), orpiment (As_2S_3), and realgar (AsS). Local mining and mineral processing at inputted approximately 3.5 tonnes of arsenic into the Moira Lake in Ontario, Canada (Azcue and Nriagu, 1995). A collection system and treatment plant were built to reduce the As loadings to both surface water and groundwater. Contaminated material was removed and on-site containment cells were established. Red mud tailings were treated with limestone, poplar trees (*Populus*) and engineered covers (CH2M HILL, 2004). Remediation was able to reduce the average daily As loadings to the nearby river from 35 kg in 1979–1982 to 6.1 kg by 1989. Other methods for arsenic tailings' stabilization are by creating an oxygen barrier or reducing water infiltration (Aubertin et al., 2016) to control the oxidation of As-bearing minerals and its leaching at the source.

In mine tailings, arsenic occurs in various forms, such as arsenopyrite (FeAsS), arsenian pyrite (As-rich FeS_2), arsenates, and association with iron oxyhydroxides. Wang and Mulligan (2006a) measured the arsenic contents of six Canadian mine tailings. ICP-MS analyses indicated that the highest arsenic concentrations reached 2200 mg/kg in tailings from a lead-zinc mine at Bathurst, NB. Subsequently, Arab and Mulligan (2020) found higher concentrations at an old gold mine. These and others are shown in Table 5.1. Many other countries, such as Thailand, Korea, Ghana, Greece, Australia, Poland, the UK (Kinniburgh et al., 2003), and the US have also experienced significant arsenic contamination associated with mining activities. McCreadie et al. (2000) have found arsenic concentrations up to 100 mg/L in the porewater extracted from tailings in the province of Ontario. Donahue and Hardy (2003) showed that dissolved arsenic concentrations within the tailings could vary from 9.6 to 71 mg/L. As with the reporting of Appelo (2008), the reports contained in Bundschuh et al. (2005) provide detailed accounting of the occurrence of natural

TABLE 5.1
Arsenic Concentrations Measured in Canadian Mine Tailings

Mine	Location	Concentration (mg/kg)	Reference
Copper Mine	Murdochville, QC	500	Wang and Mulligan, 2006a
Gold Mine	Musselwhite, ON	63	Wang and Mulligan, 2006a
Copper-zinc Mine	Val d'Or, QC	270	Wang and Mulligan, 2006a
Iron Mine	Mont-Wright, QC	<0.70	Wang and Mulligan, 2006a
Lead-zinc Mine	Bathurst, NB	2200	Wang and Mulligan, 2006a
Gold Mine	Marathon, ON	270	Wang and Mulligan, 2006a
Con Mine	Yellowknife	25,000	Ollson, 1999
Giant Mine	Yellowknife	4,800	Ollson, 1999
Giant Mine	Yellowknife	2,570	Arab and Mulligan, 2020
Negus	Yellowknife	12,500	Ollson, 1999
Rabbit Lake	Northern Saskatchewan	56 to 9,871	Moldovan et al., 2003

arsenic in groundwater in many different parts of the world, together with suggested methods for management and remediation of the problem.

5.3 RESOURCE EXTRACTION IMPACTS

It is abundantly clear that the harvesting of non-renewable natural resources discussed in this chapter is not going to satisfy "resource-sustainability" from the viewpoint of renewal or regeneration of the natural resource material. Extraction of these resources from the ground will deplete them, and will therefore fail a key sustainability issue – replenishment or renewal of supply. The primary geoenvironmental engineering concern in this respect is to (a) apply or implement engineering measures that will minimize and/or mitigate stressor impacts generated by the activities associated with harvesting of these resources, and (b) develop and implement remediation technology to restore impacted lands to pre-impact states.

The issues pertinent to sustainable practices in geoenvironmental engineering in respect to non-renewable resource extraction operations fall within the jurisdiction of land use – i.e. sustainable land use. In most instances, activities generally associated with extraction of non-renewable natural resources such as base metals begin with operations designed to extract the host ore from the ground through open-pit and/or underground mining (tunnel and shaft mining). Ores extracted from base metal mines include, in alphabetical order: Copper, gold, iron, lead, molybdenum, platinum, silver, uranium, and zinc. The predominant base metal ores are those that contain copper, iron, lead and zinc. These are primarily obtained from lode deposits using both open-pit and underground mining techniques. Beneficiation, ore dressing, and mineral extraction will generate waste materials in addition to processing wastes and discharges that find their way into the land environment. The nature of these and their impacts on the land environment and particularly with land use constitute the major concern in this chapter. The danger that one faces is the cascading or domino effect generated by these activities and discharges. In respect to the deep-underground *in situ* extraction processes illustrated in Figure 5.2, the various processes (SAGD, CSS, and hydraulic fracturing) have their proponents and opponents – vis-à-vis environmental impacts, etc. These are not germane to the discussions in this chapter. What is of importance is the identification of the stressor sources, the types of stressors produced, and their impact on the geoenvironment. In these cases, the impact problems are associated with fugitive contaminated water – i.e., water not captured as recycled-treated water used for steam injection and/or used as fracking fluid. The vulnerability of neighbouring aquifers to contamination is a key element in information gathering. These pieces of information are important because they will provide the clues for development of geoenvironmental engineering practices necessary to mitigate the adverse impacts. The emphasis on *geoenvironmental engineering practice* is necessary because there are actions that can be undertaken to reduce and even minimize adverse geoenvironmental impacts. These, however, are *operational actions* – i.e., actions that are associated with the implementation of the processes themselves.

The discussions in this section relate to four categories in this grouping of resource extraction and processing industry operations. These categories include those industries dealing with (a) ores in metalliferrous mining for recovery of various types of metals, etc. (b) ores from non-metal resource mining and processing, such as aggregates, clays, etc., (c) hydrocarbon resources extracted from the ground, such as oil sands (formerly known as tar sands), and (d) deep underground *in situ* extraction of tight oil and shale gas. These are operations mounted to process the ores or liquid/gaseous resources recovered from the various forms of mining and drilling activities. The products obtained from these industries serve as raw materials for their associated downstream industries. The associated downstream industries are discussed in detail in Chapter 4.

5.3.1 Mining-Related Industries

The first two categories mentioned previously can be considered jointly as *mining-related industries*. For these (mining-related) industries, there are typically three types of interactive contact with the geoenvironment. There are at least two different types of mining: (a) Surface mining or open pit mining, and (b) underground mining. Mining sites are characterized by piles of waste rock, huge pits, and acid mine drainage. Decommissioning procedures often must include remediation and closure strategies – requirements that have not been imposed in the past. Remediation through decontamination, isolation or immobilization may be required for contaminated areas. For closure of open pits, two simple strategies that have been employed include (a) refilling of the pit with clean material for new land use options, and (b) harmonizing with the surroundings. Before open pit filling, consultation is needed to ensure that the geological heritage is not destroyed. Maintaining formations is essential for research and teaching purposes or education of the public. Waste storage in these pits used to be a common practice. However, because the base of the quarries is often relatively permeable, groundwater contamination from waste leachates can result. If the pit is left for recreation or as a monument, slope stability must be ensured. Flooding of the pit can be quite beneficial. Waring et al. (1999) have proposed this as a method that is low cost and passive for preventing acid mine drainage by eliminating exposure of the sulphide minerals to atmosphere and hence denying oxidation of these minerals. Reservoirs can be used for fishing or wildlife or recreation or restoring biodiversity, and vegetation may also be introduced. Mines could also be sealed against seepage of rainwater and snow, and drainage systems should be developed to divert water away from the tailings and thus reduce acid mine drainage problems.

The issues of prime importance in respect to the geoenvironment and its sustainability relate to (a) the actual mining operation itself and its impact on the surficial environment, and (b) the various waste products issued from mining and resource extraction operations. Most mining wastes are associated with recovery processes of host rock material for production of natural resources, such as metal and non-metal ores and products. These include aluminium, iron, copper, gold, lead, molybdenum, silver, tungsten, uranium, zinc, coal, asbestos, gypsum, barite, syenite, potash mineral, salt mineral, bitumen (from oil sands, "tight rocks", shales, etc.) quartz, limestone, sand, gravel and stone. In addition to solid wastes generated during the mining process, liquid wastes in the form of tailings and other process liquid waste streams associated with mining and milling operations are obtained. These need to be considered as sources of stressors that could impact on the geoenvironment. Specific and detailed examples of various mining activities and milling processes can be found in mining-milling textbooks dedicated to the study of these subjects.

5.3.1.1 Discharges from Beneficiation and Processing – Stressor Sources

Procedures and processes required to extract minerals from the ores obtained in mining operations vary according to the kinds of minerals being extracted. These procedures and processes fall under the category of *beneficiation operations* and *mineral processing*. These operations are designed to physically separate the mineral of interest from the surrounding non-mineral elements commonly called gangue, and to use various means to recover the mineral of interest. Operations included in *beneficiation* and *mineral processing* are (a) ore preparation – crushing and grinding (commonly known as *comminution*), washing, (b) mineral recovery – including such processes as dissolution, filtration, calcining, roasting, leaching, concentration, separation, solvent extraction, electrowinning, and precipitation. The generalized procedure shown in Figure 5.4 describes the main elements of most of the processes used.

Variations (of procedure) from the main elements can be expected between different operating companies because of differences in the mode of operation. Bulk ore transported to the processing and extraction plant needs to be crushed to smaller fractions where specific chemicals and additives are added. The resultant mixture is then subjected to the required mineral extractions processes – as required. The points of interest in regard to interactions of the mining-milling activities and industries with the geoenvironment shown in Figure 5.5 include debris from mining, and liquid and solid waste materials from beneficiation and processing. These discharges can be considered to be sources for various kinds of stressors.

In respect to stressors and impacts on the geoenvironment, the major sources of stressors are the heaps, wastewater, and tailings' ponds. The types of stressors originating from these sources are

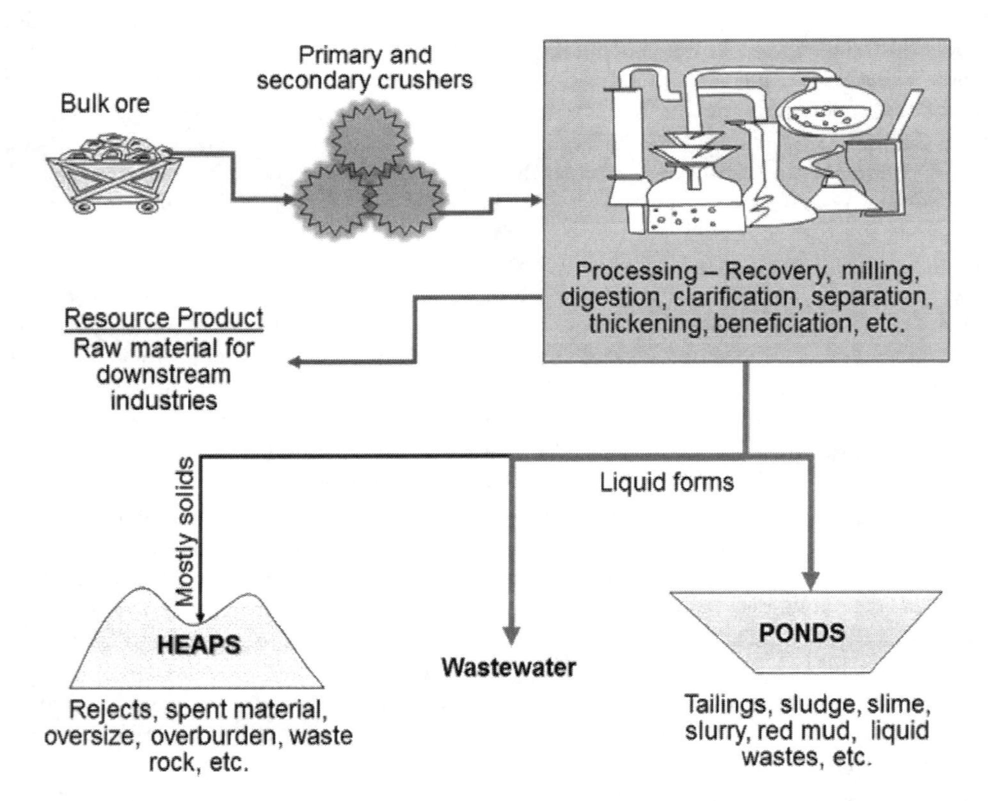

FIGURE 5.4 Generalized resource extraction-recovery process. Specific details of additives, digestion, and beneficiation, etc., will vary according to types of resources (minerals, non-minerals, hydrocarbon) being extracted. Discharges to the geoenvironment will take the general forms shown in the diagram.

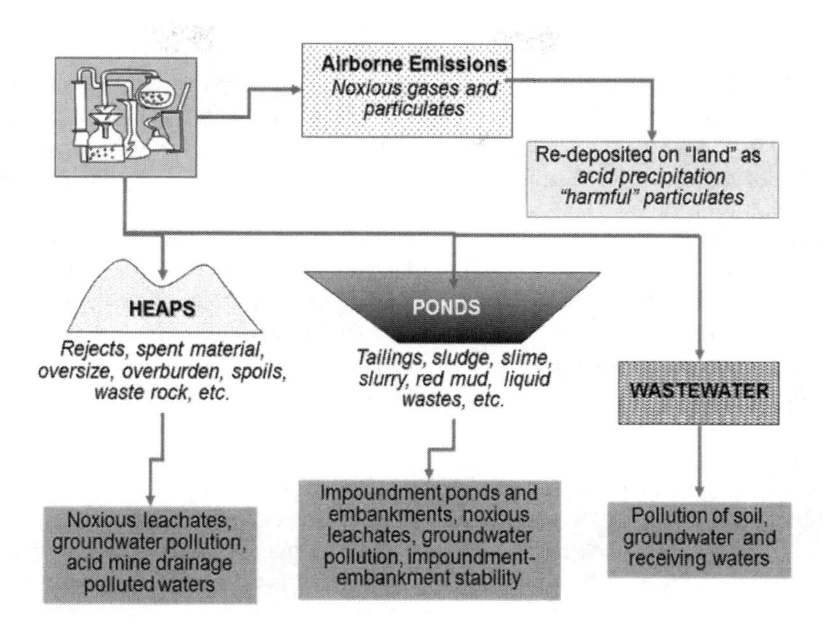

FIGURE 5.5 Discharges from resource recovery operations and some of their more significant impacts on land use and the geoenvironment. The ponds containing the slurry tailings could also be dammed-up valleys or even abandoned mine pits.

physical, chemical, physico-chemical, and biogeochemical (see Figure 5.5). Collapse and subsidence of heaps are major problems. The heaps and tailings' basins are also subjected to movement and have inflicted substantial damage in towns where the tailings have liquefied and flowed into towns burying houses and people. Toxic releases from dissolution, bacterial activity, and subsequent run-off also cause damage. The acidic water is carried into ground and surface water thereby threatening the health of humans, animals, and the local flora.

Dust inhalation by humans may also lead to cancer and other illnesses, such as asbestosis and silicosis. Bioaccumulation of mercury, lead, and other heavy metals in the food chain may also be significant. Kobayashi and Hagino (1965) reported that the run-off of cadmium from a zinc mining waste in the 1950s led to accumulation in the Jintsu River that was used for drinking water and irrigation of rice fields and that itai-itai disease afflicted the population living in the area. Secure disposal, cleanup, and management of drainage eliminated the disease.

5.3.1.2 Solid Waste Materials and Stressors

The principal sources of solid waste materials issuing from extraction of the various minerals are waste rocks, ore spoils, and overburden. There are two types of geoenvironmental impact problems from the stressors associated with the disposal of solid waste materials: (a) Physical-mechanical, and (b) chemical, physico-chemical and biogeochemical. In the first case, i.e., physical-mechanical, stability of the heaps and tailings' dams is a major concern.

Historical evidence shows numerous failures of these physical structures arising from simple slope instabilities to landslides and/or liquefaction triggered by excess porewater pressures. A chronology of some major tailings' dam and heap failures compiled by WISE (2023) can be seen in www.wise-uranium.org/mdaf.html (accessed Dec. 2023). The compilation includes a whole series of cases of dam or heap failures associated with industries extracting products such as copper, nickel, manganese, fluorite, bauxite, gold, phosphate, etc., ranging from 1960 to 2023. By and large, the major stressors responsible for the slope failures and various other types of instability are (a) excessive forces or stresses in the structures (heaps and tailings' dams) due to size and weight of the heap – inconsistent with the design stability of the structures, and (b) excess porewater pressures developed as a result of infiltrating rainwater or other sources – due to the lack of provision for slope protection facilities against infiltration and/or drainage system and pore-pressure shedding devices within the structures.

Chemical and biogeochemical stressors originating from heap and tailing dam sources can be traced to the nature of the host ores and the leaching solutions used to extract the minerals from the ores. In respect to metalliferous mining situations, sulphide minerals are found in most of the host ores. Commonly used ores for extraction of lead and zinc ores are galena (PbS) and sphalerite (ZnS). In the case of iron, the ores are in oxide and sulphide forms. These include magnetite (Fe_3O_4), hematite (Fe_2O_3), goethite ($Fe_2O_3H_2O$), siderite ($FeCO_3$), and pyrite (FeS_2). Although copper sulphide minerals are found in such minerals as chalcopyrite ($CuFeS_2$), covellite (CuS), chalcocite (Cu_2S), and bornite (Cu_5FeS_4), the most common source of copper ore exploited in present mining operations is chalcopyrite (Simons and Prinz, 1973). Oxide minerals include chrysocolla ($CuSiO_3$), malachite (Cu_2CO_3), azurite ($Cu_3(CO_3)_2(OH)_2$), tenorite (CuO), and cuprite (Cu_2O).

Using extraction of copper as an example, solid waste material discharges include the overburden material, such as soil, debris and unconsolidated material, oversize material and waste rock (rejects), spoil heaps, leach ore obtained from leaching of copper oxide mineral ores, and other solid by-products and wastes issuing from the recovery process. It is estimated that on the average, about 100 tonnes of waste material are generated in the extraction of one tonne of copper. Using the 100:1 ratio of waste material for each tonne of recovered copper as an example, we can surmise that for the various mineral extraction industries in the world, this would mean that hundreds of millions of tonnes of waste materials are generated in a year. All of these materials will need proper management and disposal. This is particularly significant because the rejects and other ores in the piles contain trace amounts of sulphide minerals. These will contribute to the generation of acid mine drainage problems. Not all the solid heaps are waste heaps. In operations where sulphur recovery is obtained, open-air storage on-site is not an uncommon procedure. Figure 5.5 shows some of the more significant impacts to the geoenvironment, and also impacts on land use. The significant impacts from the presence of solids as rejects, oversize, etc., on land use include storage, noxious leachates, acid generation (AMD) and run-offs leading to acid pollution of watercourses, groundwater, and land surface.

Alkaline waste is produced at a rate of 2 billion tonnes per year (Gomes et al., 2016). Production of steel, iron and lime, power generation from coal and chromium ore and alumina extraction (red mud) are major sources of this type of waste. Calcium, magnesium, or sodium oxides are present in these wastes, which then produce hydroxides. Leachates containing arsenic, chromium selenium and other oxyanions can be produced from rain water infiltration, particularly in old and abandoned mines. High pH and salinity are other impacts.

5.3.1.3 Liquid Waste Streams, Discharge, and Stressors

There are essentially two or three liquid waste streams associated with mining-mineral extraction industries. These include the following:

- Mine water and waste streams generated in a mining operation due to hydrologic drainage from the mining site, and percolation from waste rock and mill tailings piles, and surface run-offs. Whilst containment of mine water and generated waste streams in abandoned mines has been previously practised, present worries of acid mine generation has led to prevention of excess water entry into abandoned mine sites. Extraction pumping of mine water serves to remove a potential source for generation of acid in the mine. However, without benefit of capping and sealing of mine shafts and openings, it is inevitable that some water will be introduced into the mines – resulting ultimately in the problem identified as acid mine drainage.

- Liquid waste streams from processing plants, such as solutions or liquids containing solvent extraction sludge, spent electrolytes, spent leaching solutions, spent solvents, and used oils. In general, the liquid wastes are contained in ponds, and are often referred to as solution ponds, pregnant ponds, holding ponds, etc. Pregnant (solution) ponds are those that contain some of the minerals, and solution ponds are those that are presumably devoid of these minerals.
- Slurry tailings' discharge consisting of water and a wide range of inorganic and organic dissolved constituents obtained from the remnants of reagents used in the recovery processes and fine fractions of the host ores. Impacts on land use consist of large open containment facilities, such as dammed-up valleys and ponds containing sludges and slimes that pose health and safety threats to animals and humans, run-offs from the containment facilities, embankments for the impoundments used to contain the sludges and slimes, and leachates that contaminate ground, groundwater and receiving waters. Most of the slurry tailings' containment facilities contain slurries and slimes that will not exhibit liquid-solid separation, i.e., the solids in these facilities will not readily sediment or settle to the bottom of the containment structure. Typical of the types of tailings ponds are red-mud ponds, oil sand sludge ponds, phosphatic clay ponds, etc. Many of these ponds contain slurries and sludges that are toxic in nature. Their presence will pose potential safety and health threats to the human population within the immediate area. The strategies for maintaining and operating containment facilities are discussed in the next sub-section.
- As opposed to the discussion on tailings' dam instability in the previous section, holding ponds' embankment stability can be a problem if the containment embankments become too high and are subject to drawdown pressures when large fluctuations occur over a very short period of time in the height of the slurry or sludge in the pond. This has happened in instances when containment ponds are emptied quickly. This particular issue is a geotechnical problem that can be corrected with proper design and management techniques.

5.3.1.4 Critical Mineral Supply

Several countries have lists of critical minerals that are quite similar including the UK, South Korea, Japan, European Union, US, and Canada. Critical minerals in general have few if any substitutes, are strategic and limited, and are usually found in specific locations for extraction and processing (Government of Canada, 2022). Critical mineral demand is increasing substantially as they are required for the transition to zero carbon sustainable energy in the form of solar panels, wind turbines, and batteries. Production of solar panels, batteries, wind turbines, and semiconductors all need an increasing supply of critical minerals. The IEA (2023a) has indicated that reliable, sustainable Co, Cu, Li, and Ni supplies are essential for affordable clean energies. The market for critical minerals in the energy sector has doubled in the past 5 years (IEA, 2023b) and could increase by six-fold by 2040. For example, lithium production has grown over 202% from 2015 to 2019 (Reichl and Schatz, 2021). Copper, manganese, platinum, uranium, cobalt, nickel, indium and tellurium are some critical minerals. The Government of Canada's Critical Minerals Strategy (2022) highlights the importance of six minerals (Co, Cu, graphite, Li, Ni and rare earth minerals in addition to nine other priority minerals). Both exploration and recycling are needed to ensure the growth of these supply chains. The strategy emphasizes the need for creating nature-forward solutions with a minimal environmental footprint and using innovative ways for value recovery from residual materials (secondary sources) stream. Objectives 1 and 2 of the strategy indicate the need for a circular economy approach for extracting minerals from residues and strong environmental management through water reduction and recycling of mineral content from tailings, residues and wastewaters, responsible labour and safety practices and incorporating Indigenous practices and participation for conservation and rights protection. Practices must protect biodiversity.

5.4 CARBON CAPTURE AND STORAGE

To mitigate climate change, various methods have been developed, such as carbon sequestration. Carbon capture and storage (CCS) is the separation of carbon dioxide from industrial effluents, compressing it and bringing to a location where it can be stored long-term. Some potential locations are depleted oil and gas reservoirs, deep saline formations, and other deep geologic formations. Leakage of the carbon dioxide is a concern but proper management is needed to reduce the risk. There are still many economic and technical challenges for large-scale implementation.

In Saskatchewan, Canada, fossil fuels such as coal are used to generate electricity. This produces carbon dioxide. To prevent release into the environment carbon dioxide at the Boundary Dam, carbon capture and storage is being implemented since 2014, the first in the world. At this location, 115 MW of power is produced which is sufficient for about 100,000 Saskatchewan homes. The units are able to remove 100% of SO_2 and 90% of CO_2 emissions from the coal process. Daily average capture rate has been reported as 2674 tonnes. This resulted in an emissions intensity of 407 tonnes of carbon dioxide per gigawatt hour, which is within Canada's current federal carbon tax threshold. According to the latest report, approximately 5.6 million tonnes of carbon dioxide have been captured since 2014 (www.saskpower.com accessed Jan. 2024).

There is the potential for the reaction of calcium and magnesium in alkaline residues with carbon dioxide gas to form stable carbonates which may help to mitigate climate change by the following reaction (Gomes et al., 2016):

$$(Ca, Mg)\ SiO_3\ (s) + CO_2\ (g) \rightarrow (Ca, Mg)CO_3(s) + SiO_2\ (s) \tag{5.5}$$

High temperatures or acidic conditions would be needed to release the carbon dioxide. (Renforth et al., 2019) has estimated that there is a capture potential of up to 8.5×10^{12} kg per year. This process is not optimized but it is estimated that natural conditions have led to the capture of 100 Mt of carbon dioxide up to 2008 (Si et al., 2013). More research is needed to quantify and demonstrate full-scale potential. LCA should be performed to evaluate impacts.

5.5 TAILINGS' DISCHARGES AND MANAGEMENT

Central to the extraction of resource material contained in the ores obtained in mining is fine grinding of the ores – as the preliminary stage of the beneficiation process. The common elements of the various techniques used in the beneficiation processes include both fine-grinding and washing-flotation and finally, at some stage in the beneficiation process, separation of the resources from the finely-ground material. Flotation and other methods of resource separation (e.g., magnetic) are required. The end result of all of these processes is the generation of liquid wastes containing a suspension of the finely-ground material. This liquid waste is generally defined as *tailings' waste slurry* or *slurry tailings*. It can be well appreciated that the generous quantities of tailings are reflective of the significant amounts of water needed in the processing of the finely ground ore material for beneficiation.

5.5.1 Containment of Tailings

Slurries, slimes, sludges, and red mud ponds are all names that are given to the general class of slurry tailings discharge from beneficiation processes – the choice of which depends on the nature of the tailings being discharged. These slurry tailings cannot be directly discharged into the land environment not only because (a) they contain suspended fines, and (b) their chemical nature will most likely be toxic to the environment and public health. In short, discharge of these kinds of tailings will have detrimental impact on the land ecosystem. Some of the main reasons for containment of these slurry tailings are as follows:

- to avoid pollution of land surface environment. This is one of the principal reasons for containment, and is generally coupled with other specific disposal-containment strategies.
- to provide permanent containment of the tailings. This strategy generally includes several kinds of scenarios, ranging from *permanent ponds* to totally reclaimed solid land surfaces – e.g., a multi-stage strategy for treatment of tailings pond sludges as a land reclamation process (Yong, 1983).
- to recover water for reuse in the beneficiation processes as process recycle water or other mine-site requirements. This requires implementation of treatment of the supernatant – assuming that liquid-solids separation is effective in producing sedimentation of the suspended fines in the slurry tailings.
- to permit secondary recovery from the storage pond. This strategy presumes that some residual resource is contained in the slurry tailings waste, and that secondary recovery of this resource can be obtained when appropriate technology becomes available, and when the economic climate is favourable.

There are at least three basic types of slurry tailings impoundment facilities: (a) Abandoned, used-up, fully exploited mine pits, (b) dammed-up valleys – as illustrated in Figure 5.6, and (c) constructed ponds with containment embankments (Figure 5.7). The choice of containment facility depends on many factors, not the least of which is site conditions and company mining strategy. These are not within the scope of this book and will not be addressed.

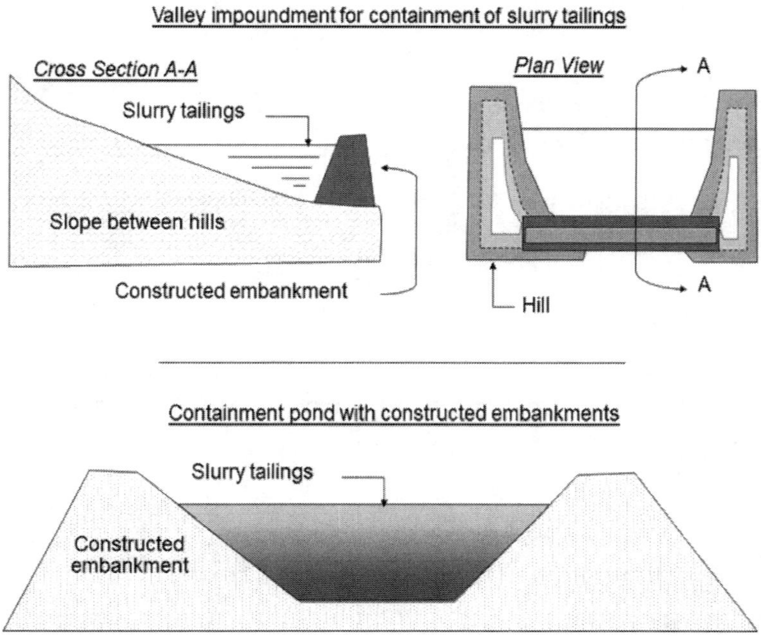

FIGURE 5.6 Illustration of valley impoundment for containment of slurry tailings (top diagram), and constructed pond (bottom diagram).

There are two distinct categories of slurry tailings contained in the containment facilities or structures:

- Containment structures that contain sedimented solids and particulates discharged from mineral extraction processes. These structures will show a "solids-sedimented" layer overlain by water as illustrated by the containment pond shown in the top diagram in Figure 5.7. For such kinds of sediment slurries, treatment of the supernatant (liquid) may or may not be necessary – depending on the chemistry and/or toxicity of the supernatant. In some mining processes, such as aggregate harvesting from transported surface soils, since water is the only agent used in the beneficiation process, the supernatant obtained is considered to be non-toxic. As an example, tin mining of placer deposits using the gravel pump and dredging method will leave behind slime ponds with well-developed sediments and clear supernatants.
- Containment structures that contain solids' suspensions. The solids in these suspensions may or may not finally sediment. Using containment ponds as an example, the bottom diagrams in Figure 5.7 show the characteristics of these kinds of ponds. The dispersion stability of these types of slurry tailings will be discussed in the next sub-section.

For suspended fines in slurry tailings that do not exhibit liquid-solid separation behaviour, some common features can be identified. If one determines the solids' concentration with the depth of the slurry tailings, one will obtain at least four distinct zones. These are (a) clear supernatant liquid, (b) a transition zone where the solids' concentration begins to register some small value that increases as one progresses in depth, (c) a stagnant zone where the solids' concentration remains relatively constant or increases imperceptibly with depth, and (d) a sediment zone that contains the

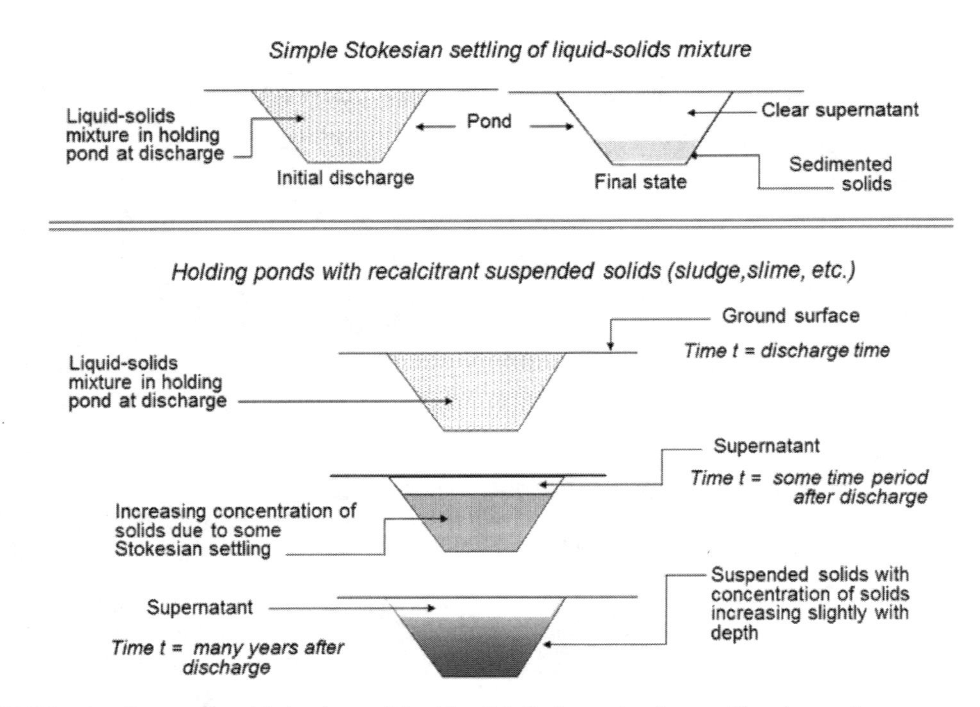

FIGURE 5.7 Two types of behaviour of liquid-solid discharge in slurry tailings' containment structures. The examples shown are containment ponds. Top diagram shows simple Stokesian settling of the solids in the liquid-solid mixture. Bottom diagram illustrates solids' behaviour over some time period, ending up with recalcitrant performance of suspended solids.

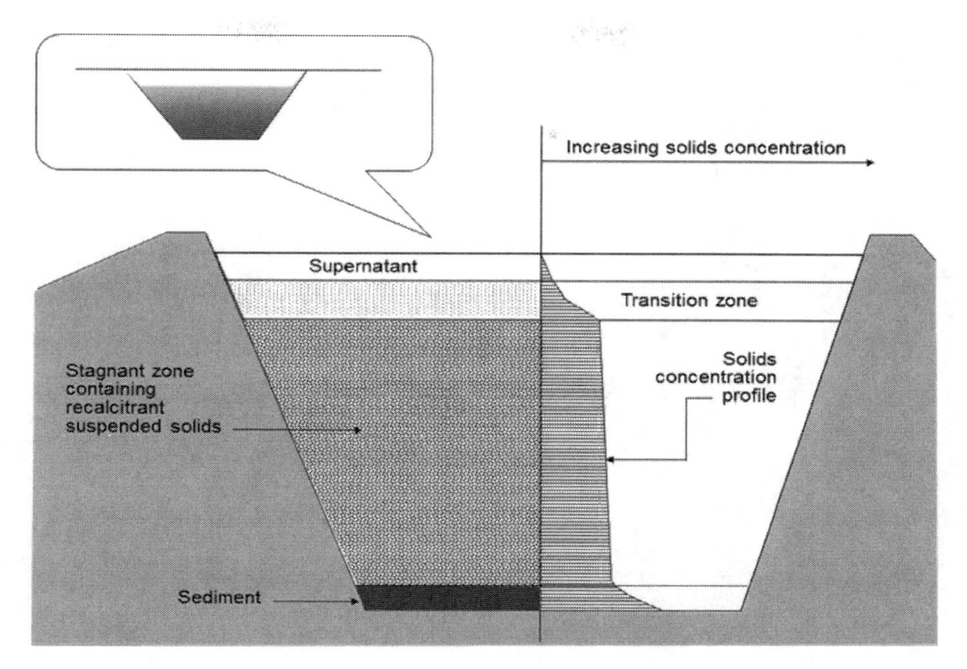

FIGURE 5.8 Illustration showing solids' concentration profile for the recalcitrant suspended solids behaviour pattern – typical of such sludges and slimes as red muds, tar sands' sludges, phosphatic, and other clay slimes.

solids that have finally settled to the bottom of the containment structure. Figure 5.8 shows a typical solids' concentration (sc) profile in the four zones that are typical of various kinds of sludges, slimes, slurries, etc. The sc values refer to the weight ratio of the suspended solids to aqueous phase that constitutes the suspension fluid. Figure 5.9 shows the solids' concentrations in the stagnant zones for various types of slurry tailings. Leaving the suspended fines in the containment structures is not an acceptable land use option. Strategies have been developed to render the material in the containment structures to a state that would not pose a threat to the immediate environment and biota (Yong, 1983a; 1983b).

5.5.2 Nature of Contained Slurry Tailings

Studies on the nature of the solids in suspensions in slurry tailings' facilities (ponds etc.) where solids remain in suspension for some considerable length of time show that the principal factors responsible for the dispersion stability of the suspended solids include (a) colloidal nature of the solid fines, (b) reactive surfaces on the fines, and (c) chemistry of the suspending fluid. The theoretical basis for the dispersion stability of the colloidal-type solid fines has been well-developed and reported (Kruyt, 1952). The diffuse double-layer (DDL) model is a good fit with the types of suspended fines (e.g., montmorillonite, kaolinite, amorphous materials) found in many slurry tailings. It essentially provides one with a basis for determining (theoretically) the maximum volume of water or fluid in the diffuse ion-layer that surrounds individual reactive suspended particles.

The intensity of the interaction forces between the two particles resulting from the inter-penetration (or overlapping) of the contiguous diffuse ion-layers is a function of (a) the extent of the overlapping or inter-penetration of the adjoining diffuse ion-layers, (b) the nature of the reactive surfaces of the particles, and (c) the chemical composition of the suspending fluid. The electrostatic interactions of the ions in the diffuse ion-layer and their relation to the surfaces of the reactive suspended particles are expressed as an electric potential ψ that decreases in an exponential manner

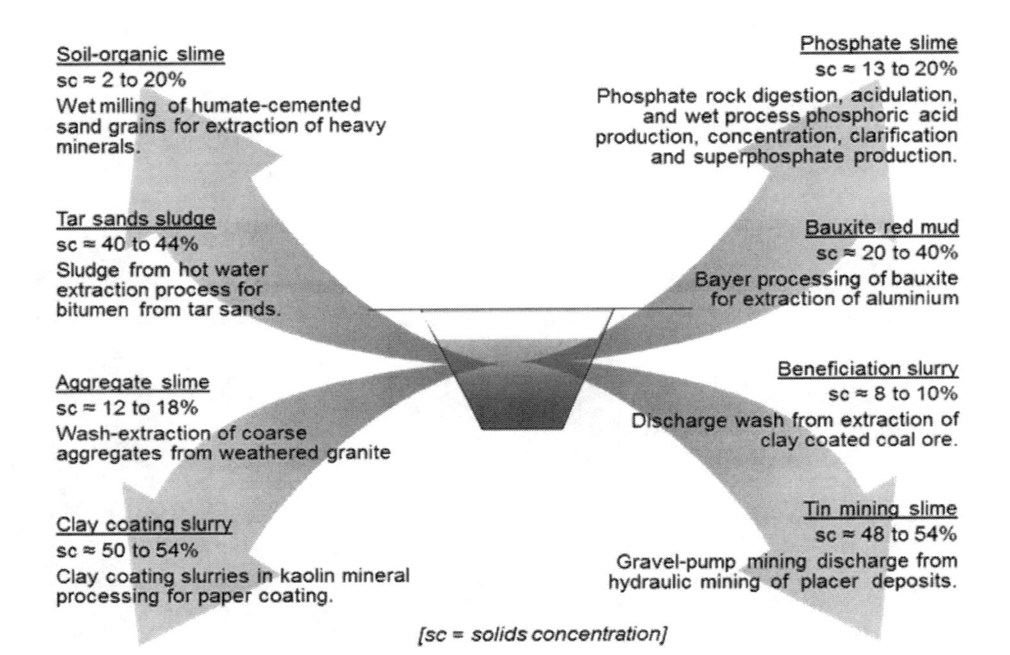

FIGURE 5.9 Some examples of slurries, slimes, and sludge found in holding ponds. The solids' concentrations (sc) are obtained from results reported by Yong and his co-workers (see, for example, Yong, 1984). Details of the compositional features of the slurries, slimes, and sludge are given in Table 5.2.

as one departs further from a particle surface. The DDL model provides one with the basis for computing the average electric potential ψ as a function of distance from the surface of the particle as follows (Yong and Warkentin, 1975):

$$\psi = -\frac{2\kappa T}{e}\ln\coth\left(\frac{x}{2}\sqrt{\frac{8\pi z_i^2 e^2 n_i}{\varepsilon\kappa T}}\right) \tag{5.5}$$

where the negative sign on the right-hand side indicates that the potential ψ decreases as one departs further away from each particle surface, and where κ = Boltzmann constant, T = temperature, e = electronic charge, n_i and z_i = concentration and valence of the ith species of ions in the bulk solution, and ε = dielectric constant. A detailed treatment of the DDL theory and models can be found in Kruyt (1952). The development and application of the DDL models to soil mineral particles such as those found in slurry tailings can be found in Yong and Warkentin (1975) and Yong (2001a).

Calculations of the volume of water associated with a gram of soil particle in equilibrium in an aqueous phase, based on the type of soil fraction and diffuse double-layer interactions can be made using the DDL models. These can be compared with measurements of equilibrium solids' concentrations obtained in soil-suspension experiments. The results of solid-suspension tests reported by Yong (1984) are shown in Table 5.2 for some typical soil solids found in slurry tailings. These results are expressed as the equilibrium volume of water per unit weight of suspended solids, and the units are given as cc/g of soil. The void ratios shown in the third column of Table 5.2 have been calculated from the measured equilibrium volumes.

Yong (1984) has shown good correlation between calculated and measured equilibrium solids' concentration for the stagnant region of a slime pond using the equilibrium volumes shown in Table 5.2. In the actual cases examined, predicted solids' concentrations were compared with actual

TABLE 5.2
Equilibrium Suspension Volumes Obtained from Soil-Suspension Tests

Suspended Solids	Equilibrium Volume, cc/g	Void Ratio
Kaolinite	1.3	3.4
Illite	3.1	8.2
Montmorillonite	21.5	57
Amorphous Fe_2O_3	20.5	82
Gibbsite	1.0	2.6
Mica	3.0	7.9
Quartz	0.14	1.12

Source: Adapted from Yong (1984).

TABLE 5.3
Composition of Slurry Tailings' Solids and Suspension Fluid in the Stagnant Zone of Slurry Tailings' Ponds

Type of Slurry Tailings	Suspended Solids' Composition	Suspension Fluid, Dominant Ions	Measured sc, %	Computed sc, %
Phosphate slime (Florida)	carbonate-flourapatite, quartz, montmorillonite, attapulgite, wavellite, feldspar, dolomite, kaolinite, illite, crandallite, heavy minerals.	Ca^{2+}, Mg^{2+}, Na^+, K^+, SO_4^{2-}, HCO_3^-	14	13.4
Aggregate slime S.E.Asia	kaolinite, montmorillonite, illite.	Ca^{2+}, Mg^{2+}, Na^+, K^+, SO_4^{2-}	14.3	10.5
Clay coating, South East USA	illite, montmorillonite, mixed-layer minerals, chlorite, quartz.	CO_3^{2-}, HCO_3^-, Cl^-, SO_4^{2-}, Na^+, Mg^{2+}, Ca^{2+}	52.7	51.8
Tin mining slime, Malaysia	kaolinite, gibbsite, mica, quartz, "other".	Na^+, Ca^{2+}, K^+, Mg^{2+}	52	50.8
Beneficiation slurry, Western Canada	montmorillonite, illite, feldspar, kaolinite, chlorite.	CO_3^{2-}, HCO_3^-, Cl^-, SO_4^{2-}, Na^+, Ca^{2+}, K^+, Mg^{2+}	8.7	9.1
Oil sands sludge, Western Canada	kaolinite, illite, chlorite, montmorillonite, mixed-layer minerals, feldspar, quartz, siderite, ankerite, pyrite, Fe_2O_3.	Not available	41.9	42.2

Source: Adapted from Yong (1984).
sc = solids' concentration

solids' concentrations obtained from samples in the stagnant zone (Figure 5.9) for phosphatic slimes, aggregate slimes, tin mining slimes, beneficiation slurry (slimes), tar sand sludges, etc. Table 5.3 shows that except for the aggregate slimes obtained from aggregate recovery of aggregate loams, a comparison of the predicted solids' concentration with actual measured values showed good accord. The ratio of predicted to measured solids concentration (predicted:measured) varied from 0.96 to 1.05. What this tells us is that colloidal dispersion of the suspended solids is responsible for the dispersion stability of the slurry tailings.

5.6 GEOENVIRONMENT IMPACTS AND MANAGEMENT

There are at least four significant kinds of geoenvironment impacts associated with resource mining and extraction operations: (a) Mining excavations, pits, underground caverns, and debris piles, waste rock, rejects, overburden, etc., (b) acid generation and/or acid mine drainage from exposure of debris piles and exposed mine cavities etc., (c) slurry tailings' containment facilities, and (d) fate of contaminated fluids in underground *in situ* extraction of hydrocarbons. The second and third concerns can be seen in Figure 5.5. Mining excavations and underground mining create situations where the excavated (empty) volumes present challenges that are beyond the scope of the material discussed in this book. The impact from debris discharge and heaping into "tips" has been briefly mentioned at the beginning of this chapter (Section 5.3.1). In addition to the sterilization of the immediate landscape surrounding the tips as a result of the leachates emanating from the tips, possible instability of the tips is a question and problem that needs attention.

5.6.1 Geoenvironmental Inventory and Land Use

By and large, a major proportion of mining and on-site resource extraction operations are initially in regions situated some distance from urban centres. Original land use, prior to the time of mining exploration in such regions would be characterized by the local physiographic features, such as those discussed in Chapter 1. An environmental inventory and more specifically a geoenvironmental inventory prior to mining operations is needed to establish a base upon which decisions regarding impacts on land use, sustainability indicators, and restorative requirements can be sensibly made. The principal features of the geoenvironmental inventory, which is a baseline descriptor of the state of the various constituents of the local geoenvironment (*ab initio* condition), include the following:

(a) regional controls, such as climate and meteorological factors,
(b) local terrain features, including linear features, physical attributes, topography, watershed, local hydrology, surface layer quality, vegetative cover, receiving waters and water quality, and
(c) sub-surface features, such as geological and hydrogeological settings, soil sub-surface system, and groundwater-aquifer regimes.

Decisions on sustainability of potential land uses or site functionality cannot be made without the inventory and without determination of the qualities of the attributes necessary for various land uses. These are highly dependent on whether (a) the mining and resource extraction operations remain isolated from urban communities, or (b) small urban communities are located contiguous to the mining site. In the first instance, i.e., mining operations remain isolated from habitable communities; return of the exploited land to original natural conditions, i.e., original site functionality, together with measures implemented to ensure protection of human health and other biota are requirements needed to satisfy sustainable land use aims. Specification of *ab initio* land environment sustainability indicators will be guided by the geoenvironmental inventory established before mining operations. Historically, geoenvironmental inventories have not been made prior to, and even during mining operations. Nevertheless, a study of the natural geoenvironment system contiguous to the mining operations will serve to provide the basis for establishment of *ab initio* sustainability indicators. Whilst return of the land to its pristine original function, as a sustainability objective, may not be absolutely possible, one can return the land to its natural state and natural site functionality. For example, one could seal shafts and openings, relevel and contour the landscape to conform to local topography to allow movement of migratory animals, add organic fertilizers, and reseed with tolerant plant species (Davies, 1999). This procedure recognizes the fact that whereas the natural resource that has been harvested no longer exists, proper mine closure allows for recovery of the functionality of the affected ecosystem.

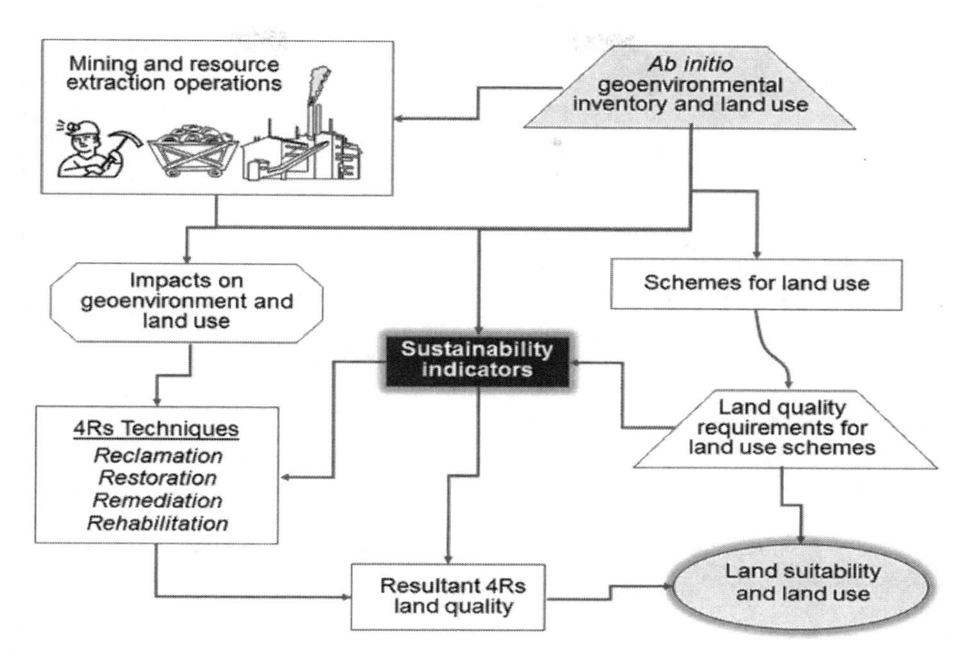

FIGURE 5.10 Basic steps for rehabilitation of lands affected by mining-extraction operations. Implementation of schemes for land use are dependent on the restored land quality. In the final analysis, land suitability is dependent on the quality of the rehabilitated land.

It is not uncommon for small urban communities to be developed within regions close to mining and resource extraction operations to service the mining operations. Land use and land restoration will depend on the requirements of the urban community and will undoubtedly be markedly different from *ab initio* land use conditions. The general procedure is to perform an assessment of impacts on land use from the mining-extraction operations, and to determine what reclamation and land restoration requirements are needed to meet the needs of the community. The basic steps shown in Figure 5.10 include the following:

(a) Determination of impacts to geoenvironment and to land use. This requires information from *ab initio* geoenvironmental inventory.
(b) Compilation of community and/or regulatory land use schemes or plans – e.g., return to *ab initio* conditions, housing estates, parklands, schools, recreation facilities, natural landscape, waste dumpsite, industrial estate, agriculture, etc.
(c) Determination of land quality requirements for land use schemes and plans.
(d) Assessment of the required 4Rs technologies – remediation, reclamation, restoration, and rehabilitation requirements to meet the required land use schemes or plans.
(e) Assessment of quality of rehabilitated lands.

Matching land quality, land suitability, and site functionality with community or regulatory land use plans and requirements for rehabilitated land use that does not necessarily mean returning the land to initial conditions, as, for example, reclamation of land for housing developments.

5.6.2 Acid Mine Drainage (AMD) Impact Mitigation

As mentioned previously, other than the presence of metal-sulphide minerals, the two main elements in the production of acidic leachates are oxygen and water. These two constitute the sources for the acid (chemical) stressors. The impacts from these stressors are seen in the contamination of receiving waters and surface land soils. Whilst it may not be possible to totally eliminate or reduce these impacts, there are two clear paths where impact mitigation can be implemented. These are (a) source control, and (b) leachate management. In the case of source control, the mitigating actions undertaken by the stakeholder include (i) control of quality of discharges by the stakeholder, and (ii) control of discharge by limiting and channelization of flow into treatment stations prior to discharge. These actions include technology and processes that are physical, chemical, geochemical and biogeochemical – and various combinations of these. In the case of leachate management, the steps taken to mitigate impact of the acid leachates are essentially remediation efforts – i.e., remediation efforts undertaken "once the horse has left the barn". These efforts have been called "passive control methods" – designed to minimize the deleterious effects of acid leachates on the environment by treating the leachates in transport through the geoenvironment. The industries involved in this type of problem, together with the pertinent governmental agencies have spent (and are still spending) some considerable effort in combating AMD. There are countless documents reporting on the various studies and efforts in "curing the problem", and the reader is advised to consult these for detailed elaboration of the various research and case studies undertaken (and also presently underway). For the discussion that follows, only the basic elements of the mitigating efforts will be addressed.

The phenomenon of acid generation from heaps with trace amounts of sulphide minerals and from exposed sulphide minerals in mined out caverns and pits, etc., has been discussed in Section 5.2. The magnitude of the sets of cascading problems attributed to the actions of acid leachate cannot be overstated. Clean-up of all the affected areas, mining sites, waterways, etc., is estimated in the billions of dollars in North America. Not only is the threat expressed in terms of acid leachates finding their way onto the land environment and into receiving waters, but also in terms of release of trace metals into the geoenvironment. The leachates and released metals will negatively affect the functionality of the various ecosystems that comprise the geoenvironment – causing them to be eventually unsustainable. In terms of sustainable land use, the obvious protection against such stressors (acid leachates) is to remove conditions and circumstances favourable for the generation of acid leachates through source control. This means denying access to oxygen and water. The principle of "*keep it dry*" is a good principle to practise at mine sites. This is an easy statement to make, but is in reality a very difficult and an almost impossible principle to adhere to. Since water is an essential element in mining-extraction processes, *keeping it dry* requires one to provide protective covers, isolation barriers, and pump-discharge operations. Implementation of these techniques is dependent on site-specific and operation-specific conditions. Adherence to the *keep it dry* principle is one of the basic requirements for minimization of impacts to land use, and to the 4Rs technique (Figure 5.10).

For conditions where acid generation has occurred and acid leachate has found its way to the land environment and its receiving waters, the two courses of required action are (a) protection of the affected land receptors and water bodies from accepting ***further*** acid leachates, and (b) treatment of the affected land and water bodies. It has been suggested that once pollution of the receiving waters, such as those described in Section 5.2 has occurred, destruction of aquatic habitat will render these waters to be bereft of aquatic life for a very long time. Nevertheless, treatment of these waters is necessary. Methods for treating polluted water have been discussed in Chapter 3.

5.6.2.1 AMD Management

Acid mine drainage (AMD) can also be managed. Active treatments that use wastewater treatment techniques are possible. They are expensive and must be maintained regularly such as the use of chemical or electrolytic treatments. The essential element here is the capture of leachate for treatment – an expense that will be ongoing for an interminable period since the source of acid generation will most likely be almost inexhaustible. Passive treatment systems include vertical flow, aerobic and anaerobic wetlands, anoxic limestone drains, open limestone channels, and alkaline leach beds may also be effective for specific flow and pH conditions. Prevention of acid mine drainage is possible by addition of limestone or by sealing the area with fly ash grouts. For example, Bulusu et al. (2005) used a grout of coal combustion by-product to reduce AMD. The grout was durable and did not exhibit signs of weathering.

Alkaline agents (calcium oxide, calcium hydroxide) and wastes (cement kiln dust, acetylene gas sludge) can be added to pools of acid mine drainage for neutralization. Alkaline trenches with limestone or soda ash are used to neutralize run-off. Bacteriocides have to be used to inhibit bacterial growth and hence pyrite oxidation. Natural treatment through carbonate formations may also be possible as in the case of cretaceous chalk underlying the coalfield slag heaps in the north of France (Chamley, 2003). Permeable reactive barriers may also be a solution. Constructed wetlands are a low-cost alternative for passive treatment as described in the next section. Anoxic limestone drains are being evaluated for use because of their ease in maintenance and operation. However, they require large areas for effective application. Skousen et al. (2019) compared the costs of six chemicals for AMD neutralization (limestone, hydrated lime, pebble quicklime, soda ash, caustic soda, and ammonia). They found that while the installation of hydrated lime systems are high, chemical costs are low. In addition, as it was effective for long-term treatment, it could offset the high initial costs. However, tests should also be performed to determine effectiveness at each site.

The essential procedures for mitigating AMD stressors impacts include (a) diversion of waters around the mine sites as part of the *keep it dry* (*keep it as dry as possible?*) strategy, (b) channelling the generated leachates through constructed aerobic and anaerobic wetlands as a neutralization procedure, and/or through permeable treatment walls – to be discussed in Chapter 10, (c) complete inundation of mined sites – to deny access to oxygen, and (d) capturing and channelling the generated leachates for active treatment before discharge (Figure 5.11). Creating an oxygen barrier or reducing water infiltration (Aubertin et al., 2016) to control AMD formation. Water covers by tailings disposal in water or by flooding the tailings can also reduce tailings reactivity.

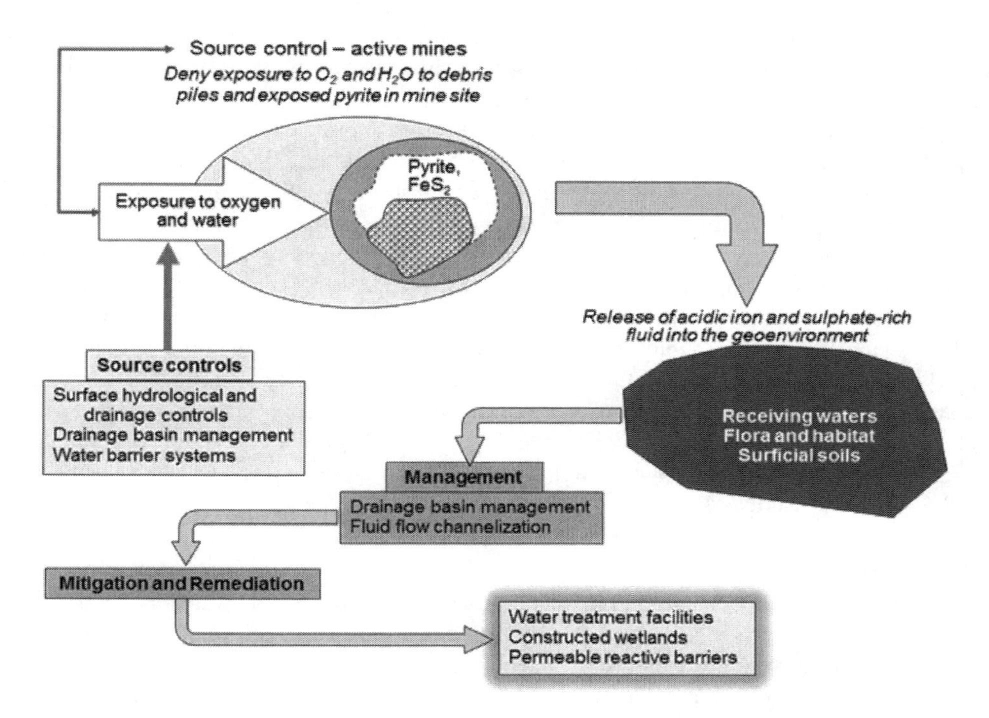

FIGURE 5.11 Basic elements of source control and passive treatment procedures for AMD stressors' mitigation.

There is also a significant capacity for natural attenuation at mining sites. A particular example of this is the Falun Copper Mine in Sweden. During its operation of more than a hundred odd years, it is estimated that a half to one megatonne of copper, lead, zinc, and cadmium were emitted into forest soils and streams in the area (Lindeström, 2003). Concentrations of 70 μg/L and 2000 μg/L were found in streams in the city of Falun after treatment of the mine water was initiated but decreased thereafter. Soils, however, in the area were able to recover substantially faster. Most of the metals concentrated in the sediments in two lakes. Copper was 120 to 130 times normal background levels while lead and zinc levels were 30 to 40 times. Cadmium accumulation was much less. The aquatic ecosystem was able to return due to the low bioavailability of the metals and possible interactions between the metals.

5.6.2.2 Natural Systems

Natural wetlands are areas of land with the water level close to the land surface, thus maintaining saturated soil conditions and vegetation that includes plants, peat, wildlife, microbial cultures, cattails (*Typha spp.*), reeds (*Phragmites spp.*), sedges (*Carex spp.*), bulrushes (*Scirpus spp.*), rushes (*Juncus, spp.*), water hyacinthe (*Eichhornia crassipes*), duckweeds (*Lemna spp.*), grasses and others (Mulligan, 2002). Algae and mosses, together with the wet areas can trap the heavy metals. Constructed wetlands have been specifically designed to include these species for the removal of BOD, suspended solids, nutrients and heavy metals for optimal performance via particle settling and metal oxidation and hydrolysis as previously shown in Figure 3.14.

Anaerobic wetlands (also called bioreactors) are deeper than aerobic wetlands (more than 30 cm). Sulphate reduction is thus promoted, which enables precipitation of metal sulphides (Neculita et al., 2007). Sorption on organic components with hydroxide precipitation can also occur. Low flow systems are most appropriate for this type of treatment.

Iron and manganese removal are often the key objectives in treatment of mine drainage. Because of possible clogging of sub-surface systems due to precipitation of iron and manganese in sub-surface systems, preference is usually given to surface systems since they can be aerated more efficiently. Since the pH typically decreases from 6 to 3 in acid mine drainage phenomena, the Tennessee Valley Authority (TVA) has developed an anoxic limestone drain (ALD) for use as a treatment tool. This consists of a high calcium limestone aggregate (20 to 40 mm) placed in a trench of 3 to 5 m wide and 0.6 m to 1.5 m in depth (Brodie et al., 1993). The anoxic conditions in the trench are ensured by backfilling with clay. A plastic geotextile is placed between the clay and limestone. The inlet of the trench is placed at the source of the acid mine drainage. However, when oxygen content in the drainage is greater than 2 mg/L or the pH is greater than 6 and the redox potential is greater than 100 mV, use of the ALD is detrimental due to the formation of oxide coatings. Installation of a sedimentation pond before the wetland treatment with or without ALD is preferred since it is easier to remove iron precipitation from the pond than the wetlands. Sanders et al. (1999) have indicated that wetland systems were used to remove Zn, Fe, Cu, Pb, and some other heavy metals from a moderate to severe acidic drainage from a mining complex in Montana. This treatment will be required for decades. Hedin et al. (2013) reported that an ALD operated for 18 years without maintenance even though clogging is possible. Long-term monitoring will be needed as shown in Table 5.4.

5.6.2.3 Biosorption

Biosorption is a potentially attractive technology for treatment of water containing dilute concentrations of heavy metals. Activated carbon is the currently recognized adsorbent for removal of heavy metals from wastewater. However, the high cost of activated carbon limits its use in adsorption. A search for a low-cost and easily available and renewable adsorbent has led to the investigation of wastes of agricultural and biological origin as potential metal sorbents (Hammaini et al., 1999). Biosorption is the ability of certain types of microbial biomasses to accumulate heavy metals

TABLE 5.4
Monitoring Requirements for an AMD Drainage Remediation Process

Category	Parameter
Physico-chemical	pH
	Redox potential
	Total dissolved solids
	Specific conductance
	Dissolved oxygen
Cationic and anionic species	Fe, Cu, Pb, Zn, Cd, Hg, As, SO_4^{2-}
Gases	O_2, CO_2, SO_2, H_2S
Flow	AMD flow rate, hydrostatic pressure
Metrological conditions	Precipitation, temperature, sunlight, wind speed

Source: Adapted from Fytas and Hadjigeorgiou (1995).

from aqueous solutions by mainly ion exchange mechanisms. A large number of micro-organisms belonging to various groups, such as bacteria, fungi, yeasts, and algae have been reported to bind a variety of heavy metals to different extents (Volesky and Holan, 1995).

The main requirement of an industrial sorption system is that the sorbent can be utilized as a fixed or expanded bed for use in a continuous process. Immobilization techniques have been developed, but the employment of immobilization procedures is expensive and complex (Liu et al., 2003). Two attempts to market two different types of immobilized microbial biomass, one by BV SORBEX and the other by the US Bureau of Mines were not commercially successful application (Tsezos, 2001). The feasibility of anaerobic granules for industrial wastewater reactors was investigated as a novel type of biosorbent for the removal of cadmium, copper, nickel, and lead from aqueous solution by Al Hawari and Mulligan (2006). Results showed that a living biomass has a higher sorption capacity than a dried biomass but due to the difficulties in maintaining the biomass, the dried biomass would be more suitable for industrial applications (Figure 5.12). Unlike most forms of biomass, immobilization or stiffening was not necessary prior to using the biomaterial. Anaerobic granules possess compact porous structures, excellent settling ability and high mechanical strength. Even under aggressive chemical environments (acidic or basic conditions), the biomass demonstrated good stability with no visible structural damage – making this biomass more advantageous over other biosorbents. The biomass was also tested for arsenic sorption and was found to favourable compare to other biosorbents (Chowdhury and Mulligan, 2011). Kim et al. (2022) reviewed various biosorbents for AMD treatment.

5.6.2.4 Other Technologies

Babaee et al. (2017) studied the effectiveness of iron/copper bimetallic nanoparticles for removal of aqueous arsenic using various sorption tests. The Fe/Cu nanoparticles had a mean diameter of 13.17 nm. Sorption capacities for As(V) and As(III) were determined as 21.32 mg g^{-1} and 19.68 mg g^{-1}, respectively, at a neutral pH. The Langmuir equation fit well, and sorption followed pseudo-second-order kinetics. No significant effect on the arsenic removal efficiency of by carbonate, phosphate and sulphate, ions was demonstrated. Overall, Fe/Cu nanoparticles showed potential for the removal of arsenic from water in acidic environments via sorption.

The Kanchan Arsenic Filter (KAF) is for treating the drinking water system used in rural areas for removing arsenic, other contaminants, and pathogens from drinking water (Nguyen and Mulligan, 2023). It is a sand filter with non-galvanized nails as a source of Fe^0 that has been adopted particularly in Nepal. However, arsenic removal efficiency is dependent on various factors. The impact

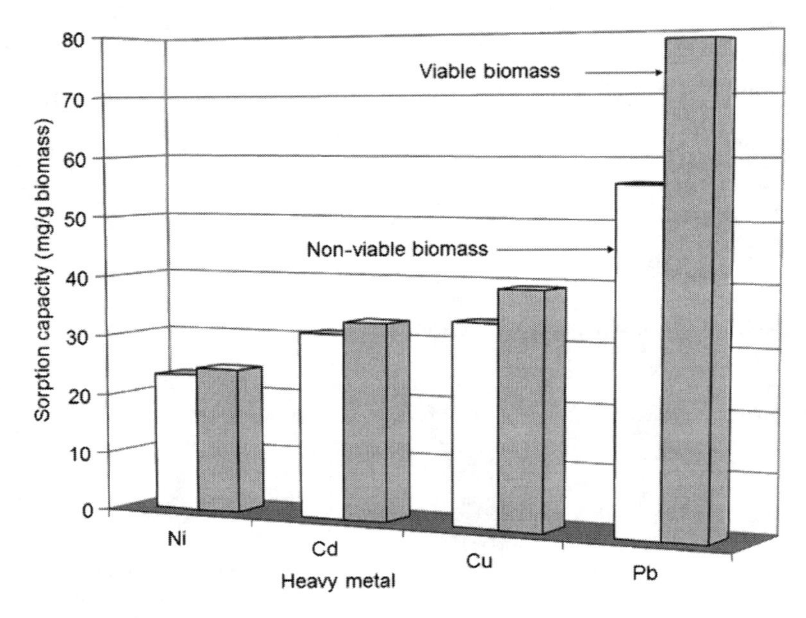

FIGURE 5.12 Comparison of a viable and non-viable anaerobic granulated biomass for the biosorption of heavy metals. (Adapted from Al Hawari and Mulligan, 2006.)

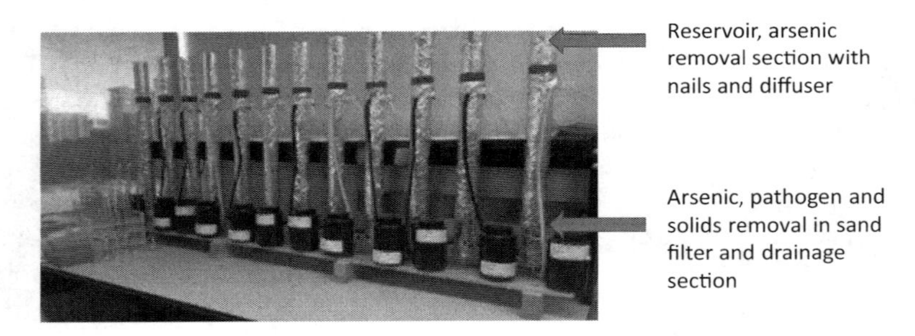

FIGURE 5.13 Photos of lab scale KAFs. (Adapted from Nguyen and Mulligan, 2023.)

of low and high levels of phosphate and hardness on iron corrosion and arsenic removal processes were studied in a lab-scale KAFs (Figure 5.13). Filters treating water with high levels of hardness showed the highest arsenic removal, reducing arsenic to under 50 ppb (Nepali and Vietnamese national accepted standard for arsenic in drinking water) from 2000 ppb. Overall, the complete process removed 85.2 to 99.4% of the arsenic while only 35.7 to 47.9% was removed by the iron nails. The findings from this study indicated that the iron corrosion affected arsenic removal efficiency less than water composition, which was not expected. Therefore, more research on these types of filters is needed.

5.6.3 Slurry Tailings' Management

Slurry tailings' ponds are by far the major type of containment facilities for slurry tailings. Their use has been discussed in the previous section. Their presence in the landscape degrades land quality and considerably reduces land use capabilities. An illustration of such kinds of ponds is shown in Figure 5.14 – for tin mining slurry ponds in S.E.Asia obtained as a result of alluvial tin mining

FIGURE 5.14 Tin mining slurry ponds obtained as a result of hydraulic dredging and pumping. Reclamation of pond is necessary to accommodate expansion of housing units seen at left of the picture (and left of the canal).

hydraulic operations using the wash-separation technique. Sand and debris are collected at the end of the sluice box – with the ponds serving as sedimentation facilities (Chow, 1998). It is not unusual for these ponds to develop a crust overlying a slime layer with solids' concentration ranging from 50 to 60%. Encroachment of housing estates onto such kinds of ponds and debris, as seen for example in Figure 5.14, will pose limits on housing and introduce safety hazards. Reclamation of the ponds is necessary to allow for utilization of the reclaimed land for further urban development and other land uses consistent with an urban ecosystem.

Sustainable land use, in the context of slurry tailings' ponds and their like, is not different in principle from the acid mine problem or mined-out caverns and pits etc. We will consider sustainability of land use in the physical landscape sense, i.e., in respect to the physical features and properties of the land and its utility. Figure 5.15 gives a summary view of the various options available for management of slurry tailings' ponds. The first option is to leave the ponds in their disposal state. This is not a generally acceptable option, but nevertheless must be considered, in the event of negligent abandonment of such ponds. Health and safety concerns require implementation of (a) monitoring procedures for these ponds, and (b) health and safety measures to protect human and wildlife population. Figure 5.15 also shows three distinct categories of sustainable land use options for the slurry tailings' ponds are available: (a) Return of land to *ab initio* physical condition, (b) reclamation of the ponds in the *keep it wet* state, and (c) reclamation of the ponds in the *take it to dry* state. Returning the land to its *ab initio* landscape condition fulfils the landscape portion of sustainability requirements.

Two basic options are available for the *keep it wet* state: (i) Fresh-water pond, and (ii) wetlands. In the *take it to dry* category, the options available depend to a large extent on (a) quality or competency of *dry* land obtained from the reclamation process, and (b) regulatory and community requirements. The basic element of all the schemes for reclamation of slurry tailings' ponds and other types of containment facilities must deal with the question of "*what to do with the stagnant layer*". Liquid-solid separation and treatment of the released water are basic requirements for any of the pond reclamation options. Physical methods for liquid-solid separation include surcharging

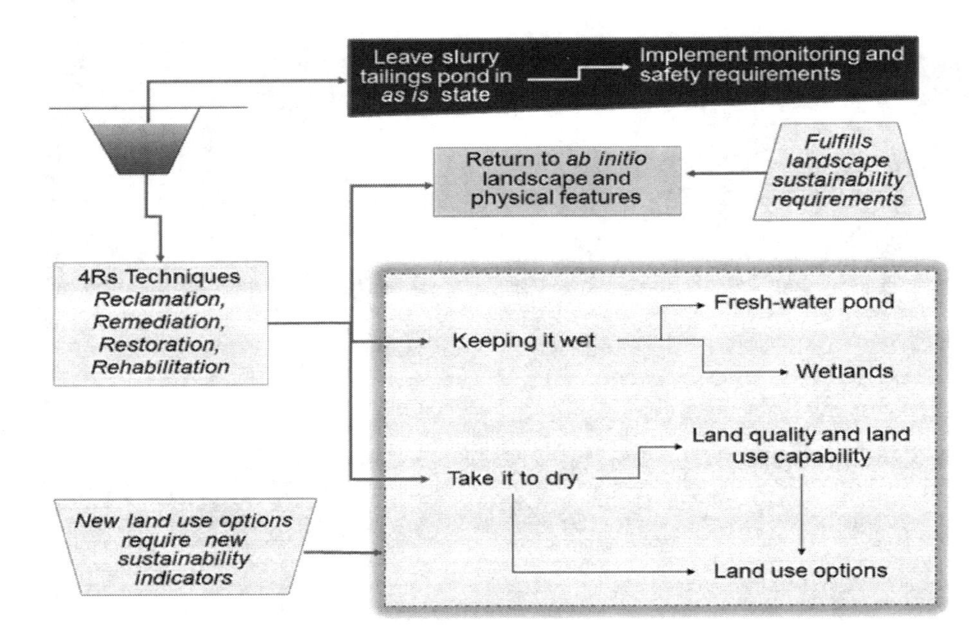

FIGURE 5.15 Sustainable land use in the context of slurry tailings pond. Options range from non-sustainable to enhanced land use and new sustainability scenarios.

the top of the stagnant layer to achieve consolidation – a geotechnical process that provides compression of the solids' skeletal matrix through the applied surcharge and drainage of the water in the skeletal structure. Other physical methods include removal of the stagnant layer for treatment, and filling of the emptied pond with new fill material.

Chemical and physico-chemical methods for increasing the sedimentation rate of the suspended solids in the stagnant layer include the use of polyacrylamides and polyelectrolytes. The basic intent of these kinds of flocculants is to overcome the domination of interparticle forces typical of colloidal interaction. Various kinds of flocculants and flocculating agents have been developed for such types of slimes and sludges (Yong and Sethi, 1982, 1983, 1989). They all have the aim of promoting aggregation of the particles into flocs, thus increasing the mass of individual groups of particles and hence allowing for gravitational forces to dominate and sedimentation to occur. Calculations performed by Yong and Wagh (1985) using two-particle collision theory to study the stability of the suspended solids in the stagnant zone, have shown the effect of aggregation on the settling velocities of the suspended solids. Confirmation of their calculations has been obtained from experiments on a pure clay mineral suspension (kaolinite) and the red mud discharge from bauxite processing (Figure 5.16). Aggregation of the red mud particles with increasing solids' concentration caused the increase in settling velocity. For the kaolinite soil suspension, increasing solids' concentration in the soil suspension served to decrease the settling velocity – probably due to the hindrance effect posed by the proximal particles.

Another method for increasing the settling velocity of the suspended particles is to increase the zeta potential ξ of the particles. This is the potential that represents the charge at the shear layer between the suspended particle surface and the suspending fluid. We recall from Equation 5.5 that the potential ψ provides us with a means for determining the electric charge distribution as a function from the particle surface. This potential has two basic components: (a) The potential at the surface of the particle, represented by ψ_0, and (b) the potential ψ_s at the Stern layer boundary (double-layer boundary) where the shear action with the liquid medium occurs. This potential, which is commonly identified as the zeta potential ξ is a function of the nature of the surface charge possessed by the

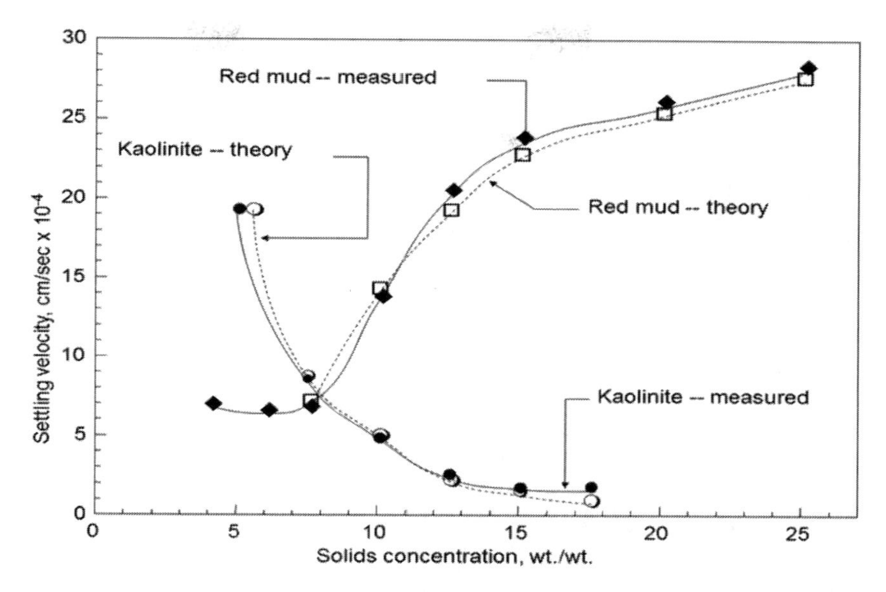

FIGURE 5.16 Calculated and measured settling velocities for a kaolinite soil suspension and bauxite red mud.

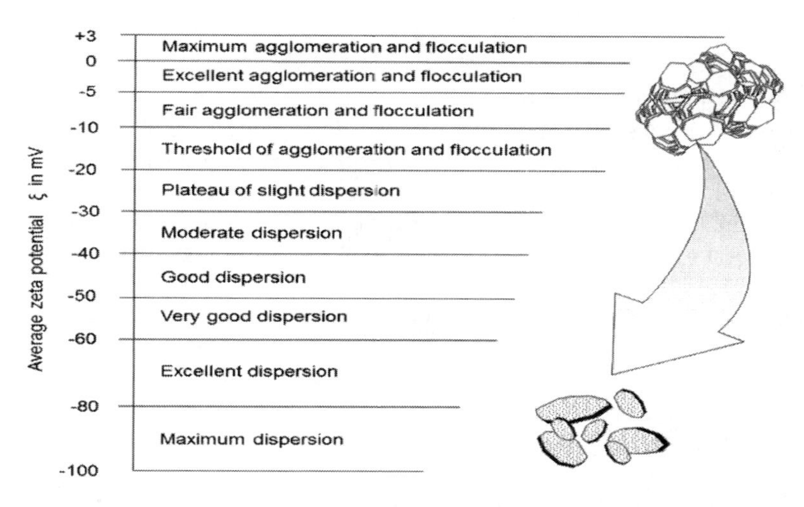

FIGURE 5.17 Relationship between zeta potential ξ, soil microstructure (flocs and aggregations of particles) and dispersion stability for clay soils.

suspended particle, the ions in the suspending fluid, and the ions in the double-layer. By changing the zeta potential, aggregation of the suspended particles in the stagnant zone can be obtained – with resultant increased settling velocities. Figure 5.17 shows the relationship between the zeta potential ξ and the dispersion stability for clay soils reported by Yong (2001b).

New additives are being developed that are more biodegradable than the synthetic flocculants. Biosurfactants such as rhamnolipids with various isolated microbial species have been shown to enhance oil sand tailings' flocculation by 2.70 times (Mulligan and Roshtkhari, 2017) (Figure 5.18). Flocculation occurs through addition of the biosurfactant, which leads to adsorption, increased hydrophobicity of the particles and bridging between particles. Faster tailings' sedimentation can reduce the pond volumes.

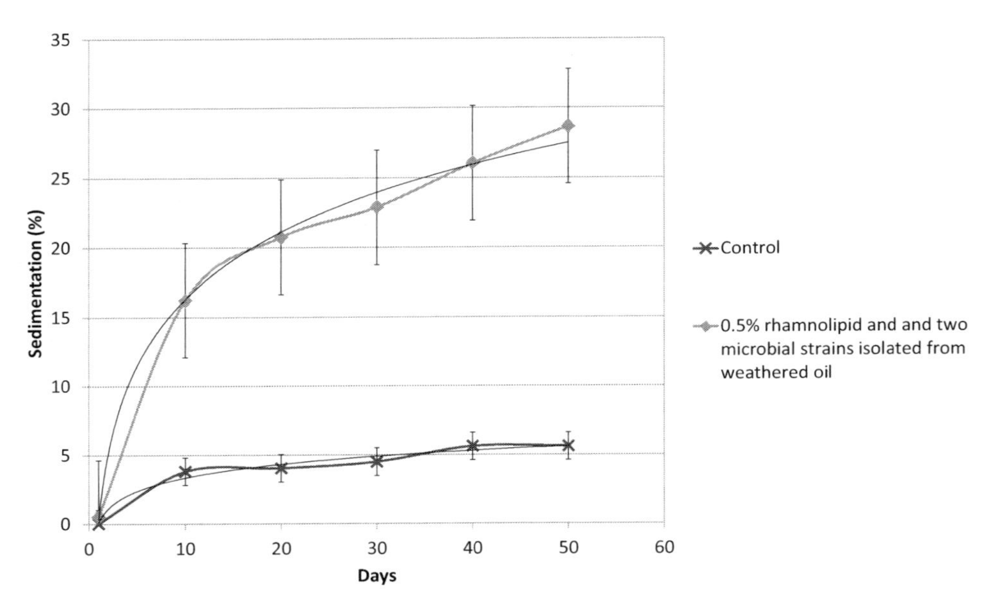

FIGURE 5.18 Sedimentation of oil sand tailings at 15 °C ± 2 °C using rhamnolipid (0.5%) and two microbial strains isolated from weathered oil over time and kinetic models for sedimentation.

5.6.3.1 Biohydrometallurgical Processes

Bacterial leaching of metals from mining ores, also called bioleaching, is a full-scale process that can be performed by slurry reactors or heap leaching (Figure 5.19). Mining wastes include low-grade ores, mine tailings, and sediments from lagoons or abandoned sites. Low pH values lead to solubilization of the metals in the mining ores. Elemental sulphur or ferrous iron may be added as bacterial substrates. Reactors such as Pachuca tanks, rollings reactors or propeller vessels have been used (Tyagi et al., 1991).

Heap leaching is more common since it allows the large volume of wastes to be treated in place (Boon, 2000). To enhance this process, aeration can be forced through the pile. Alternatively, hydrophilic sulphur compounds can be added (Tichý et al., 2000). *Thiobacilli* bacteria are responsible for the oxidation of inorganic sulphur compounds. Applications include metal dissolution of low-grade sulphide ores, generation of acidic ferric sulphate leachates for hydrometallurgical purposes, and removal of gold by oxidation of pyrite by bacterial sulphide production. The extraction of metals from low grade metals ores and refractory gold ores is a multi-billion dollar business worldwide (Rawlings, 1997). Bacterial solubilization by oxidation of the sulphide minerals, pyrite and arsenopyrite, enhances gold extraction by the traditional method of cyanidation. The solubilization mechanisms have been debated extensively, however.

It has been estimated that up to 10–15% of the global copper production and 5% of gold production are dependent on biohydrometallurgical processes (Kaksonen et al., 2018). A number of large-scale biohydrometallurgical plants have also been implemented for the treatment of water impacted by mining operations, e.g., in the Netherlands, South Africa, Dominican Republic (Paques, 2020a, 2020b, 2020c) and Canada (Parizot, 2018).

Biohydrometallurgical processes are highly efficient and cause less environmental problems than chemical methods (Torma and Bosecker, 1982). For slurry processes, oxidation rate per reactor volume, pH, temperature, particle size, bacterial strain, slurry density, ferric and ferrous iron concentrations need to be optimized. Bioleaching is very effective for recovery of gold from refractory gold pyrite and copper from chalcopyrite. According to Kaksonen et al. (2020), new directions

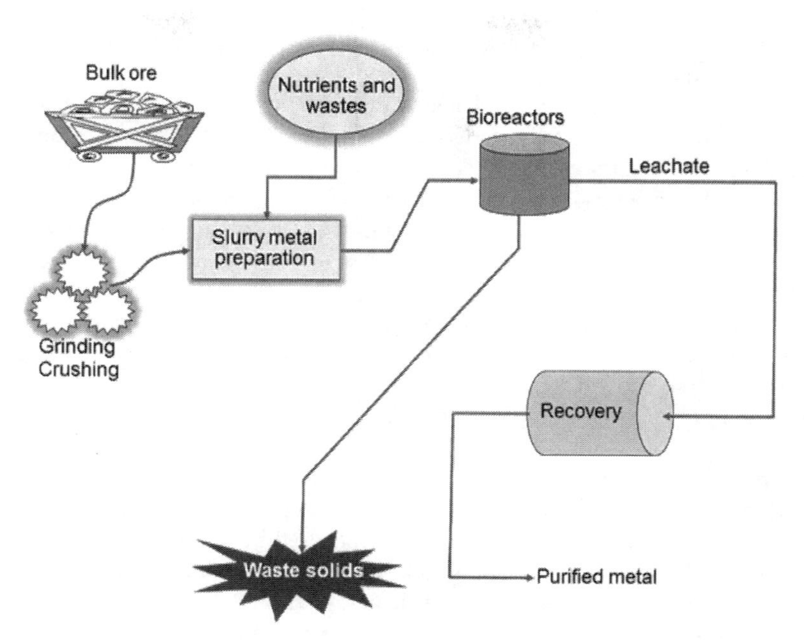

FIGURE 5.19 Flow sheet of a biometallurgical process.

are in the areas of extraction of rare earth elements and various resources from aqueous and solid wastes towards a circular economy approach. Synthetic biology, modelling, and artificial intelligent approaches will enhance new opportunities.

Rhamnolipid biosurfactants (biodegradable and of low toxicity) have also been added to mining oxide ores, to enhance metal extraction (Dahr Azma and Mulligan, 2004). Batch tests were performed at room temperature. Using a 2% rhamnolipid concentration, 28% of the copper was extracted. Addition of 1% NaOH with the rhamnolipid enhanced the removal up to 42% at a concentration of 2% rhamnolipid but decreased at higher surfactant concentrations. Sequential extraction studies were also performed to characterize the mining ore and to determine the types of metals being extracted by the biosurfactants. Approximately 70% of the copper was associated with the oxide fraction, 10% with the carbonate, 5% with the organic matter, and 10% with the residual fraction. After washing with 2% biosurfactant (pH 6) over a period of 6 days, it was determined that 50% of the carbonate fraction and 40% of the oxide fraction were removed by the biosurfactant.

Further experiments have been performed with biosurfactants and mining residues in relation to arsenic (Arab and Mulligan, 2018). Results indicated that washing the mine tailing samples with rhamnolipid or sophorolipid, the two biosurfactants were able to convert As (IV) to As (III). Rhamnolipids were more effective than the sophorolipids in mobilizing of arsenic from the mine tailings. Positive correlation between arsenic and iron mobilization were determined. Both sophorolipids and rhamnolipid were effective in the removal of arsenic and other heavy metals from soil and mine tailings. Further study with sophorolipids, was carried out at different concentrations and pH levels and at two different temperatures (15°C and 23°C), to remove arsenic and heavy metals from mine tailings (Arab and Mulligan, 2020). Increased arsenic, copper, and iron removal occurred when the temperature changed from 15 to 23°C. Different fractions of the specimens were investigated to evaluate to evaluate the effectiveness of the washing treatment by way of sequential extraction. The results showed that the arsenic removal was 0·7% from the water-soluble portion, 0·7% from the exchangeable portion, 0·6% from the carbonate fraction, 29·9% from the oxide/hydroxide fraction, 3.0% from the organic portion and 65·1% from the residual fraction. The results show the potential of sophorolipid washing as a sustainable and environmentally-friendly solution mine tailings' treatment (Figure 5.20).

Bioleaching involves *Thiobacillus sp.* bacteria, which can reduce sulphur compounds under aerobic and acidic conditions (pH 4) at temperatures between 15 and 55°C, depending on the strain. Leaching can be performed by indirect means, acidification of sulphur compounds to produce sulphuric acid, which then can desorb the metals on the soil by substitution of protons. Direct leaching solubilizes metal sulphides by oxidation to metal sulphates. In laboratory tests, *Thiobacilli* were able to remove 70 to 75% of heavy metals (with the exception of lead and arsenic) from contaminated sediments (Karavaiko et al., 1988).

Options are available for bioleaching including heap leaching and bioslurry reactors. For both heap leaching and reactors, bacteria and sulphur compounds are added. In the reactor, mixing is used and pH can be controlled more easily, leachate is recycled during heap leaching. Copper, zinc, uranium, and gold have been removed by *Thiobacillus sp.* in biohydrometallurgical processes (Karavaiko et al., 1988).

Various studies have been performed to develop a process to treat and microbially recover metals in low-grade oxide ores (Mulligan, 2002) as pyrometallurgical and hydrometallurgical techniques are either very expensive, energy intensive or detrimental to the environment. Biohydrometallurgical techniques such as those utilizing the fungus *Aspergillus niger* are thus potentially more sustainable. *A. niger* has shown potential for producing citric acid and other organic acids effective for metal solubilization (Mulligan and Kamali, 2003). Effectiveness for leaching was enhanced when sulphuric acid was added with organic acids to the medium. Different agricultural wastes such as potato peels were tested as growth substrates for the fungus. Maximum solubilization levels of 68%, 46%, and 34% were achieved for copper, zinc and nickel, respectively. Also, minimal iron dissolution was obtained (7%), which allows for further metal purification.

Another research study was to evaluate the potential for mobilization of arsenic (As) from mine tailings in the presence of natural organic matter (NOM). Humic acid (HA) was used a

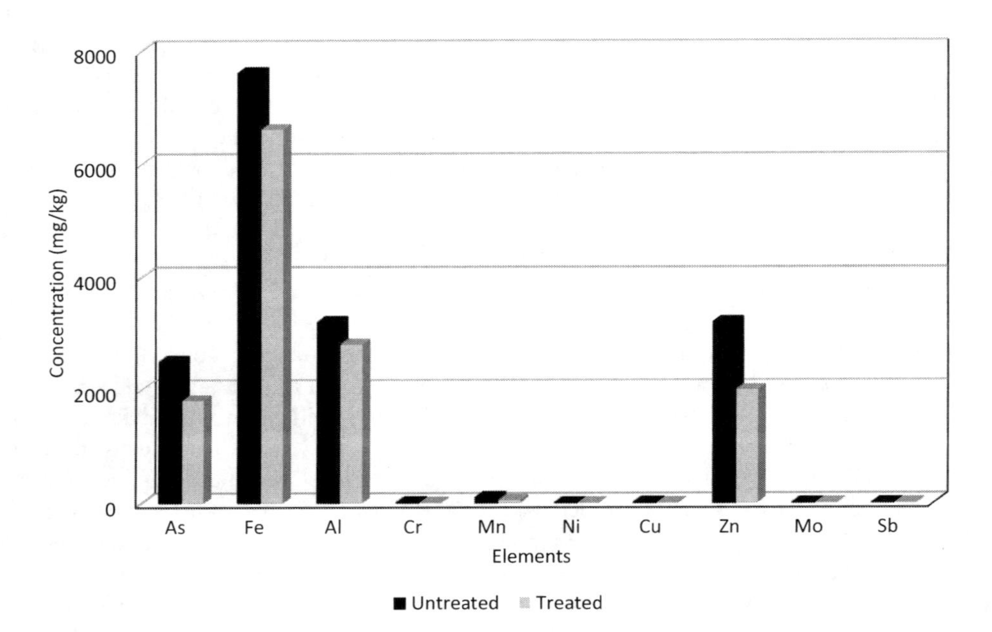

FIGURE 5.20 Concentration of elements in untreated and treated with sophorolipid mine tailing sample treated in continuous setup after 60 pore volumes. (Arab and Mulligan, 2020.)

model for NOM (Wang and Mulligan, 2006b). By introducing HA at a low mass ratio (below 2 mg HA/g mine tailings) under acidic conditions As mobilization was inhibited. However, As mobilization increased with increasing mass ratios and under alkaline conditions, As mobilization was significantly HA enhanced.

5.7 CONCLUDING REMARKS

5.7.1 Mining Activities

Harvesting of non-renewable natural resources brings with it at least three major areas of geoenvironmental impacts: (a) Depletion of the non-renewable resource, (b) mining excavations on the surface and underground, and (c) discharges from processes associated with extraction of the natural resource. The biological components of the terrestrial ecosystem are not covered in the discussions in this book. We have focused specifically on the land use aspect of the geoenvironment in mining operations as an upstream activity. The use of the raw materials obtained from such mining operations by downstream industries is covered in Chapter 7. From the viewpoint of geoenvironment sustainability, we accept that depletion of non-renewable or depleting resources will not qualify for inclusion in any category of sustainable resources. This means that if one requires land use to be sustainable, one must deal with the physical landscape problems raised by above- and below-ground mining processes, together with operations and discharges from extraction and beneficiation processes.

Underground mining excavations pose problems in two categories: (a) The physical aspect of excavations as empty chambers, tunnels, etc., together with the heaping of debris and other spoils as heaps and tips and (b) the chemical problem represented by acid generation and the resultant acid mine drainage problem – because of the interaction of the exposed sulphide minerals in the mined-out areas to oxygen and water. If sustainable land use requires one to return the land to its *ab initio* condition, this would mean filling the empty mined-out chambers and excavations. This would, or should, avoid or minimize the problem of acid generation. Since practical and economic considerations have so far militated against this course of action, attention has been directed towards mitigating acid mine and acid rock drainage problems.

5.7.2 Contaminated Water Management

The problem of water usage and "what happens to the used water" – which most often is contaminated with products issuing from extraction/beneficiation and from usage in underground *in situ* extraction of hydrocarbons (SAGD, CSS and fracking) – is perhaps one of the most critical problems that requires considerable management to avoid and/or minimize chemical stressor impacts on the geoenvironment. Two particular issues or concerns need attention: (a) Generation of acid leachates giving rise to the commonly identified problem of acid mine drainage, and (b) contamination of shallow and deep-seated aquifers from "used waters" associated with underground *in situ* hydrocarbon extraction practices. Cleaning up contaminated aquifers is a task that is almost impossible – not without some considerable expenditure and time. To avoid contamination of groundwater and aquifers, it is essential to implement monitoring of water movement in the subsurface – through the use of monitoring wells and through systematic evaluation of the hydrogeological settings and continuous hydrochemical analyses of abstracted samples from the monitoring wells. The use of appropriate and valid analytical transport and fate models that address site-specific and project-specific *in situ* hydrocarbon extraction process, whether it be the steam-assisted processes or the fracking process, would add considerable value to the monitoring schemes required to ensure that hydraulic stressors carrying contaminants are not delivered to receiving waters – above ground, and groundwater and aquifers.

Measures undertaken by industry to better protect against acid generation in present mining and extraction operations, and in underground *in situ* hydrocarbon extraction operations are continually being improved. However, for historic and abandoned mine sites, the problem of acid mine drainage remains. The direct impact on land use from acid generation problems is not only in the immediate sense of acid interactions with the land environment, but also in the ripple or cascading effect. Contamination of groundwater and receiving waters, together with contamination of shallow and deep-seated aquifers pose severe threats to native habitat and plant life and other receptors. The challenge for geoenvironmental engineering is to provide measures that will preserve and maintain the original functionality of the land component of the geoenvironment. The schematic diagrams of Figures 5.10, 5.12, and 5.14 provide a starting base for implementation of sustainable practices in geoenvironmental engineering.

5.7.3 Tailings' Discharge and Mine Closure

In regard to tailings' slurry discharges and containment facilities, sustainable land use is not different in principle from the acid mine problem or mined-out caverns and pits. Options for reclamation of the slurry tailings' ponds and containment facilities have been summarized in Figure 5.14. Much depends on both regulatory and industry requirements.

Closure of mining sites requires site restoration procedures to ensure that the restored sites may be used for other purposes and to prevent risk to the environment and humans. The reuse of mining residues, stabilization of mine areas, and neutralization of pollutants are some of the challenges. Plans for these procedures need to be initiated during the design and planning of the mine life-cycle. Numerous techniques are being investigated, such as replanting, wetland treatment, and biological treatments. As a more sustainable approach, waste products must be recycled as much as possible and should also be integrated into the treatment processes. An integrated approach for land, solids, and leachate management is highly desirable. Examples of mine rehabilitation in Australia can be found at www.mining-technology.com, accessed Feb. 2024.

REFERENCES

Al Hawari, A., and Mulligan, C.N. (2006), Biosorption of cadmium, copper, lead, and nickel by anaerobic granular biomass, *Bioresource Technology*, 97(4): 692–700.

Appelo, T., (Ed.) (2008), Arsenic in Groundwater – A World Problem. In: *Proceedings of the 2006 Utrecht Seminar*, National Committee of the IAH, Netherlands, ISBN/EAN 978-90-808258-2-6, 142 p.

Arab, F., and Mulligan, C.N. (2018), An eco-friendly method for heavy metal removal from mine tailings, *Environmental Science and Pollution Research*, 25(16): 16202–16216.

Arab, F., and Mulligan, C.N. (2020), Efficiency of sophorolipids for arsenic removal from mine tailings, *Environmental Geotechnics*, 7(3): 175–188. https://doi.org/10.1680/jenge.15.00016.

Aubertin, M., Bussiere, B. Pabst, T. James, M., and Mbonimpa. M. (2016), Review of the Reclamation Techniques for Acid-Generating Mine Wastes Upon Closure of Disposal Sites, In: A. Farid, et al. (eds.), *Geo-Chicago 2016: Geotechnics for Sustainable Energy*, Amercian Society Civil Engineering, Reston, VA, pp. 343–358. https://doi.org/10.1061/9780784480137.034.

Azcue, J.M., and Nriagu, J.O. (1995), Impact of abandoned mine tailings on the arsenic concentrations in Moira Lake, Ontario, *Journal of Geochemical Exploration*, 52: 81–89.

Babaee, Y., Mulligan, C.N., and Rahaman, Md. S. (2017), Removal of arsenic (III) and arsenic (V) from aqueous solutions through adsorption by Fe/Cu nanoparticles, *Journal of Chemical Technology & Biotechnology*, 93(1): 63–71. https://doi.org/10.1002/jctb.5320.

Barr, J. (1969), *Derelict Britain*, Penguin, London, 240 pp.

Boon, M. (2000), Bioleaching of Sulphide Minerals. In: P. Lens, and L.H. Pol (eds.), *Environmental Technologies to Treat Sulfur Pollution. Principles and Engineering,* IWA Publishing, London, pp. 105–130.

Brodie, G.A., Britt, C.R., Toamazewski, T.M., and Taylor, H.N. (1993), Anoxic Drains to Enhance Performance of Aerobic Acid Drainage Treatment Wetlands: Experiences of the Tennessee Valley Authority. In: G.

A. Moshir (ed.), *Constructed Wetlands for Water Quality Improvement*, Lewis Publishers, Chelsea, MI, pp. 129–138.

Bulusu, S., Aydilek, A.H., Petzrick, P., and Guynn, R. (2005), Remediation of abandoned mines using coal combustion by-products, *Journal of Geotechnical and Geoenvironmental Engineering,* 131: 958–969.

CH2M HILL. (2004), Deloro Mine Site Cleanup: Mine Area Closure Plan, Final Report, Ontario Ministry of Environment.

Chamley. (2003), *Geosciences, Environment and Man*, Elsevier, Amsterdam, 450 pp.

Chowdhury, Md. R.I., and Mulligan, C.N. (2011), Biosorption of arsenic from contaminated water by anaerobic biomass, *Journal of Hazardous Materials*, 190: 486–492.

Chow, W.S. (1998), Studies of slurry slime in mined-out ponds, Kinta Valley, Peninsular Malaysia, for purposes of reclamation, Ph.D. Thesis, Department of Geology, University of Malaysia, Malaysia.

Dahr Azma, B., and Mulligan, C.N. (2004), Extraction of copper from mining residues by rhamnolipids, *Practice Periodical on Hazardous Toxic And Radioactive Waste Management*, 8(3): 166–172.

Davies, C.S. (1999), Derelict Land. In: D.E. Alexander and R.W. Fairbridge (eds.), *Encyclopedia of Environmental Science*, Kluwer Academic Publishers, Dordrecht, pp. 120–123.

Donahue, R. and Hendry, M.J. (2003), Geochemistry of arsenic in uranium mine mill tailings, Saskatchewan, Canada. *Applied Geochemistry*, 18:1733–1750.

Fytas, K. and Hadjigeorgiou, J. (1995), An assessment of acid rock drainage continuous monitoring technology, *Environmental Geology*, 25: 36–42.

Gomes, H. I. , Mayes, W. M., Rogerson, M., Stewart, D. I., and Burke, I. T. (2016), Alkaline residues and the environment: A review of impacts, management practices and opportunities, *Journal of Cleaner Production*, 112: 3571–3582, https://doi.org/10.1016/j.jclepro.2015.09.111.

Government of Canada. (2022), The Canadian Critical Strategy From Exploration to Recycling: Powering the Green and Digital Economy for Canada and the World. www.canada.ca/en/campaign/critical-minerals-in-canada/canadian-critical-minerals-strategy.html.

Hammaini, A., Ballester, A., Gonzalez, F., Blazquez, M.L., and Munoz, J.A. (1999), Activated Sludge as Biosorbent of Heavy Metals. Biohydrometallurgy and the Environment Toward the Mining of the 21st Century. In: *International Biohydrometallurgy Symposium IBS'99*, 20-23 June, San Lorenzo de El Escorial, Madrid, Spain, pp. 185–192.

Hedin, R., Weaver, T., Wolfe, N., and Watlzlaf, G. (2013), Effective Passive Treatment of Coal Mine Drainage. In: *Proceedings 35th National Association of Abandoned Mine Land Programs Conference*. www.hedinenv.com/pdf/NAMLP_Effective_Passive_Treatment_Paper.pdf.

IEA. (2023a), Government Energy Spending Tracker: June 2023 update. www.iea.org/reports/government-energy-spending-tracker-2.

IEA. (2023b), Critical Minerals Market Review 2023. www.iea.org/reports/critical-minerals-market-review-2023.

Idoine, N.E., Raycraft, E.R., Price, F., Hobbs, S.F., Deady, E.A., Everett, P., Shaw, R.A., Evans, E.J., and Mills, A.J. (2023), *World Mineral Production 2017-21*. Nottingham, UK, British Geological Survey, 98pp.

IMCC (Interstate Mining Compact Commission). (1992), Inactive and Abandoned Non-Coal Mines: A Scoping Study. In: *Prepared for IMCC of Herndon, VA,* Resource Management Associates, Clancy, MT. Cooperative Agreement X-817900-01-0 (July).

Kaksonen, A.H., Boxall, N.J., Gumulya, Y., Khaleque, H.N., Morris, C., Bohu, T., Cheng, K.Y., Usher, K.M., and Lakaniemi, A.-M. (2018), Recent progress in biohydrometallurgy and microbial characterisation. *Hydrometallurgy*, 180: 7–25.

Kaksonen, A.H., Deng, X., Bohu, T., Zea, L., Khaleque, H.N., Gumulya, Y., Boxall, N.J., Morris, C., and Cheng, K. Y. (2020), Prospective directions for biohydrometallurgy, *Hydrometallurgy*, 195: 105376. https://doi.org/10.1016/j.hydromet.2020.105376.

Karavaiko, G.I., Rossi, G. Agates, A.D. Groudev S.N., and Avakyan, Z.A. (1998), *Biogeotechnology of Metals: Manual*, Center for International Projects GKNT Moscow, Soviet Union.

Kim, N., Park, D. (2022), Biosorptive treatment of acid mine drainage: A review. *International Journal of Environmental Science Technology,* 19: 9115–9128. https://doi.org/10.1007/s13762-021-03631-5.

Kinniburg, D.G., Smedley, P.L., Davies, J., Milne, C., Gaus, I., Trafford, J.M., Burden, S., Huq, S., Ahmad, N., and Ahmed, K.M. (2003), The Scale and Causes of the Groundwater Arsenic Problem in Bangladesh. In: A.H. Welch and K.G. Stollenwerk (eds.), *Arsenic in Groundwater: Occurrence and Geochemistry,* Kluwer, Boston, pp. 211–257.

Kobayashi, J., and Hagino, N. (1965), Strange Osteomalacia by Pollution from Cadmium Mining, Progress Rept. WP 00359, Okayaman University, pp. 10–24.

Kruyt, H.R. (Ed.) (1952), *Colloid Science: Irreversible Systems*, Elsevier Publishing Co., Amsterdam, 389 pp.

Lindeström, L. (2003), The Environmental History of the Falun Mine. Stiftelsen Stora Kopparberget & ÅF-Moljöforskargruppen, AB ISBN 91-631-3536-1, 110 pp.

Liu, Y., Yang, S., Xu, H., Woon, K., Lin, Y. and Tay, Y. (2003), Biosorption kinetics of cadmium(II) in aerobic granular sludge. *Process Biochemistry*, 38: 997–1102.

Manahan, S.E. (1990), *Environmental Chemistry*, 4th ed, Lewis Publishers, Boca Raton, 612 pp.

McCreadie, H., Blowes D.W., Ptacek, C.J., and Jambor, J.L. (2000), Influence of reduction reactions and solid-phase composition on porewater concentrations of arsenic, *Environmental Science and Technology*, 34: 3159–3166.

Menzie, W.D., Soto-Viruet, Y., Bermudez-Logo, O., Mobbs, P.M., Perez, A.A. Taib, M., Wacaster, S. et al. (2013), Review of Selected Global Mineral Industries in 2011 and An Outlook to 2017, U.S. Geological survey Open-File Report 2013-1091, 33 pp. www.pubs.usgs.gov/of/2013/1091.

Moldovan, B., Jiang D.T., and Hendry M.J. (2003), Mineralogical characterization of arsenic in uranium mine tailings precipitated from iron-rich hydrometallurgical solutions, *Environmental Science and Technology*, 37: 873–879.

Mulligan, C.N. (2002), *Environmental Biotreatment*, Government Institutes, Rockville, MD, 395 pp.

Mulligan, C.N., and Kamali, M. (2003), Bioleaching of copper and other metals from low-grade oxidized mining ores by *A. niger, Journal of Chemical Technology and Biotechnology*, 78: 497–503.

Mulligan, C. N., and Roshtkhari, S. (2017), Remediation of oils sands tailings ponds using biosurfactants. In: *Proceedings of the 19th International Conference on Soil Mechanics and Geotechnical Engineering, Seoul 2017.*

Neculita, C., Zagury, G.J., and Bussière, B. (2007), Passive treatment of acid mine drainage in bioreactors using sulfate-reducing bacteria, *Journal of Environmental Quality*, 36: 1–16. https://doi.org/10.2134/jeq2006.0066.

Nguyen, P. M., and Mulligan, C.N. (2023), Study on the influence of water composition on iron nail corrosion and arsenic removal performance of the Kanchan arsenic filter (KAF), *Journal of Hazardous Materials Advances*, 10: 100285. https://doi.org/10.1016/j.hazadv.2023.100285.

Ollson, C.C. (1999), Arsenic contamination of the terrestrial and freshwater environment impacted by gold mining operations Yellowknife, NWT, MEng. Thesis, Royal Military College of Canada, Kingston, Canada.

Paques. (2020a), Anglo Coal. Acidic Mine Water Treatment Without Lime. www.en.paques.nl/about-us/subpages/cases/anglo-coal-en/3, accessed Dec. 2023.

Paques. (2020b), Metal & Mining Industry – Sustainable Metal Recovery and Water Treatment (Brochure). www.en.paques.nl/mediadepot/ 184210ea7803/WEBSectorbrochureMetalMining.pdf, accessed Dec. 2023.

Paques. (2020c), The Pueblo Viejo Gold Mine. THIOTEQTM for Copper Recovery (Brochure). www.en.paques.nl/mediadepot/ 21891d2add47/WEB5800000206CaseleafletThioteqENG.pdf, accessed Dec. 2023.

Parizot, M. (2018), Teck constructs new clean water facility at Elkview. In: *CIM Magazine*, Canadian Institute of Mining, Metallurgy and Petroleum. www.magazine.cim.org/en/news/2018/teck-constructs-new-clean-water-facility-at-elkview/, accessed Dec. 2023.

Rawlings, E. (Ed.) (1997), *Biomining: Theory, Microbes and Industrial Processes*, R.G., Landes and Springer-Verlag, Heidelberg, 302 pp.

Reichl, C., and Schatz, M. (2021), *World Mining Data 2021, Volume 36*, Minerals Production, Vienna.

Renforth, P., Mayes, W.M., Jarvis, A.P., Burke, I.T., Manning, D.A., and Gruiz, K. (2012), Contaminant mobility and carbon sequestration downstream of the Ajka (Hungary) red mud spill: The effects of gypsum dosing, *The Science of the Total Environment*, 421–422: 253–259. https://doi.org/10.1016/j.scitotenv.2012.01.046. PMID: 22349140.

Renforth, P. (2019), The negative emission potential of alkaline materials, *Nature Communications*, 10: 1401. https://doi.org/10.1038/s41467-019-09475-5.

Sanders, F., Rahe, J., Pastor, D., and Anderson, R. (1999), Wetlands treat mine runoff. *Civil Engineering*, 69: 53–55.

Si, C., Ma, Y., and Lin, C. (2013), Red mud as a carbon sink: Variability, affecting factors and environmental significance, *Journal of Hazardous Materials*, 244–245: 54–59. https://doi.org/10.1016/j.jhazmat.2012.11.024.

Simons, F.S., and Prinz, W.C. (1973), Copper. In: D. A. Brobst and W. P. Pratt (eds.), *United States Mineral Resources,* US Government Printing Office, Washington, DC, p. 722. https://doi.org/10.3133/pp820.

Skousen, J.G., Sexstone, A., and Ziemkiewicz, P.F. (2000), Acid Mine Drainage and Control. In: R. I. Barnhisel, R. G. Darmody, and W. L. Daniels (eds.), *Reclamation of Drastically Disturbed Lands*, American Society of Agronomy and American Society for Surface Mining and Reclamation, Agronomy No. 41,Chapter 6. https://doi.org/10.2134/agronmonogr41.c6

Skousen, J.G., Ziemkiewicz, P.F., and McDonald, L.M. (2019), Acid mine drainage formation, control and treatment: Approaches and strategies, *The Extractive Industries and Society*, 6(1): 241–249, https://doi.org/10.1016/j.exis.2018.09.008.

Tichý, R., (2000), Treatment of Solid Materials Containing Inorganic Sulfur Compounds. In: P. Lens and L.H. Pol (eds.), *Environmental Technologies to Treat Sulfur Pollution. Principles and Engineering*, IWA Publishing, London, pp. 329–354.

Torma, A.E., and Bosecker, K. (1982), Bacterial leaching, *Progress in Industrial Microbiology*, 16: 77–118.

Tsezos, M. (2001), Biosorption of metals. The experience accumulated and the outlook for technology development, *Hydrometallurgy*, 59: 241–243.

Tyagi, R., Couillard, D., and Tran, F.T. (1991), Comparative study of bacterial leaching of metal from sewage sludge in continuous stirred tank and air-lift reactors, *Process Biochemistry*, 26: 47–54.

Volesky, B., and Holan, Z.R. (1995), Biosorption of heavy metals, *Biotechnology Progress*. 11: 235 – 250.

Wang, S., and Mulligan C.N. (2006a), Occurrence of arsenic contamination in Canada: Sources, behavior and distribution, *Science of the Total Environment,* 366: 701–721.

Wang, S., and Mulligan, C.N. (2006b). Effect of natural organic matter on arsenic release from soils and sediments into groundwater, *Environmental Geochemistry and Health*, 28: 197–214.

Waring, C., and Taylor, J. (1999), A New Passive Technique Proposed for the Prevention of Acid Drainage: GaRDS. In: *Mining in to the Next Century: Environmental Opportunities and Challenges*. Proceedings 24th Annual Environmental Workshop, Townsville, Minerals Council of Australia, pp. 527–530.

WISE (World Information Service on Energy). (2023). www.wise-uranium.org/mdaf.html.

Yong, R.N. (1983a), Method for dewatering the sludge layer of an industrial process tailings pond, U.S. Patent No. 4,399,038.

Yong, R.N. (1983b), Treatment of tailings pond sludge, U.S.Patent No. 4,399,039.

Yong, R.N. (1984), Particle Interaction and Stability of Suspended Solids. In: R.N. Yong and F.C. Townsend (eds.), *Sedimentation Consolidation Models: Predictions and Validations,* American Society of Civil Engineers Publication, ISBN 0-87262-429-3, pp. 30–59.

Yong, R.N. (1983), Treatment of tailings pond sludge, U.S. Patent 4,399,039.

Yong, R.N. (2001a), *Geoenvironmental Engineering: Contaminated Soils, Pollutant Fate, and Mitigation*, CRC Press, Boca Raton, 307 pp.

Yong, R.N., (2001b), Interaction in Clays in Relation to Geoenvironmental Engineering. In K. Adachi and M. Fukue, (eds.), *Clay Science for Engineering*, Balkema, Rotterdamn, pp. 13–28.

Yong, R.N., and Sethi, A.J. (1982), Destabilization of sludge with hydrolyzed starch flocculants, U.S. Patent No. 4,330,409.

Yong, R.N., and Sethi, A.J. (1983), Decarbonation of tailings sludge to improve settling, U.S. Patent No. 4,414,117.

Yong, R.N., and Sethi, A.J. (1989), Methylated starch compositions and their use as flocculating agents for mineral wastes, such as bauxite residues, U.S. Patent No. 4,839,060.

Yong, R.N., and Wagh, A.S. (1985), Dispersion Stability of Suspended Solids in an Aqueous Medium. In: B.M. Moudgil and P. Somasundaran (eds.), *Flocculation, sedimentation and consolidation*, National Science Foundation, New York, NY, pp. 307–326.

Yong, R.N., and Warkentin, B.P. (1975), *Soil Properties and Behaviou*r, Elsevier Scientific Publishing Co., Amsterdam, 449 pp.

6 Agricultural-Based Food Production Geoenvironment Stressors

6.1 INTRODUCTION

The discussion in this chapter is concerned with the geoenvironment stressors generated from the primary sets of activities associated with agricultural-based food production. These activities include the production and harvesting of farm products (including livestock), and the industries identified with these sets of activities. These industries are *upstream industries*. They fall under the classification of *agro industry*. By definition, upstream industries deal with the production of raw material that are later transformed to finished products by downstream industries. They are known generally as *agro processing industries*. They include industries producing food products for the consumer, textiles, and forest products, etc. Agriculture (farming and food production) is a basic activity and an essential component in the life-support system of the human population. To obtain a sustainable society, the geoenvironment natural capital must be maintained at a sustainable level – meaning that harvesting and utilization of the resources represented by the natural capital must meet the requirements of replenishment and regrowth. In respect to the vital issue of food production to satisfy the needs of a sustainable society, this means that the activities associated with food production must be managed within the context of a sustainable agroecosystem.

Agroecosystems consist of two main components: (a) Naturally occurring, and (b) human related. The naturally occurring components consist of rivers, lakes, ponds, groundwater and aquifers, and flora and fauna. The human-related group consists of components that are the result of manipulation by humans to produce food, fibres, and other products, such as cultivated pastureland, seasonal and permanent crops, cultivated forests, and livestock or animal farming. In the *seasonal crops* category, for example, the list of human-related actions include land clearing, soil tillage and planting of crops, addition of water and nutrients, weed and pest control by various mechanical and chemical aids, and harvesting when crops mature. The crops that are planted are *seasonal* in the sense that they (crops) are harvested in one seasonal growing cycle. On the other hand, the permanent crops exist for much longer periods. They consist of orchards, cotton fields, tea plantations, etc. Crops can also be used for biofuel production. There can thus be a competition between fuel and food production, depending on market conditions.

6.1.1 Food Production

Hunger and nutrient deficiencies are experienced daily by more than a billion people. Nutrient deficiency is defined as insufficient levels of food proteins and caloric energy. As can be seen in the famines in many parts of the world, nutrient deficiency is a major problem. The Food and Agriculture Organization (FAO) of the United Nations estimates that in 2021, approximately 828 million people, are currently facing hunger (FAO, 2022). This situation will only improve somewhat with nearly 670 million people still projected to face hunger by 2030 – this is despite the goal of ending hunger, food insecurity, and malnutrition under the 2030 Agenda for Sustainable Development. 2022 edition (FAO *The State of Food Security and Nutrition in the World (SOFI)*.

DOI: 10.1201/9781003407393-6

The demands for food are immense and as we have discussed previously in Chapter 1, according to the Malthusian model, food availability is linked with population growth (Malthus, 1798). Increased urbanization has decreased available agricultural land. To increase yields, the use of fertilizer and pest control chemicals has increased. Irrigated areas have expanded, and high-yield crops have been developed. Some of these agricultural activities designed to enhance crop yields have led to decreases in soil fertility and increases in soil and water contamination. Subsequent depletion of the soil resource and the presence of these contaminants could lead to a decline in per capita food production according to Meadows et al. (1992) – which could be a threat to human survival (Commoner, 1971). More than 75% of the arable land in North and Latin America, 25% in Europe, and 16% in Oceania can be considered damaged. This threatens future food supplies (Fischer Taschenbuch, 1996). North America's breadbasket (the Midwest of the USA to the Great Plains in Canada) is particularly at risk due to wind and soil erosion, and fertility loss. Loss of native habitat in North America due to farming has been significant. In North America, only about 1% of the tall-grass prairie and 20–30% of the short-grass and mixed-grass prairie remain (https://learningtheland.ca/sar/prairie/). According to the World Wildlife Fund's (WWF) 2023 Plowprint Report, 0.65 million hectares were converted to row-crops from grasslands in the US and Canada in 2021.

Food production must be increased without increasing the impact on the geoenvironment. Improper irrigation can lead to waterlogging and soil salinity problems. Use of pest control chemicals has increased pest resistance and destroyed natural species (NRC, 1991). The many aspects of agricultural engineering and soil management practices are subjects that are well studied in soil science and agronomy. Their attention on efficient food and crop production, together with research into the various issues of soil management and soil quality have the aim of providing agricultural productivity without compromising the objectives of sustainable agriculture.

6.1.2 Geoenvironment Engineering – Sustainable Issues

There are many stressor impact issues associated with agricultural-based production of food. Most, if not all, of them are issues that fall in the purview of agricultural engineering and soil science. That being said, there are some impact issues that are common to geoenvironmental engineering land management. These issues, i.e., the stressors and their impacts, constitute the focus of this chapter. The discussion in this chapter is directed toward the likely geoenvironmental impacts due to food production activities, such as pesticide use, nutrient addition, and waste management as depicted in Figure 6.1. The geoenvironment-associated problems resulting from these activities are common to those found in land disposal of wastes and other soil contamination problems encountered in geoenvironmental engineering. Discussion on management, alleviation, and mitigation of the impacts due to stressors originating from these activities will be addressed in the latter part of this book in the chapters dealing with the subject of soil contamination, management, and remediation.

The impacts from the agricultural activities depicted in Figure 6.1 that can be mitigated and alleviated fall into the category of *surface and subsoil contamination.* The stressors responsible for contamination of the surface and subsoil are (a) the various waste discharges into or onto the land surface environment, (b) the various kinds of chemicals used in pest controls (herbicides, pesticides, etc.) and as fertilizers, (c) airborne emissions from the operation of machinery and other devices in the various operations and activities, and (d) deposition of noxious airborne particulates onto the land surface.

6.2 LAND USE FOR FOOD PRODUCTION

Soil is a natural resource material and is considered a geoenvironment natural capital. Agricultural soil has a balance of inorganic and organic components. Both climate and the environment control the value of the soil. Factors such as water content, soil type (composition), soil thickness,

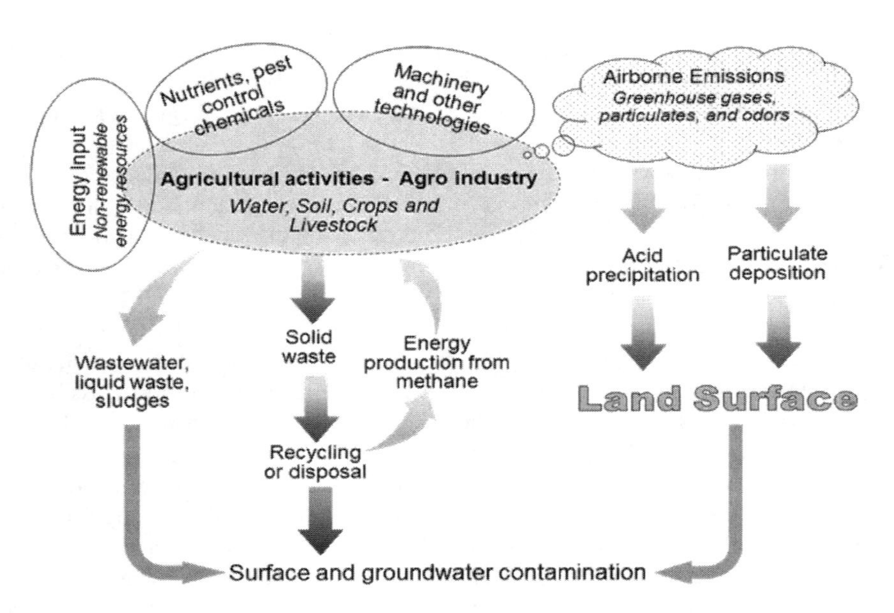

FIGURE 6.1 A summary of some of the major inputs and outputs related to agricultural activities.

salt content, and other physical, chemical and biological properties are important determinants of soil quality. In regions where winters and cold temperatures are factors, the presence or absence of permafrost will also contribute to soil value. The combination of physical, chemical, and biological agro-aids, climate, soil management, and technology has rendered the soils in Europe, and North and Central America favourable for agriculture. In contrast, the lack of many of the agro-aids, inadequate technology, poor soil management, and unfavourable climate have combined to produce nutrient-depleted soils in regions of Africa and Asia.

The requirement for sufficient food to sustain life has led humans to require substantial use of grasslands, forests, and fresh water. Factors such as excess or lack of water, drainage, soil quality and thickness, salinity, annual mean temperatures, and in cold regions, the freezing index and the presence or lack of a permafrost area determine the appropriateness of the soil for agriculture. Richards (1990) reports that there has been a five-fold increase in agricultural lands in the last 300 years. However, in the last decade, agricultural land area has remained about the same in North America (https://ourworldindata.org/grapher/agriculture-more-less-land). This needs to be balanced with the large quantities of land lost from production due to erosion, salinization from irrigation, desertification, and conversion to roads and urban uses. The capillary rise of salts (chlorides, sulphates, and carbonates) due to groundwater extraction and irrigation is threatening the US. High Plains, Canadian prairies, and Australian soils. Over the last 20 years, the US has lost more than 4.45 million hectares of agricultural land, according to a report by American Farmland Trust (https://theworld.org/stories/2020-08-07/us-lost-11-million-acres-farmland-development-past-2-decades). A senior UN official has estimated that based on current rates of soil degradation, the world's topsoil could be gone within 60 years (Arsenault, 2014). This is due to climate change, erosion resulting from deforestation, and chemical and mechanical farming techniques.

With the exception of a few high-value crops, the market value of the land for non-agricultural purposes is much higher – a factor that drives conversion of these lands to urban uses. The other consideration that is significant is the increasing productivity of agriculture in the last century – a factor that has decreased famine rates significantly (Pinstrup-Anderson et al., 1997). Innovations including high-yield crop varieties, application of fertilizers and pest control and utilization of mechanized equipment in both developed and developing countries have substantially increased agricultural productivity.

More than 50% of the land's surface is involved in one way or another with forestry, agriculture, or animal husbandry. Pastures alone make up 6 to 8% of the land. This does not include land for grazing. Agriculture, in combination with urban areas occupy up to 10 to 15% of the land (Vitousek and Mooney, 1997). More than 130,000 km^2 per year of forests were eliminated for the period 1990–2005, and of this, 98,000 km^2 per year were converted to agricultural crop use (FAO, 2006). However, the rate of conversion has slowed since 1961. Although the global population increased by 147%, agricultural land use increased by only 7% https://ourworldindata.org/deforestation. This has substantial implications in global warming since forest soils and removed trees account for recycling of much larger amounts of carbon dioxide from the atmosphere than agricultural lands. Deforestation also leads to increased risk of soil and wind erosion because of exposure to the elements. Although some forests have been allowed to regrow upon depletion of agricultural lands, reforestation rates are much lower than deforestation rates.

Ploughing disrupts: (a) Soil horizons, (b) chemical weathering, and (c) soil horizon interchanges. Ploughing compacts soil and decreases soil permeability. This inhibits evaporation processes and plant germination. Cattle and sheep herds also compact the soil. In arid climates, the soil is particularly vulnerable to erosion due to mechanization of ploughing. Overgrazing and slash and burn cultivation also reduces soil cohesion leaving many areas affected by soil degradation and loss of vegetation. More than 34% of all degraded land is due to overgrazing (FAO, 2021). Arid and semi-arid lands are particularly susceptible in Oceania and Africa.

From the geoenvironmental perspective, one could ask "how sustainable is agriculture?" Although most agree that agriculture is not sustainable under present practices, there is considerable debate concerning the means for measuring sustainability (FAO, 1995, 1996). Whilst full consensus is not available, it is agreed that the factors that need consideration are the use of genetically identical plants, irrigation water, fertilizers and pesticides, and the assortment of wastes produced. Intensive agricultural practices utilizing high levels of technology and mechanization are generally not kind to the soil environment. A simple summary of the major inputs and outputs of agricultural activities is depicted in Figure 6.1. The impact of agricultural activities on the land environment will be discussed in more detail in the next section.

6.3 STRESSOR IMPACTS ON WATER AND SOIL

Two of the most important ingredients in agricultural-based food production, other than climatic conditions, are *water* and *soil* – more specifically, adequate water supply and soil quality, i.e., soil media containing the nutrients that foster plant growth.

6.3.1 Water Utilization

The availability and quality of water are two important water factors that impact significantly on the health and welfare of a growing population, particularly in developing countries. Fresh water is a precious resource because of the limited amounts directly available for use. Agriculture is responsible for 70% of the total global water withdrawal (World Bank, 2022). Rivers are diverted to serve the needs of humans. Only 2% of the rivers in the US have not been manipulated. As an example of the impact of excessive use of water for agricultural purposes, it is reported that the water levels of the Aral Sea, located in Central Asia in the lowlands of Turan, have been substantially reduced by agricultural practices. The result of this has been (a) loss of native fish and biota, (b) creation of a source of windblown dust from the exposed salty sea bottom, (c) increase in the frequency of human diseases, and (d) creation of a drier local climate (Micklin, 1988).

The cost of energy influences one's ability to extract, pump, and irrigate abstracted groundwater. Increasing agricultural yields coupled with increasing land for agriculture result in corresponding increases in water demand. Irrigated farming produces 40% of the food and 20% of all land is irrigated (World Bank, 2022). Management for salinity and drainage is required to avoid decreases in agricultural yields. Falling water tables increase the costs of abstraction of groundwater – a factor that needs to be incorporated into management of irrigation.

Irrigation involves exploitation of rivers, aquifers, or other fresh water sources, causing a disruption of the natural hydrological cycle. If irrigation is poorly controlled, desiccation can occur between watering periods. This can lead to increasing rates of aridification and wind erosion. Over-irrigation can deplete fresh water sources and the soil can become salinated due to increased evaporation at the surface horizons. Substantial amounts of water are required for production of various crops. As shown in Figure 6.2, the growing of corn, rice, and soybeans require substantially more water than wheat. Livestock requires 100 times more water than 1 kg of vegetable protein. Overall, in 2005, less than 1% of the total amount of water is used to raise livestock (Kenny et al., 2009). It has thus been estimated that up to 15,400 L per kg are required to produce 1 kg of beef on a range, approximately 10 times the amount to produce 1 kg cereals (1644 L/kg) and 50 times that of vegetables (322 L/kg) (water footprint network) Increase in livestock and crop production required to satisfy the needs of a growing population will continue to stress water resources (Giampietro and Pimentel, 1995), in addition to extensive food wastage (up to 40% of all food produced) due to inadequate harvesting, storage, and transportation procedures (https://earth.org/what-is-food-waste/).

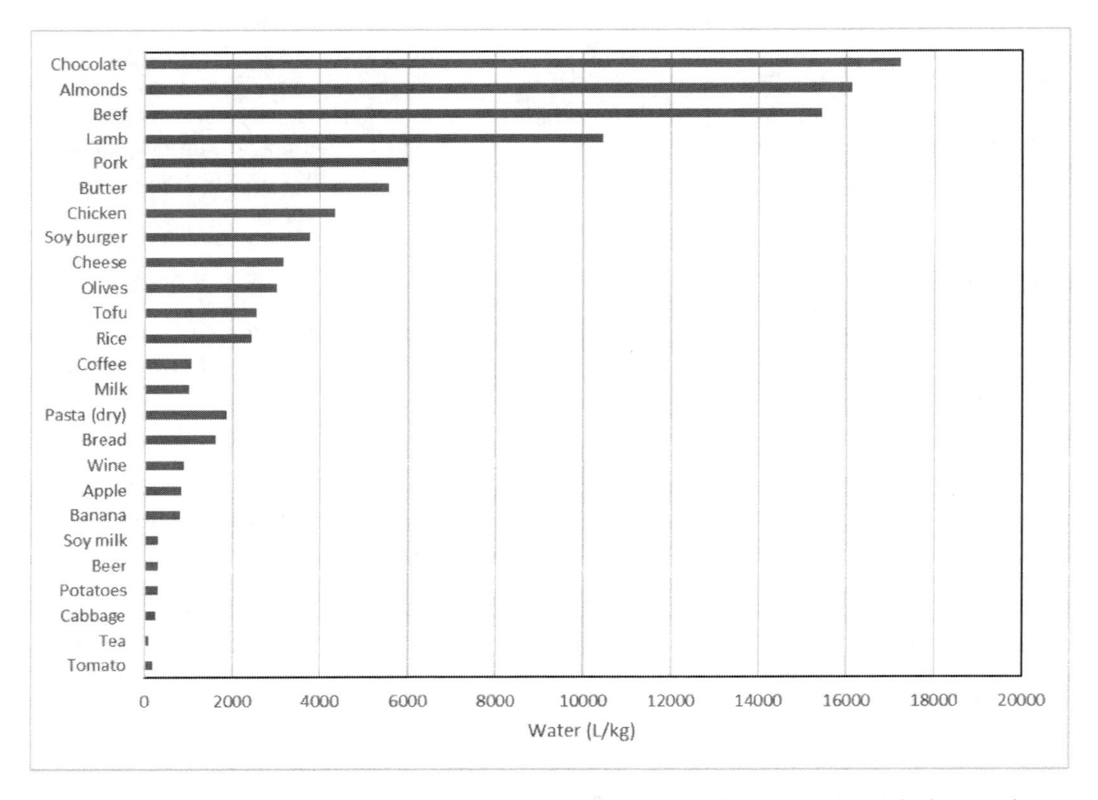

FIGURE 6.2 Amount of water required to produce a kg of food. (Data from www.watercalculator.org/water-footprint-of-food-guide/.)

Mining of groundwater is necessary because of demand for fresh water by the increasing population and the subsequent use of irrigated agricultural practices (Falkenmark, 1989). For example, from 1964 to 2003 for a well in Cook County, southwest Georgia, levels decreased by 4.6 m in a well-used for irrigation and the public (www.usgs.gov/special-topics/water-science-school/science/groundwater-decline-and-depletion). Groundwater is a valuable resource both in the US and throughout the world. Where surface water, such as lakes and rivers, are scarce or inaccessible, groundwater supplies many of the hydrologic needs of people everywhere. In the US, groundwater serves as a source of drinking water for about half the population and 190 billion litres per day for agriculture (www.usgs.gov/special-topics/water-science-school/science/groundwater-decline-and-depletion#overview).

Conflicts over water use due to irrigation have occurred in various parts of the world. The Egyptians have used the Nile for more than 5000 years while the other nations in the upper drainage basin have not (McCaffrey, 1993). Recently, however, lack of water and increasing populations have made the other nations more dependent on the Nile. Construction of a dam on the Ganges River by India has led to riots and protests in Bangladesh as the water needed for irrigation is now diverted (Kattelmann, 1990).

6.3.2 Soil and Water Quality Stressors

6.3.2.1 Nutrients

Agricultural practices can have a significant impact on groundwater and receiving waters – as summarized in Figure 6.3. Chemicals such as nitrogen (N), potassium (K) and phosphorus (P), sulphur (S), and magnesium (Mg), introduced as plant nutrients can have significant impact on water quality and can affect human health and other biotic receptors – if and when they become non-point

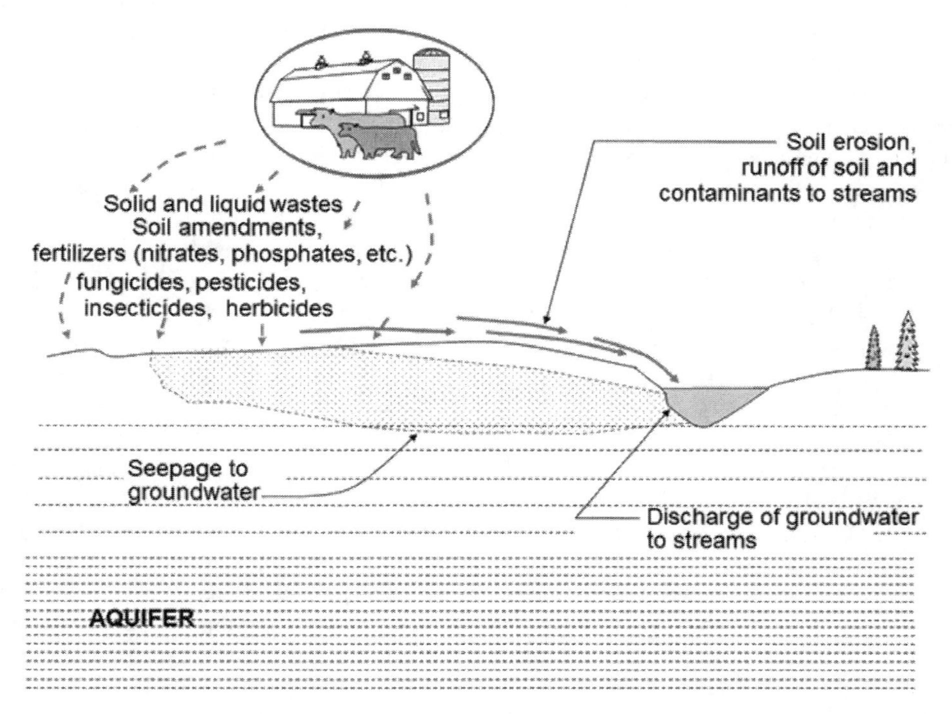

FIGURE 6.3 Schematic of contaminant transfer from agricultural activities to surface and groundwater – from a geoenvironmental perspective.

source contaminants for receiving waters and aquifers. Other sources of nutrients include inorganic fertilizers, animal manure, biosolids, septic tanks, and municipal sewages. In Canada, in 2021/2022, (www150.statcan.gc.ca/t1/tbl1/en/tv.action?pid=3210003801&pickMembers%5B0%5D=1.12&cubeTimeFrame.startYear=2015+%2F+2016&cubeTimeFrame.endYear=2019+%2F+2020&referencePeriods=20150101%2C20190101) an average of 2,160,000 tonnes of nitrogen and 733,000 tonnes of phosphorus were purchased for agricultural lands in the form of fertilizers.

The US EPA has established 10 mg/L of nitrate N as the maximum contaminant level in groundwater and a goal of 0.05 mg/L. It has also set a maximum of 0.1 mg/L for phosphate effluents that will enter a lake or reservoirs (US EPA, 1987). Nash (1993) determined that 25% of the US drinking wells had above 3 ppm levels of nitrate and other wells have nitrate levels that have reached as high as 100 ppm. According to the USGS' 2010 report, *Nutrients in the Nation's Streams and Groundwater*, 1992–2004, the median nitrate concentration in monitoring wells increased 6% from 1993–2003. During the same time period, the proportion of wells with concentrations of nitrate greater than the 10 mg/L MCL increased from 16% to 21%. Later data, showed that only three states (Oklahoma, Kansas, and Delaware) had more than 10% of the state groundwater with nitrate levels higher than 10 mg/L (1988 to 2018) (US EPA, 2023). Corn production areas in particular had higher levels of nitrates in groundwater (Garcia et al., 2017).

According to the US Geological Survey (1999), nitrate concentrations exceeded the US EPA drinking-water standard in 15% of samples collected in shallow ground water near urban and agricultural lands. Amounts were measured in agricultural streams and found to be less than 20% of the phosphorus and less than 50% of the nitrogen applied annually to the land. This is due to the greater tendency of phosphorus to attach to soil particles and move with run-off to surface water compared to nitrogen.

Soybeans, alfalfa, and other legumes fix more than 40 Tg/year (teragrams per year) of nitrogen fertilizer (Galloway et al., 1994). Depletion in nitrogen levels can impact soil fertility. Nitrate run-off from agricultural fields can decrease water quality. The nitrogen cycle can subsequently be altered through production of nitrogen oxides, which can impact human and ecosystem health (Figure 6.4). The nitrogen can travel from agricultural fields, to the rivers, streams, groundwater, and finally to the oceans. The diffusion processes for fertilizers (1 m/year) are slow. Consequently, groundwater contamination may not be seen until a decade after a large amount of fertilizer is spread (Chamley, 2003).

Nitrogen, on the other hand, is an indispensable nutrient for living organisms. The balanced circulation of nitrogen, in addition to carbon, are fundamental factors in sustainability. In this book, nitrogen and carbon cycles are covered in Chapter 12.

The nitrogen from fertilizers can be converted to nitrate before uptake by plants. When this happens, contamination of rivers, lakes, estuarine, deltaic, or coastal waters can occur – as shown for example in Figure 6.4. Toxic algal blooms (by the dinoflagellate *Pfisteria*) can result from the eutrophication of estuaries and kill fish via their toxins. Eutrophication processes also deplete the oxygen in the water, destroying aquatic ecosystems. These phenomena affect the sustainability of fisheries. Eutrophication increases costs for drinking water treatment, reduces waterfront property value, leads to amenity value and biodiversity losses, and impacts on tourism. The costs related to these are estimated at in the UK and USA as £75–114 million (Pretty et al., 2003), and $2.2 billion US, respectively (Dodds et al., 2009).

Humans can also be affected by shellfish poisoning. High nitrate levels (above 10 ppm) can also lead to methemoglobinenemia (also known as blue-baby syndrome), abortions, and increased rates of non-Hodgkin's lymphoma. Levels above 5 ppm can affect young animals (Pimentel, 1989). Algal blooms initiated by the elevated nutrient levels lead to (a) disruption of the ecology through oxygen consumption, (b) accumulation of organic content, (c) reduction of water and sediments, (d) asphyxia, and (e) biota mortality such as plankton and benthos (Chamley, 2003). Inorganic nitrogen levels in coastal areas have increased by a factor of 2.5 and phosphorus levels by 2, particularly in Western Europe. Dating of ^{210}Pb in Tahiti has allowed the determination of the accumulation

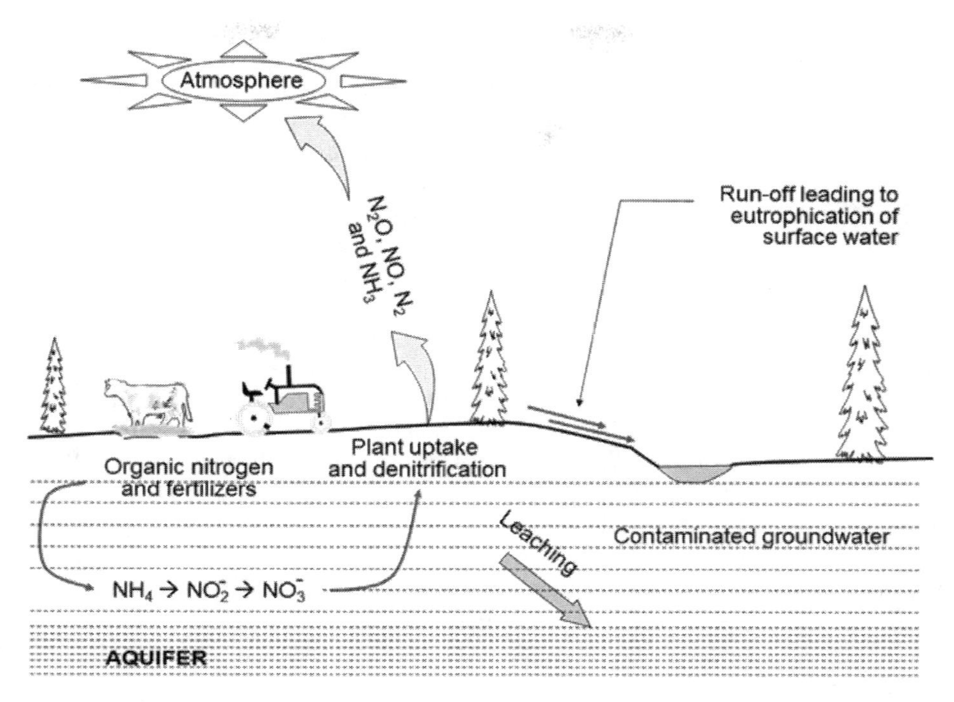

FIGURE 6.4 Illustration of the nitrogen cycle as it relates to the geoenvironment and farming practices.

of phosphorus in the sediments. The steps included exchangeable, iron-sorbed, carbonate, marine organic, and terrigenous phosphorus forms. In the 1950s, terrestrial phosphorus increased significantly due to both soil erosion and waste discharges.

The conversion of ammonia to nitrate lowers soil pH, particularly in north-eastern Canada and Scandinavia (Chamley, 2003), where the bedrock formations are siliceous. The acidity increases the mobility of the toxic components, aluminium, and manganese. This can diminish vegetation and plant growth and increase soil erosion.

Due to its charge, phosphorus binds to soils but will leach from sandy soils with low levels of clay, oxides, and organic matter. Surface run-off will most likely cause contamination of streams and lakes, whereas groundwater contamination by nitrates is more likely since these anions are held more weakly by negatively charged clayey soils and it is more soluble than phosphorus. Nitrate diffusion in the groundwater is very slow, in the order of 1 m/year. Thus, the impact from excessive fertilization of the soil may only be seen in the groundwater a decade later.

Inorganic fertilizers and other soil amendments, such as animal manure and biosolids, contribute to elevated levels of N and P in the environment. Inorganic fertilizer use increased by 20- and 4-fold from 1945 to 1980. Their use has since levelled off. Animal manure contributes 6.3 million tonnes of N and 1.8 million tonnes of P compared to 10.8 million tonnes of N and 1.8 million tons of P from inorganic fertilizers (U.S. Geological Survey Circular 1225, 1999). Nitrogen fertilizer application was about 11 billion kg in 2010 (Capel et al., 2018). Between 5 and 50% of this can run-off into streams and groundwater. About 15% of shallow groundwaters sampled beneath agricultural areas were above the acceptable levels for nitrates.

The Australian government funded a project to study the movement of phosphorus in soils supplemented with piggery effluents (Redding, 2005), soil samples were taken down to 5 cm in areas with and without the effluent amendment. Leaching was not significant if correct management procedures were used but it did occur if application rates were excessive. Significant adsorption of the phosphorus occurred in the top 5 cm and phosphorus run-off readily occurred from surface soil.

6.3.2.2 Herbicides and Pesticides

Pesticides are chemicals that have the purpose of eliminating (controlling) "pests", and are classed as *biocides*. Most of these are considered as toxic chemicals that are injurious to human health and other biotic receptors. There are several types of pesticides such as herbicides, insecticides, fungicides, rodenticides, verucides, etc., each of which is used directly for control of specific "pests". Herbicides are used for controlling unwanted plants, insecticides are used to control insects, etc. They have been used since the nineteenth century in the form of lead, arsenic, copper, zinc salts, and nicotine for insect and disease control. Since the 1930s and 1940s, with the introduction of 2,4-D and DDT, agricultural use has increased substantially. Worldwide use of pesticides for 2021 is increasing and is approximately 3.54 million metric tons (www.statista.com/statistics/1263077/global-pesticide-agricultural-use/). The *Database on Pesticides Consumption* called FAOSTAT which is maintained by the United Nations Food and Agriculture Organization (FAO, 2024) provides information on the use of specific pesticides within each country. There, pesticides use in agriculture increased by 2% from 1.74 Mt in 2020 to 1.78 Mt in 2021. The growth rate between 1990 and 2021 was 191%, with a 26% increase in the most recent decade. In 2021, the Americas imported and applied the highest levels in the world of pesticides between 1990 and 2021 per area of cropland (3.01 kg/ha), (www.fao.org/faostat/en/#data/RP).

Although pesticides have enhanced crop yields, concerns are increasing regarding their effects on the health of humans and animals and their transport in the environment. Levels of contamination of surface streams and groundwater increase with increased nutrient and pesticide use (USGS, 1999). Agricultural streams show the highest concentrations of pesticides. Herbicides are the most frequent pesticides found in agricultural streams and groundwater. Atrazine ($C_8H_{14}ClN_5$), deethylatrazine ($C_6H_{10}ClN_5$), metolachlor ($C_{15}H_{22}ClNO_2$), cyanazine ($C_9H_{13}ClN_6$), alachlor ($C_{14}H_{20}ClNO_2$), and EPTC (S-Ethyl dipropylthiocarbamate, C_9H_9NOS), are the most commonly detected chemicals – correlating well with their usage (USGS, 1999). Frequent pesticide and fertilizer use can alter the natural resistance of the plants and may also increase their resistance to parasites. This will make some soils unusable for agricultural purposes. Activities generating contaminants in agricultural areas are shown in Table 6.1. From 2000 to 2007, the application of pesticides has decreased from 0.54 to 0.5 billion kg. 80% is used for agriculture. According to Capel et al. (2018), application of pesticides in the US was about 0.3 billion kg in 2010. Overall, more than 2.3 million tonnes of pesticides (22% of the world use) are applied – resulting in the pollution of more than 10% of the rivers and 5% of the lakes in the US. Groundwater quality is also impacted with pesticide use.

TABLE 6.1
Agricultural Activities Affecting the Environment

Sources	Contaminant Transported
Fertilizer, manure, lime addition	Nitrogen (ammonia, nitrite, nitrate), phosphorus, organic nitrogen, pathogens, odours, organic carbon, nitrogen oxides
Application of herbicides	Atrazine, metolachlor, glyphosate, other chemicals
Application of insecticides	Carbaryl, DDT, permethrin, organochlorine pesticides, others
Application of fungicides	Benomyl, captan, thiabendazole, others
Animal feed additives	Tetracycline, hormones, drug residues and many other chemicals
Improper tillage, soil management, pond drainage	Suspended solids
Aquaculture	Nutrients, pesticides, salts, sediments, pathogens, metals, drug residues
Irrigation	Nutrients, pathogens, organics, salinity, soil erosion

Source: Capel et al. (2018).

Globally it has been found that 64% of agricultural land is at risk for pollution by at least one pesticide, while 31% is at high risk. The high-risk areas are in high-biodiversity regions (34%), water-scarce areas (5%) and low- and lower-middle-income nations (19%). High pesticide pollution risks were in watersheds in South Africa, China, India, Australia, and Argentina (Tang et al., 2021).

Approximately 70% of the total national use of pesticides is in agricultural areas and thus there is widespread occurrence in agricultural streams and shallow groundwater. The highest rates of detection were found for the herbicides atrazine, metolachlor, alachlor, and cyanazine. Insecticides were more frequently found in some streams draining watersheds with high insecticide use than in shallow ground water as they also tend to sorb onto soil or degrade quickly after application. As there are no US EPA aquatic-life criteria for the major herbicides, Canadian guidelines were employed and concentrations were found to be elevated, particularly for atrazine or cyanazine in 17 of the 40 agricultural streams studied. Also, currently used insecticides. The major organochlorine insecticides, (e.g., DDT, dieldrin, and chlordane), exceeded guidelines for aquatic life in at least one water sample from 18 of the 40 agricultural streams and recommended sediment-quality guidelines for protection of aquatic life at about 15% of agricultural sites. In the US, about 450 million kilograms of pesticides, including herbicides, insecticides, and fungicides, were used in the US (www.usgs.gov/centers/ohio-kentucky-indiana-water-science-center/science/pesticides).

Pesticides released into the environment for agricultural and non-agricultural purposes can contaminate surface water and groundwater, which are critical sources of drinking water. In most agricultural areas (95%) there is less than a 10% chance that the threshold-level for herbicides will be exceeded. Atrazine is currently highly used especially for sorghum and corn (Paul Stackelberg, USGS). A study by the USGS on US rivers and streams showed that during the period of 2013–2017 in 74 sites, 17 pesticides were detected at least once. Insecticides and fungicides were less frequently found than herbicides (www.usgs.gov/news/potential-toxicity-pesticides-aquatic-life-us-rivers-widespread).

Estimated mass balances of pesticides have been used to evaluate the fate and transport of the pesticides in the environment – as illustrated in Figure 6.5. These estimates exhibit substantial variation in some cases, due for example to, atmospheric drift because of weather conditions, application method, and properties of the pesticide. Unaccounted for pesticides can be due to the formation of covalent bonds with plant material or organic matter of soils (Xu et al., 2003) or biodegradation.

The interactions of organic chemicals with soil organic matter have been briefly discussed in Section 2.5 in Chapter 2 and is further discussed in detail in Chapter 9. These discussions show that the association of pesticides with the organic matter of soil can be described by the relationship: $C_{oc} = k_{oc}\, C_{aq}$, where C_{oc} is the concentration of solute contaminant sorbed onto the soil organic carbon, k_{oc} is the organic carbon-water partition coefficient and C_{aq} is the dissolved concentration of the contaminant. The Freundlich adsorption isotherm for many pesticides can be described by the following relationship: $C_{oc} = k_f C_{aq}^{\frac{1}{n}}$, where k_f and $\frac{1}{n}$ are Freundlich parameters. These parameters are known for more than 60 pesticides (Barbash, 2005).

The sorption of a herbicide, triasulfuron ($C_{14}H_{16}ClN_5O_5S$), in Molinaccio clay loam, with and without humic and fulvic acids as organic materials, determined by Said-Pullicino et al. (2004) are shown in Figure 6.6. The Freundlich isotherm values, k_f ($\mu g^{(1-1/n)} mL^{1/n}\, g^{-1}$) were 0.14 for the soil, 0.18 (with the compost), 0.20 with the humic acid and 0.15 with hydrophobic dissolved organic matter. The $\frac{1}{n}$ values were all less than 1. The high concentration of compost from municipal waste led to an increase in sorption. However, in all cases, saturation was not reached and the herbicide preferred to remain in solution and thus was highly likely to leach. The humic acids and the hydrophobic dissolved organic matter in the compost were responsible for the sorption characteristics of the compost.

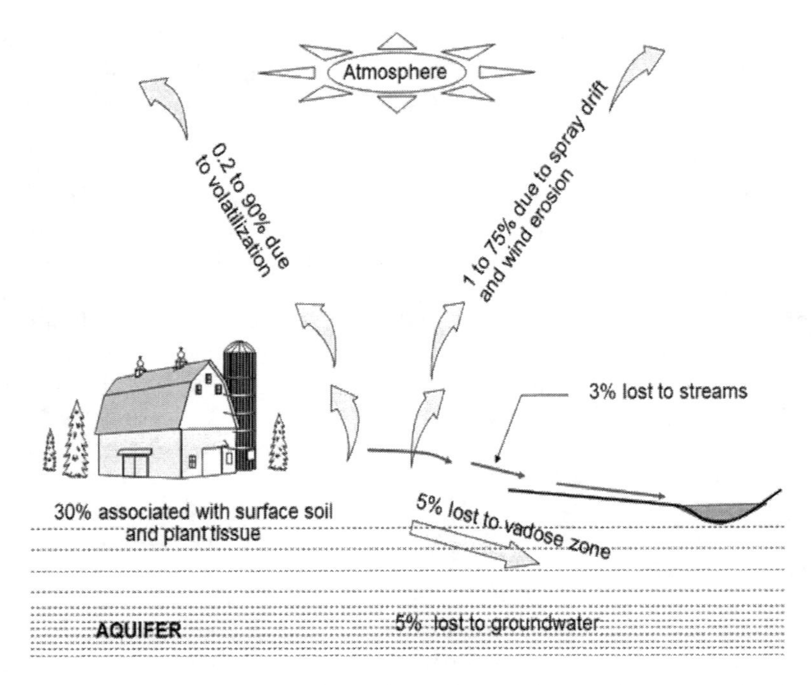

FIGURE 6.5 Estimates of the fate of pesticides after application to the soil for agricultural use (data from Barbash, 2005). Percentages of pesticide associated with plants, etc., and lost to various atmospheres, vadose zone and streams are dependent on highly climatic factors, soil management practice, type of pesticides used, and manner of application, etc.

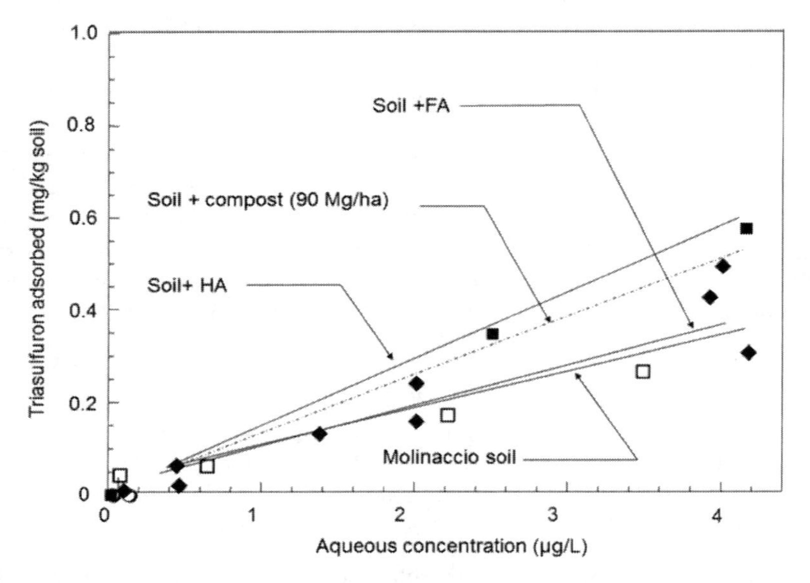

FIGURE 6.6 Adsorption of triasulfuron ($C_{14}H_{16}ClN_5O_5S$) on a Molinaccio soil, HA and FA are humic acid and fulvic acid, respectively. (Adapted from information reported by Said-Pullicino et al. 2004.)

6.3.2.3 Microbial Contaminants

Water used in agriculture can be a source of microbial contamination for food. The top four pathogenic bacteria are Shiga-toxin producing *Escherichia coli, Camphylobacter, Salmonella, Arcobacter* and *Listeria monocytogenese*. Many of the pathogens can survive on land and in run-off water for long periods of time (for weeks). Water uses include irrigation, washing of produce and equipment, and dust suppression. While there are standards by the US EPA for waters for drinking and recreational use, there are none for agricultural water. According to Bell et al. (2021), there have been few studies on pathogens in agricultural water. A notable case was in Walkerton, Ontario, Canada in 2000 where there were 2000 confirmed cases and seven deaths from *E.coli* O157:H7 and *Campylobacter jejuni* contamination of drinking water. Water was not adequately treated following heavy rainfall. Bacteria from cow manure spread for crop fertilization were transported to the shallow aquifer drawn by a well for drinking water. It was later determined that this well was too close and too shallow to the field and thus very susceptible to contamination, monitoring was inadequate, and chlorination was not sufficient. Climate change with warmer temperatures and heavier rain events may increase such incidences in the future.

The GenomeTrakr network (www.fda.gov/food/whole-genome-sequencing-wgs-program/genometrakr-network) is a database for genome sequencing for pathogen identification. Public health and university laboratories collect and share genomic and geographic data from foodborne pathogens from various sources including surface water, agricultural reservoirs, and sediments. The data is housed in public databases at the National Center for Biotechnology Information (NCBI). According to Bell et al. (2021), most submissions are from North America and western Europe with few from Africa and South America. This increased international collaboration is needed.

6.3.2.4 Greenhouse Gas Emissions

Agriculture is a major contributor to greenhouse emissions. Land degradation, and deforestation, and biodiversity loss, are contributors to climate. Globally it has been estimated that emissions from agriculture are 13–15 billion tonnes of carbon dioxide per year into the atmosphere, which represents approximately one-third of all human activities. Other GHG emissions are from rice and enteric fermentation, and nitrous oxide emissions mainly from fertilizer application that account for 50% and 75% of all emissions by humans (Tubiello et al., 2015). Overall agricultural-related GHG emissions in developing countries have increased by 32% from 1990 to 2005. As the demand for food and energy will increase, these emissions will most likely also increase particularly in developing countries.

In Germany according to the Thünen Institut (2021) (www.thuenen.de/en/thuenen-topics/climate-and-air/emission-inventories-accounting-for-climate-protection/greenhouse-gas-emissions-from-agriculture, accessed Jan. 2024), the agricultural sector accounted for GHG emissions of 56.3 million tonnes of CO_{2e} in 2021 (7.4% of the total emissions) of which 16.6 million tonnes was mainly from N_2O from soil. This is a declining trend since 2014. The N_2O emissions originate from soil amendments (animal manures, mineral fertilizers, sewage sludge and digestate from animal manures and energy crops), in addition to organic soil cultivation and crop residue degradation.

In 2009, Canada produced 56 million tonnes carbon dioxide equivalents (Mt CO_2eq) from agriculture (8% of its total emissions), while CH_4 emissions accounted for two-thirds of its total and N_2O one-third. Annual total GHG emissions increased from 1990 to 2009 from Canadian farms due to increases in the populations of swine and beef but this has stabilized recently. On the other hand, about 12 Mt CO_2eq were removed in 2009 by soil carbon. While conversion of grasslands and forests to agriculture decreases this value, since is 2000, agricultural lands have been provided a net sink for CO_2. In 2009, emissions for both livestock and crop production led to a decrease in overall total emissions (https://agriculture.canada.ca/en/environment/greenhouse-gases).

6.3.2.5 Emerging Pollutants

New pollutants have emerged in the last couple of decades. These include antibiotics, vaccines, and growth hormones. Run-off from manure on agricultural lands and aquacultures, and application of wastewater and biosolids are sources of the pollutants. Monitoring and modelling are essential for determining and mitigating impacts.

The fate and transport of pharmaceuticals have been studied. Yu et al. (2013) determined that triclosan and octyphenol, were strongly sorbed to soil, whereas others such as gemfibrozil and carbamazepine were hardly adsorbed. Clay and organic matter contents can affect sorption and biodegradation rates. Carbamazepine, therefore, could potentially leach to groundwater. A literature review on various pharmaceuticals in the soil and plant environment was performed by Gworek et al. (2021).

6.3.2.6 Aquaculture

Aquaculture, growth of fish and shellfish in water, has grown substantially since the 1980s (about 20-fold) (FAO, 2016). Developing countries are responsible for 91% of the output. This has led to increased environmental issues such as fish waste, unused fish food (fishmeal and pelletized feeds), and increased levels of antibiotics, fungicides, and anti-fouling agents. The feed contributes to higher nutrient and organic levels. For example, in Bangladesh, shrimp aquaculture produce 600 tonnes of waste per day (SACEP, 2014). Mussel farming on the other hand can clean waters through filtering. Increased production intensity by single species has led to increased pollution from antibiotics, etc.

6.4 FOOD PRODUCTION STRESSOR IMPACTS

In addition to the chemical stressors in the form of fertilizers, pesticides, and wastes from livestock, trace elements in the form of trace metals, heavy metals, metalloids, micronutrients and trace inorganics are also found in soils, and surface and groundwater water. These have been traced to application of commercial fertilizers, liming materials, irrigation waters, and biosolids. Their presence in soils and groundwater serve to reduce the quality and functionality of soils – in addition to their potential to cause injury to human health and other biotic receptors – as shown in Figure 2.2 in Chapter 2. The discussion in Chapter 10 pays special attention to the presence of these kinds of contaminants in soils and groundwater, and ways and means for minimizing the adverse impacts created by these stressors.

6.4.1 Impact on Health

The impact on public health due to activities associated with food production has been studied by many international organizations. Although agricultural yields have increased, irrigation, land conversion, and disturbances of the ecosystems have increased the rates of even older diseases such as malaria. Irrigation has increased the number of diseases, up to 30 diseases – including mosquito-borne diseases in Central and South America (WHO, 1996) and malaria and Japanese encephalitis due to rice paddy irrigation. Irrigation systems in hot climates are directly linked to schistosomiasis incidences.

Deforestation and increasing cultivation decreases the soil's ability to retain contaminants and nutrients. Chemicals such as mercury, which are normally stabilized by iron oxyhydroxide adsorption, can bioaccumulate in fish. Erosion destabilizes mercury, thus increasing the release of mercury into the water supply. Freshwater fish can contain an average of 48 μg/L of mercury, a potential health hazard (Richard et al., 2000).

The decrease in water quality due to pollution from a variety of urban and industrial sources also can contaminate crops and lead to the poor health of farm workers and consumers of the contaminated crops. Agriculture cannot be viewed in isolation to the other sectors. It has been called both a cause and victim of water pollution (Ogaji, 2005).

6.4.2 Impact on Biodiversity

One of the major impacts of land transformation for food production and other uses has been the extensive loss of biodiversity. Food production activities in the agricultural sector has led to the extinction of more species than any other sector. The rates of extinction have increased from 100 to 10,000 times the level prior to the industrial revolution. Decreases in the amounts of pollinating insects have negatively impacted yields of particular crops (Nabhan and Buchmann, 1997). The diversity of soil organisms has also decreased due to the decreased opportunity for organic decomposition and increased nutrient content. As a result, specialized predators and weeds will also develop due to the low diversity of species in agricultural lands. Changes in land use has affected the various cycles within the global system including the carbon, water, biogeochemical and biotic, to name a few. For example, lack of soil organic matter and soil organisms will impact the carbon cycle.

Biological diversity is the highest in forested areas. For example, whilst inter-tropical forests make up only about 6% of the surface area, they nevertheless make up half of all plant and animal species. Thus, conversion of these forested areas to agriculture can be highly detrimental to biodiversity. Modifying rivers and lakes for irrigation purposes can lead to extinction of the fauna, flora, and terrestrial organisms due to variations in water composition, temperature, and flow. Water pollution from run-off of pesticides, fertilizers, and salinization can influence biodiversity. Salinity levels in the Aral Sea have tripled due to cotton irrigation (WRI, 1994). This has led to extinction of 24 species of fish (Postel et al., 1996). Also, erosion can be detrimental to fish and other aquatic organisms due to siltation of breeding grounds. Mulholland and Lenat (1992) estimated that there has been a decrease of up to 50% of the species in streams affected by agriculture. Increasing the use of surface water for agriculture can also damage ecosystems in lakes and watersheds. The biodiversity losses or changes may not be evident initially but in the long-term can have significant impacts on the sustainability of agricultural practices.

New breeding methods have caused a significant loss of agricultural biodiversity. Some of these impacts are shown in Figure 6.7. The schematic diagram shows that although there is a large number of species available, only a small portion is utilized for food production. Uniformity and standardization of farming practices has led to this restricted biodiversity – a clear demonstration of the effect of human activities on biological diversity. The genetic base of crops needs to be widened to avoid dependence on a restricted genetic base that renders the world's food supply at risk to diseases, pests, and other dangers. New crops will also ease the demand for food in areas where it is scarce, and new management techniques will be required to incorporate new or underdeveloped crops to broaden the genetic base. Insofar as biodiversity is concerned, minimization of the negative impacts on natural biodiversity is paramount, together with conservation of the available genetic resources – to obtain a measure of sustainability in agricultural food production. It is useful to bear in mind that agricultural fields are not isolated from the surrounding natural environment. Current agricultural policies concentrate on product yields, demographic changes and land ownership (Fischer Taschenbuch, 1999). Genetic erosion will lead to increased risks to food security as plants will be less able to adapt to changes in the environment. Agrobiodiversity is a vital factor in agricultural management practices.

Other impacts on biodiversity have been studied in the UK. Hedgerows were used extensively in the past as field boundary markers (Nature Conservancy Council, 1986). However, more than 16,000 km of hedgerows have been removed annually in the 1960s. It has been estimated that

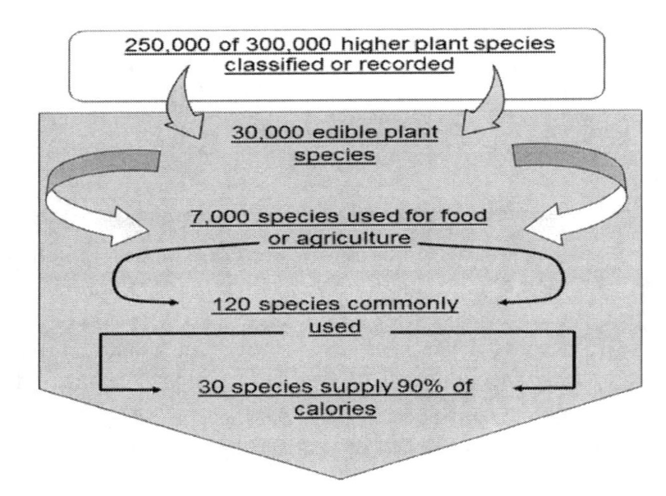

FIGURE 6.7 Risk to food security due to loss of agricultural biodiversity. (Source of data: El Bassam, 1998.)

hedgerows are important for the conservation of threatened 72 species of invertebrates, 20 species of birds, 11 species of mammals, 10 species of lichens, 5 species of reptiles and amphibians (https://hedgelink.org.uk/guidance/hedgerow-biodiversity). For example, butterflies that live in the boundaries could be threatened by the loss of the hedgerows. Biodiversity has been severely impacted due to removal of the hedgerows and other intensified agricultural production processes. Other species, such as brown hares, arthropods, insects, bees, flowers, and bats, are also threatened.

6.5 MANAGING GEOENVIRONMENT STRESSOR IMPACTS

As in the previous sections, the discussion in this section concerning agricultural practices and farm-related stressors and impacts on the geoenvironment are all structured from the geoenvironmental perspective, i.e., impacts to the geoenvironment. It is accepted that geoenvironment stressors will be generated by activities associated with agricultural food production. The section title "managing geoenvironment stressor impacts" is used to mean "implementation of measures": (a) To prevent, where possible, the various stressors from impacting on the geoenvironment, and/or (b) where it is not possible to prevent stressors from impacting on the geoenvironment, to mitigate the adverse stressor impacts on the geoenvironment. The preceding discussions have shown what the major geoenvironment stressors are and how they come about. We have seen how they impact on the geoenvironment, human health, and biodiversity. Insofar as geoenvironment sustainability is concerned, agricultural-related food production activities need to (a) eliminate (if possible) or mitigate stressor impacts on the geoenvironment, and (b) implement procedures and technology that will manage geoenvironment stressor impacts to achieve and maintain soil, surface and groundwater quality. The measures (procedure and technology) taken to manage the geoenvironment impacts can be implemented at the source, as *source management-control*, and/or impact remediation and rehabilitation procedures, i.e., procedures implemented on impacted site.

It can be argued that many of the measures for mitigating, minimizing, and even preventing impacts from contaminant loading and soil quality impairment run counter to intensive agricultural practices. From the perspective of the agroindustry, there is validity to this set of arguments. However, as has been realized for countless years, conflicts between agricultural productivity and protection of the geoenvironment have always existed. With better awareness of the sources of geoenvironment stressors, and with better tools for management of stressor impacts on the geoenvironment, these conflicts will continue to lessen as one strives towards obtaining sustainability for both agricultural-based food production and the geoenvironment.

6.5.1 Examples of Practices to Reduce Stressor Impacts

6.5.1.1 Soil Degradation Minimization

Changes in agricultural practices can lead to reduction in soil degradation rates. Most of these practices are well known in the industry and are now routinely applied in everyday agricultural engineering and farming practices. These include rotation of crops, protection of soils from erosion, increase of organic matter and nutrient contents and restructuring of the soil. It is common practice to alternate between high yield cereal or tuber plant with leguminous or fallow land planting. Adding mulches, composts, or manures can also increase the organic content. Mechanical measures including contouring, terrace cultivation, and contour hedges can reduce erosion and increase yields. In the following sections, we will look at some measures to reduce impact. They are by no means an exhaustive list but serve as examples of more sustainable practices presently being practised by farmers today.

6.5.1.2 Soil Erosion Reduction

Erosion of agricultural land can result from unsustainable methods of tillage and soil management. Overgrazing, ploughing of steep slopes, land clearing, and deforestation can increase erosion rates. It has been estimated that 10.5 megagrams (Mg) per hectare of soil erodes per year as sediment into rivers, lakes, and other water bodies (FAO, 2017a). In respect to the impact of farming practices on soil erosion, the Franklin Sustainability Project in New Zealand evaluated different methods, including raised access ways, benched headlands, silt fins and traps and contour drains for erosion minimization (Ministry of Environment, New Zealand). The New Zealand empirical erosion model (NZEEM) has been used to estimate New Zealand's annual soil erosion rate as 720 tonnes per square kilometre per year. Incorporated factors include the location of highly erodible land, average annual rainfall and land cover (Dymond et al., 2010). Grasses and sedges were effective as sediment filters and in combination with benched headlands can be very effective for trapping sediment but do not prevent erosion. Appropriate water management techniques are required to avoid subsidence and groundwater depletion. Cattle herds and flocks should be limited to avoid excessive trampling and over pasturing. Water logging should be avoided to prevent loss of the land to salinization. Adequate drainage is required to leach out the salts and remove the excess water from the soil (https://environment.govt.nz/publications/environment-aotearoa-2019/theme-2-how-we-use-our-land/).

6.5.1.3 Integrated Crop Management

The British Agrochemical Association has developed an approach called *integrated crop management* to "avoid waste, enhance energy efficiency and minimize pollution". This approach advocates (a) crop rotation, (b) selection of appropriate cultivation techniques and seed varieties, (c) minimization of fertilizers, pesticides and fossil fuels, (d) landscape maintenance, and (e) encouragement of wildlife habitats. Use of farm-produced inputs for fuels, pesticides, and fertilizers is encouraged. Soils are protected to (a) minimize energy use, (b) reduce erosion, and (c) reverse adverse effects on beetles, spiders, and earthworms. Crop production that is appropriate for the climate, soil type, and topography are maintained. Trials have shown that with such practices (a) costs are reduced by 20 to 30%, (b) pesticide use is reduced by 30 to 70%, and (c) there is a reduction in requirement for nitrogen by 16 to 25%. The results also show that biodiversity was increased and nitrate leaching, and soil erosion were reduced.

Livestock generate significant stressor impacts on the geoenvironment – such as overgrazing, erosion of the soil, river and lake pollution, desertification, and deforestation (Regenstein, 1991). Their numbers have increased substantially to the point where they now outnumber humans 3:1

(Goodland, 1998). Incentives and taxes could be used to promote good environmental practice in food and agriculture. Reduction in water use and recycling of manure can be practiced. Cattle feedlots provide the most significant impact. Sheep generate less impact since they graze on more natural grassland.

Biotechnology has been employed to increase food quality and production through the availability of transgenic plants. The main objectives are to develop plants that are resistant to bacteria, fungus, viruses, and environmental stress. At this point, conventional crop breeding is commanding more attention, due in part to the need to broaden capabilities and alternatives to the production of genetically altered plants.

Optimized uses of both fertile and fragile soils together with measures to rejuvenate production on degraded lands are procedures that can be pursued. Lower quality soils could support more species than high-yield agricultural lands (Dobson et al., 1997). Planting trees for shelter can also help to (a) reduce evaporation and transpiration by 13 to 25% (Mari et al., 1985), (b) reduce wind erosion, and (c) increase crop yields such as corn by 10 to 74% (Gregersen et al., 1989).

6.5.1.4 Water Use and Quality Improvement

The Farm Waste Management Plan of the UK (Saha, 2001) has listed the following items for pollution control:

- Delay ploughing in of crop residues.
- Reducing the use of fertilizers, manure, and sewage sludge.
- Sowing autumn crops early.
- Managing farm waste carefully.

With regard to water quality, Sagardoy (1993) listed the following items to avoid water pollution:

- Development of water quality monitoring schemes.
- Optimization of the use of farm inputs and other agricultural activities that impact wetlands.
- Establishment of water quality criteria for agriculture.
- Prevent soil run-off and sedimentation.
- Proper disposal of animal and human wastes.
- Ionization of agricultural chemicals for pest management.
- Education of the community to minimize impact on water quality and ensure food safety.

The capability of soils for retention of contaminants is a good means for reducing the threat or impact of contamination of groundwater. A case in point could be the cereal cultivation in northeastern France that has led to pollution of the Rhine Valley groundwater by nitrates (Bernard et al., 1992). These agricultural soils contained much lower organic contents than forest soils. Considering that forest soils with their higher organic contents will show greater capability in retaining the contaminants and converting the nitrates to N_2O through denitrification, it would appear that conversion of these agricultural lands through reforestation can significantly reduce groundwater pollution.

Whereas attention is normally given to soil permeability in consideration of transport of fertilizers in the ground, the attention received by the underlying geology has not been as significant. This oversight can lead to serious consequences. The example of Brittany, France, with underlying densely fractured and weathered granite rocks is a good demonstration. These densely fractured weathered rocks allow infiltration of contaminated water – in contrast to metaphoric schists that prevent seepage (Chamley, 2003). Nitrates and phosphates from farms are retained by the granite that can then contaminate areas downstream. In contrast, in schistose regions, these fertilizers do not penetrate into the underlying rock. Geologic maps and corresponding laboratory data on the rock properties are useful tools in predicting the impact of pollution from farm fertilizers. Réunion, a sloped territory with increasing agricultural activities, urban and tourist areas produced a map in the year 2000 for such purposes.

The US Soil Conservation Service (1993) developed a wetland process for treatment of agricultural runoff consisting of a wet meadow, followed by a marsh and pond with an optional vegetated polishing area. It is applicable for the removal of sediments and nutrients such as phosphorus. The wet meadow with a slope of 0.5 to 5% consists of permeable soils with cool-season grasses. The depth of the marsh with cattails varies from zero at the surface of the meadow to 0.46 m at the deep pond. The deep pond performs as a biological filter for the removal of nutrients and sediments. Fish, such as common or golden shiners should be included in the pond to feed on the plankton. Average sediment and phosphorus removal in a system for potato growing in northern Maine over two seasons were determined to be 96 and 87%, respectively (Higgens et al., 1993).

Losses to the environment of pesticides and herbicides through volatilization and run-off must also be minimized. Large quantities of agrochemicals have been found in various water bodies such as the Great Lakes in North America. Application of these chemicals during calm conditions can minimize losses due to drift. Biological methods can be used to control weeds and insects.

It has been proposed that an "Integrated Pest Management" (Janzen, 1998) strategy be utilized through chemical, biological, and cultural methods to optimize the use of pesticides. The EPA (www.epa.gov/ipm/introduction-integrated-pest-management, accessed in Dec. 2023) indicates that four steps are included: Action thresholds are set, pests are monitored and identified, pests are prevented from becoming a threat and finally, pests are controlled when other means fail.

6.5.1.5 Source Control

A major element in *source control* is reduction in use of various mineral and organic fertilizers and pesticides. These are sources for non-point contaminant stressors when they are transported in the ground. Other source control measures that can be exercised include the following:

- Strict standards and monitoring to maintain ground and surface water quality, and storage systems commonly used in industrial processes – to reduce source contaminants.
- Optimization of natural pollinators and predators to conserve species and ecosystems by maintaining natural vegetation near agricultural lands (Thies and Tscharntke, 1999).
- Optimization of the water supply and quantity by correct use of water and chemicals to protect human health and ecosystems (Matson et al., 1997).
- Use of technologies that reduce erosion, salinization, water consumption, chemical pollution, and other environmental effects.
- Minimization of tillage to prevent run-off.
- Use of intercropping and ground cover as, for example, interplanting with red clover to reduce water runoff, Wall et al. (1991) – with resultant benefits of (a) reducing water and soil loss, (b) conservation of water, (c) decreasing non-point sources of pollution, and (d) increasing water availability for plants (NGS, 1995).
- Minimization of water use through precise applications – e.g., through night application, use of surge-flow irrigation, low-pressure sprinklers, and drip irrigation (Verplancke, 1992; Goldhamer and Snyder, 1989).

6.5.2 Impact of Soil Additives

Various agricultural and industrial wastes are utilized as amendments for agricultural soils. The addition of these amendments as aids to plant growth by and large generates chemical stressors that impact on the applied soil in a way that can threaten human health and other biotic receptors – because of contamination of groundwater and receiving waters and uptake by plants. Irrigation with reclaimed sewage water is practised in arid and semi-arid areas. In Israel, irrigation for 28 years with sewage effluents has led to the accumulation of cadmium, copper, nickel, and lead in the topsoil layer in coastal plain soils (Banin et al., 1981). Significant amounts of these metals accumulated in the oxide fraction of the soils. The amounts recorded show that (a) for Cd the percentages retained varied from 9 to 20%, (b) for Cr , this varied from 5 to 11%, (c) for Cu, the variation was from 16 to 23%, (d) for Ni, this was from 10 to 14%, and (e) for Pb, the variation was from 1 to 3.9%. Approximately 20 to 45% of produced sewage sludge is added for agricultural use in the UK, Germany and US, amongst other countries (Mullins, 1990). These sludges can contain high levels of heavy metals, in addition to N, P, K and other micronutrients. Whilst extensive data are available on the distribution of copper and zinc in sludge-amended soils, the same cannot be said for such heavy metals as cadmium, chromium, nickel, and lead. After more than 10 years of application, copper, chromium, cadmium, and zinc have moved to the soil layer below the layer of application and zinc has been found below the plough layer (Han et al., 2001). Zinc has been shown to be bioavailable and has been taken up by wheat, rice soybean, and maize plants. The degree of uptake depends on the conditions of pH, soil type, and Eh. Heavy metals added via wastes can run-off or seep into the soil, potentially contaminating groundwater or impacting the quality of the food and animals within the food chain. The heavy metals may also impact the soil microorganisms. Sewage sludge and wastewater application to soils can also be a source of pharmaceuticals.

Heavy metals are a major concern in poultry and swine manure used as amendment to soils. Heavy metals in poultry litter, in particular, include As, Co, Cu, Fe, Mn, Se, and Zn (Sims and Wolf, 1994). The accumulation of heavy metals from manure amendment can occur over the long-term. Extractable Cu and Zn concentrations increase over time, in addition to the heavy metal concentrations in run-off after amendment with poultry waste. Cu concentrations have also increased in soil and plants, such as grass and corn, after the addition of swine waste amendment (Kornegay et al., 1976; Mullins et al., 1982; Payne et al., 1988).

Although beneficial, the practice of spreading of manure can negatively impact air, soil, and water quality. Besides providing nutrients, manure addition can reduce soil erosion and improve soil water holding capacity (USDA, 1992). However, excessive levels of nitrogen, phosphorus, and organic matter can accumulate in the soil if the manure is not spread properly (Figure 6.8). In addition, manure can also contain veterinary medicines, which can accumulate in the soil. For example, tetracycline from pig breeding has attained levels of 400 mg/kg in the soil (Kemper, 2008).

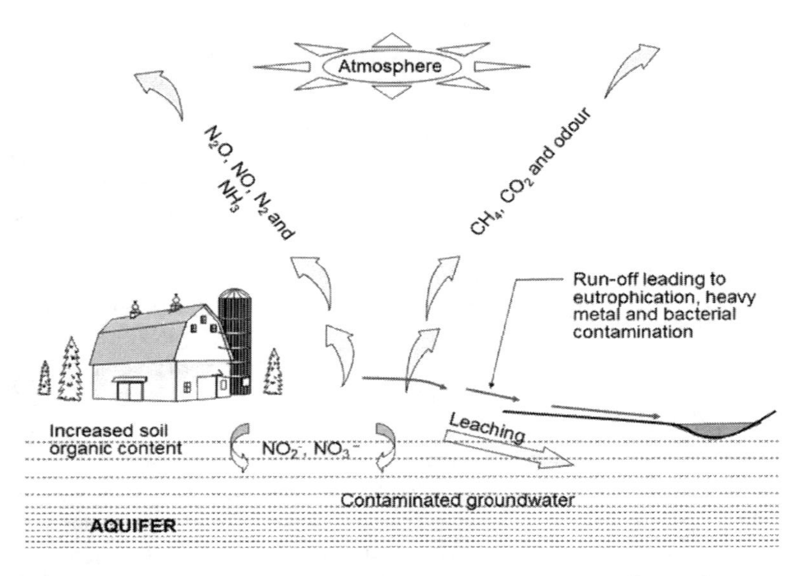

FIGURE 6.8 Illustration of the emissions and leachates due to manure spreading and storage practices.

Heavy metals, such as copper and zinc, may also accumulate in manure and subsequently in the soil due to their use as food additives. In countries, such as the Netherlands, Belgium and Germany, 15, 20, and 50%, respectively, of the groundwater have nitrate levels higher than 50 mg/L, due most likely to manure loading on the land according to the European Commission report on the Nitrates Directive). From 2000 to 2021 average nitrate concentration levels in groundwater in the EU have not changed significantly. In addition, there is no decrease in groundwater monitoring stations with greater than 50 mg/L nitrate concentrations (www.eea.eur opa.eu/en/analysis/indicators/nitrate-ingroundwater-8th-eap).

Large facilities, in particular, do not have sufficient land to apply the manure as a fertilizer. In the US, it has been reported that 1.4 billion tonnes of manure are produced each year and most of that is used on fields (Pagliari et al., 2020). About 5% of all cropland is fertilized by manure according to the US Department of Agriculture Report to Congress (USDA, 2009). Approximately, 50% of lake water and 27% of river water in the US are also contaminated with nutrients (Gleick, 1993). Run-off of the manure during rain events causes high BOD levels in surface waters. Non-point source pollution of surface waters (rivers, lakes, and oceans) and groundwater are considered to be the major source of water pollution.

Coarse-textured soils are particularly susceptible to (a) increased movement of nitrate below the surface, and (b) increased salinity levels and groundwater nitrate concentrations as manure application rates increases. More effective means of spreading fertilizers by correlating with plant needs are required. Plant uptake capacities should not be exceeded.

Han et al. (2001) reviewed the accumulation and distribution of heavy metals in soils amended with animal wastes, sewage sludges, and other wastes (Figure 6.9). For poultry litter amended soil, copper and zinc concentrations increased at a rate of 2 mg/kg-yr. Most of the copper was present in the organic (46%) and residual fractions (52%), whereas most of the zinc was found in the easily oxidizable (48%), and organic (23%) fractions. Mobility studies indicated that there was slight movement of zinc downward in the soil, and copper moved to the 40 cm depth. Zinc was particularly mobile. At depths of 60 cm, iron oxide and residual fractions were twice that of the non-amended soils.

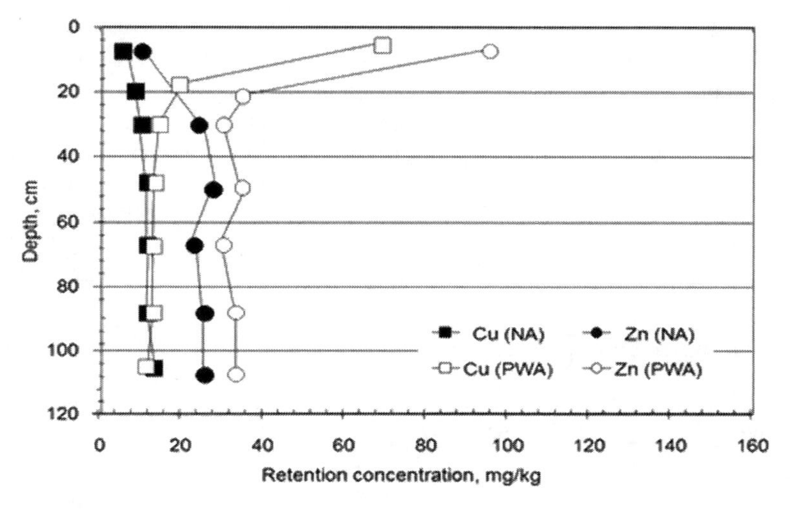

FIGURE 6.9 Retention profiles for total Cu and Zn for non-amended (NA) and poultry-waste amended (PWA). (Data from Han et al., 2001.)

6.5.3 Mitigating Manure Treatment Stressors' Impacts

6.5.3.1 Aerobic Composting

Composting can be used to stabilize manure – particularly storage. This method is preferable over other aerobic treatments that are subject to high operation costs and sludge production. It is, however, labour intensive and costly due to the requirement for aeration. The thermophilic conditions of 54 to 71 °C destroy most pathogens. Most of the ammonia is volatilized early in the process. Losses of nitrogen during the composting process however can be as high as 50% (Thomsen, 2000). Mixing with other substrates such as the bedding is usually required for bulking. Normally, straw or wood shavings are used as bulking agents with adjustments for nitrogen content to be about 1:20 or 1:40. Compared to untreated manure, volatilization and leaching of nitrogen, the risk of pathogen spreading, and odour release are reduced. Yang et al. (2003) reported that (a) greenhouse gas (GHG) emissions of compost were 1.9 times less than slurry manure and 1.5 less than stockpiled manure for dairy manure, (Table 6.2), (b) composted liquid pig manure, in comparison to liquid pig manure was more stabilized in terms of manure carbon, and (c) composted liquid pig manure produced reduced emissions of carbon dioxide and N_2O. Vervoort et al. (1998) determined that composting of poultry litter stabilizes phosphorus and reduced losses by run-off. Others such as Delschen (1999), however, found that although the addition of 2.6 tonnes of composted manure led to higher accumulation rates of soil organic matter (SOM) than untreated manure, the amounts were approximately the same if the 40 to 60% carbon loss during composting is taken into account.

Various composting systems have been described in Mulligan (2002). Production of compost for soil conditioning leads to a more stable product with practically no odour. The simpler processes may also lead to increased ammonia emissions. The selected bulking material can influence ammonia emissions. Emissions of methane and nitrous oxide depend on the aeration rates. The benefits of compost for soil conditioning are well known.

6.5.3.2 Anaerobic Digestion

The organic content of manures can be treated by anaerobic digestion. The methane produced can be used for fuel or electricity production. Treatment of the manure by anaerobic digestion can significantly reduce the impact on water resources. Greenhouse gas (carbon dioxide, in particular) emissions are reduced, and the products have improved fertilizer capability. Nitrogen and phosphorus availability for crops are increased, reducing chemical fertilizer requirements. A comparison of the mineralization of N of anaerobically stored manure with composted ruminant manure showed that anaerobic manure loses less nitrogen than during composting and therefore is a better source of inorganic N for fertilizer (Thomsen and Olesen, 2000). Only some of the nitrogen in anaerobic residues is organically bound, whereas most of the nitrogen in compost is in this form. It has been estimated that digestion can reduce greenhouse emissions by 1.4 kg of carbon dioxide per kg of volatile solids (VS) in manure. It has been reported that anaerobic treatment reduces N_2O

TABLE 6.2
Comparison of Anaerobic Digestion Residue and Compost for Nitrogen Emissions

Treatment for Waste	Global Warming (ton CO_2-equiv/year)	Acidification (ton SO_2-equiv/year)	Eutrophication (ton O_2-equiv/year)
Anaerobic digestion	654	43	688
Composting by reactor	2618	20	991

Source: Adapted from Dalemo et al. (1998).

emissions by more than 50% due to VS reduction after spring application onto soil – in comparison to untreated manure (Sommer et al., 2004). Although pathogens are reduced, the inactivation may only be about 1 to 2 log at 30 °C (Burton and Turner, 2003).

In 2010 in Germany, there are approximately 1000 anaerobic plants using agricultural substrates, now there are more than 9000 according to the Agency for Renewable Resources, FNR (2019). The combination of government and environmental factors has led to the substantial growth in the numbers used on animal farms. The EU Waste Framework Directive requires that all member states recycle 50% of their municipal waste by 2020. Although all types of manure are digestible, cow manure is more difficult to digest because of the higher fibre content – as opposed to pig and poultry manures. Typically, yields of methane are in the order of 290 L per kg of volatile solids (VS) for pig manure and 210 L of methane per kg of cattle manure (Burton and Turner, 2003). Co-substrates are often used to enhance the carbon and nutrient contents. These substrates include fodder beet and other green wastes. Cow manure in particular is well suited to co-digestion as it already contains high fibre content. The nitrogen content of the manure serves as pH buffering and as a continuous inoculum. The co-substrates also increase methane yields (Mulligan, 2002).

Lagoons have lower capital and operating costs than digestors. Organic matter is reduced while nitrogen and phosphorus remain in the end product. The use of lagoons is more frequent in warmer climates. Proper sizing and management of lagoons are required to ensure odour control. Loading rates of 60 to 90 g of VS per m^3 of lagoon per day have been suggested by the Natural Resources Conservations Service in the North East of the US. Since methane is a greenhouse gas, gas collection should be practised to avoid release into the atmosphere. Odours may also be a problem. Floating covers are becoming more popular. Temperatures of 35 °C are required to maintain optimal biogas production. Heating systems may be required, particularly in the winter. Pathogen reduction in lagoons is also minimal (Burton and Turner, 2003). Some work has been performed to reduce ammonia volatilization for ambient temperature systems such as in-storage-psychrophilic-anaerobic-digestion (ISPAD) (Madani-Hosseini et al., 2016).

6.5.3.3 Wetlands

The liquid portion of effluents from farm operations must usually be treated because of their high nutrient and organic contents. Since the wastewater contains a higher concentration of soluble ammonia than the initial feed, land spreading is a viable option since this allows for easier uptake by plants. Algal ponds or constructed wetlands (as previously discussed in Chapter 5 for mining applications) may also be used to improve water quality to allow water reuse. Constructed wetlands are operated as subsurface flow or free surface flow (Mulligan, 2002). Nutrients are removed by the plants growing in the wetlands. Subsurface systems are not as efficient for nutrient removal. Hammer et al. (1993) has reported on a two-cell surface system used for a 500 swine operation to reduce 90 kg BOD per day to 36 kg BOD per day in a 3600 m^2 wetland. Nitrogen loading rates should be from 3 to 10 kg/ha-day, and ammonia concentrations in the influents should not be higher than 100 to 200 mg/L. In general, N, P and solid reductions should be greater than 50% and BOD greater than 60% if the wetlands are not overloaded. Wetlands offer a method of enhancing biodiversity whilst treating liquid effluent discharges. Some laboratory tests have been performed on the uptake of six pharmaceuticals by *Lemna sp.* and *Spirogyra sp.*, two aquatic plants found in polishing ponds (Garcia-Rodriquez et al., 2015). The plants could uptake between 31 to 100% of these pharmaceuticals showing potential for phytoremediation but more research is needed as there is not sufficient information available.

6.5.3.4 Integrated Manure Treatment

A comparison of manure management methods is shown in Table 6.3. The choice of the most appropriate method for management of manure must include minimization of emissions and other impacts

TABLE 6.3
Comparison of the Sustainability of Manure Treatment Methods

Treatment	Energy	Emissions	Products	Costs
Land spreading	None produced, energy for spreading	Ammonia, methane, N_2O, odor, leaching of N, P, bacteria	Fertilizer	Low
Composting	Energy for mixing and mechanical separation, heat generation	Carbon dioxide, ammonia losses higher than anaerobic	Compost for soil improvement	Depends on sophistication
Anaerobic digestor	Energy produced in form of methane; heat required for the digestor	High BOD liquid effluent	Methane for electricity or energy	Depends on sophistication
Anaerobic lagoon	Energy produced in form of methane	Odors, N_2O and methane, if not covered well	Methane for electricity or energy	Low

on the environment, such as decreased water, groundwater, and soil quality. Dalemo et al. (1998) utilized a simulation model (ORWARE) for the calculation of energy and nutrient flows from soil and liquid organic wastes from restaurants. Composting and anaerobic digestion were compared. A life-cycle approach was used that included the process and the soil emissions from the product. Soil emissions were in the form of N_2O production. Since the content of ammonia is lower in the compost, emissions of ammonia in the soil are also lower in comparison to anaerobic digestion residues where the organic nitrogen is mineralized. In respect to concerns for eutrophication: (a) Emissions are mainly in the form of nitrate leaching from the soil, (b) ammonia is released during composting, and (c) NOx is released during combustion of biogas for electricity generation. In short, composting produces ammonia emissions during the composting process – in addition to NOx and SOx.

Various models were compared by Hansen et al. (2006). They included DST (Decision Support Tool, USA), IWM (Integrated Waste Management Tool, UK), IFEU (Germany), ORWARE (Sweden) and EASEWASTE (Environmental Assessment of Solid Waste Systems and Technologies (Denmark) (Varma et al., 2021). The last three are life-cycle assessment models, whereas DST considers water BOD and IWM includes reduced air emissions only. A case study comparison indicated that different assumptions, wastes, and local conditions can highly affect the obtained results. In addition, DST and IWM did not provide sufficient information to allow decision support for organic waste land application (Hansen et al., 2006).

The integrated manure management system shown in Figure 6.10 takes advantage of the less severe impact of anaerobic digestion process for primary treatment and subsequent composting for final treatment and disposal. As shown in the diagram, combined wastes (green and manure) are sent to the anaerobic digestion system for treatment. This reduces the VS content. The digestate is then removed from the digestors for solid/liquid separation by screw press or other means. The solids are sent for composting while the liquid is sent for wastewater treatment. After biological treatment (wetlands or a reactor if space is limited), the water can be stored and reused as required or spread on the land surface. The remaining solids' content is ideal for composting. The requirements for oxygen will be decreased due to the previous anaerobic digestion step. This step also requires energy that can be produced by the anaerobic digestion process. The resulting compost would be high-grade compost full of nutrients with adequate pathogen reduction. The application rates to the soil would need to be based on N and P crop requirements to avoid excess soil nutrients (Cooperbrand et al., 2002).

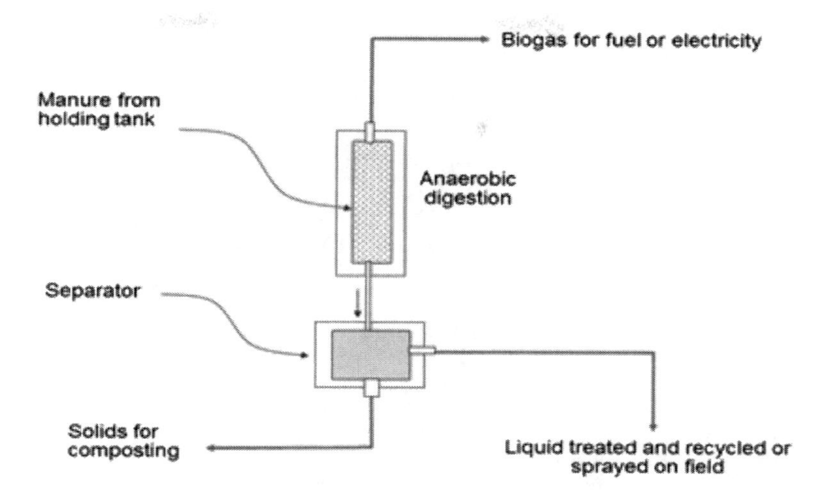

FIGURE 6.10 Schematic diagram showing flow sequence for sustainable management of manure.

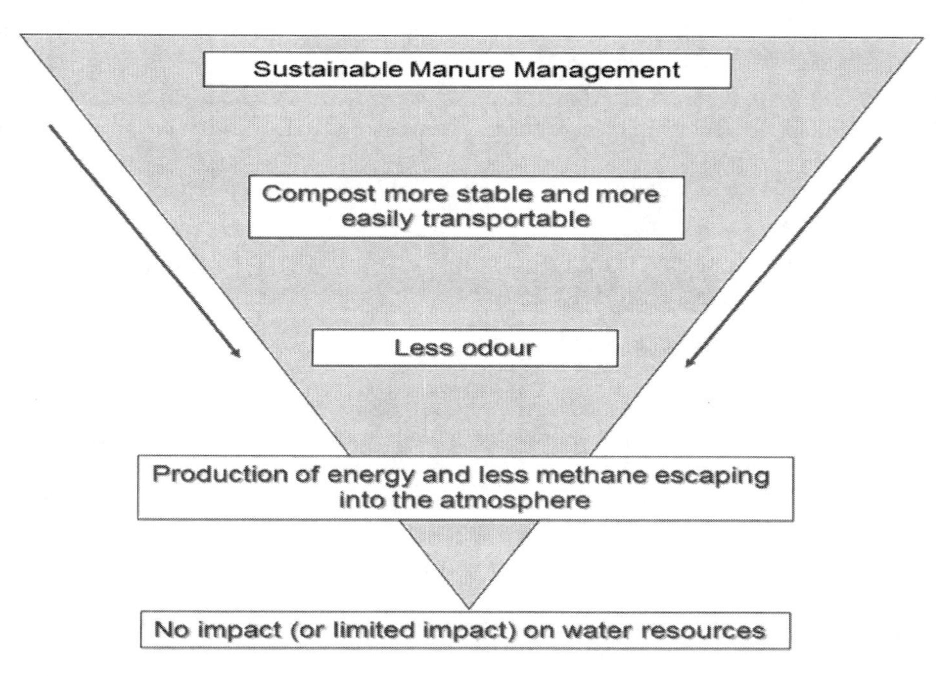

FIGURE 6.11 Impact of a sustainable manure management process on the geoenvironment.

Land spreading of manure creates numerous environmental problems in the agroecosystem. Manure treatment by anaerobic digestion is a step in the direction towards sustainable agricultural practice since it is a renewable source of energy. The main benefits are shown in schematic form in Figure 6.11. Manure treatment enables farmers to reduce (a) pathogens, odour, N_2O and carbon dioxide emissions, and (b) air, soil, and groundwater pollution caused by manure spreading. As anaerobic digestion is not a complete solution, an integrated one including composting to produce a soil conditioner will be required to ensure complete management of all aspects of manure

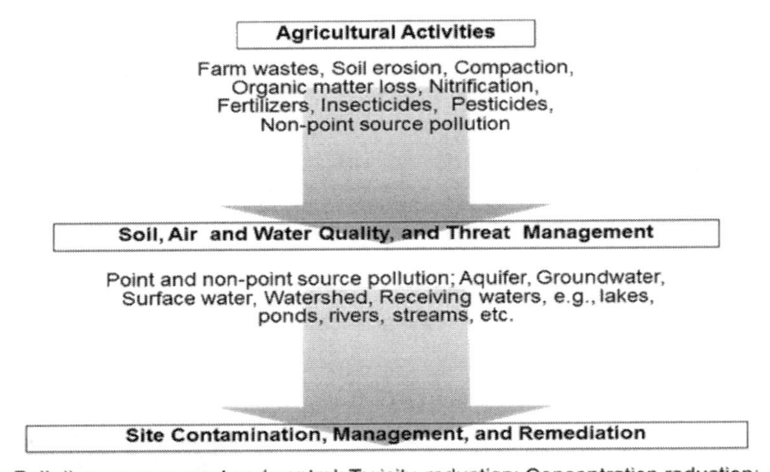

FIGURE 6.12 Summary of agricultural activities, their impacts, and minimization of the impacts.

treatment – as shown in Figure 6.12. Modelling of the emissions based on C, N, and P mass balances is an effective method to compare manure management methods. However, soil conditions, climate and other factors can significantly influence the results of the comparison. This approach is a clear example of minimization of the impact of agricultural practices shown in Figure 6.12.

6.6 TOOLS FOR EVALUATION OF GEOENVIRONMENT IMPACTS FROM FARMING STRESSOR SOURCES

6.6.1 Agricultural Sustainability

The legal definition of sustainable agriculture according to U.S. Code Title 7, Section 3103, is

> "an integrated system of plant and animal production practices having a site-specific application that will over the long-term:
>
> - Satisfy human food and fiber needs.
> - Enhance environmental quality and the natural resource base upon which the agriculture economy depends.
> - Make the most efficient use of nonrenewable resources and on-farm resources and integrate, where appropriate, natural biological cycles and controls.
> - Sustain the economic viability of farm operations.
> - Enhance the quality of life for farmers and society as a whole".

To avoid and/or mitigate risks and protect soil quality and geoenvironmental quality as a whole, we need to develop a methodology for evaluation of potential and existing risks to agricultural and geoenvironment sustainability as a result of agricultural activities. It has been estimated that agricultural practices in Europe have led to 80% of the degradation the soil (www.theconsciouschallenge.org/ecologicalfootprintbibleoverview/agriculture-soil-degradation, accessed Dec. 2023). There appears to be some significant room for improvement. One needs to (a) examine the impact of the farming practices on the quality of groundwater and other receiving waters, soil quality, and biodiversity, (b) seek methods to quantify the various kinds of emissions that take the form of stressors to the geoenvironment, and (c) determine the fate and transport pathways of the harmful emissions (stressors). By including air, water, soil, and vegetation as part of the agroecosystem, one could seek measures in an integrated approach to (a) limit the harmful discharges, and (b) curtail inefficient practices or emissions – as the first major step towards generation of sustainability. Some of these measures include (a) ecological footprint, (b) sustainable processing index (SPI), and (c) material intensity per service unit (MIPS) or land intensity per service unit (LIPS) (Quendler et al., 2002). Models and sustainability indicators are also tools that can be used to evaluate sustainability. The analysis of sustainability thus will rely on sustainability indicators, reference values, and an established evaluation method. A listing of some indicator systems has been reported by the OECD (2013).

Attempts have been made to define agricultural sustainability. According to the Brittanica, "Sustainable agriculture, a system of farming that strives to provide the resources necessary for present human populations while conserving the planet's ability to sustain future generations" (Dubey, 2023).

One of the most inclusive definitions was formulated by Christen (1996) as depicted schematically in Figure 6.13. Ethical considerations include fairness between generations. Resources and biodiversity need to be preserved without reducing production but by minimizing environmental impact. Economic viability must also be ensured both for small and large enterprise farming units. Since soil will always be used for agricultural purposes, the impact of pollution and erosion must be determined through research and field studies.

Some authors have argued that the definitions of organic farming (Quendler and Shuh, 2002, Gamage et al., 2023) follow the same principles as those used to establish sustainable agriculture. These principles include the following:

- Working with a closed system to draw on local resources.
- Maintaining the fertility of the soil over the long-term.
- Avoiding all forms of pollution.
- Producing high nutritional quality food in sufficient quantities;

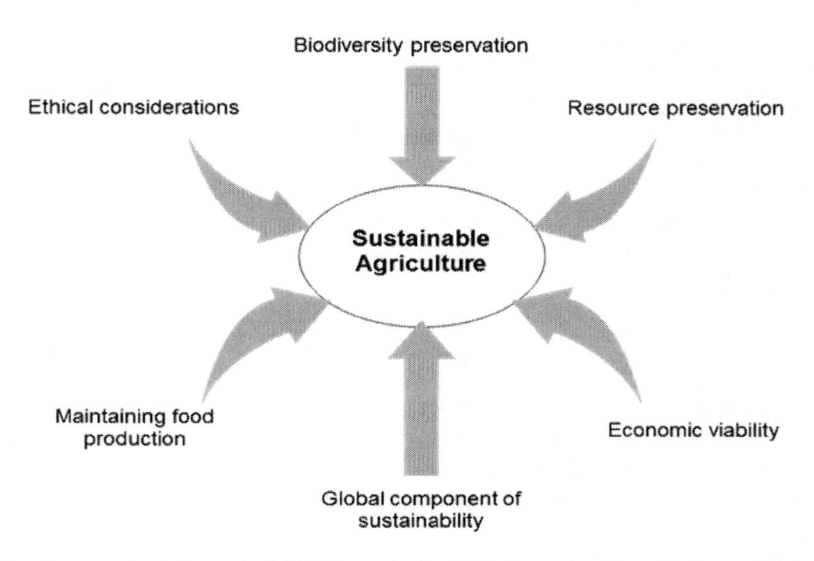

FIGURE 6.13 Components of sustainability for agriculture. (Information from Christen, 1996.)

- Reducing the use of fossil fuels to a minimum.
- Giving livestock humane conditions.
- Excluding synthetic fertilizers, hormones, chemical fertilizers, pesticides, growth hormones, and feed additives.
- Making it possible to allow agricultural producers to earn a proper living.
- Using appropriate technologies for biological systems and decentralized systems for product processing, distribution, and marketing.
- Creating aesthetically pleasing systems to all.
- Preserving and maintaining wildlife and their habitats.

The Food, Agriculture, Conservation and U.S. Trade Act of 1990 established that practice of sustainable agriculture is to

> satisfy human food and fibre needs; enhance environmental quality and the natural resource based upon which the agricultural economy depends; make the most efficient use of non-renewable resources and on-farm resources and integrate where appropriate natural biological cycles and controls; sustain the economic viability of farm operations and the quality of life for farmers and society as a whole.

Canada's International Institute for Sustainable Development agrees with this but says that organic farming is not the only way to achieve sustainable agriculture. Technology needed to achieve sustainability is minimized in organic farming.

The Sustainable Agriculture Network (SAN) (www.sustainableagriculture.eco/sustainable-agriculture-framework-2021, accessed Jan. 2024) has established a framework with ten areas of impact. Impacts and indicators are listed for each. For climate change mitigation and adaptation, SAN's sustainable agriculture approach is distributed along all sections of the SAN-SAF and focuses on building resilient agroecosystems and reducing the carbon footprint of agricultural and livestock activities, by various practices.

Related to water sustainability for agriculture, a report was prepared for the G20 Presidency of Germany by FAO (2017b). Various recommendations were made including the following:

- Modernizing schemes for irrigation.
- Improving for water supply systems.
- Improving water productivity.
- Reducing food waste and losses.
- Improving international import and export restrictions.
- Improving water governance and improving international initiatives along the water-food-energy nexus.
- Promoting data and information systems for water, including accounting and auditing systems.
- Improving communication of conditions of water scarcity to all concerned.
- Support and engage in international initiatives such as COP 22.

6.6.2 Development of Analytical Tools

A pertinent question that can be posed is: "What methods can we use to determine the impact of the practices on the geoenvironment, and how can we measure and optimize the improvements of these practices so as not to adversely impact the geoenvironment?" The development of predictive models and sustainability indicators to measure sustainability progress is an ongoing process to accomplish the task needed to answer the question posed.

We have seen in Chapter 2 that soil can provide means to retain contaminants by natural attenuation of both organic and inorganic contaminants. This subject is discussed in greater detail in

Chapter 10. For example, in Germany, levels of nitrate as high as 250 mg/L, aluminium as high as 0.64 mg/L, and potassium up to 60 mg/L have been found in the groundwater in agricultural areas (Houben, 2002). Acid rain has decreased the pH of the soil to 2.75 and that of the groundwater to 3.4, and soil buffering capacity had also been diminished. Cation exchange, autotrophic denitrification (reaction of nitrate with FeS_2), and other natural attenuation mechanisms have restricted the movement of the contaminants. Modelling, in the Houben (2002) study, together with determination of the age of the groundwater, mass balances, and reactive transport were undertaken using hydrochemical and geochemical data. The PHREEQC-2 geochemical model was used for the hydrochemical equilibrium modelling. Sorption and desorption column experiments with undisturbed samples of sandy sediments for magnesium, sodium, potassium, and aluminium ions were performed. Modelling was accurate for most ions with the exception of potassium. Competition cannot be accounted for in the mass balance approach. Due to the high velocity of fluid flow in the columns, there was insufficient time for the nitrate to react with the pyrite. The models indicated that the contaminants move a few centimetres per year. A newer version PHREEQC-3 is now available since 2021 (www.usgs.gov/software/phreeqc-version-3/, accessed Jan. 2024). New features were added from experimental results of laboratory and field studies. The code also now can be used by other software programs for calculation of chemical reactions or distributions.

A number of models have been developed to integrate scientific information to enable policy development and future management practices. Integrated assessment modelling, in particular, has been applied to determine the impacts and to predict climate change. The California Agricultural Land Evaluation and Site Assessment (LESA) is used to evaluate soil resource quality, using the indicators of the project's size, water resource availability, surrounding agricultural lands, and surrounding protected resource lands. For a given project, the factors are rated, weighted, and combined, resulting in a score that determines project's potential significance (www.nrcs.usda.gov/conservation-basics/natural-resource-concerns/land/evaluation-and-assessment, accessed Jan. 2024).

The Global Change Assessment Model (GCAM) is an integrated assessment model linking energy, agriculture, and land use with a climate model (https://gcims.pnnl.gov/modeling/gcam-global-change-analysis-model, accessed Jan. 2024). In 5-year time steps from 1990 to 2100, the model can assess various climate change policies and technology strategies for the globe over long time scales in 14 geographic regions. Emissions and atmospheric concentrations of greenhouse gases (CO_2 and non-CO_2), carbonaceous aerosols, sulphur dioxide, and reactive gases are estimated with the associated climate impacts, such as global mean temperature rise and sea level rise.

Other models focus on predicting contaminant transport and leaching in the soil. According to the U.S. Department of Agriculture, Agricultural Research Service (ARS), Great Plains Systems Research (Fort Collins, Colorado), the Nitrate Leaching and Economic Analysis Package (NLEAP) is a field-scale computer model developed to provide a rapid and efficient method of determining potential nitrate leaching associated with agricultural practices It combines basic information concerning on-farm management practices, soils, and climate and then translates the results into projected N budgets and nitrate leaching below the root zone and to groundwater supplies, and estimates the potential off site effects of leaching. The NLEAP model was designed to predict leaching of nitrate. The processes modelled include movement of water and nitrate, crop uptake, denitrification, ammonia volatilization, mineralization of soil organic matter, nitrification, and mineralization – immobilization associated with crop residue, manure, and other organic wastes. It can be used with various GIS systems in the newest version 5.0 to identify key areas and management practices to increase N use efficiency (Delgado et al. 2010) and is found at https://data.nal.usda.gov/dataset/nleap-gis-50.

The Soil and Water Assessment Tool (SWAT) has been developed to model changing land use patterns and practices on nitrogen and phosphorus movement to surface and groundwaters. SWAT+ is now available since Aug. 2023. It is found at (https://swat.tamu.edu/software/plus/) a completely revised version of the model. Even though the basic algorithms used to calculate the processes in the model have not changed, the structure and organization of both the code (object based) and the input files (relational based) have undergone considerable modification. SWAT+ provides a more flexible spatial representation of interactions and processes within a watershed. SWAT is applicable for small watersheds to river basin-scale models for the simulation of the quality and quantity of surface and groundwater and predict the environmental impact of land use, land management practices, and climate change, for assessing soil erosion prevention and controlling non-point source pollution control and regional management in watersheds. Sediment transport, crop growth, and nutrient cycling are simulated. It has also been used for pesticide transport (Rekolainen et al., 2000). It is integrated into ArcView geographic information systems (GIS) software and was developed by the USDA Agriculture Research Services.

A review by Yuan et al. (2020) accessed 14 different models including Long-Term Hydrologic Impact Assessment (L-THIA) and Nonpoint Source Pollution and Erosion Comparison Tool (N-SPECT/OpenNSPECT), Generalized Watershed Loading Function (GWLF), Loading Simulation Program C (LSPC), Source Loading and Management Model (SLAMM), and Watershed Analysis Risk Management Frame (WARMF), Agricultural Nonpoint Source pollution model (AGNPS/AnnAGNPS), Soil and Water Assessment Tool (SWAT), Stormwater Management Model (SWMM), Hydrologic Simulation Program Fortran (HSPF), Automated Geospatial Watershed Assessment Tool (AGWA), Better Assessment Science Integrating Point and Nonpoint Sources (BASINS) and Watershed Modeling System (WMS). The strengths and weaknesses of each are presented.

The Groundwater Loading Effects of Agricultural Management Systems (GLEAMS) (www.ars.usda.gov/plains-area/temple-tx/grassland-soil-and-water-research-laboratory/docs/gleams/#:~:text=Groundwater%20Loading%20Effects%20of%20Agricultural%20Management%20Systems%20%28GLEAMS%29,field%20has%20homogeneous%20land%20use%2C%20soils%2C%20and%20precipitation) (accessed Jan. 2024) was developed to determine the effect of various management practices on pesticide and nutrient leaching at, through and below the root zone. The model simulates the downwater movement of pesticides via percolated water, run-off, and sediment or upward movement of pesticides via plant uptake by evaporation and transpiration. More pesticides have been incorporated into the database and other updates have been performed.

PRZM3 is the most recent version (version 3.12.3 was released June 2006) of a modelling system developed by the EPA that includes PRZM and VADOFT, to predict pesticide transport and transformation down through the crop root and unsaturated zone (Suárez, 2005). PRZM is a one-dimensional, finite-difference model that predicts the fate of pesticides and nitrogen in the root zone of crops. It can also predict pesticide concentration in run-off water and solid particles. The latest version (PRZM3) includes the capability to simulate soil temperature, volatilization, and vapour phase transport in soils, irrigation simulation, and microbial transformation, in addition to the transport and transformation of the parent compound and up to two daughter species. The input parameters include the characteristics of the pesticide, the application of the pesticide, crop, climatic, and site information (soil and hydrological properties, agricultural practices, topography, etc.). VADOFT is a one-dimensional, finite-element code that solves Richard's equation for flow in the unsaturated zone. Several versions of the PRZM system are available on the EPA web site (www.epa.gov/ceam/przm-version-index). PRZM can also be used with the EXAMS model to simulate the fate and transport of the pesticide in water. Volatilization, sorption, hydrolysis, biotransformation, and photolysis are processes that are included in the model structure. An aseptic system module is available for evaluating nitrogen fate and transport.

LEACHM is a suite of models for simulating the leaching and fate of water and chemicals within the soil (Hutson and Wagenet, 1989) (http://datadiscoverystudio.org/geoportal/rest/metadata/item/4fd8877448d54a76bc126737eb83f21d/html#:~:text=LEACHM%20%28Leaching%20Estimation%20and%20Chemistry%20Model%29%20refers%20to,Atmospheric%20Sciences%20at%20Cornell%20University%2C%20Ithaca%2C%20New%20York, accessed Jan. 2024). Input data are similar to the PRZM-EXAMS model. One or more growing seasons can be simulated. Output includes a profile of the pesticide concentration throughout the soul and the water and pesticides fluxes in the groundwater.

The computer model Holos has been developed by Agriculture and AgriFood Canada. It is used to estimate the greenhouse gas (GHG) emissions of carbon dioxide, nitrous oxide and methane emissions. The emissions can originate from enteric fermentation and manure management, cropping systems and energy use. Carbon storage and loss from lineal tree plantings and changes in land use and management can also be estimated so that the user on the farm can identify ways to reduce farm emissions. The Holos software is continually being updated with new data and improved features. Holos 4.0, is the most recent version. (https://agriculture.canada.ca/en/agricultural-production/holos, accessed Jan. 2024).

6.7 INDICATORS OF AGROECOSYSTEM SUSTAINABILITY

Although crop yields are a measure of sustainability, they do not provide any indication of the impact on the ecosystem and the geoenvironment or land environment in particular. Monitoring of ecosystem damage must be determined, particularly to differentiate natural changes with that due to human activity. GIS, remote sensing and landscape ecology are new measuring approaches to indicate land use changes. Indicator organisms can be monitored for changes in the ecosystem. Various organizations such as the Organization for Economic Co-operation and Development, the Food and Agriculture Organization, the World Bank, and the Commission of Sustainable Development have published various indicators for agriculture related to the economic, social, environmental processes, farming practices, and environmental impacts.

For example, the UK Ministry of Agriculture, Food and Fisheries (MAFF) provides a list of pilot indicators to determine if development is improving in sustainability. The indicators are grouped in 21 families. Those related to agriculture are shown in Table 6.4. Soil quality is included as it is deemed vital for food production and an ecosystem for vital organisms. Statistical data sheets for the various indicators are found at www.gov.uk/government/statistical-data-sets/agri-environment-indicators (accessed Jan. 2024). Concentrations of organic matter, acidity, nutrient concentration (P and K), and heavy metals are the parameters included. This varies somewhat from the indicators in Canada, that were initially developed by McRae et al. (2000) and are now modified and expanded to include new indicators that are shown in Table 6.5 (Eilers et al., 2010). Indicator reporting has been done since 1993, but data has been compiled from 1981 (https://agriculture.canada.ca/en/environment/resource-management/indicators, accessed Dec. 2023). The data shows some of the environmental factors impacting agriculture and how they change over time. Some improvements, such as soil quality, are noted over time but others such as water quality are of concern due to intensification of cropping and livestock production. As an example, residual soil nitrogen levels in 1981 on farmland in Canada were in the "Desired" risk class. More recent data in 2016 showed a "Good" risk class, which is the result of nitrogen input increases by fertilizers, particularly in the 1990s.

Pesticide indicators have also been developed by the OECD (2002). Pesticide use and pesticide risks are the two indicators. Most countries have decreased the use of pesticides. Reduction in risks to human health and the environment can be achieved by reducing particular pesticides. For water use, the three indicators developed include (a) intensity of water use, (b) water volume consumed, and (c) economic value of water use. Water use is very high for many OECD countries. Technical and economic efficiency information is difficult to obtain, as well as water stress caused by diversion of water from rivers for agricultural use. The geoenvironmental impacts from agricultural practices consider impacts on (a) soil and water quality, (b) land conservation, (c) production of greenhouse gases, (d) biodiversity, (e) wildlife habitats, and (f) landscape. Soil quality was evaluated based on risk of water and wind erosion. These were considered of higher concern than soil compaction or salinization. Overall, about 10% of agricultural land is at risk to erosion. Conservation or no tillage of land, less intense crop production and retiring lands can reduce the effects of soil degradation.

TABLE 6.4
Sustainability Indicators Developed by the UK

Area	Indicator
Agriculture, rural economy, society	
Structure of the agriculture industry	Agricultural assets and liabilities
	Age of farmers
	Percentages of holdings that are tenanted
Farm financial resources	EU Producer Support Estimate (PSE)
	Agri-environment payments to farmers
	Total income from farming
	Average earnings of agricultural workers
Agricultural productivity	Agricultural productivity
Agricultural employment	Agricultural employment
Farm management systems	
Management	Adoption of farm management systems
Organic farming	Area converted to organic farming
Codes of practice	Knowledge of codes of agricultural practices
Inputs	
Pesticide use	Pesticides in rivers and groundwater
	Quantity of pesticide active ingredients used
	Area treated with pesticides
	Pesticide residues in food
Nutrients	N&P losses from agriculture
	P levels due to agriculture in soil
	Manure management
	Ammonia emissions
Greenhouse gas emissions	Emissions of methane and NOx from agriculture
Energy	Direct consumption by farms
Resource use	
Water	Trends in energy inputs to agriculture
Soil	Use of water for irrigation
	Organic matter in soil
Agricultural land	Heavy metals in topsoils
	Area of agricultural land
Non-food crops	Change in land use from agriculture to hard development
	Planting of non-food crops
Conservation value of agricultural land	
Environmental conservation	Area of agricultural land committed to environmental conservation
Landscape	Characteristic features of farmland
Habitats	Areas of cereal field margins under environmental management
	Area of semi-natural grassland
Biodiversity	Populations of key farmland birds

Source: Adapted from MAFF (2000).

The OECD reviewed all indicators worldwide and issued a compendium of indicators OECD (2013) that shows there is evidence that OECD countries from 1990 to 2010 have displayed positive signs regarding nutrient, pesticide, energy, and water management in terms of inputs per unit volume of output. Environmentally beneficial practices are being utilized more by farmers, such as conservation tillage, improved manure storage, soil nutrient testing, and drip irrigation (https://doi.org/10.1787/9789264186217-en, accessed Feb. 2024).

TABLE 6.5
Canadian Agro-Environmental Indicators (https://agriculture.canada.ca/en/environment/resource-management/indicators, accessed Jan. 2024)

Indicator Group	Sub-Group	Brief Description
Soil health	Soil organic carbon	Change in soil organic content
	Erosion risk	Soil redistribution due to tilling and cropping
	Soil salinization risk	Increases in soil salinity due to land use, hydrologic, climate and soil conditions
Water quality	Nitrogen contamination risk	Increase in nitrogen levels in water leaving farm
	Phosphorus contamination	Increase in phosphorus levels in water leaving farm
	Coliforms	Increase in coliform levels in water leaving farm
	Pesticides	Increase in pesticide levels in water leaving farm
Air quality	Agricultural greenhouse gas budget	Estimates of N_2O, CH_4, CO_2 due to agriculture in CO_2 equivalents, ammonia emissions and particulate matter in air
	Ammonia	
	Particulate matter	
Biodiversity	Soil cover	Coverage by vegetation, crop residue, or snow
	Wildlife habitat	Amount of habitat for wildlife on the farm

6.8 CONCLUDING REMARKS

The geoenvironment associated with agriculture plays a very important role in such activities and events as carbon and nutrient cycling, climate change, maintaining ecosystem biodiversity and water-groundwater quality, and pollution management. As has been stressed previously, the discussion concerning the impacts from agricultural practices in production of food are motivated primarily in respect to concerns for protection of the geoenvironment. Accordingly, the impacts and the means to mitigate impacts are all viewed from a geoenvironmental perspective. Many of these impacts and the means to alleviate and eliminate them are well known to the agricultural/farming community, and that they share the same concern as those involved with geoenvironmental management. There is validity to the thesis that many of the measures for mitigating, minimizing, and even eliminating impacts from contaminant loading and soil quality impairment run counter to intensive agricultural practices. Conflicts between agricultural productivity and protection of the geoenvironment have always existed. However, if sustainability and/or preservation of the quality of the geoenvironment are goals that have merit, it is necessary to seek sustainable practices for both agricultural production and the geoenvironment.

Whilst soils have been exploited for agricultural and livestock purposes, they are very vulnerable to environmental stresses and are renewable at very slow rates (over centuries). They are at the interface of the lithosphere, hydrosphere, atmosphere, and biosphere. Aquifers are subjected to overpumping, subsidence, and contamination. Excessive extraction leads to subsidence, erosion, aridification, and salination. Contamination from farm wastes and other activities reduces the use of the water and can cause health problems. Lack of water will have a significant impact on agriculture. Conservation of water, energy, and soil resources are required and new technologies for agricultural practices such as irrigation are needed. Recycling of crop residues and other wastes is another area where considerable development is required.

Agricultural reshaping of the land leads to erosion and other soil displacements. It changes the landscape and causes disruptions to the ecosystems and the environment overall. More than 99% of the world's food comes from the land ecosystems and more than 70% of the fresh water is used for agriculture. Lack of available and good quality water and soil limit food production today and will be even more strained as the population increases.

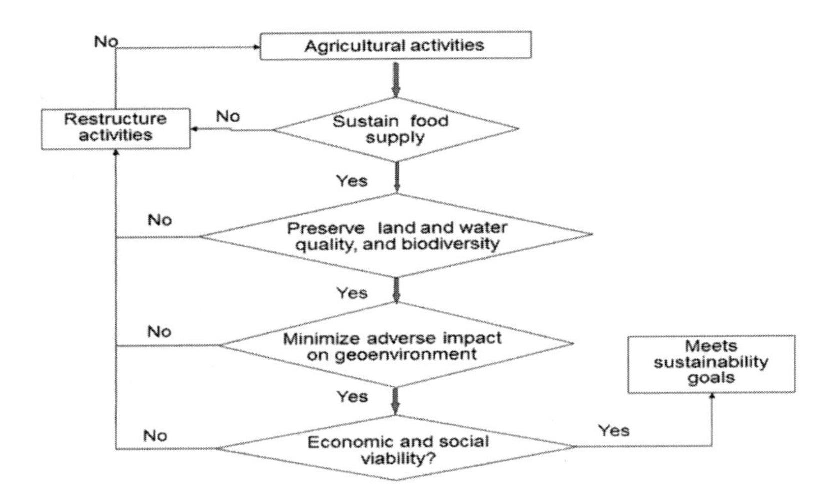

FIGURE 6.14 Evaluation of the sustainability of agricultural activities.

Many contaminants are discharged by diffuse means from farming activities and are difficult to control. Water run-off and infiltration transports the contaminants to groundwater, lakes, rivers, and marine areas. The sediments are subsequently responsible for trapping many of these contaminants. Further investigations are required to better understand the retention of the contaminants by plants and soil, and the chemical and microbial reactions governing the fate of the contaminants. Long-term monitoring is needed to develop databases for predictive modelling and to evaluate the sustainability of the various agricultural practices. An integrated approach of mitigation methods and reduction of the contaminants at the source are required to reduce the impact on the surface and groundwater. Multidisciplinary efforts and those involving the public are required to increase public knowledge and responsibility. To evaluate the sustainability of agricultural activities, a process as illustrated in Figure 6.14 will need to be followed.

If total costs involved in the production of food were to be determined, one would need to include not only the cost of physical operations in tilling and reaping the harvest, but also the cost of all the items needed to mount the effort in production. This would include the costs for production of the fuel needed to operate the machinery, production costs for the fertilizer and pesticides, etc. Expressing these all in terms of calories of energy, Wen and Pimentel (1990) calculated that more than 10 million kcal of energy are required for the operation of agricultural machines, production of fuel, fertilizer and pesticide, and for irrigation and other inputs for 1 hectare of US corn. Approximately 3–5% of total energy is used for the agricultural sector for OECD countries according to the FAO (2000). This is a useful basis for determining efficiency of operation and production of food. Government and non-governmental organizations, the scientific community, and farmers must all work together as outputs in the order of two or three times the present levels must be achieved by 2050 (NRC, 1999).

REFERENCES

Agency for Renewable Resources, FNR (2019), Bioenergy in Germany: Facts and Figures 2019, bioenergie.fnr.de, accessed Aug. 2024.

Arsenault, C. (2014), Only 60 Years of Farming Left If Soil Degradation Continues. www.scientificamerican.com/article/only-60-years-of-farming-left-if-soil-degradation-continues/, accessed Dec. 2023.

Banin, A., Navrot, J., Noi, Y., and Yoles, D. (1981), Accumulation of heavy metals in arid-zone soils irrigated with treated sewage effluents and their uptake by Rhodes grass, *Journal of Environmental Quality*, 10: 536.

Barbash, J.E. (2005), The Geochemistry of Pesticides. In: B. Sherwood Lollar (ed.), *Environmental Geochemistry*, Elsevier, Amsterdam, pp. 541–612.

Bell, R.L., Kase, J.A., Harrison, L.M., Balan, K.V., Babu, U., Chen, Y., Macarisin, D., Kwon, H.J., Zheng, J., Stevens, E.L., Meng, J., and Brown, E.W. (2021), The persistence of bacterial pathogens in surface water and its impact on global food safety, *Pathogens,* 10(11): 1391, doi: 10.3390/pathogens10111391. PMID: 34832547; PMCID: PMC8617848.

Bernard, C.C., Carbiener, R., Cloots, A.R., Groelicher, R., Schenck, Ch., and Zilliox, L. (1992), Nitrate pollution of groundwater in the Alsatian plain (France) - A multidisciplinary study of an agricultural area, *Environmental Geology & Water Science*, 20: 125–137.

Burton, C.H., and Turner, C. (eds.). (2003), *Manure Management. Treatment Strategies for Sustainable Agriculture*, 2nd ed., Silsoe Research Institute, Wrest Park, UK.

Capel, P.D., McCarthy, K.A., Coupe, R.H., Grey, K.M., Amenumey, S.E., Baker, N.T., and Johnson, R.L. (2018), Agriculture—A River Runs through It—The Connections between Agriculture and Water Quality. In: *U.S. Geological Survey Circular*, vol. 1433, 201 pp., https://doi.org/10.3133/cir1433.

Cooperbrand, L., Bollero, G., and Coale, F. (2002), Effect of poultry litter and composts on soil nitrogen and phosphorus availability and corn production, *Nutrients Cycling in Agroecosystems*, 62: 185–194.

Chamley, H. (2003), *Geosciences, Environment and Man, Development in Earth and Environmental Sciences*, Elsevier, Amsterdam, 527 pp.

Christen, P. (1996), *Konzept des "Sustainable Development" bzw. Für den landwirtschaftlichen Bereich "Sustainable agriculture"(gemäß dem erichtöur common future'bzw dem Brundtland – Report)*', Berichte über Landwirtschaft, Landwirtschaftsverlag, Münster-Hiltrup, Germany, vol. 74, pp. 66–86.

Commoner, B. (1971), *The Closing Circle, Nature, Man and Technology*, Alfred A. Knopf, New York, 326 pp.

Dalemo, M., Sonesson, U., Jonsson, H., and Bjorklund, A. (1998), Effects of including nitrogen emissions from soil in environmental systems analysis of waste management strategies, *Resources, Conservation and Recycling*, 24: 363–381.

Delgado, J., Gross, C., Lal, H., Cover, H., Gagliardi, P., McKinney, S., Hesketh, E., and Shaffer, M. (2010), A new GIS nitrogen trading tool concept for conservation and reduction of reactive nitrogen losses to the environment, *Advances in Agronomy*, 105, 117–171, https://doi.org/10.1016/s0065-2113(10)05004-2.

Delschen, T. (1999), Impacts of long-term application of organic fertilizers on soil quality parameters in reclaimed loess soils of the Rhineland lignite mining area, *Plant and Soil,* 213: 43–54.

Dobson, A.P., Bradshaw, A.D., and Baker, A.J.M., (1997), Hopes for the future: Restoration ecology and conservation biology, *Science*, 277(5325): 515–522.

Dodds, W.K., Bouska, W.W., Eitzman, J.L., Pilger, T.J., Pitts, K.L., Riley, A.J., Schloesser, J.T., and Thornbrugh, D.J. (2009), Eutrophication of U.S. freshwaters: Analysis of potential economic damages. *Environmental Science and Technology*, 43: 12–19.

Dubey, A (2023), Sustainable Agriculture. *Encyclopedia Britannica*, 22 Dec. 2023. www.britannica.com/technology/sustainable-agriculture, accessed Jan. 5, 2024.

Dymond, J.R., Betts, H.D., and Schierlitz, C.S. (2010), An erosion model for evaluating regional land-use scenarios, *Environmental Modelling and Software*, 25(3): 289–298.

El Bassam, N. (1998), Fundamentals of Sustainability in Agricultural Production Systs, and Global Food Security. In: *Proc. Of the International Conference on Sustainable Agriculture for Food, Energy and Industry (Joint Conference of FAO[Food and Agriculture Organization] Society of Sustainable Agriculture and Resource Management [SSARM] and Federal Agriculture Research Centre [FAC]), Braunschweig, Germany, June 1997*, James & James, London.

Eilers, W.D., MacKay, R., Graham, L., and Lefebvre, A. (Eds.) (2010). Environmental Sustainability of Canadian Agriculture: Agri-Environmental Indicator Report Series - Report No. 3., Agriculture and Agri-Food Canada (AAFC)/Agriculture et Agroalimentaire Canada (AAC).

Falkenmark, M. (1989), Water Scarcity and Food Production. In: D. Pimentel and C.W. Hall (eds.), *Food and Natural Resources*, Academic Press, San Diego, CA, pp. 16–191.

FAO (1995), *Livestock: A Driving Force for Food Security and Sustainable Development*, FAO, Feed Resources Group (R. Sansoucy), Rome, WAR/RMZ vol. 84/85, pp. 5–17.

FAO (1996), *Integration of Sustainable Agriculture and Rural Development Issues in Agricultural Policy*, S. Breth (ed.), FAO, May 1995 Rome Workshop, Winrock, Morrilton, AR.

FAO (2000), The Energy and Agriculture Nexus, Environment and Natural Resources Working Paper No. 4, FO, Rome. www.fao.org/docrep/003/x8054e/x8054e05.htm#P457_59563, accessed Feb. 2024.

FAO (2006), Global Forest Resources Assessment 2005 – Progress towards Sustainable Forest Management, FAO Forestry Paper 147. ISBN 92-5-105481-9. www.fao.org/forestry/fra.

FAO (2016), The State of World Fisheries and Aquaculture 2016. Contributing to food security and nutrition for all. Rome. 200 pp

FAO (2017a), *World Pollution from Agriculture: A Global Review*, Food and Agriculture Organization of the United Nations (FAO) and the International Water Management Institute on behalf of the Water Land and Ecosystems research program Colombo.

FAO (2017b), *Water for Sustainable Food and Agriculture, A Report Produced for the G20 Presidency of Germany*, Food and Agriculture Organization of the United Nations (FAO), Rome.

FAO (2021). *The State of the World's Land and Water Resources for Food and Agriculture – Systems at breaking point. Synthesis report 2021*. Rome. https://doi.org/10.4060/cb7654en

FAO (2024), Pesticides Use and Trade – 1990–2022, FAOSTAT Analytical Briefs, No. 89, Rome, https://doi.org/10.4060/cd1486en.

FAO, IFAD, UNICEF, WFP and WHO (2022). *The State of Food Security and Nutrition in the World 2022. Repurposing food and agricultural policies to make healthy diets more affordable*. Rome, FAO. https://doi.org/10.4060/cc0639en

Fischer Taschenbuch (1996), *Der Fischer Weltalmanach 1997*, Fischer Taschenbuch Frankfurt.

Fischer Taschenbuch (1999), *Der Fischer Weltalmanach 2000*, Fischer Taschenbuch Frankfurt.

Galloway, J.N., Levy II, H., and Kasibhatla, P.S. (1994), Year 2020: Consequences of population growth and development on deposition of oxidized nitrogen, *Ambio*, 23: 120–123.

Gamage, A., Gangahagedara, R., Gamage, J., Jayasinghe, N., Kodikara, N., Suraweera, P., and Merah, P. (2023), Role of organic farming for achieving sustainability in agriculture, *Farming System*, 1(1): 100005, https://doi.org/10.1016/j.farsys.2023.100005.

Garcia, V., Cooter, E.J., Hinckley, B., Murphy, M., Xing, X., and Crooks, J. (2017), Examining the impacts of increased corn production on ground water quality using a coupled modelling system, *Science of the Total Environment*, 586: 16–24, doi: 10.1016/j.scitotenv.2017.02.009.

Garcia-Rodriquez, A., Matamoros, V., Fontás, C., and Salvadó, V. (2015), The influence of *Lemna sp.* and *Spirogura sp.* on the removal of pharmaceuticals and endocrine disruptors in treated wastewaters, *International Journal of Environmental Science and Technology*, 12: 2327–2338.

Giampietro, M., and Pimentel, D. (1995), *Food, Land, Population, and the U.S. Economy*, College of Agriculture and Life Sciences, Cornell University, Ithaca, NY.

Gleick, P.H. (1993), *Water in Crisis*, Oxford University Press, New York, 473 pp.

Goldhamer, D.A., and Snyder, R.L. (eds.). (1989), *Irrigation Scheduling: A Guide for Efficient On-Water Management*, University of California, Division of Agricultural and Natural Resources, Publ. 21454, Oakland, CA.

Goodland, R. (1998), Environmental Sustainability in Agriculture: Bioethical and Religious Arguments against Carnivory, In: *Ecological Sustainability and Integrity: Concepts and Approaches*, Kluwer Academic Publishers, Dordrecht, The Netherlands, pp. 235–265.

Gregersen, H.M., Draper, S., and Elz, D. (1989), *People and Trees: The Roles of Social Forestry in Sustainable Development*, World Bank, Washington, DC.

Gworek, B., Kijeńska, M., Wrzosek, J. & Graniewska, M. (2021), Pharmaceuticals in the soil and plant environment: A review, *Water, Air, and Soil Pollution*, 232: 145, https://doi.org/10.1007/s11270-020-04954-8.

Hammer, D.A., Pullen, B.P., McCaskey, T.A., Eason, J., and Payne, V.W.E. (1993), Treating Livestock, Wastewaters with Constructed Wetlands. In: G. A. Moshiri (ed.), *Constructed Wetlands for Water Quality Improvement*, Lewis Publishers, Chelsea, MI, pp. 343–348.

Han, F.X., Kingery, W.L., and Selim, H.M. (2001), Accumulation, Redistribution, Transport and Bioavailability of Heavy Metals in Waste-Amended Soils. In I.K. Iskandar and M.B. Kirkham (eds.), *Trace Elements in Soil, Bioavailability, Flux and Transfer*, Lewis Publishers, Boca Raton, pp. 145–174.

Hansen, T.L., Christensen, T.H., and Schmidt, S. (2006), Environmental modelling of use of treated organic waste on agricultural land: A comparison of existing models for life cycle assessment of waste systems, *Waste Management & Research*, 24: 141–152.

Higgens, M.J., Rock, C.A., Bouchard, R., and Wnegrezynek, B. (1993), Controlling Agricultural Runoff by Use of Constructed Wetlands, In: G.A. Moshiri, (ed.), *Constructed Wetlands for Water Quality Improvement*, Lewis Publishers, Chelsea, MI, pp. 357–367.

Houben, G.J. (2002), Flow and transport modeling - Natural attenuation of common agricultural and atmospheric pollutants: Reactive transport of nitrate, potassium and aluminium, *IAHS Publications*, 275: 519–524.

Hutson, J.L., and Wagenet, R.J. (1989), *LEACHM. Leaching Estimation and Chemistry Model*, Version 2, Centre for Environmental Research, Cornell University, Ithaca, NY.

Janzen, D. (1998), The gardenification of the wildland nature and the human footprint, *Science*, 279: 1312–1313.

Kattelmann, R. (1990), Conflicts and cooperation over floods in the Himalaya-Ganges Region, *Water International*, 15(4): 1–5.

Kemper, M. (2008), Veterinary antibiotics in the aquatic and terrestrial environment, *Ecological Indicators*, 8: 1–13.

Kenny, J.F., Barber, N.L., Hutson, S.S., Linsey, K.S., Lovelace, J.K., and Maupin, M.A. (2009), Estimated Use of Water in the United States in 2005, USGS Circular: 1344, U.S. Department of the Interior Washington, DC U.S. Geological Survey, URL: http://pubs.er.usgs.gov/.

Kornegay, E.T., Hedges, J.D., Martens, D.C., and Kramer, C.Y. (1976), Effect on soil mineral storage and plant leaves, following application of manures of different copper contents, *Plant and Soil*, 45: 151–160.

Madani-Hosseini, M., Mulligan, C.N., and Barrington, S. (2016), Acidification of In-Storage-Psychrophilic-Anaerobic-Digestion (ISPAD) process to reduce ammonia volatilization: Model development and validation, *Waste Management*, 52: 104–111, https://doi.org/10.1016/j.wasman.2016.03.043.

MAFF (2000), *Towards Sustainable Agriculture: A Pilot Set of Indicators*, Ministry of Agriculture, Fisheries and Food, MAFF Publications, London, February, 74 pp.

Malthus, T. (1798), An essay on the principle of population, as it affects the future improvement of society with remarks on the speculations of Mr. Godwin, M. Condorcet, and Other Writers, London, Printed for J. Johnson, in St. Paul's church-yard. (HTML format by Ed Stephan, 10 Aug 1997).

Mari, H.S., Rama-Krishna, R.N., and Lall, S.D. (1985), Improving field microclimate and crop yield with low cost shelter belts in Punjab, *International Journal of Ecology & Environmental Sciences*, 11: 111–117.

Matson, P.A., Parton, W.J., Power, A.G., and Swift, M.J. (1997), Agricultural intensification and ecosystem properties, *Science*, 277: 505–509.

McCaffrey, S.C. (1993), Water, Politics, and International Law. In: P.H. Gleick (ed.), *Water in Crisis, A Guide to the World's Freshwater Resources*, Oxford University Press, New York, pp. 92–104.

McRae, T., Smith, C.A.S., and Gregorich, L.J. (eds.). (2000), *Environmental Sustainability of Canadian Agriculture – Report of the Agri-Environmental Indicator Project*, Agriculture and Agri-Food Canada, Ottawa, Ont., Canada.

Meadows, D.H., Meadows, D.L., and Randers, J. (1992), *Beyond the Limits*, Chelsea Green Publishing Co., Vermont, 299 pp.

Micklin, P. (1988), Dessication of the Aral Sea. A water management disaster in the USSR, *Science*, 241(4870): 1170–1176.

Mulholland, P. J., and Lenat., D. R. (1992), Streams of the southeastern Piedmont, Atlantic drainage. In: C.T. Hackney, S.M. Adams, and W.H. Martin (eds.), *Biodiversity of the Southeastern United States: Aquatic communities*, John Wiley and Sons, New York, pp. 193–231.

Mulligan, C.N. (2002), *Environmental Biotreatment*, Government Institutes, Rockville, MD, 395 pp.

Mullins, C.L. (1990), Physical Properties of Soils in Urban Areas. In: P. Bullick and P.J. Gregory (eds.), *Soils in the Urban Environment*, Blackwell Scientific Publications, Oxford, pp. 87–118.

Mullins, C.L., Martens, D.C., Miller, W.P., Kornegay, E.T., and Hallock, D.L. (1982), Copper availability, form and mobility in soils from three annual copper – Enriched hog manure applications, *Journal of Environmental Quality*, 11: 316–320.

Nabhan, G.P., and Buchmann, S.L. (1997), Services Provided by Pollinators. Chap. 8. In: G.C. Daily (ed.), *Nature's Services: Societal Dependence on Natural Ecosystems*, Island Press, Washington, DC, pp. 133–150.

Nash, L. (1993), Water Quality and Health. In: P. Gleick (ed.), *Water in Crsis: A Guide to the World Fresh Water Resources*, Oxford University Press, New York, pp. 29–35.

Nature Conservancy Council (1986), *Worcestershire Inventory of Ancient Woodland (Provisional)*, NCC, Peterborough, UK.

NGS (National Geographic Society). (1995), *Water: A Story of Hope*, National Geographic Society, Washington, DC.

NRC (National Research Council). (1991), *Toward Sustainability: A Plan for Collaborative Research on Agriculture and Natural Resources Management*, Panel for Collaborative Research Support for AID's Sustainable and Natural Resources Management, Academic Press, Washington, DC, 164 pp.

NRC (1999), *Our Common Journey: A Transition toward Sustainability*, Board on Sustainable Development Policy Division, National Academy Press, Washington, DC, 384 pp.

OECD (2002), *Evaluating Progress in Pesticide Risk Reduction. Summary Report on the OECD Project on Pesticide Aquatic Risk Indication*, OECD, Paris.

OECD (2013), *OECD Compendium of Agri-Environmental Indicators*, OECD Publishing, doi: 10.1787/9789264186217-en.

Ogaji, J. (2005), Sustainable agriculture in the UK, *Environment, Development and Sustainability*, 7: 253–270.

Pagliari, P., Wilson, M., and He, Z. (2020). Animal Manure Production and Utilization: Impact of Modern Concentrated Animal Feeding Operations. In: H.M. Waldrip, P.H. Pagliari, and Z. He (eds.), *Animal Manure: Production, Characteristics, Environmental Concerns and Management*, ASA Special Publication 67, ASA and SSSA, Madison, WI, pp. 1–14, https://doi.org/10.2134/asaspecpub67.c1.

Payne, G.G., Martens, D.C., Kornegay, E.T., and Lindermann, M.D. (1988), Availability and form of copper in three soils following with annual applications of copper-enriched swine manure, *Journal of Environmental Quality*, 17(4): 740–746.

Pimentel, D. (1989), Impacts of Pesticides and Fertilizers on the Environment and Public Health. In: J.B. Summers and S.S. Anderson (eds.), *Toxic Substances in Agricultural Water Supply and Drainage*, U.S. Committee on Irrigation and Drainage, Denver, CO, pp. 95–108.

Pinstrup-Anderson, P., Pandya-Lorch, R. and Rosegrant, M.W. (1997), *The World Food Situation: Recent Developments, Emerging Issues, and Long-erTm Prospects*, International Food Policy Research Institute, Washington, DC.

Postel, S.L., Daily, G.C., and Ehrlich, P.R. (1996), Human appropriation of renewable fresh water, *Science*, 271: 785–788.

Pretty, J.N., Mason, C.F., Nedwell, D.B., Hine, R.E., Leaf, S., and Dils, R. (2003), Environmental costs of freshwater eutrophication in England and Wales, *Environmental Sciences and Technology*, 32: 201–208.

Quendler, T., and Schuh, T. (2002), Sustainability, A Challenge for Future Economic and Social Policy. In: H. Wohlmeyer and T. Quendler (eds.), *WTO, Agriculture and Sustainable Development*, Greenleaf Publishing, London, pp. 193–205.

Quendler, T., Weiβ, F. and Wohlmeyer, H. (2002), Conclusions and Proposals for Solutions. In: H. Wohlmeyer and T. Quendler (eds.), *The WTO, Agriculture and Sustainable Development*, Greenleaf Publishing, London, pp. 234–251.

Redding, M. (2005), *Case Studies to Assess Piggery Effluent and Solids Application*, Department of Primary Industries and Fisheries, Australia, Feb. 2004. www.dpi.qld.gov.au/ilem.

Rekolainen, S. Guoy, V. Francaviglia, R., Eklo, O.M., and Barlund, I. (2000), Simulation of soil water, bromide and pesticide behaviour in soil with the GLEAMS model, *Agricultural Water Management*, 44(1-3): 201–224.

Regenstein, L.G. (1991), *Replenish the Earth*, Crossroad Press, New York, 305 pp.

Richards, J.F. (1990), Land Transformation. Chap. 10. In: W.C. Clark, B.L. Turner, R.W. Kates, J. Richards, J.T. Mathews, and W. Meyer (eds.), *The Earth Is Transformed by Human Action: Global Warming and Regional Changes in the Biosphere over the Past 300 Years*, Cambridge University Press, Cambridge, UK, pp. 163–178.

Richard, S., Arnoux, A., Cerdan, P., Reynouard, C., and Horeau, V. (2000), Mercury levels of soils, sediments and fish in French Guiana, South America, *Water, Air and Soil Pollution*, 124: 221–244.

SACEP (2014), Nutrient loading and eutrophication of coastal waters of the South Asian Seas – A scoping study. South Asian Co-Operative Environmental Programme (SACEP). http://hdl.handle.net/1834/34570

Sagardoy, J.A. (1993), An Overview of Pollution of Water by Agriculture. In: *Prevention of Water Pollution by Agriculture and Related Activities, Proceedings of the FAO Expert Consultation, Santiago, Chile, 20-23 Oct. 1992*, Water Report 1, FAO, Rome, pp. 19–26.

Saha, A. (2001), *Agricultural Pollution Control*. Envirospace Pollution.

Said-Pullicino, D., Gigliotti, G., and Vella, A.J. (2004), Environmental fate of triasulfuron in soils amended with municipal waste compost, *Journal of Environmental Quality*, 33: 1743–1751.

Sims, J.T., and Wolf, D.C. (1994), Poultry manure management: Agricultural and environmental issues, *Advances in Agronomy*, 52: 2–84.

Sommer, S.G., Petersen, S.O., and Moller, H.B. (2004), Algorithms from calculating methane and nitrous oxide emissions from manure management, *Nutrient Cycling in Agroecosystems*, 69: 143–154.

Suárez, L.A. (2005), *PRZM-3: A Model for Predicting Pesticide and Nitrogen Fate in the Crop Root and Unsaturated Soil Zone: Users' Manual for Release 3.12.2*, U.S. Environmental Protection Agency, Office of Research and Development, Washington, DC, EPA/600/R-05/111, September 2005.

Tang, F.H.M., Lenzen, M., McBratney, A., and Maggi, F. (2021), Risk of pesticide pollution at the global scale. *Nature Geoscience*, 14: 206–210, https://doi.org/10.1038/s41561-021-00712-5.

Thies, C., and Tscharntke, T. (1999), Landscape structure and biological control in agroecosystems, *Science*, 285: 893–895.

Thomsen, I.K. (2000), C and N transformations in ^{15}N-cross-labelled solid ruminant manure during anaerobic and aerobic storage, *Bioresource Technology*,72: 267–274.

Thomsen, I.K., and Olesen, J.E. (2000), C and N mineralization of composted and anaerobically stored ruminant manure in differently textured soils, *Journal of Agricultural Science*, 135: 151–159.

Tubiello, F.N., Salvatore, M., Ferrara, A., House, J.I., Federici, S., Rossi, S., Biancalani, R., Cóndor Golec, R.D., Jacobs, H., Flammini, A., Prosperi, P., Cardenas-Galindo, P., Schmidhuber, J., Sanz Sanchez, M.J., Srivastava, N.K., and Smith, P. (2015), The contribution of agriculture, forestry and other land use activities to global warming, 1990-2012, *Global Change Biology*, 21(7): 2655–2660, https://doi.org/10.1111/gcb.12865.

U.S. Geological Survey (USGS) (1999), The Quality of Our Nation's Waters—Nutrients and Pesticides. U.S. Geological Circular #1225. Available at water.usgs.gov

US Soil Conservation Service (1993), Nutrient and Sediment Control System, Technical Note 4, U.S. Department of Agriculture, Washington, DC, Mar.

US EPA (1987), Quality Criteria for Water 1986. EPA 440/5086-001 US Environmental Protection Agency, Office of Water Regulations and Standards, Washington, D.C.

US EPA (2023), Estimated Nitrate Concentrations in Groundwater Used for Drinking | US EPA. www.epa.gov/nutrientpollution/estimated-nitrate-concentrations-groundwater-used-drinking), accessed Feb. 2024.

USDA (2009) *US Department of Agriculture Report to Congress, Manure Use for Fertilizer and Energy*, U.S. Department of Agriculture, Washington, DC, June 2009.

USDA (United States Department of Agriculture), Soil Conservation Service. (1992), *Agriculture Waste Management Field Handbook*, USDA, Washington, DC.

USGS (United States Geological Survey). (1999), The Quality of Our Nation's Waters – Nutrients and Pesticides, U.S. Geological Circular #1225, Web Site Address: water.usgs.gov.

Varma, V.S., Parajuli, R., Scott, E., Canter, T., Lim, T. T., Popp, J., and Thoma, G. (2021), Dairy and swine manure management – Challenges and perspectives for sustainable treatment Technology, *Science of The Total Environment*, 778: 146319, https://doi.org/10.1016/j.scitotenv.2021.146319.

Verplancke, H., ed. (1992), *Water Saving Techniques for Plant Growth*, NATO ASI Series. Series E: Applied Sciences, Kluwer Academic Publishers, Amsterdam, 241 pp.

Vervoort, R.V., Radcliffe, D.E., Cabrera, M.L., and Latimore, M. Jr. (1998), Nutrient losses in surface and subsurface flow from pasture applied poultry litter and composted poultry litter, *Nutrient Cycling in Agroecosystems*, 50: 287–290.

Vitousek, P.M., and Mooney, H.A. (1997), Human domination of Earth's ecosystems, *Science*, 277(5325): 494–499.

Wall, G.L., Pringle, Q.A., and Sheard, R.W. (1991), Intercropping, red clover with sillage for soil erosion control, *Canadian Journal of Soil Science*, 71: 137–145.

Wen, D. and Pimentel, D. (1990). Ecological Resource Management for a Productive, Sustainable Agriculture in Northeast China. In T.C. Tso (ed.), *Agricultural Reform and Development in China*, IDEALS, Beltsville, MD, pp. 297–313.

WHO (World Health Organization). (1996), *The World Health Report 1996. Fighting Disease and Fostering Development*, World Health Organization, Geneva.

World Bank (2022) *Water in Agriculture*. www.worldbank.org/en/topic/water-in-agriculture.

WRI (World Resources Institute) (1994), People and the Environment, WRI Report, 4160 pp.

Xu, J.M., Gan, J., Papiernik, S.K., Becker, J.O. and Yates, S.R. (2003), Incorporation of fumigants into soil organic matter, *Environmental Science & Technology*, 37(7): 1288–1291.

Yang, X.M., Drury, C. F., Reynolds, W.D., Tan, C.S., and McKenney, D.J. (2003), Interactive effects of composts and liquid pig manure with added nitrate on soil carbon dioxide and nitrous oxide emissions from soil under aerobic and anaerobic conditions, *Canadian Journal of Soil Science*, 83: 343–352.

Yu, Y., Liu, Y. and Wu, L. (2013), Sorption and degradation of pharmaceuticals and personal care products (PPCPs) in soils, *Environmental Science and Pollution Research*, 20: 4261–4267.

Yuan, L., Sinshaw, T., and Forshay, K.J. (2020), Review of watershed-scale water quality and nonpoint source pollution models, *Geosciences* (Basel), 10(25): 1–36, doi: 10.3390/geosciences10010025.

7 Urbanization and the Geoenvironment

7.1 INTRODUCTION

Currently, 56% of the world's population live in cities compared to 20% in 2000, and more than 70% of the population will live in urban areas by the year 2050 (www.worldbank.org/en/topic/urbandevelopment/overview), making this an increasingly significant component in the global environment. Urban centres, together with their suburbs constitute what is now called the *built environment*. This built environment includes (a) the various physical structures that serve the community, (b) the resultant products and discharges associated with the various industrial, municipal, and domestic activities, such as waste piles, dumps, aeration ponds, gravel pits, etc., (c) the infrastructure, such as pipelines, transmission towers, roads, runways, bridges, etc., (c) the various utilities necessary to service the community, such as power plants, gas plants, wastewater treatment plants, reservoirs, etc., (d) the other kinds of resources associated with, and necessary to sustain, the urban population and the welfare of the community (e.g., parks, lakes, forests, recreational and sporting facilities, etc.). By its very nature, the man-made environment that defines the built environment is often in conflict with the natural environment, and in particular with the goals of sustainability of the land environment and its natural resources. The general perception is that urban centres consume significant resources and pollute the air, land, and water. Populations within the cities require clean air, clean water, sewage, and waste management systems, adequate food supply, housing, and transportation. It is estimated that 1 billion people live in cities that do not have access to clean drinking water (www.theguardian.com/environment/2023/mar/22/number-city-dwellers-lacking-access-safe-water-double-2050) and that more than 3 billion people live in cities that do not have access to solid waste collection services and facilities (https://unstats.un.org/sdgs/report/2019/Goal-11/#:~:text=Municipal%20waste%20is%20mounting%2C%20highlighting%20the%20growing%20need,according%20to%20data%20collected%20between%202010%20and%202018). It is often argued that these demands are currently not well met, and that the demand deficit will continue to escalate with time. For example, in Accra, Ghana, only 33% of the inhabitants have access to inhouse sanitation (https://cdn.who.int/media/docs/default-source/infographics-pdf/climate-change-and-health/health-and-climate-change-urban-profiles_accra_web.pdf?sfvrsn=e4c11ec1_1&download=true, accessed Aug. 2024). In Lima, Peru, 91% of the sewage is discharged directly into rivers, which reach the ocean. Some typical types of urban problems are summarized in the illustration shown in Figure 7.1.

A recent definition of urban sustainability offered at the Sustainable City Conference in Rio (2000) stated that

> The concept of sustainability as applied to a city is the ability of the urban area and its region to function at levels of quality of life desired by the community, without restricting the option available to the present and future generations and without causing adverse impacts inside and outside the urban boundary.

The 2030 Agenda for UN Sustainable Development with the adoption of the New Urban Agenda (NUA) highlight the importance of urbanization and the need for sustainability in the urban environment. Sustainable Development Goals (SDGs) 9 and 11 are related to infrastructure and cities, respectively. The related indicators and the NUA for progress measurement are discussed later in this chapter.

DOI: 10.1201/9781003407393-7

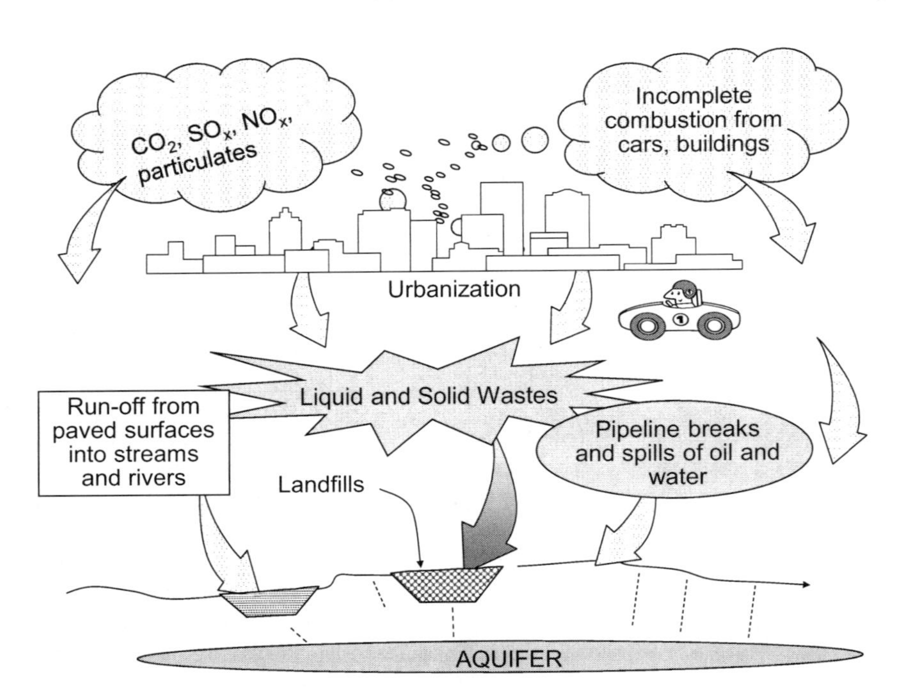

FIGURE 7.1 Urban sources of contamination and their effect on the geoenvironment.

7.2 LAND USES AND LAND USE CHANGE BY URBANIZATION

Urban development is a major consumer of land. Natural landscape areas around the cities are converted into housing estates, industrial parks, and other kinds of facilities designed to serve the community. Land is typically used for housing, businesses, industry, surface and sub-surface infrastructures, such as roads, wastewater supply, sewers, and power lines, and for recreational purposes, such as parks and playgrounds.

According to Appun (2018), Germany's agricultural land has decreased from 2000 to 2016, by 6970 km^2 as the result of urban sprawl and new infrastructure. In addition, infrastructure construction of houses or roads in Germany consumes 66 hectares every day (www.cleanenergywire.org/factsheets/climate-impact-farming-land-use-change-and-forestry-germany, accessed Jan. 2024).

Over the period of 20 years, (1992–2012) about 12.6 million hectares of farmland has been lost due to expanding urban areas (59%) and low-level residential development (41%). In addition, 62% of all development occurs using agricultural land. (www.agweek.com/business/31-million-acres-lost-development-cuts-into-u-s-farmland). According to Knutson (2018) in the US, 70 hectares of farm and ranchland are lost every hour for housing and other industries (https://globalagriculturalproductivity.org/case-study-post/americas-disappearing-farm-and-range-land/#:~:text=More%20than%2031%20million%20acres%20of%20U.S.%20agricultural,to%20make%20way%20housing%20and%20other%20industries.%201).

Urban sprawl is defined as the "rapid expansion of geographic extent of cities and towns" according the Encyclopedia Brittanica (www.britannica.com/topic/urban-sprawl, accessed Aug. 2024). Although increasing population growth is the main reason for sprawl, the desire for more space and escape from the city, and single-use zoning are also causes, among others. The increased spatial footprint correlates with an increased environmental footprint. More energy per capita is required and more wildlife habitat is destroyed. This phenomenon is not restricted to North America. In Europe, although the population increased 6% from 1980 to 2000, the spatial footprint increased 20%. Some communities, however, are developing restrictions on

TABLE 7.1
Soil Physical Properties for Various Urban Uses (Adapted from Mullins, 1990)

Requirement	Soil Properties	Application
Drainage	Hydrology, hydraulic conductivity (soil structure), porosity	Playing fields, effluent disposal
Load-bearing capacity	Bulk density, compactability, water content/ potential, penetration resistance, shear strength, compressibility, consolidation	Playing fields, foot paths, load bearing support for infrastructure, and spread footings for light structures
Plant growth medium	Drainage, air capacity, water capacity, bulk density, structure, penetration resistance	Playing fields, gardens, parks or sports fields
Prevention of erosion and runoff	Infiltration, drainage, structural stability	Foot paths, effluent disposal, parks or sports fields

construction and are implementing new planning techniques, such as smart growth, transit villages, and ecovillages.

In the urban context, land is not only lost to urban use, but surface and subsurface soils may be contaminated and degraded. They serve a variety of functions, as, for example, (a) foundation base for buildings, (b) medium for plant growth, (c) open spaces, (d) park space, (e) urban gardens, (f) bases for roads, ponds, and reservoirs, and (g) sources and sinks of pollutants. Soils in urban areas tend to be more diverse because of the introduction of additives, such as buried waste, debris, fuel ash, and other residues. Assessment of soil impairment requires specification of intended use of the soil. Particle size distribution, porosity, erodibility, structural stability, hydraulic conductivity, and rootability are some properties that need to be assessed to determine the degree of impairment of the soil. For example, if playing fields, under repeated use in wet conditions are no longer usable for playing, the situation will need to be remediated to return the land to its intended land use (Mullins, 1990). Signs of deterioration could include ponding, run-off, soil erosion, or poor grass growth. Soil properties relevant for use, soil type, and sensitivity of the properties to changes in soil use are important factors in mitigating soil damage. Some of the properties required for various uses are summarized in Table 7.1.

Derelict sites pose some unique problems. In general, these sites have become, by default, disposal sites with unauthorized disposed goods and substances that include urban garden wastes and other kinds of household wastes. Perhaps the more dominant kinds of debris found in derelict sites are those items that are the result of demolished buildings or buildings in considerable distress. The debris generally found include building materials, such as pipes, pieces of foundations, tiles, wood, plaster, rusting steel, and broken concrete slabs and structures.

7.3 IMPACT OF URBANIZATION

7.3.1 Impact on Water

In urban and suburban areas, more and more land surfaces are covered by buildings, roads, and constructed parking areas using concrete, bituminous concrete, asphalt, or other impervious coverings. Because these covered surfaces prevent infiltration of rainwater and ponded water into the sub-surface, replenishment of the underlying groundwater is denied, and groundwater levels may be consequently be lowered. Another effect of such covered surfaces is to allow surface flow (i.e., streaming) of rainwater into collecting areas. It is not uncommon to find pollutants in the surface flow water or streaming water because of noxious substances deposited onto the covered surfaces. Since these waters will eventually find their way into the receiving waters, they can be considered to be a non-point source pollution of lakes and rivers.

Urbanization has also led to a significant increase in urban flooding. According the to the Ontario Auditor General's 2022 report on urban flooding, between 2010 and 2020, cities in Ontario, Canada were impacted by damages exceeded $80 million each of seven urban flood events. The report indicates that the primary causes of increased flooding are (1) conversion of greenspace to impervious surfaces, (2) inadequate and aging stormwater infrastructure, and (3) increasing heavy rain events due to climate change. Wetlands play a significant role in mitigating flooding. Extreme rain events cannot be handled by the stormwater infrastructure, which subsequently leads to large floods. In Southern Ontario's, the wetlands are estimated to reduce flood-related damages by 38% in urban areas and 29% in rural areas. Wetlands improve water quality, inhibit erosion, promote tourism, enhance groundwater recharge, provide habitats for fish and wildlife, protect from sea level rise and reduce the intensity of droughts and high heats. This value is estimated at more than $25 billion annually. Wetlands Mitigate Flooding – Ontario Nature, (https://ontarionature.org/campaigns/wetlands/wetlands-mitigate-flooding/) Wetlands in Ontario have declined significantly over the past two centuries, with the most severe losses (over 95%) occurring in southern Ontario in urban areas. (www.ducks.ca/stories/wetlands/whats-happening-to-canadas-vanishing-wetlands/).

The US has also lost about half of its wetlands. Thus, flooding risks have also increased in many areas. Wetlands can absorb high water volumes of water and reduce direct and downstream flooding. A hectare of wetlands of 0.3 m in depth can absorb 2.77 million L of water, which is equivalent to the flooding of 26 homes by over a metre of water. A study by Ferdous (2013) showed that the combined effect of all wetlands (15% of the area of a small Canadian watershed) can lead to a 10% reduction in flood peaks. Despite the many critical benefits of wetlands, the US has lost roughly half of the wetlands since the 1780s due to urbanization, agriculture, or drainage or erosion. (www.nrdc.org/bio/ben-chou/nations-flood-risks-grow-protecting-wetlands-more-important-ever#:~:text=Research%20studies%20have%20found%20that%20wetlands%20in%20a,wetlands%20has%20been%20found%20to%20increase%20flood%20risk).

An extensive underground system of parking areas, sewers, pipes, deep building foundations, and tunnels to depths of 100 m can also significantly affect the underground terrain. Water leakage from buried and degrading water supply pipes leads to excessive water accumulation. In some cities, due to a lack of maintenance, more than 40% of the water supply is lost due to leaking pipes.

In many cities, groundwater abstraction for consumption can be excessive, particularly in regions where available surface water is difficult to access and transport, in arid regions, and in regions where the quality of surface water supply is deemed unsafe for consumption. According to Fauchon (2012), some examples are that the water table is falling in Greater Beijing, China, between 4 and 8 metres each year, and in Chennai, India, 300,000 m^3 more water per day are consumed than available. Using a land model, Li et al. (2022) investigated the change in groundwater table depth across China from 1979 to 2018. There was a significant reduction in the groundwater table depth in China by approximately 0.2 m on average, but it is higher (0.5 m) in southern China and to the north of 35° (0.8 m) (Li et al., 2022). This will lead to severe problems of ground subsidence, soil erosion, aridification, and salinization. In coastal regions particularly, high rates of groundwater abstraction can lead to land subsidence and flooding during high rainfall periods. Seawater intrusion into the aquifers can also occur as a result of groundwater abstraction from the aquifers, resulting in contamination of shallow aquifers. If vertical communication exists between shallow and deeper aquifers, contamination of the deeper aquifers will eventually occur (Yong et al., 1994, 1995). Land subsidence will also result in damage to structures with foundations affected by differential settlement (Figure 7.2). This problem has been discussed in detail in the previous chapter in respect to the impact of high rates of groundwater abstraction in regions where cities are founded on quaternary sediments.

Urban discharges can substantially degrade groundwater quality. For example, in China, phenols, cyanides, mercury, chromium, and arsenides and fluorine have been found in more than 50 urban regions. Urban effluent infiltration can also increase levels of nitrates, sulphates, and chlorine up to

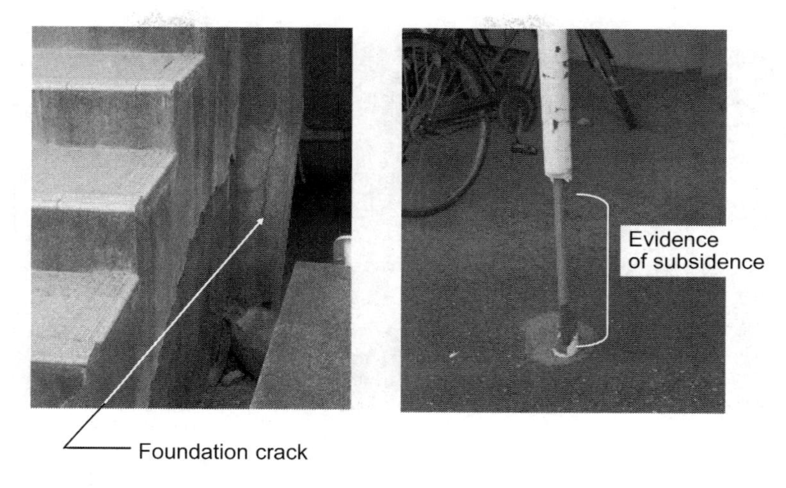

FIGURE 7.2 Subsidence and foundation impairment resulting from excessive groundwater abstraction.

90 m in depth (Chamley, 2003). Groundwater pollution causes many health problems and restricts human use.

Salinity is also increasing in the urban environment from discharges of domestic wastewater, seawater intrusion into aquifers, and contamination from road salt. Detergents, soaps, and salts are added to domestic wastewater, which is subsequently sent to wastewater treatment plants. The salts, sodium, and chloride, however, are not removed in treatment processes, resulting therefore in increased salinity of the water – as schematically illustrated in Figure 7.3. If the treated water is used for irrigation for agriculture, as for example in areas of the US, Israel, and Jordan (Vengosh, 2005), groundwater and soil quality may be impaired. The addition of road salt for de-icing also contributes to the salinization of groundwater and surface water due to run-off. Powdered road salt used for de-icing may be subject to wind dispersion, resulting in contamination of the surrounding area. Although calcium chloride can be used as a de-icing salt, it is more expensive and is not as effective as sodium chloride. Other alternatives under evaluation are beet juice, brines, coffee grounds, sand, calcium magnesium acetate, grape skin, kitty litter, and stone grit. The combination of sodium chloride with calcium chloride can reduce the sodium-to-chloride ratio. To reduce the stress on the land and water resources, processes, such as reverse osmosis and nanofiltration, will be required for desalination. Increasing road porosity or installing solar roads are being developed according to the US EPA (2020) (www.epa.gov/snep/winter-coming-and-it-tons-salt-our-roads).

7.3.2 Effect of Transportation and Energy Use

The impacts on the environment from traffic in urban areas are in the form of airborne, land, and water pollution. Understanding these problems and finding solutions for reduction of emissions are of paramount importance for the health and welfare of society and the geoenvironment. The occurrence of air pollutants from vehicular emissions at elevated levels can lead to serious health problems for the population living in high traffic areas. Pedestrians or passengers in vehicles that do not stay in the area long will not be affected but there could be significant impacts for those that live in the area. Carbon monoxide is a major pollutant from vehicles. In particular, intersections where vehicles are stationary for a period of time can substantially increase the level of carbon monoxide (CO) released into the air. The most prevalent air pollutants from vehicular emissions are nitrogen oxides (NO_x), hydrocarbons, and CO. Over the past two decades, NASA's Terra satellite has been used to measure CO atmospheric concentrations. Buchholtz et al. (2021) has reported that sine 2000,

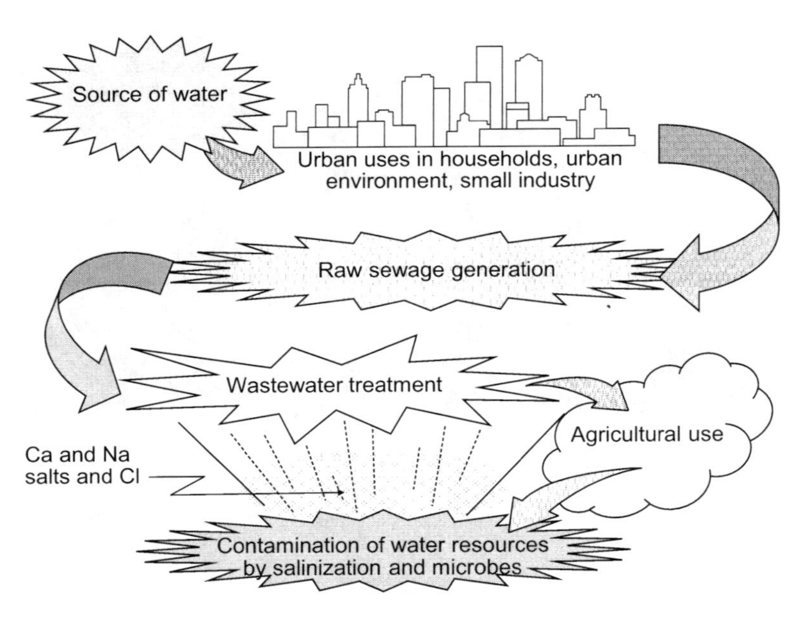

FIGURE 7.3 Cycle of salinization of water due to urban use. (Adapted from Vengosh, 2005.)

average CO concentrations decreased by 15% but that recently the rate of decline has decreased by 0.5% instead of 1% per year. Stricter regulations and cleaner burning fuels have decreased the levels.

In the past, lead in gasoline led to accumulation of lead on surface soils for many years. Page and Ganje (1970) showed that 15 to 36 μg/g over a period of 40 years accumulated in high traffic areas compared to negligible amounts in low traffic areas. Even though lead in gasoline has been eliminated in many countries, the low mobility of lead ensures that the lead remains in the ground as a contaminant for many years.

Emission standards for NO_x have been decreasing substantially until fleet averages reach 0.03 g/mile by 2025 (www.epa.gov/greenvehicles/smog-vehicle-emissions). Although emissions have decreased substantially over the years, the combination of poorly maintained cars together with more cars and other types of heavy-duty vehicles that consume more gasoline and emit higher emissions (such as sports utility vehicles (SUVs), light trucks, vans and pickup trucks) ensure that reduction in noxious emission will not be readily accomplished. Not only do vehicles emit various pollutants, but they also contribute over 20% of the carbon dioxide emissions from gasoline. Other petroleum fuels add another 12%. Reduction in vehicle use or the use of alternate fuels or other means will be required to substantially reduce the emission of this greenhouse gas.

Paved road surfaces share the same problems as covered impervious areas in cities – i.e., rainwater cannot infiltrate into the soil but will accumulate on roads and will wash away the pollutants on the streets. This can include motor oil, grease, antifreeze, metals, phosphorus, and other pollutants. These are then washed into local waterways and rivers by streaming flows, resulting in oxygen depletion and pollution of the waters – killing fish, plants, and other aquatic life. Public health is threatened when the contaminated water and contaminated fish are ingested.

Soil quality is another issue of importance. Air pollutants are deposited on the ground from precipitation passing through the airborne pollutants. These will find their way onto surface soils, and into the sub-soil and also into rivers and groundwater. Human exposure with the pollutants can be through inhalation, contact with the soil and ingestion of water and crops grown in contaminated soil. Ingestion of vegetables grown in urban gardens and children playing in exposed sand boxes and bare soil landscape are good examples of human exposure to deposited air pollutants. Whilst awareness of the potential hazards of such a form of water and soil pollution exists, the same cannot be said for information and data on pollutant concentrations and distributions from such types of deposited airborne pollutants.

Soils and sediments contiguous to roads can exhibit high levels of pollution. In Germany, levels of lead, cadmium, chromium, nickel, vanadium, and zinc in soils near roads have been found to be up to five times higher than in soils located some distance from roads. In the case of PAHs, these were found to be up to 100 times higher than in soils distant from the roads. The source for these pollutants can be traced to vehicular exhausts (Münch, 1992). Cuny et al. (2001) reports that elimination of lead in gasoline has substantially reduced lead levels in parking and rest areas near the motorways in France.

Acid rain is a direct result of precipitation through sulphur and nitrogen oxides emitted from fossil-fuel combustion in coal thermal power plants. Sulphuric and nitric acids are deposited on (a) the soil surface and into the underlying soil and perhaps into the groundwater, (b) surface water courses, and (c) other surfaces. Northeastern US and eastern Canada have experienced rain with pH values ranging from 4 to 5. Deforestation by acid rain has been significant in the nineteenth and early twentieth centuries in North America and Europe. Forests affected by acid rain, in particular, have less ability to retain water and protect against wind erosion. Heavy metals in contaminated soils may also be released by the acid rain. Fish and other aquatic organisms are susceptible to many metabolic disorders when the pH of lakes and other water bodies drops to 5 and below. Although not directly a geoenvironmental land problem, it is pertinent to note that respiratory problems in smog events are significant human health problems.

7.3.3 Implications on Health

The pathways of exposure of urban soils include inhalation and ingestion of soil and dusts through respiration, and consumption of home-grown foods and contact with the soil. Urban gardening is widely practiced. Thorton and Jones (1984) have reported on various tests conducted in the UK regarding radish and lettuces grown in typical urban gardens with soils containing different concentrations of zinc, copper, and lead. Measurable values of lead in the radish and lettuce were obtained, and it was concluded that both soil splash and foliar uptake contributed to the measured lead levels.

People living close to roads can be exposed to high levels of air pollution from traffic. The pollutants are emitted from vehicles (SUVs and trucks, in particular) fuel spills and evaporation, and tire and brake wear. Road dust, for example, containing aluminum, bohrium, lead, platinum, rhodium, vanadium, zinc, and PAHs, was found to have harmful effects on the human respiratory system, in particular, in a review by Khan and Strand (2018). Other emissions are black carbon, carbon monoxide, nitrogen dioxide, particulate and ultrafine matter, and benzene and other volatile organic compounds (VOCs). Some health issues include asthma, allergies, and other respiratory problems, childhood leukemia, adult lung cancer, and adult breast cancer (www.canada.ca/en/health-canada/services/air-quality/outdoor-pollution-health/traffic-related.html).

7.3.4 Land Degradation

Urban land can become degraded chemically and physically. As noted previously, contaminants degrade soil quality through release into the soil via spills, run-off, and other additives. Roads, sidewalk, parking lots, and other structures seal the land and reduce water infiltration into the groundwater. The installation of cables, sewers, foundation, and other underground structures disrupts the physical structure of the soil. Introduction of softer soils or wastes into the natural soil changes its characteristics.

Greenfields are lands that have not been disturbed. They have the capacity to maintain their biodiversity, ecological functions, soil quality, and can renew their groundwater resources. Brownfields, on the other hand, have been degraded by contamination from various sources – primarily industrial and manufacturing facilities, such as refineries, rail yards, gas stations, warehouses, dry cleaners,

TABLE 7.2
Urban Land Uses and Activities Leading to Contaminated Land (Adapted from Syms, 2004 and US EPA, 1999b)

Industry	Activity Leading to Soil Contamination	Type of Contaminant
Airports	De-icing and fire control run-off, servicing, fuelling	Acids/alkalis, asbestos, solvents, herbicides, PCBs, fuels, de-icing agents, fire-fighting chemicals
Animal slaughterhouses	Leaking tanks, pipework, spillages	Acids/alkalis, organic compounds, pathogens, metals, metalloids
Auto repair and refinishing	Leaking tanks, spills, sprays, solid wastes	Metals, dust, VOCs, solvent, paints and paint sludges, scrap metal, waste oils
Battery recycling and disposal	Discarded batteries	Pb, Cd, Ni, Cu, Zn, As, Cr
Incinerators	Solid wastes, gaseous emissions	Dioxin, ash, metals, wastes
Laundries and dry-cleaning	Spillage of solvents and other contaminants	Organic compounds including solvents (chloroform, TCE), PCE, PCBs, fuels, asbestos
Landfills (municipal and industrial)	Leachates and gaseous emissions	Metals, VOCs, PCBs, ammonia, methane, household products and cleaners, pesticides, wastes, hydrogen sulfide
Paper and printing works	Leakage from drums and other contaminants, may be buried on site, spillages of solvents and other materials	Metals, inorganic compounds, acids, alkalis, solvents, inks, degreasing solvents, fuels, oils, PCBs, inorganic ions
Railway yards and tracks	Maintenance and repair of tracks, engines, coal storage	Fuel oils, lubricating oils, PCBs, PAHs, solvents, ethylene glycol, creosote, herbicides, metal fines, asbestos, ash, sulphate
Sewage treatment	Disposal of sludges, stones and solid matter in landfill and other places	Metals, PCBs, PAHs, solvents, pathogens, acids/alkalis, inorganic compounds
Transport depots	Fueling, vehicle washing and maintenance, storage of tires and wastes, leaks from split drums	Metals such as Pb, Cr, Zn, Cu, vanadium, acids/alkalis, solvents, PAHs, asbestos
Waste disposal	Landfill leachates, waste transfer station spills	Inorganic chemicals, oils, metals, PCBs, PAHs, solvents, acids/alkalis, inorganic compounds detergents, asbestos

and other commercial enterprises using or storing hazardous chemicals. Table 7.2 provides a list of the various industries, activities, and contaminants that lead to land contamination, mainly in urban areas. Abandoned urban lands (brownfields) are clear indications of failure in complying with the principles of sustainable development as envisaged in the Brundtland Report (World Commission on Environment and Development, 1987) as well as that of the Club of Rome.

7.3.5 Impact of Urban Waste Disposal

Improper waste disposal has a major negative impact on the land. Wastes can be in solid forms as municipal solid waste, in liquid forms as sewage and wastewater, or in gaseous forms from vehicular emissions. Solid wastes can originate from (a) households as food and yard wastes, paper, chemicals, wood, and so on, (b) urban businesses that generate wastes similar to household wastes, (c) industrial wastes, and (d) construction wastes. Hospital wastes are discussed in more detail in Chapter 4. As there are many factors that influence production of wastes, computer models have been developed to estimate waste production rates and contaminant transport through the waste.

Various hydraulic computer models have been developed to calculate contaminant infiltration rates. Some of these models include (a) hydraulic evaluation of the landfill performance (HELP) model (Schroeder et al., 1994), (b) HYDRUS (2D/3D) based on the early work of Neuman (1972) who developed their UNSAT model, and (c) unsaturated soil and heat flow model (UNSAT-H) (Fayer, 2000). HELP has been the most frequently used model for final cover and leachate collection system design and is particularly useful for humid and semi-humid areas. Version 4 is now available (www.epa.gov/sites/default/files/2020-10/help_4.0.1.zip). UNSAT-H is more appropriate for arid and semi-arid regions for landfill cover design (version 3 is at https://github.com/pnnl/unsat_h). MIGRATEv9 (www.gaea.ca/migrate.php) can be used for sorption, radioactive and biological decay, and transport through fractures, in landfills, buried waste deposits, spills, or disposal ponds. POLLUTEv8 is a contaminant migration analysis program and requires little computational effort for landfill design from simple systems to those with composite liners and multiple barriers and aquifers and site remediation (www.gaea.ca/pollute.php). The WHI UnSat Suite Plus combines SESOIL, VLEACH, PESTAN, VS2DT, and HELP in a revolutionary graphical environment specifically designed for simulating one-dimensional groundwater flow and contaminant transport through the unsaturated zone, such as a landfill (www.waterloohydrogeologic.com/products/other-software-offerings/).

Wastes can also be classified as demolition, non-hazardous (municipal), and hazardous wastes. Inert or non-leachable wastes include many types of construction wastes such as soil, bricks, concrete, tiles, and gypsum board. As long as they are not contaminated, these materials can be reused as backfill material, sub-grade and road materials, and even as building materials. Organic and inorganic wastes have been discussed previously in Section 2.3.

Wastes can be classified according to the physical state, origin, degree of hazard, or ability to be recycled or transformed. Wastes are classified as *hazardous* if they have any of the following characteristics:

- Ignitability-potential for fire hazard during storage, transport or disposal under standard temperatures and pressures.
- Corrosivity – potential for corrosion of materials in contact with candidate waste, resulting in environmental and health threats due to a pH less than 2.0 or greater than 12.5.
- Reactivity – potential for adverse chemical reactions when in contact with water, air, or other wastes.
- Toxicity – as per the Toxicity Characteristics Leaching Procedure (TCLP). The leaching test system for application of the TCLP procedure is shown in Figure 7.4 and the regulatory levels of the compounds in the leachate are given in Table 7.3.

In common with wastes generated from the various industries and manufacturing facilities, municipal solid wastes also will ultimately find their way into one or all of three disposal media: (a) Receiving waters, (b) atmosphere, and (c) land. Land disposal of waste products and waste streams appears to be the most popular method for waste containment and management. The various impacts arising from this mode of disposal and containment include degradation of land surface environment and ground contamination by pollutants. Because of the inhomogeneous nature of wastes, stability problems on the surface of the landfill, slope stability, and settling of the landfill can occur. Chemical and biochemical reactions within the landfill and water infiltration will inevitably change the mechanical properties of the landfill.

Both the wastes themselves and the products they produce (such as leachates and emissions) are health and geoenvironmental threats. Disposal of wastes in the ground, illicit dumping, leaking underground storage tanks and others are all causes for concern. A sampling of backyards in the

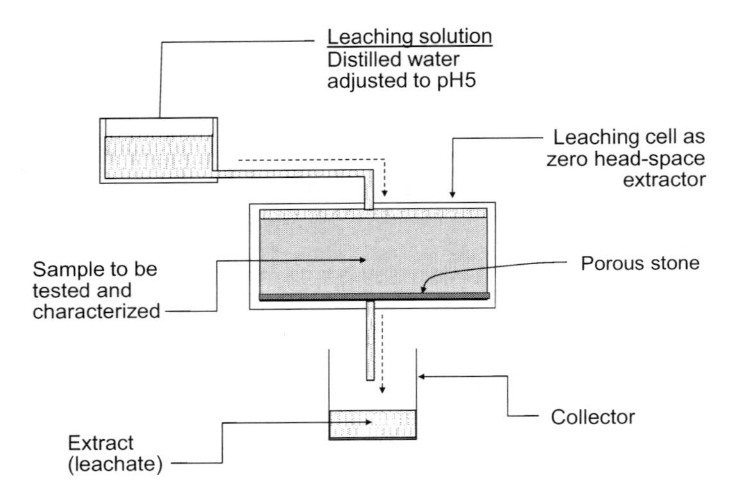

FIGURE 7.4 Typical leaching cell used as a zero-head extractor for application of TCLP procedure. The results of chemical analysis of the extract (leachate) should be compared with the regulatory values shown in Table 7.3. (Adapted from Yong, 2001.)

TABLE 7.3
TCLP Compounds and Regulatory Levels in Extract

Compound	Level (mg/L)	Compound	Level (mg/L)
Arsenic	5.0	Hexachloro-1,3-butadiene	0.5
Barium	100.0	Hexachloroethane	3.0
Benzene	0.5	Lead	5.0
Cadmium	1.0	Lindane	0.4
Carbon tetrachloride	0.5	Mercury	0.2
Chlordane	0.03	Methoxychlor	10.0
Chlorobenzene	100.0	Methyl ethyl ketone	200
Chloroform	6.0	Nitrobenzene	2.0
Chromium	5.0	Pentachlorophenol	100.0
o-Cresol	200.0	Pyridine	5.0
m-Cresol	200.0	Selenium	1.0
p-Cresol	200.0	Silver	5.0
1,4-Dichlorobenzene	7.5	Tetrachloroethylene	0.7
1,2-Dichloroethane	0.5	Toxaphene	0.5
1,1-Dichloroethylene	0.7	Trichloroethylene	0.5
2,4-Dinitrotoluene	0.13	2,4,5-Trichlorophenol	400.0
Endrin	0.2	2,4,6-Trichlorophenol	2.0
Heptachlor	0.008	2,4,5-TP (Silvex)	1.0
Hexachlorobenzene	0.13	Vinyl Chloride	0.2

Montreal area where wastes had been previously dumped indicated elevated levels of the heavy metals – lead and zinc (Figure 7.5).

Water entry into waste piles in landfills, together with dissolution processes result in the generation of waste leachates. A liner and leachate collection system such as that shown in Figures 1.11 and 3.12 in Chapters 1 and 3, respectively, is required to protect the groundwater from fugitive

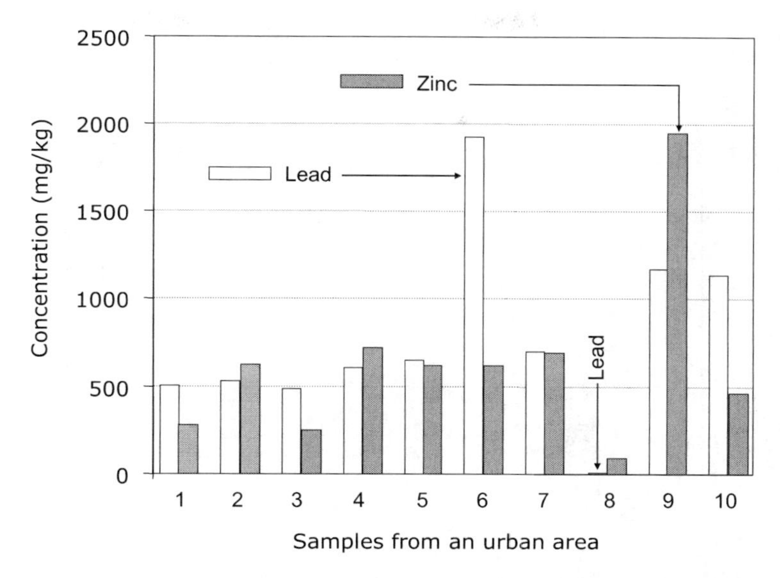

FIGURE 7.5 Lead and zinc concentrations of soil samples taken from various locations in the Montreal area (seven samples from communal garden and three samples from locations near it). (Data from Huang 2005.)

leachates. A cap system is required also to prevent rain from entering the waste pile and producing leachates. Typical leachates from landfills include organic chemicals (e.g., BTEX, PAHs, phthalates, ketones, dioxins, phenols, pesticides, solvents, etc.) and inorganic components (e.g., mercury, cadmium, chromium, copper, zinc, lead, nickel, etc.). The composition of leachates varies significantly depending on the age and type of waste and landfill technology used to contain the waste. The groundwater level and the nature of the soil under the landfill are important factors in managing leachate pollution risks to the groundwater. Concentrations of both organic and inorganic components can be high as shown in Figure 7.6.

In the past, quarries and other pits without proper barrier and liner systems were used for waste disposal – as for example the Gloucester landfill in Canada (Lesage et al., 1990). From 1969 to 1980, hazardous wastes, including laboratory organic chemicals, were disposed in this landfill, which was on glacial outwash deposits with a semi-confined aquifer. The chemicals reacted with the explosive charges that were also in the waste pile, and the leachate plume of more than 300 m in the aquifer in 1990 was found to contain the unaltered and transformed chemicals from the waste pile.

Gaseous emissions from biodegradation of organic materials in landfills also occur. These gases generally contain both carbon dioxide and methane. More than 350 m^3 of biogas are produced by a tonne of municipal solid waste (MSW) (Genske, 2003). Volatile solvents, paints, and other chemicals, such as toluene, phenol, ethylbenzene, naphthalene, vinyl chloride, methylene chloride, xylene, and chloroform, will also evaporate. Dissolution processes involve chemical reactions between the various constituents in a waste pile, the end result of which will be transformed products and leachates. Hazardous waste dumping in the Love Canal in Niagara Falls, US in the 1940s and 1950s and subsequent use of the site as a hazardous waste landfill resulted in a situation (La Grega et al., 1994) where the public was exposed to dioxin and other chemical fumes. Investigation of the many illnesses arising from the exposure showed that the chemicals causing the illnesses were neurotoxins, teratogens, fetotoxins, carcinogens, pulmonary toxins, and hepatoxins (Bridges, 1991).

Another case of illegal dumping of industrial wastes occurred on the island of Teshima in Japan where a low population of residents was living (Takatsuki, 2003). From 1978, a company was allowed to dispose of non-hazardous wastes. However, residents complained of disposal of oil, shredder waste, and open burning of wastes. Finally, as the result of an investigation, the company

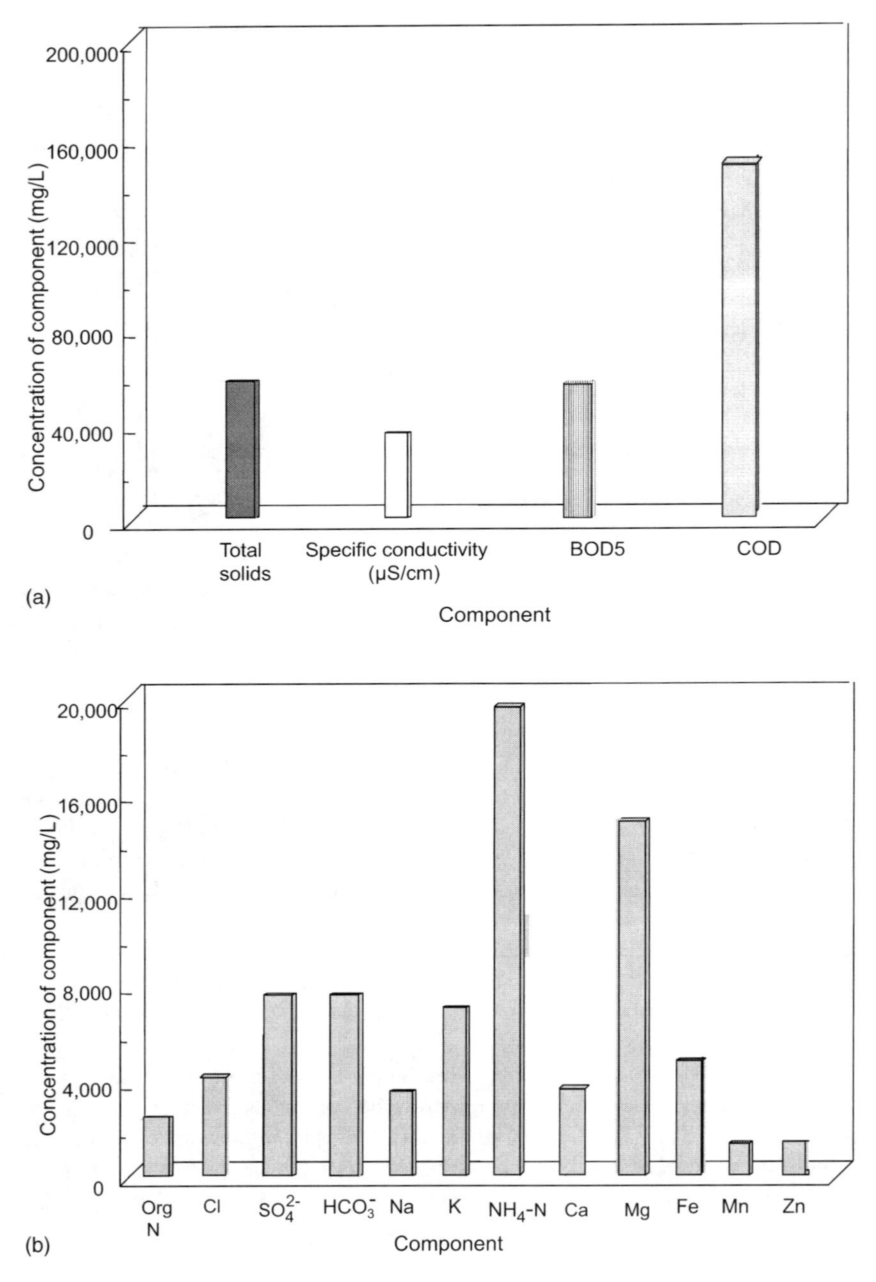

FIGURE 7.6 (a, b) Range of selected components found in landfill leachate. (Adapted from data reported by Kjeldsen et al., 2002.)

stopped doing business on the island but left large amounts of waste at the site. Leaching of the contaminants from the landfill and proper disposal of the contaminants (more than 600,000 tonnes) had to be addressed. Analysis showed contamination in the soil and leachate, surface and groundwater (Table 7.4). Lead, benzene, and PCBs, in particular, were notable. An impermeable wall was constructed on the north coast of the disposal area to prevent polluted water from flowing out to sea. The industrial waste was then excavated (Figure 7.7) and brought to Naoshima Island where a

TABLE 7.4
Average Concentrations of Hazardous Contaminants in the Waste Leachate, Groundwater, and Surface Water at Teshima (Takatsuki, 2003)

Component	Leachate (mg/L)	Ground Water (mg/L)	Surface Water (mg/L)
Arsenic	0.048	0.004	0.02
Benzene	1.2	1.1	ND
Cadmium	0.022	ND	ND
1,3 dichloropropylene	0.54	ND	ND
Chromium (VI)	0.1	ND	0.03
Dioxins (ng TEQ/L)	8.0	0.031	0.18
Lead	6.1	0.804	0.12
Mercury	0.0017	ND	ND
PCBs	0.016	ND	0.0005
Tetrachloroethylene	0.19	0.0041	ND
Trichloroethylene	ND	0.003	ND

ND: Not detected

FIGURE 7.7 Two views of the Teshima waste site with retaining wall where the waste and contaminated soil are contained before transfer to the treatment plant.

melting furnace was constructed. The slag was to be reused as aggregate in concrete. Comprehensive environmental preservation measures and full environmental monitoring of sewage and emission gas are planned in Teshima and Naoshima. The cleanup cost was in the order of 30 billion yen (approximately $300 million US).

7.3.6 Greenhouse Gas Emissions

Activities associated with urban living contribute significantly to the production of greenhouse gases. The combustion of fossil fuels is a major source of carbon dioxide production. Wood burning is another significant source. Methane released from anaerobic decomposition of organic waste in dumps and other sources contributes about 15% of the greenhouse effect. Chlorofluorocarbons (CFC) used as refrigerants in air conditioners and as propellants, were produced in the 1950s. Their use has decreased significantly in developed countries, with the exception of developing countries, after the introduction of measures for their reduction. Concentrations of ozone, produced in internal combustion engines, have also increased. Levels have tripled in Europe and North America in the troposphere. The evidence indicates that human activities during urbanization have clearly increased these levels (greenhouse gases and CFCs), and that it is very likely that these will affect the geoenvironment in the future. If global warming comes to pass, sea levels will rise as a result of ice cap melting and flooding of coastal regions will occur. This will disrupt coastal ecosystems and habitats. Permafrost melting will impact arctic ecosystems and will also cause considerable distress to physical structures because of terrain instabilities.

Due to the targets set for GHG reductions, efforts are being made to determine the emissions in cities. The CO_2-USA project (Mitchell et al., 2022) has constructed a dataset for a Canadian and several US cities for carbon dioxide (CO_2), methane (CH_4), and carbon monoxide (CO) using several techniques. The data is easily accessible at www.unidata.ucar.edu/software/netcdf/, accessed Feb. 2024.

7.3.7 Impact on Ecosystem Biodiversity

Globally, humans have impacted three-quarters of the land surface, adversely impacted two-thirds of the oceans, and destroyed over 85% of wetlands. These impacts have the numbers of species by more than 20%, and another 1 million species are at risk of imminent extinction (IPBES, 2019). The multitude of microorganisms and their diversity in the soil ecosystem is essential in maintaining and developing a healthy soil. Organic matter and nutrient recycling, mineralization and decomposition are essential processes that have not been studied extensively in urban soils, particularly when compared to agricultural soils (Harris, 1990). Soil organisms include microbiota (bacteria, fungi, algae, protozoa), mesobiota (arthropods, nematodes, springtails, etc.), and macrobiota (earthworms, mollusc, larger arthopods, enchytraeids, etc.).

The impact of urbanization on soil properties and attributes includes great variability, compaction, bare topsoil that is often water repellent, altered pH conditions and restricted aeration and drainage, altered nutrient cycling of the soil organisms, presence of other materials and contaminants in the soil and altered temperature profiles (Craul, 1985). These changes all significantly impact on the soil organisms. For example, trampling by humans in urban forests can significantly reduce the numbers of earthworms. More information is needed, however, on species diversity and numbers in urban soils and the impact of recreation, disturbance, compaction, and contamination on these numbers. This microbial community can be used as an indicator of soil quality.

Chemicals from wastes, leaks, and emissions enter the soil ecosystem. They can accumulate in organisms via bioaccumulation as higher animals on the food chain eat lower contaminated organisms. As the contaminants increase in the species, the species may become compromised, causing an imbalance in the whole system. Anecdotal evidence indicates that fewer organisms, less biomass and fewer species of organisms are found in urban soils. Assuming the evidence to be valid, this would be an indication of the stress on these organisms caused by the impact of human activities on urban soils. Whilst direct measures of the level of degradation or impairment of the quality of urban soils are not available, an argument could be made in support of the use of sensitive soil fauna and microflora as indicators for urban soil quality.

7.3.7.1 Impact on Resources

Material consumption by cities is extensive. In 2010, it was estimated by the UN that more than 40 billion tonnes of materials are consumed by cities, which will more than double by 2050 to 90 billion tonnes (UN Environment, 2018). This is due both to increased population and consumption per capita. More efficient use of resources is highly needed.

7.4 IMPACT AVOIDANCE AND RISK MINIMIZATION

As cities grow, the problems and stresses created by human-urban activities on the land environment will escalate. The requirements to satisfy the classical "food, shelter, clothing, and recreation" needs of the community, impose several demands and stresses on the urban environment and its contiguous regions. To a very large extent, the main challenges for an urban society are linked to (a) sensible land management, (b) energy utilization-consumption and management, (c) non-energy resource (food, minerals, building materials, etc.) consumption and management, (d) waste management, (e) reduction of pollution to air and water, and (e) water resources management.

7.4.1 Waste Management

7.4.1.1 Pollution Management and Prevention

Waste generation (urban and industrial) is a large problem (https://datatopics.worldbank.org/what-a-waste/trends_in_solid_waste_management.html). New land must be acquired for classical waste disposal techniques involving landfills since older landfills become filled and are de-commissioned. Modern requirements for environmentally sensitive waste disposal and treatment require considerable resources. The essence of these requirements is embodied in the 4Rs – recycle, reuse, reduce, and recover. Reduction of waste entering landfills is the stated objective of almost, if not all communities and municipalities. However, municipal waste is projected to increase from 2 billion tonnes in 2016 to 3.4 billion by 2050 (statista.com). Since 2000, Europeans (Eu-27) on the average generate 520 kg of household waste per capita, which has stabilized despite economic growth until 2008 (European Environment Agency, Municipal waste generation CSI016/waste 001, published Dec. 2011). Annual municipal waste generation in Denmark is 845 kg per capita. Annual municipal waste generation in Germany is 53.7 tonnes. Japan also produces a steady amount (1990 to 2007) at 400 kg/capita whilst Canada continues to increase production (777 kg per capita) twice the amount of Japan. Since 2005, the US is starting to decrease its production (Conference Board of Canada, Jan 2013). In 2018, the US and Canada produced 2.58 kilograms and 2.33 kilograms per day, respectively (statistia.com). The United States and China are the largest producers of municipal waste in the world, generating annually over 200 million metric tons. In contrast, municipal waste generation in the UK was approximately 31 million tons in 2020 (statista.com).

Other types of wastes are also significant and growing. For example, e-waste has increased by 60% from 2010 and 2019 when 53.6 million metric tons was produced and is expected to reach 75 million metric tons by 2030 (statista.com). Technology is advancing quickly; demand increased and life cycles of the product are short. China is the world's largest e-waste producer but produces three times less e-waste per capita than Europe. Despite the large quantities, only 17.4% was collected and recycled. Large amounts are shipped to developing countries without proper waste management. Components, like arsenic, mercury, and flame retardants, can leach into the ground from improperly managed e-waste. This creates health and environmental problems in countries like Ghana.

Another waste that is growing is plastic waste. It was globally estimated at 353 million metric tons in 2019, which is double that of 2000 (statista.com). Only 9% was recycled, half was landfilled and 23% was mismanaged and can be found in rivers, oceans, and other waterways. This has severe impacts on marine life and ecosystems, particularly as plastics are very difficult to degrade (taking up to 500 years).

Food waste is also a substantial component of the waste stream, particularly in China and India, producing 92 million and 69 million metric tons every year, respectively (statista.com). Approximately 17% of total food is wasted. Households are the main sources, followed by the food service sector. Food waste is costly, expends precious resources and produces large volumes of GHG on degradation, which in 2015 totalled 17.9 billion metric tons of CO_2 equivalent. Food waste per capita is quite similar among developed and developing countries. Highest per capita food waste production is in Western Asia and Sub-Saharan Africa. In Switzerland, a family of four generating 100 tonnes of household waste would require 200 m^3 of landfill volume. Reduction in land required waste landfill can be obtained through implementation of the other 3Rs (recycling, reuse, and recover) together with incineration. Historically, landfills used to be located on the outskirts of cites. However, with the expansion of cities and suburbs in particular, these landfills are now part of the urban-suburban landscape. Although incineration is cleaner and energy efficient and can substantially reduce the volume of waste, incinerator facilities without benefit of modern burning systems and efficient scrubber units can emit pollutants into the air and produce toxic materials and ash that require disposal. Hazardous wastes, such as solvents, oils, medical wastes, and highly contaminated soils are often treated by incineration. With present capabilities and efficiencies in incineration and scrubber systems that can clean properly capture pollutants before emission discharge, incinerators are being considered as the disposal system of choice since they can be located in or near cities. The system in the city of Vienna, Austria is an example.

Pollution prevention is a method to reduce damage to the environment for future generations. Practices such as pesticide addition to lawns, city parks, etc., should be modified or eliminated. Some municipalities are now banning their use. New methods to control pests should be introduced, such as biological control. Many cities have banned the use of pesticides for private use. The use of less harmful of hazardous products can reduce pollution through substitution. A key example where this should have been practiced is the case of CCA-treated wood. Whilst it has only recently been banned for use in many countries, wood previously treated still exists in service. Chromated copper arsenate (CCA) is a major wood preservative that was used in North America for many years for lumber treatment against insects and microorganisms because (a) it was inexpensive, (b) it left a dry, paintable surface, and (c) it bonded well, and was thus relatively leach-resistant. Primary utilization was for decks, playground equipment. Even today, it would not be unusual to find that outdoor wooden facilities with CCA treatment. CCA associated with playground equipment is particularly problematic since children are directly exposed through physical contact and subsequent oral contact by ingesting food with hands having previously been in contact with CCA contaminated wood. The U.S. Consumer Product Safety Commission (CPSC) has stated that this might even cause cancer in children (Green Building News, Feb. 2003 issue). There is increasing concern about potential environmental contamination from leaching of Cu, Cr, and As from treated wood in service and from wood removed from service and placed in landfills and composted material. Open burning of this type of wood is also potentially hazardous. Proper incineration procedures are necessary. The life-cycle of treated wood is estimated to be about 25 years and the wood is then discarded as waste. By 1995, more than 90% of 67 million kg of utilized waterborne preservatives was CCA-treated (Solo-Gabriele and Townsend, 1999). The quantity of removed treated wood from services was estimated to increase from 6 million cubic metres by the Forest Products Laboratory (FPL) in Madison, Wisconsin, to 16 million cubic metres by the year 2020 (Cooper, 1993).

Moghaddam and Mulligan (2008) tested CCA treated Gray Pine species wood for leaching using a modified TCLP method to determine the leaching of the three metals under various conditions. To obtain the results, samples of ground wood were soaked in acetic acid (0.1N) at pH values of 3, 4, or 5, temperatures were 15 °C, 25 °C, or 35°C and leaching time was 5 days, 10 days, and 15 days. The amounts of chromium as chromium oxide (III) (CrO_3), copper as copper oxide (CuO) and arsenic as arsenic oxide (III) (As_2O_3) leached from the wood were determined to be 49% CrO_3, 34% As_2O_3, and 17% CuO. The study also examined the effects of pH and temperature on the leaching of the three metals, from the wood, for a 5-day period, and found that measurable amounts of chromium, copper, and arsenic in the leachates. Arsenic was found to be the least resistant metal to leaching when the temperature increased, and chromium was the most resistant. In addition, there was more leaching of all three elements as the pH decreased. The effect is shown for pH 4 in Table 7.5. The results of the study showed that there is the risk of soil, water, and environmental contamination by chromium, copper, and arsenic, wherever chromated copper arsenate treated wood is used or disposed in a landfill. Chromium was leached the least despite being present in the greatest proportion. Disposal must be in a lined landfill to avoid contamination of the groundwater.

Since CCA-treated wood was exempt from the Toxicity Characteristic Leaching Procedure (TCLP) developed by the US EPA, few reports concerning tests on these substances are available.

TABLE 7.5
Results of Leaching Tests with Acetic Acid Under Various Conditions (Adapted from Moghaddam and Mulligan, 2008)

Conditions	Chromium (mg/L)	Copper (mg/L)	Arsenic (mg/L)
TCLP regulatory level	5.0	Not on the list	5.0
Experimental data at 35°C, pH 4			
5 days	0.6	4.9	4.5
10 days	1.1	5.9	5.7
15 days	1.4	6.4	5.5

Wilson (1997) tested CCA-treated wood and found that it failed for arsenic and barely passed for chromium. Since the ash also failed the test, one would conclude that CCA-treated wood cannot be burnt in incinerators since the metals will remain in the ash. Arsenic can vaporize and be captured by air control equipment or escape into the atmosphere, but chromium and copper will stay in the ash. Mixing the CCA wood with mulch for landscape purposes is also problematic because of the potential for arsenic leaching. Wood-burning power plants cannot accept this type of wood if they use their ash for application onto agricultural fields. The ash becomes hazardous if 10.7% or more of the wood is CCA treated. Whilst landfilling in municipal landfills is an option, it is not uncommon to find the waste (ash) sent to unlined *construction and demolition (C&D)* landfills – a practice that leads to contamination of groundwater. Recently, an alternative method was examined involving acid digestion followed by ethanol production by fermentation (Moghaddam, 2010). A slightly lower amount of ethanol from CCA treated than untreated wood was produced (6 g/L and 7 g/L, respectively). In general, it suggests that production of ethanol from a hazardous waste (CCA-treated wood) could assist in the disposal of CCA-treated wood while generating a clean fuel as a source of energy.

Attempts in removal of the CCA treatment in the wood before reusing the wood have not been very successful. However, disposal of these products in the future is still uncertain. As landfilling in lined landfills is the only current option, the production of CCA-treated wood will be limited and was phased out from consumer application at the end of the year 2003 in the US – as an agreement with manufacturers and the US EPA. European countries had already banned this type of treated wood. Other alternatives such as ammonium copper quaternary (ACQ) and copper boron azole (CBA), alkaline copper quaternary compounds (ACQ), copper azole (CuAz), ammoniacal copper zinc arsenate (ACZA), copper citrate, and copper HDO (CuHDO) preservatives could be used.

7.4.1.2 Waste Reduction

Reduction of the wastes or reduction of the source of waste is the key to reducing emissions from landfills and other waste management techniques. Life-cycle analysis (LCA) has been identified as a tool to help achieve sustainable consumption as it accounts for the emissions, and resource uses during production, distribution, use, and disposal (ISO, 1997). Three steps are involved: (a) The processes of the life-cycle, (b) the environmental pressures of the processes (Figure 7.8), and (c) the environmental impact of the use – including the use of impact indicators. Whilst ISO 14040:2006 (reviewed and confirmed in 2010) defines the inventory analysis and impact assessment steps, the other steps involving definition of the process and the interpretation of the results are not necessarily simple "steps". Several databases such as EcoInvent (version 3.0) and SimaPro are now available for some materials, but data is lacking. GaBi (now Sphera Product Sustainability Software (https://sphera.com/product-sustainability-software/), Sima Pro (https://simapro.com/), CcaLC version 3.3 (www.ccalc.org.uk/) and Ecoinvent (https://ecoinvent.org/, version 3.10), the US LCA Commons (www.lcacommons.gov have been developed and enable more users to perform the LCA analysis

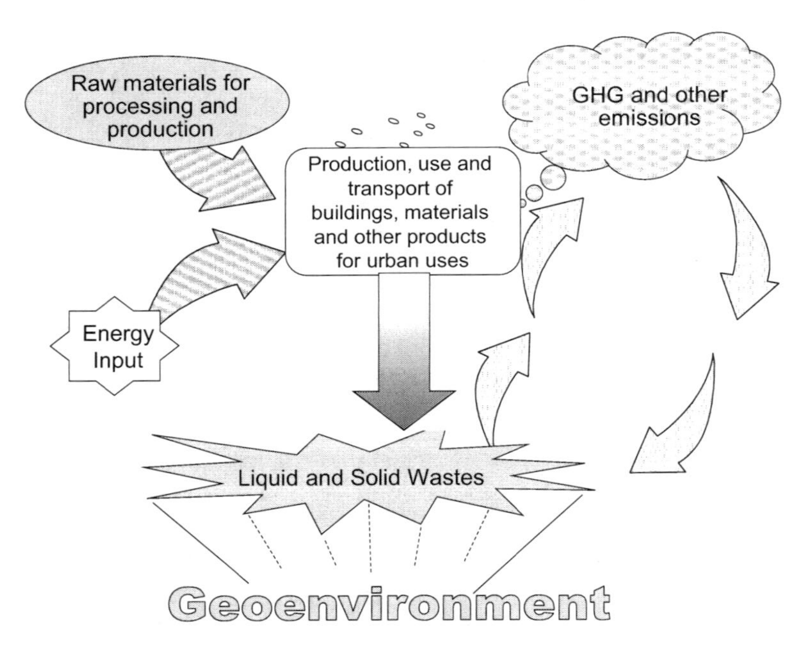

FIGURE 7.8 Elements of a life-cycle assessment.

as they include background data (secondary data for materials and energy). Primary data although preferable may be difficult to obtain. This ultimately will enable improvement in the design of products, processes, and services.

A *Life-Cycle Initiative, Joint Initiative of UNEP* and the Society for Environmental Toxicology and Chemistry (SETAC) was launched in 2002 and was formed to promote the development of standard methods and exchange of information (UNEP, 2004) (www.lifecycleinitiative.org/, accessed Feb. 2024). For example, waste management experts can determine which wastes are causing problems and inform the public on alternatives or methods for waste reduction, however most of the current information for LCA methods has been in regard to energy analysis (Hertwich, 2005). In the third phase of activity (2017–2022), one of the main objectives is towards mainstreaming the use of life-cycle approaches. Some currently available tools are as follows:

- Bilan Produit, which helps to assess and reduce the environmental impact of a product, (https://base-empreinte.ademe.fr/).
- Carbonostics, an online, carbon accounting and product life-cycle assessment for the food sector.
- OGIP, a tool to achieve an architecturally and ecologically optimised overall project within a budget. (www.dora.lib4ri.ch/empa/islandora/object/empa%3A8255, accessed Aug. 2024).
- ECODESIGN+, a tool for supporting product carbon footprint (PCF). (www.ecodesign-company.com/en/ecodesign-plus-en/about-en).
- WRATE (Waste and Resources Assessment Tool for the Environment), software which determines the environmental impact of different municipal waste management systems. (https://fellowsenvironmental.com/wrate-life-cycle-analysis/), accessed Aug. 2024.
- Eco, the USai and Eco-bat software tools have been merged to facilitate the work of professionals in the building sector. The result is the first software that makes it possible to analyse the U value, thermal inertia, and life-cycle assessment of a construction (homogeneous and inhomogeneous) Eco-Sai a tool for quickly performing a thorough life cycle impacts assessment of a building. (https://eco-sai.com/).

- EQUER, a LCA tool based on the building simulation by Pleiades, it is approved for BREEAM credits. (www.izuba.fr/logiciels/outils-logiciels/analyse-du-cycle-de-vie-acv/)
- LEGEP, a calculation tool for integrated design of sustainable buildings. (https://legep.de/?lang=en)

7.4.1.3 Waste Recycling

Recycling, reuse of wastes and incineration reduce the volumes of waste entering landfills. It is clear that Canada whose per capita waste generated is continually increasing (from 510 kg/capita in 1980 to 2.33 kilograms per day in 2018) must increase efforts through recycling and waste to landfill reduction. Countries such as Finland and Japan have initiated aggressive waste management plans. Application of waste limits could be implemented as a "*polluter pays*" mechanism. Whilst recycling and reduction of the wastes and thus the sources of pollution has made inroads, particularly in advanced countries, a lot more progress is required. Although difficult to achieve, 100% recycling should be the goal for all materials including those for construction (asphalt, concrete, wood, etc.). Collection of paper, plastics, glass, trees, and other materials for recycling is practised to reduce energy, material requirements, and landfill spaces. Backyard composting can substantially reduce organic waste transport, collection, processing and disposal, while providing a fertilizer for home gardens.

Sewage sludge or biosolids is produced by wastewater treatment plants. The amount of sludge depends on the amount of water treated by the city, town or municipality. The sludge is typically treated to remove water and stabilization is by thickening, digestion, conditioning, dewatering, drying, and incineration. After drying/incineration, the sludge can be reused as a powder for manufacturing bricks or mixed with Portland cement and stabilized ash. Land application for agricultural uses or for land reclamation is also possible. Pathogen and heavy metal concentrations must be removed to protect groundwater quality and public safety.

Sewage sludge has been used as an amendment to urban and agricultural gardens for organic carbon and nutrients. The sludges generally contain about 45% organic matter, 2.0% nitrogen, 0.3% phosphorus, and 0.2% potassium (Bridges, 1991). These sludges have been known to contain up to 3000 ppm of zinc, 2000 ppm of chromium, 1400 ppm of copper, 385 ppm of nickel, 240 ppm of lead, 60 ppm of cadmium and 60 ppm of arsenic. Repeated application of this sludge leads to an accumulation of heavy metals in the soil. More recently, the metal content of biosolids was evaluated in Canada (Table 7.6). Although the levels are lower, they still can be problematic for Cd, Co, Cu, and Se, which are higher than Quebec regulations for fertilizer use.

TABLE 7.6
Heavy Metals from Biosolids from a Canadian Wastewater Treatment Plant in 2007

Heavy Metals	Concentration (mg/kg)
As	6
Cd	9
Co	19
Cr	131
Cu	477
Hg	1
Mo	7
Ni	35
Pb	105
Se	9
Zn	596

Glass is dense and takes up significant landfill space. Increasing recycle rates is required. However, separation of colored glasses is difficult and must not be contaminated with other materials, such as ceramics. Crushed glass can be processed to reach the characteristics of gravel or sand. It can thus replace aggregate in backfill, road construction, and retaining walls. The use of mixed glass as glassphalt is another possibility to replace aggregate in asphalt. More engineering tests (compaction, durability, skid resistance) will need to be performed on the properties of glassphalt and other materials. TCLP tests would need to be undertaken to determine the leachability of heavy metals from the glass. Lead has been shown to leach below the US EPA levels (CWC, 1998).

A recent LCA study (Tushar et al., 2023) on crushed glass use for road materials showed that recycled glass aggregate produced by the crushing process reduces carbon emissions by 46.67% compared to natural sand obtained from a quarry. However, CO_2 emissions increased by 89.9% compared to natural sand extraction if both the washing and crushing processes are applied to produce recycled aggregates. If clearer glass is collected separately, then simple crushing only is needed. Regardless of the type of waste glass collected, recycling into crushed glass aggregate produces a significant reduction of the environmental impacts compared to sending it to landfills; this was quantified to be approximately 10.62 kg CO_2-eq per ton of recycled crushed glass (RCG). If the RCG is to be used in asphalt and concrete mixtures, the compressive strength and tensile strength must be determined to ensure that increasing proportions of RCG does not affect the engineering properties. Between 15% to 30% RCG by mass has been recommended by different road authorities.

According to the UNEP (www.unep.org/interactives/beat-plastic-pollution/) only 10% of the plastic waste of 7 billion tonnes annually is recycled. The remaining 90% is disposed of in a variety of ways, including landfilling, dumping, burning or even shipped to other destinations for disposal. More effective management of the waste is urgently needed. For example, the UNEP, indicates that plastic waste floating in the Mekong River is largely due to household littering and dumping. Meijer et al. (2021) has indicated that rivers (>1000) are a major source (>80%) of plastics in the oceans. Plastics comprise around 10% of municipal solid waste but because it is not very biodegradable, it can remain there for very long periods of time. Recycling can be achieved to form new products from recycled polypropylene and polystyrene by injection moulding, blow moulding, and foam moulding. Products can be redesigned to include higher contents of recycled materials and for multiple uses. Many of these single use plastics are banned in many cities. The Government of Canada has a plan for zero plastic waste by 2030. A few municipalities, however, such as Montreal, Edmonton and Moncton, N.B., have local bylaws banning either plastic shopping bags or single-use plastics. Replacing the fossil fuel plastics by biodegradable plastics are another alternative that is increasing in use.

Scrap tires are environmental threats: (a) As mosquito hazards from accumulation of rainwater, and (b) as fire hazards, since they are difficult to extinguish, burning for months, creating substantial pollution in the air and ground. Landfilling is difficult since landfills (a) are not compactable, (b) require hundreds of years to decompose, and (c) occupy substantial amounts of space. Contaminants may also leach into the soil. Heavy metals, PAHs, and TPHs have leached out under various conditions. This has limited the reuse of tires as artificial reefs and other environmental applications. The type of tire shredding can substantially affect the leaching results. However, as new laws prohibiting tire disposal in landfills came into effect, the Environmental Protection Agency indicates that about 85% of (or roughly 340 million) scrap tires are now recycled each year (https://autosphere.ca/tires/2022/03/11/tire-recycling-effective-and-efficient-with-dr-ben-chouchaoui/). Various applications for scrap tires are shown in Table 7.7. Since there will be restriction for tire-derived fuel in the future, the engineering applications for tires are the most promising. Shredded tires, known as tire derived aggregate (TDA), have many civil engineering applications, such as a backfill for retaining walls, fill for landfill gas, trench collection wells, cover material in landfills, backfill for landslide repair projects and vibration damping material for railway lines. Ground and crumb rubber, also known as size-reduced rubber, can be used in both paving type projects and in mouldable products.

TABLE 7.7
Applications of Recycled Scrap Tires (Adapted from information reported in Sharma and Reddy, 2004)

Tire Form	Application
Tire chips	Embankments, aggregate materials retaining walls, bridge abutment backfills
	Reduction in frost penetration due to low thermal conductivity
	Adsorption of organic chemicals in leachate collection systems, for gas migration control trenches, gas collection and venting layer in caps, leachate recirculation trenches, drainage layer in covers in landfills
Whole tires	Low height retaining structures
	Highway applications for stabilization of road shoulders or noise barriers
Rubber	Rubberized asphalt concrete with long life and decreased thermal cracking
Crum rubber	Rubber is moulded to form panels for railway crossings that fit between tracks instead of timber crossings
Tire shreds	VOC movement reduction in groundwater by addition to bentonite slurry walls, sub-grade fill and embankments, back-fill for walls and bridge abutments, sub-grade insulation for roads, landfill projects, septic system drain fields
Ground tires	Sorption of VOCs from wastewater in wastewater treatment plants

Other waste materials include demolition and concrete from demolition and construction work. According to the United Nations Agenda 21 (1993), the promotion of environmentally sound waste disposal and treatment methods for construction debris is highly desirable and is one of the programmes. Concrete can be recycled into aggregates of concrete, base foundation, new pavement, road shoulders, or backfill. Scrap wood can be used for landscaping (wood chips, mulch, groundcover), wood-based geotextiles for landfills as fuel or in building products (fibreboard products, rigid boards, plastic lumber). Asbestos, lead-based paints, PCBs, CCA, and pentachlorophenol (PCP) leaching from treated wood may be a concern. Cardboard, dry wall, rubble can be used as aggregate. The US EPA estimated that the amount of building-related waste in the United States was approximately 170 million tonnes in 2003, which is an increase of 25% compared to 1996. This amount is still increasing. In 2018, 600 million tons of C&D debris were generated (90% due to demolition and less than 10% due to construction). Of this, 455 million tonnes of C&D debris were reused while 145 million tonnes were sent to landfills (www.epa.gov/smm/sustainable-management-construction-and-demolition-materials, accessed Jan. 2024). According to the US EPA resources and energy can be saved avoided by reusing materials compared to new materials. Some examples include the following:

- Donation or use during the rebuilds of doors, hardware, appliances, and fixtures.
- Wood chipping on site for use as mulch or groundcover.
- Crushed and de-papered gypsum can be used as a soil amendment in moderate quantities.
- Brick, concrete, and masonry for use as fill, sub-base material, or driveway bedding on site.
- Excess exterior wall insulation in interior walls use for noise attenuation.
- Remix of paints for garage or storage areas, or as a primer.
- Reuse or recycling of packaging materials.

7.4.1.4 Composting and Anaerobic Digestion of Organic Wastes

Composting is one of the simplest processes that can be used for treatment of organic wastes. It can also be very sophisticated. At the present time, it is used mainly for food wastes (Torsha and

Mulligan, 2024). However, composting materials can include garden and vegetable cuttings, paper and cardboard, garbage and any decomposable organic matter. It is applicable for homeowners, individual institutions, and even communities. It has increased in popularity significantly due to consumer awareness in decreasing the amount of wastes going to landfills. Moisture and oxygen levels, pile temperature, and odour must be monitored throughout the process. Carbon to nitrogen ratios are the other important factors that must be optimal to ensure the success of the process. Composted products can be used as soil conditioners. The use of in-vessel systems or biofilters with aerated piles will reduce odour levels considerably.

Anaerobic digestion is the microbial stabilization of organic materials without oxygen to produce methane, carbon dioxide, and other inorganic products. It is the most common biological treatment method used for high strength organics in the world. At present, the technique is used for industrial, commercial, and municipal sludges (Mulligan, 2002). It will continue to be popular since it produces methane that can be recovered and used for energy. Solid concentrations up to 35% can be processed. In addition, less carbon dioxide is produced in comparison to aerobic processes. Solids from anaerobic digestion can be composted. Supernatants from the sludge often have high organic contents and need to be treated further. Anaerobic digestion of municipal solid waste (MSW) organics is performed at solid concentrations of 4 to 10% (WEF, 2017). Higher concentrations must be diluted. Gas production is 1.5 to 2.5 m^3/m^3 reactor or 0.25 to 0.45 m^3/kg of biodegradable volatile solids. The biogas contains 65% to 70% methane (CH_4) and 25% to 30% carbon dioxide (CO_2), with trace amounts of other gases. Retention times are approximately 15 to 20 days. Most reactors operate under mesophilic conditions (WEF, 2017).

Aerobic digestion is a highly stable process that can be operated so that nitrification can also occur. Because of aeration requirements, it is more suitable for small- and medium-sized plants. Capital costs are less than costs for anaerobic digestion, and operation is safer since there are no explosive gases produced. Supernatants are also of higher quality. Thermophilic digestion is becoming increasingly popular due to pathogen destruction, low space requirements, high sludge treatment rates, and organic disposal regulations. A recent review of autothermal aerobic digestion (ATAD) was performed by Zhang et al. (2022).

7.4.2 Water Resource Management

As up to 90% of human sewage in developing countries is untreated, this is a growing threat to the health of surface water and groundwater. The capacity of rivers to absorb these pollutants without irreparable impairment of water quality is being exceeded. A combination of regulations and consumer incentives is needed to remedy the situation. Infrastructures must also be improved to reduce leakage from water supply systems. Since water quality monitoring is non-existent in many countries, the impact of urbanization on water quality and water ecosystems is largely unknown – a situation that must be remedied.

The lack of clean water, on the one hand, and flooding, on the other hand, are problems that beset many cities worldwide. Development of dams upstream instead of policies for water sustainability, has led to water supply problems for downstream cities. We have previously discussed some of these problems in Chapter 3. Reuse of wastewater to reduce contaminant discharges into water systems may also be beneficial. Introduction of taxes or tax credits can also promote reduction in water use. Various other approaches have been proposed by König (1999) including the use of plant roofs to adsorb roof run-off with a combination of seepage or delayed drainage areas to allow rainwater to return to the groundwater. Roof run-off could also be collected for use in toilets as shown in Figure 7.9.

In 2002, water consumption in Canada per capita was approximately 1500 m^3 per capita as compared to less than 197 for Denmark (www150.statcan.gc.ca/n1/daily-quotidien/210817/cg-c001-eng.htm). Through national efforts, a water conservation programme was introduced including leak detection and repair, changing water taxation, and pricing introducing metering and water-saving

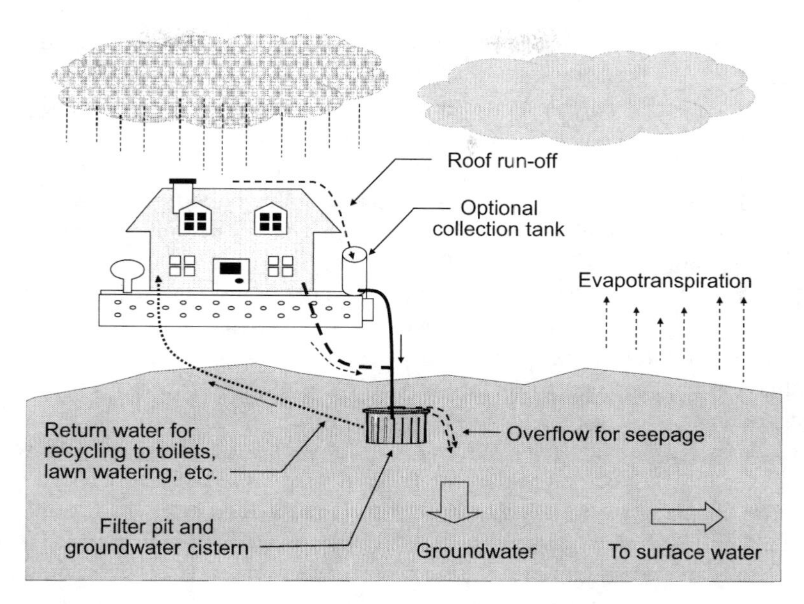

FIGURE 7.9 Recycling of roof and wastewater to replenish groundwater and provide non-potable water for housing. (Adapted from information in König, 1999.)

devices, and public education, water consumption decreased from 1997 to 2006 by 18.5% (Danish Water and Wastewater Association, 2007).

7.4.3 Reduction in Climate Change Impacts

As the burning of wood, coal, petroleum, and natural gas leads to substantial production of greenhouse gases and promotion of acid rain, it is obvious that reduction in the use of these energy sources is essential – as a step to fulfilling the target of net zero by 2050, which many countries have pledged to work towards. Whilst many different other steps are needed to meet the requirements, and whereas this is an essential step in the right direction, it is recognized that combustion of fossil fuels is not a sustainable practice. Other sources of energy, such as solar, wind, geothermal, hydrothermal, and hydroelectricity produce less greenhouse gases than the above-mentioned fuels. In the case of acid rain, the choice of fuels with lower sulphur and nitrogen oxide contents can also minimize acid fallout and damage to soil, water, fauna and people. Nuclear energy is a clean energy. However, the problems of spent fuel rods and other radioactive waste materials are substantial and have yet to be fully resolved. A recounting of the repository disposal of high-level nuclear wastes and the technological tools for managing such wastes can be found in Pusch and Yong (2005). Biogas from landfills should be recovered and used for heating purposes.

For example, the St. Michel Environmental Complex, in Montreal, Canada recovers, at the landfill, approximately 20 megawatts of electricity, which is enough to power 12,000 homes (https://montreal.ca/en/articles/parc-frederic-back-unique-metamorphosis-18997). Other metrics are for material reuse during the development of the former landfill into a park. This includes (in metric tonnes) excavated soil (200,000), sand (40,000) gravel (10,000), composted leaves on site (24,000), and wood chips for tree pruning (10,000). The park grounds are composed of soil from construction sites, wood chips, leaf compost, and sand. In addition, benches were made from felled ash trees, and new roads and trails were constructed from cliff rockfalls. More than 525 trees, 17,800 bushes and perennials and over five types of herbaceous plants were planted to promote biodiversity and have a low ecological footprint. Little water is required, and the root systems are shallow to not impact the landfill covers.

Reforestation and increases in forest growth will increase carbon dioxide sink capacity. Sustainable energy generation and low-energy consuming practices for transportation, heating, and so on, must be practiced, although they are not without political implications. Green buildings for houses and commercial use to reduce energy use are now being constructed. Eco-building codes and architecture will need to be developed and improved. Santa Monica, California, for example, replaced city building electricity with geothermal electricity in 1999 and Austin, Texas offers a "GreenChoice" option for consumers that includes renewable electricity by wind energy (https://austinenergy.com/green-power/greenchoice). In 2022, by 28,000 choosing this option, carbon dioxide emission reduction is equivalent to the amount that 201,000 hectares of US forests can remove in one year, or the emissions that the average gasoline-powered passenger vehicle emits over 1,677,000,000 km in a year.

Biodiesel is gaining popularity as a renewable fuel. It is made from the oil of various crops, including palm, mustard, rapeseed (canola), sunflower, and soybeans (Weeks, 2005). Rapeseed, which produces the highest yield, is the most common source in Europe, whereas soybean is the most common in the US. Waste frying oil can also be used, which diverts waste products from the landfill. All emissions (with the exception of NO_x) are reduced by up to 80% when biodiesel is burnt in comparison to petroleum diesel. Spills are of less impact since biodiesel is less toxic and degrades faster in the environment than diesel. In addition, wastewater and solid waste generation rates are lower.

7.4.4 Green Spaces

The encroachment of cities into natural landscape areas has a detrimental impact on biodiversity. To mitigate this impact, one needs to have access to a base inventory of the natural places and biodiversity existing within the target areas so as to develop a basis for determination of the nature and extent of potential impacts. This base inventory also forms the basis for planning for urban expansion that is sensitive to the need to (a) prevent introduction of invasive species, (b) protect endangered species, and/or (c) protect the natural habitats. Cities such as San Francisco, California are looking at ways to understand biodiversity, protect and restore the natural ecosystems, protect sensitive plants and animals through education, purchasing green spaces, developing pest management programmes, and enforcing environmental regulations (Portney, 2003). Planting flora, fruits and veggies native to the area or nectar-producing wildflowers, staying on walking paths or hiking trails, and building bee boxes are ways homeowners can support biodiversity (https://sustainability.yale.edu/blog/6-ways-preserve-biodiversity). Cities can maintain green spaces, such as gardens, parks, and green roofs. Incorporating nature-based solutions has many benefits for the environment, biodiversity, human wellness, air pollution reduction, education and recreation (www.iucn.org/story/202305/embracing-biodiversity-paving-way-nature-inclusive-cities). Reducing destruction of wetlands is another avenue to preserve biodiversity and serve as flood barriers.

7.4.5 Alternative Forms of Transportation

The 2020 UN Sustainable Development Goals Report states only half the world's urban population has convenient access to public transportation, according to 2019 data from 610 cities in 95 countries. Transport accounts for substantial percentages of global oil consumption (64%), energy use (27%), and carbon dioxide emissions (23%) (www.iisd.org/articles/deep-dive/road-sustainable-transport). Therefore, much progress is needed. The UN Secretary-General's High-Level Advisory Group on Sustainable Transport defines sustainable transport as

> the provision of services and infrastructure for the mobility of people and goods—advancing economic and social development to benefit today's and future generations—in a manner

that is safe, affordable, accessible, efficient, and resilient, while minimizing carbon and other emissions and environmental impacts.

HLAG-ST, 2016, p. 10

Whilst some might argue otherwise, there is a consensus that automobiles have facilitated transport and contributed to economic progress. However, there are many environmental and sustainability issues related to transportation that need to be addressed. The method of delivery of urban transportation needs (buses, cars, trucks etc.) require modification if ozone production is to be reduced. Cities should be planned in an optimal manner to minimize transportation requirements. Bicycles and walking should be facilitated instead. Many cities have introduced extensive bike lane networks and bike share programmes as a major focus. Walking is a completely sustainable transportation mode. Copenhagen was one of the earliest cities to promote pedestrian streets (Vega, 1999). Numerous cities within Europe are promoting pedestrian areas and others such as Amsterdam are promoting cycling. Public transportation will also need to be improved. Due to the COVID pandemic, it was found that transitioning from in-person to virtual conferencing can substantially reduce the carbon footprint by 94% and energy use by 90% (Tao et al., 2021).

Electric cars are another alternative as no emissions are produced. More charging stations will be needed. Electric scooters, electric-assist bicycles, and electric mopeds are now available in over 600 cities across more than 50 countries (www.iisd.org/articles/deep-dive/road-sustainable-transport). Electric buses and trains are also being introduced. The transition for electric in heavy-duty trucks, aviation, and shipping has been much slower.

Initiated during COVID, several European cities closed streets to cars, created pedestrian malls, and expanded bicycle lanes to enhance quality of life. From January 2021, China, has put significant funding into electrification of its public transit has more than 400,000 electric buses, about 99% of the world's total (Margolis, 2019). The multitude and breadth of innovations is encouraging for the future of sustainable transport. Electric vehicles could further expand with the addition of charging stations. However, many developing countries still need access to reliable and low carbon electricity sources (www.iisd.org/articles/deep-dive/road-sustainable-transport).

7.4.6 Brownfield Redevelopment

New policies are being developed in various countries to reduce greenfield consumption. In the UK, regulations require that 60% of all housing must be on brownfields by 2015. Germany is aiming to reduce greenfield consumption by 75% by 2020 (Genske, 2003). The US EPA made special efforts since 1998 to encourage brownfield redevelopment – in line with the goals of sustainability (US EPA, 1999a). The US EPA Brownfields and Land Revitalization Program has created many benefits such as the leveraging of more than 197,600 jobs nationwide, cleaned up more than 2400 properties, increased residential property values by 5–15.2% near brownfield sites and increased local tax revenues after completion of the cleanup (www.epa.gov/brownfields/). The programme aims to reduce uncertainties related to the environmental, economic, and social challenges of using contaminated sites. Seed money and technical assistance are provided to encourage land reuse and attract additional funding.

An overview of the action plan devised by the US EPA (2017) regarding brownfield development is shown in Figure 7.10. Before starting down the road map, reuse goals should be set, the regulations and liability aspects must be understood, the community should be engaged, and funding and professional assistance should be identified. The first part of the evaluation is the collection and assessment of the environmental information for the site. The data will determine what cleanup may be required depending on the intended use of the land. These data will need to be compared with regulatory levels to determine if the contaminant levels in the site are above or below regulatory limits. Risks to the local community should be assessed. Then cleanup options should be investigated and the most

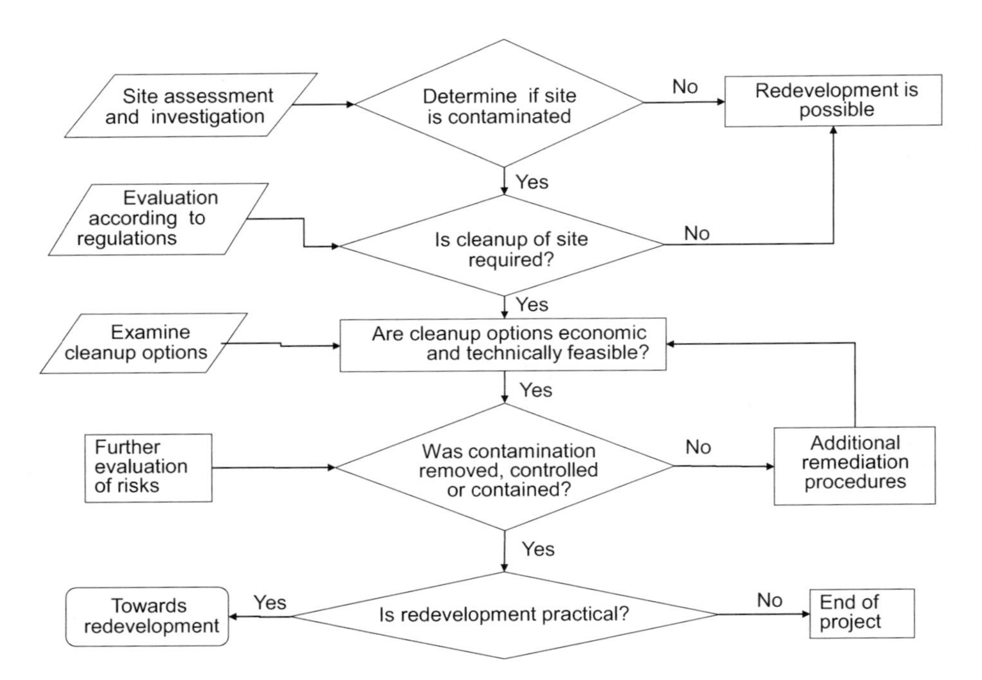

FIGURE 7.10 Evaluation of a brownfield project. (Adapted from US EPA 2017.)

appropriate selected. Greener cleanup practices should be implemented. These will be discussed further in Chapter 9. Once the cleanup option is selected and is implemented, best practices for the cleanup should be optimized. The technology will be implemented and monitored to determine if cleanup levels are reached. Additional contamination may be found that was not initially discovered. Long-term monitoring, particularly for natural attenuation, may be required. The site can only be developed when the regulatory levels have been achieved. Additional contamination discoveries may lead to a revision of the remediation options. Once all contamination is contained, removed, and controlled, the site reuse can be implemented.

The US EPA (1999a) organized various elements into a model framework – illustrated in Figure 7.11. Many of the principles of sustainability are incorporated into this process to enable the project to be sustainable for the community in the future. The framework is designed to assist municipalities, planners, and developers to undertake brownfield projects. The technologies used should incorporate resource conservation, materials reuse, public safety and mobility, and information availability. Factors to measure sustainability should be incorporated into the future and projects should be reevaluated every few years to monitor progress and failures.

Success stories are highlighted at www.epa.gov/brownfields/success-stories. For example, a Harvesting Arts and Culture Center has been established at a Former Brownfield site in Council Bluffs, Iowa. In 2003, the city worked to purchase the International Harvester Company property and begin planning for redevelopment. Two $200,000 Brownfields Assessment Grants were obtained, one in 2005 and one in 2008. Contaminants on the property were identified and site cleanup requirements were determined. Contaminants in the soil, including PAHs, benzene, lead and volatile organic compounds. Lead paint and asbestos were found in the building. In 2009, a $200,000 Brownfields Cleanup Grant was obtained from the US EPA for remediation of the site. The decontamination involved removing approximately 1700 tons (tonnes) of material, including the upper 1 m of soil in contaminated areas. The total remediation cost was $240,000. The Pottawattamie Arts, Culture and Entertainment (PACE) organization and Iowa West Foundation needed a space for an arts and culture centre. More than $27 million was raised

FIGURE 7.11 Elements to be incorporated into a sustainable brownfields' redevelopment. (Adapted from US EPA, 1999a.)

from various sources. Some elements of the building such as the large windows in the entry were retained for the new construction. Cleanup of the site took place from 2012–2013 and the site was rehabilitated from 2016 to 2020. The Hoff Family Arts and Culture Center was opened in February 2020.

7.4.7 Sustainability Indicators for Urbanization

Indicators are useful for determining progress and comparisons with other cities or practices. There has been considerable effort in recent years to create indicators as a measure of the sustainability of cities. Most of the indicators presently available deal with easy-to-obtain data such as recycling information. Cities such as Seattle (Portney, 2003) use a wide range of indicators such as air quality, biodiversity, energy, climate change, ozone depletion, food and agriculture, hazardous materials, human health, parks and open space, economic development, environmental justice, education, etc. (Figure 7.12).

Environmental indicators are generally associated with elements or parameters, such as water, energy, solid wastes, air, and land use. For example, in regard to waste, one is interested in quantities: (a) Generated, (b) disposed, and (c) recycled. Other indicators regarding urban planning are (a) areal size of informal settlements as a percent of city area, (b) green area (hectares) per 100,000 population. Water indicators track (a) wastewater treatment facilities with primary, secondary, and tertiary treatment, (b) water consumption, (c) water loss, and (d) water service interruption per household. For energy, the indicators are generally concerned with (a) use of energy and its form (less polluting energy sources are favored), and (b) the amount of energy used, (c) power interruptions (d) % of city population with electrical service. Air quality indicators seek to provide controls on (a) PM10 concentration, (b) GHG emissions due to transportation, industry and the energy sector. A water quality index, greenhouse gas index and total energy use indices are under development. Others for land quality and use should also be included.

Table 7.8 shows an index value system developed for Amsterdam (Jonkhoff and van Eijnatten, 2012). The indicators are valued equally and use 2010 as the base, which is indexed at 100. As the value decreases, this indicates a more sustainable city, which was occurring as a preliminary basis in 2011. www.amsterdam.nl/en/policy/sustainability/, accessed Feb. 2024.

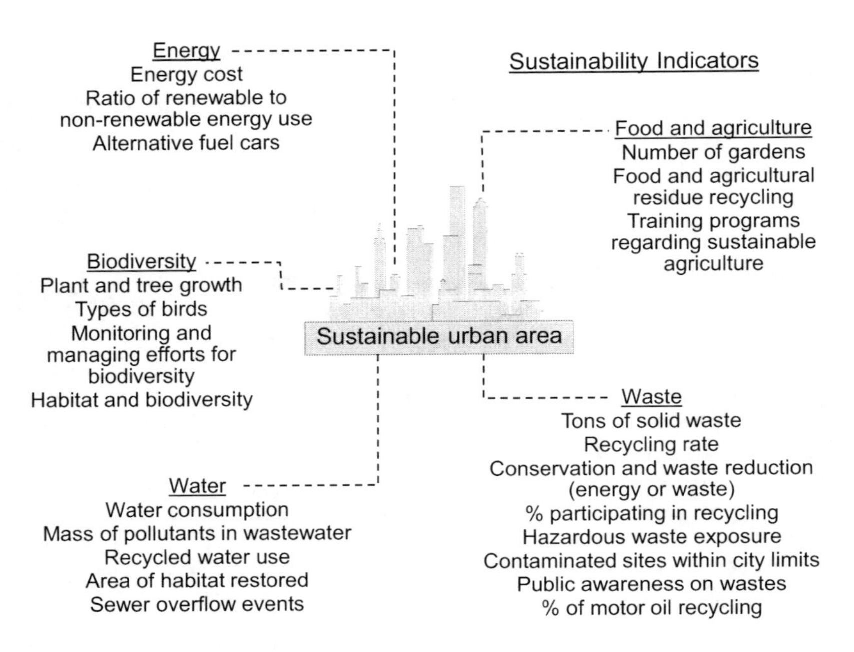

FIGURE 7.12 Selected geoenvironment relevant city sustainability indicators.

TABLE 7.8
Sustainability Indicators for Amsterdam (Adapted from Jonkhoff and van Eijnatten, 2012)

Type of Indicator	Indicator
General indicators	CO_2 emissions per inhabitant (tonnes) NO_x emissions per inhabitant (μg/m^3)
Climate and energy	Energy use (households) (GJ) per inhabitant Sustainable energy production (GJ) per inhabitant (inverse)
Sustainable mobility and air quality	Bicycle share in (%) of total of bicycles, mopeds, motorbikes and cars Share clean trucks and lorries (%) with Euro 4 or cleaner engine
Climate and energy	Attractiveness of Amsterdam for new companies (according to ECM ranking) Energy use per added value (MJ/€)
Sustainable innovative economy	Amount of residual household waste (kg) per inhabitant Liveability indicator (x/10) given by inhabitants when asked how satisfied they were with their neighbourhood on a scale of 1 to 10

Rating systems such as Envision or BREEAM Infrastructure (CEEQUAL) can assist in the sustainability assessments of infrastructure projects through goals and indicators. These are under continual development. Envision (ISI 2023) is used to evaluate the sustainability of infrastructure projects by designers, constructors, community groups, owners, and policy makers. Bronze, silver, gold, and platinum levels can be obtained according to the points reached. Institute for Sustainable Infrastructure (ISI) was founded by the American Public Works Association (APWA), ASCE and the American Council of Engineering Companies (ACEC). Envision can be used for planning, design, construction, and operation of various types of infrastructure projects related to airports, bridges, dams, roads, landfills, and water treatment systems among others. A quick assessment can be done with a checklist for the early stages of a project. A certified evaluator must do the full assessment. In Canada, the tool is promoted by Envision Canada in partnership with the Canadian Society for Civil Engineering (CSCE).

The rating system is a comprehensive framework of 60 sustainability criteria that addresses environmental, social, and economic impacts to sustainability. These criteria, known as credits, are placed in the following categories: Quality of life, leadership, resource allocation, natural world, and climate and risk. The tool can help stakeholders understand the sustainability of the project.

An example of a Canadian project is the 2nd Concession, which is a major north-south arterial corridor under the jurisdiction of the Regional Municipality of York (York Region) (envisioncanada.com). It is the second project in Ontario to earn an Envision award for sustainability from the ISI. Here are some of the features of how the project succeeded.

- An elevated wooden boardwalk through wetlands and marshes connecting forests with growing residential communities.
- Benches, bike racks, and a flagstone meeting area.
- A wooden-clad pedestrian bridge in the Rogers Reservoir overlooks the historic canal and lock system, which the project protected and preserved.
- Bridges spanning an active rail corridor and the Holland River.
- The new retaining wall for the road is both functional and aesthetically pleasing as it incorporates tree patterns that improve community attractiveness for residents and trail users.
- Dedicated cycling tracks (the first in the region).

7.5 MITIGATION AND REMEDIATION OF IMPACTS

7.5.1 Mitigation of Impact of Wastes

Proper daily cover of wastes deposited in active landfills can mitigate odour, dust, fires, and pests. A common technique is to use a granular-type soil fill material cover that is compacted on the daily load of waste brought to the landfill. Techniques for placement of wastes in landfills, using compartments, for example, have been developed and are well illustrated in many handbooks dealing with disposal of wastes in landfills. The details of the interactions between wastes and the land environment are discussed in Chapter 9.

For landfill surface closure, a waterproof cover system as shown previously in Figure 1.11 in Chapter 1, will prevent water infiltration and reduce the requirement for treatment of the leachates. This is the technique of landfill construction and closure that is called a "dry garbage bag" system. The intent of this system is to keep the material entombed in the land with liners surrounding the entire waste that is designed to be impermeable to water. Since water is the carrier for contaminants, i.e., contaminants cannot be transported into the surrounding ground without water, there is a school of thought that argues that denial of water will not only obviate dissolution processes but will also deny production of leachates. Along this line of reasoning, a dry garbage bag system will presumably keep the contained wastes in a dry state "forever". However, reality forces one to accept the fact that engineered liners and barriers have a life span for secure containment that sometimes will have flaws, the result of which will admit water to the system and ultimately generate leachates that will find their way into the surrounding ground. In recognition of that fact, and in support of the thesis that if one could generate a bioreactor system with water entering the waste pile, leachate recycling into the waste pile would accelerate dissolution of the waste material in the landfill. The outcome of this kind of strategy is to obtain faster dissolution of the wastes in the landfill, and a quicker return of the landfill to more fruitful and beneficial land use.

For closures that are designed to prevent moisture entry into the waste pile, a vegetative cover on top of the impermeable barrier can assist in reducing dust generation. The roots of the plants need to be short to avoid penetrating the waste that would inhibit their growth. Biogas at active landfills should be collected for energy recovery – as done, for example, in the landfill in the city of Montreal. Other gases generated in the waste pile will need to be collected to avoid problems – as seen in the Love Canal problem discussed previously.

Wastes are treated in many ways according to the regulations in the country where it is generated. In many ways, unsorted municipal solid wastes can be treated as a resource containing discarded bottles, paper, glass, and cans that can be recovered and recycled. Construction wastes contain many items that are recyclable – as, for example, broken concrete elements crushed and used as granular fill, and scrap lumber and wood chips that can be converted to particle board and paper into cardboard. Organic wastes can be converted via composting or anaerobic digestion (Mulligan, 2002). One tonne of green waste produces 0.3 tonne of compost and 0.12 tonnes of biogas.

Mixed wastes can be very difficult to treat and thus may require incineration. However, this reduces the weight of the waste to one-third and the volume to one-fifth of the original volume. In addition, one tonne of household waste can produce an equivalent energy value of 120 L of heating oil or 200 kg of coal (Genske, 2003). Despite these positive aspects, wastes generated from incinerators include gaseous emissions (74%), bottom (23%) and fly ashes (3%) and process water. These ashes must be stored and properly disposed since exposure to acidic solutions enhances metal leaching that will find its way into aquifers or surface water. If these ashes are recycled, the heavy metals must be immobilized. Volatile substances such as dioxin are produced during incineration and can accumulate in animals, plants, and humans. Filtration and chemical methods are needed to reduce emissions of dioxins and other hazardous components such as sulphur compounds that lead to acid rain, and carbon and nitrogen oxides that can contribute to the greenhouse gas effect. The calorific value of the waste also must be sufficient to run the incinerator in a sustainable manner.

Treatment of liquid wastes is by physical and chemical means. Deep underground injection and storage is a method that has been used. However, in addition to possible compaction and ground subsidence, groundwater contaminant also may result. Modelling and experimental investigations to predict the consequences of long-term storage are required.

7.5.1.1 Fresh Kills Urban Dump, New York City, New York, USA

Fresh Kills Landfill is 7,500 ha in area and was established in 1948 on Staten Island for the disposal of waste for the 20 million people living in New York (Chamley, 2003). Although more than 21,000 tonnes were dumped daily in the 1980s, this decreased to 13,000 tonnes per day by the 1990s, mainly due to sorting and recycling. The landfill was the largest in the world when it closed in 2001 with the exception of the waste from the rubble from the September 2001 World Trade Centre disaster. As the site was originally a wood, meadow, and saline-soil mixed coastal landscape, plans are ongoing to restore the site. Local vegetation has been planted on a new 100 m hill, with an ultimate objective of transformation of the site into a nature and leisure place. It will be a 890-hectare park that is about three times the area of Central Park and will emphasize sustainability and ecological restoration. The first part of the park (Schmul Park) was opened in Oct. 2012 (FreshkillsParkNewsletter-Winter/Spring 2013, www.nycgovparks.org/park-features/freshkills-park). Storm water management and reduced water and energy consumption are practiced. As 90% of the wetlands in New York-New Jersey Harbor Estuary were lost, an effort is underway to restore the wetlands. The process consisted of using goats to graze on invasive species and installing biodegradable coir (coconut fibre) fibre logs with mussel shells and native wetland plants to reduce erosion and dissipate wave energy. Freshkills Park played a role in absorbing and slowing Hurricane Sandy in 2012. Fresh Kills is now named Freshkills. The park has been opening in phases including: Owl Hollow Soccer Fields which opened in 2013, the New Springville Greenway which opened in 2015 and North Park Phase 1 which is the latest opened (2023). The South Park Anchor Park project is currently being designed and will include creeks, wetlands, meadows, and views of New York City. The rest will open by 2036. Since then, various layers of soil and infrastructure has been engineered. The park is now a home for wildlife, and various recreational, scientific, educational, and artistic activities (https://freshkillspark.org/).

7.5.2 Vertical Barriers and Containment

In Chapter 3, we have seen how vertical barriers can be used to isolate and contain the contaminants in groundwater. When a landfill is in contact with the groundwater or when a landfill leachate contaminates the groundwater, an available treatment option is to install vertical barriers to confine the contaminants. A low permeability base is required to key in the cutoff walls. If this does not exist, an artificial base will have to be added. For older landfills where liners have degraded or do not exist, other means may be required to reduce the potential for contaminant leaching into the groundwater, as shown in Figure 7.13. Silicate gels, pozzolanics, or cements can be injected into the soil under the landfill to form an impermeable layer thus immobilizing the plume. Groundwater levels may also need to be lowered by pumping to maintain their level below the landfill and the seal-off area. Pump and treat systems for treating the groundwater may also be necessary. Land settlement and decreased plant growth may result from lowering of the water table. Long-term monitoring is required to ensure that the system performs according to requirements specified the indicators prescribed for safe performance.

7.5.3 Excavation

Excavation, partially or totally, to remove offending landfills can be a solution to decreasing contaminant generation and allow land reuse. The waste can be recycled, incinerated, or dumped in a safe landfill. Handling and transport of these wastes is not without risk. Corroded drums are not easy to handle. The excavated landfill can be filled with clean or recycled soil and subsequently used for construction of buildings. Numerous measures have to be instituted to ensure worker safety and protection of the environment during the excavation work. Health and geoenvironmental threats arise from breaking bags, dust, emission of harmful gases or liquids, and unstable waste.

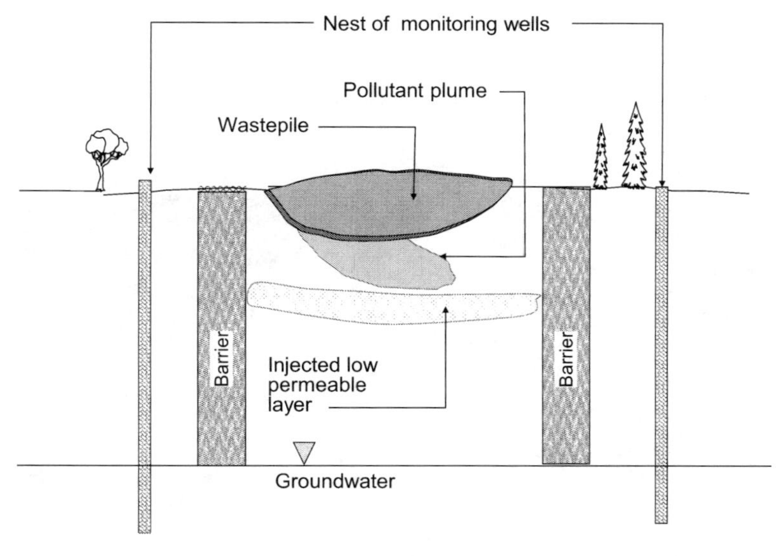

FIGURE 7.13 Vertical barriers and injections for an artificial low permeable layer under a landfill. (Modified from Genske, 2003.)

7.5.4 Landfill Bioreactor

Landfills traditionally have been operated without addition of liquid, i.e., they have been operated under the "dry garbage bag" concept. Increasing the moisture content of the waste, however, can increase waste degradation and methane production since most of the conversion processes are anaerobic. This concept is known as a landfill bioreactor and is becoming increasingly used worldwide. Experience with this type of landfill shows that gas recovery can be increased up to 90%, and that wastes can be stabilized within 10 years instead of 30 to 100 years (Block, 2000). Landfills can be designed to be aerobic, anaerobic, or anaerobic-aerobic bioreactors. Anaerobic bioreactors maintain moisture contents of 10 to 20% to optimize anaerobic degradation conditions. Aerobic reactors require injection of air or oxygen with vertical or horizontal injection wells. Although wastes can be stabilized in 2 years, by aerobic landfills, injection of the air can be costly, and can lead to fires and thus is not practised often. The hybrid or anaerobic-aerobic bioreactor takes advantage of aerobic and anaerobic bacteria. The upper layer of waste is aerobically treated before burial and treatment with anaerobic bacteria. In all cases, moisture control is the most important parameter. Other factors include pH, waste pre-treatment, such as shredding, nutrient addition, settlement, cellulose content, leachate quantity and quality, and temperature control. Monitoring of these factors is essential. There is also the potential for mining the dredged waste for humic material and other recyclables. Figure 7.14 shows the major elements of a waste landfill bioreactor system. In addition to recycling of the collected leachate back into the waste pile, the option of addition of inoculum, nutrients, and other dissolution aids is provided. Gas generated from the waste pile is collected in a collection system for treatment.

In the US according to the US EPA subtitle D rule (www.epa.gov/landfills/bioreactor-landfills, accessed Feb. 2024), landfill bioreactors can only be used for landfills with composite liners such as 0.61 m of clay covered with a high-density polyethylene (HDPE) liner – as shown in Figure 7.14. Most installations increase the moisture content through recirculation of the leachate. Recirculation assists biodegradation by enhancing the transport of nutrients through the waste, redistribution of methane bacteria, buffering, dilution of inhibitory components and retention of the constituents of the leachate in the landfill. It can be accomplished by spray application, infiltration ponds, horizontal or vertical injection wells or trenches. Accumulated amounts of leachate will depend on (a) regulatory

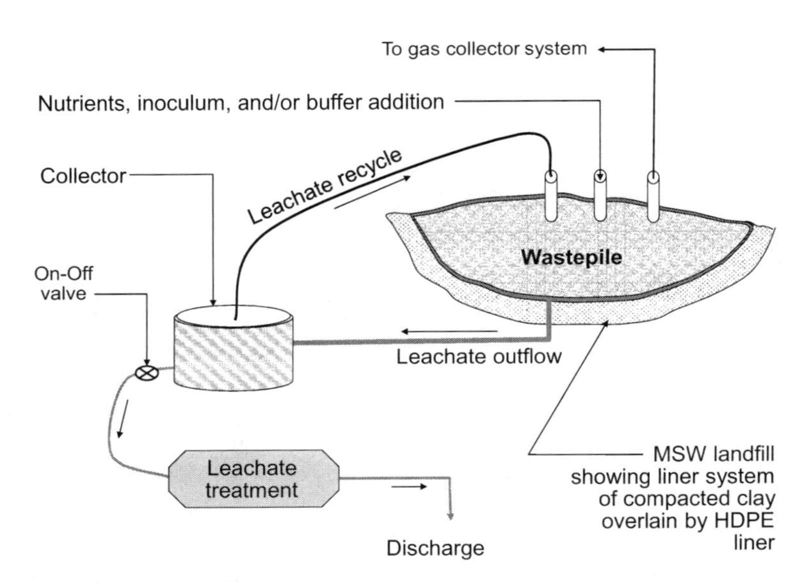

FIGURE 7.14 Schematic of a landfill bioreactor. Leachate can be directly fed back to the waste pile in the landfill without treatment or after treatment in the "leachate treatment" facility. The use of on-off valves will facilitate choice of treated or untreated leachate for recycling. (Adapted from Mulligan, 2002.)

requirements, (b) how much is needed for each landfill compartment, and (c) minimization of quantity of leachate requiring disposal. Older areas in the landfill can serve as seeds for new areas. Ideally, other liquids will have to be added to increase the moisture content from 30 to 40%. However, a balance must be made such that organic acids or other components do not accumulate and inhibit methane production (Pohland and Kim, 2000). Leachate and gas monitoring are integral to the success of the process. In terms of heavy metal concentration in the leachate, pH is a major factor in mobility of the metals. Mobility of the metals increases as the pH decreases. In addition, organic and inorganic agents can serve as ligands, promoting metal transport. However, the mechanisms of precipitation, encapsulation and sorption ensure that the heavy metals are captured in the waste.

Solubility, volatility, hydrophobicity (k_{ow}), biodegradability, and toxicity influence the behaviour of organic contaminants in the landfill. Compounds such as dibromomethane, TCE, 2-nitrophenol, nitrobenzene, pentachlorophenol (PCP), and dichlorophenol tend to be highly mobile and thus are found in the leachate and gas phase (Pohland and Kim, 2000). They will also be biologically altered by reduction, complexation, complete or partial degradation. Other compounds such as hexachlorobenzene, dichlorobenzene, trichlorobenzene, lindane and dieldrin are more hydrophobic and thus will remain in the waste and are available for biodegradation (Table 7.9).

The US EPA (2007) reviewed the literature and evaluated five full-scale operating bioreactor landfills. They found that the bioreactors complied with existing solid waste regulations and technical guidance and that the addition of leachate and other liquids require appropriate design such that the injection systems to distribute the moisture uniformly within the waste. The design of leachate collection systems was able at all sites to maintain leachate levels of under 30 cm of head on the liner. Slope stability was not problematic, and any issues were easily corrected but proper design and operations were necessary to provide for slope stability. As there was more potential for fires in aerobic landfills than anaerobic, monitoring and liquid addition without delay are essential. Fires or “hot spots” appear to have greater potential in aerobic landfills but can be managed with good monitoring of temperature and aeration and subsequent addition of liquids. Issues for the anaerobic bioreactors were similar to regularly operated landfilled.

In summary, the development of landfill bioreactors will continue due to its advantages over conventional landfills. Further efforts will be necessary to optimize leachate recirculation, gas generation, and removal of recalcitrant compounds. There are still many challenges, including regulator

TABLE 7.9
Conversion or Transport of Inorganic and Organic Components in a Landfill Bioreactor (Adapted from Mulligan, 2002)

Component	Transport or Conversion Mechanism
Heavy metals (Cd, Cu, Cr, Fe, Hg, Ni, Pb, Zn)	Reduction of Fe, Cr, Hg
	Complexation with organic or inorganic components and mobilization
	Precipitation as hydroxide (Cr) or sulphides (Cd, Cu, Fe, Hg, Ni, Pb and Zn) after sulphate reduction
	Sorption and ion exchange with waste
	Precipitation under alkaline conditions
Halogenated aliphatics including PCE, TCE, dibromomethane	Volatilization and mobilization in leachate due to high vapour pressure and solubility
Chlorinated benzenes such as hexachlorobenzene, trichlorobenzene and dichlorobenzene	Volatilization and sorption on waste due to low solubility and high k_{ow}
Phenols and nitroaromatics such as dichlorophenol, nitrophenol and nitrobenzene	Low volatility, vapour pressure and k_{ow}, with high solubility in leachate
PAHs and pesticides (lindane and dieldrin)	Low volatility and mobility due to low vapour pressure and high k_{ow}

reluctance, ability to wet the waste uniformly, and availability of design criteria. Slope stability and settlement will differ from traditional landfills due to the increased moisture contents and degradation rates and thus will need to be properly monitored.

7.5.6 Natural Attenuation

Although we discussed natural attenuation and its application for groundwater in Chapter 3 and will be discussing this in greater detail in Chapter 9, we want to turn our attention here to the use of monitored natural attenuation (MNA) of soil to reduce the toxicity and concentration of contaminants in a soil-water system. This subject has been treated in detail in Yong and Mulligan (2019). This concept of passive remediation has gained acceptance by many jurisdictions and regulatory agencies. To ensure effectiveness, monitoring guidelines and criteria have been (or are being established. The soil and the contaminants must be compatible to ensure the reactions of effective and optimum partition of the contaminants with the soil solids and the reactions and interactions to reduce the toxicity of the contaminants. Target concentrations must not be exceeded to protect the environment and biotic receptors.

To a certain extent this is true because the soil attenuation layer of an engineered barrier system is often composed of soil materials that are chosen for their attenuation capability – a designed soil-water system, which is generally called an *engineered clay barrier*. Figure 7.15 shows a general view of an engineered barrier system used for containment of a waste pile. The details of the filter, membrane, and leachate collection system and the nature and dimensions of the pollutant attenuation layer are specified by regulatory *command and control* requirements or by performance requirements.

The geomembrane consists of polymeric materials ranging from 0.25 to 3.5 mm thick that must be compatible with the chemicals in the landfill. They are designed to restrict leachate transmission. Synthetic polymers consist of ethylene propylene diene monomers (EPDM), high-density polyethylene (HDPE), linear low-density polyethylene (LLDPE), polyvinyl chloride (PVC), or polypropylene (PP). They are typically joined by welding techniques and must resist tearing, punctures, and indentations. They must reduce chemical vapor or liquid transmission. Geotextiles are made of

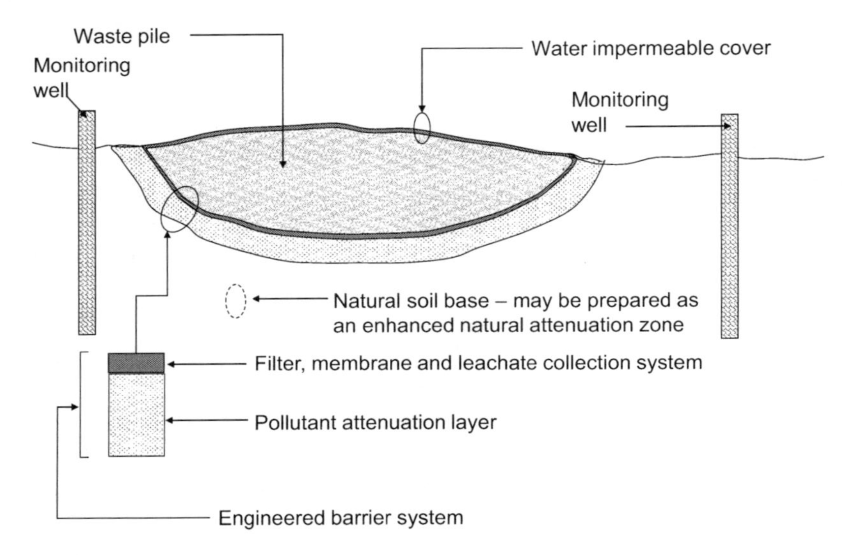

FIGURE 7.15 Pollutant attenuation layer constructed as part of an engineered barrier system. The dimensions of the attenuation layer and the specification of the various elements that constitute the "filter, membrane, and leachate collection system" are generally determined by regulations or by performance criteria. (From Yong and Mulligan, 2019.)

polypropylene fibres in either woven or non-woven forms and allow water to pass while retaining solids. They serve as a separate layer between the sub-base soil and the waste aggregate. The main issue to be considered is clogging. Geosynthetic clay liners (GCLs) can be used instead of conventional compacted clay liners. GCLs consist of low-hydraulic conductivity processed sodium bentonite in between two layers of geotextiles. Geosynthetic clay liners are resistant to physical and chemical deterioration in landfills. They are easier to install than clay barriers and more environmentally friendly as one truckload of GCL replaces 200 truckloads of clay.

The nature of material comprising the engineered clay barrier that underlies the synthetic membrane is determined on the basis of a maximum permissible hydraulic conductivity performance expressed in terms of the Darcy permeability coefficient k. There is an implied understanding (not always well founded) that the minimum specified hydraulic conductivity is somehow related to the attenuation capability of the engineered soil.

The basic idea in the design details of the engineered barriers is that if leachates inadvertently leak through the high-density polyethylene membrane (HDPE) geomembrane and are not captured by the leachate collection system, the pollutants in the leachate plumes will be attenuated by the engineered clay barrier. The engineered clay barrier serves as the second line of defence or containment. Even though the specifications refer only to a maximum permissible k value (generally in the order of 10^{-9} m/s) for the engineered clay barrier material, it is prudent to conduct additional tests of the material. These tests, which determine the pollutant assimilation capability of the clay material, are part of the protocol in the *evidence of Engineered Natural Attenuation (EngNA) capability* that assesses the capability of the engineered clay barrier material to attenuate the pollutants in the leachate plume. This will be discussed in the next section.

The EngNA as a direct link from NA is also used as the foundation base for the double-liner barrier system for landfill containment of hazardous waste. There are several options for the foundation base seen in the diagram. Since a fully compacted foundation base is a standard requirement – to provide support for the material contained above – one has the option of working with the native material if it has the proper assimilative potential, or with imported fill material. Once again, the purpose of the foundation base is to provide attenuation of pollutants should leakage of pollutants through the double-liner system occur. This in essence constitutes a third line of defence against pollutant transport into the subsurface soils.

One of the significant benefits in applying the protocols for *evidence of EngNA capability* is the determination of the required thickness of the pollutant attenuating layer and the engineered clay barrier shown in Figure 7.16. This specification can be obtained from the calculations from fate and transport models using the results of supporting laboratory research on partitioning and other attenuating phenomena. Since the dimensions for the engineered clay barriers specify "*greater than or equal to*" designations, application of the protocols for *evidence of EngNA capability* will likely provide dimensions that should satisfy regulatory requirements.

Water quality in the vicinity must remain acceptable in spite of contaminant dilution, degradation, and sorption processes. According to the Action Plan for *Switzerland Towards a Sustainable Development* (FOEFL, 1997), these controlled leachate landfills fall within the concept of sustainable development since few resources are consumed. Contact with the waste is not necessary. As with all bioreactor-type landfills, a proper monitoring programme is necessary – to ensure that fugitive leachates are captured and treated effectively with natural attenuation processes. To date, although little hard documentation exists regarding the use of monitored natural attenuation (MNA) for remediation of fugitive leachates from landfill sites, the limited options available favour its use. Examples of its use are shown in Table 7.10. Christensen et al. (2000) has suggested that there are five critical factors including the following:

- Local hydrogeological conditions.
- Size of the landfill and the variable nature of the leachate plume or plumes.

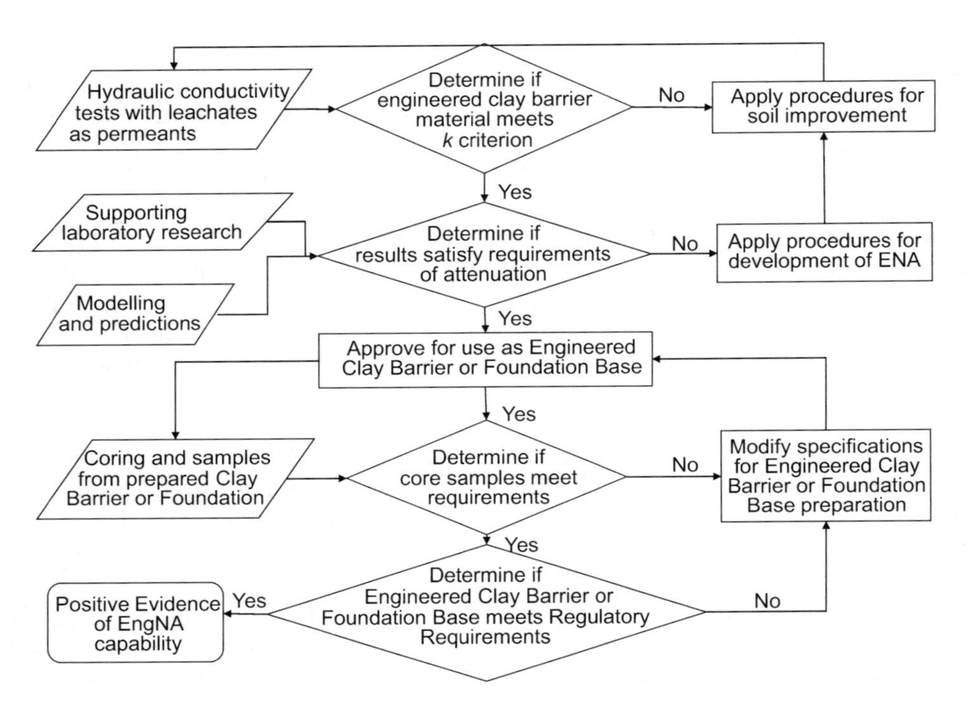

FIGURE 7.16 Protocol for determination of *positive evidence of EngNA capability as* required for use of engineered clay barriers or as foundation base material for double-liner systems. (Yong and Mulligan, 2019.)

TABLE 7.10
Application of Natural Attenuation at Landfill Sites

Location	Geology	Chemicals	Electron Donors	Microbial Process	Studies
Farmington, NH landfill (1995 to present)	Silty sand (up to 22 m below surface) Bedrock	TCE, DCE, VC, trace ethane, DCM, TEX, ketones	TX, DCM, ketones	Acetogenesis, methanogenesis. Cometabolic oxidation	NA investigation in groundwater, laboratory studies
Niagara Falls, NY (1994)	Overburden, fractured bedrock	TCE, DCE, VC, DCA, CA, CT, CF, DCM, CM, ethene ethane	Landfill leachate, other chemicals	Methanogenesis Sulfate reduction Aerobic oxidation	NA investigation in groundwater
Cecil County, MD (1995–1996)	Sand and fill over fractured saprolitic bedrock	VC release	VC	Anaerobic oxidation	NA investigation in groundwater
Norman, OK (2000–2003)	Sand, silt/clay lenses, poorly sorted gravel sediments	VOCs including n-propylbenzene, i-propylbenzene	Dissolved organic carbon in landfill leachate	Sulfate reduction, iron reduction	Field experiments at landfill in groundwater

Source: Adapted from Sharma and Reddy (2004) and Christenson et al. (2003).

- Complexity of the leachate plumes.
- Long time frame for evaluation of the attenuation capacity of the soil.
- Demonstrating the effectiveness of natural attenuation based on a mass reduction basis.

More recently, there has been strong evidence for the presence of PFAS in groundwater originating from landfill leachates (Moore, 2022). For example, in New Hampshire, the US EPA found that 77% of groundwater samples exceed the Health Advisory Level (Doherty and Schlosser, 2020) and that 92% of the groundwater impacts are from unlined landfills. A USGS report indicated that 60% of public water wells and 20% of private ones in the eastern US contained at least one PFAS (McMahon et al., 2022). MNA is viewed as an energy-efficient, sustainable, and cost-efficient method of dealing with contaminants and is being investigated by the Defence Department. Since PFAS biodegradation is slow, enhanced MNA is being evaluated by adding colloidal activated carbon (CAC) (Regenesis Patents, 2024), which enhances sorption of PFAS (McGregor, 2018).

7.5.7 Remediation of Urban Sites

There are numerous benefits to restoring contaminated urban and brownfield sites. They included reducing sprawl, providing tax revenue, improving land and public health by improving air quality, removing threats to safety and reducing greenhouse gas emissions (NRTEE, 1998). The UK has national efforts in place for brownfield redevelopment. In addition, transportation costs can be reduced by up to $66,000 per hectare per year if brownfields are redeveloped compared to greenfields by reducing urban sprawl (NRTEE, 2004). Also 4.5 hectares of *greenland* can be preserved for every hectare of brownfield restored. Land use is more compact, thus increasing city competitiveness. The Revi-Sols programme in Montreal and Québec, Canada has led to the cleanup and development of 153 projects for a total of 220 hectares of land by the year 2004. The programme was extended to 2005. The tax revenues in Montreal increased by $25.6 million and 3400 new housing units were established. At the provincial level, the ClimatSol-Plus programme continues to support reusing contaminated land and is available for municipalities and private landowners.

One example of the restoration of a former contaminated site is the Angus Shops in Montreal (NRTEE, 2004). Between 1904 and 1992, the site was used as an area for railway and military maintenance and the construction of new equipment. Approximately one-third of the sites (309 acres out of 1240 acres) were contaminated with heavy metals, petroleum hydrocarbons, and PAHs. All hazardous wastes were disposed of off-site. Recyclable materials, all debris and contaminated soil were removed. Backfill consisted of clean on-site soil. The cleanup cost a total $12 million. 500 houses, and a supermarket and industrial mall have been built, with a biotechnology centre under construction. Property taxes have increased to $2.2 million a year and more than $391 million has been invested by private parties.

To restore urban contaminated sites, various parameters need to be considered. These include the characterization of the soil (mineral, texture, geochemical characteristics), the factors influencing fate and mobility of the contaminants such as dilution, sorption/desorption, biodegradation and transformation, and other factors such as climate, hydrology, and microorganisms present. As a first step in the remediation of land for redevelopment, an investigation will be required to determine underground heterogeneity and hidden object and obstacles. This can be accomplished by studying historical records and conducting geophysical surveys. Sewers, cables, underground tanks, pipelines, foundations, etc., must all be identified. They should be grouped according to their ease of extraction. For example, large foundations cannot be easily extracted and may require blasting, whereas tanks and other similar objects are not difficult to remove. Another factor is the type of material. Wood piles can be cut off, brick foundations may be extracted but concrete foundations require special procedures. Various strategies could also be used to reduce redevelopment costs and promote sustainability. These are summarized in Figure 7.17.

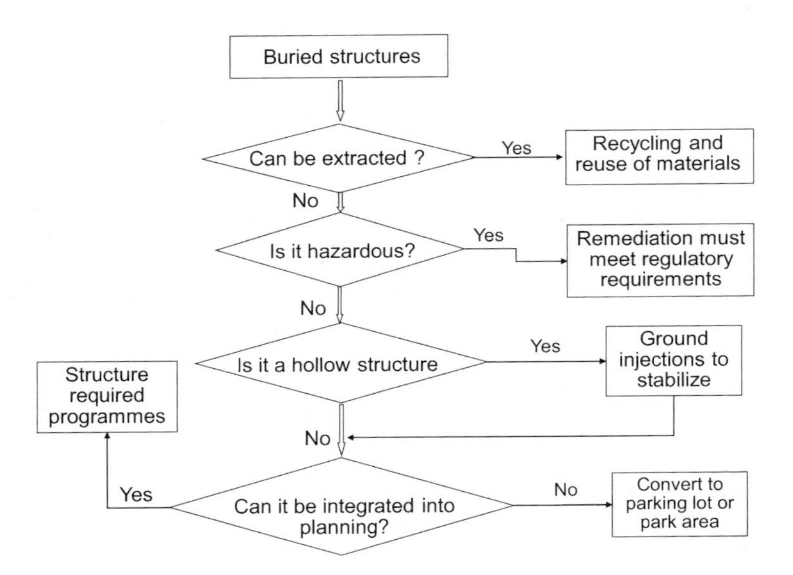

FIGURE 7.17 Schematic protocol for managing underground structures in a redevelopment plan. (From Genske, 2003.)

FIGURE 7.18 Biopile contaminated soil treatment site with soil preparation (A) and aeration system (B).

As will be discussed in Chapter 11, a variety of remediation techniques exist. They include excavation, contaminant fixing or isolation, incineration or vitrification, and biological treatment processes. *In situ* processes such as (a) bioremediation, air or steam stripping or thermal treatment for volatile compounds, (b) extraction methods for soluble components, (c) chemical treatments for oxidation or detoxification, and (d) stabilization/solidification with cements, limes, resins for heavy metal contaminants. Phytoremediation is a developing technique (Mulligan, 2002). The most suitable types of plants must be selected based on pollutant type and recovery techniques for disposal of the contaminated plants.

In Copenhagen, a soil treatment plant of 45,000 tonnes per year was recently established – mainly for oil-contaminated soils, with requirement for reduction of levels of 700 to 2000 mg/kg to below 50 mg/kg (Cooper, 1999). The treatment procedures call for green waste to be added to the soil in a 50:50 ratio and placed in a 1.5 m high windrow. Turning every week is required, to maintain the

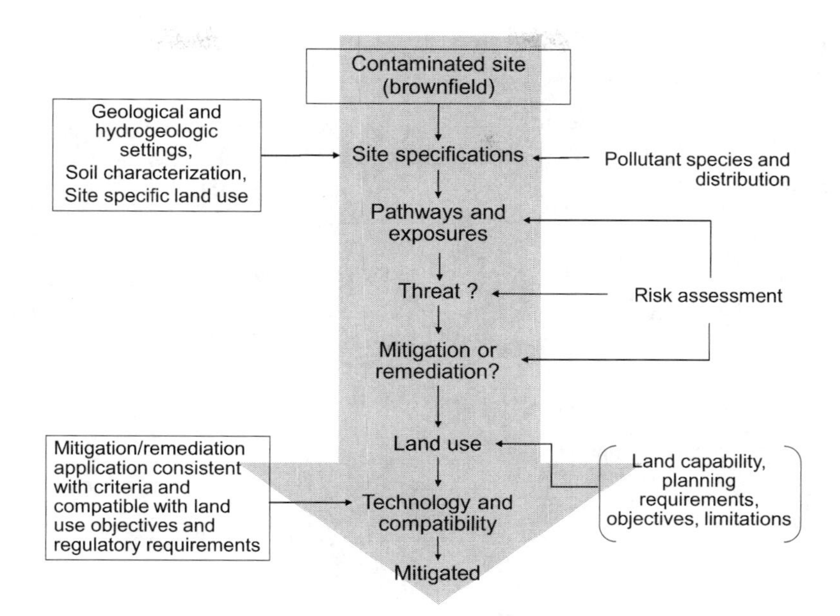

FIGURE 7.19 Simple protocol for rehabilitation of a contaminated site. (Adapted from Yong, 2001.)

temperature above 15 °C even in the winter. Because of better air permeability characteristics, sandy soils are much easier to treat than clay soils. If the treatment is successful, the remediated soil can be used to grow vegetation in embankments or in construction projects. If the treatment fails or does not meet the required criteria, the soil must be sent to a special landfill.

In the provinces of Quebec and Ontario, treatment centres for contaminated soil serve various clients including real estate developers (for example, conversion of a gas station to condos), excavation contractors, environmental consultants, government agencies, industries, and manufacturing companies. The process is an *in situ* biopile that uses microorganisms to breakdown the contaminants while controlling aeration, humidity and nutrient levels. The process is shown in Figure 7.18. Treated contaminants include petroleum hydrocarbons, volatile organic contaminants, PAHs, PCP, creosote, and phenol. Water and air emissions from the process can also be treated biologically via an aerated wastewater treatment system, wetlands, and a biofilter. Depending on the remaining contaminant levels, the treated soil may be able to be returned to the original site conditions.

Regardless of the origin of the contaminants and pollutants in the area, an evaluation of the threats to human health and the environment must be undertaken before the remediation process. Both the potential exposure time and level must be considered. Figure 7.19 gives a flow chart that illustrates a simple procedure for evaluation and treatment. Techniques such as selective sequential extraction are useful in determining the likelihood that the heavy metals are mobile. Selective sequential extraction studies were performed on nine soil samples (Huang, 2005). Figures 7.20A and B show the results for lead and zinc, respectively. It can be seen that both lead and zinc have different affinities towards different soil fractions. Both Pb and Zn have higher affinities towards the soil fractions of organic matter and oxides. Only a small fraction of both metals is associated with the exchangeable fraction. Metals bound to the exchangeable fraction of soil are mostly physically adsorbed (by electrostatic force) to the soil surfaces, and thus the bonding is weaker compared to other binding mechanisms. The moderate-to-high degree of leaching by rainfall and the competition from other cations present in the leachate solution possibly explains why only a limited amount of Pb and Zn were retained by this soil fraction. There is a high degree of association of Pb and Zn with soil oxides and organic matter. The metals associated with oxides are

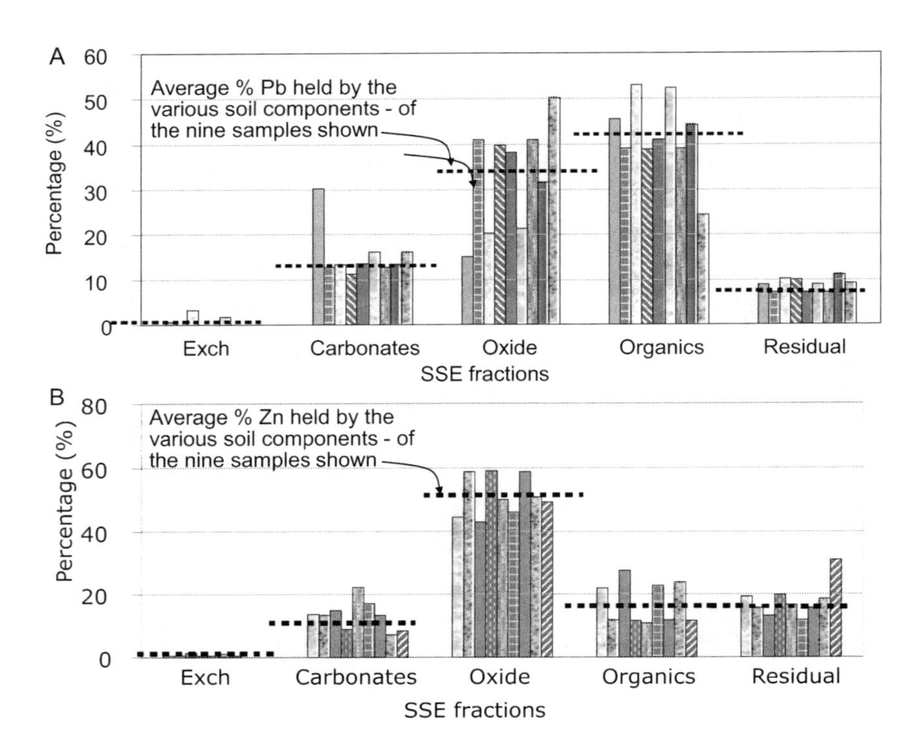

FIGURE 7.20 Selective sequential extraction characterization of nine different soil samples for Pb and Zn retention by the different soil components and mechanisms (data from Huang, 2005). Note that *Exch* represents Pb or Zn held by cation exchange mechanisms, and *carbonates, oxides* and *organics* represent soil components responsible for retaining the metals. *Residual* represents "remaining" in the soil matrix.

particularly susceptible to oxidation-reduction reactions and solubilization upon a decrease in pH by acid rain.

Sequential extraction was performed on the untreated soil from a site slated for rehabilitation and construction (Okoro, 2006). The results showed that copper existed mainly more in the organic fraction (50%), while zinc was mainly in the oxide fraction (36%) and nickel more in the exchangeable and carbonate fractions (50%). Some soil washing tests were performed with a series of five washings with biosurfactants (Figure 7.21), which enabled the oxide fraction of zinc, organic fraction of copper, exchangeable and carbonate fractions of nickel to be substantially reduced, compared to the control and the untreated soil. These results indicated the feasibility of removing heavy metals and the total petroleum hydrocarbon content of a mixed contaminated soil with the anionic biosurfactants tested.

7.6 CASE STUDY OF A SUSTAINABLE URBAN AREA

The Federation of Canadian Municipalities (FCM) provides various awards each year for a variety of concerns including innovative sustainability practices. In 2016, the Region of Waterloo won the award for a project of duration from 2009 to 2015. Benzene, hydrocarbons, and zinc contaminated soil and groundwater in an abandoned industrial area within Kitchener. The site was remediated, which reduces health risks, and the land redeveloped to include renovated historical buildings and a LEED GOLD energy and water-efficient building. The heavy industrial machinery was removed by a recycling company in exchange for the values of the materials. Jobs have been created by

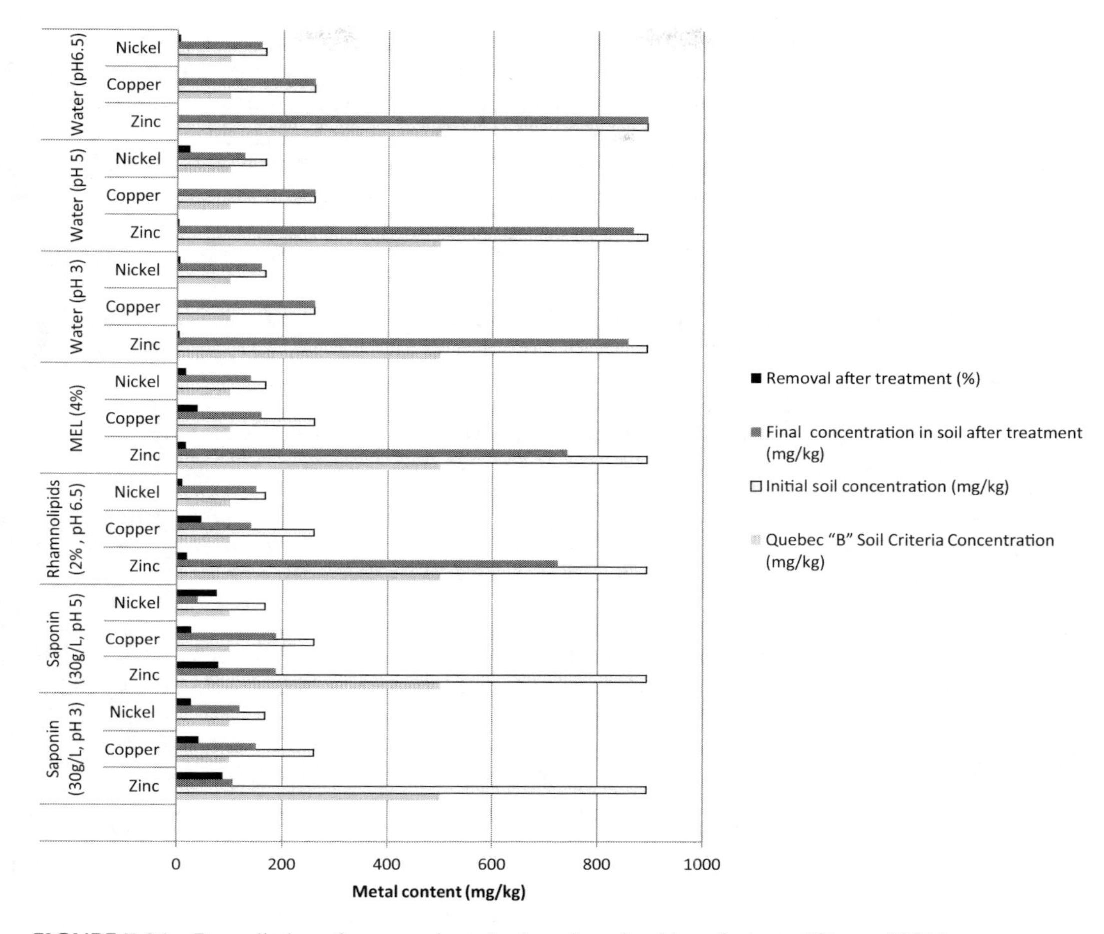

FIGURE 7.21 Remediation of a contaminated urban site using biosurfactants. (Okoro, 2006.)

tenants such as Google. Sprawl is reduced due to use of existing infrastructure. A new transportation hub with light rail helps to reduce GHG emissions. New green spaced have been created (https://greenmunicipalfund.ca/case-studies/2016-fcm-sustainable-communities-awards-brownfields-project).

7.7 CONCLUDING REMARKS

Cities have a significant impact on the geoenvironment. Increasing urban population will increase pressures on the geoenvironment in the years to come. All aspects of the geoenvironment need to be protected. Optimal design of landfills is one of the methods to protect the urban environment. Natural attenuation of the landfill leachate plumes is a sustainable method of remediation. The operation of a landfill bioreactor has potential to increase methane production rates and waste degradation rates, which subsequently increases the life of the landfill. Although waste reduction and recycling programmes exist in many cities, substantial improvements are needed to decrease wastes going into landfill and increase recycling rates. Life-cycle analysis of products can be used to assist in determining which products have less impact on the environment. However, more data (particularly regarding emissions and leachates during production and disposal) is required to facilitate the comparison process.

Cities need to redevelop brownfields through remediation and other planning initiatives. This will substantially reduce the need for greenfields for future development projects. Few efforts have been made to protect the diversity of plants, animals, and microorganisms (i.e., biodiversity). Substantial efforts can be made to conserve resources such as energy. Electricity from geothermal and other renewable sources can be chosen. Buildings and their materials must also be energy efficient. Efforts must also be made regarding transportation to reduce its impact on the geoenvironment.

REFERENCES

Appun, K. (2018), Climate Impact of Farming, Land Use (Change) and Forestry in Germany, Factsheet, Oct. 30, 2018. www.cleanenergywire.org/factsheets/climate-impact-farming-land-use-change-and-forestry-germany.

Block, D. (2000), Reducing greenhouse gases at landfills, *BioCycle*, 41(4): 40–46.

Bridges, E.M. (1991), Waste Materials in Urban Soils. In P. Bullock and P.J. Gregory, (eds.), *Soil in the Urban Environment*, Blackwell, Oxford, pp. 28–46.

Buchholz, R. R. Worden, H.M., Park, M., Francis, G., Deeter, M.N., Edwards, D.P., Emmons, L.K., Gaubert, B., Gille, J., Martínez-Alonso, S., and Tang, W. (2021), Air pollution trends measured from Terra: CO and AOD over industrial, fire-prone, and background regions, *Remote Sensing of Environment,* 256: 112275.

Chamley, H. (2003), *Geosciences, Environment and Man,* Elsevier, Amsterdam, 450 p.

Christensen, T.H., Bjerg, P.L., and Kjeldsen, P. (2000), Natural attenuation: A feasible approach to remediation of groundwater pollution at landfills? *GWMR*, 20(1): 69–77.

Cooper, P.A. (1993), Disposal of Treated Wood Removed from Service: The Issues, In: *Proceedings, Environmental Considerations in Manufacture, Use and Disposal of Preservative Treated Wood,* Carolinas-Chesapeake Section, Forest Products Society, Madison, WI, pp. 85–90.

Cooper, J. (1999), Solid Waste Management in Copenhagen. In: A. Atkinson, J.D. Davila, E. Fernandes and M. Mattingly, (eds.), *The Challenge of Environmental Management in Urban Areas*, Ashgate, Aldershot, pp. 139–150.

Craul, P.J. (1985), A description of urban soils and their desired characteristics, *Journal of Arboriculture,* 11: 330–339.

Cuny, D., Van Haluwyn, C., and Pesch, R. (2001), Biomonitoring of trace elements in air and soil compartments along the major motorway in France. *Water, Air and Soil Pollution*, 125 (1–4): 273–289.

CWC. (1998), Crushed Glass Cullet Replacement of Sand in Topsoil Mixes, Final Report No. GL-97-10, February, CWC, Seattle, Washington.

Danish Water and Wastewater Association. (2007), *Water in Figures 2007*, Skanderborg, DANVA, 4 p.

Doherty, A.T., and Schlosser, K.E.A. (2020), Sources of PFAS Impacts to New Hampshire Groundwater, Presented at: Northeast Waste Management Officials' Association Webinar Series, July 29, 2020, New Hampshire Department of Environmental Services.

Fayer, M. J. (2000, June), *UNSAT-H Version 3.0: Unsaturated Soil Water and Heat Flow Model Theory, User Manual, and Examples Pacific Northwest National Laboratory*, US. Department of Energy, National Technical Information Service, Springfield, VA.

Fauchon, L. (2012), *Cities and Water Security Sustainable Cities, Building Cities for the Future*, Green Media Ltd., Long, pp. 117–118.

Ferdous, A. (2013), Cumulative hydrologic impact of wetland loss: Numerical modeling study of the Rideau River Watershed, Canada, *Journal of Hydrologic Engineering*, 19(3). https://doi.org/10.1061/(ASCE)HE.1943-5584.0000817.Table

FOEFL. (1997), *Sustainable Development: Action Plan for Switzerland. Report of the Federal Office of Environment*, Forests and Landscapes, Bern.

Genske, D.D. (2003), *Urban Land, Degradation, Investigation and Remediation*, Springer, Berlin, 331 pp.

Harris, J.A. (1990), The Biology of Soils in Urban Areas. In: P. Bullick and P.J. Grogory, (eds.), *Soils in the Urban Environment*, Blackwell Scientific Publications, Oxford, pp. 139–152.

Hertwich, E.G. (2005), Life cycle approaches to sustainable consumption: A critical review, *Environmental Science and Technology*, 39(13): 4673–4684.

Huang, Y-T. (2005), Heavy Metals in Urban Soils, M.A.Sc. Thesis, Concordia University, Montreal, Canada.

IPBES (Intergovernmental Science-Policy Platform on Biodiversity and Ecosystem Services). (2019), *Global Assessment Report on Biodiversity and Ecosystem Services of the Intergovernmental Science-Policy Platform on Biodiversity and Ecosystem Services*. Edited by E. S. Brondizio, J. Settele, S. Díaz, and H. T. Ngo, IPBES Secretariat, Bonn, Germany.

ISI. (2023), Envision Overview. www.sustainableinfrastructure.org/about-isi/, accessed, Feb. 2024.

ISO. (1997), *ISO 14040: Environmental Management-Life Cycle Assessment-Principles and Framework*, International Organization for Standardization, Geneva.

Jonkhoff, E., and van Eijnatten, W. (2012), *Measuring Sustainability: The Amsterdam Sustainability Index Sustainable Cities, Building Cities for the Future*, Green Media Ltd., Long, pp. 27–28.

Khan, R.K., and Strand, M.A. (2018), Road dust and its effect on human health: A literature review, *Epidemiology and Health*, 40: e2018013. https://doi.org/10.4178/epih.e2018013.

Kjeldsen, P., Barlaz, M.A., Rooker, A.P., Baun, A., Ledin, A., and Christensen, T.H. (2002), Present and long-term composition of MSW landfill leachate- A review. *Critical Reviews in Environmental Science and Technology*, 32: 297–336.

Knutson, J. (2018), 31 Million Acres Lost: Development Cuts into U.S. Farmland, Agweek, May 09, 2018 at 10:01 AM. www.agweek.com/business/31-million-acres-lost-development-cuts-into-u-s-farmland, accessed Jul. 2024.

König, K.W. (1999), Rainwater in Cities: A Note on Ecology and Practice. In: T. Inoguchi, E. Newman and G. Paoletto, (eds.), *Cities and the Environment: New Approaches for Eco-Societies,* United Nations University Press, Tokyo, pp. 203–215.

LaGrega, M.D., Buckingham. P.L., and Evans, J.C. (1994), *Hazardous Waste Management*, McGraw-Hill, New York, 1146 p.

Lesage, S., Jackson, R.E., Priddle, M.W., and Riemann, P.G. (1990), Occurrence and fate of organic solvent residues in anoxic groundwater at the Gloucester Landfill, Canada, *Environmental Science and Technology,* 24: 559–566.

Li, M., Wu, P., Ma, A., Lv, M., Yang, Q., and Duan, Y. (2022), The decline in the groundwater table depth over the past four decades in China simulated by the Noah-MP land model, *Journal of Hydrology*, 607: 127551, https://doi.org/10.1016/j.jhydrol.2022.127551.

Margolis, J. (2019), China Dominates the Electric Bus Market, But the US is Getting on Ooard. The World. www.pri.org/stories/2019-10-08/china-dominates-electric-bus-market-us-getting-board, accessed Jul. 2024

McGregor, R. (2018), In situ treatment of PFAS-impacted groundwater using colloidal activated Carbon, *Remediation Journal*, 28(3): 33–41. doi:10.1002/rem.21558.

McMahon, P.B., Tokranov, A.K., Bexfield, L.M., Tokranov, A.K., Bexfield, L.M., Lindsey, B.D., Johnson, T.D., Lombard, M.A., and Watson, E. (2022), Perfluoroalkyl and polyfluoroalkyl substances in groundwater used as a source of drinking water in the Eastern United States, *Environmental Science & Technology*, 56(4):2279–2288. https://doi.org/10.1021/acs.est.1c04795.

Meijer, L.J. J., van Emmerik, T., van der Ent, R., Schmidt, C., and Lebreton, L. (2021), More than 1000 rivers account for 80% of global riverine plastic emissions into the ocean, *Science Advances*, 7(18): eaaz5803. https://doi.org/10.1126/sciadv.aaz5803.

Mitchell, L.E., Lin, J.C., Hutyra, L.R. Bowling, D.R., Cohen, R.C., Davis, K.J., DiGangi, E., Duren, R.M., Ehleringer, J.R., Fain, C., and Falk, M. (2022), A multi-city urban atmospheric greenhouse gas measurement data synthesis, *Scientific Data*, 9: 361. https://doi.org/10.1038/s41597-022-01467-3.

Moghaddam, A. (2010), Development of a Sustainable Method for the Disposal of Chromated Copper Arsenate (CCA) Treated Wood, Ph.D. thesis, Concordia University, Montreal, Canada, December.

Moghaddam, H.A., and Mulligan, C.N. (2008), Leaching of heavy metals from chromated copper arsenate treated wood, *Waste Management,* 28: 628–637.

Moore, R. (2022), Enhanced natural attenuation: A sustainable solution to manage PFAS in groundwater at landfills, *Waste Advantage*, Mar. 30, 2022. www.wasteadvantagemag.com/enhanced-natural-attenuation-a-sustainable-solution-to-manage-pfas-in-groundwater-at-landfills/, accessed Feb. 2024.

Mulligan C.N. (2002), *Environmental Biotreatment,* Government Institutes, Rockville, MD, 395 pp.

Mullins, C.E. (1990), Physical Properties of Soils in Urban Areas. In: P. Bullick and P.J. Gregory, (eds.), *Soils in the Urban Environment*, Blackwell Scientific Publications, Oxford, pp. 87–118.

Münch, D. (1992), Soil contamination beneath asphalt roads by polynuclear aromatic hydrocarbons zinc lead and cadmium, *Science of the Total Environment*, 126(1-2): 49–60.

National Round Table for the Environment and Economy (NRTEE). (1998), *Greening Canada's Brownfield Sites*, NRTEE, Ottawa.

National Round Table for the Environment and Economy (NRTEE). (2004), *Cleaning up the Past, Building the Future, A National Brownfield Redevelopment Strategy for Canada*, NRTEE, Ottawa.

Neuman, S.P. (1972), Finite Element Computer Programs for Flow in Saturated-Unsaturated Porous Media, Second Annual Report, Project No. A10-SWC-77, Hydraulic Engineering Lab., Technion, Haifa, Israel.

Okoro, C. (2006), Biosurfactant Enhanced Remediation of a Mixed Contaminated Soil, MASc thesis, Concordia University, Montreal, Canada.

Page, A.L., and Ganje, T.J. (1970), Accumulation of lead in soils for regions of high and low motor vehicle traffic density, *Environmental Science and Technology*, 4: 140–142.

Pohland, F.G., and Kim, J.C. (2000), Microbially mediated attenuation potential of landfill bioreactor systems, *Water Science and Technology*, 4 (3): 247–254.

Portney, K.E. (2003), *Taking Sustainable Cities Seriously*, MIT Press, Cambridge, MA.

Pusch, R., and Yong, R.N. (2005), *Microstructure of Smectite Clays and Engineering Performance,* Taylor and Francis, London, 335 pp.

Regenesis Patents. (2024), REGENESIS Remediation Solutions. www.regenesis.com/en/patents/, accessed Feb. 2024.

Schroeder, P.R, Lloyd, C.M., and Zappi, P.A. (1994, Sept.), The Hydraulic Evaluation of Landfill Performance (HELP) Model: User's Guide for Version 3, EPA/600/R-94/168a, Office of Research and Development, US EPA, Washington, DC.

Sharma, H.D., and Reddy, K.R. (2004), *Geoenvironmental Engineering*, John Wiley and Sons, Hoboken, NJ.

Solo-Gabriele, H., and Townsend, T. (1999), Disposal – End Management of CCA – Treated Wood. In: *95th Annual Meeting of the American Wood Preservers' Association.* Lauderdale, Florida, May 16–19, pp. 65–73.

Syms, P. (2004), *Previously Developed Land, Industrial Activities and Contamination*, 2nd ed, Blackwell Publishing, Oxford.

Takatsuki, H. (2003), The Teshima Island industrial waste case and its process towards resolution, *Journal of Material Cycles and Waste Management,* 5: 26–30.

Tao, Y., Steckel, D., Klemeš, J.J., and You, F. (2021), Trend towards virtual and hybrid conferences may be an effective climate change mitigation strategy, *Nature Communications*, 12: 7324 https://doi.org/10.1038/s41467-021-27251-2.

Thorton, I., and Jones, T.H (1984), Sources of Lead and Associated Metals in Vegetables Grown in British Urban Soils: Uptake from the Soil Versus Air Deposition. In: D.D. Hemphill, (ed.), *Substances in Environmental Health-XVIII*, University of Missouri, Columbia, Missouri, pp. 303–310.

Torsha, T., and Mulligan, C.N. (2024), Anaerobic treatment of food waste with biogas recirculation under psychrophilic temperature, *Waste*, 2: 58–71. https://doi.org/10.3390/waste2010003.

Tushar, Q., Salehi, S., Santos, J., Zhang, G., Bhuiyan, M. A., Arashpour, M., and Giustozzi, F. (2023), Application of recycled crushed glass in road pavements and pipeline bedding: An integrated environmental evaluation using LCA, *Science of The Total Environment*, 881: 163488. https://doi.org/10.1016/j.scitotenv.2023.163488.

UNEP. (2004), *UNEP/SETAC Life Cycle Initiative*, United Nations Environment Programme, Paris, Vol. 2004.

UN Environment. (2018), *The Weight of Cities: Resources Requirements of Future Urbanization*, Paris, France.

United Nations Agenda 21 Department of Public Information. (1993), *Agenda 21: Programme of Action for Sustainable Development, Rio Declaration on Environment and Development*, United Nations Publication, New York.

US EPA. (1998), *Office of Mobile Sources*, U.S. Environmental Protection Agency, Washington, DC. www.epa/oms.

US EPA. (1999a), A Sustainable Brownfields Model Framework, United States Environmental Protection Agency, Office of Solid Waste and Emergency Response, Washington, DC, EPA500-R-99-001. EPA Jan. 1999.

US EPA. (1999b) Brownfield Road Map to Understanding Innovative Technology Options for Brownfields Investigation and Cleanup, 2nd ed., U.S. Environmental Protection Agency, Office of Solid Waste and Emergency Response, Washington, DC, EPA 542-B-99-009.

US EPA. (2007), *Bioreactor Performance*, EPA530-R-07-007, August 15.

US EPA. (2009), *Estimating 2003, Building-related Construction and Demolition Materials Amounts*, USEPA EPA530-R-09-002, March.

US EPA. (2017), Brownfield Road Map to Understanding Options for Site Investigation and Cleanup, 6th ed., EPA, 542 R-17-003, United States Environmental Protection Agency, Office of Land and Emergency Response, Washington, DC.

Vega, M. (1999), The Concept and Civilization of an Eco-Society: Dilemmas, Innovations and Urban Dramas. In: T. Inoguchi, E. Newman and G. Paoletto, (eds.), *Cities and the Environment: New Approaches for Eco-Societies*, United Nations University Press, Tokyo, pp. 47–70.

Vengosh, A. (2005), Salinization and Saline Environments. In: B. Sherwood Lolar (ed.), *Environmental Geochemistry*, Vol. 9, Elsevier, Amsterdam, pp. 333–366.

WEF. (2017), Fundamentals of anaerobic digestion, fact sheet, WSEC-2017-FS-002—Municipal Resource Recovery Design Committee—Anaerobic Digestion Fundamentals.

Weeks, J. (2005), Building an energy economy on biodiesel, *Biocycle*, 46(7): 67–68.

Wilson, A. (1997), Disposal: The Achilles' heel of CCA-treated wood, *Environmental Building News*, 6(3) , March. www.buildinggreen.com/features/tw/treated_wood.html.

World Commission on Environment and Development. (1987), *Our Common Future*, Oxford University Press, Oxford, 400 p.

Yong, R.N., Turcott, E., and Gu, D. (1994), Artificial Recharge Subsidence Control, IDRC Canada Report File No. 90-1020-1, Bangkok, Thailand.

Yong, R.N., Turcott, E., and Maathuis, H. (1995), Groundwater Abstraction-Induced Land Subsidence Prediction: Bangkok and Jakarta Case Studies. In: F.B.J. Barends, F.J.J. Brouwet, and F.H. Shröder, (eds.), *Land Subsidence,* Vol. 234, IAHS Publication, pp. 89–97.

Yong, R.N. (2001), *Geoenvironmental Engineering: Contaminated Soils, Pollutant Fate and Mitigation,* CRC Press, Boca Raton, 307 pp.

Yong, R.N., and Mulligan, C.N. (2019), *Natural and Enhanced Attenuation of Contaminants in Soils*, Lewis Publishers, Boca Raton.

Zhang, M., Tashiro, Y., Ishida, N., and Sakai, K. (2022), Application of autothermal thermophilic aerobic digestion as a sustainable recycling process of organic liquid waste: Recent advances and prospects, *Science of The Total Environment*, 828: 154187. https://doi.org/10.1016/j.scitotenv.2022.154187.

8 Coastal Marine Environment Sustainability

8.1 INTRODUCTION

For many different reasons, the coastal marine environment can be considered to be an important part of the geoenvironment. It is the recipient of (a) liquid discharges from surface run-offs, rivers, and groundwater, and (b) waste discharges from land-based industry, municipal, and other anthropogenic sources. It is also a vital element in the geoenvironment that provides the base for life-support systems. The combination of these two large factors, with their direct link to human population, makes it (the coastal marine environment) an integral part of the considerations on the sustainability of the geoenvironment and its natural resources.

8.2 COASTAL MARINE ENVIRONMENT AND IMPACTS

The coastal marine environment is a significant resource. A healthy coastal marine ecosystem ensures that aquatic plants and animals are healthy and that these do not pose risks to human health when they form part of the food chain. In this chapter, we will discuss (a) the threats to the health of the coastal sediments resulting from discharge of contaminants and other hazardous substances from anthropogenic activities, (b) the impacts already observed, and (c) the necessary remediation techniques developed to restore the health of the coastal sediments.

8.2.1 Geosphere and Hydrosphere Coastal Marine Environment

The discussion in Section 1.2 in Chapter 1 on the inclusion of many elements of the hydrosphere in the geoenvironment, points out that the coastal marine environment is included in the discussions on the receiving waters of the geoenvironment. This is because it is impacted by the outcome of anthropogenic activities. The sea provides the habitat for many living organisms and is an important ecosystem in the geoenvironment. Seawater is one of the major supply sources for our water resources through evaporation of the seawater and subsequent deposition on land as rainfall. Evapotranspiration constitutes the other major source of water for rainfall. Circulation and recycling of water on land is a very important process for preservation of the global environment and life support for all living organisms. This is not only because no living organisms can survive without water, but also because many substances required for the preservation of the sea environment are transported with the circulation of water.

Nutrients, such as phosphorus, nitrogen, silicate, etc., are produced during the decomposition of plants by the actions of organisms and are transported into the sea by water. Phosphorus and nitrogen are produced from the decomposition of withered leaves, whilst silicate originates primarily from inorganic soils. These nutrients are essential to the organisms in the sea, and are basic elements of the ecosystems in the sea. In general, there are fewer nutrients in shallow coastal zones, in comparison to deep seawater. This makes deep seawater more attractive for fish farming and for creating fishing grounds. To increase the amount of nutrients in the fishing grounds, fishermen have begun to plant broad-leaved trees in mountainous areas as a means to increase production of phosphorus and nitrogen as decomposition products for eventual rainfall (land surface flow) transport

DOI: 10.1201/9781003407393-8

into the sea. Recognition of the high mineral and nutrient values that can be obtained from deep seawater has led to harvesting of deep seawater for extraction of salt and other products.

8.2.2 Impacts of Climate Change on the Marine Environment

Increased levels of carbon dioxide lead to warmer ocean temperatures. Ocean levels are rising as ice bergs, glaciers, and sea water ice melt. The rate of the rise over the last three decades has increased from 2.1 mm/year to 4.5 mm/year. This increase will have a direct effect on flooding, coastal erosion, and landslides. Increased carbon dioxide levels decrease the pH. It has been estimated that the ocean water has absorbed 90% of the heat and 30% of the GHG emissions (https://unfccc.int/news/urgent-climate-action-is-needed-to-safeguard-the-world-s-oceans). Also, higher ocean temperatures will decrease oxygen levels. The US EPA has identified ocean heat, ocean acidity, and sea surface temperatures as the main indicator of climate change (https/epa.gov/climate-indicators, accessed Feb. 2024). Heat waves have doubled according to the IPCC since the 1970s. These changes impact marine life in various ways. For example, warmer water can cause bleaching or the even death of coral (www.epa.gov/coral-reefs/threats-coral-reefs, accessed Feb. 2024). A 1.5 °C increase will destroy 70 to 90% of the coral worldwide while a 2 °C global increase will destroy about 99%. Other species have declined in numbers or have shifted their patterns, migrating to cooler temperaturs. Plankton, which are at the bottom of the food chain, are very sensitive to oceanic changes. This can then impact the entire food chain and the ecosystem (www.noaa.ogov/education/resource-collections/marine-life/aquatic-food-webs, accessed Feb. 2024).

8.2.3 Sedimentation

The sea bottom is the interface between the seawater as the hydrosphere and the sediments and rock as the geosphere. Sediments are formed from substances deposited in the hydrosphere or produced in the hydrosphere itself. Because of the concentration of cations in seawater, suspended clay particles can aggregate more easily and settle faster than in fresh water. In addition, most sediment solids have a specific gravity greater than that of seawater – which explains why most of the non-aggregated solids will finally settle to the bottom of the sea. The settlement or sedimentation of solids is probably one of the strongest agents responsible for the purification of seawater, because

(a) various (harmful and noxious) substances sorbed or attached to the sedimenting substances (particles) will be sedimented with the particles – resulting in a measure of purification of the seawater,
(b) turbidity is reduced and transparency is promoted.

8.2.4 Eutrophication

In some closed sea areas, increased concentrations of nutrients can be found. This phenomenon is called *eutrophication*. This can happen naturally. More often than not, this phenomenon is developed as a result of the input of additional nutrients due to anthropogenic sources. This is sometimes called anthropogenic eutrophication. The main sources of these nutrients are sewage effluents, nutrients washed out of farmland, golf courses, lawns, and deposition of nitrogen from nitrous oxide emissions. Low to moderate eutrophication is beneficial because it enhances production of microscopic plants that live in the ocean called phytoplankton. Because they (phytoplankton) are bait for zooplankton, they are the basis for the marine food chain and their increased presence means a better food supply for the fish that relies on them as their source of food. However, when eutrophication is high, an excessive amount of phytoplankton will be produced, and the resultant phytoplankton bloom will contribute to the reduction in the amount of dissolved oxygen in the immediate region – creating problems for the fish population, as explained later in this section as red tide.

Resuspension of decayed algae and inorganic and organic particles will contribute to the turbidity of water, adding to the sunlight shading effect. Turbidity affects growth of sea plants that need sunlight. Decay of algae by bacteria removes large amounts of oxygen from the water, and may kill living organisms. Oxygen deficiency will cause sediments to change from an aerobic to anaerobic state. When this occurs, sulphides, such as pyrite, can be formed (Fukue et al., 2003). Hydrogen sulphide is a hazardous substance for fish. The combination of oxygen depletion with hydrogen sulphide makes the seawater look blue or greenish-blue – a condition sometimes called *blue tide*.

Accelerated production of phytoplankton algae will lead to algae blooms – an overpopulation of certain types of algae that are readily distinguished on the water surface as patches of bloom because of their high population density. The red algae bloom known as *red tide* is a vivid demonstration of such an occurrence involving certain types of algae that contain red pigment. Some, but not all, of the red algae species are toxic. The prominent ones are the dinoflagellate *Alexandrium tamarense* and the diatom *Pseudo-nitzschia australis.* (Their production of neurotoxins makes them harmful to fish and other aquatic life forms – and even humans.) Remediation treatment is needed to maintain a healthy environment in the coastal marine ecosystem to counter the effects of the preceding.

Many countries, states, and communities have set up programmes for monitoring of surface water quality. The term *surface water* is used here in the overall defining sense to mean all water that is naturally exposed to the atmosphere. This includes rivers, lakes, reservoirs, ponds, streams, impoundments, seas, estuaries, and all springs, wells, or other collectors directly connected to surface water. In general, the quality of surface water has a direct influence on its sediments, benthic organisms and plankton. Data obtained on surface water quality are used to structure measures to counter eutrophication and contamination of surface water.

8.2.5 Food-Chains and Bioaccumulation

Some organic matter, such as phytoplankton, deposited onto the sea bottom or suspended in seawater will be decomposed by microorganisms or will be consumed by small benthic animals called zooplankton. Phytoplankton is the first level in the marine food chain and zooplankton are the second trophic level. Decomposition of the phytoplankton and other organic matter will produce detritus. This is a good source of nutrients. The food-chain starting from detritus is often called the *detritus food-chain*, in contrast to the food-chain starting from phytoplankton in seawater. These food-chains are the most important process for the preservation of marine environment and living organisms, such as fish, shellfish, sea plants, etc. These are important sources of protein and minerals for humans. However, the food source for these marine living organisms can contain toxic and hazardous substances bioaccumulated through uptake, as for example, through the gills of fishes.

Bioaccumulation refers to the uptake and storage of a contaminant – generally a toxic substance – by an organism in the food-chain. Bioaccumulation in fishes presents health problems to humans when these are ingested. Whilst uptake of a toxic substance is an important measure, we need to account for possible excretion of the substance and also metabolic transformation of the substance as factors that will reduce the uptake amount of the substance. Hence, storage becomes an important consideration. The literature shows that the term *bioconcentration* is quite often used in place of bioaccumulation. Bioconcentration is related to bioaccumulation, but is more specific in that it expressly refers to the uptake and storage of toxic substances from water. The term bioconcentration has also been used to indicate the condition when concentrations of a substance in a particular biota are higher than in the surrounding medium. For this discussion, we use the first meaning of the term bioconcentration – uptake and storage of contaminants from water. When uptake and storage occurs from water and also from food, the more general term bioaccumulation is used. Available evidence (Chiou, 2002) indicates that there is a correlation between bioaccumulation and the octanol-water coefficient k_{ow} (see Section 2.63) for many of the organic chemical contaminants.

8.2.6 Contamination of Sediments

Historical discharge of contaminants and other toxic substances into the environment and the receiving waters from industrial development is a matter of record. This discharge became more significant during and after the industrial revolution at around the eighteenth century. Many of the discharges included substances, such as heavy metals, PCBs, dioxins, polycyclic aromatic hydrocarbons, tributyltin, triphenyltin, etc. – all of which are known toxicants. When these substances find their way into the ocean environment, some of them will dissolve. However, most of them will find their way onto the sea bottom through eventual sedimentation or attachment to suspended solids – a process that is considered as "dispersion" of noxious substances in the sea. Their effects on the human food chain can be deduced from Figure 8.1. This figure shows the food-chain beginning with phytoplankton anchoring the lowest trophic level and progressing upward through the zooplankton, fish and other marine aquatic species, and finally to humans. Mineral particles and organic matter settling in seawater can adsorb toxic and hazardous elements and compounds. The record shows that bioaccumulation of such elements and compounds as polychlorinated biphenyls (PCB), polychlorinated dibenzo-p-dioxins/dibenzo furans/dioxins, tributyltins (TBT), heavy metals, etc., in living organisms, such as seaweed, sea turtles, shellfish, fish etc. (Jensen et al., 2004; Gardner et al., 2003; Green and Knutzen, 2003). In some coastal areas and bays near urban centres, studies show that the sediments are heavily polluted with heavy metals and other hazardous substances (Jones and Turki, 1997; Kan-Atireklap et al., 1997; Fukue et al., 1999; Ohtsubo, 1999; Cobelo-García and Prego, 2003; Romano et al., 2004; Selvaraj et al., 2004). Makiya (1997) reports an intake of dioxins in the range of 60 to 70 % from a one day ingestion of fish and shellfish. To eliminate the root of the human food-chain problem requires one to decontaminate the contaminated sediments. This will eliminate the food source for the lower trophic

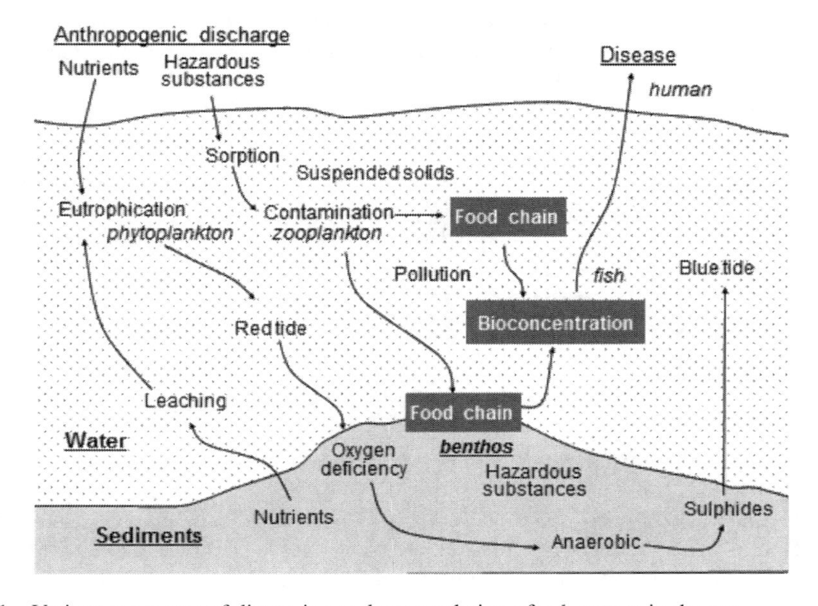

FIGURE 8.1 Various processes of dispersion and accumulation of substances in the sea.

levels. Until such is achieved, the danger of ingesting fish and shellfish that have bioaccumulated toxic and hazardous substances will always be present.

8.2.6.1 Some Case Studies of Sediment Contamination

Evidence of sediment contamination from land-based anthropogenic activities can be found in both marine coastal regions and also at the bottom of rivers, lakes, and other bodies of receiving waters. At a site in Germany, a lignite seam was found to accumulate aliphatic and aromatic chlorinated hydrocarbons downstream from a chemical plant (Dermietzel and Christoph, 2002). An initial fast desorption occurred from the outer surface of the sediment – followed by a slower diffusion-controlled released from the interior of the sediment.

Sediment samples from Lake Harwell, SC, were taken at five places in 1998, to determine if natural attenuation of polychlorinated biphenyls (PCBs) was occurring (Pakdeesusuk et al., 2005). From an analysis of the mole percentage of each congener of PCB and/or the total of meta, para, and ortho chlorines and total chlorines per biphenyl, it was determined that solubilization and desorption were negligible according to mass balances since 1987. *In situ* dechlorination was occurring though after an initial rapid rate, followed by a slow rate since 1987. Microcosm studies supported the findings. There was a lack of information on organic matter and electron acceptors, such as nitrate, sulphate, iron, and manganese, which made it difficult to predict optimal dechlorination conditions. To reduce the risk of bioaccumulation in fish, capping with fresh sediment may need to be increased.

In 1982, trichloroethene contamination in the groundwater was first detected at a Michigan National Priorities List site (An et al., 2004). Since then, samples were taken in 1991, 1992, 1994, 1995, and 1998, 100 m from the shore and later, 3 m from the shore. Anaerobic degradation was indicated as the products of dichloroethene (DCE), vinyl chloride (VC), ethene, and methane were found. Analysis of the water within the lake sediments indicated natural attenuation.

At the Columbus Air Force Base, the fate and transport of jet fuel contaminants was evaluated in 60 sediment samples (Stapleton and Sayler, 1998). 10^7 to 10^8 organisms per gram of sediment were found by DNA probes, compared to 10^4 to 10^6 organisms per gram by traditional methods. There was evidence of the degradation of BTEX and naphthalene, particularly after 5 to 7 days. Without nutrient addition, more than 40% of these ^{14}C-labeled compounds were mineralized in the sediments. Correlations of laboratory assay and field analyses are required.

At the Dover Air Force Base, which was contaminated with chlorinated ethenes, low biomass levels ($<10^7$ bacteria per g sediment) were found (Davis et al., 2002). However, mineralization of vinyl chloride and cis-DCE was found to be occurring. The 16 S rRNA gene sequence indicated the presence of anaerobic microorganisms capable of anaerobic-halorespiration and iron reduction. It was concluded that microorganisms were the major mechanism for reductive and oxidative attenuation of the chlorinated ethenes.

A study of heavy metals in fish and shellfish was conducted in Port Philip Bay, Australia (Fabris et al., 1999) with the objective of determining the partitioning of heavy metals in dissolved and particulate species in the bay waters. Concentrations in the near-shore and estuarine areas were not higher than in the coastal marine waters despite a flushing time of 10 to 16 months in the bay. The mechanisms for partitioning were related to co-precipitation of iron and manganese oxyhydroxides with dissolved heavy metals. A strong correlation of iron with chromium, nickel and zinc was seen in the particulates. Contrary to the metals, arsenic concentrations (as As (III)) increased in depth in the sediments and thus did not seem to be the result of anthropogenic activity. Near the surface layer of sediments, arsenic is oxidized to As(V) and leaves the sediments while Fe (III) can co-precipitate some of the arsenic and become trapped in the sediments. More recent data (2013–2021) has indicated good water quality. However elected levels of pesticides, metals and *E. coli* were found in the bay in Jan. 2023 after record rainfalls in Oct. and Nov. 2022 (Lenton-Williams, 2023). While industrial discharges have been controlled, pollution from animal waste, garden waste, and vehicles is an ongoing problem and is exacerbated by heavy rainfalls.

8.2.6.2 Sediment Quality Criteria

Table 8.1 gives an indication of the different criteria or definitions structured by some countries for characterization of sediment quality. Although many countries and jurisdictions have established guidelines for water quality, and especially for drinking water, very few countries have set up sediment quality guidelines. The importance of available guidelines and criteria can be seen in the need to protect pollution of the sea from dumping and indiscriminate discharge of hazardous wastes – all of which will eventually find their way onto the sea bottom and ultimately into the human food chain. The 1954 International Convention on the Prevention of Pollution of the Sea by Oil, together with the 1958 Geneva Convention on the High Seas were the earliest formal attempts to regulate and control discharge of hazardous substances into the sea. These have been reinforced with more attention paid to discharge of various kinds of hazardous wastes, especially wastes originating from land sources.

8.3 LONDON CONVENTION AND PROTOCOL

The London Convention and Protocol consists of the original London Convention 1972, which expanded the Oslo Convention for North-East Atlantic to cover marine waters worldwide, except for the inland waters of the various 80 States that were signatories to the Convention. The Oslo Convention came into force in 1974 and the expanded Oslo Convention (to worldwide marine waters), which became the London Convention that came into force in 1975. With this Convention, elimination of future marine pollution from deliberate discharge of industrial and other wastes is to be achieved through regulation of dumping of wastes at sea. The wastes of concern included the original oily wastes from the 1954 and 1958 Conventions, dredging spoils and wastes, and industrial wastes (i.e., land-based generated wastes).

Adoption of the *1996 Protocol to the Convention on the Prevention of Marine Pollution by Dumping of Wastes and Other Matter, 1972* essentially strengthened the London Convention with the *Precautionary Approach* and the *Polluter Pays Principle.* With the precautionary approach, the

TABLE 8.1
Outlines of Sediment Quality Guidelines

Country or Organization	Criterion or Standard for Guideline
USEPA	Screening concentrations for inorganic and organic contamination
	SQCoc:draft sediment quality criteria; oc = organic carbon
	SQALoc: sediment quality advisory levels
	ERL: Effects: Range-Low
	ERM: Effects: Range-Median
	AET-L: Apparent effects threshold-low
	AET-H: Apparent effects threshold-high
	TELs: Threshold effects levels
	PELs: Probable effects Levels
US-NOAA*	ERL: Effects: Range-Low
	ERM: Effects: Range-Median
Canada	ISQs: Interim sediment quality guideline
	PEL: Probable Effects Levels
Australia	ISQG-low: Interim sediment quality guidelines – low
	ISQG-high: Interim sediment quality guidelines – high
Netherlands	Target Value
	Intervention value

* U.S. National Oceanic and Atmospheric Administration

burden of responsibility for determining whether a waste designated for ocean dumping is potentially hazardous is now borne by the originator of the waste. The Annex in the 1996 protocol, "Assessment of wastes or other matter that may be considered for dumping" states that "acceptance of dumping under certain circumstances shall not remove the obligations under this Annex to make further attempts to reduce the necessity for dumping." The wastes or other matter that may be considered for dumping in accordance with the objectives of the Protocol and the precautionary approach include the following:

1 dredged material,
2 sewage sludge,
3 fish waste, or material resulting from industrial fish processing operations,
4 vessels and platforms or other man-made structures at sea,
5 inert, inorganic geological material,
6 organic material of natural origin; and
7 bulky items primarily comprised of iron, steel, concrete and similarly non-harmful materials for which the concern is physical impact, and limited to those circumstances where such wastes are generated at locations, such as small islands with isolated communities, having no practical access to disposal options other than dumping.

Of particular interest in the Annex is the following:

> For dredged material and sewage sludge, the goal of waste management should be to identify and control the sources of contamination. This should be achieved through implementation of waste prevention strategies and requires collaboration between the relevant local and national agencies involved with the control of point and non-point sources of pollution. Until this objective is met, the problems of contaminated dredged material may be addressed by using disposal management techniques at sea or on land.

Apparently, dredged material from the sea bottom has always occupied a special position under the Convention – to a large extent because dredging is an important requirement and a necessity for keeping navigation routes open, and also for ports and harbours. The volume of dredged materials is considerable. However, there is incontrovertible evidence to show that a significant portion of the dredged material ports, harbours and coastal regions is highly polluted. This realization has now energized many countries to begin considering dredged sediments as contaminated-polluted, and to insist that proper disposal of dredged materials be obtained.

8.4 QUALITY OF MARINE SEDIMENTS

Changes in seawater quality occur quickly over a very short time period because of the effect of currents and dilutions. Whilst these effects may lead to low concentrations of contaminants in seawater, the record shows that contaminants can be (and will be) adsorbed onto suspended and sedimentary particles. This is especially true when the settling particles are of biological origin – since they have the capability for sorbing heavy metals. This sorption process is known as biosorption, and the sorbent is called a biosorbent. Generally speaking, a biosorbent refers to the capability of a biomass to sorb heavy metals from solutions. Of particular importance is the fact that biosorbents have a combination of functional groups such as those described in Section 2.5.2 in Chapter 2 and in Figure 2.12 for organic chemicals. The combination of functional groups endows the combined group with significantly enhanced biosorbent capability to sorb various kinds of heavy metals – as opposed to mono-functional groups. Microalgae, for example, are well-known biosorbents (Wilde and Benemann, 1993). The study reported by Inthorn et al. (2002) on 52 strains of microalgae and their capabilities for removal of Pb, Cd, and Hg from various solutions showed that both green algae and blue-green algae functioned well in removing the heavy metals. Whilst deposition of the contaminant-associated settling particles onto the seabed may serve to remove a proportion of the water-borne contaminants from seawater itself, the accumulation of contaminated particles in the sediment presents a significant problem for the benthic population.

Determination of the quality of marine sediments is required for at least two different purposes:

- Preservation of the ecosystem. Knowledge of the types and nature of the contaminants in the sediments is essential, to enable one to structure the necessary measures for remediation and management of the quality of the sediments in conformance with established guidelines for sediment quality.
- Safety assurance for the lower trophic levels in the marine-derived human food-chain. Since phytoplankton and zooplankton are the first and second trophic levels, respectively, in this marine-derived human food-chain, preservation of the quality of these trophic levels means eliminating or reducing the concentrations of contaminants in the sediments – inasmuch as these can be bioaccumulated or bioconcentrated by the organisms that occupy these trophic levels. The record shows that organisms, such as benthos, fish, and mammals have been more or less contaminated through the food-chain and through bioconcentration (Kavun et al., 2002; Do Amaral et al., 2005; Moraga et al., 2002). Whilst guidelines have yet to be established, it is evident that in the absence of detailed records and tests, specification of allowable concentrations will likely be severely conservative, in the interest of eliminating health threats to the human population.

8.4.1 Standards and Guidelines

8.4.1.1 Guidelines

Guidelines have been established in some countries for the purpose of evaluating sediments contaminated with toxic chemicals. The aim of these guidelines is to limit the concentration of toxic chemicals through the use of various criteria – such as those shown in Table 8.1. As with all criteria based on observed effects on human health, differences exist between the guidelines used by the various countries. To a large extent, this is because of the different means for determination of the effective levels, and also in the perception of what constitutes an acceptable risk. For example, since apparent effects threshold (AET) values are essentially determined by a single result (i.e., the highest non-toxic sample) as opposed to the entire distribution of results (e.g., as with threshold effect levels TEL or probable effects level PEL), the final AET values used by the regulatory agency may vary substantially depending on the outcome of their analyses. A considerable amount of work remains to be done in this area. The use of interim values as preliminary values at the present time recognizes the fact that additional technical work on individual AET values together with reliability analyses and discussions with other involved agencies are required.

8.4.1.2 Chemicals

The guidelines issued by the various countries and agencies for environmental quality for sediments include trace and heavy metals and different types of organic chemical compounds. There is no definitive common listing of elements and chemicals between the agencies and countries, and no common agreement as to criteria used to evaluate and target the listed elements and chemicals. An example of sediment quality guidelines can be seen in the ISQG (interim sediment quality guidelines) and PEL (probable effect level) values given in the 2003 Canadian Environmental Quality guidelines (Table 8.2). The US National Oceanic and Atmospheric Administration (NOAA) on the other hand has established sediment quality guidelines that contain nine trace metals, 13 individual polycyclic aromatic hydrocarbons (PAHs), three classes of PAHs, and three classes of chlorinated organic hydrocarbons. As an example, for lead the ERL (effects: range – low) is 46.7

TABLE 8.2
Environmental Quality Guidelines for Sediments

	Fresh Sediments		Marine Sediments	
Substance	**ISQG (μg/kg)**	**PEL (μg/kg)**	**ISQG (μg/kg)**	**PEL (μg/kg)**
Arsenic	5,900	17,000	7240	41,600
Cadmium	600	3500	700	4,200
Chlordane	4.5	8.87	2.26	4.79
Chromium	37,300	90,000	52,300	160,000
Copper	35,700	197,000	18,700	108,000
DDD(2,2-Bis(p-chlorophenyl)-1,1-dichloroethane; Dichloro diphenyl dichloroethane)	3.54	8.51	1.22	7.81
DDE(1,1-Dichloro-2,2-bis(p-chlorophenyl)-ethene Diphenyl dichloroethylene)	1.42	6.75	2.07	3.74
DDT(2,2-Bis(p-chlorophenyl)-1,1,1-trichloroethane, Dicholoro diphenyl trichloroethane)	1.19	4.77	1.19	4.77
Dieldrin	2.85	6.67	0.71	4.3
Endrin	2.67	62.4	2.67	62.4
Heptachlor (Heptachlor epoxide)	0.6	2.74	0.6	2.74
Lead	35,000	91,300	30,200	112,000
Lindane(Hexachlorocyclohexane)	0.94	1.38	0.32	0.99
Mercury	170	486	130	700
Polychlorinated biphenyls(PCBs)	34.1	277	21.5	189
Arochlor 1254	60	340	63.3	709
Polychlorinated dibenzo-p-dioxins/dibenzo furans (PCDD/Fs)	0.85 ngTEQ/kgdw	21.5 ng TEQ/kgdw	0.85 ng/TEQ/kgdw	21.5 ngTEQ/kgdw
Acenaphthene	6.71	88.9	6.71	88.9
Acenaphthylene	5.87	128	5.87	128
Anthracene	46.9	245	46.9	245
Benzo(a)anthracene	31.7	385	74.8	693
Benzo(a)pyrene	31.9	782	88.8	763
Chrysene	57.1	862	108	846
Dibenzo(a,h)anthracene	6.22	135	6.22	135
Fluoranthene	111	2,355	113	1,494
Fluorene	21.2	144	21.2	144
2-Methylnaphthalene	20.2	201	20.2	201
Naphthalene	34.6	391	34.6	391
Phenanthrene	41.9	515	86.7	544
Pyrene	53	875	153	1,398
Toxaphene	0.1	No PEL derived	0.1	No PEL derived
Zinc	123,000	315,000	124,000	271,000

Source: Adapted from the Canadian Environmental Quality Guidelines (2003).

ppm dry wt., ERM (effects: range – medium) is 218 ppm dry wt., and the incidence of effects is 90.2% when the concentration is higher than the ERM. When concentration exceeded the ERM values, the incidence of adverse effects increased from 60% to 90% for most trace metals and 80% to 100% for most organics. However, the reliability of the ERMs for nickel, mercury, DDE, total DDTs, and total PCBs are much lower than those for other substances.

8.4.2 Background and Bioconcentration

8.4.2.1 Background Concentration

As discussed in Section 1.5 in Chapter 1, the elements shown in Table 1.3 are known to exist naturally in the environment, generally in the form of compounds and minerals, such as sodium chloride, copper carbonate (azurite $Cu_3(CO_3)_2(OH)_2$ and malachite $Cu_2CO_3(OH)_2$), magnesite (for magnesium), and in food sources, such as spinach and nuts (for magnesium). Not all the elements shown are toxic or totally harmful to human health. In the listing shown in Table 1.3, both lead (Pb) and cadmium (Cd) are known as toxic elements and can be safely identified as contaminants. There are no acceptable daily intake values for these. The rest of the elements shown in Table 1.3 are known to be essential elements, and the lack of any of these can be harmful to human health. However, ingestion of concentrations of these essential elements in excess of acceptable daily intake (ADI) can be harmful to human health. Table 8.3 shows some of the effects to human health for some of the essential elements when ingested concentrations are deficient (*lack of*) or in excess (*toxic*). All the harmful effects shown in the last two columns are not meant to be totally definitive. They should be considered as *potential effects* since very few totally-controlled studies have been conducted to fully isolate the noted *harmful effects*.

The record shows that background concentration of many of these elements and several other known toxicants exist in the environment and especially in the coastal marine environment and sediments – naturally derived and more likely due to anthropogenic discharges. Exposure to concentrations of these elements and toxicants that are higher than the PEL (probable effects level) or ERM (effects: range – median) raises questions relating to safety and risks to human health.

TABLE 8.3
Average Daily Intake of Some Inorganics in Typical North American Adults Compared with Typical Dosages in a Common Dietary Supplement

Element	Daily intake (mg/day)* *Typical average value for adult*	Typical *dosage* in dietary pills mg/day	Possible effects from deficiency**	Possible toxic effects**
Potassium	3750	32	hypokalemia, muscle weakness, abnormal heart rhythms	diarrhea, nephrotoxicity, hyperkalemia, muscle fatigue, cardia arrhythmia
Calcium	420	530	loss of calcium from bone, muscle spasms, leg cramps,	calcium deposition in soft tissue, kidney stones
Sodium	5660		muscle cramps	high blood pressure
Phosphorus	1500	400	weakness, rickets, bone pain	kidney/liver damage
Magnesium	375	100	electrolyte imbalance of Ca and K	muscle weakness
Zinc	13	22.5	reduced appetite and growth	irritability, nausea
Iron	19.5	12	anaemia	gastrointestinal irritation
Chromium	0.115	0.027	atherosclerosis	tubular necrosis of the kidney
Fluoride	3		possible osteoporosis	dental fluorosis, possible osteoscierosis
Copper	1.7	2	anaemia, loss of pigment, reduced growth, loss of arterial elasticity	disorder of copper metabolism, hepatic cirrhosis

Source: Adapted from Yong (1996).
RSC = relative source contribution
* Information from Lappenbusch (1988).
** Deficiency and toxicity effects are *probable effects*, and are very much dependent on initial health, diet, local environment, cultural attitudes, body features, physiology, etc.

Since the PEL and ERM type of criteria and guidelines have been provided for determination of the direct toxic effects on aquatic life, the potential risk to human health resulting from bioconcentration (of toxicants) have yet to be fully evaluated. Excessive bioconcentration in marine species constituting seafood for humans will pose problems to human health, and may well result in chronic or acute poisoning. The Minamata disease originated from methyl mercury pollution of fish. This is discussed in greater detail in a later section.

Target concentrations reflecting maximum allowable background values for sediments are necessary to provide the necessary protection for both aquatic species and human health. These have yet to be set because (a) available data on present backgrounds are not available, (b) the links between these and bioconcentration in aquatic species and human health are also not available, and (c) the highly variable nature of background concentrations of a substance. The variability and changing nature of background concentrations are functions of many factors, such as sediment characteristics, local geology, mineralogical aspects of soils near the deposited place, and discharges from land-based industries. Isolating discharges from land-based industries allows one to consider the deviation of the background concentrations with sediment type as an indication of a range of tolerable intake.

8.4.3 Sulphide and Its Effects on Marine Life

8.4.3.1 Toxic Sulphide

An excess of sedimentary organic matter might lead to high bacterial sulphate reduction rates, oxygen depletion, and the subsequent release of toxic hydrogen sulphide, especially during the warm season (Magni et al., 2008). The effects of eutrophication on marine benthic communities have been well documented (Pearson and Rosenberg, 1978). A significant reduction in survival time is obtained when individual macrobenthic species are subjected to hypoxia and sulphidic conditions (Rosenberg et al., 2001; Gamenick et al., 1996). In general, sulphide combined with hypoxia is more toxic to benthic animals than hypoxia alone (Diaz and Rosenberg, 1995). Effects on infaunal benthos are also more acute than on epifauna because of their frequent burrowing activities and, hence, exposure to severe hypoxia and sulphidic conditions (Hagerman, 1998).

In the Seto Inland Sea, in Japan, the qualities of seawater and sediment were degrading up to 1970. Since the Act on Special Measures Concerning Conservation of the Environment of the Seto Inland Sea was issued in 1973, the quality of seawater has been considerably improved. However, to date, organic matter and nutrients have accumulated in the sediments, resulting in the production of organic matter in the sediments, and toxic hydrogen sulphide. In Japan, this situation is similar for other sea areas, including ports and harbors.

For example, the sulphide concentrations for the Harima-nada sediments are plotted against ignition loss (Figure 8.2). The original data were obtained by the Ministry of Land, Infrastructure, Transport and Tourism (MLITT). In Figure 8.2, solid circles and squares show sandy sediments, and the triangles indicate silt-clay sediments. The figures show no significant relationship between sulphide content and ignition loss, because of the varied nature of the organic content and sediment age.

The details are explained as follows: New organic sediments are possible under aerobic conditions, because they initially contact dissolved oxygen. An excess of sedimentary organic matter leads to high bacterial sulphate reduction rates, oxygen depletion, and the subsequent release of toxic hydrogen sulphide, from summer to autumn (Magni, 2008; Rosenberg et al., 2001). At this stage, the ignition loss tends to decrease and the sulphide content will increase (Yamamoto et al., 1997). The trend is indicated by the arrow in Figure 8.2.

The concentration of the sulphide in sediments depends on many factors, such as the type of sediment, organic content, pH, redox potential, dissolved oxygen, water and sediment temperature, etc. The physical and chemical properties of Seto Inland Sea sediments, reported by Yamamoto et al. (1997), are shown in Figure 8.3 plotted in terms of redox potential versus pH values for the period between October 1993 and June 1994 for various sites in the Seto Inland Sea.

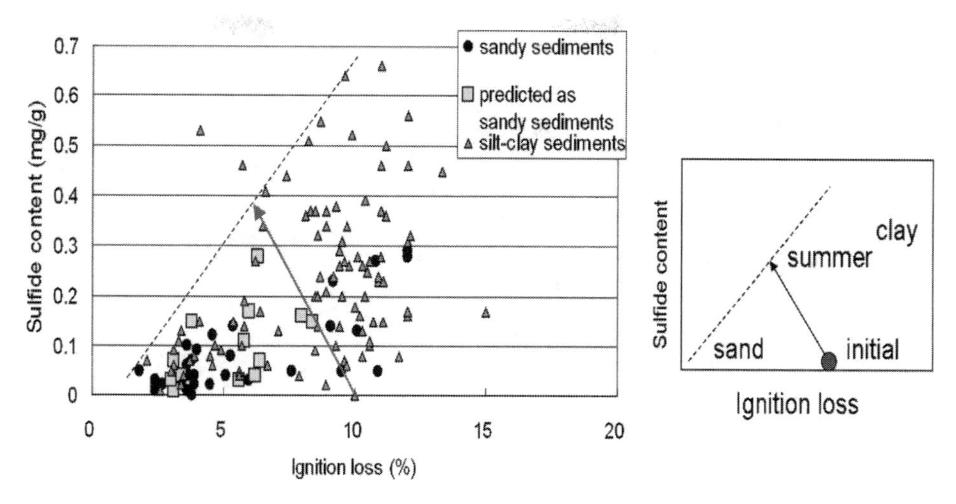

FIGURE 8.2 Sulphide content versus ignition loss for the Harima-nada sediments. (Data from MLITT.)

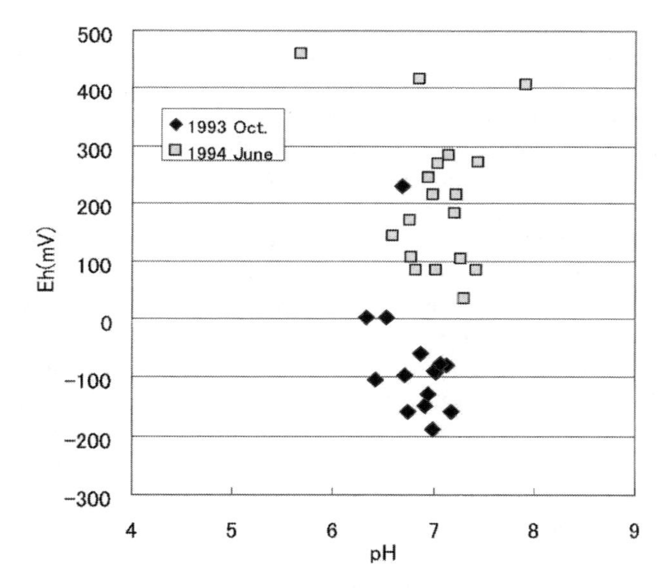

FIGURE 8.3 Comparison between redox potentials in October 1993 and June 1994. (Data from Yamamoto et al., 1997.)

Figure 8.4 shows the changes in sulphide concentration obtained from Harima-nada in the month of Oct. between 2000 and 2008. The concentration and its fluctuation are unpredictable. The data for site 10 were obtained from fine sand, which had a relatively low level (less than 0.1 mg/g) of sulphide (between 2000 and 2006) and a rapid increase to 0.2 mg/g after that.

8.4.3.2 Guideline of Sulphide for Surface Water and Sediments

The Quality Criteria for Water (US EPA, 1976) concluded that the hazard from hydrogen sulphide to aquatic life is often localized and transient. Available data indicate that water containing concentrations of 2.0 μg/L undissociated H_2S would not be hazardous to most fish and other aquatic wildlife, but concentrations in excess of 2.0 μg/L would constitute a long-term hazard.

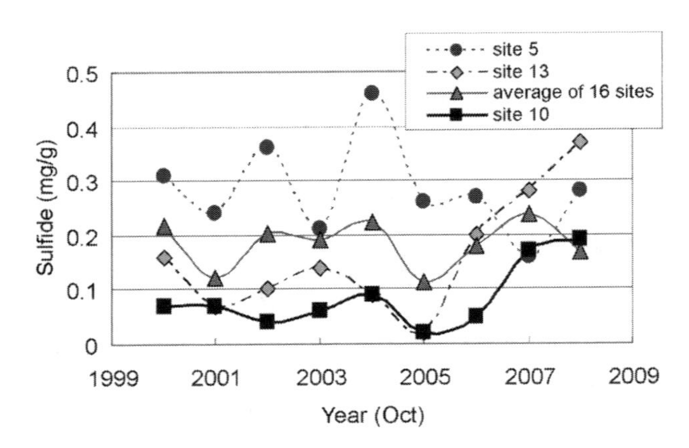

FIGURE 8.4 Annual changes in sulphide content for the Harima-nada sediments. (Data from MLITT.)

Hydrogen sulphide has an adverse impact on aquatic life, especially to crustacea. The standard level provided by the Japanese sediment guideline for fish is 0.2 mg/g. Herein, the sulphide content is defined by the percentage of sulphide/dry weight of sediment. If the pore space of sand is the habitat of marine life, the conventional definition of sulphide should be modified for an aestivating fish, such as sand lance.

The comparison between both the guidelines for water and sediments, above mentioned, indicates that the value of the criterion for sediments will be somewhat higher than that of water, even if the difference between the densities of sediments and water are considered.

8.4.3.3 Connecting Problems of Geoenvironment and Bioenvironment

It is well understood that the marine geoenvironment is directly connected to the bioenvironment through the food-chain. Eutrophication of marine geoenvironment often results from the blooming of phytoplankton – with the latter event (phytoplankton blooming) being the result of human activities.

Guidelines and/or criteria are necessary requirements in addressing environmental problems. It is important to realize that in respect to the bioenvironment, environmental change is often irreversible, with no prospect for recovery or rehabilitation. A specific example of adverse effects of eutrophication most likely responsible for extinction of marine life in the bioenvironment is explained below.

Japanese sand lance (*Ammodytes personatus*), also called sand eel, is one of about 18 species of marine fish of the family *Ammodytidae* (order Perciformes). Sand lances are slim, elongated, usually silver fish that are especially abundant in northern seas. Individuals range from about 8 to 20 cm in length. They have a forked tail, a long head, a long dorsal fin, and peculiar skin folds on the lower sides, as shown in Figure 8.5. They live in schools, often under the sand below the surf (aestivation) in summer, which means that they can be affected by sediment quality. They lay their eggs in sand, and the eggs hatch in winter. The fry become a target of fishery for about one month in the spring.

The haul of sand lance in the Seto Inland Sea has continued to decrease since 1970. This is considered to be mainly due to the decrease in sand beach and tideland. Sand was harvested from the bottom of the Seto Inland Sea for use as aggregates for concrete. This also decreased the habitat for sand lance and their egg-laying sites. According to a report by the Ministry of Economy, Trade, and Industry and the Ministry of Land, Infrastructure, Transport and Tourism, about 10,000 to 20,000 m^3 of sand were harvested annually from the bottom of the Seto Inland Sea from 1971 to 2004. During this period, a large coastal area was reclaimed from the sea, as shown for example by the reclamation

FIGURE 8.5 Japanese sand lance.

of 16 km of coast in the Okayama and Kagawa Prefecture facing the middle of Seto Inland Sea. This has resulted in the loss of sand lance habitat. The haul of sand lance, which has not recovered since harvesting of sand at the bottom was prohibited, continues to decrease, with large fluctuations. There has been no explanation for this phenomenon. Because of anthropological activities, various kinds of solutes and solids have been discharged into the Seto Inland Sea. In particular, organic matter has caused eutrophication. The organic matter discharged from land has spread over the areas of the Seto Inland Sea, except for straits with a strong current. In this situation, the clean sand of the sea bottom has changed to sand with clay or silt fractions and organic matter.

The sedimentary rate in the Seto Inland Sea determined using the ^{210}Pb method varies between 0.11 and 1.13 g/cm^2 (Hoshika and Shiozawa, 1987). The sedimentary rate is relatively high in the central parts of Osaka Bay, Harima-nada and Hiuchi-nada. These areas are also the habitat of sand lance. Water quality might be one of the governing factors for the health of the population of sand lance. This area had the worst water quality around 1970. Whilst water quality has been improved since 1970, red tide still occurs. The biodegradation of dead bodies of phytoplankton accumulating on the bottom and penetrating into the pore spaces of sediment renders the sediment anoxic. This creates a critical condition for sand lance because they must aestivate in the sand during the summer.

The fishery and haul of sand lance are now strictly controlled by the local governments. Even under management and control, the population and haul of sand lance are still decreasing, as shown in Figure 8.6. However, no explanation could account for the abrupt drop of the haul. Note that sand lance is a key member of the food-chain in the Seto Inland Sea.

The rapid decrease to one tenth of the haul in 2017 from 2016 may indicate a difficulty in population recovery. It is apparent that the lowest haul in 2020 was due to the COVID-19 pandemic. If the sand lance disappears from the sea area, fish diversity will be completely changed, because the sand lance belongs to a lower level of the food-chain in the sea. It will also damage the food culture in the anthrosphere through the biosphere and geoenvironment. Sand lance is a raw material, which is a well-known famous processed food, so called caramelized sand lance (*kugi ni*) in Japan.

In order for fishery resources in the Seto Inland Sea to recover, it is necessary to increase the population of sand lance. The detailed study showed that the decrease in population of sand lance in the Seto Inland Sea was due to the sulphide concentration in the pore mud water of sandy sediments (Fukue et al., 2012a), where the pore mud water means the organic mud water in sandy sediments. This was examined by converting the sulphide concentration of sediments to that of the pore mud water of the sediments. Figure 8.7 shows the conversion of sulphide concentration from sediment to pore mud water. The figure shows that aestivating sand lance will be killed or damaged by the toxic sulphide in summer to autumn because the concentration of converted sulphide content (in pore mud water) is considerably higher than the Japanese criteria for fish. For the countermeasure for this problem, the resuspension technique can be applied (Fukue et al., 2012b). The technique will be discussed in Section 8.5.4 in this chapter.

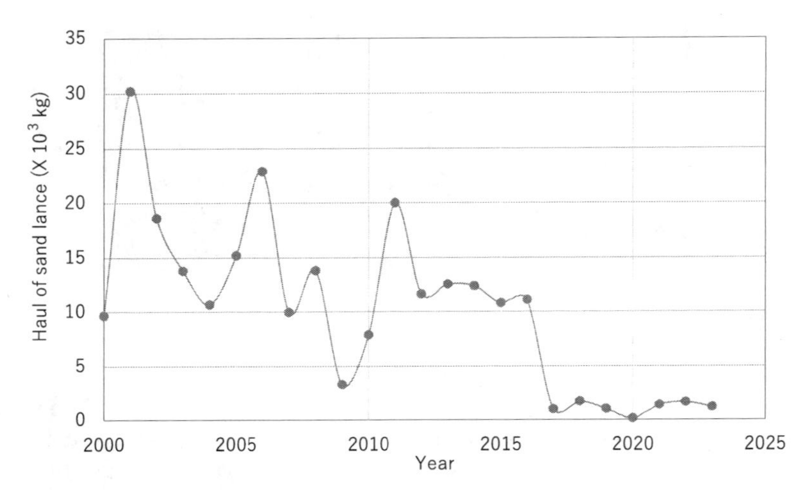

FIGURE 8.6 Fluctuation of haul of sand lance as published by Hyogo Prefectures. (Data published by the Japan Ministry of Agriculture, Forestry and Fisheries, on Sept. 15, 2023.)

www.maff.go.jp/j/tokei/kouhyou/kaimen_gyosei/index.html.

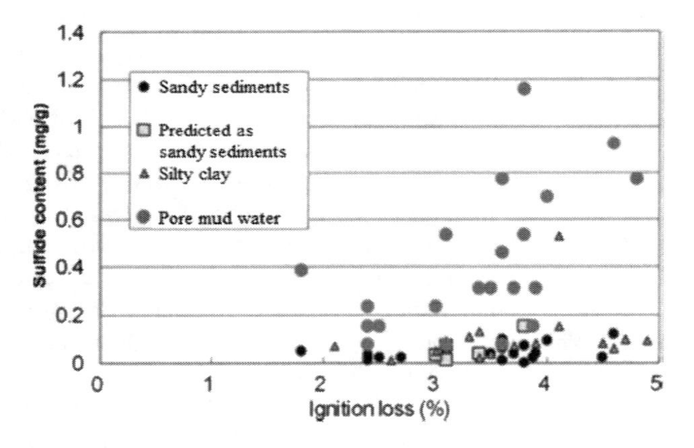

FIGURE 8.7 Conversion of sulphide concentrations from sandy sediments to the pore mud water. (Data from Fukue et al., 2012a.)

8.4.4 Heavy Metals

Many heavy metals are trace metals. These are metals that exist in extremely small quantities and are almost at the molecular level. They reside in, or are present in, animal and plant cells and tissue, and are a necessary part of good nutrition – as shown for example in Table 8.3. Because excess intake of heavy metals may cause damage to human health, it is necessary to take into consideration the background concentrations in sediments in structuring safe limits for ingestion of aquatic species. We define the *background concentrations* (values) as the concentrations of substances under natural condition without any significant input or effect from human activities.

Background values of metals were found to agree well with average values for gneisses rocks (Carral et al., 1995). However, the background values for copper, zinc, lead, and cobalt do not necessarily agree with previously obtained values. Various techniques have been used to reduce data scattering and to allow for a more accurate statement of background values. Fukue et al. (1999) used carbonates as the normalizing substance in their measurements. Cobelo-García and Prego (2003)

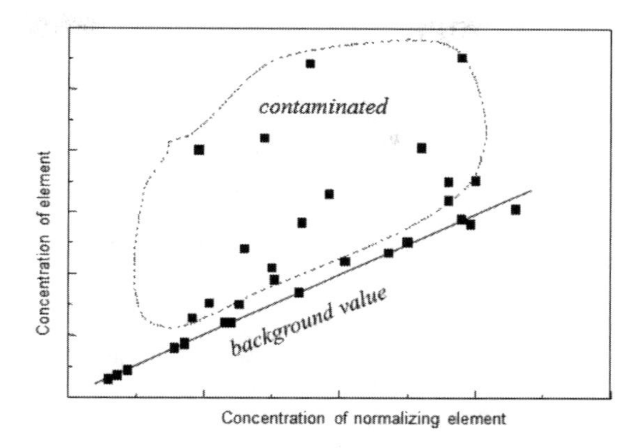

FIGURE 8.8 Concept of a normalization technique for obtaining a background value.

obtained the baseline relationships between the concentration of iron and contaminant, whilst Din (1992), Cortesão and Vale (1995), and Santschi et al. (2001) used aluminium to normalize heavy metals. Titan has also been used as a normalizing element. Fukue et al. (2006) calculated the specific surface area of sediment particles assuming them to be spheres, and that the specific surface areas will be related to the background values of heavy metals. This is consistent with the thesis that the amount of metals sorbed onto particles is to a large extent dependent on the particle size. The results reported by Fukue et al. (2006) using these normalization characteristics were found to be reasonable. Santschi et al. (2001) reported that concentrations and fluxes of most trace metals found in sediment cores recovered from Mississippi River delta, Galveston Bay and Tampa Bay in the US, when normalized to Al, were typical for uncontaminated Gulf Coast sediments. Similar results can be cited with other normalizing elements. The concept for the normalization technique is shown in Figure 8.8. The drawback of these methods is that non-contaminated sediment samples are required so that relationships between normalized substances and the objective concentrations can be constructed. If the depth and extent of sampling are insufficient, normalization characteristics cannot be obtained.

Many contaminants in water are sorbed on sorbates that may be inorganic and organic particulate materials. These particulates, together with the sorbed contaminants eventually settle to form sediments on the seabed. The properties and surface characteristics of the soil solids that comprise the particulate materials have been discussed briefly in Sections 2.4 and 2.5 in Chapter 2 in respect to contaminant-particle interactions. Some of the more pertinent surface activities include parameters, such as charge density, CEC, specific surface area, the equilibrium (natural) concentration of contaminants, etc. A more detailed explanation of all of these surface properties and their interactions with contaminants can be found in Yong (2001). The background concentration values of sediments will be directly related to the inherent properties and characteristics of the constituents.

There are two concepts of background concentration: (a) concentration in the mother rocks, and (b) mother rock concentrations together with sorbed substances under natural conditions during transport process and deposition. The second concept is more appropriate when background concentrations are used to evaluate the risk to human health, especially when bioconcentration is factored into risk evaluation.

8.4.4.1 Profile of Heavy Metal Concentration

Most profiles of metal concentration for sediments show trends similar to those portrayed in Figure 8.9. The total concentration consists of the background and contributions from the discharges originating from land-based industries. Beginning from the bottom of the profile shown in Figure 8.9, the total concentration consists only of the background values. As one progresses upward, toward the top of the sediment, the total concentration begins to include contaminant discharge

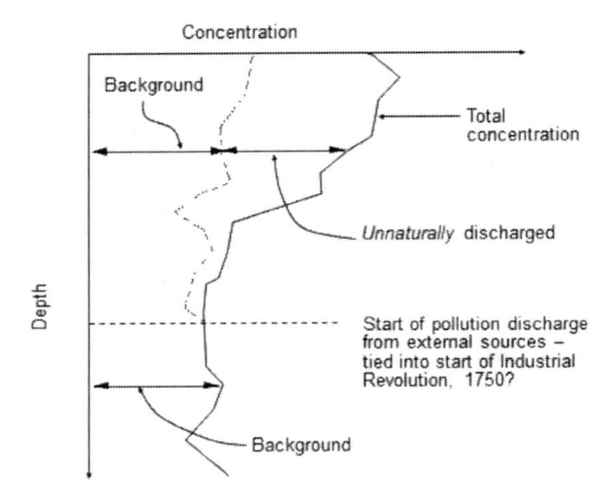

FIGURE 8.9 Concept of background value of metal concentrations in sediments. The *unnaturally* discharged concentrations are likely due to discharges from land-based industries and land-based non-point pollution sources.

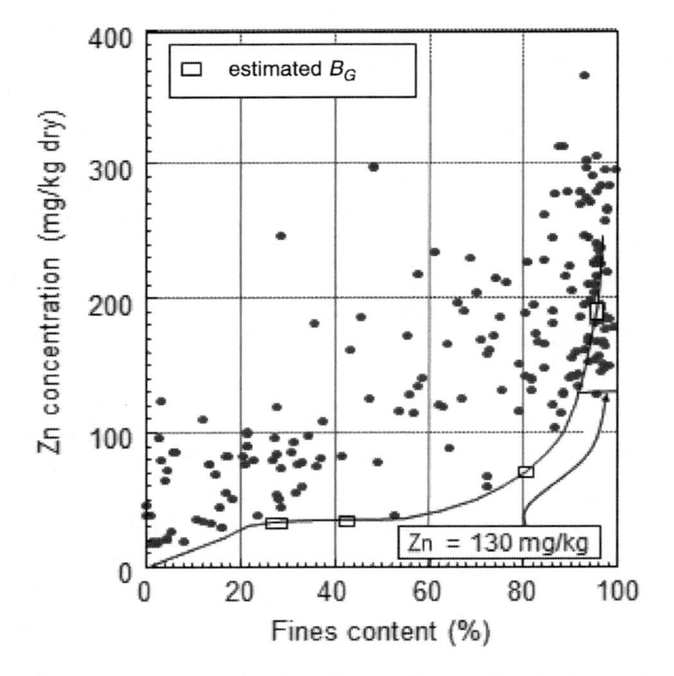

FIGURE 8.10 Deviation of zinc concentration in various sediments. B_G = background concentration.

contributions from land-based industries. The beginning point shown in the diagram assumes that this can be tied to the start of the Industrial Revolution in about 1750. The total concentration increases as one progresses upward in the sediment, indicating increased discharge of contaminants into the marine environment. As an example, Figure 8.10 shows zinc (Zn) concentrations in various sediments obtained from coastal regions in Japan. Because the sediment samples were obtained from various depths in the total sediment layer, the concentration levels were not uniform, and some of the shallower samples were contaminated. The background concentrations can be obtained

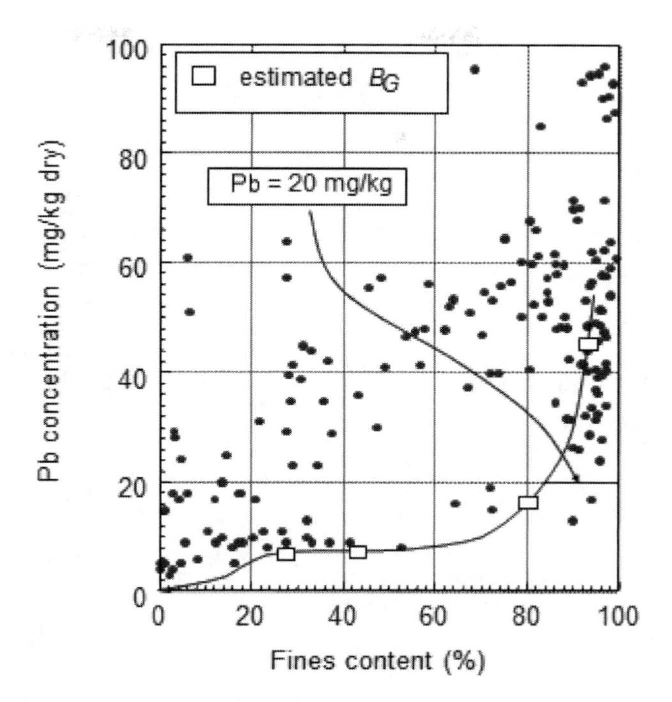

FIGURE 8.11 Deviation of lead (Pb) concentration in various sediments.

using a normalizing factor. The procedure adopted for the zinc concentrations shown in Figure 8.10, was to use the fines content as the normalizing factor. The fines content refers to the fine particle sizes in the sediment, generally clay and silt, and is defined as the content of clay and silt fractions <0.075 mm. The line representing the approximate lower limit shown in Figure 8.10 can be defined as the background zinc concentrations for the sediments. The detailed theoretical approach to obtain the background line has been reported by Fukue et al. (2006).

Similar relationships with lead (Pb) and copper (Cu) have been obtained – as shown in Figures 8.11 and 8.12. Fukue et al. (2006) have used relationships between calculated specific surface area and fine content, and also limitation of sorption of fine particles under a relatively low equilibrium concentration. This is due to the limited concentrations of the substance in nature. The background values for Zn, Pb, and Cu shown in the figures are lower than the values presented as the ISQGs (interim sediment quality guidelines) in Table 8.2. When the background concentration B_G is known, the degree of pollution, P_d will be obtained as

$$P_d = (C - B_G)/B_G \tag{8.1}$$

where C is the current concentration of a specific substance or element (Fukue et al., 2006).

As was stated in the previous section and in Chapter 1, and shown in Table 8.3, there is a risk that the intake of hazardous substances beyond the PEL (probable effect level) values may be harmful to human health.

8.4.5 Minamata Disease

In the1950s, a significant number of people succumbed to a disease, later identified as Minamata disease, which was traced to ingestion of poisoned fish and shellfish in Minamata Bay in Kyushu Island (Japan). Measurements of methylmercury chloride showed very high concentrations, up to

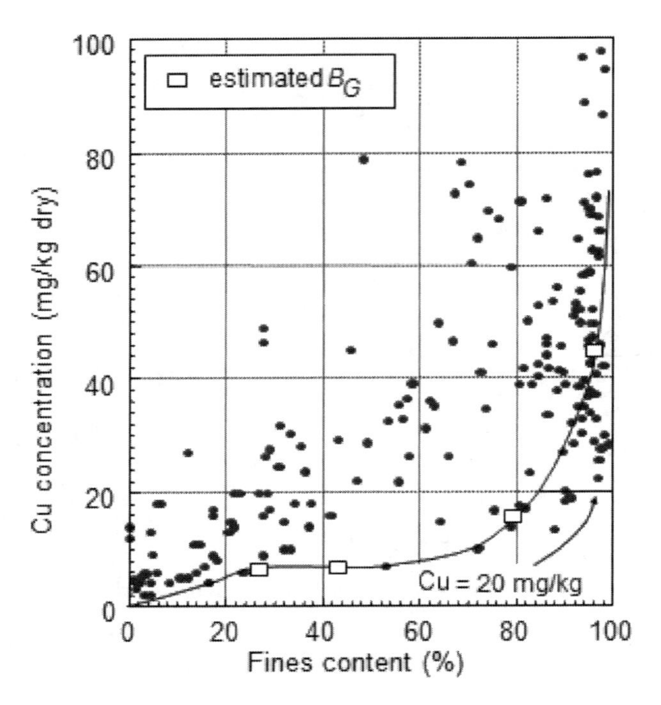

FIGURE 8.12 Deviation of copper (Cu) concentration in various sediments.

approximately 50 ppm in fish and 85 ppm in shellfish from the contaminated areas in Minamata Bay. The more than 100 people that ingested the contaminated fish and shellfish showed initial symptoms of numbness of the limbs and areas around the mouth, sensory disturbance, lack of coordination, weakness and tremor, speech and audio-visual difficulties. Progressive worsening of all of these with time led to general paralysis, convulsions, brain damage, and death. Traces of mercury poisoning were also found in animals living around the bay. Many of these animals also died. More details of this terrible demonstration of the impact of polluted sediments can be found at www.env.go.jp/en/index.html (accessed Jul. 2024).

It is useful to note that whilst Minamata disease is well documented and serves as a dramatic demonstration of the chain of effects originating from polluted sediments, discovery and diagnosis of the health problems and source of the health problem took some considerable time, effort, and tracking. There are undoubtedly countless cases of poisoned aquatic animals in the first level of the food-chain serving the human population, and unreported or undiagnosed (under-diagnosed?) poisoning of the affected population.

8.4.6 Organic Chemical Contaminants

8.4.6.1 Organotins

Other than from direct and indirect discharge of land-based industrial waste contaminants and leachates from waste piles, some hazardous substances enter the marine environment through direct contact and use, such as organotins used as antifoulants for ships, quays, buoys, etc. Organotins are compounds of tributyltin (TBT). They are highly toxic chemicals comprising of tin combined with organic molecules. They are used not only as antifoulants, but also as wood preservatives, slimicides, and biocides. In the context of a coastal marine environment, organotins are used essentially as biocides to prevent the build-up of barnacles and algae. They are self-polishing co-polymers and generally have a service life between 3 to 5 years – meaning that they have to be reapplied to the marine structures at the end of their service lives. They are poisonous to marine life including whales, dolphins, seals, fish, and sea birds. Linley-Adams (1999) reports that concentrations in bottlenose dolphin liver from the US Atlantic and Gulf Coasts found in the period between 1989 and 1994 ranged from 110 to 11340 ng/g wet weight. He further reports on the presence of organotins in sea otters in California waters and harbour porpoises in Turkish coastal waters in the Black Sea. TBT has been reported to result in the development of imposex, a pseudo-hermaphroditic condition in female gastropods (snails) at ng/L levels of concentration (Horiguchi et al., 1994). The UK guideline provides 0.002 μg/L of TBT for seawater quality, and 0.008μg/L for seawater quality for triphenyltin (TPT) guideline. The Canadian environmental guideline for fresh water quality is 0.022 μg/L for TBT (no guideline exists for marine water).

TBT can enter the marine environment through (a) leaching of the antifoulant paints, (b) during application of TBT as an antifoulant, and (c) when the paint is removed from the pieces of equipment painted with the antifoulant. The degradation of TBT is relatively slow in sediments. The half-life of TBT in sediments has been reported to be approximately 2.5 years (de Mora et al., 1995), whilst it is only a week or so in marine waters (Seligman et al., 1988). Since 1990, their use has been banned for all vessels in certain countries (Japan, Australia, and New Zealand), and in some other countries, their use have been restricted to vessels with lengths greater than 25 m. The IMO (International Maritime Organisation) in 2001 adopted the International Convention on the Control of Harmful Anti-Fouling Systems – a Convention that prohibits the use of harmful organotins as antifoulants. The Convention came into effect at the beginning of 2003 for all ships and was expanded to include all floating platforms, floating units, floating production, and storage units at the beginning of 2008.

8.4.6.2 Chlorinated Organic Microcontaminants

Chlorinated organic microcontaminants have been accumulating in sediments and sea animals for many years. These contaminants are highly toxic and direct or indirect ingestion of these will be life-threatening. The particular group of chemicals known as CDDs (chlorinated dibenzo-p-dioxins) is perhaps the one that has gained the most attention and concern, since it is known to enter the geoenvironment and into the marine environment from many sources. Of the more than 70 chemicals that make up the CDDs, the one known as 2,3,7,8-TCDD or more simply as TCDD, (2,3,7,8-tetrachlorodibenzo-p-dioxin) is considered to be one of the most toxic. The WHO (World Health Organization) has designated this as a known human carcinogen. In general, CDDs find their way into the land and marine environment as waste discharges from some processes that employ chlorine, such as paper mills that use chlorine bleaching processes (see Section 7.4 of Chapter 7), manufacturing of chlorinated organic intermediates, and water treatment plants. Generically, CCDs are known as dioxins. They also occur naturally and are discharged when materials containing such substances are degraded and combusted – as in forest fires. Dioxins released as atmospheric emissions from forest fires, combustion of fossil fuels, and from incinerators serving industrial plants, municipal and hazardous waste facilities, will subsequently be deposited onto land and water bodies at points distant from their sources. In water bodies, they will be attached to the suspended and sedimenting particles and will form part of the sediment. A limit of 3×10^{-5} micrograms of 2,3,7,8-TCDD per litre of drinking water (30 picograms per litre, i.e., 30 pg/L) has been set by the EPA.

Dioxins generally occur in the environment with chlorinated dibenzofurans (popularly known as furans). They are ubiquitous, and have been found in all media: Air, water, soil, sediments, animals, and food (Johnson, 1992). They fall into the class of POPs (persistent organic pollutants) and are bioaccumulative. Since they have a strong affinity for soil and sediment particles, their presence in the media containing these POPs pose significant health threats. To determine the risks resulting from exposure to these POPs, many environmental regulatory bodies, such as the US Environmental Protection Agency (EPA) have adopted an evaluation technique that uses *toxicity equivalency factors* (TEFs). This technique compares the toxicity of designated dioxins and furans to that of the most toxic dioxin, i.e., the 2,3,7,8-TCDD. With this technique, the TCDD establishes the height of the toxicity bar and is given a TEF value of one, and all other organic chemical compounds being scrutinized as assigned TEF values according to how toxic they are perceived to be. Determination of toxicity intake (or uptake) of a particular organic chemical contaminant requires the use of *toxicity equivalents* (TEQs). A TEQ for a particular organic chemical contaminant ingested is determined by the product of the contaminant weight (in grams) and its assigned TEF, and the units are grams TEQ. As an example, Makiya (1997) stated that 60 to 70% of daily intake of dioxins (approximately 3.5 TEQ pg/kg/day) is obtained from fish and shellfish. The tolerable daily intake (TDI) established in Japan is 4 TEQ pg/day per kg weight. The Japanese standard value for soils contaminated with dioxins is 1000 pg-TEQ/g or less, whereas the corresponding standard value for sediments is 150 pg-TEQ/g. This value drops to 1 pg-TEQ/L or less for water – in accordance with the Ministry of Environment (Japan) standards. Sediments containing dioxins that are more than the standard value are required to be treated. The classical remediation techniques used are dredging and disposal on land – procedures that are most often costly and sometimes prohibitive when dredged materials have to be treated before land disposal.

8.4.6.3 Micro and Nanoplastics

According to Plastics Europe (2020), the largest consumers of plastics are the building, construction and packaging industries. Polypropylene (PP), polyethylene (PE), and polyvinyl chloride (PVC) are some of the most common. Despite initiatives to improve collection and recycling, plastics end up in the environment and are transported to the oceans via rivers. Substantial amounts can sink to river and lake beds and the ocean floor. It has been estimated that 139 Mt of plastic has accumulated in the ocean as of 2019, which will double by 2060 (OECD, 2023). These plastics can last for very long periods of time since they do not decompose easily and fragment into smaller pieces. For example, high-density polyethylene (HDPE) milk bottles and low-density polyethylene (LDPE) plastic products like LDPE plastic bags and HDPE milk bottles have estimated half-lives of 5 to 250 years on land and 3 to 58 years in marine environments (Chamas et al., 2020). Marine species can get entangled in the plastic debris or ingest the macro and microplastics. Accumulation of the microplastics in species such as mussels and fish can lead to human consumption (Lusher et al., 2017).

Fish et al. (2023) has reviewed the physical, chemical, and biological methods that exist for removal of plastics in the environment. However, it is indicated that there are no standard and cost-effective methods presently. A system approach of reducing plastic introduction into the environment and its removal is needed.

8.5 REHABILITATION OF COASTAL MARINE ENVIRONMENT

Several techniques have been tried and used in the rehabilitation of coastal marine environments. Table 8.4 gives a short summary of some of the recent procedures used to treat contaminated seawater and sediments, and also measures taken to create coastal marine ecosystems. The methods used include (a) removal of the contaminants and hazardous substances by various techniques, (b) immobilization and isolation of these substances, and (c) neutralization and detoxification. The techniques for treatment of seawater include aeration, filtration, adsorption, accumulation, isolation, etc. Common treatment techniques for contaminated sediments include dredging and capping. Other lesser-used techniques include cultivation and lime strewing. The essence of the various techniques, together with present capabilities, economics of operation and requirements are also briefly noted in the table. The ultimate aim of sediment and seawater treatment procedures is to remove the threat posed by contaminants and hazardous substances in the water and sediments.

TABLE 8.4
Various Techniques Developed for Remediation of Contaminated Seawater and Sediments

Technology	Name	Feature	Example of Use	Economics	Durability	Maintenance	Comments
			Seawater				
Aeration	Air bubble curtain	Aeration	Developing	Needs electricity	Durability of hardware	Clogging due to organisms	
		Aerobic condition					
		Convection of seawater					
	Microbubbling	Aeration	Oyster farms				
		Very small bubbles					
	Aerator	Aeration	Developing	Needs electricity			
		Mixing of seawater					
	Special seawall	Natural bubbling	Developing				
		Mixing of seawater					
	Air mixing flow	Aerobic condition of deep seawater	Developing				
		Mixing of seawater					
Filtration	Filtration	Control and management of eutrophication	Many examples	Exchange of filter materials	Depends on instruments	Exchange of filter materials	Enclosed water area
		Removal of suspended solids		Removal of filtered materials			
Gravel oxidation	Control and management of eutrophication	Developing					
	Removal of suspended solids						
	Actions of microorganisms						
Inclined wall	Control and management of eutrophication	Used in rivers		Exchange of filters	Removal of sediments		
		Removal of suspended solids					
		Filtration					
	Artificial leaves	Control and management of eutrophication	Used in sediment ponds			Removal of sediments	
		Removal of suspended solids					

(continued)

TABLE 8.4 (Continued)

Technology	Name	Feature	Example of Use	Economics	Durability	Maintenance	Comments
		Sedimentation by aggregation agents					
Adsorption	Clays	Control and management of eutrophication	Many examples			Removal of sediments	Muddiness due to spraying agent Impact on living organisms
	Aggregation agent	Control and management of eutrophication	Many examples	Polymer		Removal of sediments	Impact on environment from aggregation agents
		Promotion of sedimentation					
Lime strewing	Adsorption of sulfide Increase in pH	Many examples		Short term effect	Strew often when no effects are observed	Fossil limestone	Fossil limestone
Sorption/ accumulation	Accumulation of plants	Reduction of eutrophication	Lakes			Recovery of plants	Management of aquatic plants
	Bioaccumulation of shellfish	Removal of suspended solids	Well known				
Isolation	Silt fence			\$500–2000/m			
	Oil fence	Prevention of diffusion of suspended solids	Many examples	\$50/m	About 5 years	Treatment of adhesive organisms	
		Promotion of settlement of suspended solids					
			Sediments				
Cultivation		Oxidation		Simple and easy			
In situ solidification		Solidification of seabed	Established	Expensive			
Dredging and disposal		Removal of contaminants	Many examples	\$300/m^3			Disposal using bags

Trench	Collection of mud in trench with about 2 m depth	Needs experience		Influenced by waves	Removal of sediments	Need disposal sites
Capping of sand	Capping of sand	Many examples	Easy task with a simple machine			Not for navigation route of ships
Artificial tidal land	Sand beach	Many examples	Lack of sand materials	Influenced by waves		
Lime strewing	Adsorption of sulfide Increase in pH	Many examples		Short term	Strew often when no effects are observed	
Phyto-remediation	Heavy metals, TBT, etc.	Developing				
		Need seawater purification				
		Creation of Ecosystem				
Artificial wet lands	Biological diversity	Many examples	High cost			Difficulty in evaluation
Seagrass	Biological diversity, Increase in haul of fish and shellfish					
Tidal pool	Recreation in seaside					Easy
Artificial coral reef	Biological diversity					
Artificial fish compartment	Biological diversity, Increase in haul of fish and shellfish					

8.5.1 Removal of Contaminated Suspended Solids

8.5.1.1 Confined Sea Areas

In confined sea areas, accumulation of hazardous substances and eutrophication are significant problems. A major factor in the constitution of seawater quality is the amount of suspended solids in the seawater. These solids include mineral particles, plankton, organic matter, and others kinds of particulate matter. The sorption characteristics and properties of suspended solids make them useful tools for sorption of hazardous substances, such as heavy metals, PAHs. chlorinated organic compounds, etc., in the seawater. Bacteria such as colon bacilli may also be found on the surfaces of suspended solids. Table 8.5 provides an example of the concentration of heavy metals adhering to some typical suspended solids. Removal of the contaminated suspended solids will be a step towards obtaining better seawater quality. Together with the process of bioaccumulation, these suspended solids can be removed from the seawater using pumping and filtration techniques, e.g., released phosphorus (into the sea) initially taken by phytoplankton and algae will be removed when the phytoplankton and algae are themselves subsequently removed. Figure 8.13 gives a schematic illustration of this simple concept. The ultimate aim is to ensure that the concentrations of the hazardous substances are lower than the allowed values in the guidelines. Improved seawater quality

TABLE 8.5
Examples of Metal Concentrations in the SS of a Coastal Sea Area (mg/kg dry)

Element	Concentration	Element	Concentration	Element	Concentration
Fe	8,000–30,000	Cu	240–500	P	2,600–3,000
Al	14,000–44,000	Pb	26–97	Mg	9,600–14,000
Ti	1,000–32,000	Cd	2.6–3.0	Ca	9,000–90,000
V	40–53	Ni	2.8–36	K	9,500–15,000
Zn	350–940				

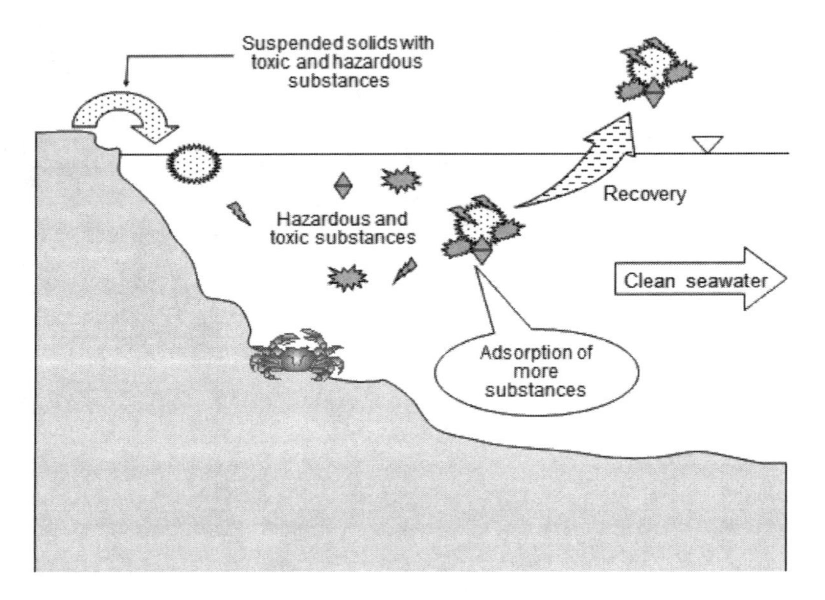

FIGURE 8.13 Concept of removal of suspended solids (SS).

provides for (a) greater transmission of light to sea bed, (b) acceptable water quality for leisure use by the local coastal community, (c) reduction of potential for eutrophication and reduced capability for development of red tide, and (d) better sediment quality and also further reduction in potential for eutrophication, benthic pollution, and development of blue tide. Technology and procedures for removal of suspended solids in closed sea areas using filtering techniques should be determined by the targeted final requirements and results of site investigation.

8.5.1.2 Large Bodies of Water

To treat large bodies of seawater, filter units consisting of beach sand and steel slag have been used successfully to remove the suspended solids. These units were installed on a vessel (approximately 2500 tonnes), as shown in Figure 8.14. The case study conducted with thirty eight filter units (1.5 m × 3.6 m) for purification of seawater in closed sea area achieved a purification capacity of about 6000 m^3/day. The quality of the treated (purified) seawater satisfied the regulatory requirements for allowable SS, COD, pH, DO, etc. (Fukue et al., 2004). A comparison of the nature of the suspended solids showed that these contained large amounts of substances – as much as the underlying sediments. Figure 8.15 shows the results of removal of the suspended solids. The left-hand plastic cylinder with a height of 30 cm contains the seawater with suspended solids. That is why no circle lines on the bottom can be seen. The right-hand picture shows that removal of almost 100% of suspended solid from the contaminated seawater produced transparent seawater that to all intents and purposes was devoid of suspended solids and all the substances adhering to the suspended solids. Studies show that this technique can be used for purification treatment water in various kinds of marine applications including dredging, and treatments to prevent development of red tide and blue tide.

8.5.1.3 Continuous Removal of Suspended Solids

In some closed sea areas, land-based hazardous substances are continuously discharged and delivered by rivers etc. to the coastal marine waters. Filter units similar to those used in the purification vessel, mounted on semi-permanent pier fixtures can function well to remove the hazardous substances adhering to the suspended solids.

FIGURE 8.14 Purification vessel to remove suspended solids from seawater.

FIGURE 8.15 Plastic cylinders showing (a) non-filtered and (b) filtered seawater obtained with purification units in the purification vessel.

8.5.2 Sand Capping

Containment and isolation of nutrients and hazardous substances in sediments will discourage eutrophication and prevent uptake of hazardous substances. Two methods presently in use are (a) sand capping, and (b) dredging removal and dumping on land. Sand capping is a procedure that places a bed of clean sand on the contaminated sediment. It may be the easiest and one of the most economical ways to restore the health of the bottom environment. *In situ* capping refers to placement of a covering or cap over an *in situ* deposit of contaminated sediment (Palermo et al., 1998). The cap may be constructed of clean sediments, sand, gravel, or may involve a more complex design using geotextiles, liners and multiple layers, as used for the capping of landfill waste disposal sites. This capping isolation procedure not only prevents resuspension of polluted fines and other sediment particles, but also reduces the flux of contaminants into the water column above.

A sand cap can provide a new sand beach and tidal flats that will serve as a habitat for various kinds of organisms – from bacteria in pore spaces to macro-benthos such as bivalves, sea cucumber, and seaweed. These organisms consume a large amount of nutrients. When sand is obtained from regions distant from the capping site, organisms and small animals may exist in the transported sand that are not native to the capping site. This is especially true when the source of the sand is land-based. Newly created sand cap beaches can function as physical purification units, i.e., filtration, through the action of waves and tides. Removal of suspended solids from the seawater and decomposition of organic matter adhering to the suspended solids by microorganisms in the pores are additional benefits obtained in sand capping. Some case studies are summarized in the following report (https://astswmo.org/files/policies/CERCLA_and_Brownfields/2013-04-Sediment_Remedy_Effectiveness_and_Recontamination.pdf, accessed Feb. 2024).

For sand capping to be robust and successful in providing the necessary isolation capability, the following must be accounted for in the design and construction of the cap:

(a) Nature and type of currents, waves, flood, etc., that would cause drifting of the sand during and after emplacement. The record shows that hurricanes and typhoons are capable of dramatically altering the near-shore bottom conditions.
(b) Possible movement of some benthic organisms through the sand cap.
(c) Disruption of the food-chain.

Table 8.6 shows a sampling of some major sand capping projects. The benefits from sand capping are shown in Figure 8.16. Data reported by the Port of Kanda, Ministry of Land, Infrastructure and Transport, Japan, for COD, T-N, T-S, and T-P following sand capping indicate environmental benefits such as less COD, less sulphide, and a lower amount of nutrients in the sand cap, in comparison to

TABLE 8.6
Purification Projects Using Sand Caps

Project Location	Contaminants	Site Conditions	Cap Design	Construction Methods	Reference
Kihama Inner Lake, Japan	Nutrients	3700 m^2	Fine sand, 5 and 20 cm		
Akanoi Bay,	Nutrients	20,000 m^2	Fine sand, 20 cm		
Denny Way, Washington	PAHs, PCBs	1.2 ha nearshore with depths from 0.6 to 18.3 m.	Avg. 80 cm of sandy sediment	Barge spreading	Sumeri, 1995
Simpson Tacoma, Washington	Creosote, PAHs, dioxins	6.8 ha nearshore with varying depth	1.2 to 6.1 m of sandy sediment	Hydraulic pipeline with "sandbox"	Sumeri, 1995
Hamilton Harbor Ontario	PAHs, metals, nutrients	10,000 m^2 portion of large, industrial harbour	0.5 m sand	Tremie tube	Zeman and Patterson, 1996
Eitrheim Bay, Norway	Metals	100,000 m^2	Geotextile and gabions	Deployed from barge	Instanes, 1994

that of the surrounding silty seabed. From the viewpoint of biological diversity, there is every indication that the sand cap is better than the surrounding silty seabed.

8.5.3 Removal of Contaminated Sediments by Dredging

8.5.3.1 Dredging

Dredging has historically been used (a) for beach reclamation, (b) to remove bottom sediments to deepen channels, waterways and harbours and ports, (c) for maintenance of desired water depths for sea routes, and (d) to obtain materials for reclamation. Regardless of the type of equipment used – cutter wheels, augers, bucket wheels, pumping, suction hopper, self-propelled, automatic, etc. – dredging is a very invasive and destructive procedure. When used for removal of contaminated sediments, total disruption of the bottom ecosystem occurs. To return the benthic layer to full functionality, replacement of the food supply for benthic organisms is necessary. Removal of the first trophic level in the food-chain will have severe consequences on the higher trophic levels. In that regard, planning for dredge removal of the contaminated bottom layer must include restoration of the functionality of the benthic layer.

8.5.3.2 Treatment of Dredged Sediments

Two options are available for disposal of dredged contaminated sediments: (a) Disposal in a secure landfill, and (b) treatment of the contaminated sediments and reuse of the treated sediments. Option (a) is not an option that has many proponents. Treatment of contaminated sediments (option (b)) can be an expensive procedure, especially when the quantities are large. An expedient procedure is to perform gravity separation of the contaminated sediment and to remove the coarse fractions for treatment and reuse as construction material. A useful technique for sediments that do not contain much organic matter is to form larger particles by promoting aggregation of the fines with lime. Contaminated fine fractions can be treated or disposed in secure settling ponds. These settling ponds are not unlike those obtained in natural resource extraction processes (Chapter 5). Techniques for dewatering and hastened sedimentation of the suspended fines that constitute the fine fractions of the

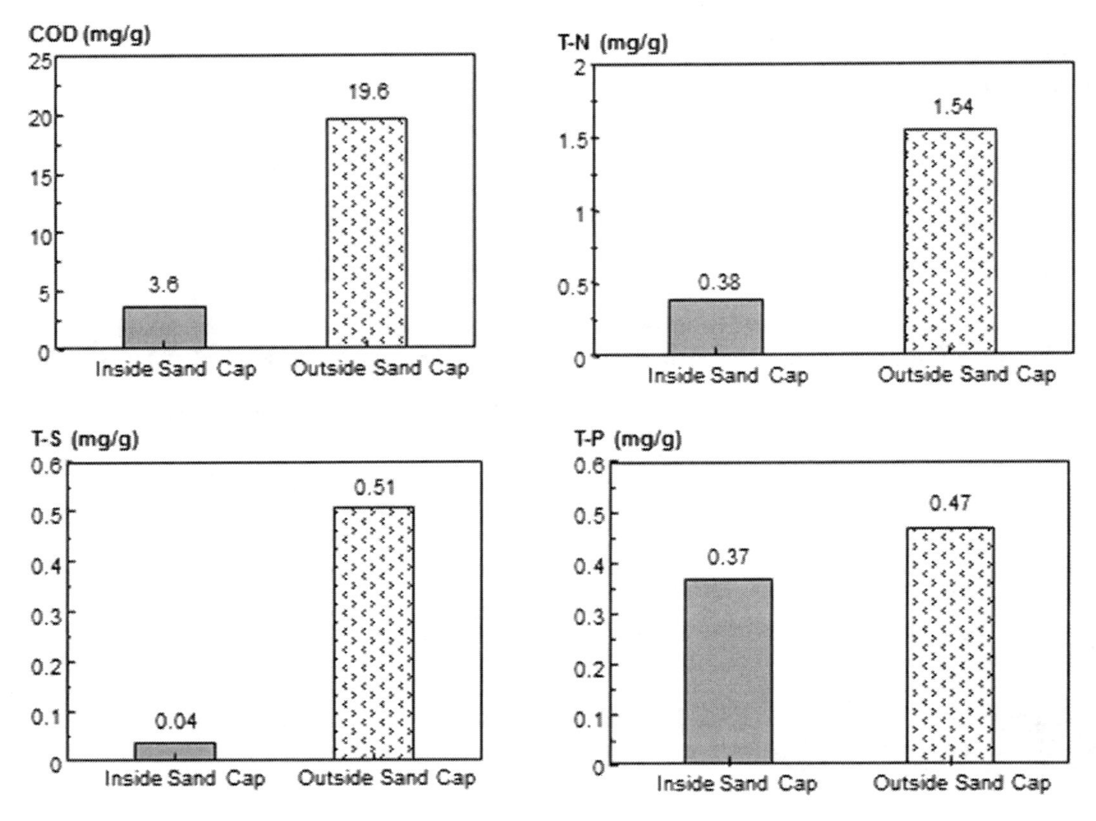

FIGURE 8.16 Effects of a sand cap (Port of Kanda, the Ministry of Land, Infrastructure and Transport, Japan). COD = chemical oxygen demand, T-N, T-S, and T-P = total nitrogen, total sulphides, and total phosphorus, respectively.

sediment have been discussed in Chapter 5. In the case of the fines in sediments, solidification and compression by filter pressing can be used (Yamasaki et al., 1995).

8.5.4 Removal of Contaminated Sediments by Resuspension

Dredging of the sediments is necessary to facilitate the passage of the vessels or remediate contaminated sediments. However, disturbance of the contaminated sediments can spread the pollution. Alternatively, a two-stage resuspension method is suggested as a new *in situ* method for remediation of highly contaminated sediments (Fukue et al., 2012b; see Figure 8.17). In the first stage, air jets are used in the confined water column to create a strong turbulent flow to suspend the sediments. After a certain period of time, the coarser sediments are allowed to settle while the finer sediments remain suspended. In the second stage, the fine suspended sediments are then pumped from the water and filtered. Recently a pilot test in Japan was evaluated in an area of 3000 m^2 with a high organic content (30% loss on ignition) with hydrogen sulfide production. About 11 tonnes of wet resuspended solids were removed during the resuspension pilot test. In the test, 50 cm of the bottom sediment layer was resuspended and 3% of the organic and smaller particles were removed. Analyses showed that full-scale implementation would enable the removal of about 10% of the resuspended solids, and reduce COD by 95%, T-P by 50%, T-N by 100% and sulfide by 75% for the redeposited sediments compared to the untreated sediments. Further tests were performed in Canada on samples obtained from a harbour (Pourabadehei and Mulligan, 2016). In total, 17 samples (surface and core samples) were taken at selected stations near the dock and the breakwater

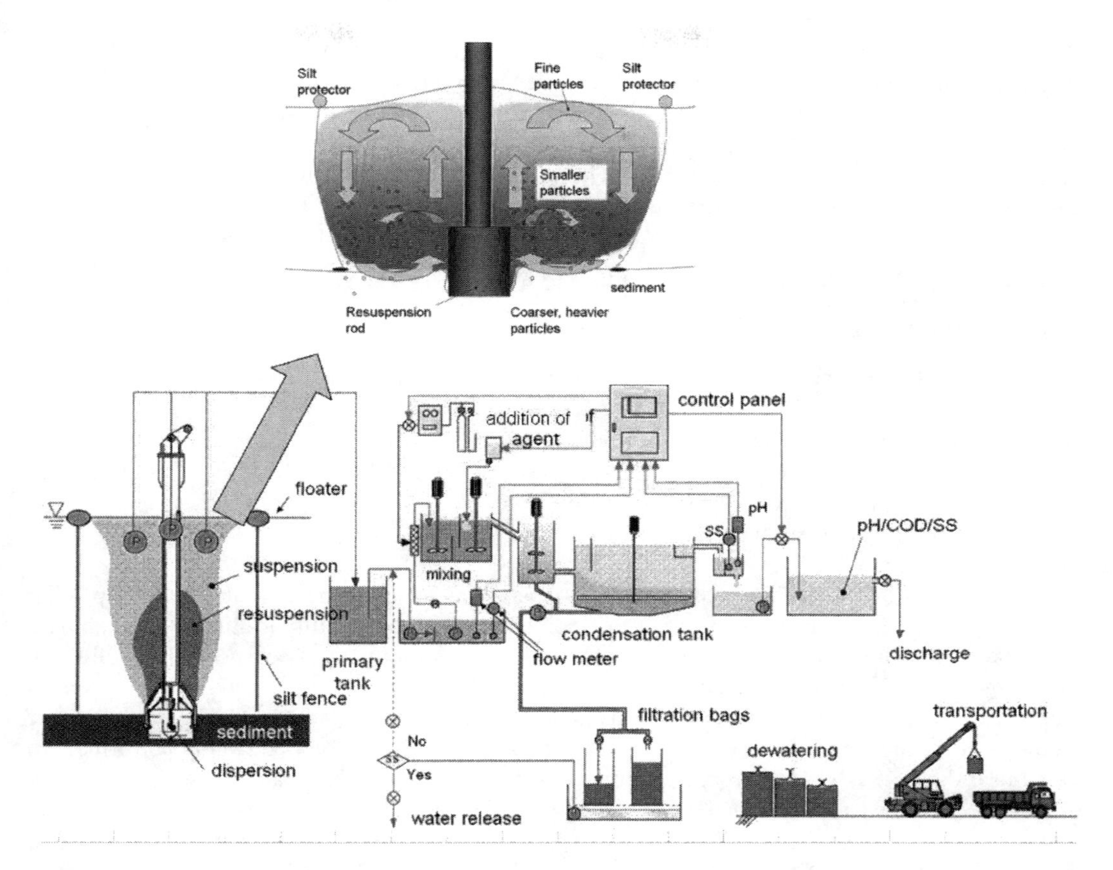

FIGURE 8.17 Schematic of the entire resuspension process. P denotes pumps. Insert indicates resuspension principle.

area. Sediment size distribution, water content, and loss of ignition (LOI) were determined, and the concentrations of the heavy metals were measured before and after the tests. Final results indicate the improvement of sediment quality. Aeration reduced the concentration of pollutants (mostly organic) and by removing the fine sediments, the heavy metal concentrations were decreased for some metals.

8.5.5 Cleanup of Oil Spills

Marine oil spills can occur due to a variety of human activities. These include oil tanker or oil rig accidents, pipeline leaks, ship spills, and discharges. The oil can evaporate, react with sunlight, disperse naturally, biodegrade, and/or reach shorelines where major ecosystem damage can occur. Tourism and marine resource extraction industries, such as fisheries, can be highly impacted. Cleanup strategies vary according to conditions.

Controlled burning can reduce the amount of oil in water, but should be done in low wind and may cause air pollution. Skimming and booms that surround the oil requires calm waters. On the shoreline beaches, beach raking, vacuums, shovels and other road equipment can be used to pick up oil. Sorbents can also be used to absorb the oil and small droplets. A recent review by Bi et al. (2023) compares dispersants, surface washing agents and solidifiers for shoreline cleanup. Solidifiers change the spilled oil to a solid, semi-solid, or a rubber-like material that floats on water for easy cleanup. Dispersants can be used to break up oil masses by transferring it into the water column. Dredging for oils dispersed with detergents and other oils denser than water may be required. Dispersed oil droplets can be lethal to coral. Corexit is the main surface washing agent that has gained regulatory approval. However, it has been found to be toxic to corals and other organisms. Other less toxic dispersants are required. Saborimanesh and Mulligan (2018) have applied sophorolipids for enhanced dispersion of weathered biodiesel, diesel, and light crude oil-contaminated water under various salinities, temperatures, and pHs. The dispersion by sophorolipids was the highest for weathered biodiesel, and followed by diesel, and light crude oil in seawater.

8.6 CREATION OF A NATURAL PURIFICATION SYSTEM

8.6.1 Creation of Sand Beaches and Tidal Flats

Artificial sand beaches and tidal flat are created for one of the following purposes:

(a) formation of a clean beach for resort areas and parks,
(b) farming for shellfish,
(c) recovery of a beach following reclamation, and
(d) rehabilitation of coastal marine environment.

Sand beaches and tidal flats possess natural capabilities for cleaning seawater under repeating waves and tides. This capability arises from a combination of their ability to filter a large amount of suspended solids (mostly organic matter), and the dissolution of the suspended solids by microorganisms. Whilst the organic matter entrapped in the sand pores is food for microorganisms and benthic animals, there are no easy means to quantify the process and its benefits. Evaluation of the impacts arising from construction of the tidal flats and beaches cannot be readily performed. In part, this is due to the dynamic processes initiated by the actions of currents and waves. Stabilization of the new beaches and tidal flats will be a long-term process. The use of breakwaters on beaches brings problems of decreasing redox in the region due to the dead organisms and excrements. Figure 8.18 shows one of the three tidelands (Kasai Rinkai Park, 270,837 m^2), created artificially by the Tokyo Metropolitan Government in 1965, at a time when Tokyo Bay was losing its valuable natural environment. The area incorporates vast tidelands, which were once the breeding areas for birds and were also once abundant with fish and shellfish.

FIGURE 8.18 An example of creation of an artificial tideland.

8.6.2 Creation of Seaweed Swards

Eelgrass (*Zostera marina*) is a water plant with long grass like leaves. Figure 8.19 shows a dense sward of eelgrass. There are many different species of eelgrass. Shallow intertidal-water eelgrass has shorter and narrower leaves, whereas deeper subtidal-water eelgrass has longer and wider leaves. They tend to grow in tidal creeks, sandy bays, estuaries, and on silty-sandy sediments and are a vital part of the food web chain for the coastal marine ecosystem. In dense swards of eelgrass, silt and clay particles tend to be deposited with organic matter. Decomposition of organic matter will render the seabed anaerobic, and the colour of the sediments will become black because of the effect of sulphide.

The eelgrass family is one of the few flowering plants that lives in salt water, and the long grass blades are home to various kinds of small marine plants and animals. They are the breeding ground and habitat for all kinds of marine animals, including crabs, scallops, and other kinds of shell fish. They not only serve to foster and stabilize the benthic habitats, but they also have the potential for phytoremediation. Eelgrass can absorb trace metals and organotins (Brix and Lyngby, 1982; Fransois et al., 1989).

Table 8.7 gives a comparison of concentrations of various heavy metals in sediments with and without eelgrass. The sediments that were taken from a small eelgrass sward at an estuary of Kasaoka Bay in Seto Inland Sea, Japan, consisted of a number of small communities – with bare parts between the communities. The total area of the eelgrass sward was 1491 m^2. The sampling points from A to H were located in the bare parts and communities. The results show that the sediments with eelgrass contain a lesser amount of heavy metals – most likely attributed to heavy metal absorption (uptake) by the eelgrass. Since eelgrass grows from spring to summer, and their dead leaves drift upward to the sea surface at the end of their growing season, collection of the dead leaves can be simply implemented. This means that if eelgrass is used for phytoremediation, the absorbed heavy metals can be harvested with the dead seagrass leaves.

Reclamation and other near-shore industrial activities can negatively impact the coastal habitat, particularly on the seagrass beds that form the seaweed fields. Reduction of seaweed fields not only decreases the habitat of marine living things but will also result in a marked decrease in the haul of inshore fish. For example, in Japan, approximately 6000 ha of seaweed field have disappeared since 1978, and about one third of this is due to the impact of reclamation projects.

FIGURE 8.19 Dense sward of Eelgrass (*Zostera marina L.*) in a coastal region.

TABLE 8.7
Comparison of Heavy Metal Concentrations in Sediments With and Without Welgrass (Seto Inland Sea, Japan)

Sampling Point		Heavy Metal (mg/kg)		
		Cu	Pb	Zn
A	with eelgrass	11	20	76
B	without eelgrass	13	10	83
C	with eelgrass	11	25	69
D	without eelgrass	14	130	90
E	with eelgrass	15	17	77
F	with eelgrass	16	18	82
G	without eelgrass	27	17	110
H	with eel grass	17	18	83
	average with eelgrass	14	19.6	77.4
	average without eelgrass	18	52.3	94.3

8.7 RECOVERY OF DRIFTING SEA DEBRIS

Dumping or discharging land-based industrial waste into the sea is essentially prohibited with the burden of responsibility resting on the waste generator to ensure that any waste material entering the sea must be non-toxic and non-hazardous. Section 8.2 has discussed the strict prohibitions articulated in the London Convention and Protocols. Since many countries and jurisdictions with restricted land areas do not have sufficient land space for land disposal of waste, controlled and regulated discharge of municipal and industrial wastes into the sea remains as the option of last resort. When such a need arises, waste disposal sites in the sea must be constructed to meet safety and health protection requirements. Isolation of the waste from contact and dispersion into the sea is a prime requirement. In some countries, artificial islands have been constructed for the principal purpose of emplacing secure disposal facilities. These island-based disposal sites must conform to all the regulations that attend land-based disposal sites – with the strict requirement for monitoring and control, to ensure no escape of leachate into the sea.

A lot of garbage has washed ashore on the coast. These are caused by natural disasters that cause plants to be washed into the sea by rivers, illegal dumping of garbage containing plastic, or the death of plants produced in the sea. It is not uncommon for them to cross internationally and wash ashore on the shores of foreign countries.

Under this situation, in Japan, the Act on the Promotion of Disposal of Coastal Debris Related to the Conservation of the Coastal Landscape and Environment and the Marine Environment in order to Protect Beautiful and Abundant Nature (Coastal Debris Disposal Promotion Act: Parliamentary Legislation) was enacted (www.env.go.jp/water/marine_litter/law.html).

Table 8.8 shows an example of a series of activities by the prefecture under the law which gives a short summary of the recovery of sea wastes on the beaches in the Kanagawa-prefecture, Japan.

8.8 COASTAL EROSION

Coastal erosion is a problem that has confronted coastal communities for a long time. Some will argue that this is a natural phenomenon and that the "*problem*" arises when the human population occupies the coastal regions and imposes various requirements on the coast, such as building structures, infrastructures, wharfs, etc. Alteration of the coastline is a natural phenomenon, and

TABLE 8.8
Change in the Recovery of Sea Wastes from the Coasts in Kanagawa-Prefecture, Japan

Year	2015	2016	2017	2018	2019	2020	2021	2022
Combustible	2126	1307	1456	1416	1471	1028	1422	991
Non-combustible	314	276	368	357	326	203	240	220
Seaweed	2386	974	79	691	462	239	176	208
Total (tonnes)	4826	2557	2617	2464	2259	1470	1838	1419

Extract from Koueki zaidan Houjin Kanagawa Kaigan Bika-zaidan: (www.bikazaidan.or.jp/kaigangomi/suii/.)

FIGURE 8.20 An erodible coast in Japan facing the Pacific Ocean.

erosion is a major factor in the alteration process. This alteration process becomes a problem because of the requirements and expectations imposed by the human population. Erosion arises not only from the aggressive action of waves and currents, but also from diminished sediment recharge from streams and rivers feeding into the sea. Flood control dams and other river restoration projects can curtail the flow of suspended particulates and sediments. Without this sediment recharge and with erosive forces acting on the coastal plains, erosion becomes a considerable issue. To counter the erosive forces and to prevent beach erosion, armouring of the beach with revetments and seawalls is a procedure that has been adopted by many coastal regions. Figure 8.20 shows an erosive beach on the coast of Japan facing the Pacific Ocean. A huge number of concrete blocks have been installed to protect the coastline and to gather sand. Figure 8.21 shows soil dumping as a countermeasure for erosion. The record shows that some period of time after levelling of the dumped soil, erosion of the slope occurred. To overcome this, geotextile tubes can be used to protect coastline from waves and tides – as has been utilized in some countries. The tubes, which can be installed along the coastline, are a few metres in diameter and a few kilometres in length can be filled with dredged soils (Figure 8.22). They can also serve as a breakwater for man-made islands and wetlands.

FIGURE 8.21 Countermeasure for coastal erosion by dumping and the action of erosion.

FIGURE 8.22 Geotextile tube used for protection and creation of a coast.

8.9 REMEDIATION OF CESIUM-CONTAMINATED DEPOSITS

8.9.1 Examples of Removal of Cs-Contaminated Deposits

There are several isotopes of cesium Cs, such as Cs-133, Cs-134, and Cs-137, etc. In nature, there is only Cs-133, which does not emit radiation. Cs-134 and Cs-137 are substances generated when nuclear fission occurs, and these are called artificial radionuclides, in contrast to natural radionuclides such as uranium, radium, and strontium. The characteristics of the isotope Cs are shown in Table 8.9.

In addition to the isotopes of Cs listed in Table 8.9, there are Cs-129, 131, 132, and 135. With the exception of Cs-135, the half-lives of Cs-129, 131, and 132 are between 1 and 2 weeks. On the other hand, Cs-135 has a half-life of 2,300,000 years, but its precursor Xe-135 often absorbs neutrons to become Xe-136, and its fission yield is negligible in the reactor. Furthermore, the half-life of the other isotopes of Cs is 2–3 s to 1/10 s, and their fission yield is also negligible. Therefore, with Cs pollution, both Cs-137 and Cs-134 have been the subjects of remediation (Falciglia et al., 2017; Kosaka et al., 2012; Lee et al., 2022; Li et al., 2022; Liu et al., 2021; Wang et al., 2023; Yasunari et al., 2011).

The effects of Cs-137 and Cs-134 on the human body differ between external exposure and internal exposure. External exposure is further divided into whole body exposure and local exposure. In this case, the effect may be on the exposed area. In addition, the intensity of the impact varies depending on the type of organ exposed.

In the case of internal exposure, the exposure dose is high in the organs and tissues where radioactive substances tend to accumulate. If this easily accumulating organ or tissue is highly radiosensitive, there is a high risk of radiation effects. For details, refer to related informative publications.

Cs-137 and Cs-134 emitted into the atmosphere have different paths and fates depending on the characteristics of sites of the deposition (Buesseler et al., 2011; Koma et al., 2017; Yasunari et. al., 2011). Cs accumulated on the soil is adsorbed by soil particles or transported by rainfall, and those that flow into rivers flow out from the estuary to the sea. In addition, some reach lakes and agricultural reservoirs. During these processes, Cs reacts with other substances or is adsorbed by plankton, dissolved organic matter and soil particles (Buesseler et al., 2012; Ikenoue et al., 2022; Kambayashi et al., 2017; Koma et al., 2017).

Plankton with Cs possibly can be ingested by fish. As inorganic substances with Cs settle on the bottom, the food-chain from plankton may cause the accumulation and diffusion of Cs. Fortunately, Cs fixed on soil particles is not very mobile in sediments, unless the soil surfaces are disturbed or transported (Liu et al., 2021; Nemoto et al., 2020). Therefore, regardless of whether it is on land or at the bottom of water, the removal of Cs is made possible by the removal of Cs adsorbed to the soil. In order to do so, it is important to efficiently remove the soil to the appropriate layer and to store and manage it without causing secondary contamination.

TABLE 8.9
Isotopes of Cs

Isotope	Number of protons	Number of neutrons	Properties	Half-life (years)
Cs-133	55	78	stable	-
Cs-134	55	79	radioactive	2.1
Cs-137	55	82	radioactive	30

8.9.2 Outline of the Fukushima Nuclear Power Plant Accident

The Fukushima Daiichi Nuclear Power Plant (FDNPP) consisted of Units 1 to 6. Of these, Unit 1 was the oldest, and started operation in Mar. 1971. At the time of the accident, it was one of the oldest nuclear power plants in Japan.

The FDNPP disaster occurred due to the Tohoku-Pacific Ocean Earthquake and tsunami that occurred at 14:46:18.1 on Mar. 11, 2011. The following is a revised version of the chronological situation of the accident compiled by the Japan Nuclear Safety Institute (old Japan Nuclear Technology Association).

The original information was based on the information from the Prime Minister's Office Press, the Nuclear and Industrial Safety Agency Press, and the Tokyo Electric Power Company (TEPCO Press).

Mar. 11, 2011 14:46: Turbines and reactors shut down automatically (Units 1–3).
15:42: Total AC power lost (Units 1–3) (Article 10 of the Nuclear Disaster Countermeasures Act).
16:36: Water injection not possible (Units 2 and 3).
19:03: A nuclear emergency declared.
20:50: Evacuation order from within a 2 km radius.
21:23: Evacuation from within a radius of 2 km, indoor evacuation order within a radius of 3–10 km.
Mar. 12, 01:20: PCV pressure abnormal rise (Article 15 of the Nuclear Disaster Countermeasures Act).
05:44: Evacuation order from within a 10 km radius.
06:50: Pressure control instruction in PCV (Nuclear Reactor Regulation Act).
10:17: Venting starts (Unit 1).
16:17: Abnormal increase in site boundary radiation dose (>500 μSv/h) (Article 15 of the Nuclear Disaster Countermeasures Act).
18:25: Evacuation order from within a 20 km radius.
20:05: Support for injecting seawater into nuclear reactors (Nuclear Reactor Regulation Act).
20:20: Started injecting seawater and boric acid into the reactor via the fire extinguishing system line (Unit 1).
Mar. 13, 05:10: ECCS water injection failure (Unit 3) (Article 15 of the Nuclear Disaster Countermeasures Act).
08:41: Venting starts (Unit 3).
08:56: Abnormal increase in site-boundary radiation dose (>500 μSv/h) (Article 15 of the Nuclear Disaster Countermeasures Act).
09:25: Started injecting fresh water and boric acid into the reactor by the fire extinguishing system line (Unit 3).
11:00: Venting starts.
13:12: Injecting fresh water and boric acid into the reactor on the fire extinguishing system line (Unit 3).
14:15: Abnormal increase in site-boundary radiation dose (>500 μSv/h) (Article 15 of the Nuclear Disaster Countermeasures Act).
Mar. 14, 01:10: Seawater injection suspended (Units 1 and 3)
03:20: Seawater injection resumes (Unit 3).
03:50: Abnormal increase in site-boundary radiation dose (>500 μSv/h) (Article 15 of the Nuclear Disaster Countermeasures Act).
04:08: SFP water temperature rising to 84°C (Unit 4).
04:15: Abnormal increase in site-boundary radiation dose (>500 μSv/h) (Article 15 of the Nuclear Disaster Countermeasures Act).
05:20: Venting starts (Unit 3).
06:10: D/W pressure rises to about 460 kPa (Unit 3).
07:44: Abnormal PCV pressure rise (Unit 3) (Article 15 of the Nuclear Disaster Countermeasures Act).

09:27: Abnormal increase in site-boundary radiation dose (>500 μSv/h) (Article 15 of the Nuclear Disaster Countermeasures Act).
11:01: R/B panel release (impact of explosion) (Unit 2), hydrogen explosion (Unit 3).
13:25: Loss of cooling function (Unit 2) (Article 15 of the Nuclear Disaster Countermeasures Act).
16:34: Started injecting seawater into the reactor (Unit 2).
18:22: Furnace water level -3700 mm, fuel exposed (Unit 2).
21:37: Abnormal increase in site-boundary radiation dose ((>500 μSv/h) (Article 15 of the Nuclear Disaster Countermeasures Act).
22:50: Abnormal PCV pressure rise (Unit 2) (Article 15 of the Nuclear Disaster Countermeasures Act).
March 15, 00:22: Venting started (Unit 2).
06:10: Abnormal noise in the S/P room, S/P pressure drop (Unit 2).
06:14: Smoke was release (Unit 3). A part of the wall was damaged due to a noise (Unit 4).
06:51: Abnormal increase in site-boundary radiation dose (>500 μSv/h) (Article 15 of the Nuclear Disaster Countermeasures Act).
08:11: Abnormal increase in site boundary radiation dose (>500 μSv/h) (Article 15 of the Nuclear Disaster Countermeasures Act).
08:25: White smoke from near the 5th floor (Unit 2).
09:38: Fire broke out near the 3rd floor (Unit 4).
10:22: Dose of 400 mSv/h confirmed in the vicinity (Unit 3).
10:30: Support for early injection of water into the reactor and D/W venting (Units 2 and 4) (Reactor Regulation Act).
10:59: Evacuation order to off-site centre, evacuation to Fukushima Prefectural Office.
11:00: Indoor evacuation order within a radius of 20–30 km.
12:25: Fire extinguishment confirmed (Unit 4).
16:17: Abnormal increase in site boundary radiation dose (>500 μSv/h) (Article 15 of the Nuclear Disaster Countermeasures Act).
22:00: Instructions to inject water into SFP (Nuclear Reactor Regulation Act).
23:05: Abnormal increase in radiation dose at the site boundary (Article 15 of the Nuclear Disaster Countermeasures Act).
The chronological recording continued until 15:00 on Jul. 11, 2011.

As a result, five disasters occurred in which a large amount of radioactivity was released into the atmosphere (Stohl et al., 2012). Since then, many surveys and simulations have been conducted to verify radiation damage (Aoyama et. al., 2019; Buesseler et al., 2011, 2012; Stohl et al., 2012).

8.9.3 Cesium Contamination in Water

Since Cs-137 is in cationic form, which can bond negative charges, then, in comparison to other radioactive nuclides, Cs-137 may be easier to recover from water. In water, at any moment Cs-137 exists in many forms, such as solutes, deposits to dissolved organic matter, adsorbents to soil particles (organic or inorganic matter), and single or complex compounds, as described in Chapter 8.9.1. However, the position and form of those substances, the concentration of Cs and the amount of substances changes in time (Tatsuno et al., 2023). The change in time is complicated with many factors, such as physical phenomena, chemical reactions, and biological actions.

For example, as a physical phenomenon, the disintegration and a half time of radioactive Cs are important terms. In addition, absorption into microorganisms, clay minerals, and organic matter may occur (Akai et al., 2013). The chemical reaction may include the formation of single and complex Cs compounds (Koma et al., 2017; Yasunari et. al., 2011). If the compounds are settled on the water bottom, this is physical behaviour.

Cs solutes and Cs adsorbed to dissolved organic matter in water absorb well to zooplankton (Ikenoue et al., 2022; Tateda et al., 2013). Since the plankton is at the bottom of the food chain, radioactive Cs can be taken up into the food chain (de With et al., 2021; Nemoto et al., 2020; Wada et al., 2023). However, even Cs-137, the most troublesome of the radioactive Cs, can be decontaminated due to its high adsorption properties of Cs depending on the place of deposition.

Because of high adsorption characteristics of radioactive Cs, various adsorbents to remove radioactive Cs have been examined (Li et al., 2022). On the other hand, various suspended materials act as adsorbents or absorbents of radioactive Cs in water. It was found that the removal of Cs-134 and Cs-137 in pond water by resuspension and resettling increased markedly after 24 hours if the water column was mixed with sediments (Kosaka et. al., 2012). Akai et al. (2013) found that interactions with Cs ions, i.e., such as adsorption to clay minerals, adsorption/desorption to organic matter and intake by microorganisms. The removal of radioactive Cs from soils is dealt with in Chapter 11.4.9.

Kurikami et al. (2016) studied sediments and Cs-137 discharge from the reservoirs during various rainfall events. As a result, we have obtained very interesting data. One is the relationship among the concentration (mg/L) of suspended solids, particle size, and time. Another is the relationship between the concentration (Bq/L) of Cs-137 adsorbed or transferred by adsorbing to soil particles with the same time frame. These two relationships show similar trends over time. This is because the amount of soil particles classified for each compartment adsorbs/attaches the amount of Cs-137 that matches its characteristics. These two relationships, which are thought to be related to adsorption capacity were determined by the specific surface area, are related to the amount of sand, silt, clay and the overall transferred Cs-137.

Radioactive Cs released by a sudden accident generally forms a thin contaminated layer in underwater sedimentation, because of high adsorption and low remobilization. Therefore, in reservoirs, a method of removing this thin layer is considered as the most ideal method to remove most of the Cs.

A survey of the contamination of Sr and Cs from the Fukushima Daiichi nuclear accident was conducted by Kavasi et al. (2023). In the study, Sr-90 and Cs-137 active concentrations taken from the Fukushima exclusion zone were measured by radiometry in 76 soil samples (soils, litters, gutters, sediments, and roadside sediment samples). From the activity ratio values, Sr-90/Cs-137, they concluded that the Sr-90 released to the atmosphere was only around 0.0003–0.02 PBq which is negligible compared to the Chernobyl accident (~10 PBq) or other nuclear accidents. However, from the standpoints of radioecology and radiation safety, Cs-137 remains the primary pollutant of the FDNPS accident.

Ikenoue et al. (2022) concluded as follows. At least since 2017, the higher Cs-137/Cs-133 ratios in zooplankton than in seawater are probably not due to the presence of highly radioactive particles or resuspended sediment in zooplankton samples but are related to suspended particles enriched with Cs-137 that are continuously entering the ocean from land via rivers or directly discharged from the FDNPP. Future work is required to elucidate the types and inventories of suspended particles with low radioactivity per particle.

In the Matsukawaura Lagoon, the largest lagoon in Fukushima, the activity concentration of radioactive cesium and resuspended particles in surface sediments were monitored (Kambayashi et al., 2017). Their results suggest that the radioactive cesium distribution in surface sediments changed spatiotemporally and irregularly due to the tsunami. Also, the observed half-life was significantly shorter than the expected half-life. Therefore, the cesium release in the lagoon resulted in some benefits. Observations of sediment traps revealed that resuspended particles were transported to the sea from sediments, suggesting that the suspending of particles in the lagoon and their transport to the ocean by seawater exchange was an important process for radioactive cesium transport out of the lagoon. Furthermore, it was shown that the seawater exchange process contributes to the diffusion of radioactive cesium to the ocean.

Wada et al. (2023) analyzed Cs-137 activity concentrations in fish (15 species, $n = 164$) and water collected from the Maeda River (3.3–8.9 km from the nuclear power plant) and Shimo-Fukasawa Pond (2.9 km) in 2017 to elucidate the Cs-137 contamination level and mechanism of fish living in the river and pond environment near FDNPP. In addition, an 8-week breeding experiment was conducted using Japanese hornwort fed with non-contaminated pellets and pond water (average concentration of Cs-137 of 2.0 Bq/L), and the accumulation of Cs-137 from water to fish was evaluated.

The Cs-137 concentration of Japanese dace, *Pseudaspius hakonensis* is the only species collected at five sampling sites from the mouth of the Maeda River to the upper reaches of the river. This correlated with the ambient air dose rate and fish size, showing large variations of 16.5-2.6 × 10^3 Bq/kg-wet. On the other hand, the dissolved Cs in river water increased from upstream to downstream (0.025–0.28 Bq/L), and the water-body concentration ratio (CR) (60.0–35700 L/kg-wet) varied greatly. These CRs (geometric mean 3670 L/kg-wet) were much higher than the steady-state CRs (9.7 L/kg-wet) of domesticated fish, indicating that river fish ingest Cs-137 from aquatic and riparian prey.

A statistically significant negative correlation was detected between K^+ concentrations in water and CR in river fish, and CR tended to decrease from upstream to estuary. These results suggest that the variability of Cs-137 concentration and CR in river fish varies greatly in river basins, in relation to the food habitats and the size effects of fish in the air dose rate, K^+ concentration, and estuarine process heterogeneity in brackish water habitats. In contrast, pond fish had a Cs-137 concentration (4.3–14.6 kBq/kg-wet) higher than river fish. The CR of pond fish was always high, but its range was smaller (1010–3440 L/kg-wet) and greater in fish with higher trophic levels. These results suggest that biological accumulation in the pond is the main cause of Cs-137 contamination of pond fish.

8.9.4 Challenges of Remediation at Large Depths

As was mentioned in the earlier sections, the sediments may contain more Cs-137 in the reservoirs near FDNPP. If it is a shallow reservoir, the remediation of the bottom sediments can be achieved. However, if the water depth is greater than 10 m, larger equipment is required, and there may be access issues. The cost of increasing the size is likely to be very high. In such cases, other methods may be employed. In a large-scale dam reservoir, monitoring or prevention of the spread of Cs due to dredging is also issue.

8.9.5 Selective Removal Technology of Polluted Sediments

8.9.5.1 Removal of Cs Ions from Turbid Water by Soil Particles

It is known that in water, Cs ions adsorb or absorb to dissolved organic matter, zooplankton, soil particles such as minerals like clay and silt, and organic matter (Akai et al., 2013; Buesseler et al., 2012; Ikenoue et al., 2020; Kambayashi et .al., 2017; Kavasi et al., 2023; Kurikami et al., 2014; Lim et al., 2014; Liu et al., 2021; Park et al., 2021; Wang et al., 2023; Zhang et al., 2022). Furthermore, Cs is also absorbed into microorganisms. This implies that the sediments are the adsorbents of Cs and play an important role in preventing Cs dispersal. On the other hand, as Cs-rich sediments are radioactive, they have to be removed promptly, because there are risks for Cs mobilization by the food-chain, floods and erosion due to natural disasters and remobilization of radiocaesium from bottom sediments to the water column (Kurikami et al., 2014; Funaki et al., 2022).

Muslim et al. (2023) conducted a batch adsorption experiment to remove Cs-137 from radioactive wastewater generated during the 1999 Gulf War. The adsorbent used consisted of natural clay minerals (bentonite, attapulgite, kaolinite) deposited in the western desert of Iraq. As a result of the experiment, adsorption equilibrium was achieved after 2 hours, and the removal efficiency of Cs-137 was 98%, 97%, and 75% for bentonite, attapulgite, and kaolinite, respectively. They concluded that these adsorbents are a promising material for the removal of Cs-137, which is available and effective at low cost.

For removal of Cs-rich sediments in reservoirs, the following are the requirements.

(a) The technology must be capable of not dispersing Cs-137.
(b) The methodology must be based on scientific evidence for reduction in removed sediment volume.

(c) Efforts to reduce the exposure dose must reach as much as possible to below the permissible radiation dose.

Dredging is the most common method for remediation of polluted deposits in reservoirs. However, with regard to radioactive Cs, ordinary dredging diffuses Cs-rich sediments, and, as mentioned previously, there is a sufficient risk of run-off due to rainfall. Therefore, it is required to dredge in a completely closed area, and volume reduction is required because it is necessary to store under control.

Adsorption and absorption of Cs are due to the interface phenomena, which are affected by the surface area of adsorbent (soil particles). The surface area of soil particles per a given soil mass or soil volume can be presented by the specific surface area of particles of soil (SSA), which may be defined by the average value of the total particles. However, SSA decreases with increasing particle size, as shown in Figure 8.23, which indicates that the average SSA of sand is 5.8 mm^2/mm^3, 185 mm^2/mm^3 for silt and 6000 mm^2/mm^3 for clay. In soils, it is known that SSA is proportional to CEC. In addition, Cs adsorption correlates well with CEC (Park et al., 2021).

The sediment profile in terms of Cs concentration attributed to the fallout of FDNPP accident is dependent on the transfer process from the fallout until deposit in the reservoir. However, without knowing the process, the measurement results in the sediment will show at least the grain size distribution and Cs concentration for sediment depth in the reservoir. The measurements give information on whether remediation of the reservoir is required.

The classification depends on the particle size, but the names differ based on the nature of the soil. For example, "granular" or "cohesive" means sand and silt or clay minerals or silicate, clay or silt. However, even if the grain size is the same, the properties differ depending on whether it is a regional soil (for example, volcanic ash). Therefore, the classification differs slightly from country to country.

8.9.6 Segregation of Sediments Due to Resuspension

When turbid water containing Cs intrudes into reservoirs and settles on the bottom, the segregation of particles occurs. The greater the particle size and the density of particles, the higher the settling

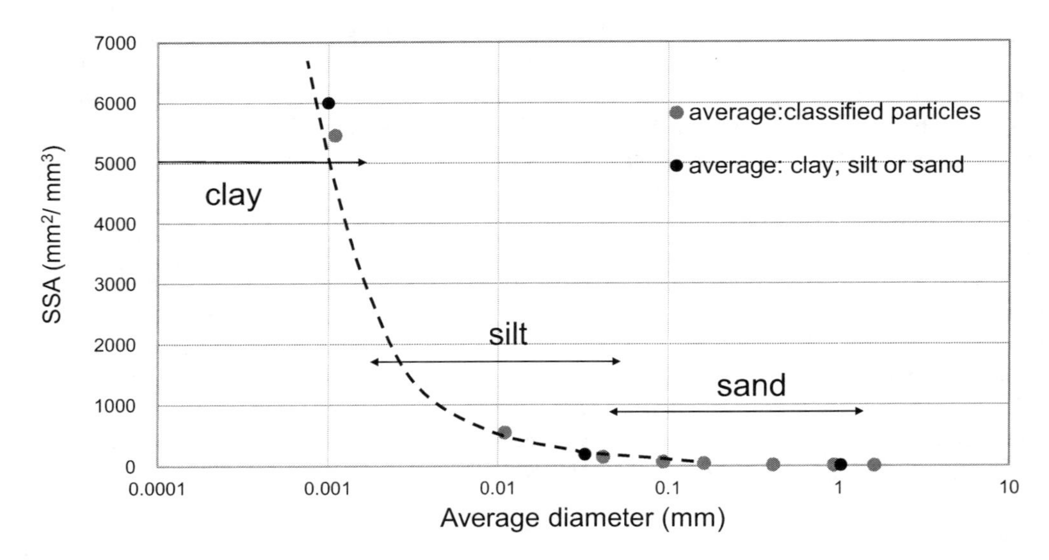

FIGURE 8.23 Specific surface area of spherical particles as a function of grain size. The classification of soil particles is presented in Table 8.10. (Jahn et al., 2006.)

TABLE 8.10
Particle-Size Classification

Fine Earth Particles	≤ 2 mm
Sand	>63 µm – ≤ 2 mm
Very coarse sand	>1250 µm – ≤ 2 mm
Coarse sand	>630 µm – < 1250 µm
Medium sand	>200 µm – ≤ 630 µm
Fine sand	>125 µm – ≤ 200 µm
Very fine sand	>63 µm – ≤ 125 µm
Silt	>2 µm – ≤ 63 µm
Coarse silt	>20 µm – ≤ 63 µm
Fine silt	>2 µm – ≤ 20 µm
Clay	≤ 2 µm
Coarse clay	>0.2 µm – ≤ 2 µm
Fine clay	≤ 0.2 µm

Source: Jahn et al. (2006).

velocity. This is explained by Stokes' law, and is used for the classification of all fine particles in soils. It can be found as "Grain size analysis of soils" in Jahn et al. (2006).

For the determination of particle (grain) size, D, Stokes' law is used in the following form.

$$v_t = C_p D^2 \tag{8.2}$$

where v_t is terminal velocity and C_p is expressed by

$$C_p = \frac{(\rho - \sigma)g}{18\eta} \tag{8.3}$$

where ρ is the density of sphere, σ is the density of liquid, η is the viscosity of liquid, and g is the acceleration of gravity. The terminal velocity of spheres in liquid is proportional to D^2. Therefore, when various sized spheres settle in water, the larger spheres will be deposited first. Theoretically, when a sphere is put in the upward flow with a velocity of v_t, the sphere is suspended theoretically. Furthermore, in upward flow with a velocity of 2 v_t, the sphere moves upward. If the diameter is 2D, the terminal velocity is given by $4 \times Cp\ D^2$. Thus, if the diameter becomes twofold, the terminal velocity will be fourfold.

As was demonstrated, there exists an upward flow velocity to suspend a given size of particles. If an upward flow velocity is higher than the terminal velocity, the particle can be elevated. Assuming that D^* is the particle's diameter at the terminal velocity, any particles having D values greater than D^* go down in the upward flow velocity equal to the terminal velocity. On the other hand, any particles with smaller than D^* will rise under the same upward flow velocity. In a suspension containing various sized particles, an upward flow makes segregation of particles going up or down by size. In this case, the terminal velocity is different depending on the particle size, greater particles settle faster, and smaller particles rise faster and a few particles stay in place.

This means that the upward flow velocity can control the behaviour of particles, because the terminal velocity is a function of the particle's size. Thus, the behaviour of particles, up or down, can be controlled by the upward flow velocity. The illustration of segregation of particles by different

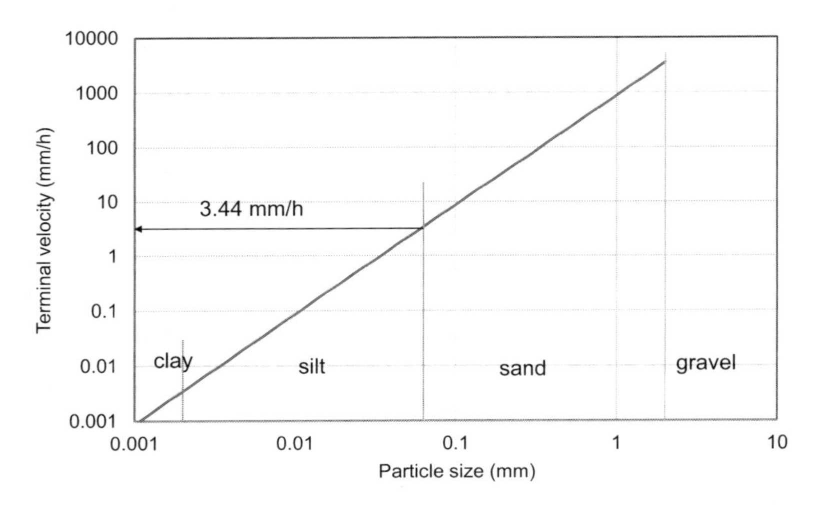

FIGURE 8.24 Illustration of segregation by different diameters of initially suspended particles (spheres) due to advection in a gravitational field.

diameters is presented in Figure 8.24. In fact, since the initial and boundary conditions are not considered, the concept is not applicable, at which the initial and final conditions have a strong effect, when the advection distance is short. If an analysis is made, the ranges in advection velocity and terminal velocity will be affected by particle size, which vary very widely.

8.9.6.1 Removal of Cs Ions and Cs-Rich Sediments by Segregation

Ivanic et al. (2020) investigated relationships among grain size, CEC and SSA on fine silt-coarse clay obtained from the Eastern Adria Sea. The results showed that the decrease in grain size increased CEC and SSA, when organic matter was removed before the measurements. At the same time, relatively more Cs is adsorbed on the surfaces of fine particles. When turbid water contains more fine particles, a higher amount of Cs is retained by the particles. Similar trends were obtained by many researchers (Akai et al., 2013; Muslim et al., 2023; Siroux et al., 2021).

As described earlier, Cs-137 ions are adsorbed strongly on some minerals such as those with higher CEC and greater SSA. As was shown in Figure 8.24, the smaller the soil particles, the higher the SSA. It is apparent that the area of particle surfaces per mass of soil particles increases abruptly with the size of particles. In geotechnical engineering, physicochemical properties of soils result from the particle size, which can control the interfacial phenomena. The reason why clay particles have cohesion is because of the interfacial phenomenon. The permeability of soils is also dependent on the specific surface area, which can control the friction loss. At one time, the hydraulic conductivity of sand was expressed as a function of 10% of the particle size (called effective diameter). This is thought to mean that the interfacial phenomenon of the entire sand could be expressed by representing the specific surface area of the 10% mass particle group of the finer fraction. Note that a 10% mass of particles from the finer side have more than 80% of surface area of the total sand particles. Since adsorption of Cs-137 on soil particles is a common interfacial phenomenon, there are ways to devise methods that can be expected to have that effect. Accordingly, the methodology for effective removal of Cs-rich sediments is to remove the smaller particles.

8.9.7 Case Studies of Reservoirs in Fukushima Prefecture

In Section 8.9.2, an outline of the Fukushima Nuclear Power Plant Accident was described. Japan, the Ministry of Agriculture, Forestry and Fisheries (MAFF), in cooperation with the Fukushima

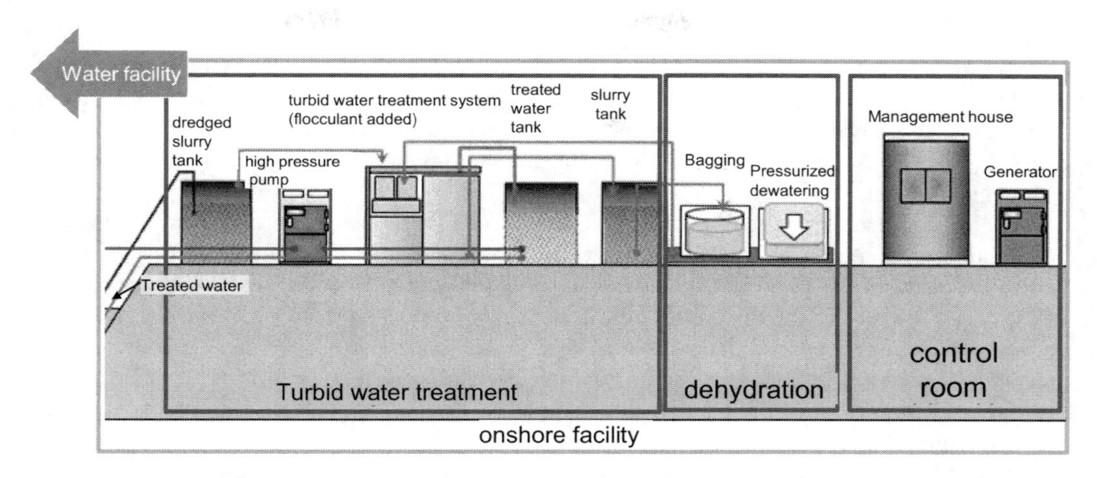

FIGURE 8.25 Illustration of remediation equipment consisted of three systems, except pump-dredging system.

Prefecture and the National Institute of Rural Engineering of the National Agriculture and Food Research Organization, has been working to determine the actual situation of radioactive substances in irrigation ponds and other agricultural irrigation facilities, their impacts, and to demonstrate technical measures to reduce the impact through the "Demonstration Project for Measures to Prevent the Spread of Pollution in Irrigation Ponds". In particular, with regard to measures against radioactive substances in reservoirs, Fukushima Prefecture conducted technical demonstrations at 42 locations and 15 construction types, and data and knowledge have been accumulated in the period of 2012–2015. Based on the data and knowledge accumulated new projects were conducted in the period of 2016–2019.

The case study presented in this section was conducted by Aomi Construction Co. LTD. Proprietary in 2016–2019, on remediation of 11 reservoirs, which were polluted with Cs-137 after the FDNPP accident. This section introduces a multi-functional remediation technology for Cs contaminated sediments, which is based on separation and grading by resuspension of sediments with water jets and effective recovery of smaller particles by pumping.

The remediation equipment used consisted of pump-dredging, turbid (contaminated) water treatment, dehydration treatment, pressurized dewatering and remote-control systems, which were installed at different positions. Except for the pump-dredging system, three other systems are roughly illustrated in Figure 8.25.

8.9.7.1 Separation of Sediments by Resuspension and Pump Dredging

It was preliminarily confirmed in the earlier Decontamination Technology Demonstration Test Project that the Cs ions adsorbed to soil particles are difficult to desorb from the soils. On the other hand, the adsorption properties and behaviour of Cs ions have been studied by many researchers (Latrille and Bildstein, 2022; Zhang et al., 2022). The adsorption of Cs ions to soil particles mainly depends on the concentration of Cs and the type of soils according to cation exchange capacity (CEC), and specific surface area. Thus, the Cs adsorption capacity varies according to many factors.

Therefore, to design the remediation construction, information is needed on the sediment profiles in terms of Cs content (concentration), dry density, and grain size distribution. The more detailed the information, the better to avoid difficulties from the viewpoint of pollution control. The remediation depth can be determined from the contaminated depth and Cs content. In general, the contaminated depth of Cs can be a limiting factor as it is difficult to remove deposited sediments with adsorbed Cs ions.

The Cs contaminated sediments consist of various sized particles. As the smaller soil particles have a larger specific surface area (m^2/g), the cation exchangeable capacity (CEC) of particles is usually higher for smaller particles. In general, clay minerals have a grain size less than 2 μm and a high CEC. Consequently, more Cs ions can be retained by clay minerals per mass of particles. This suggests that the Cs amounts adsorbed by sediment particles can be segregated by the grain size of sediment particles.

Considering that the circulation of water has a velocity v_c, and soils (to be suspended) are added at the bottom of circulation system as shown in Figure 8.26. The movement of soil particles is dependent on the circulation velocity and the mass of each particle. Assuming appropriate conditions as explained in Figure 8.24, there must be a critical upward flow velocity such that sand and gravel particles cannot be suspended. If it is achieved, sand and gravel particles remain at the bottom, and clay and silt particles can be pumped out from the circulation system. In fact, the size and shape of circulation system are determined, and also the upward flow velocity varies with time and position in the circulating system. Therefore, the observation of the advected materials and other experimental results required are important to design the equipment. What is even more important is whether the amount of recovered contaminated sediments are as expected.

The terminal velocity varies mainly with the particle size. From Equations (8.2) and (8.3), the only variable is grain size D and the others can be assumed to be constant. For general conditions, the density of soil particles is 2.7 t/m^3 for silt and clay, the density of liquid (water containing suspended matter) is assumed to be 1.1 t/m^3, the viscosity coefficient of water is 0.001005 Pa s (20 °C), and the acceleration of gravity 9.807 m/s^2 were used for the following calculations. The relationship between terminal velocity v_t and particle size D is obtained from Equations (8.2) and (8.3), using the above constants.

From Equation (8.3)

$$C_p = 8.667 \times 10^2 \text{ (1/ms)} \tag{8.4}$$

According to Equation (8.2), the terminal velocity is

$$v_t = 8.67 \times 10^2 D^2 \text{ (m/s)} \tag{8.5}$$

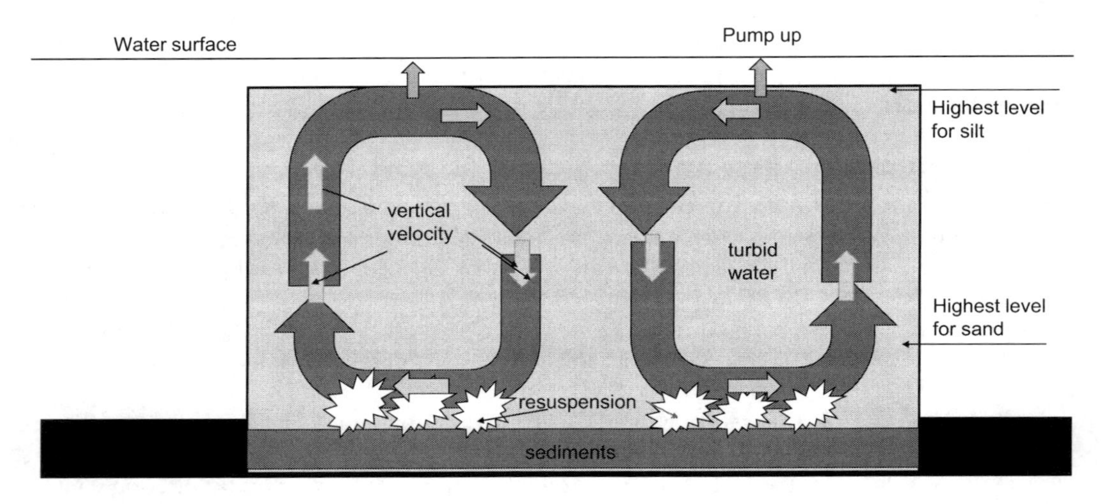

FIGURE 8.26 A simple model of resuspension and circulation system for removal of Cs contaminated sediment particles by water jets.

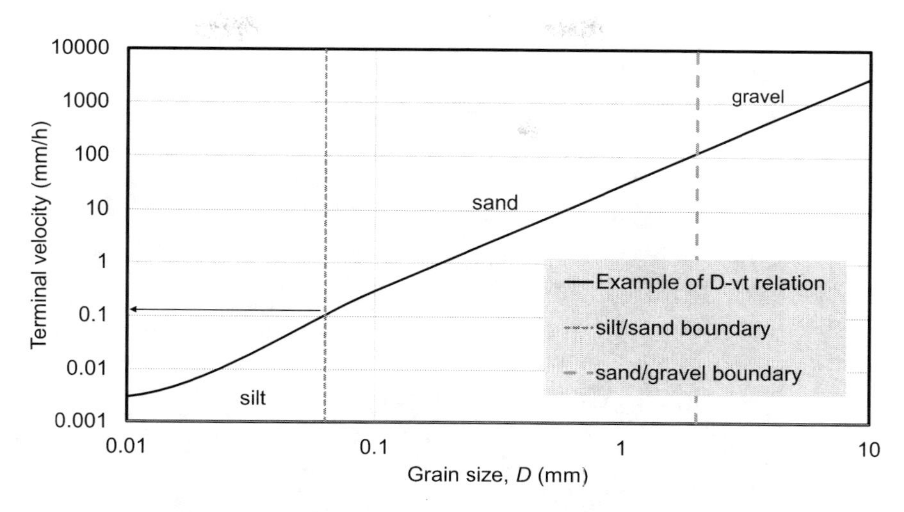

FIGURE 8.27 Example of terminal velocity v_t as a function of particle diameter D.

Equation (8.4) can be expressed as the quadratic functions of D. The relationship is expressed by a logarithmic graph, which can show a wide range of v_t-D relationships, as shown in Figure 8.27.

A terminal velocity corresponding to the maximum grain size 0.063 mm for silt is 0.12 mm/h. Therefore, when the v_t is greater than 0.12 mm/h at the upward flow velocity is nearly as high as v_t, theoretically most particles greater than silt have to move downward from the present suspension levels. In turn, smaller particles move by the action of upward flow. Thus, the upward flow velocity can be used to control the removal particle size, where smaller particles are pumped up from the top level of the circulation flow. In an actual case, theoretical v_t values are not important, but the empirical solution should be significantly used.

It is noted that the large particles cannot be contaminated with Cs and they ultimately will remain at the bottom. It is noted that organic matter has low density, so that they can behave like silt and clay particles. It means that most organic matter can also be removed. The segregation system by water jet was developed by Aomi Construction Co. Ltd. Ultrasonic and cavitation systems were also used to completely disperse aggregated particles.

8.9.7.2 Decontamination Equipment and Technology

The detailed pump-dredging technique was developed by Aomi Construction Co. Ltd., using various types of nozzles for cleanup of a thin polluted sediment layer, as illustrated in Figure 8.28. The dredger includes a number of water jets, cavitation and ultrasonic wave nozzles, respectively. To avoid the loss of contaminated sediments, the resuspension of sediments is made in a closed space. As shown in Figure 8.28, the polluted surface layer is resuspended by the injection from the 28 water jet nozzles. The aggregates suspended are disaggregated by the actions of ultrasonic waves and cavitation generated respective nozzles during circulating turbid water. After recovery of the particles of the silt and clay fractions by a pump, the circulation of turbid water is stopped and the particles greater than silt fractions settle to the bottom. It is noted that to avoid the pollution outside the dredging area, the pump is kept running, until the barge is moved to a new position. Therefore, the connection between dredgers and the land tanks is always open (Figure 8.29).

The dredged materials can be easily separated into turbid water and sediments. Therefore, the turbid water is recovered in the turbidity water tank and sediments are moved into the sediment tank. Note that a high-pressure pump is needed to move the sediments. Then, flocculants are added in the turbidity water tank to separate suspended particles and water. The concentrated flocculants were

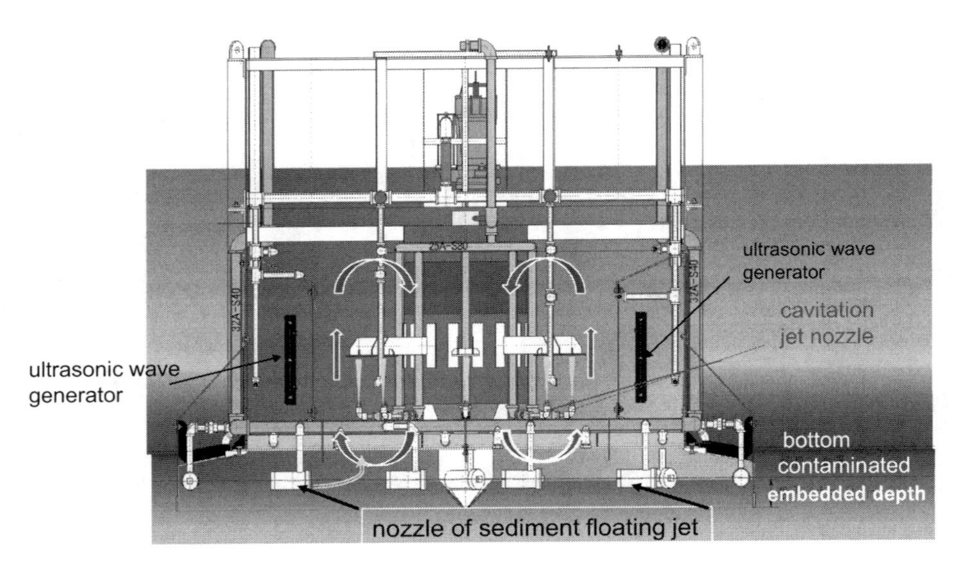

FIGURE 8.28 Illustration of recovering mechanism of Cs-rich fine sediments by resuspension of Cs contaminated sediments.

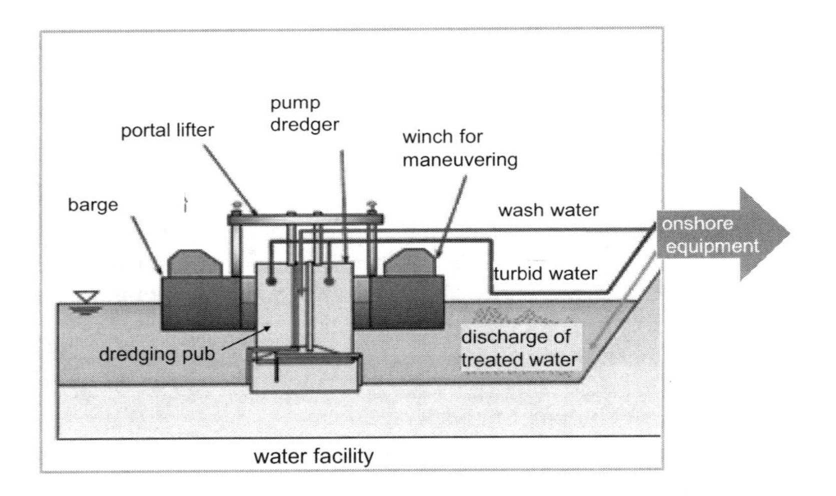

FIGURE 8.29 Connections between the pump-dredging system and other onshore systems.

dehydrated using loading. After the drainage under the self-weight, further drainage was allowed by loading when required.

8.9.7.3 Remote Controlled Dredging Equipment

Two RTK-GNSS devices are mounted on the barge to measure the position and direction of the barge in real time. By manoeuvring the barge so that the designated position is separated by the mesh overlaps with the barge position projected on the screen, the operator can guide the ship to the dredging position with high accuracy (Figure 8.30). In addition, for each mesh, the underwater topography before and after dredging is displayed at the dredging position by elevation, and it can be reliably dredged while confirming that the planned dredging depth has been reached. The management of dredging operations is programmed by setting modes, such as moving up and

FIGURE 8.30 (a) Panoramic view of the installation status of construction equipment, and (b) the dredging facility.

down the dredger bucket, starting and stopping pumps, managing the flow rate of pipelines, and displaying measured values of various sensors. Therefore, the operator needs only to perform the up and down operation of the dredger bucket and the mode switching operation of the dredger bucket while looking at the control screen, and it is close to an automatic operation. Thus, together with the positioning guidance system, the dredger can be remotely controlled by a single operator. Furthermore, the data system on the measured Cs was also controlled in real time in the control room, and it was reflected in the dredging operation at real time. Figure 8.31 shows the dose rates measured on the water bottom before and after the operation.

8.9.7.4 Results

A new technology using the remote-controlled pump dredging was applied to remove Cs-137 from the reservoirs that were contaminated by the FDNPP accident. The results are shown in Table 8.11.

As presented in Table 8.11, by using the resuspension pump dredging method, the contaminated sediments of 13,463 m^3 of Cs-137-rich deposits were removed from 11 reservoirs in the Fukushima prefecture from 2016 to 2023. The total remediation area of the 11 reservoirs was 73,266 m^2, and the average removal thickness was 184 mm.

Prior to the work, the quality of sediments before and after the operation was investigated in the demonstration project, by the Ministry of the Environment, Government of Japan, 2014. The average Cs concentration was 135,000 Bq/kg dry before the remediation, and after the average was 2995 Bq/kg dry. The Cs-137 removal rate was 97.8 %. Note that in other projects, similar results to the demonstration project were obtained. The results were satisfactory because the threshold concentration presented by the authorities was 8000 Bq/kg-dry. Thus, it can be assured that this new remediation technology can be applied to remove Cs contaminated sediments.

8.10 CONCLUDING REMARKS

The health of coastal marine environments is vital to the productivity of the marine aquatic resources in the ecosystems within this environment. The oceans and the coastal marine environments are significant resource bases, and are essential components of the life-support system for the human population. The two groups of events that pose significant threats to the health of this environment are distinguished on the basis of *natural* and *man-made*. In the first instance, the natural events include phenomena, such as coastal erosion resulting from aggressive current, waves, and tidal action. Man-made events result in fouling of the coastal marine environment

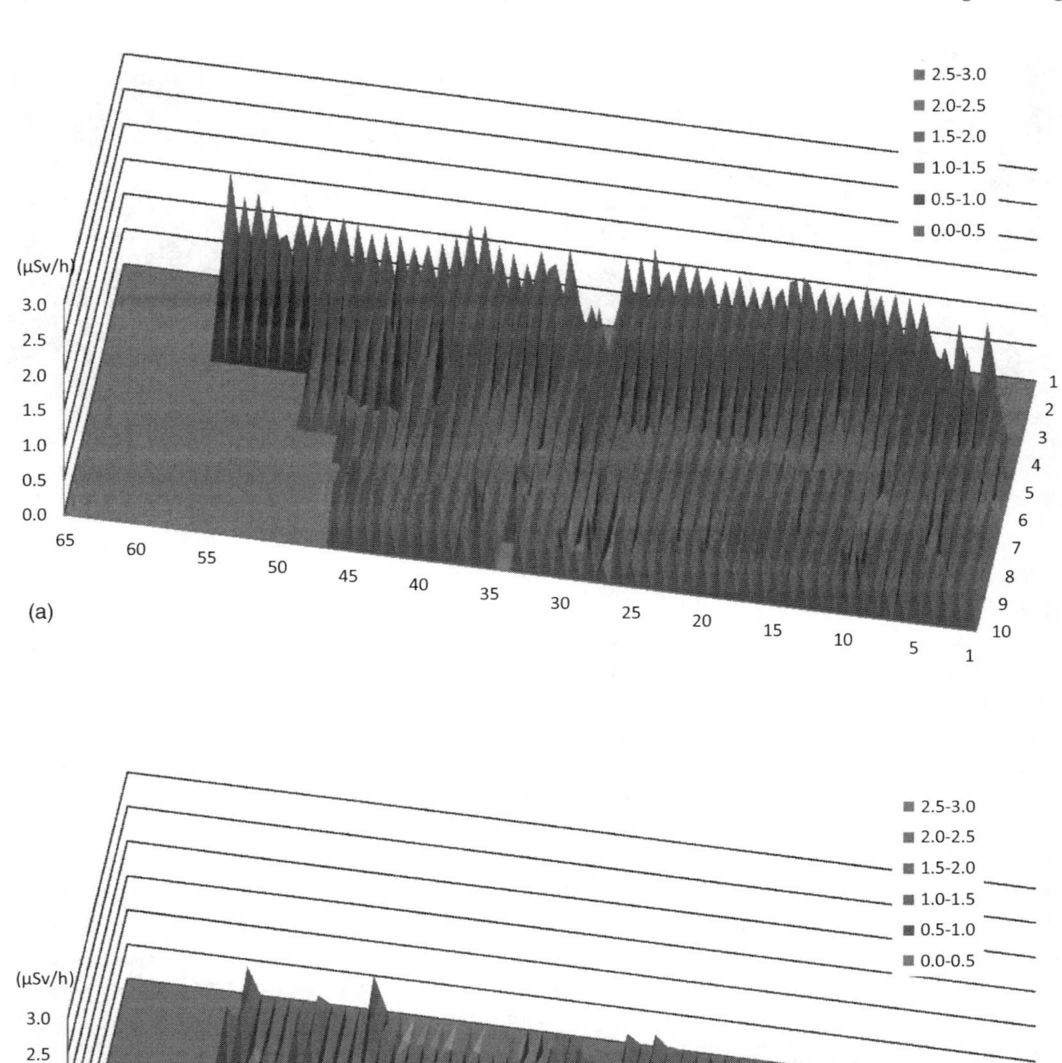

FIGURE 8.31 Image of dose rates measured on the in-situ bottom, (a) before and (b) after the operation.

and of the seas through discharge of wastes from ocean vessels and from land-based industries and activities. It has been argued that coastal erosion is also a victim of man-made events, according to the thesis that *man-generated variations* in sea level contribute to the aggressive actions of the currents, waves, and tidal actions. The thrust of this thesis is that global warming

TABLE 8.11
Performance and Quality for Remediation Work by the Resuspension Technique (Aomi Construction Co. Ltd.)

	Remediation Performance on Contaminated Sediments for 11 Reservoirs		
2016–2019	removal area	(m^2)	73,266
	removal thickness	(mm)	184
	removal sediment volume	(m^3)	13,463
	operation term	(d)	869
	dredging capacity	(m^2/d)	84.3
	Demonstration project, Ministry of Environment, Government of Japan		
2014	Cs concentration (in advance)	(Bq/kg)	135,000
	Cs concentration (after)	(Bq/kg)	2995
	reduction rate	(%)	97.8

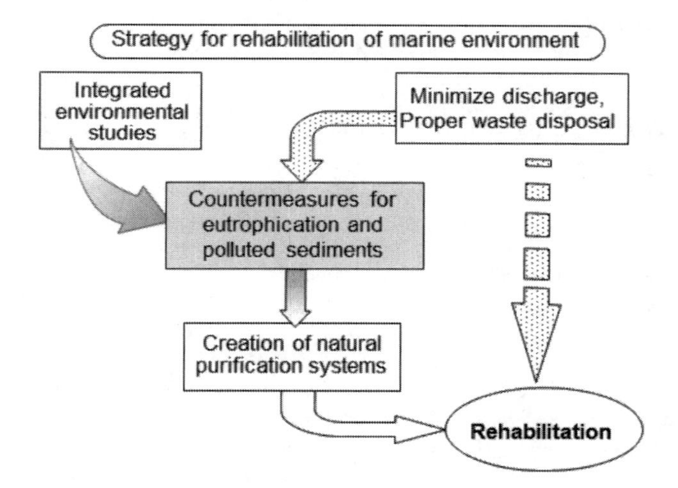

FIGURE 8.32 A strategy for rehabilitation of the marine environment.

is in part responsible for the variations in sea level not only through ice-melting, but also through changes in the seabed levels. it is important to point out that the many facets of geoenvironmental sustainability are affected by global warming.

Degradation of the coastal marine environment and of the seawater affects both seawater and sediments. Contaminants are accumulated in the sediments, whilst others are put into circulation through the food-chain and bioconcentration – ultimately posing health threats to the human population. Two basic problems exist in the coastal marine environment and the seawater: Eutrophication and concentration of toxic and hazardous substances. Figure 8.32 gives an illustration of a simple strategy for rehabilitation of the coastal marine environment. Hazardous substances and nutrients have been discharged into the coastal marine environment for many decades. These have to be collected and removed as seen in the example of the caesium removal in the Fukushima Prefecture. A balance in the amount of nutrients removed is needed. On the one hand, sufficient removal of excess nutrients is needed to avoid eutrophication, and on the other hand, sufficient nutrients must be available for the aquatic animals that rely on these nutrients for their food supply. A sustainable coastal marine environment requires a natural purification system. Research and development into how this can be achieved would contribute substantially to the goals of a sustainable coastal marine environment. In the final analysis, since degradationof the marine coastal environment and the sea is in large measure attributable to the land-based activities and industries, proper control and management of the land pollution would go a long way towards mitigating this.

REFERENCES

Akai, J., Nomura, N., Matsushita, S., Kudo, H., Fukuhara, H., Matsuoka, S., and Matsumoto, J. (2013), Mineralogical and geomicrobial examination of soil contamination by radioactive Cs due to 2011 Fukushima Daiichi Nuclear Power Plant accident, *Physics and Chemistry of the Earth*, 58–60: 57–67, http://dx.doi.org/10.1016/j.pce.2013.04.010.

An, Y.-J., Kampbell, D.H., Weaver, J.W., Wilson, J.T., and Jeong, S.-W. (2004), Natural attenuation of trichloroethene and its degradation products at a lake-shore site, *Environmental Pollution*, 130: 325–335.

Aoyama, M., Inomata, Y., Tsumun, D., and Tateda, Y. (2019), Fukushima radionuclides in the marine environment from coastal region of Japan to the Pacific Ocean through the end of 2016, *Progress in Nuclear Science and Technology*, 6: 1–7.

Bi, H., Mulligan, C. N., Zhang, B., Biagi, M., An, C., Yang, X., Lyu, L., and Chen X. (2023), A review on recent development in the use of surface washing agents for shoreline cleanup after oil spills, *Ocean and Coastal Management*, 245(106877): 245, DOI 10.1016/j.ocecoaman.2023.106877.

Brix, H., and Lyngby, J.E. (1982), The distribution of cadmium, copper, lead, and zinc in eelgrass (*Zostera marina* L.), *The Science of the Total Environment*, 24: 51–63.

Buesseler, K., Aoyama, M., and Fukasawa, M. (2011), Impacts of the Fukushima nuclear power plants on marine radioactivity, *Environmental Science & Technology*, 45: 9931–9935, dx.doi.org/10.1021/es202816c.

Buesseler, K.O., Jayne, S.R., Fisher, N.S., Rypina, I.I., Baumann, H., Baumann, Z., Breier, C.F., Douglass, E.M., George, J., Macdonald, A.M., Miyamoto, H., Nishikawa, J., Pike, S.M., and Yoshida, S. (2012), Fukushima-derived radionuclides in the ocean and biota off Japan, *Proceedings of the National Academy of Sciences*, 109(16): 5984–5988. www.pnas.org/cgi/doi/10.1073/pnas.1120794109.

Carral, E., Villares, R., Puente, X., and Carballeira, A. (1995), Influence of watershed lithology on heavy metal levels in estuarine sediments and organism in Galicia (north-west Spain), *Marine Pollution Bulletin*, 30: 604–608.

Chamas, A. Moon, H., Zheng, J., Qiu, Y., Tabassum, T., Jang, J.H., Abu-Omar, M., Scott, S.L., and Suh, S. (2020), Degradation rates of plastics in the environment, ACS *Sustainable Chemistry & Engineering*, 8(9): 3494–3511, https://doi.org/10.1021/acssuschemeng.9b06635.

Chiou, C.T. (2002), *Partition and Adsorption of Organic Contaminants in Environmental Systems*, John Wiley & Sons, Inc., New York, 257 pp.

Cobelo-García, A., and Prego, R. (2003), Land inputs, behaviour and contamination levels of copper in a ria estuary (NW Spain), *Marine Environmental Research*, 56(3): 403–422, doi: 10.1016/S0141-1136(03)00002-3.

Cortesão, C., and Vale, C. (1995), Metals in sediments of the Sado Estuary, Portugal, *Marine Pollution Bulletin*, 30(1): 34–37.

Davis, J.W., Odom, J.M., DeWeerd, K.A., Stahl, D.A., Fishbain, S.S., West, R.J. Klecka, G.M., and DeCarolis, J.G. (2002), Natural attenuation of chlorinated solvents at Area 6, Dover Air Force Base: Characterization of microbial structure. *Journal of Contaminant Hydrology*, 57: 41–59.

de Mora, S.J., Stewart, C., and Phillips, D. (1995), Sources and rate of degradation of tri(n-butyl)tin in marine sediments near Auckland, New Zealand, *Marine Pollution Bulletin*, 30: 50–57.

de With, G., Bezhenar, R., Maderich, V., Yevdin, Y., Iosjpe, M., Jung, K.T., Qiao, F., and Perianez, R. (2021), Development of a dynamic food chain model for assessment of the radiological impact from radioactive releases to the aquatic environment, *Journal of Environmental Radioactivity*, 233, 106615, https://doi.org/10.1016/j.jenvrad.2021.106615.

Dermietzel, J., and Christoph, G. (2002), The release of contaminants from aged field sediments, *IAHS Publications*, 147–151.

Diaz, R.J., and Rosenberg, R. (1995), Marine benthic hypoxia: A review of its ecological effects and behavioural responses of marine macrofauna, *Oceanography and Marine Biology*, 33: 245–303.

Din, Z. (1992), Use of aluminium to normalize heavy metals data from the estuarine and coastal sediments of the strait of Melaka, *Marine Pollution Bulletin*, 24: 484–491.

Do Amaral, M.C.R., Rebelo, M. de F., Torres J.P.M., and Pfeiffer, W.C. (2005), Bioaccumulation and depuration of Zn and Cd in mangrove oysters (*Crassostrea rhizphorae*, Guilding, 1828) transplanted to and from a contaminated tropical coastal lagoon, *Marine Environmental Research*, 59(4): 277–285.

Fabris, G.J., Monahan, C., and Batley, G.E. (1999), Heavy metals in waters and sediments of Port Philip, Australia, *Marine and Freshwater Research*, 50: 503–513.

Falciglia, P.P., Romano, S., and Vagliasindi, F.G.A. (2017), Stabilization/solidification of 137Cs-contaminated soils using novel high-density grouts: g-ray shielding properties, contaminant immobilization and a gRS index-based approach for in situ applicability, *Chemosphere*, 168: 1257e1266, http://dx.doi.org/10.1016/j.chemosphere.2016.10.068.

Fish, K.E., Clarizia, L., and Meegoda, J. (2023), Microplastics in aquatic environments: A review of recent advances, *Journal of Environmental Engineering and Science*, 18(4): 138–156, https://doi.org/10.1680/jenes.23.00018.

Fransois, R., Short, F.T., and Weber J.H. (1989), Accumulation and persistence of tributyltin in Eel grass (*Zostera marina L.*) tissue, *Environmental Science and Technology*, 23(2): 191–196.

Fukue, M., Nakamura, T., Kato, Y., and Yamasaki, S. (1999), Degree of pollution for marine sediments, *Engineering Geology*, 53: 131–137.

Fukue, M., Sato, Y., Inoue, T., Minato, T., Yamasaki, S., and Tani, S. (2004), Seawater Purification with Vessel-Installed Filter Units. In: R.N. Yong and H.R. Thomas (eds.), *Geoenvironmental Engineering: Integrated management of Groundwater and Contaminated Land*, Thomas Telford, London, pp. 510–515.

Fukue, M., Sato, Y., Yamashita, M., Yanai, M., and Fujimori, Y. (2003), Change in microstructure of soils due to natural mineralization, *Applied Clay Science*, 23: 169–177.

Fukue, M., Sato, S., Yasuda, K., Kato, K., and Ono, S. (2012a), Evaluation of adverse effects of anoxic sediments on aestivating sand lance (*Ammodytes personatus*), Contaminated Sediments, 5th Volume: Restoration of Aquatic Environment, STP 1554, ASTM, pp. 159–174. STP104222, DOI:10.1520/STP104222

Fukue, M., Uehara, K., Sato, Y., and Mulligan, C.N. (2012b), Resuspension technique for improving organic rich sediment-water quality in a shallow sea area, *Marine Georesources and Geotechnology*, 30: 222–233.

Fukue, M., Yanai, M., Sato, Y., Fujikawa, Y., Furukawa, Y., and Tani, S. (2006), Background values for evaluation of heavy metal contamination in sediments, *Journal of Hazardous Materials*, 136(1): 111–119, https://doi.org/10.1016/j.jhazmat.2005.11.020.

Funaki, H., Tsuji, H., Nakanishi, T., Yoshimura, K., Sakuma, K., and Hayashi, S. (2022), Remobilization of radiocaesium from bottom sediments to water column in reservoirs in Fukushima, Japan, *Science of the Total Environment*, 812: 152534, https://doi.org/10.1016/j.scitotenv.2021.152534.

Gamenick, I., Jahn, A., Vopel, K., and Giere, O. (1996), Hypoxia and sulphide as structuring factors in a macrozoobenthic community on the Baltic Sea shore: Colonization studies and tolerance experiments, *Marine Ecology: Progress Series*, 144: 73–85.

Gardner, S.C., Pier, M.D., Wesselman, R., and Arturo, J. (2003), Organochlorine contaminants in sea turtles from the Eastern Pacific, *Marine Pollution Bulletin*, 46 (9): 1082–1089,

Green, N.W., and Knutzen, J. (2003), Organohalogens and metals in marine fish and mussels and some relationships to biological variable at reference localities in Norway, *Marine Pollution Bulletin*, 46(3): 362–374.

Hagerman, L. (1998), Physiological flexibility: A necessity for life in anoxic and sulphidic habitats, *Hydrobiologia*, 373/376: 241–254.

Horiguchi, T., Shiraishi, H., Shimizu, M., and Morita, M. (1994), Imposex and organotin compounds in *Tais clavigera* and *T. bronni* in Japan, *Journal of the Marine Biology Association UK*, 74: 651–669.

Hoshika, A., and Shiozawa, T. (1987), personal communication.

Ikenoue, T., Yamada, M., Ishii, N., Kudo, N., Shirotani, Y., Ishida, Y., and Kusakabe, M. (2022), Cesium-137 and 137Cs/133Cs atom ratios in marine zooplankton off the east coast of Japan during 2012–2020 following the Fukushima Dai-ichi nuclear power plant accident, *Environmental Pollution*, 311: 119962, https://doi.org/10.1016/j.envpol.2022.119962.

Instanes, D. (1994), Pollution control of a Norwegian Fjord by use of geotextiles. Fifth International Conference on Geotextiles, Geomembranes and Related Products, Singapore, 5–9 September.

Inthorn, D., Sidtitoon, N., Silapanuntakul. S., Aran, and Incharoensakdi, A. (2002), Sorption of mercury, cadmium and lead by microalgae, *Science Asia*, 28: 253–261.

Ivanic, M., Durn, G., Skapin, S.D., and Sondi, I. (2020), Size-related mineralogical and surface physicochemical properties of the mineral particles from the recent sediments of the Eastern Adriatic Sea, *Chemosphere*, 249: 126531, https://doi.org/10.1016/j.chemosphere.2020.126531.

Jahn, R., Blume, H.P., Asio, V., Spaargaren, O., and Schád, P. (2006), *Guidelines for Soil Description*, 4th ed., Physical properties, Food and Agriculture Organization of the United Nations, Rome. www.fao.org/soils-portal/data-hub/soil-properties/physical-properties/en/.

Jensen, H.F., Holmer, M., and Dahllöf, I. (2004), Effects of tributyltin(TBT) on the seaweed *Ruppia maritima*, *Marine Pollution Bulletin*, 49(7-8): 564–573.

Johnson, B.L. (1992), *Public Health Implications of Dioxins, Congressional Testimony*, Subcommittee on Human Resources and Intergovernmental Relations, Committee on Government Operations, US House of Representatives, Washington, DC, June 10, 1992.

Jones, B., and Turki, A. (1997), Distribution and speciation of heavy metals in surficial sediments from the Tees Estuary, North-east England, *Marine Pollution Bulletin*, 34-10: 768–779.

Kambayashi, S., Zhang, J., and Narit, H. (2017), Spatial assessment of radiocaesium in the largest lagoon in Fukushima after the TEPCO Fukushima Dai-ichi Nuclear Power Station accident, *Marine pollution Bulletin*, 122: 344–352, http://dx.doi.org/10.1016/j.marpolbul.2017.06.071.

Kan-Atireklap, S., Tanbe S., and Sanguansin, J. (1997), Contamination by butyltin compounds in sediments from Thailand, *Marine Pollution Bulletin*, 34-11: 894–899.

Kavasi, N., Arae, H., Aono, T., and Sahoo, S.K. (2023), Distribution of strontium-90 in soils affected by Fukushima dai-ichi nuclear power station accident in the context of cesium-137 contamination, *Environmental Pollution*, 326: 121487. https://doi.org/10.1016/j.envpol.2023.121487.

Kavun, V.Ya, Shulkin, V.M., and Khristoforova, N.K. (2002), Metal accumulation in mussels of the Kuril Islands, north-west Pacific Ocean, *Marine Environmental Research*, 53: 219–226,

Koma, Y., Shibata, A., and Ashida, T. (2017), Radioactive contamination of several materials following the Fukushima Daiichi Nuclear Power Station accident, *Nuclear Materials and Energy*, 103: 5–41, http://dx.doi.org/10.1016/j.nme.2016.08.015.

Kosaka, K., Asami, M., Kobashigawa, N., Ohkubo, K., Terada, H., Kishida, N., and Akiba, M. (2012), Removal of radioactive iodine and cesium in water purification processes after an explosion at a nuclear power plant due to the Great East Japan Earthquake, *Water Research*, 46 (14, 15): 4397–4404.

Kurikami, H., Funaki, H., Malins, A., Kitamura, A., and Onishi, Y. (2016), Numerical study of sediment and 137Cs discharge out of reservoirs during various scale rainfall events, *Journal of Environmental Radioactivity*, 164: 73–83, http://dx.doi.org/10.1016/j.jenvrad.2016.07.004.

Kurikami, H., Kitamura, A., Yokuda, S.T., and Onishi, Y. (2014), Sediment and 137Cs behaviors in the Ogaki Dam Reservoir during a heavy rainfall event, *Journal of Environmental Radioactivity*, 137: 10–17, https://doi.org/10.1016/j.jenvrad.2014.06.013.

Lappenbusch, W.L. (1988), *Contaminated Waste Sites, Property and Your Health*, Environmental Health Inc., Virginia, 360 pp.

Latrille, C., and Bildstein, O. (2022), Cs selectivity and adsorption reversibility on Ca-illite and Ca-vermiculite, *Chemosphere*, 288: 132582, https://doi.org/10.1016/j.chemosphere.2021.132582.

Lee, H.K., Jun, B.M., Kim, S.-Il., Song, J.-S., Kim, T.-J., Park, S., and Chang, S. (2022), Simultaneous removal of suspended fine soil particles, strontium and cesium from soil washing effluent using inorganic flocculants, *Environmental Technology & Innovation*, 27: 102467, https://doi.org/10.1016/j.eti.2022.102467.

Lenton-Williams, G. (2023), Water Quality in Port Philip Bay Still Recovering after Victoria's Record Floods. www.abc.net.au/news/2023-01-05/victoria-beach-bay-water-quality-epa-melbourne-floods/101826148?utm_campaign=abc_news_web&utm_content=link&utm_medium=content_shared&utm_source=abc_news_web, accessed Mar. 2024.

Li, X., Xu, G., Xia, M., Liu, X., Fan, F., and Dou, J. (2022), Research on the remediation of cesium pollution by adsorption: Insights from bibliometric analysis, *Chemosphere*, 308: 136445, https://doi.org/10.1016/j.chemosphere.2022.136445.

Lim, Y.S., Kim, J.K., Kim, J.W., and Hong, S.S. (2014), Evaluation of suspended-sediment sources in the Yeongsan River using Cs-137 after major human impacts, *Quaternary International*, 344(10): 64–74, https://doi.org/10.1016/j.quaint.2014.05.033.

Linley-Adams, G. (1999), The Accumulation and Impact of Organotins on Marine Mammals, Seabirds and Fish for Human Consumption, WWF-UK Report, WWF-UK Project No. 90854, 26 pp.

Liu, W., Li, Y., Yu, H., Saggar, S., Gong, D., and Zhang, Q. (2021), Distribution of 137Cs and 60Co in plough layer of farmland: Evidenced from a lysimeter experiment using undisturbed soil columns, *Pedosphere*, 31(1): 180–190, doi:10.1016/S1002-0160(19)60837-4.

Lusher, A., Hollman, P., and Mendoza-Hill, J. (2017), Microplastics in fisheries and aquaculture: Status of knowledge on their occurrence and implications for aquatic organisms and food safety, FAO Fisheries and Aquaculture Technical Paper. No. 615, Rome, Italy.

Magni, P., Rajagopal, S., van der Velde, G., Fenzi, G., Kassenberg, J., Vizzini, S., Mazzola, A., and Giordani, G. (2008), Sediment Features, Macrozoobenthic Assemblages and Trophic Relationships ($d^{13}C$ and $d^{15}N$ Analysis) following a Dystrophic Event with Anoxia and Sulphide Development in the Santa Giusta Lagoon (Western Sardinia, Italy), *Marine Pollution Bulletin*, 57: 125–136.

Makiya, K. (1997), Current situation concerning government activities associated with dioxins, *Waste Management Research*, 8(4): 279–288 (Text in Japanese).

Moraga, D., Mdelgi-Lasram, E., Romdhane, M.S., El Abed, A., Boutet, I., Tanguy, A., and Auffret, M. (2002), Genetic response to metal contamination in two clams: *Ruditapes decussates* and *Ruditapes philippinarum*, *Marine Environmental Research*, 54: 521–525.

Muslim, W.A., Al-Nasri, S.K., Talib, M., and Albayati, T.M. (2023), Evaluation of bentonite, attapulgite, and kaolinite as eco-friendly adsorbents in the treatment of real radioactive wastewater containing Cs-137, *Progress in Nuclear Energy*, 162: 104730, https://doi.org/10.1016/j.pnucene.2023.104730.

Nemoto, Y., Oomachi, H., Saito, R., Kumada, R., Sasaki, M., and Takatsuki, S. (2020), Effects of 137Cs contamination after the TEPCO Fukushima dai-ichi nuclear power station accident on food and habitat of wild boar in Fukushima Prefecture, *Journal of Environmental Radioactivity*, 225: 10634, https://doi.org/10.1016/j.jenvrad.2020.106342.

OECD. (2023), Plastic Leakage and Greenhouse Gas Emissions are Increasing – OECD. www.oecd.org/environment/plastics/increased-plastic-leakage-and-greenhouse-gas-emissions.htm#:~:text=In%202019%20alone%2C%206.1%20Mt%20of%20plastic%20waste,1.7%20Mt%20flowing%20into%20the%20ocean%20in%202019, accessed Jan. 2023.

Ohtsubo, M. (1999), Organotin compounds and their adsorption behaviour on sediments, *Clay Science*, 10: 519–539.

Pakdeesusuk, E, Lee, C.M., Coates, J.T., and Freedman, D.L. (2005), Assessment of natural attenuation via in situ reductive dechlorination of polychlorinated biphenyls in sediments of the Twelve Mile Creek Arm of Lake Hartwell, SC., *Environmental Science and Technology*, 39: 945–952.

Palermo, M., Maynord, S., Miller, J., and Reible, D. (1998), Guidance for in-situ subaqueous capping of contaminated sediments, EPA 905-B96-004, Great Lakes National Program Office, Chicago, IL. www.epa.gov/glnpo/sediment/iscmain/one.html, accessed Dec. 14, 2023.

Park, C.W., Kim, S.-M., Kim, I., Yoon, I.-H., Hwang, J., Kim, J.-H., Yang, H.-M., and Seo, B.M. (2021), Sorption behavior of cesium on silt and clay soil fractions, *Journal of Environmental Radioactivity*, 233: 106592, https://doi.org/10.1016/j.jenvrad.2021.106592.

Pearson, T.H., and Rosenberg, R.(1978), Macrobenthic succession in relation to organic enrichment and pollution of the marine environment, *Oceanography and Marine Biology*, 16: 229–311.

Plastics Europe. (2020), Plastics – The Facts 2020. An Analysis of European Plastics Production, Demand and Waste Data. Plastics Europe, Brussels, Belgium, http://plastics.org/knowledge-hub/plastics-the-facts-2020/, accessed Feb. 2024.

Pourabadehei, M., and Mulligan C.N. (2016), Resuspension of sediment, a new approach for remediation of contaminated sediment, *Environmental Pollution*, 213: 63–75.

Romano, E., Ausili, A., Zharova, N., Magno, M.C., Pavoni, B., and Gabellini, M. (2004), Marine sediment contamination of an industrial site at Port of Bagnoli, Gulf of Naples, Southern Italy, *Marine Pollution Bulletin*, 49: 487–495.

Rosenberg, R., Nilsson, H. C., and Diaz, R.J. (2001), Response of benthic fauna and changing sediment redox profiles over a hypoxic gradient, estuarine, *Coastal and Shelf Science*, 53: 343–350.

Saborimanesh, N., and Mulligan, C.N. (2018), Investigation on the dispersion of weathered biodiesel, diesel, and light crude oil in the presence of sophorolipid biosurfactant in seawater, *Journal of Environmental Engineering*, 144(5), https://doi.org/10.1061/(ASCE)EE.1943-7870.0001369, Article Number: 04018028.

Santschi, P.H., Presley, B.J., Wade, T.L., Garcia-Romero, B., and Baskaran, M. (2001), Historical contamination of PAHs, PCBs, DDTs, and heavy metals in Mississippi River Delta, Galveston Bay and Tampa Bay sediment cores, *Marine Environmental Research*, 52: 51–79.

Seligman, P.F., Valkirs, A.O., Stang, P.M., and Lee, R.F. (1988), Evidence for rapid degradation of tributyltin in a marina, *Marine Pollution Bulletin*, 19(10): 531–534.

Selvaraj, K., Mohan, V., Ram, and Szefer. P. (2004), Evaluation of metal contamination in coastal sediments of the Bay of Bengal, India: Geochemical and statistical approaches, *Marine Pollution Bulletin*, 49: 173–185.

Siroux, B., Latrille, C., Beaucaire, C., Petcut, C., Tabarant, M., Benedetti, M.F., Pascal, E., and Reiller, P.E. (2021), On the use of a multi-site ion-exchange model to predictively simulate the adsorption behaviour of strontium and caesium onto French agricultural soils, *Applied Geochemistry*, 132: 105052, https://doi.org/10.1016/j.apgeochem.2021.105052.

Stapleton, R.D., and Sayler, G.S. (1998), Assessment of the microbiological potential for the natural attenuation of petroleum hydrocarbons in a shallow aquifer system, *Microbial Ecology*, 36: 349–361.

Stohl, A., Seibert, P., Wotawa, G., Arnold, D., Burkhart, J. F., Eckhardt, S., Tapia, C., Vargas, A., and Yasunari, T. J. (2012). Xenon-133 and caesium-137 releases into the atmosphere from the Fukushima Dai-ichi nuclear power plant: determination of the source term, atmospheric dispersion, and deposition, Atmospheric Chemictry and Physics, 12, 2313–2343, https://doi.org/10.5194/acp-12-2313-2012.

Sumeri, A. (1995), Dredged Material Is Not Spoil - A Status on the Use of Dredged Material in Puget Sound to Isolate Contaminated Sediments, 14th World Dredging Congress, Amsterdam, The Netherlands, 14–17 Nov.

Tateda, Y., Tsumune, D., and Tsubono, T. (2013), Simulation of radioactive cesium transfer in the southern Fukushima coastal biota using a dynamic food chain transfer model, *Journal of Environmental Radioactivity*, 124: 1–12, http://dx.doi.org/10.1016/j.jenvrad.2013.03.007.

Tatsuno, R., Waki, H., Kakuma, M., Nihei, N., Takase T., Wada, T., Yoshimura, K., Nakanishi, T., and Ohte, N. (2023), Effect of radioactive cesium-rich microparticles on radioactive cesium concentration and distribution coefficient in rivers flowing through the watersheds with different contaminated condition in Fukushima, *Journal of Environmental Management*, 329, 116983, https://doi.org/10.1016/j.jenvman.2022.116983.

US EPA. (1976), Quality Criteria for Water, EPA 440-9-76-023, Office of Water and Hazardous Materials, Washington, DC.

Wada, T., Hinata, A., Furuta, Y., Sasaki, K., Konoplev, A., and Nanba, K. (2023), Factors affecting 137Cs radioactivity and water-to-body concentration ratios of fish in river and pond environments near the Fukushima Dai-ichi Nuclear Power Plant, *Journal of Environmental Radioactivity*, 258: 107103, https://doi.org/10.1016/j.jenvrad.2022.107103.

Wang, W., Shi, L., Wu, H., Ding, Z., Liang, J., Li, P., and Fan, Q. (2023), Interactions between micaceous minerals weathering and cesium adsorption, *Water Research*, 238: 119918, https://doi.org/10.1016/j.watres.2023.119918.

Wilde, E.W., and Benemann, J.R. (1993), Bioremoval of heavy metals by the use of microalgae, *Biotechnology Advances*, 11(4): 781–812.

Yamamoto, T., Imose, H., Hashimoto, T., Matsuda, O., Go, A., and Nakaguchi, K.J. (1997), Results of seasonal observations on sediment quality in the Seto Inland Sea, *Journal of the Faculty of Applied Biological Science, Hiroshima University*, 36: 43–49 (text in Japanese with English summary).

Yamasaki, S., Yasui, S., and Fukue, M. (1995), Development of solidification technique for dredged sediments, dredging, remediation, *Containment for Dredged Contaminated Sediments*, 1293, 136–144.

Yasunari, T.J., Stohl, A., Hayano, R. S., Burkhart, J.F., Eckhardt, S., and Yasunari, T. (2011), Cesium-137 deposition and contamination of Japanese soils due to the Fukushima nuclear accident, Edited by James E. Hansen, *Proceedings of the National Academy of Sciences*, 108(49): 19530–19534. www.pnas.org/cgi/doi/10.1073/pnas.1112058108.

Yong, R.N. (1996), Waste Disposal, Regulatory Policy and Potential Health Threats. In: S.P. Bentley (ed.), *Engineering Geology of Waste Disposal*, Geological Society of London Special Publication No. 11, pp. 325–340.

Yong, R.N. (2001), *Geoenvironmental Engineering: Contaminated Soils, Contaminant Fate, and Mitigation*, CRC Press, Boca Raton, FL, 307 pp.

Zeman, A.J., and Patterson, T.S. (1996), Preliminary Results of Demonstration Capping Project in Hamilton, Harbour, NWRI Contribution No. 96-53, National Water Research Institute, Burlington, Ontario.

Zhang, K., Li, Z., Qi, S., Chen, W., Xie, J., Wu, H., Zhao, H., Li, D., and Wang, S. (2022), Adsorption behavior of Cs(I) on natural soils: Batch experiments and model-based quantification of different adsorption sites, *Chemosphere*, 290: 132636, https://doi.org/10.1016/j.chemosphere.2021.132636.

9 Contaminants and Land Environment Sustainability Indicators

9.1 INTRODUCTION

We use term *land environment* instead of *geoenvironment* because we do not include the coastal marine environment (considered in the previous chapter) in our discussions in this chapter. As we have seen and discussed in the previous chapters – in respect to the land environment – other than depletion of natural resources and habitat destruction, ground contamination by all kinds of contaminants and pollutants from anthropogenic activities poses one of greatest threats to the sustainability of the natural capital of the geoenvironment. The term *pollutants* is used to remind the reader that these are contaminants that are deemed by regulatory agencies to be injurious to human health. In the treatment of the subject of contaminants and pollutants in this chapter, the encompassing term *contaminants* will be used. When emphasis is needed, the term *pollutants* will be used. Within the context of the land environment, the term *natural capital* refers to the land ecosystem, which includes (a) the receiving waters in the land surface environment, such as rivers, lakes, ponds, and streams, (b) the solid land surface and the underlying soil-water system, (c) the natural resources, such as forests, mineral, and carbon resources, and (d) the various biotic species and the biodiversity of the ecosystem. The presence of contaminants in the ground affect not only soil and water quality, but also those living elements and biota that more or less depend on soil and water for their well-being. This would include forests, agricultural production, habitats, and the host of biotic species contributing to the biodiversity of the land ecosystem.

Anthropogenic activities associated with production of goods and services, such as resource exploitation, agricultural production, harvesting of forest and carbon resources, urbanization, industrial production, and manufacturing, are by far the greatest contributors to the generation of waste and contaminants that ultimately find their way onto and into the land environment. Sustainability of the land environment natural capital cannot be easily or readily defined for many of the individual components that make up the natural capital of interest because of their dynamic nature. This is particularly true for the *living* components. It is easier to define sustainability goals for the natural capital as part of the life-cycle assessment process discussed in Chapter 4. This task is facilitated if one can establish an acceptable natural capital baseline.

9.2 INDICATORS

The general understanding of sustainability of a natural capital of the land environment can be stated as follows: "*To ensure that each natural capital maintains its full and uncompromised functioning capability without loss of growth potential*". Clear baseline values for the various components that constitute the land environment natural capital are needed to define or establish sustainability requirements of the natural capital in quantitative terms. Since absolute sustainability is not always attainable, the use of indicators is a means to establish benchmarks or targets that point the way towards sustainability of specific components of the natural capital and of the natural capital itself.

DOI: 10.1201/9781003407393-9

9.2.1 Nature of Indicators

Section 1.4 of Chapter 1 has discussed, very briefly, the nature of indicators. They are essentially "signs" or "markers" used daily by ordinary people and also by professionals. The example cited earlier of vehicular traffic lights at an intersection is a very good example of one's everyday encounter with indicators. In terms of day-to-day living, personal events such as prolonged headaches, stomach aches, muscle pain, toothaches, etc., are used as indicators of some form of health distress and that one should visit one's health care specialist. Numbers, statistical information, events, etc., can all be used as indicators. Unemployment rates, gross domestic product (GDP), and financial indices such as those used for tracking the ups and downs of various equities, are all indicators of the health of the economy. In a sense, indicators have been used by humans for as long as there has been human life on the planet.

From the geoenvironmental perspective, the indicators discussed in this chapter are considered to be markers or benchmarks specified or prescribed by individuals and/or organizations interested in tracking: (a) The progress of events, (b) operations and performance of systems, (c) the outcome of actions on something specific, and (d) the status of a particular set of events, actions, process, activity, and situation – all of which are with specific reference to the land environment. The indicators can also be used (i) for performance assessment of various systems, processes, actions, etc., and (ii) to set goals and/or targets for operation and performance of systems, processes, activities, etc. The example of the use of soil functionality indices (SFIs) shown in Section 1.4 of Chapter 1 and in Figure 1.7 is a good demonstration of the usefulness of indicators for tracking operations of systems.

There are many varieties and types of indicators that can be used to determine or track the sustainability of the geoenvironment. For convenience, we can group all of these into two major groups: (a) System status indicators, and (b) material performance indicators. System status indicators refer to the status of a system – an ecosystem for example – at any time period. They are essentially performance indicators. This is because the system status at any one time period is the result of the performance of the system over the time period under consideration. In that same sense, material performance indicators are designed to provide information on the performance of materials at any specified or required time period – quantitatively or qualitatively. Since material properties and characteristics may be functions of system processes, it follows that material performance indicators are also material properties' indicators. When the term *indicators* is used, it is understood that this refers to both system status and material performance indicators. The greatest benefit of status indicators comes when they are used in relation to some predetermined or defined criterion, target, or objective.

Not all indicators are sustainability indicators. A good example is the measurement of acid precipitation onto the ground where the sources of SO_x and NO_x are known. The acidity measurements themselves are material indicators, i.e., they are acidity indicators. This begs the question of "so what?" Four courses of action come to mind: (1) Do absolutely nothing, (2) implement a ground surface neutralization programme using perhaps a surface spray technique, (3) monitor target components such as soil and water quality, environmental mobility of contaminants in the affected area, and (4) determine the source of the gaseous discharge and rectify the situation through implementation of better process technology and more effective scrubber systems. Figure 9.1 illustrates this example from a land environment perspective.

Figure 9.1 shows that *indicators* (a) inform us on "where we are" in respect to specific indices and prescribed parameters, (b) provide specific information to be used to compare with (or contrast with) baseline values or other target values, (c) provide monitoring and tracking information for use in determining the transient state of items being monitored, and (d) tell us if we are on target or "off-track". The "SO_x, NO_x, PM_{10} and TOMPs" shown in the top right-hand corner of the diagram are *material indicators*. They provide information on the noxious gases and other airborne contaminants that are responsible for the chemistry of the airborne deposition of contaminants on the land surface.

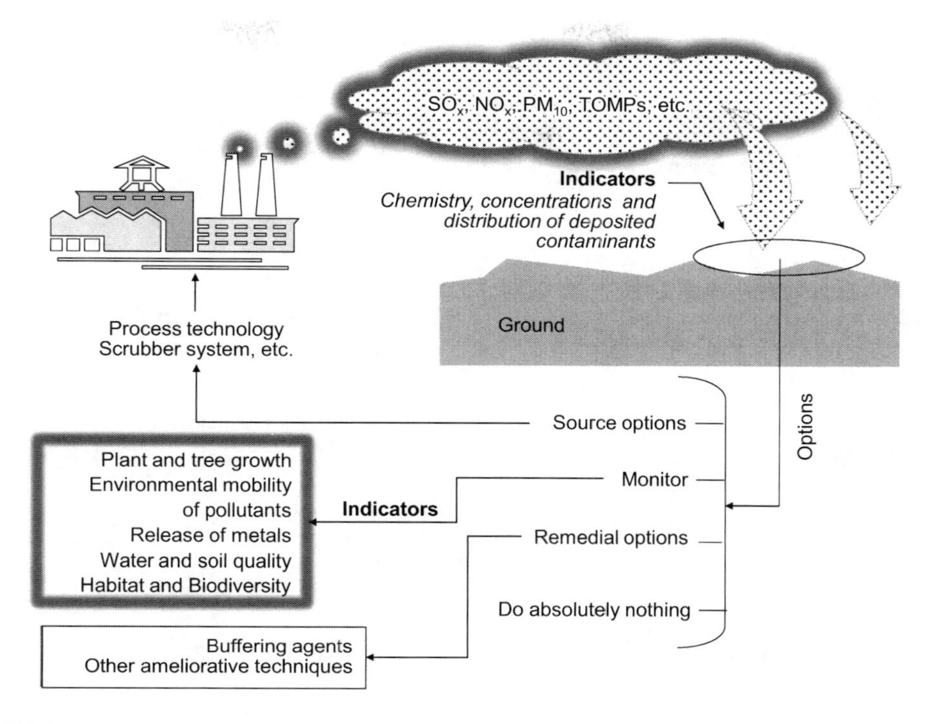

FIGURE 9.1 Illustration of the role of indicators in a typical precipitation situation in an atmosphere with noxious gaseous and other airborne contaminants – in the context of impacts on the land surface. The indicators shown do not cover the entire impact problem. Note that changes to process technology and scrubber system do not fall within the purview of this book – and neither do plant-tree growth, habitat, and biodiversity. Note that PM_{10} = particulate matter less than 10 μm in size, and TOMPs = toxic organic micropollutants such as obtained PAHs, PCBs, dioxins and furans from incomplete combustion of fuels.

This information gives one the opportunity to exercise any one of four different options – as shown in the diagram. Depending on the nature of the contaminants (chemistry, concentration, distribution, etc.) being deposited on the land surface, the two distinct courses of available action are (a) actions undertaken at the industrial plant itself, and (b) actions taken to address the impact of the deposition of the airborne discharges. *Action (a)* is essentially *source control*, e.g., improve the scrubber system in the industrial plant shown in the top left-hand corner of the diagram, whilst *action (b)* is *contaminant management.*

For the second option, i.e., contaminant management, information from material indicators is required for one to determine the management course of action – as shown in Figure 9.1. The actions are (a) do nothing, (b) monitor, and (c) undertake remedial actions. In the case of the "monitor" option, the associated indicators are *status indicators.* These indicators essentially track the status of the various items, such as water and soil quality at various times, to determine whether negative changes occur. More importantly, these system status indicators are important markers along the road to sustainability, and where appropriate, they may be called sustainability indicators. *Sustainability indicators* (SIs) are those indicators that point the way towards achievement of the goals and objectives of sustainability. For these indicators to have value and meaning, it is necessary to provide or specify the sustainability goals or objectives. Because true sustainability may never be attained, as for example in the case diminishing non-renewable natural resources, the sustainability indicators provide us with a measure of non-renewable natural resource depletion (depletion rate).

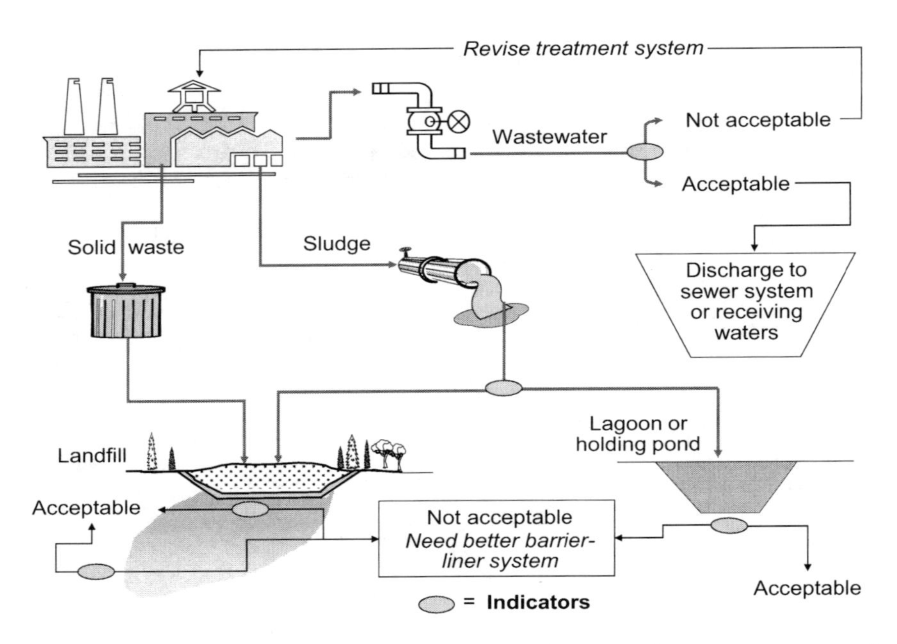

FIGURE 9.2 The role of indicators in assessing status and sustainability of operations in a plant discharging liquid and solid wastes. The indicators in the diagram are shown as shaded ovals.

By targeting a depletion rate that would allow for a longer time before exhaustion of the resource, SIs can alert us as to whether we are on target or whether drastic remedial or corrective steps need to be taken. However, as will be evident in further discussions on indicators, SIs are really a collection of many intermediate indicators situated as markers along the road to sustainability – with the final set of indicators located at the terminal point of the road. The various intermediate indicators should set the requirements and conditions of performance at stages along the way, as shown for example in Figure 9.2. The combination of these indicators can be considered as "requirements to satisfy sustainability objectives".

9.2.2 Contaminants and Geoenvironment Indicators

An example of impacts from the discharge of liquid waste and/or sludge and solid waste is shown in Figure 9.2. Not all the indicators shown in the diagram (depicted by the shaded ovals) are SIs. These indicators have several functions. In the top right-hand corner, the various pieces of information from the wastewater discharge from the industrial plant, which are wastewater chemistry indicators, tell us whether the wastewater discharge meets the regulatory requirements permitting discharge into the sewer system or receiving waters. The wastewater chemistry indicators, which are material indicators, could include, for example, suspended solids, alkalinity, metals, fats, oils, grease, organic chemicals, etc. Failure of the wastewater chemistry to meet discharge standards will require a re-engineering of the treatment system.

Information obtained from analyses of the sludge being discharged (from the discharge end of the sludge pipe, i.e., the sludge discharge indicators) can be used to determine whether the toxicity and sludge characteristics meet lagoon (holding pond) capabilities or whether the sludge should be contained in a secure landfill. This assumes that lagoons and holding ponds do not generally have the same type of secure impervious barrier-liner systems that are required for landfills. In this instance, the sludge discharge indicators are not sustainability indicators but are status indicators that provide one with the information required to make the necessary judgment for disposal of the sludge.

The contaminant indicators located below the lagoon and the landfill serve several purposes. In the lagoon case, assuming that the barrier-liner system for the lagoon is not as secure as the landfill system, the contaminant indicators will tell us whether the contaminants escaping from the lagoon are in the range of acceptable concentrations. One needs to rely on the attenuation characteristics and properties of the sub-soil strata to further ameliorate the concentration of contaminants as the contaminant plume travels further into the soil sub-stratum. If the contaminant indicators show unacceptably high concentrations of contaminants immediately under the lagoon, regulations would require corrective action to be taken.

The same situation applies to the contaminant indicators immediately adjacent to, or under the barrier-liner system of the landfill. Corrective action is needed if the contaminant indicators exceed specified concentration trigger levels. Acceptable concentration levels reported by the contaminant indicators immediately under the barrier-liner system do not mean that the other contaminant indicators located further away from the landfill would report favourably. Much depends on the nature of the contaminants and the transport processes. Contaminant transport modelling, using the information from the first set of indicators under the barrier-liner system, can be useful. Predicted transport values can be used to compare with the second set of contaminant indicators. The contaminant indicators are seen to be markers that show progress of the contaminant front (i.e., trackers of transport), and if design specifications and expectations of attenuation are correct, the indicators should accord well with predictions – provided that the transport models accurately predict performance. The transport of contaminants has been briefly discussed in Chapter 2. A further detailed discussion of these transport and fate processes will be found in Section 9.3.

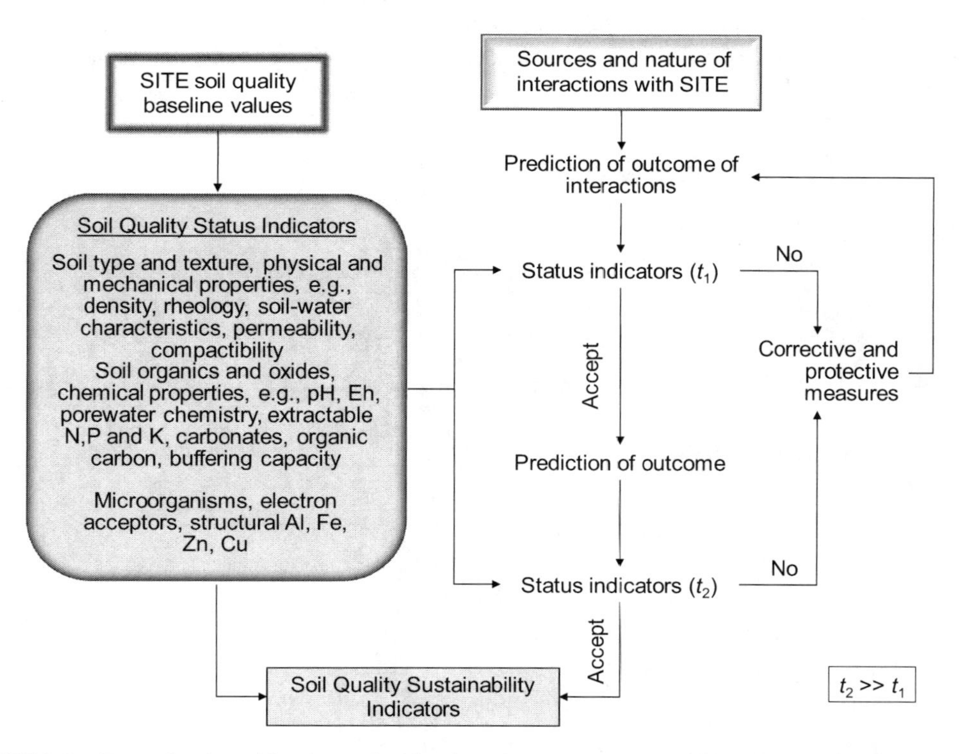

FIGURE 9.3 Example of specification and utilization of system status and material performance (properties) indicators using sustainability of soil quality as a goal. Note that the physical, chemical, and biological indicators listed in the *soil quality status indicators* are meant to portray the kinds of indicators that might be used. (t_1) and (t_2) indicate two different times.

9.2.3 Prescribing Indicators

Where and how are indicators prescribed? To a large degree, material performance, and system status indicators are specified or determined on the basis of how or what one needs to know and undertake to meet the specific targets identified in *sustainability status indicators.* There are two starting points for delineation of indicators:

(1) Starting with the objective itself. In this instance, as an example, we could begin with a particular item in a natural capital component, e.g., maintaining the quality of the soil in a tract of land or a particular site. This is important because of the life-support role of the specific tract of land. In this instance, one begins by (a) defining or establishing what indicators are needed as sustainability status indicators, e.g., specifics of soil quality as shown, for example, in Figure 9.3(b) determining the sources and nature of interactions with the tract of land and their impacts, and (c) establishing the required actions to ameliorate, mitigate, avoid and protect the desired soil quality. Material and macro status indicators are established to determine and track the results of the corrective and protective measures. Figure 9.3 shows the use of status indicators in tracking anticipated or predicted outcomes from analyses or modelling of the processes initiated by the corrective actions. Failure to meet tracking results from the status indicators requires one to decide: (a) to ignore indicators, or (b) modify or add or correct the actions previously prescribed to manage the impacts.
(2) Starting with the external sources of interactions with the land environment itself. With this starting point, one follows a reverse sequence. Figure 9.4 shows a typical protocol to determine or evaluate sustainability capabilities for the actions or impacts resulting from a specific project, activity, or industry. The status indicators (1) and (2) are indicators that may refer to different time periods, intervals, circumstances, or locations. Prediction of the outcome of

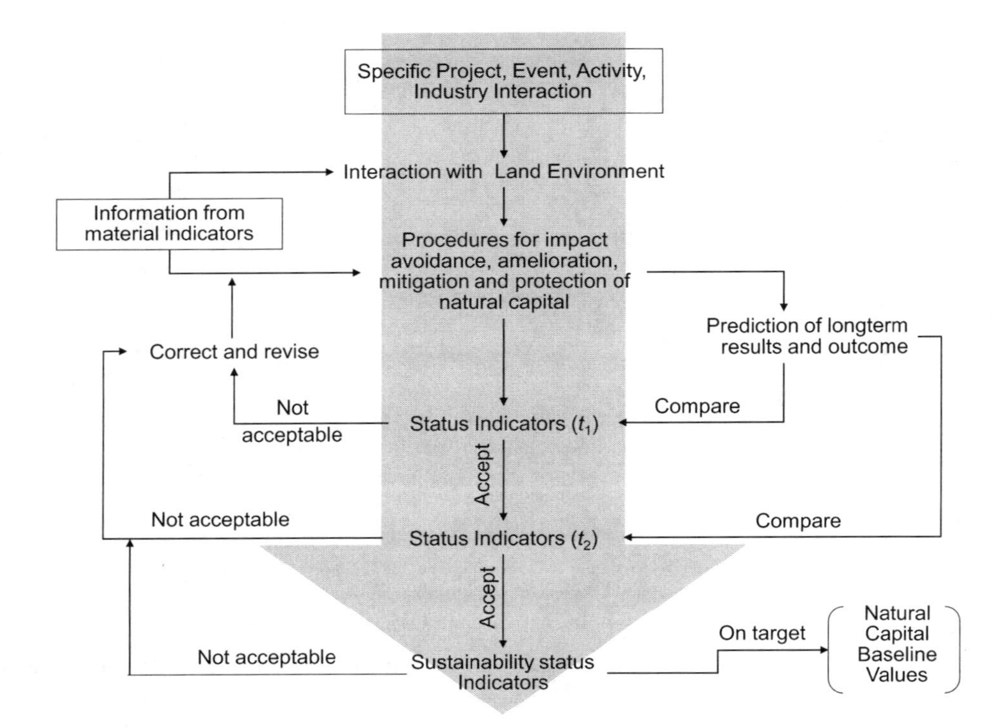

FIGURE 9.4 Prescription of material performance (properties) and system status indicators for determination of procedures for impact avoidance, amelioration, mitigation and protection of natural capital.

preventative or ameliorative actions (corrective actions) is generally obtained via modelling of the processes involved in the impact and corrective interactions – as also in the case of the actions shown in Figure 9.3.

The importance of system status and material performance indicators is evident in the *soil quality indicators* shown in the left-hand side of Figure 9.3. The list of physical, chemical, and biological indicators shown in the diagram are not meant to be comprehensive. They are prescribed according to the specifics of the natural capital component under consideration. The status indicators (1) and (2) are prescribed in accordance with tracking or monitoring requirements. These are both spatial and temporal in nature and can include more than the numbers shown in the diagrams. Even though true sustainability may not be attained, it is necessary for the objectives and goals for sustainability to be properly articulated. These will serve to establish what, where, how many, and how often status indicators will be used or required.

9.3 ASSESSMENT OF INTERACTION IMPACTS

Figure 9.5 shows many of the pollutants and contaminants found as discharge waste from activities associated with urbanization and industrial manufacturing and production. Knowledge of the geoenvironment impact interactions with specific projects and industrial manufacturing and production processes is required in structuring defensive and mitigation actions. In the case of contaminants in the land compartment of the geoenvironment, this means that knowledge of the transport and fate of contaminants in the sub-soil must be obtained. The aim of the outcome of defensive and mitigation actions is to provide the best course of action required to meet the objectives for protection and conservation of the particular natural capital component under consideration. Contaminant fate and transport modelling is a commonly used tool to predict the outcome of these actions. The basic elements of fate and transport processes and modelling have been discussed in Chapter 2. We will continue the discussion in this section.

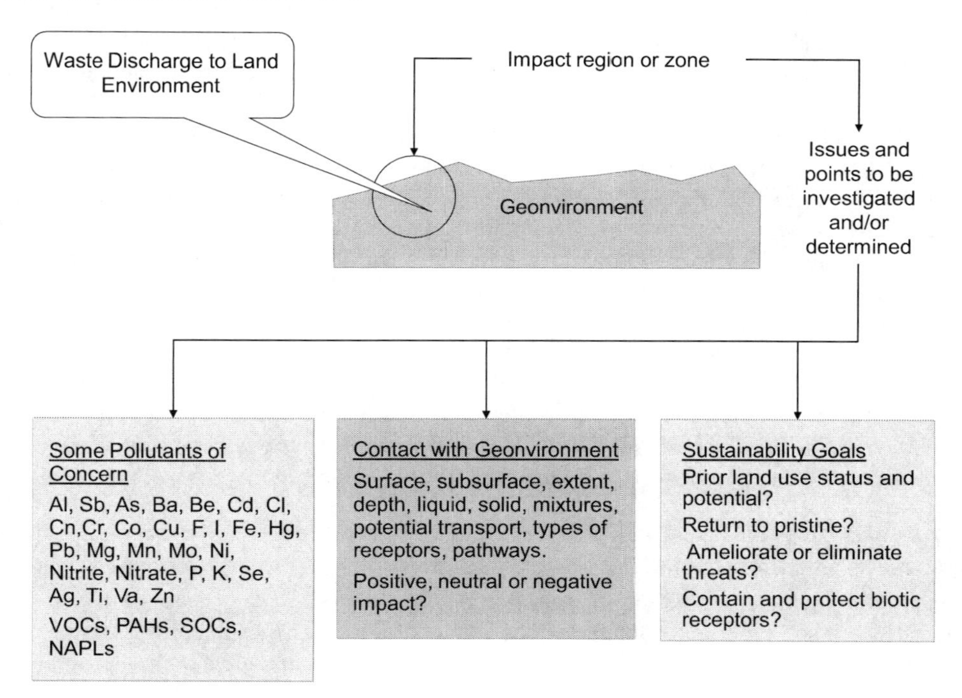

FIGURE 9.5 Simple schematic showing first sets of determination for assessment of impact of waste discharge to the land environment.

9.3.1 Sustainability Concerns

The previous chapters have shown that the major sets of interactions between the various external agents (activities, projects, industries, etc.) that have the potential to severely impact the land environment are the liquid and solid discharges (spills and other discharge forms) from these agents. The necessary pieces of information and knowledge required to generate preventive and mitigation solutions (Figures 9.3 and 9.4) relate to the local geological and hydrogeological settings, the types and nature of spills and discharges, the various environmental and biotic receptors, and most importantly, the status and sustainability indicators. A useful rule-of-thumb in prescribing sustainability status indicators (the last set of indicators shown in Figures 9.3 and 9.4) is to set objectives and targets that do not yield negative impact results, i.e., interaction and preventive-mitigation impacts should not diminish the natural capital component under consideration. Whilst activities associated with upstream and downstream industries may inevitably generate impacts that cannot be fully ameliorated, minimized, avoided, or totally remediated, it is nevertheless necessary to set status indicators that target sustainability of the natural capital of land environment.

9.3.2 Surface Discharge – Hydrologic Drainage, Spills, and Dumping

Surface discharge or liquid/solid contaminant loading of land surface occurs under circumstances that include (a) hydrologic drainage occurring at a mine site at intersections of mine openings with the water table, (b) percolation from waste rock and mill tailings piles and surface run-offs, (c) inadvertent spills or deliberate dumping, (d) leaking pipelines, (e) deposition of airborne noxious contaminants from rainfall and snow loads, (f) application of agricultural chemicals (including pesticides) and irrigation, and (g) designed land farming treatment of contaminated materials. Figure 9.6 shows a schematic of some of the situations leading to contaminant and contaminant loading of the ground surface. The bottom-left diagram in the figure concerns the problems discussed in Chapter 6. The land farming scheme shown in the bottom right of the diagram assumes that a prepared impermeable base is in-place before placement of the material to land farmed. Techniques for land farming will be discussed in a later section when mitigation and treatment alternatives are addressed.

The principal issues for contaminant loading of the land surface relate to the transport of contaminants from the contaminant source. By and large, spills and deliberate dumping of waste materials involve small surficial areas (i.e., small areas on the land surface). These will serve as point-sources for transport of contaminants into the ground. Surface run-offs from these regions tend to be small. On the other hand, surface run-offs for other land application of contaminants resulting from deposition of airborne noxious substances and use of agricultural chemicals (including pesticides) can be serious issues – because of the extent of the land surface affected. The surface run-offs of contaminants arise if rainfall occurs or continues before the noxious substances and agricultural chemicals have a chance to effectively infiltrate or leach into the subsurface. Since the receiving sites for surface run-offs are topographically lower than the source location, these sites will likely be lowlands, wetlands and receiving waters. Contamination of these regions (from surface run-offs) is commonly identified as *non point-source contamination*.

Contamination of the sub-surface material (sub-surface geologic material) and the underlying aquifers from contaminant sources occurs because of the transport of contaminants in the subsurface material. Laboratory studies on the transport of contaminants in soils have been conducted for a large variety of soil types and contaminants. Field studies have also been performed in support of remediation projects and as due diligence work. Procedures and analytical-computer models have

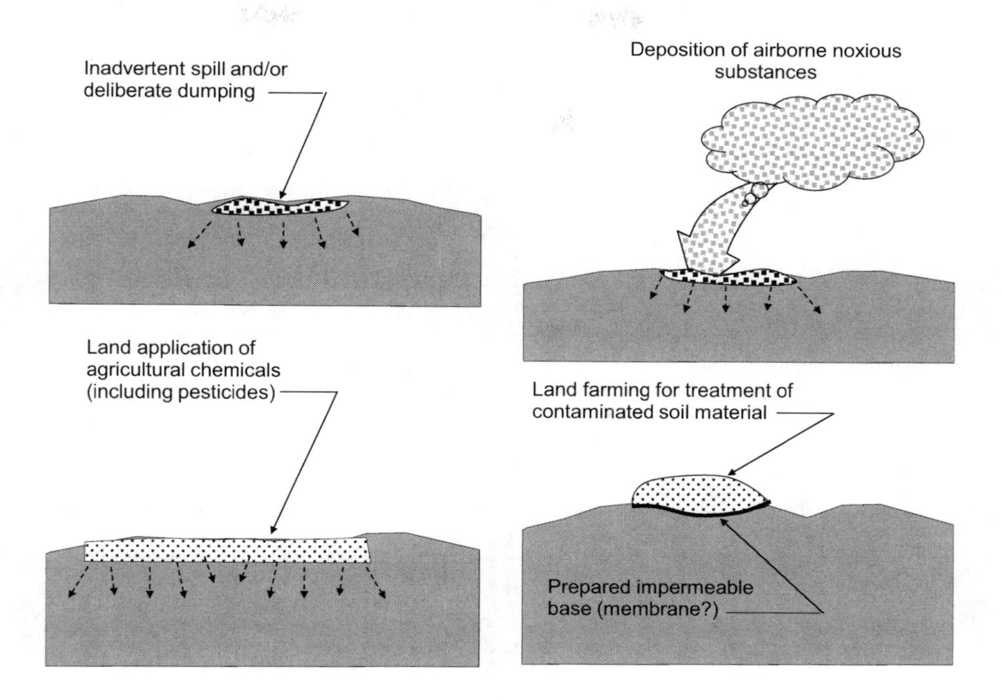

FIGURE 9.6 Scenarios showing contaminant loading on land surface of geoenvironment. For small spills and dumps, point source contamination is assumed. For precipitation and land application of control agents, etc., non-point source contamination is generally assumed.

been developed to provide one with the capability to determine and/or predict the movement, distribution and concentration of contaminants in the subsurface soil. A brief summary of these can be found in Section 2.5.4 in Chapter 2. As also emphasized previously, detailed treatment of these subjects can be found in textbooks dedicated to the study of contaminant fate and transport in soils. The discussion of contaminant transport processes and predictions will be found in the latter portion of this section.

The essential elements required for assessment or evaluation of the impacts from surface discharge phenomena, such as spills from industrial operations and systems, such as pipelines and vehicular traffic transporting liquid wastes or heavy oils and bitumen, dumping, etc. are shown in Figure 9.7. In addition to determination of the size of affected area and quantity of material spilt, discharged and/or dumped, one is required to determine the following:

(1) Nature and composition of the discharged material.
(2) Hydrogeological setting in the region.
(3) Sub-soil profile, material composition and properties.
(4) Transport processes involving the types of contaminants found in the compositional analyses – from laboratory tests.
(5) Transport and fate of contaminants (emanating from the surface discharge) in the sub-soil – using predictive models and test information on partitioning of contaminants and transmissivity characteristics of the contaminants.
(6) What are the environmental and biotic receptors, and how will these be impacted?
(7) Source-path-receptor relationships.

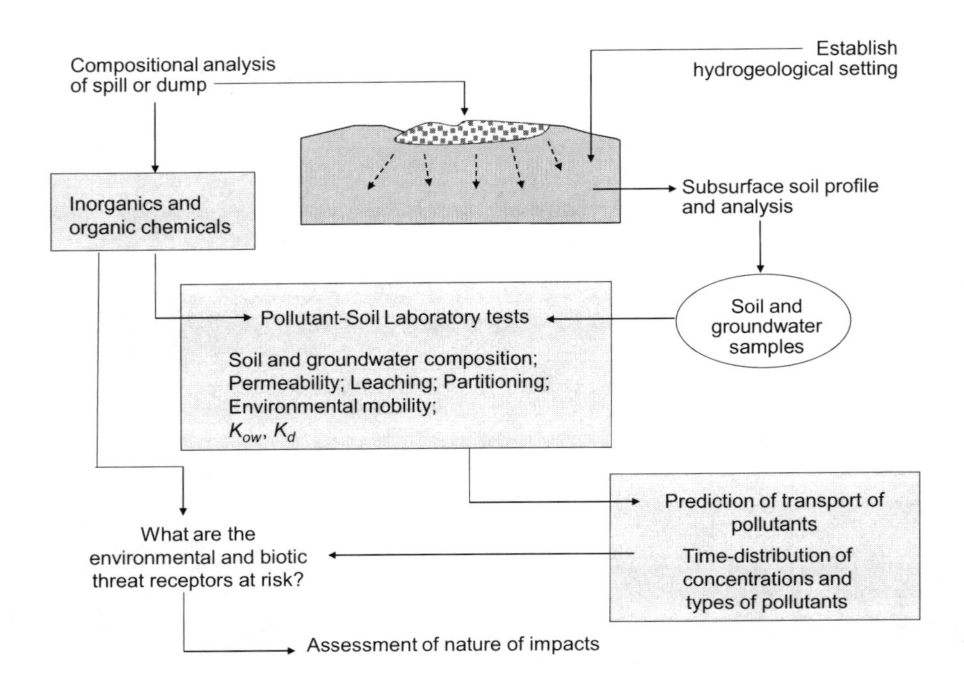

FIGURE 9.7 Schematic diagram showing procedures, factors, tests, and analyses required to begin the process for determination of consequence of surface discharge.

9.3.3 Sub-Surface Discharges

Sub-surface discharges of contaminants arise when leaks or breakages in underground liquid/gaseous storage facilities and underground liquid or gaseous pipelines occur. Sub-surface discharges of contaminants also occur when leachates escape from waste landfills (Figure 9.8). The left-hand portion of the schematic in Figure 9.8 shows the leachate plume escaping at the bottom of the constructed waste landfill. When properly designed, with adequate barrier and liner systems underlying and surrounding the sides of the landfill, and leachate collection intercepts placed in the bottom barrier system, leakages are not expected. Furthermore, with proper impermeable caps placed on the top of the landfills, water is not permitted to enter into the system. This subject is considered further in the next chapter. A comprehensive discussion of waste landfill barrier systems can be found in textbooks dealing with the subject of landfills and barrier systems (e.g., Yong et al., 2010).

Leachates emanating from landfills result from (a) landfills not properly constructed, (b) breaks in the liner and barrier system, (c) deterioration of barrier system, (d) incompatible interactions between barrier material and leachate chemistry resulting in failure of barrier to perform effectively, and (e) historic (old) landfills built without any real site preparation, and therefore not constructed with attention to requirements for proper and secure containment with barrier and liner systems. Typical constituents in municipal solid waste (MSW) leachates consist of various metals and salts, such as Cd, Fe, Pb, Zn, Mn, Na, Ca, Mg, K, and other inorganics including Cl^-, P, $CaCO_3$, and SO_4^{2-}. Biological and other parameters characterizing MSW leachates include biochemical oxygen demand (BOD_5), chemical oxygen demand (COD), total suspended solids (TSS), total dissolved solids (TDS), pH, total nitrogen (N), and total organic carbon (TOC). Typical organics found in MSW leachates include phenols, volatile acids, organic nitrogen, tannins, lignin, oil and

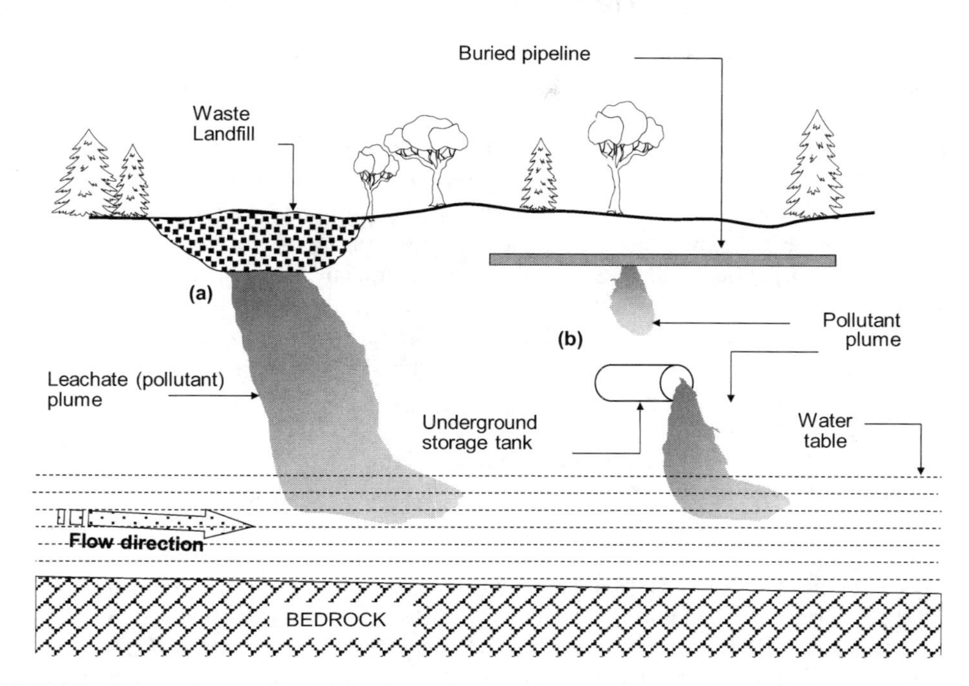

FIGURE 9.8 Schematic showing (a) leachate plume with contaminants, (b) contaminant plumes from a leaking underground storage tank and leaking buried pipeline.

grease, and chlorinated hydrocarbons. One of the many problems encountered when various kinds of organic and inorganic chemicals and elements are combined in a mixture is the reactions that can occur between some of the constituents. For example, reactions between halogenated organics and metals will generate heat in a leachate stream. If this occurs in a relatively dry environment, fire will result. The same occurs when saturated aliphatic hydrocarbons and phenols interact with oxidizing mineral acids.

The constituents of leachate streams in hazardous waste landfills are intimately associated with the composition of the hazardous waste contained in the landfill. The contaminants of concern in the leachate streams are both inorganic and organic in nature. Heavy metals constitute the primary inorganic contaminants whereas the organic contaminants can cover a whole range of organic chemicals – from VOCs (volatile organic chemicals) to SOCs (synthetic organic chemicals). For both the MSW and hazardous waste landfills, the principal sets of problems, outside of the control and remediation-type problems, concern determination of the transport, and fate of the contaminants. Knowledge or prediction of where, when, and extent of leachate penetration into the subsurface surroundings will provide one with (a) the requirements for a proper and effective sub-surface monitoring programusing monitoring wells and other sensing devices, and (b) the possible scenarios for various management options and remedial actions. The procedures and requirements shown in Figure 9.7 in respect to surface discharge problems also apply to sub-surface discharges. A summary of the various transport processes involving sorption and partitioning of inorganic and organic contaminants has been given in Chapter 2 together with a brief mention of the widely used transport relationship. A detailed treatment of these subjects can be found in Yong (2001) and Yong et al. (2010).

9.4 CONTAMINANT TRANSPORT AND FATE

Potential sources of contamination of sub-surface water or groundwater (porewater and aquifers) other than inadvertent spills and deliberate dumping of hazardous materials, include landfills, underground storage tanks, waste piles and waste sites, underground injection wells, unplugged oil and gas wells, various kinds of surface impoundments and settling ponds, lands treated with pesticides, insecticides, fertilizers, and pipelines transporting carbon resources. Figure 9.8 shows many of these potential sources. The types of contaminants, their concentrations and proportions, and the transport of these in the sub-soil and their fate are critical to the structuring of protective measures necessary for the protection of public health and the health of the land environment. The need for predictive tools is obvious.

Questions posed by regulators, investigators, containment facility designers, site remediation technologists, and all others working with containment and remediation of site contamination are summarized as follows:

- Source of contaminants? Nature of contaminant plume?
- Size of contaminant plume? Concentrations and distribution of the various contaminants in contaminant plume? Dominant toxic elements and contaminants in the contaminant plume?
- Where are the receptors? Paths to receptors? Source-pathways-receptors linkage?
- Rate of transport of contaminants? Can we predict its rate and extent of advance?
- Fate and/or persistence of the contaminants in the plume? Can we predict?
- Will contaminant plume threaten or contaminate water resources? Will contaminants pose threats to environment and health of biotic receptors?
- Measures for risk management? Risk tolerance?

9.4.1 Analytical and Predictive Tools

Analytical and predictive tools dealing with the fate and transport of contaminants must account for the following:

- concentrations of the various target contaminant species,
- hydraulic conductivity of the subsurface material (soil),
- diffusive capabilities of the target contaminants,
- hydrogeologic setting,
- partitioning potential of the target contaminants,
- solubility of the target contaminants,
- speciation, complexation and products formed,
- abiotic and biotic reactions and transformations.

The factors and elements to be considered fall conveniently into two groups: (a) transport, and (b) reactions. Two types of analytical-computer models have been developed: (a) models dealing with fate and transport of contaminants, and (b) models that take into account geochemical reactions and their products. By and large, the models dealing with fate and transport of contaminants are non-reactive models. This means to say that other than using the partitioning coefficients to account for sorption of contaminants from the porewater, no attempt is made to account for the chemical reactions in the soil-water system. In particular, speciation and complexations are not included in the structuring of the basic functions. Attempts have been made, (and are being made), to develop reactive fate and transport models. In the second type of models, we have geochemical models that pay attention to geochemical speciation equilibria between the various phases (solids, liquid, and gaseous) in the sub-surface setting. These include the dissolved and adsorbed elements in the various phases.

Assessment and prediction of the transport and fate of contaminants commonly rely on analytical and/or numerical (computer) models designed to take into account the various processes, site contamination situations, and properties of the contaminants and sub-surface materials. These models are useful for regulatory agencies in risk management and performance assessment of target sites and situations, for operators of landfills, and for those involved in site remediation (Figure 9.9). The quality of models, i.e., how accurately their predictions accord with real performance, depends on how well they represent the real problem situation and processes involved. This requires not only proper and accurate problem conceptualization, but also the capability to render these into the appropriate mathematical relationships.

Problem conceptualization includes the following:

(a) *Problem recognition.* This encompasses more than an adequate description of the site and interacting elements. To structure the output requirements for the model and the manner in which the results need to be expressed, knowledge of the end purpose (use) of the results is required. The three different areas of application of models shown at the top of the illustration shown in Figure 9.9 (*Risk, Design, Forensic*) will not have the same input and output requirements inasmuch as the decision-making process and the decisions required are different between them. Taking the problem of the contaminant plume emanating from the landfill, shown in the top left-hand corner of the figure, the differences between the outputs obtained from analysis (or prediction) of plume advance in respect to risk management, design of barrier systems, or determination of source-pathway-receptor are obvious.

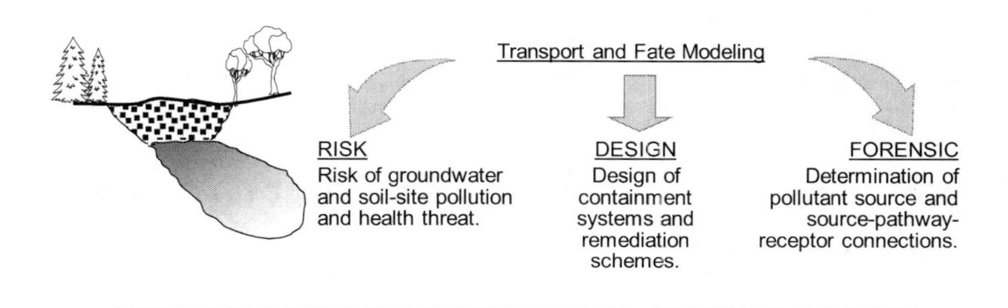

(1) Problem Conceptualization

Knowledge of problem and the various processes controlling persistence and fate of pollutants. Assumptions and simplifications.

(2) Model and Code Development

Analytical description of system. Conservation laws, continuity conditions, constitutive relationships, solution algorithms, numerical approximations.

(3) Calibration and Verification

Input data sets and parameters, reliabilities, uncertainties, relevance, parameter assignment, sensitivities, input-output flow, verifiable situations, benchmarking, analytical solutions.

(4) Validation

Availability of long term platform for validation of performance? 50 years? 100 years? Acceleration of chemical and biogeochemical processes in laboratory tests and simulations?

FIGURE 9.9 Principal objectives, issues, and requirements for contaminant transport and fate modelling.

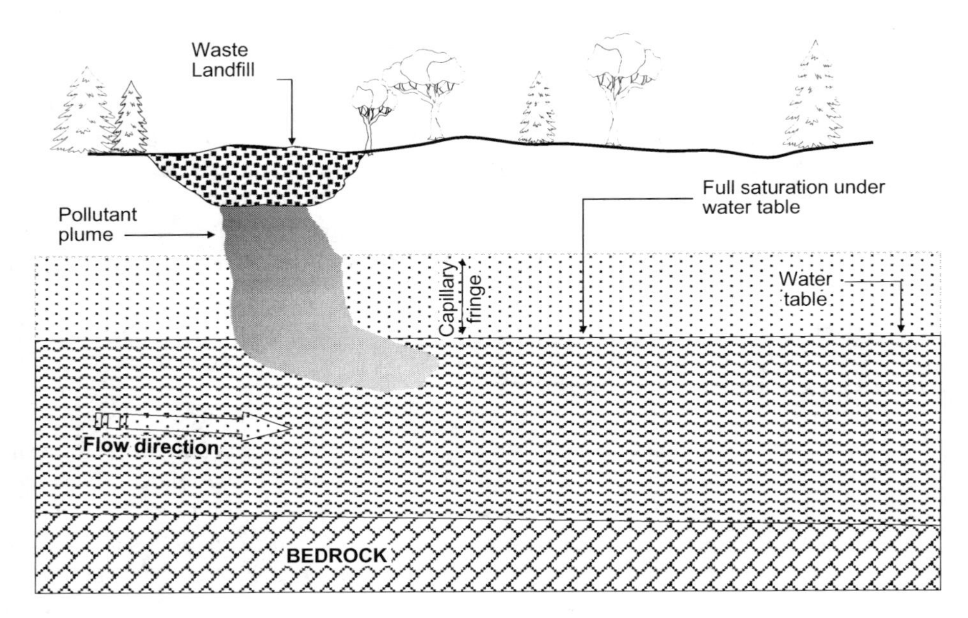

FIGURE 9.10 Schematic showing leachate plume as a contaminant plume spreading downwards toward the water table and spreading in the direction of flow of the groundwater.

(b) *Elements involved in the total system within the problem domain.* In the example shown in Figure 9.10, the contaminant plume emanating from the landfill in Figure 9.10 passes through the capillary fringe before meeting the saturated zone under the water table. The elements involved in the example problem include not only the interacting elements typified by the soil fractions, porewater and contaminants, and microorganisms, but also the problem setting – i.e., the capillary fringe and water-saturated zone.

(c) *Mechanisms and processes of interactions between participating elements, initial and boundary conditions, and output requirements.*

9.4.2 Basic Elements of Interactions between Dissolved Solutes and Soil Fractions

Interactions occurring between contaminants in the porewater, which are essentially dissolved solutes, and reactive soil particle surfaces are responsible for the transfer of these solutes from the porewater to the soil particle surfaces (partitioning). Molecular interactions governing sorption of contaminants are essentially electrostatic in nature. They are Coulombic interactions between nuclei and electrons. Of particular importance are the interatomic bonds, such as the ionic, covalent, hydrogen, and van der Waals. Ionic forces are Coulombic forces. These are forces between positively and negatively charged atoms and the bonds formed are called ionic or electrovalent bonds. The simplest example of ionic bonding is between a sodium atom and a chlorine atom – resulting in the formation of NaCl. The strength of the attractive forces, and hence the strength of the ionic bonds decrease as the square of the distance separating the atoms.

Another example of ionic bonding is the bond established between the oxygen from a water molecule to the oxygen on a clay particle's surface. This is due to the hydrogen atom, which can attract two electronegative atoms, and the ionic bond formed is called the *hydrogen bond.* In comparison to other bonds between neutral molecules, the hydrogen bond is a strong bond. Hydrogen bonding between two oxygen atoms, which are electronegative, is important in bonding layers of clay minerals together, in holding water at the clay surface, and in bonding organic molecules to clay particle surfaces.

van der Waals forces of attraction can be categorized into three components: (a) Keesom forces developed as a result of dipole orientation, (b) Debye forces developed due to induction, and (c) London dispersion forces. Adsorption of organic anions onto clay particle surfaces can be in the form of (a) anion associated directly with cation, or (b) anion associated with cation via a water bridge – referred to as a *cation bridge*. The process consists of replacement of a water molecule from the hydration shell of the exchangeable cation by an oxygen or an anionic group, e.g., carboxylate or phenate of the organic polymer. Hydrogen bonding to the oxygens of siloxane (mica-type) surfaces of clay particles are generally weak bonds. Adsorption of the organic anion is readily reversible by exchange with chloride or nitrate ions.

Cation exchange in soils refers to the exchange of positively charged ions associated with clay particle surfaces. The process is stoichiometric and electroneutrality at the clay particle surfaces must be satisfied. Cations will be attracted to the reactive soil particle surfaces in accordance with the relationship, $\frac{M_s}{N_s} = \frac{M_o}{N_o} = 1$, where M and N represent the cation species and the subscripts s and o represent the surface and the bulk solution. *Exchangeable cations* are cations that can be readily replaced by other cations of equal valence, or by two of one-half the valence of the original one. Thus, for example, if a clay containing sodium as an exchangeable cation is washed with a solution of calcium chloride, each calcium ion will replace two sodium ions, and the sodium can be washed out in the solution.

The quantity of exchangeable cations held by the soil is called the *cation-exchange capacity (CEC)* of the soil, and is expressed as milliequivalents per 100 g of soil (meq/100 g soil). One milliequivalent is equal to 6.023×10^{20} cation exchange sites in the soil. The CEC is a measure of the amount of negative sites associated with the soil fractions. The predominant exchangeable cations in soils are calcium and magnesium, with smaller amounts of potassium and sodium. The valence of cations plays a significant role in the exchange process. Higher valence cations will show greater replacing power. The higher the charge, the higher is its attraction to exchange sites. The converse also holds true, i.e., higher valence cations at the surfaces of clay particles will be more difficult to replace. The replacing power or the strength of attraction of cations the soil particle surfaces is given by the lytropic series. An example of some typical cations and the replacing power is given as follows:

$$Th^{4+} > Fe^{3+} > Al^{3+} > Cu^{2+} > Ba^{2+} > Ca^{2+} = Mg^{2+} > Cs^{+} > K^{+} = NH_4^{+} > Li^{+} > Na^{+}$$

Exchange-equilibrium equations can be used to determine the proportion of each exchangeable cation to the total cation-exchange capacity (CEC) as the outside ion concentration varies. The simplest of these is the Gapon relationship:

$$\frac{M_e^{+m}}{N_e^{+n}} = K \frac{\left[M_o^{+m}\right]^{\frac{1}{m}}}{\left[N_o^{+n}\right]^{\frac{1}{n}}} \tag{9.1}$$

where m and n refer to the valence of the cations and the subscripts e and o refer to the exchangeable and bulk solution ions. The constant K is dependent on the effects of specific cation adsorption and the nature of the clay surface. K decreases in value as the surface density of charges increases.

The adsorption of ions due to the mechanism of electrostatic bonding is called *physical adsorption* or *non-specific adsorption*. The ions involved in this type of process are identified as *indifferent ions*. The other mechanism of ion adsorption is a chemical reaction, which involves covalent bonds and activation energy in the process of adsorption. This type of adsorption process is generally identified as *chemisorption* or *specific adsorption*, and occurs at specific sites. *Specific cation adsorption* refers to the situation where the ions penetrate the coordination shell of the structural atom and are bonded by covalent bonds via *O* and *OH* groups to the structural cations. It is useful to note that when the energy barrier is overcome by the activation energy (in the chemisorption process), desorption of the ions will not be easily accomplished since desorption energy requirements may be prohibitively large. This has considerable significance in the evaluation of contaminant-soil interaction, especially with respect to the environmental mobility of sorbed contaminants.

Chemisorption and adsorption on soil particle surfaces involving siloxane cavities are generally confined to the surface layer and the monolayer next to the electrified interface. To obtain a better picture of the adsorption processes at the surface monolayer and beyond, we need to look more closely into the various energies of interaction developed. In short, the net energy of interaction due to adsorption of a solute ion or molecule onto the surfaces of the soil fractions is the result of both short-range chemical forces, such as covalent bonding, and long-range forces, such as electrostatic forces. Furthermore, sorption of inorganic contaminant cations is related to their valences, crystallinities, and hydrated radii.

9.4.3 Elements of Abiotic Reactions between Organic Chemicals and Soil Fractions

Abiotic adsorption reactions or processes involving organic chemicals and soil fractions are governed by (a) the surface properties of the soil fractions, (b) the chemistry of the porewater, and (c) the chemistry and physical-chemistry of the contaminants. Mechanisms pertaining to ion exchange involving organic ions are essentially similar to those between inorganic contaminants and soil fractions. The adsorption of the organic cations is related to the molecular weight of the organic cations. Large organic cations are adsorbed more strongly than inorganic cations by clays because they are longer and have higher molecular weights. Organic chemical compounds develop mechanisms of interactions at the molecular level that include the following:

- London-van der Waals forces – these have been described previously, and include the three types: Keesom, Deybe and London dispersion forces.
- Hydrophobic reactions and bonding – organic chemical molecules bond onto hydrophobic soil particle surfaces because this requires the least restructuring of the original water structure in the pore spaces of the soil. For soil organic matter (SOM), the fulvic acids are by and large hydrophilic. These kinds of organic matter have the least influence on the structuring of water. On the other hand, the SOM humins are highly hydrophobic. These have considerable influence on the restructuring of the water structure.
- Hydrogen bonding and charge transfer – hydrogen bonding is a special case of charge-transfer complex formation. These are complexes formed between electron-donor and electron-acceptor. The bonding between the aromatic groups in soil organic matter and organic chemicals is an example of charge transfer.
- Ligand and ion exchange – the bonding process requires that the organic chemical possesses a higher chelating capacity than the replaced ligand.
- Chemisorption – for soils and contaminants, the process of chemical adsorption involves chemical bonding between the contaminant molecule or ion in the porewater and the reactive soil particle surfaces. The process is sometimes called *specific adsorption* and the bonds are covalent bonds.

The functional groups for organic chemical compounds (organic chemical contaminants) are either acidic or basic. The characteristics and properties of these groups in organic molecules, such as shape, size, configuration, polarity, polarizability, and water solubility are important in the adsorption of the organic chemicals by the soil fractions. The various functional groups associated with

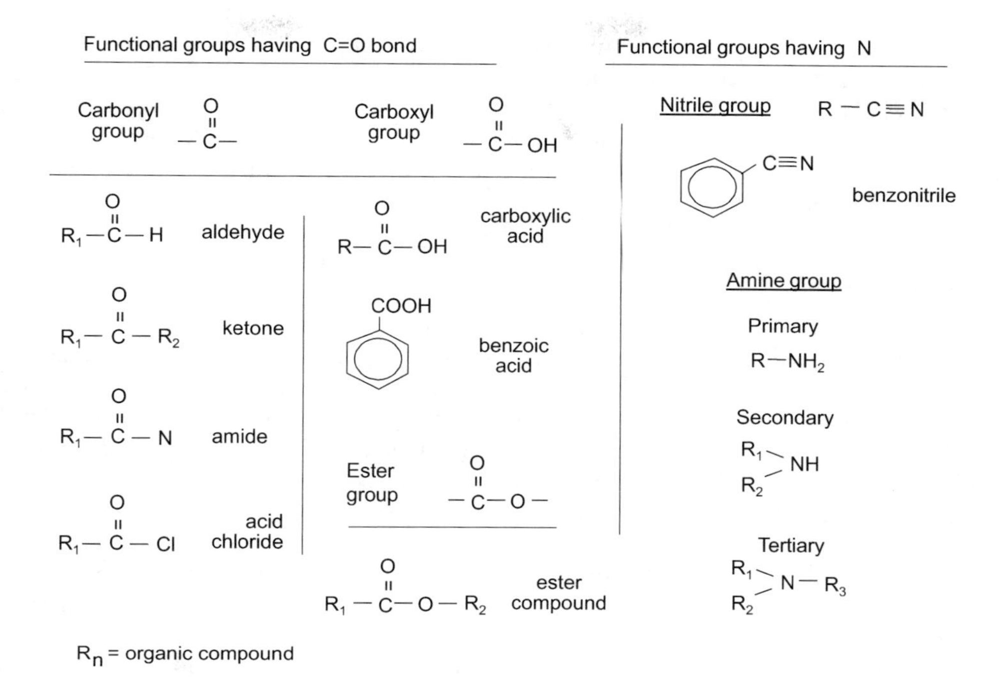

FIGURE 9.11 Some common functional groups for organic chemical contaminants.

organic chemical contaminants have been described in Section 2.5.2. A summary of some of these groups is shown in Figure 9.11. These include the hydroxyl group (alcohols and phenols), the carboxyl group (COOH), carbonyl (CO), and the amines (primary, secondary, and tertiary). The carbonyl group technically includes the carboxyl group, and is considered to be the most important functional group in organic chemistry. Most of the organic chemical contaminants found in the ground contain the carbonyl group. These chemicals are associated with production of pharmaceuticals, synthetic chemicals, and synthetic materials.

Not shown in the illustrations in Figure 9.11 are the hydroxy (OH) compounds. These are the compounds that contain the hydroxyl functional group. The two main groups are (a) aliphatic, and (b) aromatic. The aliphatic compounds are the alcohols and the aromatics are the phenols. Alcohols are hydroxyl alkyl compounds (R–OH), with a carbon atom bonded to the hydroxyl group. The more familiar ones are CH_3OH (methanol) and C_2H_5OH (ethanol). Adsorption of the hydroxyl groups of alcohol can be obtained through hydrogen bonding and cation-dipole interactions. Most primary aliphatic alcohols form single-layer complexes on the negatively charged surfaces of the soil fractions, with their alkyl chain lying parallel to the surfaces of the soil fractions. Phenols, on the other hand, are compounds that possess a hydroxyl group attached directly to an aromatic ring. As with the carbonyl functional group, the various hydroxy compounds are widely used and can be found in the manufacture of industrial products and pharmaceutical agents. Both alcohols and phenols can function as weak acids and weak bases.

Functional groups exert considerable influence on the characteristics of organic compounds, and contribute significantly to the processes that control accumulation, persistence and fate of the organic chemical compounds in soil. Organic chemicals with C=O bond functional groups and nitrogen-bonding functional groups (see Figure 9.11) are fixed or variable-charged organic chemical compounds. They can acquire a positive or negative charge through dissociation of H^+ from or onto the functional groups depending on the dissociation constant of each functional group and the pH of the soil-water system. An increase in the pH of the soil-water system will cause these functional groups (i.e. groups having a C=O bond) to dissociate. The outcome of the release of H^+ is a development of negative charges for the organic chemical compounds. Charge reversal (i.e., from positive to negative charges) could lead to the release of organic chemical contaminants held originally by cation bonding to the negatively-charged reactive surfaces of the soil particles. The phenomenon is a particular case of environmental mobility of previously sorbed contaminants.

9.4.4 Reactions in Porewater

Since contaminants consist of both inorganic and organic chemicals, it is more convenient to use the Brønsted-Lowry concepts of acids and bases to describe the various reactions and interactions occurring in a soil-water-contaminant system. In the Brønsted-Lowry concept, an *acid* is a substance that has a tendency to lose a proton (H^+), and conversely, a *base* is a substance that has a tendency to accept a proton. With this acid-base scheme, an *acid* is a *proton donor.* It is a *protogenic* substance. Similarly, a *base* is a *proton acceptor,* i.e., it is a *protophillic* substance. Water is both a *protophillic* and a *protogenic* solvent, i.e., it is *amphiprotic* in nature. It can act either as an acid or as a base. It can undergo self-ionization, resulting in the production of the conjugate base OH^- and conjugate acid H_3O^+. The self-ionization of water is called *autoprotolysis. Neutralization* is the reverse of autoprotolysis. Substances that have the capability to both donate and accept protons such as water and alcohols are called *amphiprotic* substances.

Chemical reactions in the porewater include (a) acid-base reactions and hydrolysis, (b) oxidation-reduction (redox) reactions, (c) speciation and complexations. Acid-base reactions and equilibrium in the porewater have important consequences on the partitioning and transport of contaminants in the soil. Acid-base reactions are *protolytic* reactions resulting from a process called *protolysis* – i.e., proton transfer between a proton donor (acid) and a proton acceptor (base). To assess the bonding and partitioning relationships between heavy metals and soil solids, it is useful to use the Lewis (1923) concept of acids and bases. This concept defines an acid as a substance that is capable of accepting a pair of electrons for bonding, and a base as a substance that is capable of donating a pair of electrons. This means that *Lewis acids* are electron acceptors, and *Lewis bases* are electron donors. All metal ions M^{nx} are Lewis acids. The Lewis acid-base concept permits us to treat metal-ligand bonding as acid-base reactions. Hydrated metal cations can act as acids or proton donors, with separate pk values for each. The dissociation constant k is a measure of the dissociation of a compound. This constant k is generally expressed in terms of the negative logarithm (to the base 10) of the dissociation constant, i.e., $pk = -\log(k)$. The smaller the pk value, the higher is the degree of ionic dissociation, and the more soluble is the substance. A comparison of the various pk values between compounds tells which compound would be more or less soluble in comparison to a target compound.

Oxidation-reduction reactions involve the transfer of electrons between the reactants, and the activity of the electron e^- in the chemical system plays a significant role. There is a link between redox reactions and acid-base reactions since the transfer of electrons in a redox reaction is accompanied by proton transfer. Redox reactions involving inorganic solutes result in a decrease or increase in the oxidation state of an atom. Organic chemical contaminants on the other hand show the effects of redox reactions through the gain or loss of electrons in the chemical. Biotic redox reactions are of greater significance than abiotic redox reactions. These reactions are significant factors in the processes that result in the transformation, persistence and fate of organic chemical compounds in soils.

The stability of inorganic solutes in the porewater is a function of such factors as pH, the presence of ligands, temperature, concentration of the inorganic solutes and the *Eh* or *pE* of the porewater. *Eh* is the redox potential and *pE* is a mathematical term that represents the negative logarithm of the electron activity e^-. The redox potential Eh is a measure of electron activity in the porewater, and is described by the following relationship:

$$Eh = E^0 + \left(\frac{RT}{nF}\right)\ln\frac{a_{i,ox}}{a_{i,red}} \tag{9.2}$$

where E^0 is the standard reference potential, n is the number of electrons, R is the gas constant, T is the absolute temperature, F is the Faraday constant, a_i is the activity of the ith species, and the subscripts ox and red refer to the oxidized and reduced ith species. At a temperature of 25 °C, the relationship between Eh and pE will be obtained as $Eh = 0.0591\ pE$. Figure 9.12 (from Yong, 2001) shows the pE-pH diagram for Fe and water with a maximum soluble Fe concentration of $10^{-5}M$. As can be seen, the valence state of the reactants is a function of the pH-pE status.

Speciation and complexations are central to the processes that control the fate of heavy metals in soils. Speciation refers to the formation of complexes between heavy metals and ligands in the aqueous phase (porewater). In a soil-water system, speciation provides competition between the ligands and the reactive soil solids for sorption of heavy metals. Various dissolved solutes in the porewater participate in the aqueous and surface complexation that are characteristic of the interactions between the solutes and the reactive soil particles. These interactions impact the predictions of contaminant transport directly, and especially modelling procedures, which rely on the use of simple partition coefficients. For example, Cl^- ions, sulphates, and organics can form complexes with heavy metals. The end result of this is seen as a lesser amount sorbed onto the soil particles – i.e., a lower adsorption isotherm performance, and a larger amount of the target heavy metal transported in the porewater. In other words, the environmental mobility of heavy metals is enhanced with speciation and complexation. Studies on Cd adsorption by kaolinite soil particles indicate that the Cd that were not adsorbed by the soil were in the form of $CdCl_2^0$, $CdCl_3^-$, and $CdCl_4^{2-}$ (Yong and Sheremata, 1991). The amount of Cd not adsorbed by the kaolinite soil particles in the presence of Cl^- was due to (a) a decrease in activity due to the presence of NaCl, (b) competition from Na^+ for adsorption sites, and

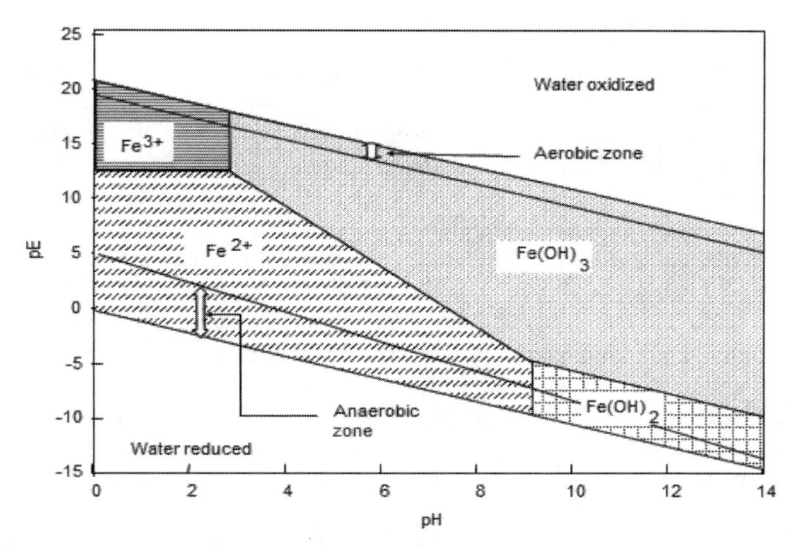

FIGURE 9.12 pE-pH chart for Fe and water with maximum soluble Fe concentrations of $10^{-5}\ M$. The zone sandwiched between the aerobic and anaerobic zones is the transition zone. (From Yong, 2001.)

(c) complexation of Cd^{2+} as negative and neutral chloride complexes. With increasing pH values, there is a tendency for Cd to be removed from the porewater as hydroxides, and if the pH is above the precipitation pH of Cd, precipitation of the Cd onto clay particle surfaces is likely. At pH values higher than the isoelectric point (iep), Cd is more likely to remain in solution in the presence of Cl^- than in the presence of ClO_4^-. The is the result of competition of Cl^- with OH^- for formation of complexes with the Cd^{2+} that were not sorbed by the kaolinite soil.

9.5 SURFACE COMPLEXATION AND PARTITIONING

As opposed to speciation and complexation in the aqueous phase (porewater), surface complexation refers to the complexes formed by the inorganic contaminants and the reactive sites on the soil particle surfaces. Surface complexation includes several mechanisms of solute-particle surface interaction, described broadly as sorption mechanisms. These have been discussed previously as non-specific adsorption, specific adsorption, and chemisorption. The processes involved have been described as Coulombic molecular interactions, with bonds formed that include ionic, covalent, and van der Waals.

The result of surface complexation is partitioning. We describe *partitioning* of contaminants as the transfer of contaminants in the porewater to the soil solids as a result of sorption mechanisms between the two. This is also called mass transfer (of contaminants). Section 2.5.3 has considered the partitioning of contaminants in a very general manner. Partitioning, as a phenomenon, includes the transfer of both inorganic and organic chemical contaminants. This is an important phenomenon since this is the outcome of one of the fundamental processes that determines the persistence and fate of contaminants. In the case of organic chemicals, the processes involved include London-van der Waals forces, hydrophobic reactions, hydrogen bonding and charge transfer, ligand and ion exchange, and chemisorption. In this section, we will be examining partitioning of both inorganic contaminants and organic chemicals in greater detail. Whilst there may be some similarity in mass transfer mechanisms in the partitioning of inorganic solutes and organic chemicals, it is generally more convenient to consider these separately.

9.5.1 Partitioning of Inorganic Contaminants

A popular measure of partitioning of inorganic and organic chemical contaminants is the partition coefficient k_p. Chapter 2 has given a very brief description of the general types of partition coefficients. In brief, partition coefficients describe the relationship between the amount of contaminants transferred onto soil particles (sorbed by the soil particles) and the equilibrium concentration of the same contaminants remaining in the porewater (Figure 9.13). The popular relationships such as Langmuir and Freundlich are shown in Figure 2.15 in Chapter 2. To distinguish between partition coefficients obtained using different laboratory techniques, the term *distribution coefficient* k_d is often used to denote partitioning of contaminants obtained with batch equilibrium adsorption isotherm procedures. By and large, the distribution coefficient is the partition coefficient most commonly used to describe partitioning of heavy metals and other inorganic solutes.

What is the significance of partitioning? It is necessary to bear in mind that *partitioning* is the result of mass transfer of contaminants from the porewater onto the soil particles. How one determines the results of mass transfer (i.e., partitioning) can be a contentious issue, and can severely affect one's prediction of the transport and fate of contaminants under consideration. A quantitative determination of partitioning, such as the distribution coefficient k_d, is needed in many mathematical relationships structured to evaluate the fate of contaminants transporting in a soil system. For example, Equation (2.2) in Chapter 2 uses the distribution coefficient k_d as a key parameter in the relationship used to predict the transport of contaminants in a saturated soil.

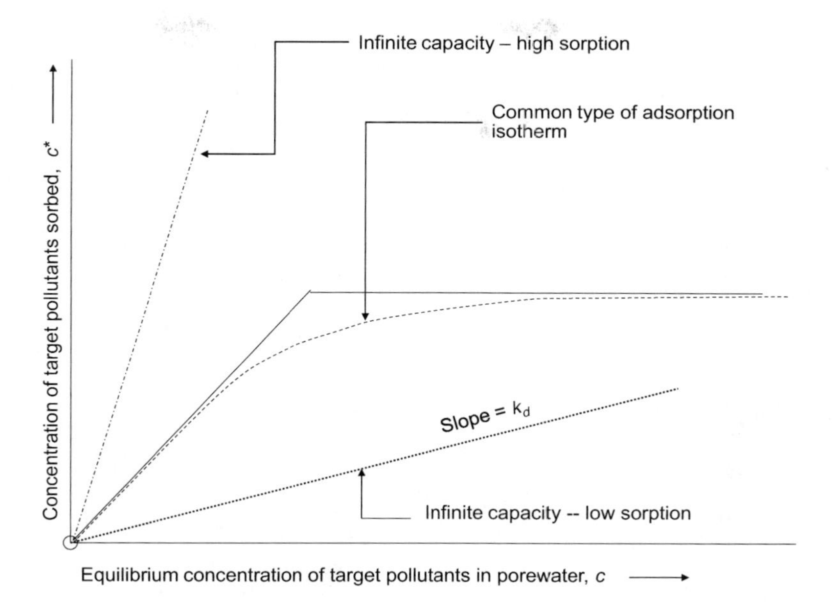

FIGURE 9.13 Illustration of partitioning of target contaminants between contaminants in porewater and contaminants sorbed by soil particles. Note that unless limits are placed on the maximum sorption capacity of the soils, specification of constant k_d values means infinite sorption capacity by the soil particles in the soil-water system.

There are at least two broad issues regarding the determination and use of the distribution coefficient k_d, namely, (a) types of tests used to provide information for determination of k_d, and (b) range of applicability of k_d in transport and fate predictions. We will discuss the former in this section and leave the latter discussion for a later section where the problem of prediction of transport and fate of contaminants is addressed. Laboratory tests used to provide information on the mass transfer of contaminants from the porewater onto soil solids are the most expedient means to provide one with information on the partitioning of contaminants. By and large, these tests provide only the end result of the mass transfer, and not direct information on the basic mechanisms responsible for partitioning.

The distribution coefficient k_d is determined from information gained using batch equilibrium tests on soil solutions. Ratios of 10 parts or 20 parts solution to one part soil are generally used, and the candidate or target contaminant is part of the aqueous phase of the soil solution. In many laboratory test procedures, the candidate soil is used in the soil solution, and the candidate or target contaminant is generally a laboratory-prepared contaminant, e.g., $PbNO_3$ for assessment of sorption of Pb as a contaminant heavy metal. Since the soil particles are in a highly dispersed state in the soil solution, one would expect that all the surfaces of all the particles are available for interaction with the target contaminant in the aqueous phase of the soil solution. By using multiple batches of soil solution where the concentration of the target contaminant is varied, and by determining the concentration of contaminants sorbed onto the soil solids and remaining in the aqueous phase, one will obtain characteristic adsorption isotherm curves such as those shown in Figures 9.13 and 2.15. The slope of the adsorption isotherm defines k_d.

Consider the illustration shown in Figure 9.14. This schematic compares the loose structure or dispersed state of soil particles in a soil suspension (bottom illustration) with an aggregation of a multitude of particles constituting a microstructural unit that is representative of a soil structure in a natural soil sub-surface environment (top illustration). The schematic shows that whilst the two

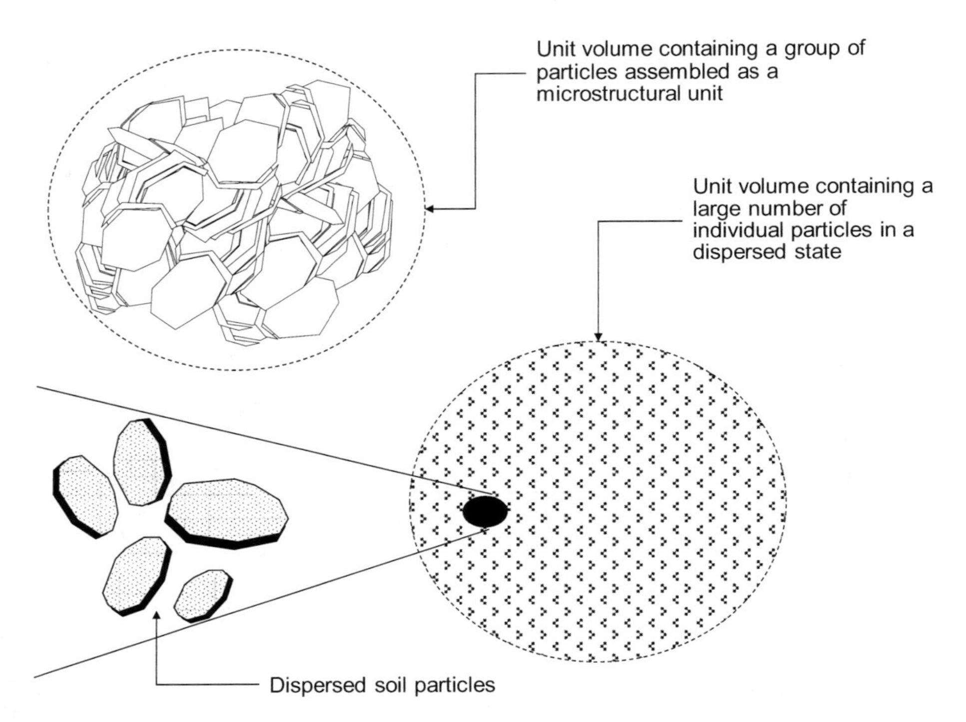

FIGURE 9.14 Illustration of influence of soil structure on availability of exposed reactive soil particle surfaces for sorption of contaminants. The dispersed particles are typical of soil suspensions used for batch equilibrium test for determination of adsorption isotherms, whilst the microstructural unit is typical of soil structures in subsurface soils.

identical unit volumes do not necessarily contain the same number of particles, a relatively small proportion of the reactive particle surfaces in the microstructural unit is available for interaction with the contaminants in the porewater.

Distribution coefficients k_d obtained from adsorption isotherms using the batch equilibrium with soil solutions and prepared target contaminants are very useful in that they define the upper limit of partitioning of the target contaminant. Problems or pitfalls arising from the application of k_d values reported in the literature for use in models to predict actual fate and transport of contaminants in the natural subsoil can be traced to the following:

- Inappropriate use of the coefficient, i.e., using the reported k_d value to represent partitioning effects in a natural compact subsoil. This can arise from a lack of appreciation or knowledge of the particulars of the batch equilibrium tests. If the tests were conducted to provide the upper limit of partitioning, as has been described, the model would over-predict sorption and there under-predict transport.
- Major differences in the composition of the leachate and contaminants in the leachate plume being modelled. Unless the batch equilibrium tests were conducted with actual leachates from the site under consideration, it is inappropriate to use results from single-species contaminant tests to represent multi-species behaviour. In any event, it needs to be remembered that the k_d values would be upper limit values. Competition for sorption sites, preferential sorption, and speciation-complexation are some of the major factors that would directly affect the nature of the adsorption isotherm obtained.

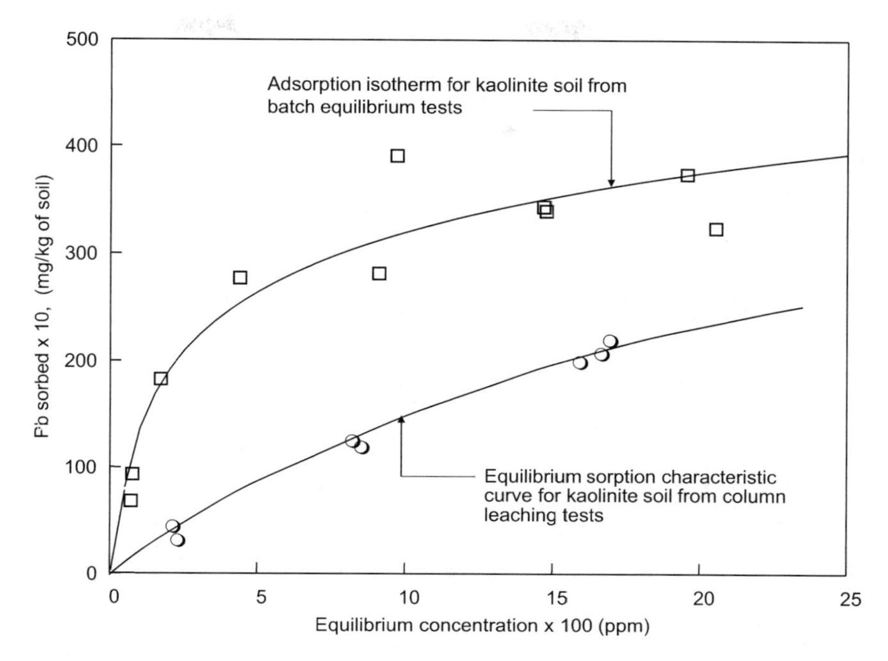

FIGURE 9.15 Comparison of Pb sorption curves obtained from batch equilibrium and column leaching tests for kaolinite soil. (From Yong, 2001.)

For assessment of partitioning using soils in their natural compact state, it is necessary to conduct column-leaching or cell-diffusion tests. In these kinds of tests, the natural soil is used in the test cell or column, and either laboratory-prepared candidate contaminants or natural leachates are used. The partition coefficient deduced from the test results is not the distribution coefficient identified with the adsorption isotherms obtained from batch equilibrium tests. Instead, the partition coefficients obtained from column-leaching or cell-diffusion tests need to be properly differentiated from the traditional k_d. Yong (2001) has suggested that these partition coefficients be called *sorption coefficients* – to reflect the sorption performance of the soils in their natural state in the column or cell. The disadvantages in conducting column-leaching and cell-diffusion tests are (a) the greater amount of effort required to conduct the tests, (b) the much greater length of time taken to obtain an entire suite of results, and (c) inability to obtain exact replicate soil structures in the companion columns or cells. The results indicate that the characteristic curves obtained from column-leaching tests, for example, are much lower than corresponding adsorption isotherms. Figure 9.15 gives an example.

9.5.2 Organic Chemical Contaminants

The partitioning of organic chemical contaminants is a function of several kinds of interacting mechanisms between the organic chemicals and the soil solids in the natural soil-water system that constitutes the sub-soil. A key factor in the development of the kinds of interaction mechanisms is the type or class of organic chemicals. The degree of water miscibility of the organic chemical appears to be a key element. A good example of this is the difference between non-aqueous phase liquids (NAPLs) and water miscible alcohols. The family of NAPLs include those that are denser and lighter than water. The DNAPLs (dense NAPLs) include the organohalides and oxygen-containing organic compounds, and the LNAPLs (light NAPLs) include gasoline, heating oil, kerosene, and aviation

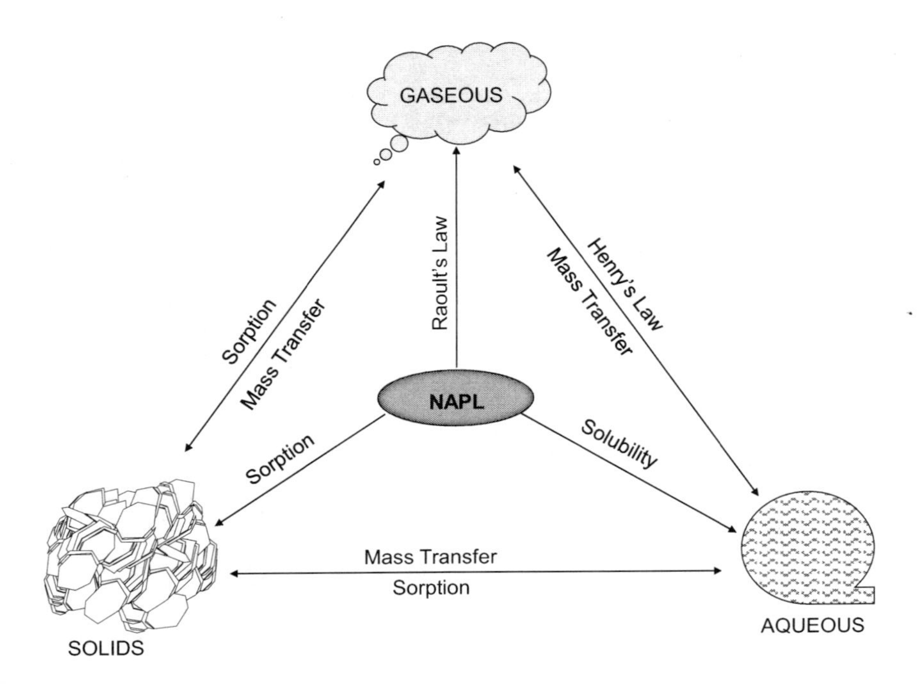

FIGURE 9.16 Processes involved in partitioning and fate of NAPLs.

fuel. Most NAPLs are partially miscible in water. Consideration of the transport of NAPLs in the saturated zone requires attention to two classes of substances: (a) miscible or dissolved substances, and (b) immiscible substances.

The basic processes involved in transport and fate of NAPLs are demonstrated in Figure 9.16. The chemical properties that affect NAPL transport and fate include (1) volatility, (2) relative polarity, (3) affinity for soil organic matter or organic contaminants, and (4) density and viscosity. The higher the vapour pressure of the substance, the more likely it is to evaporate. Movement in the vapour phase is generally by advection. At equilibrium between NAPLs and the vapour phase, the equilibrium partial pressure of a component is directly related to the mole fraction and the pure constituent vapour pressure – as described by Raoult's law. Designating P_i as the partial pressure of the constituent, x_i as the mole fraction of the constituent, and P_i^0 as the vapour pressure of the pure constituent, Raoult's law states that when equilibrium conditions are obtained, and when the mole fraction of a constituent is greater than 0.9, $P_i = x_i P_i^0$.

As Figure 9.16 shows, an organic chemical compound in the soil may be partitioned between the soil water, soil air, and the soil constituents. The rate of volatilization of an organic molecule from an adsorption site on the solid phase in the soil (or in solution in the soil water) to the vapour phase in soil air and then to the atmosphere is dependent on many physical and chemical properties of both the chemical and the soil, and on the process involved in moving from one phase to another. The three main distribution or transport processes involved are

(a) compound in soil ↔ compound in solution,
(b) compound in solution ↔ compound in vapour phase in soil air,
(c) compound in vapour phase in soil air → compound in atmosphere.

Partitioning of a chemical among the three phases can be estimated from either vapour-phase or solution-phase desorption isotherms. The process by which a compound evaporates in the vapour phase to the atmosphere from another environmental compartment is defined as volatilization.

This process is responsible for the loss of chemicals from the soil to the air, and is one of the factors involved in the persistence of an organic chemical. Determination of volatilization of a chemical from the soil to the air is most often achieved using theoretical descriptions of the physical process of volatilization based on Raoult's law and Henry's law. The rate at which a chemical volatilizes from soil is affected by soil and chemical properties, and environmental conditions. Some of the properties of a chemical involved in volatilization are its vapour pressure, solubility in water, basic structural type and the number, nature and position of its basic functional groups. Other factors affecting volatilization rate include adsorption, vapour density and water content of the soil in the sub-surface.

Adsorption impacts directly on the chemical activity by reducing it to values below that of the pure compound. In turn, this affects the vapour density and the volatilization rate since vapour density is directly related to the volatilization rate. Vapour density is the concentration of a chemical in the air, the maximum concentration being a saturated vapour. The role of water content is seen in terms of competition for adsorption sites on the soil. Displacement of non-polar and weakly polar compounds by water molecules can occur because of preferential sorption (of water). Hydrates, i.e., hydration layer on the soil particle surfaces, will increase the vapour density of weakly polar compounds. If dehydration occurs, the compound sorbs onto the dry soil particles. This means that the chances for volatilization of the organic chemical compound are better when hydrates are present.

When a vapour is in equilibrium with its solution in some other solvent, the equilibrium partial pressure of a constituent is directly related to the mole fraction of the constituent in the aqueous phase. Once again, designating P_i as the partial pressure of the constituent, X_i as the mole fraction of the constituent in the aqueous phase, and H_i as Henry's constant for the constituent, Henry's law states that $P_i = H_iX_i$. By and large, so long as the activity coefficients remain relatively constant, the concentrations of any single molecular species in two phases in equilibrium with each other will show a constant ratio to each other. This assumes ideal behaviour in water and the absence of significant solute-solute interactions and also absence of strong specific solute-solvent interactions.

Partitioning of organic chemicals is most often described by the partition coefficient k_{ow}. This is the octanol-water partition coefficient and has been widely adopted in studies of the environmental fate of organic chemicals. The octanol-water partition coefficient is sometimes known as the *equilibrium partition coefficient,* i.e., coefficient pertaining to the ratio of the concentration of a specific organic contaminant in other solvents to that in water. Results of countless studies have shown that this coefficient is well correlated to water solubilities of most organic chemicals. Since *n*-octanol is part lipophilic and part hydrophilic (i.e., it is amphiphilic), it has the capability to accommodate organic chemicals with the various kinds of functional groups. The dissolution of *n*-octanol in water is roughly eight octanol molecules to 100,000 water molecules in an aqueous phase. This represents a ratio of about one to twelve thousand (Schwarzenbach et al., 1993). Since water-saturated *n*-octanol has a molar volume of 0.121 L/mol as compared to 0.16 L/mol for pure *n*-octanol, the close similarity permits one to ignore the effect of the water volume on the molar volume of the organic phase in experiments conducted to determine the octanol-water equilibrium partition coefficient. The octanol-water partition coefficient k_{ow} has been found to be sufficiently correlated not only to water solubility, but also to soil sorption coefficients. In the experimental measurements reported, the octanol is considered to be the surrogate for soil organic matter.

Organic chemicals with k_{ow} values less than 10 are considered to be relatively hydrophilic – with high water solubilities and small soil adsorption coefficients. Organic chemicals with k_{ow} values greater than 10^4 are considered to be very hydrophobic and are not very water-soluble. Chiou et al. (1982) has provided a relationship between k_{ow} and water solubility S as follows:

$$\log k_{ow} = 4.5 - 0.75 \log S \text{ (ppm)} \tag{9.3}$$

Aqueous concentrations of hydrophobic organics such as polyaromatic hydrocarbons, PAH, in natural soil-water systems are highly dependent on adsorption/desorption equilibrium with sorbents present in the systems. Studies of compounds which included normal PAHs, nitrogen and sulphur heterocyclic PAHs, and some substituted aromatic compounds suggest that the sorption of hydrophobic molecules (benzidine excepted) is governed by the organic content of the substrate. The dominant mechanism of organic adsorption is the hydrophobic bond established between a chemical and natural organic matter in the soil. The extent of sorption can be reasonably estimated if the organic carbon content of the soil is known (Karickhoff, 1984) by using the expression: $k_p = k_{oc} f_{oc}$, where f_{oc} is the organic carbon content of the soil organic matter, k_{oc} is the organic content coefficient, and k_p is the linear Freundlich isotherm obtained for the target organic chemical. This approach works reasonably well in the case of high organic contents (e.g., $f_{oc} > 0.001$). Relationships reported in the literature relating k_{ow} to k_{oc} show that these can be grouped into certain types of organic chemicals. For PAHs, the relationship given by Karickhoff et al. (1979) is

$$\log k_{oc} = \log k_{ow} - 0.21 \tag{9.4}$$

For pesticides, Rao and Davidson (1980) report that

$$\log k_{oc} = 1.029 \log k_{ow} - 0.18 \tag{9.5}$$

For chlorinated and methylated benzenes, the relationship given by Schwarzenbach and Westall (1981) is:

$$\log k_{oc} = 0.72 \log k_{ow} + 0.49 \tag{9.6}$$

The graphical relationship shown in Figure 9.17 uses some representative values reported in the various handbooks (e.g., Verscheuren, 1983; Montgomery and Welkom, 1991) for log k_{ow} and log

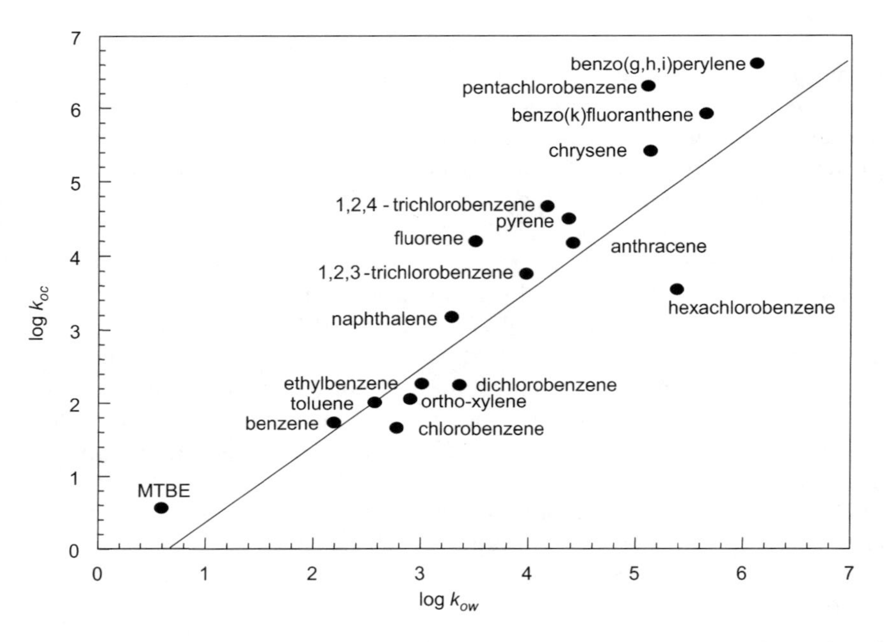

FIGURE 9.17 Relationship of log k_{oc} with log k_{ow} for several organic compounds. (Adapted from Yong and Mulligan, 2019.)

k_{oc}. The values used for log k_{ow} are essentially mid-range results reported in the handbooks and in many studies. Not all log k_{oc} values are obtained as measured values. Many of these have been obtained through application of the various log k_{oc}-log k_{ow} relationships reported in the literature, e.g., Kenaga and Goring (1980) and Karickhoff et al. (1979). The linear relationship shown by the solid line in Figure 9.17 is given as

$$\log k_{oc} = 1.06 \log k_{ow} - 0.68 \tag{9.7}$$

This graphical relationship is useful in the sense of partitioning of the organic chemical compounds shown in the diagram. Yong and Mulligan (2019) have discussed some of the pertinent correlations, stating, for example, in regard to the k_{oc} values shown in Figure 9.17 for dichlorobenzene, that they indicate that it partitions well to sediments, and particularly to the organic fractions (SOM, soil organic matter). Because of its resistance to anaerobic degradation, it is very persistent.

9.6 PERSISTENCE AND FATE

One of the many necessary actions needed along the path towards land environment sustainability is to return contaminated land to its uncontaminated state. This requires one to clean up contaminated sites. The problem of concern in remediation of contaminated sites is the persistence of contaminants in the site. In many instances, natural attenuation can be relied upon as a remediation tool – a sort of natural self-remediation process that is intrinsic to the site properties – as will be discussed in a later chapter. There exists however, several kinds of contaminants that are not easily "self-remediated" with natural attenuation processes. These contaminants fall under the general class of persistent pollutants. The term *persistence* has been defined as "*the continued presence of a contaminant pollutant in the substrate*". The persistence of inorganic and organic contaminants differs in respect to meaning and application. The persistence of heavy metals refers to their continued presence in the subsurface soil regime in any of their individual oxidation states and in any of the complexes formed. Organic chemical contaminants, on the other hand, can undergo considerable transformations because of micro-environmental factors. We define *transformation* to mean the conversion of the original organic chemical contaminant into one or more resultant products by processes which can be abiotic, biotic, or a combination of these. The intermediate products obtained from transformation of organic chemical compounds by biotic processes along the pathway towards complete mineralization are generally classified as degradation products. Transformed products resulting from abiotic processes in general do not classify as being intermediate products along the path to mineralization. They are, however, not easily distinguished because some of the abiotic transformed products themselves may become more susceptible to biotic transformations. When this occurs, the process is known as a combination transformation process.

A characteristic term used to describe organic chemicals that persist in their original form or in altered forms is *persistent organic chemical pollutants*, *POPs* (persistent organic pollutants). These include dioxins, furans, pesticides and insecticides, polycyclic aromatic hydrocarbons (PAHs), and halogenated hydrocarbons. The persistence of organic chemical contaminants in soils is a function of at least three factors: (a) the physico-chemical properties of the organic chemical contaminant itself, (b) the physico-chemical properties of the soil, and (c) the microorganisms in the soil. Resultant abiotic reactions and transformations are sensitive to factors (a) and (b), and all factors are important participants in the dynamic processes associated with the activities of the microorganisms in the biologically mediated chemical reactions and transformation processes.

9.6.1 Biotransformation and Degradation of Organic Chemicals and Heavy Metals

The various types of organisms and microorganisms responsible for the biotransformation (including degradation) of organic chemical compounds can be classified under the Whittaker

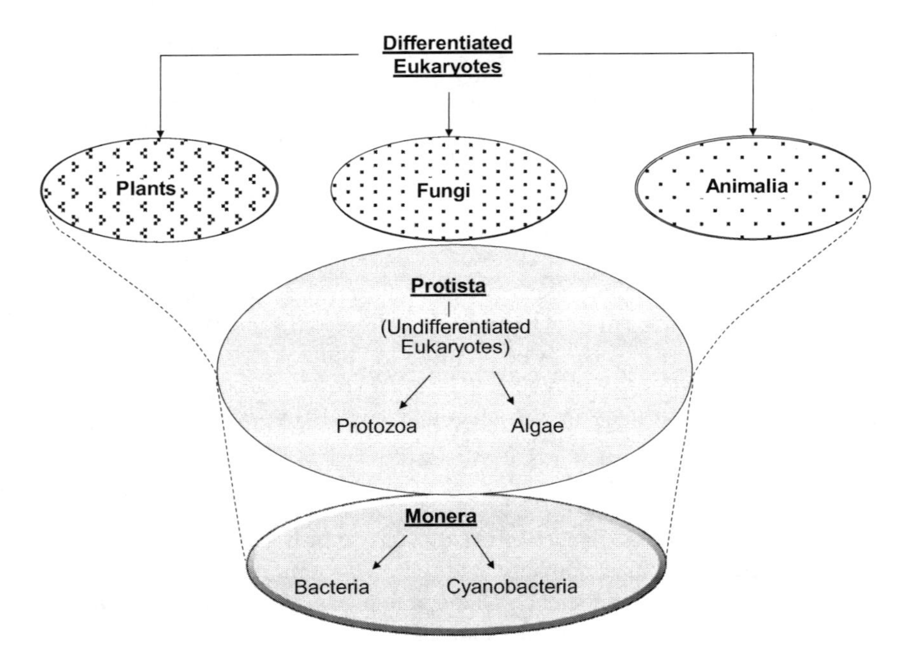

FIGURE 9.18 Organisms and microorganisms participating in natural bioremediation processes grouped according to the Whittaker 5-kingdom hierarchial system. (From Yong and Mulligan, 2019.)

(1969) five-kingdom classification scheme shown in Figure 9.18. The reader should consult the regular textbooks on microbiology for detailed descriptions of these.
The descriptions given by Yong and Mulligan (2019) are summarized as follows:

- Protozoa include pseudopods, flagellates, amoebas, ciliates, and parasitic protozoa. Their sizes can vary from 1 to 2000 mm. They are aerobic, single-celled chemoheterotrophs, and are eukaryotes with no cell walls. They are divided into four main groups: (the Mastigophora which are flagellate protozoans, (b) the Sarcodina which are amoeboid, (c) the Ciliophone which are ciliated, and (d) the Sporozoa which are parasites of vertebrates and invertebrates.
- Fungi are aerobic, multicellular, eukaryotes and chemoheterotrophs that require organic compounds for energy and carbon. They reproduce by formation of asexual spores. In comparison to bacteria, they (a) do not require as much nitrogen, (b) are more sensitive to changes in moisture levels, (c) are larger, (d) grow more slowly, and (e) can grow in a more acidic pH range (less than pH 5). Fungi mainly live in the soil or on dead plants and are sometimes found in fresh water.
- Algae are single-celled and multi-cellular microorganisms that are green, greenish tan to golden brown, yellow to golden brown (marine), or red (marine). They grow in the soil and on trees or in fresh or salt water. Those that grow with fungi are called lichens. Seaweeds and kelps are examples of algae. Since they are photosynthetic, they can produce oxygen, new cells from carbon dioxide or bicarbonate (HCO_3^-), and dissolved nutrients including nitrogen and phosphorus. They use light of wavelengths between 300 and 700 nm. Red tides are indicative of excessive growth of dinoflagellates in the sea. The green color in a body of lakes and rivers is eutrophication due to the accumulation of nutrients such as fertilizers in the water.
- Although viruses are smaller than bacteria and require a living cell to reproduce, their relationship to other organisms is not clear. In order for them to replicate, they have to invade various

kinds of cells. They consist of one strand of DNA and one strand of ribonucleic acid (RNA) within a protein coat. A virus can only attack a specific host. For example, those that attack bacteria are called bacteriophages.

- The most significant animals in the soil are millimetre-sized worms. Nematodes are cylindrical in shape and are able to move within bacterial flocs. Flatworms such as tapeworms, eel worms, roundworms, and threadworms, which are nematodes, can cause diseases such as roundworm, hookworm and filarisis.
- Bacteria are prokaryotes that reproduce by binary fission by dividing into two cells, in about 20 minutes. The time it takes for one cell to double, however, depends on the temperature and species. For example, the optimal doubling time for *Bacillus subtilis* (37°C) is 24 minutes and for *Nitrobacter agilis* (27°C) is 20 hours. Classification is by shape, such as the rod-shaped bacillus, the spherical-shaped coccus and the spiral-shaped spirillium. Rods usually have diameters of one-half to one micron and lengths of 3 to 5 microns. The diameter of spherical cells varies from 0.2 to 2 microns. Spiral-shaped cells range from 0.3 to 5 microns in diameter and 6 to 15 microns in length. The cells grow in clusters, chains or in single form and may or may not be motile. The substrate of the bacteria must be soluble. In most cases, classification is according to the genus and species (e.g., *Pseudomonas aeruginosa* and *Bacillus subtilis)*. Some of the most common species are *Pseudomonas, Arthrobacter, Bacillus, Acinetobacter, Micrococcus, Vibrio, Achromobacter, Brevibacterium, Flavobacterium* and *Corynebacterium*. Within each species, we will have various strains. Each can behave differently. Some strains can survive in certain conditions that others cannot. The ones that are better adapted will survive. For survival, strains, called mutants, originate due to problems in the genetic copying mechanisms. Some species are dependent on other species, though, for survival. Degradation of chemicals to an intermediate stage by one species of bacteria may be required for the growth of another species that utilizes the intermediate.

9.6.1.1 Alkanes, Alkenes, and Cycloalkanes

Alkanes, alkenes, and cycloalkanes, amongst others (PAHs, asphaltenes etc.) are components of petroleum hydrocarbons (PHCs). Low molecular weight alkanes are most easily degraded by microorganisms. As the chain length increases from C_{20} to $C_{40,}$ hydrophobicity increases and both solubility and biodegradation rates decrease. Alkenes with a double bond on the first carbon may be more easily degradable than those alkenes with the double bond at other positions (Pitter and Chudoba, 1990). Cycloalkanes are not as degradable as alkanes due to their cyclic structure, and their biodegradability decreases as the number of rings increase.

9.6.1.2 Polycyclic, Polynuclear Aromatic Hydrocarbons (PAHs)

As with cycloalkanes, the compounds become more difficult to degrade as the number of rings of PAHs increases. This is due to decreasing volatility and solubility and increased sorption properties of these compounds. They are degraded one ring at a time in a manner similar to single ring aromatics.

9.6.1.3 Benzene, Toluene, Ethylbenzene, and Xylene – BTEX

Benzene, toluene, ethylbenzene, and xylene (BTEX) are volatile, water soluble components of gasoline. Aromatic compounds with benzene structures are more difficult to degrade than cycloalkanes. Aerobic degradation of all components of BTEX occurs rapidly when oxygen is present. Aromatic compounds can also be degraded under anaerobic conditions to phenols or organic acids to fatty acids to methane and carbon dioxide (Grbic-Galic, 1990). Degradation is less assured and is slower than under aerobic conditions.

9.6.1.4 Methyl Tert-Butyl Ether – MTBE

Methyl tert-butyl ether (MTBE), which is an additive to gasoline, is highly resistant to biodegradation. It is reactive with microbial membranes.

9.6.1.5 Halogenated Aliphatic and Aromatic Compounds

Halogenated aliphatic compounds are pesticides such as ethylene dibromide (DBR) or $CHCl_3$, $CHCl_2Br$, and industrial solvents including methylene chloride and trichloroethylene. Halogenated aromatic compounds are also pesticides, and they include such pesticides as DDT, 2,4-D and 2,4,5-T, plasticizers, pentachlorophenol, and polychlorinated biphenyls. The presence of halogen makes aerobic degradation of the halogenated aliphatic compounds difficult to achieve – due to the lower energy and the higher oxidation state of the compound. Anaerobic biodegradation is easier to achieve. This is particularly true when the number of halogens in the compound increases – making aerobic degradation more difficult. In the case of the halogenated aromatic compounds, the mechanisms of conversions include hydrolysis (replacement of halogen with hydroxyl group), reductive dehalogenation (replacement of halogen with hydrogen), and oxidation (introduction of oxygen into the ring causing removal of halogen).

PCBs and dioxins are very stable in the environment. A review by Urbaniak (2013) examined the physical and chemical properties of dioxins and PCBs and how they can be broken down in the environment. Microbiological transformation is discussed with reference to aerobic, anaerobic, and sequential anaerobic-aerobic conditions. Physical transformations include photochemical and thermal degradation. Different phytoremediation processes also can affect environmental degradation of dioxins and PCBs.

9.6.1.6 Heavy Metals

Microbial cells can accumulate heavy metals through ion exchange, precipitation, and complexation on and within the cell surface containing hydroxyl, carboxyl, and phosphate groups. Processes involving bacterial oxidation-reduction will alter the mobility of the heavy metal contaminants in soil. An example of this is the reduction of Cr(IV) in the form of chromate (CrO_4^{2-}) and dichromate ($Cr_2O_7^{2-}$) to Cr(III). Conversion can also be indirect by microbial production of Fe(II), sulphide and other components that reduce chromium. Oxidation of selenium in the four naturally occurring major species of selenium [selenite (SeO_3^{2-}, IV), selenate (SeO_4^{2-}, VI), elemental selenium (Se (0)) and selenide (-II)] can occur under aerobic conditions. Transformation of selenate can occur anaerobically to selenide or elemental selenium, and methylation of selenium detoxifies selenium for the bacteria by removing the selenium from the bacteria. Some microorganisms (bacteria and fungi) and natural processes can transform mercury in the environment from one form to another, usually under anaerobic (oxygen-deficient) conditions. The most common organic mercury compound that these microorganisms and natural processes generate from other forms is methylmercury.

Per- and polyfluoroalkyl substances (PFAS). Perfluorooctanesulfonic acid (PFOS) and perfluorooctanoic acid (PFOA) compounds are used as firefighting foams, cleaners, nonstick surfaces, and food packages. Over time, various health impacts have originated from exposure to these compounds, such as increased risk of cancers and diseases to liver and kidneys. They are highly water soluble and very low in volatility (ATSDR, 2009). Long-chain perfluorocarboxylic acids (PFCs) have low vapour pressures (Environment Canada, 2010). These compounds are very stable and resistant to biodegradation, atmospheric photooxidation, direct photolysis, and hydrolysis (OECD, 2002).

The fluorine-carbon bonds of PFAS make them very persistent in the environment (Buck et al., 2011; ASTSWMO, 2015; Jia et al., 2019). Soil contamination with PFAS can occur at firefighting training sites, composts, farms, and parks with biosolids addition and landfills. The microbial transformation of polyfluorinated precursors yield perfluorinated compounds in aerobic environments (Liu and Avendaño, 2013; Dasu and Lee, 2016). While some studies have found that perfluorinated compounds such as PFOS and PFOA are resistant to microbial degradation transformation in aerobic environments. Fewer studies under anaerobic conditions have been performed (Liu and Avendaño, 2013).

9.6.1.7 Micro and Nanoplastics

Plastics have gained attention recently as they are abundant, difficult to biodegrade and of high toxicity. They are found frequently in marine environments in fragments smaller than 5 mm as described in Chapter 8. Light, UV radiation, and mechanical weathering cause the fragmentation. Much more research has been performed in aquatic environments and only in the last decade or so, has microplastic pollution been found as a potential health issue in soils (Rilling, 2012). The behaviour of soils has thus been much less studied than in water (He et al., 2018).

Sources of microplastics are discharges from water treatment plants, waste disposal, compost, sludge, and manure addition to soils and irrigation water. For example, in Canada it has been reported that microplastic removal is 98% after secondary treatment (Gies et al. 2018) while in Shanghai, removal rates are much lower (62.6%) (Jia et al., 2019). Recycling rates of plastic are still relatively low and thus disposal to the environment is common. While the microplastics can be transported in the soils, they can accumulate in the soil (He et al., 2018).

In a review by Yang et al. (2021), it was indicated that microplastics can change the soil pH electrical conductivity and can affect the decomposition of soil organic matter. Other pollutants can sorb onto the microplastics, such as heavy metals, antibiotics, and pesticides. These processes can enhance the mobility of the pollutants in the soil and potentially their bioavailability. The microplastics can be adsorbed into crops or microorganisms or soil organisms such as earthworms, thus changing the soil biodiversity.

The microplastics can be transported through irrigation practices, rainfall, and bioturbation. Aggregation of the microplastics affects the rate of transport (Dong et al., 2022). The ionic strength of pore water, other colloids and organic matter, and the microplastic properties (density, size, shape, functional groups, surface properties) can have significant influence on the aggregation.

In summary, significant research is needed to understand the fate of transport of nano and microplastics. Dong et al. (2022) summarized that more field work is needed, the transformation during transport needs to be better understood, including secondary transport of other contaminants, the different forms of the microplastics needs to be considered and fate and transport models need to be developed.

9.7 PREDICTION OF TRANSPORT AND FATE OF CONTAMINANTS

A key factor in the decision-making process in structuring sustainability objectives for the land environment is to have knowledge of the movement and spread (transport) of contaminants in the ground. Apart from procurement of field information required to delineate the parameters of the contamination problem at hand, this requires development of techniques for prediction of the transport and fate of the contaminants in the ground. The main elements in mass transport and mass transfer of contaminants in a soil-water system in the ground are shown in Figure 9.19.

Prediction of the transport and fate of contaminants in the sub-soil requires consideration and incorporation of these elements in the analytical and mathematical analyses. Mass transport refers to transport of the dissolved solutes by advective, diffusive and dispersion forces. Mass transfer of contaminants in the soil refers to chemical mass transfer processes. These have been discussed in the previous sections. They include sorption, dissolution and precipitation, acid-base reactions and hydrolysis, oxidation-reduction (redox) reactions, speciation-complexation, and biologically mediated transfer. For aspects of design, containment of high level nuclear waste, where prediction of the transport of radionuclides is of particular concern, radioactive decay is one of the important chemical transfer processes that need to be considered. Finally, biological mediated transfer of

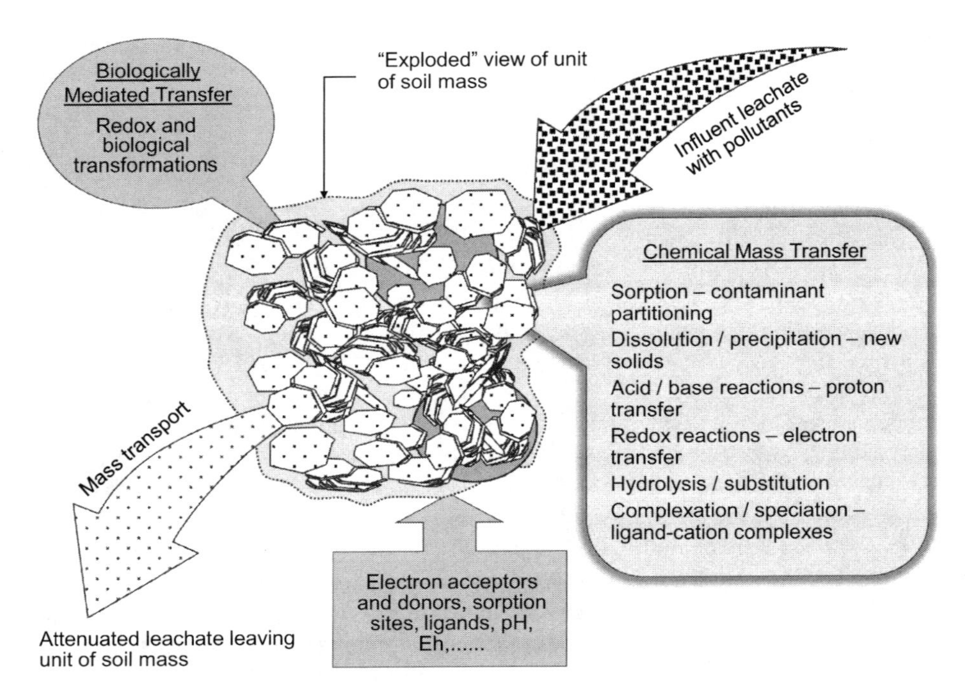

FIGURE 9.19 Elements of basic mass transport and mass transfer in attenuation of contaminants in leachate transport through a soil element.

contaminants completes the three main categories of contaminant transfer from porewater to soil solids. Transformation and degradation of the organic chemical contaminants require as much attention as sorption and transport of the organic chemicals.

Many of the chemical mass transfer processes are kinetic processes, and reactions are not instantaneous. However, in a majority of the analytical considerations, most of the reactions are considered as instantaneous and chemical equilibrium is immediately obtained. This leads to modelling of transport of contaminants as a non-reactive process – an outcome that is not always appreciated by users of developed models. Consideration of transport and fate as a reactive process requires incorporation of the many transfer mechanisms that are time dependent, and also the transformations and degradations of the chemical. This is not easily accomplished. In the recent attempts to incorporate the reactive processes, incorporation of geochemical models into the "standard" transport models has been attempted – with differing degrees of complexity and success.

9.7.1 Mass Transport

In the conventional treatment of mass transport of contaminants, the three mechanisms for mass transport include advection, diffusion, and dispersion. Advection refers to the flux generated by the hydraulic gradient and is given in terms of the advective velocity v. transport of dissolved solutes in the porewater solely by advective means will progress in concert with the advective velocity. In situations where the solutes possess kinetic energy and demonstrate Brownian activity, diffusion of these solutes will occur. Using the porewater as the carrier, the solutes will combine their diffusive capability with advective velocity to produce the combined mass transport. This permits the dissolved solutes to move ahead of the advective front, i.e., the general tendency is for the diffusion front to precede the advection front. In situations where tortuosity and pore size differences combine with pore restrictions to create local mixing in the movement of the dissolved solutes, dispersion results. The degree of dispersion and the resultant effect on mass transport is not readily quantified.

The diffusive movement of a particular solute (contaminant) is characterized by its diffusion coefficient. This diffusion coefficient D_c is most often considered as being equivalent or equal to the effective molecular diffusion coefficient. In dilute solutions of a single ionic species, the diffusion coefficient of that single species is termed as the *infinite solution diffusion coefficient* D_o. The infinite solution diffusion coefficients are dependent on such factors as ionic radius, absolute mobility of the ion, temperature, viscosity of the fluid medium, valence of the ion, equivalent limiting conductivity of the ion, etc. A useful listing of these coefficients for a range of solutes and for various sets of conditions can be found in many basic handbooks and other references (e.g., Li and Gregory, 1974; Lerman, 1979). From a theoretical point of view, the studies of molecular diffusion by Nernst (1888) and Einstein (1905) show the level of complex interdependencies that combine to produce the resultant coefficient obtained. From studies dealing with the movement of suspended particles controlled by the osmotic forces in the solution, the three expressions most often cited, are

$$\text{Nernst-Einstein} \quad D_o = \frac{uRT}{N} = uk'T \tag{9.8}$$

$$\text{Einstein-Stokes} \quad D_o = \frac{RT}{6\pi N \eta r} = 7.166 \times 10^{-21} \frac{T}{\eta r} \tag{9.9}$$

$$\text{Nernst} \quad D_o = \frac{RT\lambda^o}{F^2 |z|} = 8.928 \times 10^{-10} \frac{T\lambda^o}{|z|} \tag{9.10}$$

where D_o is the diffusion coefficient in an infinite solution, u = absolute mobility of the solute under consideration, R = universal gas constant, T = absolute temperature, N = Avogadro's number, k' = Boltzmann's constant, λ^o = conductivity of the target ion or solute, r = radius of the hydrated ion or solute, η = absolute viscosity of the fluid, z = valence of the ion, and F = Faraday's constant. A large listing of experimental values for λ^o for major ions can be found in Robinson and Stokes (1959).

We define a dimensionless Peclet number as $P_e = v_L\, d/D_o$, where v_L is the longitudinal flow velocity (advective flow). From the information reported by Perkins and Johnston (1963), it is seen that for Peclet numbers less than one, ($P_e<1$), diffusive transport of the contaminant solutes in a contaminant plume travels faster than the advective flow of water. For $P_e > 10$, advective flow constitutes the dominant flow mechanism for the movement of solutes. In between the values of 1 and 10, there is a gradual change from diffusion-dominant to advection-dominant transport (Figure 9.20). The longitudinal diffusion coefficient D_L consists of both the molecular diffusion coefficient D_m and the hydrodynamic (mechanical) dispersion coefficient D_h. This is written as: $D_L = D_m + D_h = D_o\tau + \alpha v$ where D_m = molecular diffusion = $D_o\tau$; $D_h = \alpha v$ and α = dispersivity parameter, τ = tortuosity factor.

The tortuosity factor is introduced to modify the infinite solution diffusion coefficient to acknowledge that we do not have an infinite solution, and that diffusion of a single solute species in a soil-water system is subject to constricting pore volumes and non-linear paths. Figure 9.20 shows that in the diffusion-dominant transport region, we can safely neglect the v_L term since v_L is vanishingly small. Under those circumstances, the diffusion-dominant transport region, we will have $D_L = D_o\tau$. In the advection-dominant transport region, if we consider that diffusion transport is negligible, then $D_L = v_L$. In the transition region, the relationship for D_L will be given as: $D_L = D_o\tau + \alpha v_L$.

The significance of a correct choice or specification of a diffusion coefficient cannot be overstated. Figure 9.21 is a schematic illustration showing the variation of D (or D_L) coefficients calculated using Equation (2.2) in Section 2.5.4 of Chapter 2, and using the concentration profiles shown at

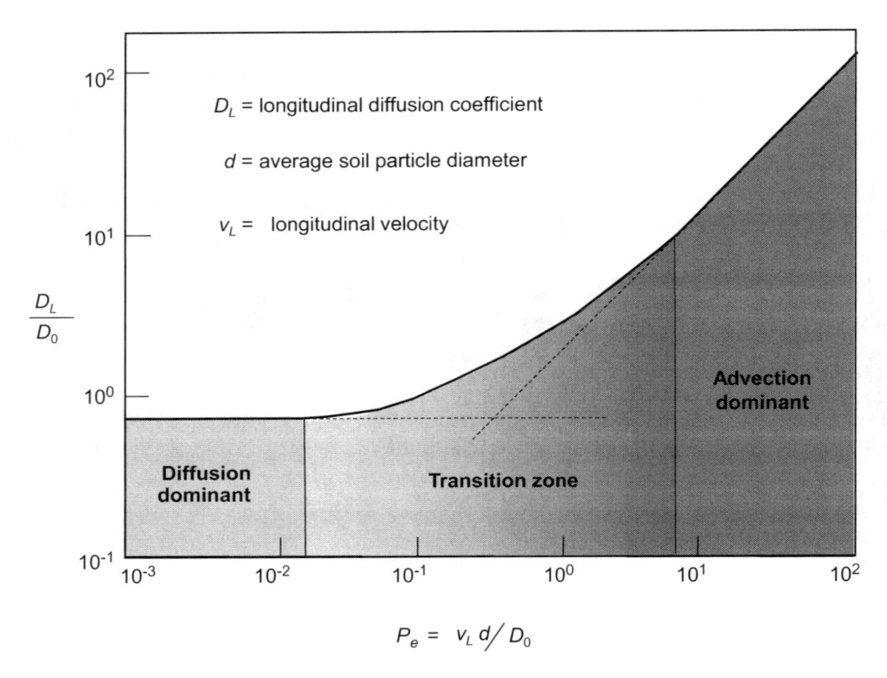

FIGURE 9.20 Diffusion and advection dominant flow regions for solutes in relation to Peclet number. (Adapted from Perkins and Johnston, 1963.)

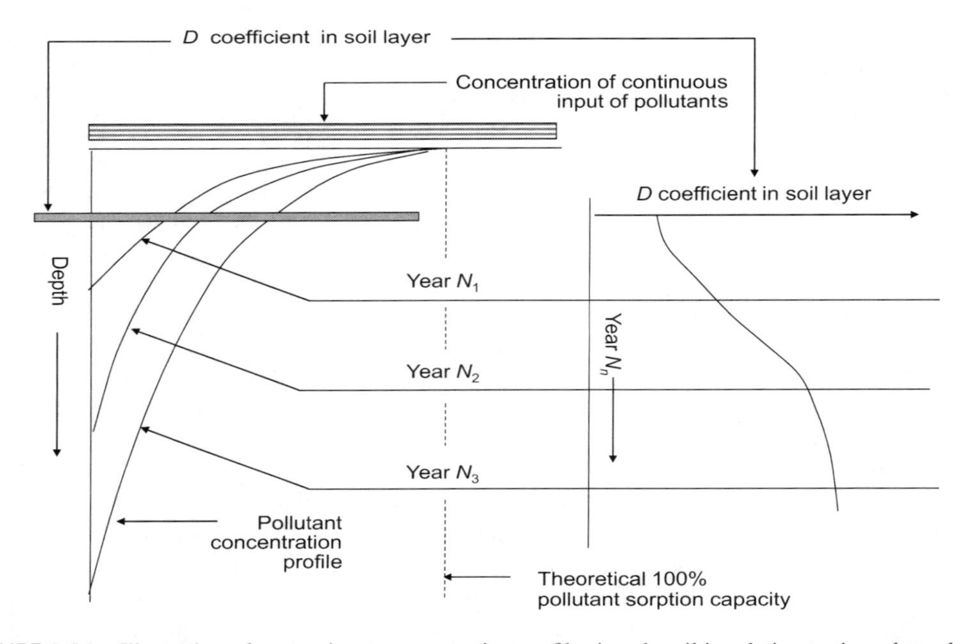

FIGURE 9.21 Illustration of contaminant concentration profiles in sub-soil in relation to time elapsed given as years N_n. Diffusion coefficient profile (right-hand curve) is obtained from calculations reflecting changes in contaminant concentration with time.

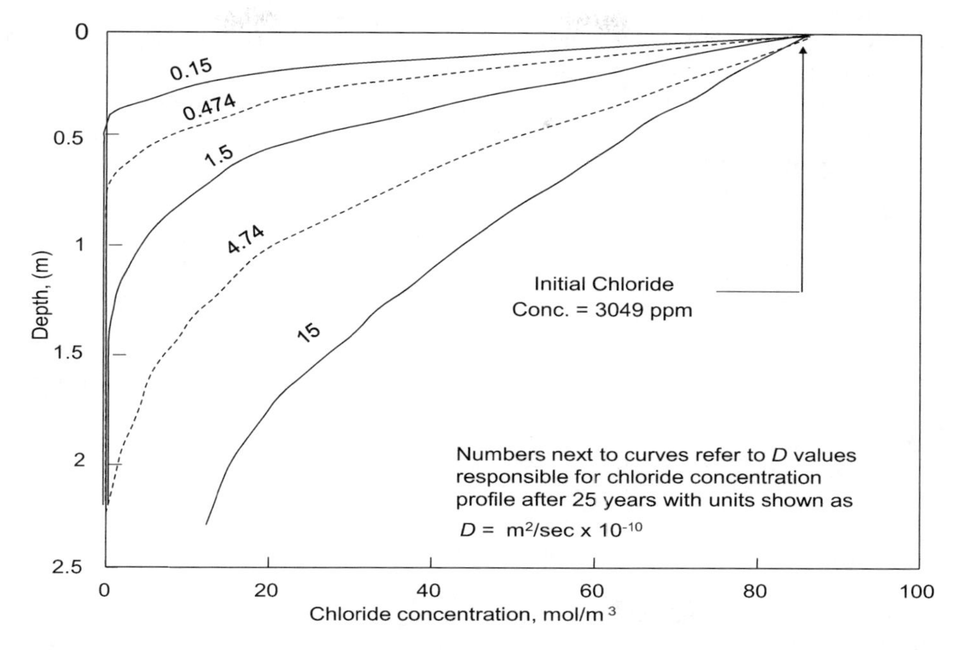

FIGURE 9.22 Effect of different values of coefficient of diffusion D on chloride concentration profile after 25 years of continuous input of chloride at 3049 ppm.

the left-hand side of the diagram. The Ogata and Banks (1961) solution of the transport equation similar to the one shown in Equation (2.2) for an initial chloride concentration of 3049 ppm as the input source (Figure 9.22), shows the different chloride concentration profiles obtained in relation to variations in the D value used in the calculations. The differences are not insignificant.

9.7.2 Transport Prediction

For contaminants that can be partitioned in the transport process in the sub-soil system, it is not uncommon to use transport relationship given as Equation (2.2) in Chapter 2. For some situations, such as equilibrium-partitioning processes, this is an adequate method for predicting the transport and fate of contaminants that can be partitioned. The assumption is generally made that the rate of reactions is independent of the concentration of contaminants – i.e., a zero-rate reaction process. However, for many other situations relating to contaminants that are partitioned during transport, such as non-equilibrium partitioning, this relationship needs to be knowledgeably applied and perhaps modified to meet the conditions of partitioning. The relationship given as Equation (2.2) in Chapter 2 can be written in its original expanded form as

$$\frac{\partial c}{\partial t} = D_L \frac{\partial^2 c}{\partial x^2} - v \frac{\partial c}{\partial x} - \frac{\rho}{n\rho_w} \frac{\partial c^*}{\partial t} \tag{9.11}$$

where c = concentration of contaminants of concern, t = time, D_L = diffusion coefficient, v = advective velocity, x = spatial coordinate, ρ = bulk density of soil media, ρ_w = density of water, n = porosity of soil media, and c^* = concentration of contaminants adsorbed by soil fractions (see ordinate in graph shown in Figure 2.15 in Chapter 2). We recall that the adsorption isotherms portrayed in

Figures 2.14 and 2.15 are derived from batch equilibrium tests with soil solutions, and we further recall the discussion in the previous Section 9.5.1 and Figure 9.15 that the distribution coefficient k_d obtained from the adsorption isotherms refers directly to a maximum reactive surface reaction process. Equation (2.2) is obtained when c^* is assumed to be equal to $k_d c$, i.e., if a linear adsorption isotherm is assumed. Substituting for c^* in Equation (9.11) gives us

$$\frac{\partial c}{\partial t} = D_L \frac{\partial^2 x}{\partial x^2} - v\frac{\partial c}{\partial x} - \frac{\rho}{n\rho_w}\frac{\partial (k_d c)}{\partial x} \tag{9.12}$$

Collecting terms and defining R as the *retardation* $= \left[1 + \frac{\rho}{n\rho_w}k_d\right]$, the equation previously seen as Equation (2.2) is obtained:

$$R\frac{\partial c}{\partial t} = D_L \frac{\partial^2 c}{\partial x^2} - v\frac{\partial c}{\partial x} \tag{9.12}$$

If the partition isotherms obtained from actual laboratory tests do not show linearity, the obvious solution is to use the proper function that describes c^*, e.g., the Freundlich or Langmuir (see Figure 2.15) or some other equivalent relationship. In this instance, the opportunity to properly reflect the partitioning process through a compact soil mass should be taken. Instead of using the non-linear adsorption isotherm obtained from batch equilibrium tests on soil solutions, the sorption relationship obtained from column leaching tests through compact soil should be used. This has been discussed in Section 9.5.1 and illustrated in terms of adsorption curves' differences in Figure 9.15.

Equations of the type shown as Equations (2.2) or (9.12) have been described as non-reactive transport relationships. This description has been applied to such equations to reflect the observation that biotic and abiotic chemical reactions in the aqueous phase have not been factored into the structuring of the relationship. The chemical reactions discussed in Section 9.4 cannot be ignored since they not only compete with the soil solids for partitioning of the contaminants, but they may also transform the contaminants – especially the organic chemicals. The reactions and transformations will not only change the character of the contaminants, but they will also change the distribution of contaminants in the zone of interest. The problem is magnified in field situations because of multi-species contaminants and a mixture of both inorganic and organic chemicals. Unlike laboratory leaching column and batch equilibrium tests, real field situations provide one with a complex mix of contaminants transporting through an equally complex subsoil system.

9.7.2.1 Chemical Reactions and Transport Predictions

To meet the objectives of sustainability of the land environment, proper prediction of transport and fate of contaminants requires knowledge of how the abiotic and biotic reactions affect the long term health of the terrain system – especially the sub-soil system. From the myriad of possibilities in handling the complex problem of chemical reactions and reaction rates, and transformations, there exist at least four simple procedures that provide some accounting, to a greater or lesser degree, of the various processes controlling transport. These include (a) the addition of a reaction term r_c in the commonly-used advection-diffusion equation given as Equations (2.2) or (9.12), (b) accounting for the contaminant adsorption-desorption process, (c) use of first or second-order or higher-order reaction rates, and (d) combining transport models with geochemical speciation models. None of these appear to handle biotransformations and their resultant effect on the transport and fate processes.

Addition of a reaction term r_c to Equation (9.12) is perhaps the most common method used to accommodate a kinetic approach to fate and transport modelling. The resultant formulation is a linearly additive term to Equation (9.6) as follows:

$$R\frac{\partial c}{\partial t} = D_L\frac{\partial^2 c}{\partial x^2} - v\frac{\partial c}{\partial x} + r_c \tag{9.13}$$

The last term in Equation (9.13) can be expressed in the form of a general rate law as follows:

$$r_c = -k\vartheta|A|^a\,|B|^b \tag{9.14}$$

where r_c in this case is the rate of increase in concentration of a contaminant of species A, k is the rate coefficient, ϑ represents the volume of fluid under consideration, A and B are the reactant species, and a and b are the reaction orders.

The use of an adsorption-desorption approach to fate and transport modelling recognizes that in field situations, desorption (or displacement) occurs as part of the ion exchange process. Determination of the transport and fate of contaminants using the partitioning approach typified by the advection-diffusion relationship with adsorption-desorption consideration requires that one can write a relationship for c^* in Equation (9.11). Curve fitting procedures are commonly used to deduce information obtained from batch equilibrium and/or flow-through (leaching column) tests. The Freundlich and Langmuir curves for example (see Figure 2.15) are specific cases of such procedures.

Prediction and modelling for biotransformation and biodegradation and their effects on fate and transport require a different approach. Yong and Mulligan (2019) have provided an accounting of some of the more popular analytical-computer models used in application of natural attenuation schemes. These in effect are fate and transport models since biotransformation and biodegradation are the primary attenuation processes – a principal feature in fate and transport of organic chemicals. For example, the analytical model BIOSCREEN (Newell et al., 1996) developed for the Air Forces Centre for Environmental Excellence by Groundwater Services Inc. (Houston, TX) assumes a declining source concentration with transport and biodegradation processes for the soluble hydrocarbons that include advection, dispersion, adsorption, aerobic, and anaerobic degradation. Most of the available analytical-computer models developed to handle biotransformation and biodegradation in fate and transport modelling have the essential items contained in BIOSCREEN (www.epa.gov/water-research/bioscreen-natural-attenuation-decision-support-system). The principal distinguishing factors between the available computational packages such as BIOPLUME III (www.epa.gov/water-research/bioplume-iii) (Rifai et al., 1997), MODFLOW and RT3D (Sun et al., 1996), BIOREDOX (Carey et al., 1998, http://porewater.com/software_bioredox.html), and BIOCHLOR (Aziz et al., 2000, www.epa.gov/water-research/biochlor-natural-attenuation-decision-support-system), include (a) structure of the outputs, (b) manner in which degradation is handled, such as order of degradation and degradation rates, (c) types of organic chemicals, (d) inclusion of heavy metals and some other inorganics, (e) availability and types of electron acceptors, and (f) adsorption-desorption. MT3D (www.usgs.gov/software/mt3d-usgs-groundwater-solute-transport-simulator-modflow) is a groundwater solute transport model for complex transient and steady-state flows, anisotropic dispersion, first-order decay and production reactions to include da, and sorption (linear and nonlinear) that works with MODFLOW (Bedekar et al. 2016) MODFLOW 6 is the latest version (www.usgs.gov/software/modflow-6-usgs-modular-hydrologic-model) and can translate other versions including MODFLOW-2005, MODFLOW-NWT, or MODFLOW-LGR (Version 2) model.

Footprint is based on the Domenico model (1987), which was released in 2008 by the EPA (Martin-Hayden and Robbins, 1997; USEPA, 2008). It can be used to estimate the extent of a plume using either a zero-order or first-order decay rate or can simulate the biodegradation of BTEX and/or ethanol. The Remediation Evaluation Model for Chlorinated Solvents (REMChlor) released in 2007 is used to simulate the first-order sequential decay and production of several species of chlorinated solvents (www.epa.gov/water-research/remediation-evaluation-model-chlorinated-solvents-remchlor). It is applicable for simulating contaminants in groundwater source and remediation whereas REMFuel, released in 2012, is applicable for fuel hydrocarbons (www.epa.gov/water-research/remediation-evaluation-model-fuel-hydrocarbons-remfuel).

Solution of the transport relationships shown as Equations (9.12) and (9.13) and other similar relationships can be achieved using analytical or numerical techniques. For well-defined geometries, initial and boundary conditions, and processes, analytical techniques provide exact solutions that can further insight into the processes involved in the problem under consideration. Numerical techniques such as finite difference, finite element and boundary element are useful and are perhaps the techniques favoured by many because of their capability to handle more complex geometries and variations in material properties and boundary conditions.

9.7.3 Geochemical Speciation and Transport Predictions

Abiotic reactions and transformations, together with the biotic counterparts, form the suite of processes that are involved in the transport and fate of contaminants in the subsoil. The reactions between the chemical species in the porewater and also with the reactive soil particle surfaces discussed in the previous sections and chapters constitute the basic platform. Because individual chemical species have the ability to participate in several types of reactions, the equations to describe the various equilibrium reactions can become complicated – particularly since one needs to be assured that all the reactions are captured.

Geochemical modelling provides a useful means for handling the many kinds of calculations required to solve the various equilibrium reactions. Specific requirements are a robust thermodynamic database and simultaneous solution of the thermodynamic and mass balance equations. Appelo and Postma (1993) provide a comprehensive treatment of the various processes and reactions, together with a user guide for the geochemical model PHREEQE developed by Parkhurst et al. (1980). As with many of the popular models, the model is an aqueous model based upon ion-pairing, and includes elements and both aqueous species and mineral phases (fractions).

Other available models include the commonly used MINTEQ (Felmy et al., 1984) and the more recent MINTEQA2 version 4.03 released in 2006 that includes PROFEFA2 (Allison et al., 1991, (www.epa.gov/ceam/minteqa2-equilibrium-speciation-model), a preprocessing package for developing input files, GEOCHEM (Sposito and Mattigod, 1980), HYDROGEOCHEM (Yeh and Tripathi, 1990), and WATEQF (Plummer et al., 1976). GEOCHEM-EZ (Shaff et al., 2010) is a multi-functional chemical speciation program, designed to replace GEOCHEM-PC. HYDROGEOCHEM 2 is the newer version of HYDROGEOCHEM V1.0 (Yeh and Tripathi,1990). The modification includes replacement of the EQMOD chemical equilibrium subroutines by a mixed chemical kinetic and equilibrium model (KEMOD) to deal with species whose concentrations are controlled by either thermodynamics or kinetics. HYDROGEOCHEM 2 is a coupled model of hydrologic transport and geochemical reaction in saturated-unsaturated media. HYDROGEOCHEM 6.1 is a more recent series of models that is used for simulation of geothermal systems, radioactive waste and sequestration of CO_2 (Yeh and Tsai, 2015) . The newer version of the chemical speciation program WATEQF is WATEQ4F which is maintained by the Chemical Modeling and Thermodynamic Data Evaluation Project of the USGS and mainly applicable for large numbers of water analyses (Ball and Nordstrom, 2001) (www.brr.cr.usgs.gov/projects/GWC_chemtherm/software.htm).

PHREEQC version 3 is now available (www.usgs.gov/software/phreeqc-version-3/) works with various models such as the Lawrence Livermore National Laboratory model and WATEQ4F), a Pitzer specific-ion-interaction aqueous model, and the SIT (specific ion interaction theory) aqueous model. The PHREEQC model can perform

(1) speciation and saturation-index calculations,
(2) batch-reaction and one-dimensional (1D) transport calculations,
(3) inverse modelling.

By and large, most of the geochemical codes assume instantaneous equilibrium, i.e., kinetic reactions are not included in the calculations. In part, this is because reactions such as oxidation-reduction, precipitation-dissolution, substitution-hydrolysis and to some extent, speciation-complexation, can be relatively slow. To overcome this, some of the models have been able to provide analyses that point towards possible trends and final equilibria. The code EQ3/6 (Delaney et al., 1986) does however provide for consideration of dissolution-precipitation reactions. Transformations however are essentially not handled by most of the codes, although more models are including them.

9.8 CONCLUDING REMARKS

The concern in this chapter is for land environment sustainability, as it pertains to the effects of anthropogenic discharges of wastes in the land environment, have focused attention on developing concepts that consider the natural capital of land environment, and have clearly specified the objectives for sustainability: "*To ensure that each natural capital component maintains its full and uncompromised functioning capability without loss of growth potential*".

For the objectives to be fulfilled, actions, reactions, and management techniques require specification of *indicators* that mark the path towards sustainability of the natural capital. Whilst we recognize that absolute sustainability is not generally attainable, one should nevertheless adopt and implement strategies and technological/engineering means that point towards sustainability objectives.

The first half of this chapter has examined the nature of indicators and has distinguished between system status and material performance indicators. This distinction is necessary since various situations demand a proper accounting of the relationship between the two. Figure 9.4 provides the protocols that assist in this type of accounting.

Contaminants from anthropogenic activities are perhaps the largest sources and types of geoenvironment stressors. The manner in which they are handled in respect to land environment impact will, to a large extent, greatly dictate whether sustainability or "near sustainability" of the land environment and/or its natural capital can be achieved. The impact of these (contaminants) and the implementation of indicators as a technique for assessment need proper consideration. Figure 9.5 provides the sustainability goals in respect to waste discharge onto the land environment, and the subsequent figures provide examples. To some extent, we can learn how to prescribe the necessary indicators to achieve the objective. Obviously, real field situations are both site specific and industry specific.

In the assessment of waste impacts, it is clear that this cannot be achieved without an understanding of both the interactions with the sub-soil system and the goals of sustainability. Central to the various issues is the problem of gaining a proper knowledge of the health status of the sub-soil system. This requires one to be able to predict the transport and fate of pollutants in the subsoil system. The problem of prediction is not a simple problem that can be handled with one set of tools. Analytical-computer modelling is perhaps the most common technique used to provide information that allows one to predict system behaviour.

For pollutants that can partition between the aqueous phase and the soil solids in the subsoil system, we have well-developed advection-diffusion transport models that can address the problem. The pitfalls in implementation of such models include the availability of appropriate and realistic input parametric information (especially partition and distribution coefficients), and chemical reactions that affect the status of the contaminants in the system.

The use of geochemical speciation modelling allows one to determine these reactions. However, since kinetic reactions are not readily handled in some of the present available geochemical models, and since many of these models are not coupled to the regular transport models, much work remains at hand to obtain a reactive prediction model that can tell us about the fate and transport of contaminants in the subsoil system. Present research into coupling between geochemical models and advection-dispersion models has identified the complex and highly demanding computational requirements for a coupled model. Nevertheless, a realistic reactive coupled model is needed if we are to reach the stage where knowledge of the fate and persistence of contaminants in the sub-soil system is to be obtained.

REFERENCES

Allison, J.D., Brown, D.S., and Novo-Gradac, K.J. (1991), MINTEQA2/PRODEFA2, A Geochemical Assessment Model for Environmental Systems, Version 3. 0 user's manual, USEPA, United States.

Appelo, C.A.J., and Postma, D. (1993), *Geochemistry, Groundwater and Contamination*, Balkema, Rotterdam, 536 pp.

Agency for Toxics and Disease Registry (ATSDR), Division of Toxicology and Environmental Medicine. (2009), Toxicological Profile for Perfluoroalkyls, U.S. Government Printing Office. www.atsdr.cdc.gov/toxprofiles/tp.asp?id=1117&tid=237.

ASTSWMO (Association of State and Territorial Solid Waste Management Officials). (2015), *Perfluorinated Chemicals (PFCs): Perfluorooctanoic Acid (PFOA) & Perfluorooctane Sulfonate (PFOS) — Information Paper*, Washington DC, 68 pp.

Aziz, C.E., Newell, C.J., Gonzales, J.R., Haasm, P.E., Clement, T.P., and Sun, Y. (2000), BIOCHLOR Natural Attenuation Decision Support System, User's Manual, Version 1.1, EPA/600/R-00/008, USEPA Office of Research and Development.

Ball, J.W., and Nordstrom, D.K. (2001), User's Manual for WATEQ4F, with Revised Thermodynamic Data Base and Text Cases for Calculating Speciation of Major, Trace, and Redox Elements in Natural Waters, Open-File Report 91-183, U.S. Geological Survey, Menlo Park, California, 1991. Revised and reprinted – April, 2001.

Bedekar, V., Morway, E.D., Langevin, C.D., and Tonkin, M. (2016), MT3D-USGS version 1: A U.S. Geological Survey release of MT3DMS updated with new and expanded transport capabilities for use with MODFLOW: U.S. Geological Survey Techniques and Methods 6-A53, 69 pp. http://dx.doi.org/10.3133/tm6A53.

Buck, R.C., Franklin, J., Berger, U., Conder, J.M., Cousins, I.T., de Voogt, P, Jensen, A.A., Kannan, K., Mabury, S.A., and van Leeuwen, S.P. (2011), Perfluoroalkyl and polyfluoroalkyl substances in the environment: Terminology, classification, and origins, *Integrated Environmental Assessment & Management,* 7(4): 513–554.

Carey, G.R., van Geel, P.J., Murphy, J.R., McBean, E.A., and Rover, F.A. (1998), Full-scale field application of a coupled biodegradation-redox model BIOREDOX, In: G.B Wickramanayake., and R.H., Hinchee, (eds.), *Natural Attenuation of Chlorinated Solvents*, Batelle Press, Columbus, OH, pp. 213–218.

Chiou, G.T., Schmedding, D.W., and Manes, M. (1982), Partition of organic compounds on octanol-water system, *Environmental Science and Technology*. 16: 4–10.

Dasu, K., and Lee, L.S. (2016). Aerobic biodegradation of toluene-2,4-di (8:2 fluorotelomer urethane) and hexamethylene-1,6-di (8:2 fluorotelomer urethane) monomers in soils. *Chemosphere*, 144: 2482–2488.

Delaney, J.M., Puigdomenech, I., and Wolery, T.J. (1986), Precipitation Kinetics Option of the EQ6 Geochemical Reaction Path Code, UCRL-56342, Lawrence Livermore National Laboratory Report, Livermore, California, 44 pp.

Domenico, P.A. (1987), An analytical model for multidimensional transport of a decaying contaminant species, *Journal of Hydrology,* 91: 49–58.

Dong, S., Yu, Z., Huang, J., and Gao, B. (2022), Chapter 9 – Fate and Transport of Microplastics in Soils and Groundwater, In: Bin Gao (eds.), *Emerging Contaminants in Soil and Groundwater Systems*, Elsevier, pp. 301–329. https://doi.org/10.1016/B978-0-12-824088-5.00001-X.

Einstein, A. (1905), Uber die von der Molekularkinetischen theorie der Warme Geoforderte Bewegung von in Rubenden Flussigkeiten Suspendierten Teilchen, *Annalen der Physick*, 4: 549–660.

Environment Canada. (2010), *Risk Management Scope for Perfluorooctanoic Acid (PFOA), and Its Salts, and Its Precursors and Long Chain (C9 – C20) Perfluorcarboxylic Acids (PFCAs) and Their Salts and Their Precursors*, Government of Canada.

Felmy, A.R., Girvin, D.C., and Jeene, E.A. (1984), MINTEQ – A Computer Programme for Calculating Aqueous Geochemical Equilibria, PB84-157148, EPA-600/3-84-032, February.

Gies, E.A., Lenoble, J.L., Noël, M., Etemadifar, A., Bishay, F., Hall, E.R., and Ross, P.S. (2018), Retention of microplastics in a major secondary wastewater treatment plant in Vancouver, Canada, *Marine Pollution Bulletin*, 133: 553–561. https://doi.org/10.1016/j.marpolbul.2018.06.006.

Grbic-Galic, D. (1990), Methanogenic transformation of aromatic hydrocarbons and phenols in groundwater aquifers, *Journal of. Geomicrobiology*, 8: 167–200.

He, D.F., Luo, Y.M., Lu, S.B., Liu, M.T., Song, Y., and Lei, L.L. (2018), Microplastics in soils: analytical methods, pollution characteristics and ecological risks, *TrAC Trends in Analytical Chemistry*, 109: 163–172. https://doi.org/10.1016/j.trac.2018.10.006.

Jia, Q.L., Chen, H., Zhao, X., Li, L., Nie, Y.H., and Ye, J.F. (2019), Removal of microplastics by different treatment processes in Shanghai large municipal wastewater treatment plants. *Environmental Science*, 40(09): 4105–4112 (in Chinese with English abstract). https://doi.org/10.13227/j.hjkx.201903100.

Karickhoff, S.W. (1984), Organic pollutants sorption in aquatic system, *Journal of Hydraulic Engineering*, 110: 707–735.

Karickhoff, S.W., Brown, D.S., and Scott, T.A. (1979), Sorption of hydrophobic pollutants on natural sediments, *Water Resources*, 13: 241–248.

Kenaga, E.E., and Goring, C.A.I. (1980), Relationship between water solubility, soil sorption, octanol-water partitioning and concentration of chemicals in biota, *ASTM-STP,* 707, pp. 78–115.

Lerman, A. (1979), *Geochemical Processes*: *Water and Sediment Environments*, John Wiley and Sons., New York, 481 pp.

Lewis, G.N. (1923), *Valences and the Structure of Atoms and Molecules*, The Chemical Catalogue, New York.

Li, Y.H., and Gregory, S. (1974), Diffusion of ions in sea water and in deep- sea sediments, *Geochemica et Cosmochimica Acta*, 38: 603–714.

Liu, J., and Avendaño, S.M. (2013), Microbial degradation of polyfluoroalkyl chemicals in the environment: A review, *Environment International*, 61: 98–114.

Martin-Hayden, J., and Robbins, G.A. (1997), Plume distortion and apparent attenuation due to concentration averaging in monitoring wells, *Ground Water*, 35(2): 339–346.

Montgomery, J.H., and Welkom, L.M. (1991), *Groundwater Chemicals Desk Reference,* Lewis Publishing, Ann Arbor, 640 pp.

Nernst, W. (1888), Zur Kinetik der in Losung befinlichen Korper, *Zeitschrift fur Physikalishe Chemie*, 2: 613–637.

Newell, C.J., McLeod, R.K., and Gonzales, J. (1996), BIOSCREEN Natural Attenuation Decision Support Systems, Report EPA/6000/R-96/087, August.

Organization for Economic Cooperation and Development (OECD) Environment Directorate. (2002), Hazard Assessment of Perfluorooctane Sulfonate (PFOS) and Its Salts. www.oecd.org/chemicalsafety/risk-assessment/2382880, pdf.

Ogata, A., and Banks, R.B. (1961), A Solution of the Differential Equation of Longitudinal Dispersion in Porous Media, U.S. Geological Survey Paper, US Government Printing Office, Washington, DC, 411-A.

Parkhurst, D.L., Thorstenson, D.C., and Plummer, L.N. (1980), PHREEQE – A computer programme for geochemical calculations, *US Geological Survey Water Resources Investigatio*n, 80–96: 210.

Perkins, T.K., and Johnston, O.C. (1963), A review of diffusion and dispersion in porous media, *Journal Society of Petroleum Engineering,* 17: 70–83.

Pitter, P., and Chudoba, J. (1990), *Biodegradability of Organic Substances in the Aquatic Environment*. CRC Press, Boca Raton, 167 pp.

Plummer, L.N., Jones, B.F., and Truesdell, A.H. (1976), WATEQF – A FORTRAN IV version of WATEQ, a computer code for calculating chemical equilibria of natural waters, *US Geological Survey Water Resources Investigation*, 76–83, 61.

Rao, P.S.C., and Davidson, J.M. (1980), Estimation of Pesticide Retention and Transformation Parameters Required in Nonpoint Source Contamination Models. In: M.R. Overcash and J.M. Davidson (eds.), *Environmental Impact of Nonpoint Source Contamination*, Ann Arbor Sciences, Ann Arbor, MI, pp. 23–27.

Rifai, H.S., Newell, C.J., Gonzales, J.R., Dendrou, S., Kennedy, L., and Wilson, J., (1997), *BIOPLUME III Natural Attenuation Decision Support System, Version 1.0 User's Manual prepared for the US Air Force Center for Environmental Excellence*, Brooks Air Force Base, San Antonio, TX.

Rillig, M.C. (2012), Microplastic in terrestrial ecosystems and the soil? *Environmental Science and Technology*, 46(12): 6453-6454. doi: 10.1021/es302011r. Epub 2012 May 31. PMID: 22676039.

Robinson, R.A., and Stokes, R.H. (1959), *Electrolyte Solutions,* Butterworths, London, 571 pp.

Schwarzenbach, R.P., and Westall, J. (1981), Transport of non-polar organic compounds from surface water to groundwater: Laboratory sorption studies, *Environmental Science and. Technology*, 15(11):1360–1367.

Schwarzenbach, R.P., Gschwend, P.M., and Imboden, D.M. (1993), *Environmental Organic Chemistry,* John Wiley and Sons, New York, 681 pp.

Shaff, J.E., Schultz, B.A., Craft, E.J., Clark, R.T., and Kochian, L.V. (2010), GEOCHEM-EZ: A chemical speciation program with greater power and flexibility, *Plant and Soil*, 330: 207–214.

Sposito, G., and Mattigod, S.V. (1980), *GEOCHEM: A Computer Programme for the Calculation of Chemical Equilibria in Soil Solutions and Other Natural Water Systems*, Deparment of Soils and Environment Report, University of California, Riverside, 92 pp.

Sun, Y., Petersen, J.N., Clement, T.P., and Hooker, B.S. (1996), A Monitoring Computer Model for Simulating Natural Attenuation of Chlorinated Organics in Saturated Groundwater Aquifers. In: *Proceeding of the Symposium Natural Attenuation of Chlorinated Organic in Groundwater*, Dallas, TX, RPA/540/R-96/509.

Urbaniak, M. (2013), Biodegradation: Engineering and Technology. In: R. Chamy and F. Rosenkranz (eds.), *Intech Open Science*, ISBN: 978-953-51-1153-5, Chapter 4, pp. 73–100. doi:10.5772/50829

USEPA. (2008), FOOTPRINT (A Screening Model for Estimating the Area of a Plume Produced From Gasoline Containing Ethanol, Publication No. EPA/600/R-08/058.

Verscheuren, K. (1983), *Handbook of Environmental Data on Organic Chemicals*, 2nd ed, van Norstrand Reinhold, New York, 1310 pp.

Whittaker, R.H. (1969), New concepts of kingdoms or organisms: Evolutionary relations are being represented by new classification than by the traditional two kingdoms, *Science*, 163(863): 150–160.

Yang, L., Zhang,Y., Kang, S., Wang, Z., Wu, C. (2021), Microplastics in soil: A review on methods, occurrence, sources, and potential risk, *Science of The Total Environment*, 780: 146546, https://doi.org/10.1016/j.scitotenv.2021.146546

Yeh, G.T., and Tripathi, V.S. (1990), *HYDROGEOCHEM, a Coupled Model of HYDROlogic Transport and GEOCHEMical Equilibria in Reactive Multicomponent Systems*, Oak Ridge National Laboratory, Oak Ridge, TN.

Yeh, G.-T. & Tsai, C.-H. (2015). HYDROGEOCHEM 6.1 A Two-Dimensional Model of Coupled Fluid Flow, Thermal Transport, HYDROGEOCHEMical Transport, and Geomechanics through Multiple Phase Systems Version 6.1 (A Two Dimensional THMC Processes Model) Theoretical Basis and Numerical Approximation. https://doi.org/10.13140/RG.2.1.1766.2483.

Yong, R.N., and Sheremata, T.W. (1991), Effect of chloride ions on adsorption of cadmium from a landfill leachate, *Canadian Geotechnical Journal*, 28: 84–91.

Yong, R.N. (2001), *Geoenvironmental Engineering: Contaminated Soils, Pollutant Fate, and Mitigation,* CRC Press, Boca Raton, 307 pp.

Yong, R.N., and Mulligan, C.N. (2019), *Natural and Enhanced Attenuation of Contaminants in Soils*, Lewis Publishing, Boca Raton, 324 pp.

Yong., R.N., Nakano, M., and Pusch, R. (2010), *Containment of High-Level Radioactive and Hazardous Solid Wastes with Clay Barriers*, Spon Press, Taylor and Francis, London, 468 pp.

10 Geoenvironment Impact Mitigation and Management

10.1 INTRODUCTION

10.1.1 GEOENVIRONMENTAL IMPACTS

As we have seen from Chapter 1, impacts on the geoenvironment come from stressors whose sources include (a) natural events, such as earthquakes, tornadoes, hurricanes, typhoons, floods, drought, etc., (b) activities associated with life-support systems for humans, such as mining for resources, drilling, and fracking for extraction of oil and gas, farming, manufacturing, etc., (c) disasters and failures associated with man-made structures and activities, such as collapse of dams and holding ponds, failure of pipelines carrying crude oil and bitumen, derailment of trains carrying dangerous goods, etc., and (d) inadvertent and deliberate stressor impact actions, such as application of chemical aids in control of pests and dumping of hazardous wastes. The adverse impacts generated by geoenvironmental stressors result in diminishing the natural capital of the geoenvironment – thereby reducing the capability of the geoenvironment to provide the wherewithal to provide for the future needs of society. This, in essence, is the picture of an unsustainable geoenvironment.

10.1.1.1 Types of Stressors

Both Chapters 1 and 2 show that regardless of the sources of stressors on the geoenvironment, the stressors generated by the various sources can be grouped or classified according to the type of action or mechanism/process involved. The most useful way of classifying stressors is to group them according to the types of action that would result from the application of the stressor in question. Such a classification of stressors would be as follows: (a) thermal, (b) hydraulic, (c) mechanical, (d) chemical and geochemical, and (d) biological and biological mediated. This kind of classification scheme allows one to examine the processes or actions involved, and in conjunction with the appropriate knowledge of the geoenvironment landscape, one would be able to determine the outcome of the impact on the geoenvironment. Since the discussions in this book are directed towards maintaining the health of the soil ecosphere (see Figure 1.1 in Chapter 1), the types of geoenvironment landscape information required refer particularly to soil properties and behaviour.

10.1.1.2 Impact Mitigation and Management

Section 2.1.1 in Chapter 2 has provided some detailed information of the kinds of impacts generated from the various types of stressors. One can obviate stressor impacts on the geoenvironment by eliminating the direct stressor sources associated with the activities, etc., of humans – i.e., exercising source control. Source control is, however, not an option for natural events, such as earthquakes, etc. Experience has shown however that source control can rarely fully eliminate stressors delivered to the geoenvironment landscape. We need to develop strategies and technologies to mitigate the stressor impacts and to undertake remediation of the impacted sites to maintain the health of the geoenvironment. This holds true for geoenvironment stressors generated from both natural and anthropogenic sources.

We cite two recent examples of "incidents" (or "perceived" incidents) to illustrate the importance of impact mitigation and management. The first geoenvironment impact "incident", which stems from the voiced objections to the building of pipelines across virgin terrain carrying bitumen

DOI: 10.1201/9781003407393-10

and other similar products, claiming that pipeline failures (ruptures, leaks, etc.) will discharge large quantities of the product into the geoenvironment (i.e., onto the ground and also into receiving waters). Other than using more robust materials and prudent engineering practice in the construction of the pipeline as source control measures, impact mitigation is this case requires one to develop strategies and technologies that would (a) minimize or perhaps even eliminate future failure events, (b) provide strict monitoring of pipeline integrity through instrumentation that sense pressure drops or other performance factors, and (c) establish a disaster response protocol for effective corrective action in the case of a failure whilst implementing immediate remediation of the contaminated region.

The second example is one which seeks to develop remediation schemes for an unexpected disaster – namely the fall-out of radioactive cesium from the destruction of the nuclear power plants in Fukushima. The details of contamination of contiguous land surface by radioactive cesium from this unexpected tragedy will be discussed in a later chapter. For now, it is sufficient to point out that implementation of remediation of the land surface to pre-contamination levels is required – if original land use activities is to be restored, i.e., if *original site functionality* is to be restored.

10.2 SITE FUNCTIONALITY AND RESTORATION

The restoration of sites impacted by natural causes or by anthropogenic activities requires one to establish restoration goals or objectives. The central question to be addressed is "what does one want to have as a restored site"? Two distinct choices are available (a) restoration of the impacted site to its pre-impacted state, or (b) restoration of the site to an altered (i.e., different from pre-impacted) state that meets the requirements of regulatory agencies and stakeholders. Both of these choices require one to establish the functionality of the site – i.e., the functional purpose or purposes that have been served or can be served by the site.

10.2.1 Site Functionality

Much like the concepts of soil quality and soil functionality, introduced in the first two chapters of this book, we can apply the same conceptual model to a site or even a region (assuming a site to be on a smaller scale than a region) facing the threat of stressors from natural and anthropogenic sources. Site functionality essentially tells one about the capability of the site, i.e., the present site usage and more importantly, what the usage of the site could be if one makes full use of all the attributes (properties and characteristics) of the site. There are at least three reasons for determining or articulating the functionality of a particular site: (a) *pre-impact status* – establishing the *initial conditions, status and usage* of the site, (b) *design status* – stating the type, manner, and usage as the outcome of planned human intervention, and (c) *optimum status* – defining the potential (best benefit) usage and status of the site.

To establish *pre-impact status* site functionality, one needs to determine the present usage of the site, i.e., present status of the site. Examples of present status would include landscape or terrain scenarios, such as natural virgin land, farmland, pastureland, developed land, natural, or cultivated forest, etc. Of the preceding scenarios, except for natural virgin land, all the other land or terrain scenarios are land-use scenarios. This means to say that *natural virgin land* is a category by itself, and that all the other "present usage" scenarios fall into the *land use* category. By definition, land use indicates a land status, which has had, or does have, human intervention. Farmlands, cultivated forests, developed lands, etc., are all the result of human intervention.

Both *design status* and *optimum status* site functionality options classify under the category of restoration of site to an *altered status*. *Design status* site functionality may or may not coincide with *optimum status* site functionality. Obviously, this depends on whether the planned (design) land use takes advantage of the full potential of the site – i.e., meaning that all the attributes, properties, and site characteristics are exploited at their maximum potential.

10.2.1.1 Choice and Use of Attributes

What are *attributes*? In this particular case, one should ask "what are geoenvironment attributes"? *Attributes* are characteristics of a particular item. In the case of the geoenvironment and in view of the subject of interest in this book – the soil component of the landscape – we are interested in the various elements or parameters that characterize the land or ground. Simply put, the geoenvironment attributes of particular interest and concern are those that govern soil health and behaviour. These include soil composition (chemical, mineralogical, biological, etc.), properties and characteristics and the various factors and elements that constitute a soil and the ground.

The decision on the attributes that need to be determined to characterize site functionality can be simple or arduous, depending on (a) the nature and extent of the geoenvironment impacts involved in the site, and (b) the aims or objectives of the planned site restoration, i.e., the land use plans. In the example of the type of protocols used to assess changes in site functionality in Figure 10.1, we show the importance and use of attributes in assessment process.

The partial list of attributes seen in the second upper-right box of Figure 10.1 is a small example of the kinds of properties, characteristics, data, and settings that one might choose to gather. Obviously, not the entire vast array of information is required for each site or region under consideration. The choice that one makes must take into account the type of planned project or activity contemplated. Also seen in the right-hand group of boxes is the specification of required attributes – i.e., attributes required to meet such requirements as geoenvironment sustainability objectives and design or planned project objectives.

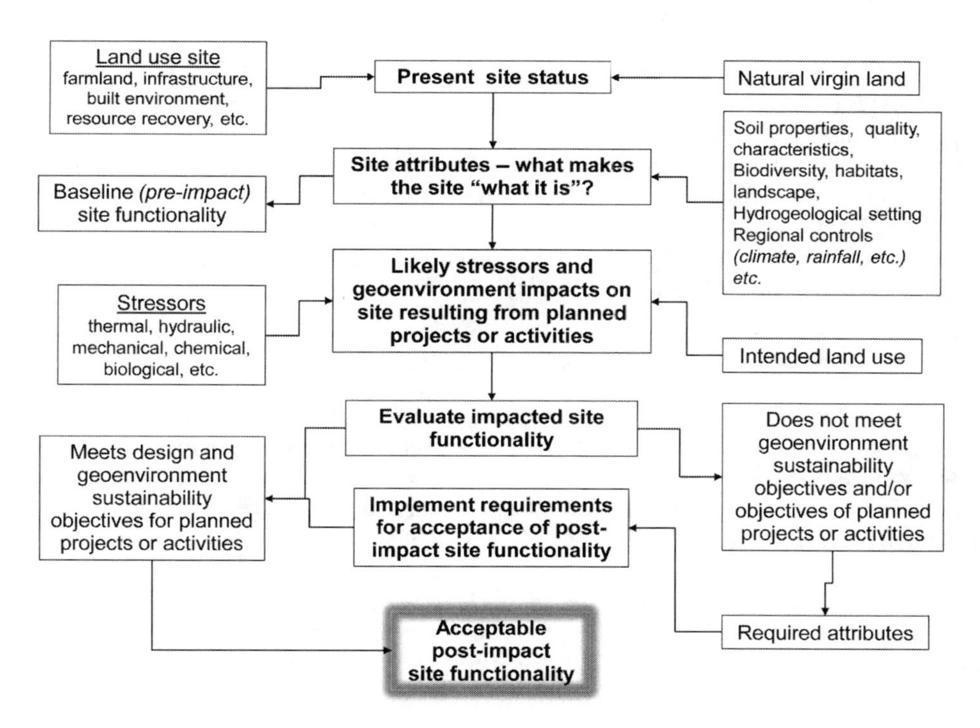

FIGURE 10.1 Example of protocol used to assess changes in site functionality as a result of planned changes or activities to a specific site.

10.2.2 Site Restoration

Site restoration means *restoring the impacted site to pre-impact state*. A good example of site restoration to pre-impact state is the planned rehabilitation of farmlands contaminated by the fallout of radioactive nuclides from the disastrous Fukushima nuclear power plant explosions following the 2011 East Japan earthquake (Nakano and Yong, 2013). Rehabilitation of the contaminated farmlands to pre-contaminated productive farm status is the present site restoration goal.

It should be noted that there are many cases and reasons why restoration to pre-impact state may not be the prime objective of site restoration. The following considerations are important in arriving at decisions and objectives concerning the technical details and implementation of site restoration plans:

- The pre-impact state of the site may be such that it does not meet the site functionality requirements such as that represented, for example, by a derelict site – i.e., a site that has served no useful purpose because of the lack of positive site attributes – and hence would require restoration to some level of positive site functionality.
- The objectives for restored site functionality by the stakeholder undertaking the restoration do not coincide with the pre-impact state site functionality – meaning that the stakeholder plans to develop the impacted site and has to remediate the site before implementing site development. A good example of this is the remediation of a brownfield site so as to be able to provide for development of housing projects. (A brownfield site – generally in an urban setting – is one which has been impacted by various kinds of chemical stressors.)
- The extent of site restoration means remediating the impacted site to a state that may or may not meet the pre-impact state level – depending on the interests of the stakeholder conducting the restoration, and more importantly, depending on regulatory acceptance of site restoration plans. This restoration plan assumes that site restoration will be implemented to a level that is on a lower level of site functionality as compared to pre-impact levels. Under such conditions, regulatory permission is the key to implementation of site restoration plans.
- The type of restored site, applied technology and extent of site restoration plans will be decided in accord with regulatory requirements and public acceptance – with the latter being of utmost importance. In instances where the impacted site directly involves the public at the impacted site, it is not a surprise to expect that the affected public would want to "have a say" in what type of site restoration scheme is being considered, and in particular, how the restored site will impact their daily lives. Under such circumstances, combined regulatory-public hearings are required.

10.3 STRESSOR IMPACTS AND MITIGATION

The nature and types of stressors have been discussed in the preceding chapters and have been classified or grouped according to their actions or impacts in the geoenvironment (Section 2.1.1) – e.g., thermal, hydraulic, mechanical, chemical, geochemical, and biological-mediated. Included in the discussion in Section 2.1.1 are some of the main sources of stressors and the encountered impacts from the various kinds of stressors. Mitigation of stressor impacts is a course of action that assumes (expects?) that source control is not available. Such is the case for most, if not all, natural geo-disasters such as earthquakes, tsunamis, tornadoes, hurricanes, etc. For geo-disasters that occur as a result of human activities, source control can minimize the number and intensity (concentration, strength, magnitude, etc.) of stressors from the source.

10.3.1 Geo-Disaster Mitigation and Protection

Disasters occurring in the land compartment of the geoenvironment have often been termed as *geo-disasters*. There are two distinct causative sources for geo-disasters: (a) Naturally occurring events such as earthquakes, tornadoes, hurricanes, drought, etc., resulting in geo-disasters, and (b) human related, i.e., disasters occurring directly and indirectly due to human activities such as failure of embankments and retaining walls, foundation failures due to failure in supporting capacity of the ground, contamination of ground by non-point sources and by fallout from smokestack emissions. Regardless of, or independent of the causative sources, there are many reasons that geo-disasters happen in the constructed environment. One of the main reasons is inadequate consideration or accounting of: (a) the nature and intensity of potential geoenvironment stressors, (b) the nature, extent and magnitude of their impacts on constructed facilities, and (c) the capability of the land compartment to fulfil its design site functionality requirements.

10.3.1.1 Naturally Occurring Events

Naturally occurring events provoking or resulting in geo-disasters fall into two categories: (a) weather-related or weather-provoked, and (b) non-weather-related events. Included in the first category of weather-related events are

- Seasonal winter-summer cycles: Stressor impacts include frost heave in winter, ground collapse in summer due to thawing of ice lenses formed in winter, flooding due to melting of snowpacks, avalanches, seasonal drought conditions, etc.

By and large, knowledge of these kinds of stressors and their impacts is well established in civil and geotechnical engineering practice, and preventative design and construction procedures have been developed to avoid development or mitigation of geo-disasters resulting from these stressors. Examples of some of these include the use of frost-free soil material underpinning foundations, restricting water inflow to underpinning foundations in frost susceptible regions, alleviation of excessive pore pressures in slopes to minimize or eliminate slope instability and slope failures due to build-up of excessive porewater pressures, application of avalanche control procedures, construction of flood-protective barriers along river banks, construction of diversion ditches and canals, etc.

- Hurricanes, typhoons, and tornadoes: Whilst the common feature to these weather-related events is *wind,* the results of the actions represented by these events differ somewhat. The distinguishing feature is *water.* The resultant stressors from hurricanes and typhoons passing over open water are both hydraulic and mechanical – taking the form of flooding of shorelines and low-lying areas and wind-forces acting on exposed facilities and objects.

As with the previous naturally occurring seasonal events, knowledge of the extent of the effects of these events are well-appreciated. Predictive analytical-computer models have been developed to provide for the ability to forecast the advent and magnitude of such events – to a greater or lesser degree of accuracy depending on the availability of on-site data. Flood control levees and embankments form one of the major mitigating or preventative measures against the hydraulic (water and flooding) stressors and their impacts. Wind-resistant design of structures will minimize damage to structures, and most importantly, the construction and use of robust storm shelters is perhaps the best means for withstanding the impacts from the mechanical (wind force and tornado forces) stressors.

Outside of ensuring robust ground stability and support capability to mitigate the mechanical and hydraulic geoenvironment stressor impacts, mitigation of stressor impacts from the more commonly reported type of naturally occurring events provoking geo-disasters such as earthquakes, hurricanes, tornadoes, and floods are technically not "mitigation" but are "protection" against stressor impacts. Examples of these are storm shelters, earthquake-resistant design of constructed facilities, flood control dikes and embankments, avoidance of earthquake zones, and other natural geo-disaster zones or regions, etc. To be considered a geo-disaster, one needs to make a distinction between (a) a disaster that happen to an above-ground facility due solely to failure of the facility itself due to impacts from stressors, and (b) failure of the above-ground facility due to fail of the ground to provide durable support. The latter disaster is considered a geo-disaster. Most of the types of stressors involved in the preceding natural geo-disaster types are generally mechanical and hydraulic (see Section 2.1.1). Civil (structural, hydraulic, and geotechnical) engineering capabilities have developed technology that will respond to the need for protection against these types of geo-disasters.

Ground improvement is one of the techniques used to provide stable supporting platforms and competent soil capable of resisting excessive ground motion under mechanical and hydraulic stressors. In addition to the more traditional geotechnical engineering methods for ground improvement, a new innovative and sustainable method for ground improvement has recently been developed. This will be discussed in detail in Chapter 12.

10.3.1.2 Anthropogenic Actions

A major group of geo-disasters occurring as a result of anthropogenic actions are chemical in nature – as has been demonstrated in the previous chapters – more often than not resulting in threats to the health of biotic receptors. Inadequate or deficient foundation or subgrade or soil stability considerations under mechanical and/or hydraulic stressors – leading to ground failure – constitute another group. The means for mitigating or countering stressor impacts for this second group are similar to those mentioned in the previous sub-section discussion relating to *natural events.*

By and large, a significant proportion of geo-disasters of a chemical nature are really geo-hazards – i.e., they present threats to the health of biotic receptors and also to the geoenvironment. In most instances, the line separating *geo-hazards* from *geo-disasters* is a function of several factors: (a) a matter of scale (size and intensity), (b) direct consequence and level of threat (toxicity, lethality, harm, exposure), (c) magnitude of physical and geoenvironmental damage, and (d) economic consequence. Geo-hazards and/or geo-disasters involving chemicals – in one form or another – occur as a result of the following:

- *Source control* – inadequate or insufficient measures of control of plant operations involving fugitive and planned discharges in operational processes.

  Mitigation of stressor impacts consist of corrective action by management to reduce or eliminate chemical stressors and their sources, and to install or implement *capture and treat* systems – i.e., systems to capture fugitive and planned discharges for treatment prior to final discharge from the plant.
- *Direct application of chemical aids* – use of chemical aids (a) for control of pests and unwanted plant species (e.g., pesticides, herbicides, fungicides, insecticides, etc.), and (b) as soil amendments (fertilizers) or as control agents (e.g., de-icing compounds). These direct actions give rise to non-point source contamination of the soil and receiving waters.

  Mitigation of stressor impacts for contaminants in the soil consists of reduction or elimination of chemical stressor toxicity – concentration and toxic level – before the contaminant reaches any biotic receptor. This is the subject of discussion in the rest of this chapter and the next chapter.
- *Consequence of industrial operations* – resource extraction (minerals, hydrocarbons, aggregates, etc.) and farming (cattle and other livestock) provide the major sources for chemical stressors. The discussions on the types of stressors and their sources have been reported in the previous chapters.

  Mitigation of stressor impacts begin with source control and implementation of "collect and treat" systems. For chemical stressors (contaminants) that find their way into the sub-soil, the techniques for mitigation will be discussed in this and the next chapter.
- *Leachates, spills, accidents during hazardous material transport, illegal dumping, etc.* – leachates leaking from containment systems, spills, accidents of trains during transport

resulting in fires or leakage of hazardous materials and illegal dumping of chemical-type liquid and solid wastes are some of the major sources of chemical stressors found in the subsurface soil. These are, by and large, point-source chemical stressors.

The discussion on procedures for mitigating the impacts from the stressors will be found in the remainder of this chapter and continued in the next chapter as options in remediation.

10.4 CHEMICAL STRESSORS – CONTAMINANTS

Regardless of the cause of the geo-disaster, the primary concern arising therefrom is in respect to the health of the geoenvironment and the biotic receptors in the presence of contaminants in the subsoil. Contaminants are chemical stressors that can severely affect the quality of water and groundwater resources, and soil quality. These (contaminants) include non-point-source contaminants, such as herbicides, pesticides, fungicides, etc., spread over large land surface areas, and point-source contaminants from effluents, waste treatment plants, and liquid discharges as wastes and spills from industrial plants (e.g., heavy metals and organic chemicals). The previous chapters dealing with urbanization and industries have shown that liquid and solid waste discharges, together with rejects, debris, and inadvertent spills in the plants, all combine to create significant threats to the health of biotic receptors and also the environment. To demonstrate the magnitude of the problem, we can cite the example of sites contaminated with hazardous wastes and other material discards. The USEPA (2004) has estimated that there are thousands of contaminated sites in the US that would need to be cleaned up over the next 30 years.

Methods and procedures for mitigating some of the major impacts from contaminant stressors, together with treatment and remediation options will be discussed in the later sections in this chapter. The discussion in this chapter recognizes that the impact from the presence of contaminants in the ground need to be mitigated and managed – as a beginning step towards protection of the resources and the natural capital in the geoenvironment as a necessary step towards achievement of a sustainable geoenvironment. The emphasis will be on using the properties and characteristics of the natural soil-water system as the primary agent for such purposes. The motivation for this is not because of the high expenditures incurred with the use of various technological remediation schemes and processes, but because it allows one to address pollution sources that encompass the range from point-source to non-point-source. Managing the impact from non-point-source pollution with technological solutions can be prohibitive because of the extent of the source (if such is known), and the extent of pollution resulting from such a source. A good example of this is the transport of contaminants in the ground and on the ground surface in conjunction with pesticides and herbicides use. One needs to pay more attention to the impacts, from both atmospheric-based and land-based non-point-sources of contamination, on the health of both soil and water resources. If transport of contaminants is to receiving waters, such as streams and rivers, how does one use technological aids and engineered systems to manage and control the advance of the contaminants? Erecting barriers that run a certain length of the stream can be prohibitively costly. A good practical solution is to invoke the properties of the natural soil system as a partner in mitigation-management. Amending and enhancing the properties to make it more effective as a control tool would also be a good tactic since this allows the subsoil to remain in place as a mitigation-management tool. This tactic is now being used in a limited way in passive remediation-treatment of contaminated sites.

10.5 SOILS FOR CONTAMINANT IMPACT MITIGATION AND MANAGEMENT

We use the term *contaminant* as a more encompassing term that includes both contaminants and pollutants. This means to say that the interactions and relationships established between contaminants and soil particles pay no attention to whether it is a contaminant or a pollutant. The designation of *pollutant* is a "human thing", made necessary to ensure protection of public health from contaminants that threaten the well-being of humans if and when they are exposed to such contaminants – by direct contact, inhalation, ingestion, etc.

The latter part of Chapter 2 dealt with the nature and basic properties of soils as they relate to the transport and fate of contaminants (pollutants) in soil. To be factually correct, we will use the term *contaminant* in all the discussions to follow and reserve the use of the term pollutant when this is factually required. In this section, we will be dealing with the aspects of soils for mitigation of impacts from containments. The role of soil as a resource material for management of contaminants is due to its physical, mechanical, chemical, and biological properties. These properties constitute the basic tools for the many different strategies and measures available for passive and aggressive management of the land and water resources in the geoenvironment. The short discussion of these tools in Chapter 3 referred to the total actions of the various soil properties in management of contaminant waste streams as the *natural attenuation* process of soils. In this section, we will examine the basic properties and attributes of soils in respect to "why and how" they can function as tools for mitigation and management of contaminant waste streams in soils.

The properties of soil directly involved as a contaminant mitigation and control tool are as follows:

(a) Those that refer directly to the soil solids themselves. These are primarily the physical and mechanical properties of the soil, and also the surface properties of the soil solids responsible for sorption of contaminants. These include the density, macro- and microstructure, porosity and continuity of void spaces, permeability, exposed surface areas in the void channels, specific surface area (SSA), cation-exchange capacity (CEC), and functional groups associated with the soil solids.
(b) Those that depend on the interactions between the contaminants and the soil solids and the chemical constituents in the porewater. In this respect, the properties of the complete soil-water system become more important. These would be the chemistry of the porewater, the presence and types of inorganic and organic ligands in the porewater, pH and *Eh* or *pE*, exchangeable ions, SSA, and CEC. All these properties, together with the biological properties of the soil-water system will define the initial state of the soil-water system and hence the capability of the soil to react with incoming waste leachate streams and contaminants.

10.5.1 Physical and Mechanical Properties

In Chapter 2, we pointed out that the two primary types of interactions between the soil particles (soil solids) and liquid waste streams and contaminants in transport in the soil sub-surface are physical and chemical in nature. For soil conditions both in the *in situ* state and in the prepared state (i.e., engineered soil barriers, for example), we need to have good mechanical and hydraulic performance characteristics from the soil. The physical, hydraulic, and mechanical properties useful for mitigation and control of liquid and solid waste substances and contaminants are those that impede and/or prevent the flow or passage of liquid and solid substances. We need to distinguish between: (a) the natural *in situ* soil condition where the soil is in the landscape as a surface and subsurface soil, and (b) the situation where human intervention and manipulation of the soil is possible, i.e., placement of a prepared soil in the ground as an active mitigation and/or treatment tool. In the case of natural soils in the landscape and subsurface, transport of liquid waste streams, leachates, and contaminants are controlled by the *in situ* physical properties of the sub-surface soil. Without human intervention, it becomes a case of "*getting what the natural situation dictates*". Changes in the physical hydraulic, and mechanical properties of the sub-soil that will likely occur because of chemical and physico-chemical interactions with contaminants and pollutants will be discussed in the next sub-section.

In the case where prepared soil is used as a sole treatment tool or as one of the tools in a designed mitigation-treatment process, control on the soil physical, hydraulic, and mechanical properties can be exercised. At this stage, the design physical, hydraulic, and mechanical properties of the soil to be

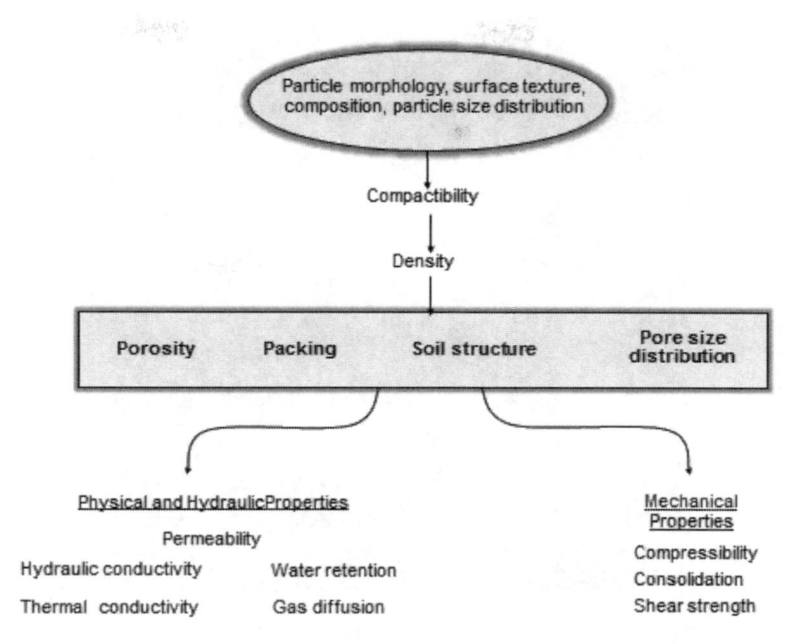

FIGURE 10.2 Control of physical and mechanical properties by soil composition and particle properties and characteristics.

used are important factors in the mitigation and prevention of contaminant transport. The soil properties of significance include physical and hydraulic properties, such as density, permeability, porosity, and mechanical properties, such as compactibility, compressibility, consolidation, and strength. All of these properties depend on the texture, grain (particle) morphology, particle size distribution, and composition. These all combine to control the packing of the particles and density of the compacted material. Figure 10.2 shows the relationship between all of these and the physical, hydraulic, and mechanical properties obtained in relation to the compactibility of the soil material. The mechanical properties are of importance in liner, buffer, and barrier systems, and the physical and hydraulic properties feature prominently in fluid and gas transport through the soil. The common assumption that transport of contaminants and pollutants is halted when fluid transport is stopped is wrong since transport mechanisms for contaminants and pollutants are via diffusive processes. So long as there is water in the soil barrier system, diffusive transport will occur. The water in the soil, even if it is immobile, serves as the carrier for the contaminants.

10.5.1.1 Soil Microstructure Controls on Hydraulic Transmission

Figure 2.11 in Chapter 2 illustrates the main points of physical interactions between a liquid waste stream permeating amongst the soil particles. The physical and hydraulic properties of the soil immediately involved in defining the nature of the fluid permeation are known as the transmission properties of the soil. These are essentially linked to the permeability of the soil to aqueous and gaseous phases – as shown in the bottom left compartment in Figure 10.2. Considering only fluid flow, the factors that affect hydraulic conductivity can be conveniently divided into three distinct categories: (a) External environmental factors, such as hydraulic head and temperature, (b) fluid phase factors, and (c) soil structural factors. The fluid phase factors of significance in the rate of fluid movement in the soil include the density, viscosity, and chemistry of the solutes contained in the fluid phase. Soil structural features are very influential in controlling flow rate and partitioning of contaminants. These include the microstructure and the micropores in the soil, the pore size distribution and the continuity of pores. All of these are functions of the density of the soil.

FIGURE 10.3 Scanning electron microscopy (SEM) picture showing typical aggregate grouping of particles forming the structure of a clay soil. The black band in the bottom middle portion of the picture represents a scale of 10 μm. Note the variety and sizes of voids, ranging from micro voids in the aggregate group (cluster, ped,) to the macro voids between aggregate groups of particles.

For any given density of soil, there is an almost infinite number of arrangements of soil particles in a typical unit volume of soil. The sketches shown in Figures 2.9 and 2.10 in Chapter 2 indicate that individual particles acting as single units are rarely found – except for granular soils. Figure 10.3 shows a scanning electron microscopy (SEM) picture of a typical clay soil unit composed of aggregate groups of clay particles, and depending on the sizes of these groups, they are generally called domains, clusters, peds, or microstructural units. Because of the variety in sizes, we will use the general term *microstructural units* in our discussion of the aggregate groups that make up the microstructure of the soils. The importance of soil structure in defining flow through a soil is evident from Figure 10.3.

Since flow occurs through void spaces that are connected, the nature of the void spaces and how these spaces are connected will be influential in determining the flow rate of the liquid waste and contaminants. Greater densities of soil will show smaller void spaces. Note that the micropores in the microstructural units will not show the same characteristics of flow as found in the macropores – i.e., the pores between peds. Because of the infinite variations in sizes and types of microstructural units and their distribution, it would not be surprising to find that soils with similar compositions can have different densities and correspondingly different hydraulic conductivities. If soil is to be used as a tool to control flow and distribution of contaminants, i.e., to mitigate pollution, it is important to determine what key factors are involved in controlling contaminant partitioning and distribution in the soil. Soil permeability to liquids is determined as the hydraulic conductivity of a soil. This holds true for saturated soils. However, for unsaturated soils, movement of liquids is generally identified as diffusive flow, even though this may not be exactly correct. The permeability of a soil is generally expressed in terms of a permeability coefficient, k. The common technique is to perform laboratory permeability tests – as shown for example in Figure 10.4, which depicts a constant head permeability test. Procedures for conducting permeability tests using constant head and falling head techniques, and also with flexible wall permeameters have been written as Standards, e.g., ASTM D5084-16a (2016). Whilst the double-Mariotte tube system shown in Figure 10.4 is not the prescribed or specified system for administering the constant hydraulic head for permeameter tests,

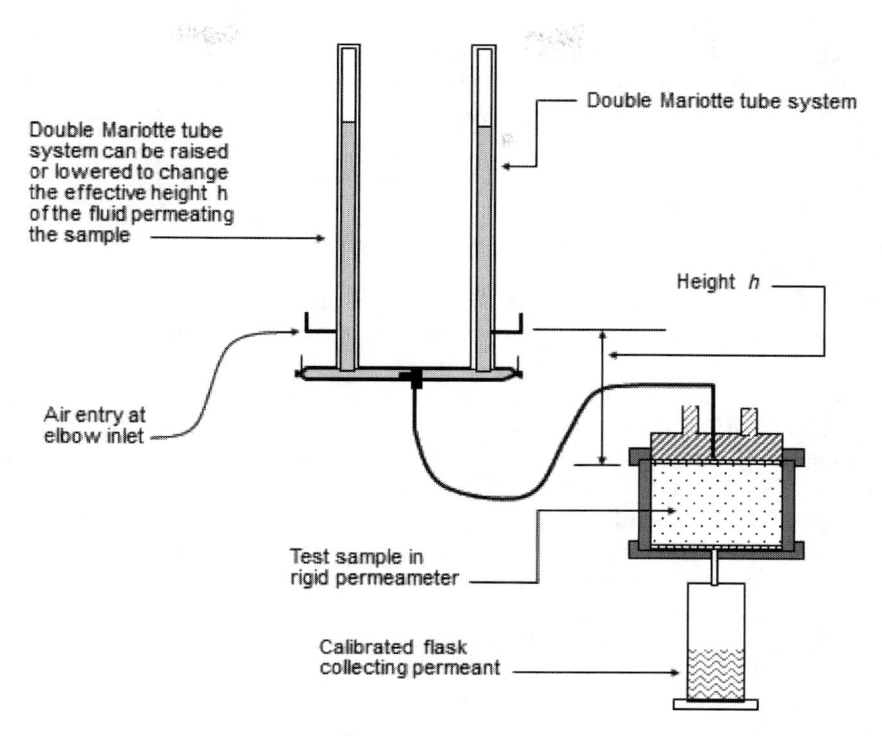

FIGURE 10.4 Water entry experiment with constant hydraulic head h for permeating fluid in a rigid permeameter.

it is nevertheless a proper and useful system to use. It permits flexibility in adjusting the hydraulic head required for constant head permeation. The Darcy coefficient of permeability k is obtained from the relationship: $Q = kiA = k(\Delta h/\Delta L)A$. The hydraulic gradient i is the ratio of the hydraulic head Δh and ΔL, the spatial distance, and A is the cross-sectional area of the test sample.

Since the Darcy model for determination of the permeability coefficient k from experiments, such as those shown in Figure 10.4 does not consider the properties of the permeant and the microstructure of the soil, Yong and Mulligan (2004) have proposed a relationship that uses a modification of the combined form of the Poiseuille and Kozeny-Carman relationships. This takes into account the influence of the properties of pore channels defined by the structure of a soil, and the fact that the wetted soil particles' surface area is controlled by the microstructure of the soil. The relationship obtained is shown as

$$v = k^* i = \frac{C_s n^3 \gamma}{\eta T^2 S_w^2} \frac{\Delta \psi}{\Delta l} \tag{10.1}$$

where

- k^* = PKC (Poiseuille-Kozeny-Carman) permeability coefficient which considers permeant and soil microstructure properties, $= \dfrac{C_s n^3 \gamma}{\eta T^2 S_w^2}$
- C_s = shape factor – with values ranging from 0.33 for a strip cross-sectional face to 0.56 for a square face. Yong and Warkentin (1975) have suggested that a value of 0.4 for C_s may be used as a standard value – with a possible error of less than 25% in the calculations for an applicable value of k^*.

- i = hydraulic gradient = ratio of the potential or hydraulic head difference $\Delta\psi$ between the entry and exit points of the permeant, and the direct path length Δl of the soil mass being tested.
- T = tortuosity = ratio of effective flow path Δl_e to thickness of test sample Δl and which is quite often taken to be $\approx \sqrt{2}$.
- γ and η = density and viscosity of the permeating fluid, respectively.
- n = porosity of the unit soil mass.
- S_w = wetted surface area per unit volume of soil particles.

Equation (10.1) uses the soil property parameters C_s, T, and S_w in structuring the relationship that describes permeability of a soil. These soil property parameters are dependent on soil composition and soil structure. Assuming that the physical properties of a leachate permeant are not too far distant from that of water at about 20 °C, and further assuming a tortuosity T value of $\sqrt{2}$, and $C_s = 0.4$, the graphical relationships shown in Figures 10.5 and 10.6 will be obtained. These graphs show the relationship between the PKC permeability coefficient k^* and the amount of surface area wetted in fluid flow through the soil. A comparison of the calculated wetted surface areas S_w for the soils shown in Figure 10.6 show that the wetted surface areas vary from about 3% to 7% of the specific surface area of the soils. This indicates that micro-structural units such as those shown in Figure 10.3 encompass a large number of soil particles – to the extent that the effective surface areas presented to a contaminant leachate stream represents only a small fraction of surface areas present. For soils that are prepared for use as barriers and liners, the sizes and distribution of micro-structures are significant factors in determining the effectiveness of the barriers and liners.

10.5.1.2 Microstructure, Wetted Surfaces, and Transport Properties

Hydraulic conductivity through a soil engineered barrier is greatly facilitated when interconnected voids and their connecting channels are large. Large voids in a compact soil generally mean large grain sizes and/or large microstructural units. In addition to the advantage of larger flow paths, large, interconnected voids generally mean that the surfaces presented as the surrounding surfaces

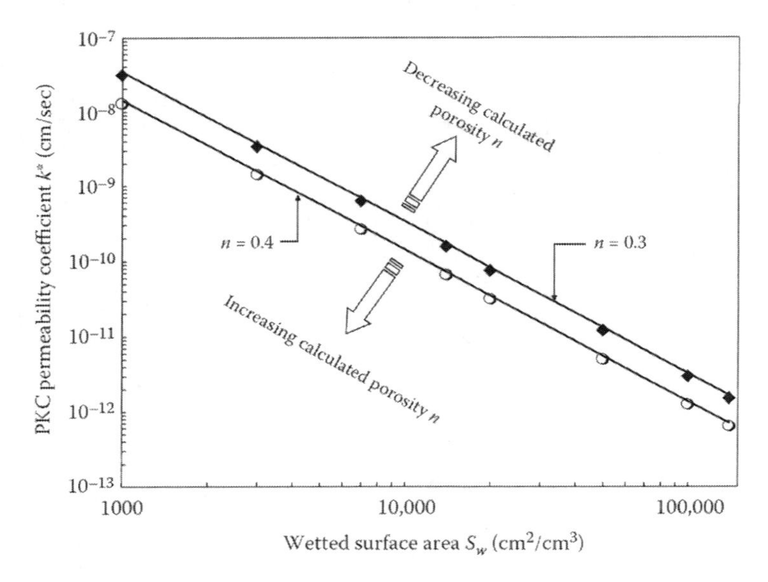

FIGURE 10.5 Variation of PKC permeability coefficient k^* with wetted surface area S_w and calculated porosity n.

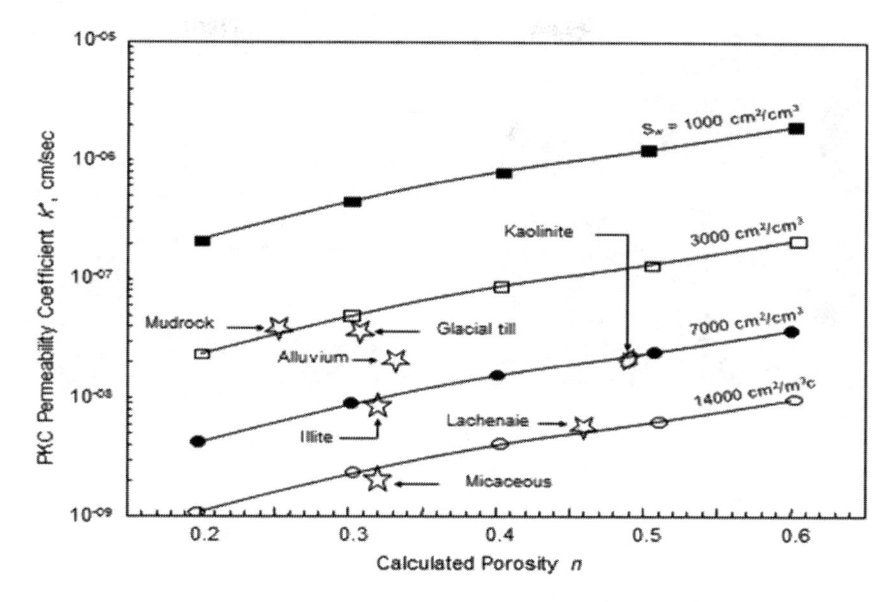

FIGURE 10.6 Variation of PKC permeability coefficient k^* with calculated porosity – in relation to wetted surface area S_w.

in the void spaces are lesser than if the void volumes were smaller. The surrounding flow path contact surfaces are important factors because they (a) offer drag or boundary resistance to flow, and more importantly (b) provide the surface areas and corresponding functional groups for chemical reactions that include sorption, ion exchange, and complexation. We will discuss this latter aspect in greater detail in the next section.

Figure 10.7 shows two scanning electron micrographs of the same clay soil. The clay soil was obtained as a core sample from a recently cut slope in northern Québec, Canada, where the winters can be cold and harsh. The picture on the left shows the structure of the soil obtained from a core sample to consist of small micro-structural units apparently uniformly distributed in the cross-section. The right-hand picture shows the same soil in the thawed state after 32 cycles of freezing and thawing (Yong et al., 1984). The dramatic increase in sizes of the micro-structural units is obvious, testifying to the marked decrease in surface areas presented to a permeating fluid, and also testifying to the significant increase in void spaces. Transmission or transport through the soil material shown on the right-hand side will be considerably facilitated by the freeze-thaw effect. The lesson to be learnt from the pictures in Figure 10.7 is that environmental effects can alter the initial conditions to such an extent that design of mitigation and treatment procedures must anticipate such events.

10.5.2 Chemical Properties

The chemical properties of significance include those that promote ion exchange, sorption and precipitation of solutes in the fluid phase (including porewater) in the soil, and complexation. These are properties that are more appropriately defined as soil-water system properties. These have been discussed briefly in Chapters 2 and 9 in respect to partitioning processes involving heavy metals. To fully utilize soil as a resource material for management of waste leachate streams and contaminants, a broader discussion of the important chemical properties and interactions between contaminants and soil particles or fractions is needed.

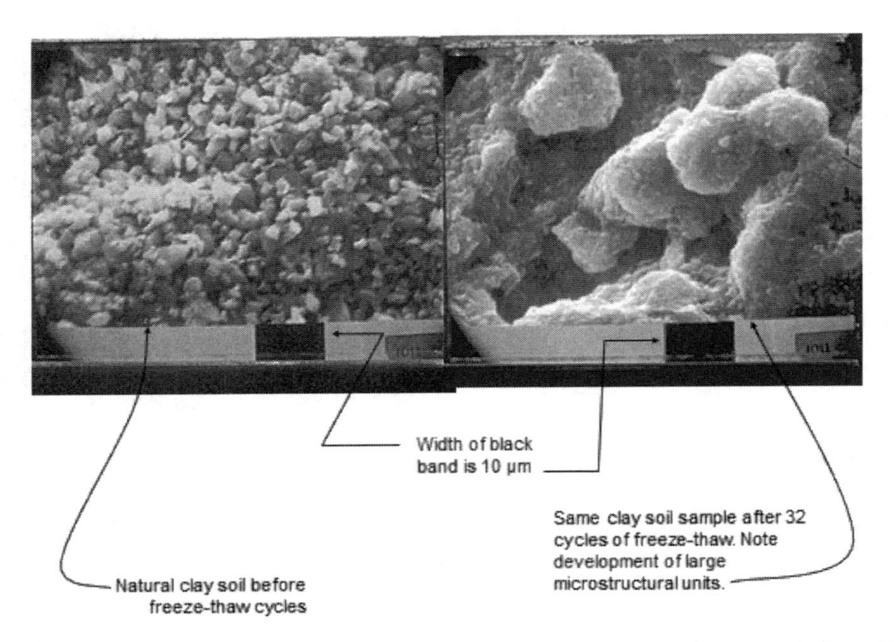

FIGURE 10.7 SEM picture showing formation of large microstructural units in a natural clay soil after 32 cycles of freezing and thawing.

10.5.2.1 Sorption

As discussed in the previous chapter, sorption processes involving molecular interactions are (a) Coulombic in nature, (b) interactions between nuclei and electrons, and (c) essentially electrostatic in nature. The major types of interatomic bonds are ionic, covalent, hydrogen, and van der Waals. Ionic forces hold together the atoms in a crystal. The various types of bonds formed from various types of forces of attraction include (a) ionic bonds, i.e., electron transfer between the atoms which are subsequently held together by the opposite charge attraction of the ions formed, (b) covalent bonds developed as a result of electron sharing between two or more atomic nuclei, and (c) Coulombic bonds developed from ion-ion interaction.

For interactions between instantaneous dipoles, we have the three types of *van der Waals* forces: Keesom, Debye, and London dispersion forces. Bonding developed by van der Waals forces is, by and large, the most common type of bonding between organic chemicals and mineral soil fractions. Electrical bonds can be formed between negatively charged organic acids and positively charged clay mineral edges. Sorption of organic anions can occur if polyvalent exchangeable cations are present. The polyvalent bridges formed will be due to (a) anion associated directly with cation, or (b) anion associated with cation in the form of a cation bridge (water bridge).

10.5.2.2 Cation Exchange

Cation exchange involves those cations associated with the negative charge sites on the soil solids, largely through electrostatic forces. Ion exchange reactions occur with the various soil fractions, i.e., clay minerals and non-clay minerals. This process, which has been discussed in detail in Chapter 9, is set in motion because of the need to satisfy electroneutrality and is stoichiometric. Calculations or determinations of the proportion of each type of exchangeable cation to the total cation exchange capacity of the soil can be made using exchange equilibrium equations such as the Gapon relationship shown in Equation (9.1).

From the electrostatic point of view, physical adsorption (or sorption) of contaminants in the porewater (or from incoming leachate) by soil fractions is due to the attraction of positively charged contaminants, such as the heavy metals to the negatively charged surfaces of the soil fractions. This type of adsorption is called non-specific adsorption. By definition, we can refer to *non-specific adsorption* when ions are held by the soil particles primarily by electrostatic forces. This distinguishes it from *specific adsorption*, which is another way of identifying *chemisorption*, a process that involves covalent bonding between the contaminant and the soil particle (generally mineral) surface. Examples of non-specific adsorption are the adsorption of alkali and alkaline earth cations by the clay minerals. By and large cations with smaller hydrated size or large crystalline size would be preferentially adsorbed.

10.5.2.3 Solubility and Precipitation

The contaminants affected by solubility and precipitation processes are mostly heavy metals. The pH of the soil-water system plays a significant role in the fate of heavy metal contaminants because of the influence of pH on the solubility of the heavy metal complexes. According to Nyffeler et al. (1984), the pH at which maximum adsorption of metals occurs varies according to the first hydrolysis constant of the metal (cationic) ions. When the ionic activities of heavy metal solutes in the porewater of a soil exceed their respective solubility products, precipitation of heavy metals as hydroxides and carbonates can occur. The two stages in precipitation are nucleation and particle growth. This will generally be under slightly alkaline conditions. The precipitate will either form a new separate substance in the porewater or will be attached to the soil solids. Gibbs phase rule restricts the number of solid phases that can be formed.

Factors involved in formation of precipitates include soil-water system pH, type and concentration of heavy metals, presence of inorganic and organic ligands, and the individual precipitation pH of heavy metal contaminants. In the solubility-precipitation diagram shown in Figure 10.8 for a metal hydroxide complex, the left-shaded area marked as *soluble* identifies the zone where the metals are in soluble form with positively charged complexes formed with inorganic ligands. The right-shaded *soluble* area contains the metals in soluble form with negatively charged compounds. The *precipitation region* in-between the two shaded areas contains various metal hydroxide species.

Figure 10.9 shows heavy metal precipitation information using data reported by MacDonald (1994). Transition from soluble forms to precipitate forms occurs over a range of pH values for the three heavy metals. The onset of precipitation can be as early as pH of about 3.2 in the case of the single heavy metal species (Pb). The process of precipitation is a continuous process that begins with onset at some early pH and finally concludes at somewhat higher pH value – generally around pH 7 for most metals. The influence of other metal species in the precipitation process is felt not only in terms of when onset pH occurs, but also in the rate of precipitation in relation to pH change. Figure 10.9 shows that the onset of precipitation of Zn as a single species is about pH 6.4, and that reduces to about pH 4.4 when other metals are present. Given that the experiments were conducted with equal amounts of each of the three heavy metals, it is expected that the concentrations of the other metals would also have an effect on modification of the onset pH. We must note that the precipitation boundaries are not distinct separation lines, and that transition between the two regions or zones occurs in the vicinity of the boundaries throughout the entire pH range.

The role of pH in the soil-water system is important because of the various complexes formed in relation to pH. For example, when a heavy metal contaminant solution such as a $PbCl_2$ salt enters a soil-water system at pH values below the precipitation pH of Pb, a portion of the metals will be adsorbed by the soil particles. The ions remaining in solution would either be hydrated or would form complexes such as Pb^{2+}, $PbOH^+$, and $PbCl^+$. These would be contained in the left-shaded area of Figure 10.8. When the pH is raised to the pH levels shown in the right-shaded area of Figure 10.8, one would form complexes such as PbO_2H^- and PbO_2^{2-} that would reside in the right-shaded area.

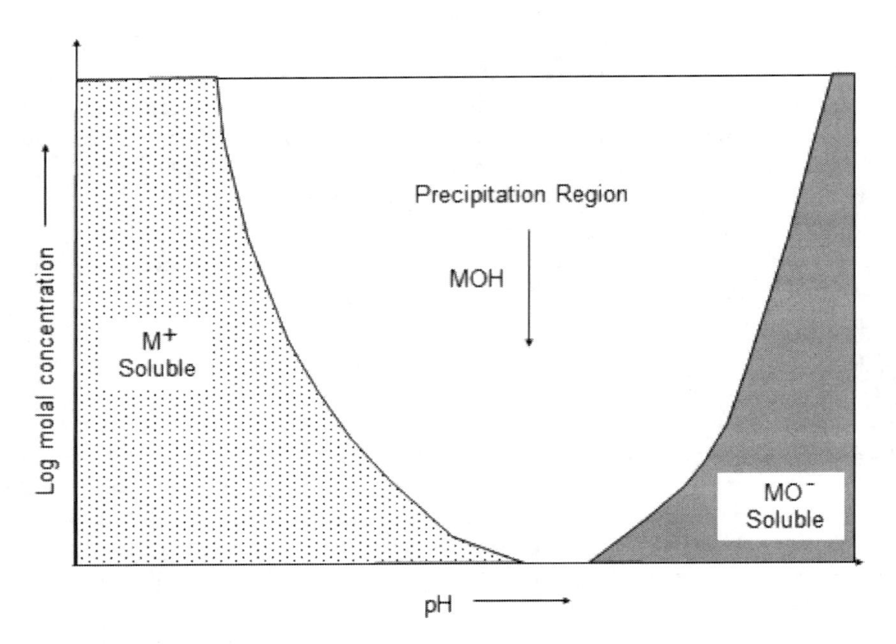

FIGURE 10.8 Solubility-precipitation chart for a metal hydroxide complex.

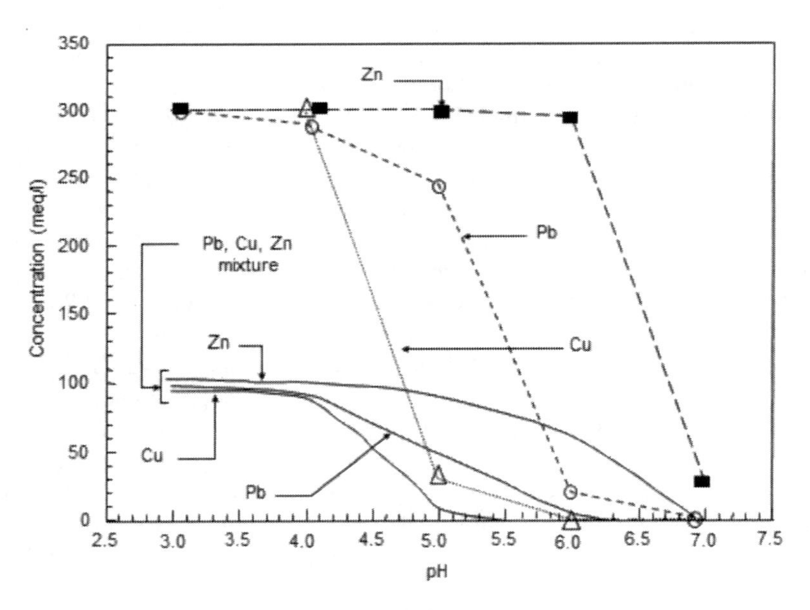

FIGURE 10.9 Precipitation of heavy metals Pb, Cu, and Zn in aqueous solution. Bottom curves are precipitation of individual metal from a mixture of Pb, Cu, and Zn in equal proportions of metal nitrate solution (100 meq each). Top curves are for single solutions of individual metals at 300 meq/L concentration. (Data from MacDonald, 1994.)

10.5.2.4 Speciation and Complexation

The processes associated with speciation and complexation apply primarily to inorganic contaminants in the liquid waste streams, particularly the heavy metals. The term *speciation* refers to the formation of complexes between heavy metals and ligands in the aqueous phase. Ligands are defined as anions that can form coordinating compounds with metal ions. Inorganic and organic ligands include, for example, $CO_3^{2-}, SO_4^{2-}, Cl^-, NO_3^-, OH^-, SiO_3^{2-}, PO_4^{3-}$, and humic and fulvic acids. For the complexes formed between heavy metals and humic and fulvic acids, these would generally be chelated compounds. From the preceding, we note that the chemistry of the porewater in the soil-water system is an important factor in mitigation and control of contaminant transport in the soil-water system – through competition between the ligands and the soil solids for sorption of the heavy metals.

10.5.3 Biological Properties

Biological properties of soils (soil-water systems) are very important factors in the passive and aggressive treatment and management of organic chemical contaminants in the subsurface soil regime. These properties are determined by the large variety of microorganisms that reside in the soil-water system. These microorganisms consist of viruses, bacteria, protozoa, fungi, and algae. Microorganisms are the key to biological treatment and management of contaminants and contaminants.

10.5.3.1 Protozoa

Protozoa are aerobic, single-celled chemoheterotrophs. They are classified as eukaryotes with no cell walls and with sizes that vary from 1 to 2000 μm. They include pseudopods, flagellates, amoebas, ciliates, and parasitic protozoa. The four primary groups of protozoa are (a) Mastigophora – flagellate, (b) Sarcodina – amoeboid, (c) Ciliophone – ciliated, and (d) Sporozoa – parasites of vertebrates and invertebrates. Protozoa are found in water and soil and feed on bacteria and need water in order to move. Although they do not generally biodegrade contaminants, they are useful in reducing bacterial numbers near injection wells that become clogged due to excessive growth. Soil protozoa are heterotrophic and although their general food source is bacteria, they are known to feed on soluble and even insoluble organic material. They mineralize nutrients and release excess nitrogen as NH_4^+ which is beneficial to plants and others on the food web.

10.5.3.2 Fungi

Fungi include slime moulds (filamentous fungus), yeasts and mushroom, and are aerobic, multicellular, eukaryotes and chemoheterotrophs that require organic compounds for energy and carbon. They are larger than bacteria, and do not require as much nitrogen. They grow more slowly and in a more acidic pH range than bacteria, and are more sensitive to changes in moisture levels. Yeasts are unicellular organisms that are larger than bacteria, and are shaped like eggs, spheres or ellipsoids.

10.5.3.3 Algae

Algae are single-celled and multi-cellular microorganisms, and are considered to be the abundant photosynthetic microorganisms in soils. According to Martin and Focht (1977): (a) The availability of inorganic nutrients such as C, N, P, K, Fe, Mg, and Ca are said to be responsible for their (soil algae) abundance, and (b) their principal functions in soil are nitrogen fixation, colonization of new rock and barren surfaces, supplying organic matter and nitrogen for humus formation, weathering of rocks and minerals, and binding of soil particles through surface bonding.

10.5.3.4 Viruses

Viruses are the smallest type of microbe and can be 10,000 times smaller than bacteria. They require a living cell to reproduce. It is said that their primary function is to reproduce, and they do it well by taking over a host cell. They have a direct influence on bacterial abundance, and through lysis (cell destruction) and transduction, i.e., transfer of viral DNA from one cell to another through viruses that attack bacteria (bacteriophages), they can alter bacterial genetic diversity. Beyond their direct attack on the various microbial cells and their influence on community composition, their other functions in soil are not too well known or established.

10.5.3.5 Bacteria

Bacteria are single-celled microorganisms that vary in size and shape from very small spheres to rods that can vary from 1 μm to a few microns in length and width. There are literally many thousands of different bacterial species co-existing in the soil. With favourable conditions of temperature and nutrient availability, it is reported that bacterial population in soil can be in the order of 10^8 to 10^{10} per gram of soil. They are both autotrophic and heterotrophic. Most bacteria used for bioremediation treatment of organic chemicals are chemoorganotrophs and heterotrophs. Those requiring organic substrates for energy are called chemoorganotrophs and those using organics as a carbon source are called heterotrophs. Those that use inorganic compounds as an energy source are named chemolithotrophs. Nitrifying bacteria (*Nitrosomonas* and *Nitrobacter*) that use carbon dioxide as a carbon source instead of organic compounds are called autotrophs. Nitrifying bacteria produce nitrite from ammonium ion, which is then followed by conversion to nitrate.

10.6 NATURAL ATTENUATION CAPABILITY OF SOILS

By definition, the reduction of toxicity and concentration of contaminants in a contaminant plume during transport in the subsurface soil is called *contaminant attenuation*. We use the general term contaminants to include pollutants and all other kinds of hazardous substances in the fluid phase of the soil-water system. If the various processes responsible for contaminant attenuation are naturally-occurring, the attenuation process is said be the result of the *natural attenuation capability* (i.e., *assimilative capacity)* of the subsurface soil. What are these naturally-occurring attenuation processes? These are the physical, chemical, and biological properties discussed in the previous section. They all contribute to the assimilative capacity of soil, i.e. the capacity of the soil to "cleanse itself" through attenuation of the flux of contaminants by means of processes that include physical, chemical and biologically mediated mass transfer, and biological transformation.

The American Society for Testing and Materials (ASTM) (2015) defines *natural attenuation* as the "reduction in mass or concentration of a compound in groundwater over time or distance from the source of constituents of concern due to naturally occurring physical, chemical, and biological processes, such as; biodegradation, dispersion, dilution, adsorption, and volatilization." The USEPA (1999) on the other hand considers natural attenuation specifically in the context of a monitored scheme for treatment of polluted sites. Accordingly, they use the term monitored natural attenuation and define it as

> the reliance on natural attenuation processes (within the context of a carefully controlled and monitored site cleanup approach) to achieve site-specific remediation objectives within a time frame that is reasonable compared to that offered by other more active methods. The 'natural attenuation processes' that are at work in such a remediation approach include a variety of physical, chemical, or biological processes that, under favorable conditions, act without human intervention to reduce the mass, toxicity, mobility, volume, or concentration of contaminants in soil or groundwater. These in-situ processes include biodegradation; dispersion; dilution; sorption; volatilization; radioactive decay; and chemical or biological stabilization, transformation, or destruction of contaminants.

In the specific context of contaminants and transport of liquid wastes, leachates etc. in the sub-soil, the reduction and detoxification of all of these contaminants from processes associated with natural attenuation is termed as *intrinsic remediation*. More specifically, reduction in

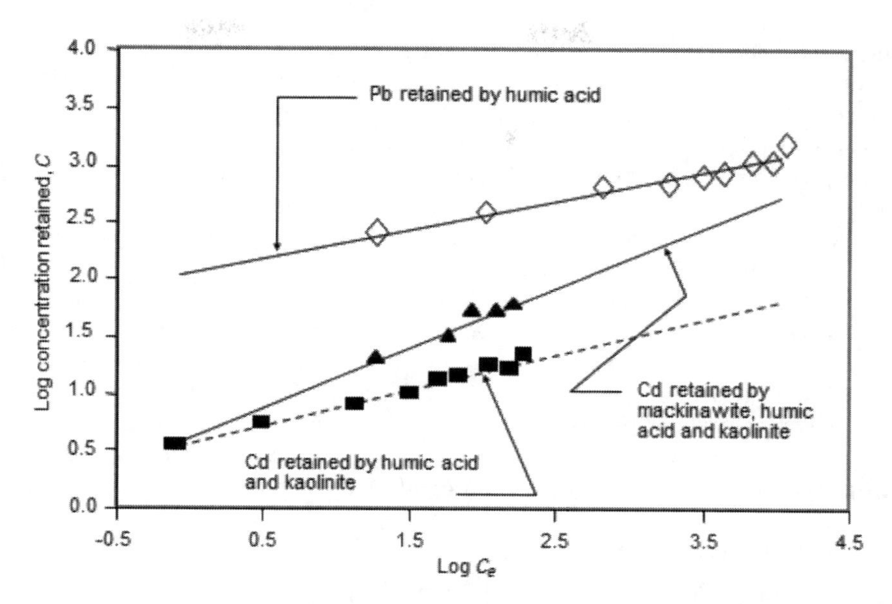

FIGURE 10.10 Freundlich-type adsorption isotherms are showing the retention of Pb humic acid, and Cd by kaolinite and humic acid, and mackinawite $(FeNi)_9S_8$, humic acid and kaolinite. Solutions of Pb and Cd at pH 4.3 were obtained as $PbCl_2$ and $CdCl_2$, respectively (data from Coles and Yong, 2004). C_e represents the equilibrium concentration.

concentration of contaminants is by processes of partitioning and dilution, and reduction in toxicity of the contaminants is generally achieved by biological transformation (of organic chemicals) and sequestering of the toxic inorganic contaminants. The test results reported by Coles and Yong (2004) in Figure 10.10 show the importance of soil composition on the retention of lead (Pb) and cadmium (Cd). In the particular case shown in the figure, humic matter in the form of fulvic acid, a sulphide mineral called mackinawite $(Fe,Ni)_9S_8$ and kaolinite were used as control soil material.

Natural attenuation of contaminants can be considered as being a set of positive processes that mitigate the impact of contaminants in the ground through a reduction of their intensity – as measured in terms of concentration and toxicity of the contaminants. In the past, the use of natural attenuation processes had been considered almost exclusively in connection with remediation of contaminated sites – and more specifically with sites contaminated with organic chemical contaminants, as shown for example in the definition provided by the USEPA. Little distinction was made between intrinsic remediation and intrinsic bioremediation. More recently, with a better appreciation of the assimilative capacity of soils and especially in view of a growing body of research information on contaminant-soil interactions, more attention is being paid to the use of natural attenuation as a tool for mitigating and managing the transport and fate of contaminants in the ground. A contributing factor has also been the accelerating costs for application of aggressive remediation techniques to treat contaminated sites. The reader is reminded that *contaminants* include all the polluting and health-threatening elements entering into the ground, such as contaminants, toxicants, leachates, liquid wastes, hazardous substances, etc.

10.6.1 Natural Attenuation by Dilution and Retention

We have, up to now, considered natural attenuation of contaminants as being due to the processes associated with physical, chemical, and biological properties of soil. We have considered that these properties contribute directly to the partitioning of contaminants, i.e., the transfer of contaminants in the porewater to the surfaces of the soil solids. Strictly speaking, there is another set of processes that arguably can be considered as part of the natural attenuation capacity of soils – except that we would now have to refer to this as the natural attenuation capacity of soil-water systems. Whilst the groundwater and porewater aspects of the soil-water system have heretofore been considered only in respect to their physical and chemical interactions with the soil solids and contaminants, they attenuate contaminants through processes of dilution. Thus, in addition to the processes previously described reduction in concentrations of contaminants can be accomplished by dilution through mixing of the contaminants with uncontaminated or less contaminated groundwater. In total, natural attenuation of contaminants in soils includes (a) dilution, (b) interactions and reactions between contaminants and soil solids resulting in partitioning of the contaminants between the soil solids and pore water, and (c) transformations that reduce the toxicity threat posed by the original polluting contaminants (contaminants). The likelihood of only one mechanism being solely responsible for attenuation of contaminants in transport in the soil is very remote. In all probability, all the various processes or mechanisms will participate to varying degrees in the attenuation of contaminants – with perhaps partitioning being by far the more significant factor in attenuation of contaminants.

10.6.1.1 Dilution and Retention

In the context of contaminant transport in soils, *dilution* refers to the reduction in concentration of contaminants in a unit volume as a result of a reduction of the ratio of number of contaminants n_c to the volume V of the host fluid. An example of this would be when the original contaminant load is given as 100 ppm, and dilution with groundwater reduces this to 50 ppm, the singular process responsible for the decrease or reduction in concentration is dilution, and transport in the subsoil will likely be consistent with the advective velocity of the groundwater. Except for physical controls on groundwater flow, no other soil properties are involved in the dilution process.

Retention refers to the retention of contaminants by the soil solids through partitioning processes that involve physico-chemical and chemical mass transfer. The result of retention is a decrease in the concentration of contaminants in a leachate plume or liquid waste stream as one progress away from the source. Using the same numerical example as above, one would see a reduction from 100 ppm to 50 ppm in a contaminant transport stream at a point further downstream from the contaminant source. The difference between this and the previous example is that when reduction of concentration is obtained through retention processes, the contaminants retained will not be readily available for transport. In the case of dilution as a means for reduction in concentration of contaminants, there are no contaminants held by the soil particles. All the contaminants will be delivered downstream in due time. Figure 10.11 illustrates the differences using assumed ideal bell-shaped concentration distribution pulses. In the case of dilution, the diagram shows that eventually, the total contaminant load will be delivered downstream. The areas of the assumed bell-shaped dilution pulses are all constant and equal to the original rectangular distribution shown on the ordinate. In contrast, the assumed bell-shaped retention pulses will show decreasing areas. These retention pulses will diminish to zero so long as the assimilative capacity of the soil is not exceeded. This is a significant point of consideration in the use of the natural attention capacity of soils for mitigation and management of impacts from liquid waste and contaminant discharges.

10.6.2 Biodegradation and Biotransformation

The common perception is that biological activities in the sub-soil will degrade organic chemical compounds. Not always understood or perceived is whether this does in fact contribute to attenuation of the concentration and toxicity of the contaminants in the subsoil. Figure 10.12 provides the overall view of various attenuating mechanisms in the soil. The top left-hand corner of the diagram shows the biologically mediated transfer mechanisms participating in the attenuation process. Not well illustrated or sometimes not fully acknowledged is the redox reactions (this includes both

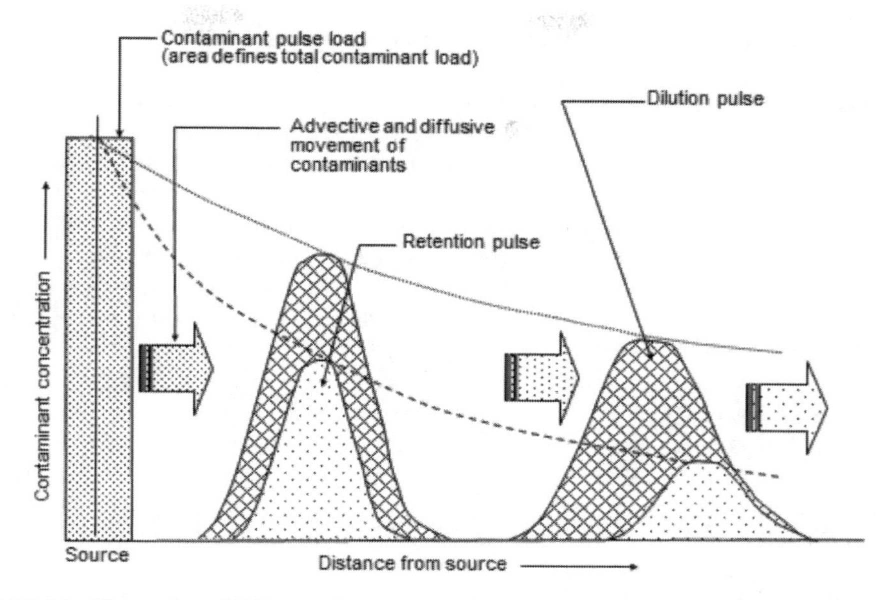

FIGURE 10.11 Illustration of difference between retardation and retention of contaminants during transport of a contaminant pulse load. Note that the areas of the assumed bell-shaped dilution pulses are constant and that they are equal to the original contaminant pulse load. The areas of the assumed bell-shaped retention pulses diminish as one progresses further away from the source.

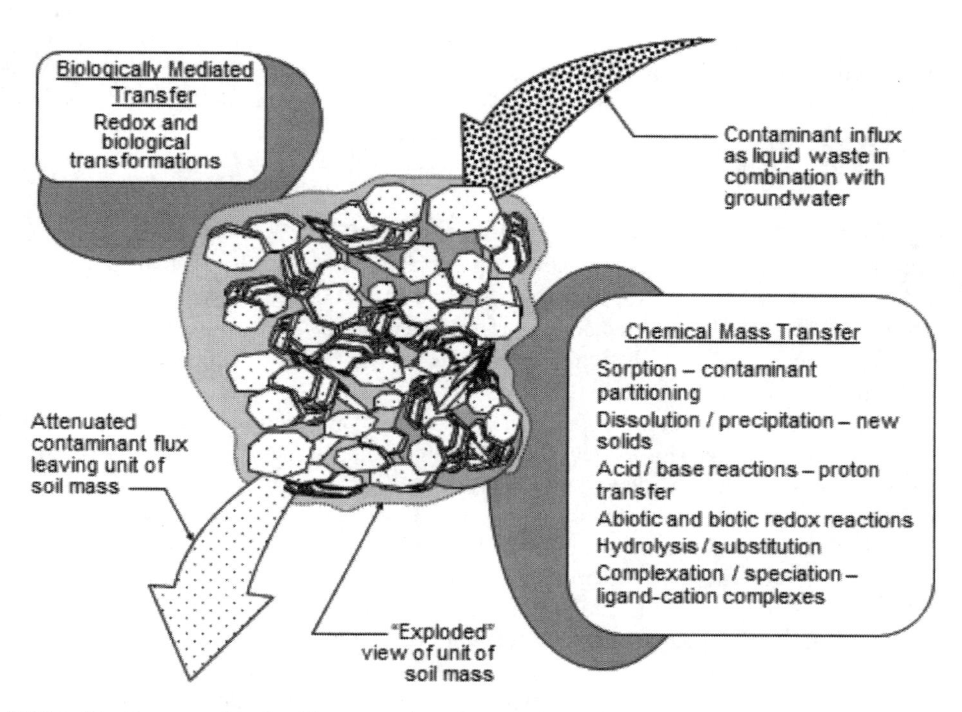

FIGURE 10.12 Processes involved in attenuation of contaminants in leachate transport through a soil element. Dilution of contaminants is not included in the schematic illustration. (Adapted from Yong and Mulligan, 2019.)

abiotic and biotic redox reactions) listed under the *Chemical Mass Transfer* box in the bottom-right corner of the diagram.

The discussions in the latter sections of this chapter will provide case histories dealing with treatments using various procedures and techniques. Amongst these will be the use of intrinsic bioremediation in management of contaminant transport. Before embarking on these discussions, it is useful to clarify the use of the term substrate in the discussions to follow. In the context of biological activities in the subsoil, the term *substrate* is used to mean the food source for microorganisms. This term should not be confused with the term substrate used in some soil mechanics and geotechnical engineering literature to mean sub-surface soil stratum. The use of *substrate* as a term here is consistent with its usage in microbiology and is meant to indicate that it (*substrate*) serves as a nutrient source and also a carbon or energy source for microorganisms.

By definition, *biodegradation* refers to the decomposition of organic matter by microorganisms. The end result of the metabolic and enzymatic processes is seen in terms of smaller compounds, and in respect to organic contaminant remediation goals, ultimately as $CO_{2,}$ $CH_{4,}$ and H_2O. The organic wastes and chemicals of interest in this book are contaminants such as organic compounds that contain carbon, and organic waste matter. Strictly speaking, biodegradation is a particular form of biotransformation, since *biotransformation* (or biological transformation) means the conversion of a chemical substance into another chemical substance (generally a metabolite) by enzymatic action or other biological processes. The use of the term *biodegradation* implies that the biological transformation process reduces the original organic chemical compound into smaller fragments, with the presumed conclusion that these smaller fragments are less toxic or less threatening to the receptors. To some extent, this is probably valid. However, the well-documented and well-reported example of anaerobic "degradation" of C_2Cl_4 (perchloroethylene, PCE) to C_2H_3Cl (vinyl chloride) shows that the latter "degraded" compound is more of a threat than the original, because C_2H_3Cl is more toxic and volatile than PCE, and does not partition well in the soil. In any event, the transformation of original organic chemical compounds in the sub-surface soil into smaller units occurs by oxidation and reduction mechanisms (redox reactions) resulting from the metabolic activities of the microorganisms in the soil. In the next few subsections, we will show examples from Yong and Mulligan (2019) of biotransformation and biodegradation of some organic chemical compounds. Greater details of the many kinds of transformations and conversions can be found in specialized texts dealing with bioremediation.

10.6.2.1 Petroleum Hydrocarbons – Alkanes, Alkenes, and Cycloalkanes

Petroleum hydrocarbon consists of various compounds such as alkanes, cycloalkanes, aromatics, polycyclic aromatic hydrocarbons, asphaltenes, and resins. Their biodegradability in the sub-soil ranges from very biodegradable to recalcitrant. This is because of the varying degrees of branching, chain lengths, molecular sizes, substitution with nitrogen, oxygen or sulphur atoms. Many of the alkanes found in petroleum are branched.

Alkanes (C_2H_{2n+2}) are aliphatic compounds. Low molecular weight alkanes are the most easily degraded by microorganisms. However, as the chain length increases from C_{20} to $C_{40,}$ hydrophobicity increases while solubility and biodegradation rates decrease. Conversion of alkanes leads to the formation of an alcohol using a monooxygenase enzyme. This is followed by oxidation to an aldehyde and then to a fatty acid (Pitter and Chudoba, 1990). Further oxidation (β-oxidation) of the fatty acid yields products less volatile than the original contaminants. Anaerobic bacteria such as sulphate-reducing bacteria are capable of degrading fatty acids via this step (Widdel, 1988). Although bacteria that are capable of degrading n-alkanes cannot degrade branched ones (Higgins and Gilbert, 1978), *Brevibacterium ethrogenes, Corynebacteria sp., Mycobacterium fortuitum, Mycobacterium smegmatis*, and *Nocardia* sp. have been shown to grow on branched alkanes. The first degradation step is the same as for the unbranched alkanes. However, the β-oxidation is more difficult and less efficient (Pirnik, 1977). In addition, in the presence of *n*-alkanes, the metabolism of the branched alkanes will be repressed, which will cause difficulties during the degradation of mixtures such as petroleum.

Alkenes with a double bond between carbons and have not been extensively studied for biodegradation. Those containing the double bond on the first carbon may be more easily degradable than those alkenes with the double bond at other positions (Pitter and Chudoba, 1990). The products of oxidation of 1-alkenes can be either diols or the methyl group.

Because of their cyclic structure, cycloalkanes are not as degradable as alkanes, and they become less degradable as their number of rings increase. Pitter and Chudoba (1990) attribute some of this to their decreasing solubility. Species of *Nocardia* and *Pseudomonas* are able to use cyclohexane as a carbon source. Oxidation of the cycloalkanes with the oxidase enzyme leads to production of a cyclic alcohol and then a ketone (Bartha, 1986).

10.6.2.2 Gasoline Components BTEX and MTBE

Benzene, toluene, ethylbenzene, and xylene, (BTEX), are volatile, water soluble, hazardous components of gasoline. Aerobic degradation of all components of BTEX occurs rapidly with available oxygen. Under anaerobic conditions, degradation is less reliable and is slower than under aerobic conditions. Bacterial metabolism proceeds through a series of steps depending on the availability of electron acceptors.

The gasoline additive methyl tert-butyl ether (MTBE) is believed to be highly resistant to biodegradation since it is reactive with microbial membranes. Some believe that it is slowly biodegraded (Borden et al., 1997) whilst others believe that it partially degrades to tert-butyl alcohol (TBA), a health hazard (Landmeyer et al., 1998). More recently, it has become generally accepted that MNA of MTBE is an acceptable remediation scheme. Sorption and volatilization are limited. However, mechanisms such as uptake by plants, and abiotic degradation by oxidation and hydrolysis are likely. In addition, biodegradation and its by-products are biodegradable under aerobic and anaerobic conditions (Bradley et. al., 2001). Guidance documents have been prepared by the API for the natural attenuation of MTBE (Zeeb and Wiedemeier, 2007).

10.6.2.3 PAHs – Polycyclic Aromatic Hydrocarbons

Polycyclic aromatic hydrocarbons (PAHs), $C_{4n+2}H_{2n+4}$, are components of creosote. As with cycloalkanes, they are difficult to degrade and as the number of rings increases, the compounds become more difficult to degrade – a result of their decreasing volatility and solubility, and increased sorption. They are degraded one ring at a time. As an example, the pathway for biodegradation of anthracene is from anthracene *cis*-1,2-dihydrodiol to salycilate with at least six intermediates beginning with 1,2-dihydroxy anthracene onwards to 1-hydroxy-2-naphthoic acid as the last intermediate before salycilate.

10.6.2.4 Halogenated Aliphatic and Aromatic Compounds

Halogenated aliphatic compounds include (a) pesticides such as ethylene dibromide (DBR) or $CHCl_3$, $CHCl_2Br$ and (b) industrial solvents such as methylene chloride and trichloroethylene. Because of the presence of halogen, the lower energy, and higher oxidation state makes aerobic degradation more difficult to achieve than anaerobic biodegradation. Methylene chloride, chlorophenol, and chlorobenzoate are the most aerobically biodegradable. Removal of the halogen and replacement by a hydroxide group is often the first step of the degradation process, particularly when the carbon chain length is short. An example of this is methylene chloride, with formaldehyde, 2-chloroethanol and 1,2-ethanediol as intermediates and carbon dioxide as the final product (Pitter and Chudoba, 1990).

Biodegradation of chlorinated ethenes involves formation of an epoxide and hydrolysis to carbon dioxide and hydrochloric acid. Reductive dehalogenation can occur anaerobically and involves replacement of the halogen with hydrogen or formation of a double bond when two adjacent halogens are removed (dihalo-elimination). This is the particular case for PCE. As discussed previously, perchloroethylene (PCE) and TCE can be reduced to form vinylidene and vinyl chloride (VC) that are more toxic and volatile than the original compound. Oxidation of vinyl chloride to carbon dioxide and water occurs under aerobic conditions. Induction of monooxygenase or dioxygenase enzymes can lead to the co-metabolism of TCE by methanotrophs (Alvarez-Cohen and McCarty, 1991). However, molecular oxygen, and a primary substrate (methane, ethene, phenol, toluene, or other compounds) must be available for natural attenuation by this mechanism.

Halogenated aromatic compounds include pesticides such as DDT, 2,4-D and 2,4,5-T, plasticizers, pentachlorophenol, polychlorinated biphenyls. Although PCBs have been banned since the 1970s, the record shows that they are still found in aqueous and sediment systems. Congeners containing fewer chlorines are degraded more quickly than those with more than four chlorine atoms (Harkness et al., 1993). Soluble forms are much more likely to biodegrade through natural attenuation than those sorbed to solids or entrapped in NAPLs. Mechanisms involved in transformations and conversions halogenated aromatic compounds include biodegradation, hydrolysis (replacement of halogen with hydroxyl group), reductive dehalogenation (replacement of halogen with hydrogen), and oxidation (introduction of oxygen into the ring causing removal of halogen). As the number of halogens rise, reductive dehalogenation will occur. In addition, ring cleavage could occur before oxidation, reduction or substitution of the halogen.

Bacterial strains of *Pseudomonas sp.*, *Acinetobacter calcoaceticus*, and *Alkaligenes eutrophus* have been able to degrade aromatic halogenated compounds by oxidizing them to halocatechols followed by ring cleavage (Reineke and Knackmuss, 1988). Cleavage for chlorobenzene, for example, can occur either at the ortho position to form chloromuconic acid or at the meta position to form chlorohydroxymuconic semialdehyde. Subsequent dehalogenation can be spontaneous (Reineke and Knackmuss, 1988). As reported by Yong and Mulligan (2004), chlorinated benzoates (Suflita et al., 1983), 2,4,5-T pesticides, PCBs (Thayer, 1991), 1,2,4-trichlorobenzenes (Reineke and Knackmuss, 1988) are known to undergo reductive dehalogenation under anaerobic conditions. *Rhodococcus chlorophenolicus* (Apajalahti and Salkinoja-Salonen, 1987), and *Flavobacterium sp.* (Steiert and Crawford, 1986) can aerobically biodegrade pentachlorophenol while anaerobic degradation of 3 -chlorobenzoate and PCBs has been identified by methanogenic consortia (Nies and Vogel, 1990).

10.6.2.5 PFAS

Polyfluorinated precursors can be microbially transformed in to perfluorinated compounds, in mainly aerobic conditions (Dasu and Lee 2016, Liu and Avendaño 2013). A few aerobic studies have indicated that perfluorinated compounds (PFOS and PFOA) are resistant to microbial degradation or transformation. Anaerobic studies are even fewer (Liu and Avendaño 2013).

10.6.2.6 Pharmaceuticals

Pharmaceuticals have gained attention due to their endocrine-disrupting potential. Yu et al. (2013) determined that pharmaceutical biodegradation and sorption in soil were major mechanisms. Other pharmaceuticals, such as caffeine, acetaminophen, sulfamethoxalone, naproxen, and carbamazepine were studied to evaluate their biodegradability (Martinez-Hernandez et al., 2017). Acetaminophen and caffeine were very degradable while carbamazeprine was the least. However, concentration levels affected these biodegradation rates. Sulfamethoxalone and carbarmazeprine produced intermediate products.

10.6.2.7 Nitroaromatics

Nitroaromatics in the form of 2,4,6-trinitrotoluene (TNT), nitrobenzene, nitrophenol isomers, 2,6- and 2,4- isomers of dinitrotoluene, 1,3-dinitrobenzene, and nitroaniline can be found at contaminated sites from explosive manufacture and use (Wujcik et al., 1992). While the nitrogen group on an aromatic ring can make biodegradation very difficult, degradation by two different pathways has been demonstrated: Nitrogroup reduction (Preuss et al., 1993), or oxidation of the nitrophenol (Spain and Gibson, 1991). The conditions are not well understood for natural attenuation for TNT and RDX (Coleman et al., 1998; Esteve-Nunez and Ramos, 1998). A study by Amin et al. (2017) showed that TNT and pentaerythritol tetranitrate (PETN) could be anaerobically degraded.

10.6.2.8 Metals and Metalloids

It has long been assumed that biological transformation and degradation applied primarily to organic chemical compounds. More recently, however, research has shown that microbial conversion of metals occurs. The following short account summarizes the discussion from Yong and Mulligan (2019). Microbial conversion includes bioaccumulation, biological oxidation/reduction, and biomethylation (Soesilo and Wilson, 1997). Microbial cells can accumulate heavy metals through ion exchange, precipitation, and complexation on and within the cell surface containing hydroxyl, carboxyl, and phosphate groups. Bacterial oxidation/reduction could be used to alter the mobility of the metals. For example, some bacteria can reduce Cr (IV) in the form of chromate (CrO_4^{2-}), and dichromate ($Cr_2O_7^{2-}$) to Cr(III), which is less toxic and mobile due to precipitation above pH 5 (Bader et al., 1996).

The rhamnolipid biosurfactant has been demonstrated for the removal and reduction of hexavalent chromium from contaminated soil and water in batch experiments (Ara and Mulligan, 2015). A sequential extraction study was used on soil before and after washing to determine from what fraction the rhamnolipid removed the chromium. The exchangeable and carbonate fractions accounted for 24% and 10% of the total chromium, respectively. The oxide and hydroxide portions bound 44% of the total chromium in the soil. On the other hand, 10% and 12% of the chromium was associated with the organic and residual fractions. Rhamnolipid was able to remove most of the exchangeable (96%) and carbonate (90%) portions and some of the oxide and hydroxide portion (22%) but from the other fractions. This information is important in designing the appropriate conditions for soil washing and for potential aspects of natural attenuation in the presence of biosurfactant producing microorganisms.

Mercury can be found as Hg(II), volatile elemental mercury (Hg(0)), methyl- and dimethyl forms. Metabolism occurs through aerobic and anaerobic mechanisms through uptake, conversion of Hg(II) to Hg(0), methyl and dimethylmercury or to insoluble Hg(II) sulphide precipitates. Although volatilization or reduction during natural attenuation would still render mercury mobile, Hg(II) sulphides are immobile if sufficient levels of sulphate and electron donors are available.

Arsenic can be found as the valence states As(0), As(II), As(III), and As(V). Forms in the environment include As_2S_3, elemental As, arsenate (AsO_4^{3-}), arsenite (AsO_2^-) and other organic forms such as trimethyl arsine and methylated arsenates. The anionic forms are mobile and highly toxic. Microbial transformation under aerobic conditions produces energy through oxidation of arsenite. Other mechanisms include methylation, oxidation, and reduction under anaerobic or aerobic conditions.

Selenium, which is a micronutrient for animals, humans, plants, and some microorganisms can be found naturally in four major species, selenite (SeO_3^{2-}, IV), selenate (SeO_4^{2-}, VI), elemental selenium (Se (0)) and selenide (-II) (Frankenberger and Losi, 1995; Ehrlich, 1996). Oxidation of selenium can occur under aerobic conditions while selenate can be transformed anaerobically to selenide or elemental selenium. Methylation of selenium detoxifies selenium for the bacteria by removing the selenium from the bacteria. Immobilization of selenate and selinite is accomplished via conversion to insoluble selenium. Due to the many forms of selenium, selenium decontamination by microorganisms is not promising.

10.6.2.9 Nitrogen

Bacteria are able to remove nitrogen from the soil and releases it into the atmosphere in a process called dissimulative nitrate reduction. Denitrifying bacteria are always facultative and thus can function in the presence and absence of oxygen. Another process that is anaerobic occurs when both ammonium and nitrite/nitrate ions are present and is called annamox (Caschetto et al., 2018), and is shown in Equation (10.1):

$$NH_4^- + NO_2^- \rightarrow N_2 + 2H_2O \tag{10.1}$$

10.6.2.10 Sulphur

Sulphate-reducing bacteria can convert sulphate to hydrogen sulphide, H_2S, or sulphur, S. Only about 12 species of bacteria are sulphate-reducing. Sulphur bacteria that are facultative can process sulphur, hydrogen sulphide, thiosulphate and organic sulphides or organic materials for energy. Some reactions are as follows:

$$\begin{aligned} &H_2S + 2O_2 \rightarrow SO_4^{2-} + 2H^+ \\ &S^0 + H_2O + \frac{3}{2}O_2 \rightarrow SO_4^{2-} + 2H^+ \\ &S_2O_3^{2-} + H_2O + 2O_2 \rightarrow 2SO_4^{2-} + 2H^+ \end{aligned} \tag{10.2}$$

The production of sulphate from hydrogen sulphide occurs vias several steps. The species *Beggiatoa* is often found in sediments with high contents of hydrogen sulphide (Eweis et al., 1998).

Metal (Me) sulphides are formed by sulphate- reducing bacteria (SRB) according to the following:

$$\begin{aligned} &CH_3COOH + SO_4^{2-} \rightarrow 2HCO_3 + HS^- + H^+ \\ &H_2S + Me^{2+} \rightarrow MeS + 2H^+ \end{aligned} \tag{10.3}$$

The presence of an electron donor such as methanol, sulphate, and low redox conditions are required. An increase in pH can stimulating sulphate reduction and allow metal hydroxide and oxide formation that precipitate and inhibit migration in soils and groundwater.

10.6.3 Oxidation-Reduction (Redox) Reactions

It is useful to recall that (a) the chemical reaction process defined as *oxidation* refers to a removal of electrons from the subject of interest, and (b) *reduction* refers to the process where the "subject (electron acceptor or *oxidant*)" gains electrons from an electron donor (*reductant*). By gaining electrons, a loss in positive valence by the subject of interest results and the process is called a reduction. Oxidation-reduction (redox) reactions have been briefly discussed in Chapter 9. Biological transformation of organic chemical compounds results from biologically mediated redox reactions. Bacteria in the soil utilize oxidation-reduction reactions as a means to extract the energy required for growth. They are the catalysts for reactions involving molecular oxygen and organic chemicals (and also soil organic matter) in the ground. Oxidation-reduction reactions involve the transfer of electrons between the reactants. The activity of the electron e^- in the chemical system plays a significant role. Reactions are directed toward establishing a greater stability of the outermost electrons of the reactants, i.e., electrons in the outermost shell of the substances involved. The link between redox reactions and acid-base reactions is evidenced by the proton transfer that accompanies the transfer of electrons in a redox reaction. Manahan (1990) gives the example of the loss of three hydrogen ions that accompanies the loss of an electron by iron(II) at pH 7 resulting in the formation of a highly insoluble ferric hydroxide, as indicated by the following:

$$Fe(H_2O)_6^{2+} \rightarrow Fe(OH)(H_2O)_5^{2+} + H^+ \tag{10.4}$$

It is not easy to distinguish between abiotic and biotic (biologically mediated) redox reactions. To a large extent, it is not always possible to eliminate or rule out involvement of microbial activity in abiotic redox reactions. There does not appear to be a critical need to distinguish between the two in reactions that concern organic chemical contaminants, since it is almost certain that with all the microorganisms in the subsoil, some measure of microbial activity would be involved. In any event, the number of functional groups of organic chemical contaminants that can be oxidized or reduced under abiotic conditions is considerably smaller than those under biotic conditions (Schwarzenbach et al., 1993).

The two classes of electron donors of organic chemical contaminants are (a) Electron-rich B-cloud donors, which include alkenes, alkynes, and the aromatics; and (b) lone-pair electron donors which include the alcohols, ethers, amines and alkyl iodides. Similarly, in the case of electron acceptors, we have (i) electron-deficient π-electron cloud acceptors which include the π-acids; and (ii) weakly acidic hydrogens such as s-triazine herbicides and some pesticides.

A measure of the electron activity in the porewater of a soil-water system is the *redox potential Eh*. It provides us with a means for determining the potential for oxidation-reduction reactions in the contaminant-soil-water system under consideration, and is given as:

$$Eh = pE\left(\frac{2.3RT}{F}\right) \tag{10.5}$$

where E is the electrode potential, R is the gas constant, T is the absolute temperature and F is the Faraday constant. The electrode potential E is given in terms of the half reaction:

$$2H^+ + 2e^- \Leftrightarrow H_2(g) \tag{10.6}$$

When the activity of $H^+ = 1$ and the pressure H_2 (gas) =1 atmosphere, we obtain $E = 0$.

10.7 NATURAL ATTENUATION AND IMPACT MANAGEMENT

The natural attenuation capacity of soils in the substratum has long been recognized and described by soil scientists as the assimilative capacity of soils. The discussions at the beginning of the previous section and in the earlier chapters of this book show that this is now a tool that can be used as a passive treatment process in the remediation and management of sites contaminated by organic chemicals. The USEPA has wisely coupled the requirement for continuous on-site monitoring of contaminant presence whenever natural attenuation is to be used as a tool for site remediation – as seen in the definition provided in the first part of Section 10.4. The procedure for application of this attenuation process is called *monitored natural attenuation (MNA)*. Guidelines and protocols for application of MNA as a treatment procedure in remediation of contaminated sites have been issued. Since site specificities differ from site to site, the prudent course of action is to adapt the guidelines and protocols for site specific use. A general protocol, from Yong and Mulligan (2019) for considering MNA as a remediation tool is shown in Figure 10.13. A very critical step in the application of MNA as a site remediation tool is to have proper knowledge of (a) *lines of evidence* indicating natural or intrinsic remediation, (b) contaminants, soil properties and hydrogeology, and (c) regulatory requirements governing *evidence of success* of the MNA remediation project. Lines of evidence and evidence of success will be discussed in the latter portion of this section.

The data and information inputs shown on the left-hand-side of the diagram in Figure 10.13 tell us what is required to satisfy site-specific conditions, and whether the *indicators* for natural bioremediation are sufficient to proceed with further examination to satisfy that the use of MNA is a viable treatment option. Negative responses from the first two decision steps will trigger technological

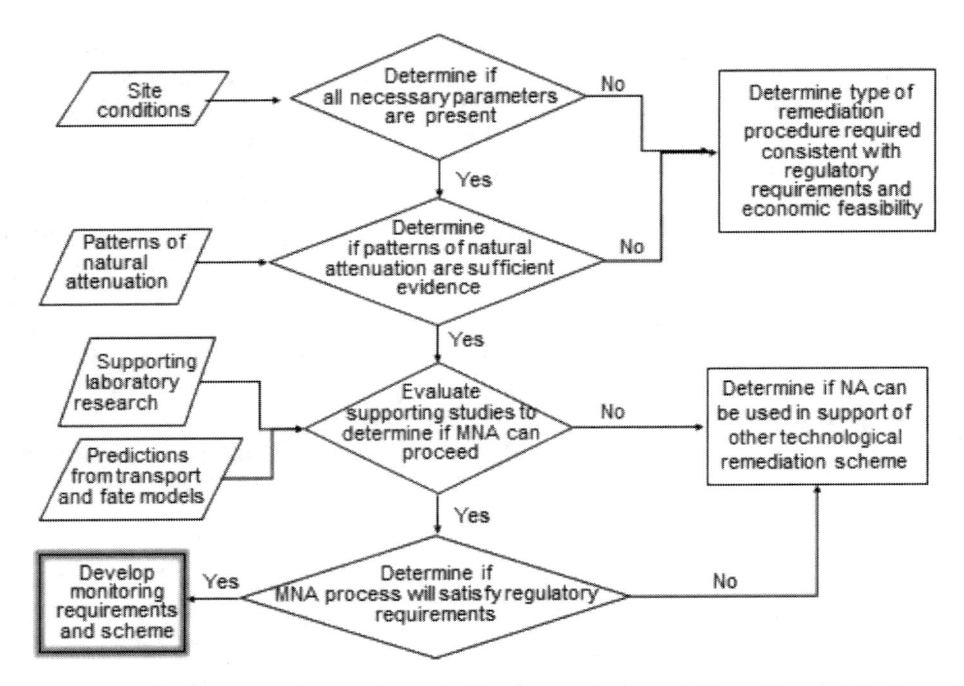

FIGURE 10.13 General protocol for considering MNA as a remediation tool. (From Yong and Mulligan, 2019.)

and/or engineered solutions to the remediation problem. As will be seen, laboratory research and transport and fate modelling are needed to inform one about the ability of the site materials and conditions to attenuate the contaminants.

Adoption of natural attenuation (NA) as an active tool in the management of contaminant impact and transport, as opposed to the use of MNA as a passive tool, has been hampered because of insufficient knowledge of the many processes that contribute to the natural attenuation process. The designation of MNA as more of a passive tool as opposed to an active tool is based on the fact that except for the monitoring requirement, the use of natural attenuation processes as existent *in situ* is essentially a "do-nothing" solution. The "do-nothing" part refers to human contribution to the processes resulting in natural attenuation of contaminants. As a clarification, we should point out that the acronym "NA" is used to denote natural attenuation as a process tool. When we wish to discuss the processes that result in the natural attenuation of contaminants, we will use the complete term *natural attenuation*. To make NA an active tool, we can (a) enhance the processes that contribute to natural attenuation capability, (b) incorporate NA as part of a scheme to mitigate and manage the geoenvironmental impacts from discharge and/or containment of waste products and contaminants.

10.7.1 Enhancement of Natural Attenuation Capability

Successful application of waste discharge and ground contamination impact mitigation procedures reduces and/or minimizes the damage done to the land environment and its inhabitants. The main objective of such procedures is to achieve reduction or elimination of health threats to the biotic receptors through elimination or minimization of ground contamination, and prevention of contamination of receiving waters and groundwater. The objectives or desired end points sought in contaminant impact mitigation fit very well with the capabilities of the processes in soils that contribute to natural attenuation. Accordingly, natural attenuation can be used as tool to provide impact mitigation and management since the basic processes involved in natural attenuation result in reduction of concentration of contaminants and toxicity. To increase the capability of natural attenuation as a process tool – i.e., the process tool NA – one can consider enhancement of the natural assimilative capacity of the soil-water system. These can take the form of geochemical and biogeochemical aids, bioaugmentation and biostimulation. Successful enhancement of NA as a process tool will produce a subsoil with properties that can be considered as *enhanced NA capability.*

10.7.1.1 Soil Buffering Capacity Manipulation

The capability of a soil to accept and retain inorganic and some organic contaminants can, in some instances, be assessed by determining its chemical buffering potential – particularly if the reactions in the soil-water system result in changes in the pH of the system. The chemical buffering system contributes significantly to the carrying capability of the soil, i.e., the capability of the soil barrier or sub-soil to accept and retain contaminants. The main issue is the ability of the soil-water system to maintain a natural pH level (within acceptable limits) in spite of input of acidic or alkaline contaminant leachates. *In situ* soil pH manipulation for the purpose of contaminant impact mitigation requires introduction of buffering agents generally through leaching methods or via injection. In contaminant-soil interaction, the chemical buffering system describes the capability of the system to act as chemical barrier against the transport of contaminants.

The buffering capacity of a soil determines the potential of a soil for effective interaction with leachate contaminants and is more appropriate for inorganic soils and inorganic contaminant leachates. The principal features that establish the usefulness of buffering capacity assessment centre around the "*acidity*" or "*alkalinity*" of the initial soil-water system and the solutes in the leachate. Soil conditioning in respect to changes in the natural soil buffering capacity is usually considered in terms of addition of buffering agents, much in the same manner as solution chemistry. In the *in situ* soil conditioning case, however, addition of buffering agents needs to be effectuated through injection wells or through leaching – as for example by adding lime to the surface as the leach source to increase the pH of the soil. In a site contaminated with heavy metals, raising the pH of the soil-water system would precipitate the heavy metal contaminants and thus make them less environmentally mobile and less bioavailable. However, we must recognize that this is not a permanent solution because if the pH of the system is subsequently reduced by environmental forces or external events, the same heavy metals will become mobile again. To avoid subsequent solubilization, the precipitated heavy metals should be removed from the contaminated site.

10.7.1.2 Biostimulation and Bioaugmentation

Chapter 3 introduced the use of biological aids, *biostimulation* and *bioaugmentation,* as part of the available tools for groundwater management. *Biostimulation* occurs when stimuli, such as nutrients and other growth substrates, are introduced into the ground to promote increased microbial activity of the microorganisms existing at the site. The intent is to obtain improved capabilities of the microorganisms to more effectively degrade the organic chemical contaminants in the soil. The addition of nitrates, Fe (III) oxides, Mn (IV) oxides, sulphates and CO_2 for example, will allow for anaerobic degradation to proceed. Biostimulation is perhaps one of the least intrusive of the methods of enhancement of the natural attenuation capacity soils.

Bioaugmentation denotes the process whereby exogenous microorganisms are introduced *in situ* to aid the native or indigenous microbial population in degrading the organic chemicals in the soil. The reason one would use bioaugmentation is presumably because the microorganisms in the soil are not performing up to expectations. This could be because the concentrations of microorganisms are insufficient, or maybe because of inappropriate consortia. The function of the exogenous microorganisms is to augment the indigenous microbial population such that effective degradative capability can be obtained. Frequently, biostimulation is used in conjunction with bioaugmentation. There is the risk that (a) use of microorganisms grown in uncharacterized consortia, which include bacteria, fungi, and viruses, can produce toxic metabolites (Strauss, 1991), and (b) the interaction of chemicals with microorganisms may result in mutations in the microorganisms themselves, and/or microbial adaptations.

10.7.1.3 Biochemical and Biogeochemical Aids

Introduction of geochemical aids in-situ utilize the same techniques employed to introduce the various kinds of growth substrate, nutrients and exogenous microorganisms for biostimulation and bioaugmentation. Manipulations of pH and *pE* or *Eh* using geochemical aids can increase the capability of the soil to mitigate the impact of some toxic contaminants. A good case in point is the changes in toxicity for chromium and arsenic because of changes in their oxidation state. Chromium (Cr) as Cr (III) is an essential nutrient that helps the body use sugar, protein and fat. On the other hand, chromium as Cr (VI) has been determined by the World Health Organization (WHO) to be a human carcinogen. Cr (III) can be oxidized to Cr (VI) by dissolved oxygen and quite possibly with manganese dioxides. If such a possibility exists in a field situation, management of the potential impact can take the form of in-situ geochemical and/or biogeochemical intervention to create a reducing environment in the subsurface. A useful procedure would be to deplete the oxygen in the subsurface to create a reduced condition in the soil. The danger or risk of manipulation of the *Eh* of the soil-water regime is incomplete knowledge of all the elements in the subsoil that are vulnerable to such manipulation. The case of arsenic in the ground is a good example. It is known that arsenic, (As), as As (III) is more toxic than As (V). If ground conditions show that arsenic is present as As (V), creating a reducing environment to prevent oxidation of chromium to the more toxic oxidation state would create the reverse effect on As (V). Reduction of As (V) to As (III) would increase the toxicity of arsenic.

Manipulation of pH can change the nature of the assimilative capacity soils – as stated previously. It addresses the precipitation of heavy metals in solution (pore water) or dissolution of precipitated heavy metals. Changes in pH of the soil-water system will produce changes in the sign of surface electrostatic charges for those soil materials with amphoteric surfaces, i.e., surfaces that show pH-dependency charge characterization. Subsoils containing oxides, hydrous oxides and kaolinites are good candidates for pH-dependent charge manipulation. Changes in surface charge characterization can result in increased bonding of metals or release of heavy metals from disruption of bonds. Both pH and *Eh* changes will have considerable influence also on acid-base reactions and on abiotic and biotic electron transfer mechanisms. Abiotic transformations of organic chemical contaminants due to acid-base and oxidation-reduction reactions are minor in comparison to biotic transformations.

10.7.2 NA Treatment Zones for Impact Mitigation

Treatment zones are regions in the subsoil that utilize the NA process tool and more specifically, enhanced NA, to attenuate the impact of contaminants during transport in the subsoil. Since it is rare to find source discharges conveniently located directly in a region where the various elements contributing to natural attenuation capabilities are available, the usual procedure is to provide a treatment zone that would capture the contaminant plume during transport. A good knowledge of site hydrogeology is essential for this type of mitigation procedure to function properly. Figure 10.14 shows a simple SPR (source-pathway-receptor) problem that uses a treatment zone to mitigate the impact of the contaminant plume generated by the pollution source shown at the bottom left of the diagram. To determine if the treatment zone is effective in mitigating the impact of the generated contaminant, the monitoring scheme shown in Figure 10.15 is recommended. By this means, determination of reduction in concentration and toxicity of the contaminants can be obtained. This procedure also allows one to establish evidence of success of the treatment zone. It can be argued that one should remove the contamination source as part of the mitigation procedure. Assuming that the source is an industrial facility, the obvious course of action is to (a) implement operational procedures in the facility that will reduce output of contaminating items, and (b) establish or improve treatment of

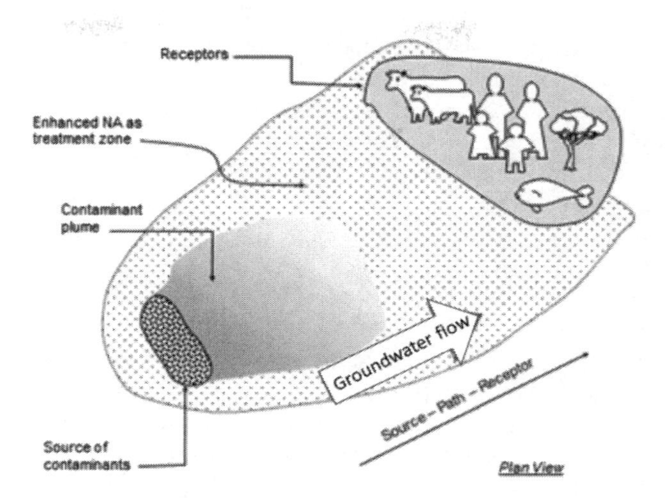

FIGURE 10.14 Simplified plan view of a treatment zone established with enhancement of the natural attenuation capability of the site sub-soil – to mitigate impact of contaminants on land environment and to protect *Receptors*.

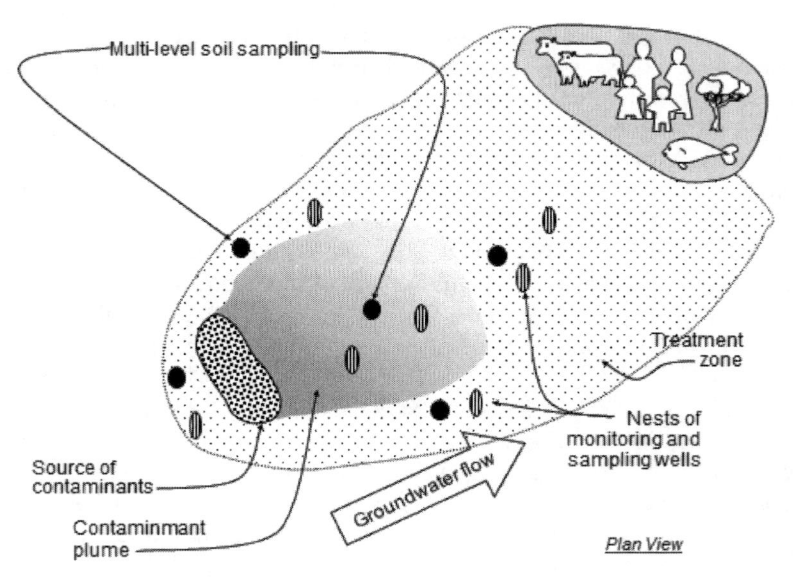

FIGURE 10.15 Plan view of distribution of monitoring wells and soil sampling boreholes for verification monitoring of treatment zone effectiveness and long-term conformance monitoring.

discharges to capture all the noxious substances discharged into the land environment. The combination of the treatment zone with removal of contaminant source will ensure short-term mitigation and longer-term elimination of the threats posed by the contamination plume – a positive step towards geoenvironmental sustainability.

10.7.2.1 Permeable Reactive Barriers and NA

Permeable reactive barriers (PRBs) are engineered material-barriers constructed and placed in the ground to intercept contaminant plumes. The material in these barriers generally consists of permeable soil material containing various elements designed to react with the kinds of contaminants entering the barrier. The whole intent of PRBs is to provide a chemical-physical sieve or filter that would capture the contaminants as they pass through the barrier. In a sense therefore, PRBs are barriers with highly engineered and efficient attenuation properties and characteristics. They are sometimes also known as treatment walls.

The major contaminant capture, and immobilization processes needed for the engineered materials in the PRBs to function effectively include (a) sorption, precipitation, substitution, transformation, complexation, oxidation, and reduction for inorganic contaminants, and (b) sorption, biotic, and abiotic transformations, and degradation for organic chemical contaminants. The types of reagents, compounds, and micro-environment in the PRBs include a range of oxidants and reductants, chelating agents, catalysts, microorganisms, zero-valent metals, zeolite, reactive clays, ferrous hydroxides, carbonates and sulphates, ferric oxides and oxyhydroxides, activated carbon and alumina, nutrients, phosphates, soil organic materials. The selection of engineered materials in the PRBs, such as reagents and compounds, and the manipulation of the pH-*pE* micro-environment in the treatment walls will need to be made on the basis of site-specific knowledge of the nature of the contaminants.

The success of PRBs in mitigating contaminant impacts depends on the following:

(a) Effectiveness of types of engineered material in the PRB: This depends on a proper knowledge of the contaminants, barrier material, and the kinds of processes (interactions and bonding mechanisms) resulting from contaminant-barrier material interactions.
(b) Sufficient residence time of the contaminant plume in the PRB: There must be sufficient residence time in the PRB for contaminant-material interactions and reactions to be fully realized. This is a function of both the permeability of the barrier itself and the kinds of reaction times needed between contaminants and the barrier material. It would be useless if the contaminant passed through the barrier at high rates – rates that would not permit reactions to be completed. Conversely, it would be useless if the contaminant would not penetrate the barrier, hence denying any opportunity for reactions to occur.
(c) Proper intercept of contaminant plume advance: A thorough knowledge of site hydrogeology is required to allow one to place the barrier for optimum intercept of the contaminant plume. Some knowledge also of the advective velocity is also required.

For more effective use of PRBs, an enhanced NA treatment zone can be used and placed ahead of the PRB. Such a case has been shown in Figure 3.11 in Chapter 3. The treatment zone in the diagram is shown as an optional tool. Contaminant plumes can be channelled to flow through reactive walls by shepherding the plume with a "*funnel-gate*" technique. In this technique, the plume is essentially guided to the intercepting reactive wall by a funnel constructed of impermeable material, such as sheet pile walls, and placed in the contaminated ground to channel the plume to the PRB (Figure 10.16). Other variations of the funnel-gate technique exist – obviously in accordance with site geometry and site specificities.

10.8 LINES OF EVIDENCE

Lines of evidence is a term that is associated with the use of NA as a tool for mitigation and management of impacts from waste and contaminant discharges to the land environment. *Lines of evidence* (LOE) refers to the requirement to determine whether a soil has capabilities for *in situ* attenuation of contaminants. This requirement originates from procedures associated with the use of monitored natural attenuation (MNA) as a treatment procedure. This is a prudent course of action since there is need to determine how effective a particular soil will be in attenuating contaminants. The types of information and analyses required for LOE indicators are shown in Figure 10.17. Site and problem

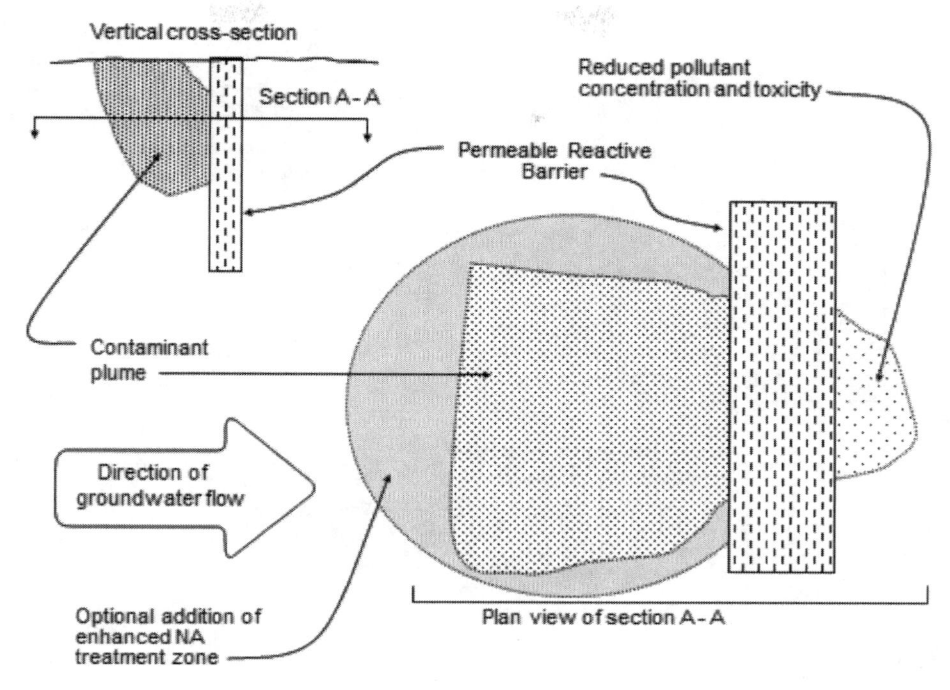

FIGURE 10.16 Cross-section and plan view of permeable reactive barrier (PRB). Plan view shows leachate plume entering the PRB with contaminants and leaving the PRB with reduced contaminant concentration and toxicity. If required, an enhanced NA treatment zone can be situated in front of the PRB – to increase the effectiveness of the contaminant impact mitigation scheme. (Adapted from Yong and Mulligan, 2019.)

specificities will dictate how many pieces of information and what specific kinds of analyses will be needed. The type of information needed to define the site characteristics is shown in the top right-hand corner of the diagram. The physical (geologic and hydrogeologic) setting sets the parameters of the problem to be resolved. Whether the natural attenuation capability of the sub-soil is capable of mitigating and managing the contaminant plume anticipated within the site boundaries will be established by the other two categories: Patterns of natural attenuation and supporting laboratory tests and analyses.

Knowledge of the patterns of natural attenuation identified in the central box is essential. What is sought in this category is evidence of previous natural (intrinsic) remediation of contaminants. To determine this, it is necessary to recall the various mechanisms and processes that establish retention and transformation of the various kinds of contaminants generally found in the sub-soil. In addition, one needs to determine or assess the environmental mobility of the contaminants in the site under consideration. These are necessary pieces of information for prescription of *indicators* for the lines of evidence (LOE).

10.8.1 Organic Chemical Compounds

The previous chapters have shown that organic chemicals such as organic solvents, paints, pesticides, oils, gasoline, creosotes, and greases, etc., are responsible for many of the chemicals found in contaminated sites. These chemicals are known generally as xenobiotic compounds. It is not possible to categorize them all in respect to how they would interact in a soil-water system. The more common organic chemicals found in contaminated sites can be grouped into three broad groups:

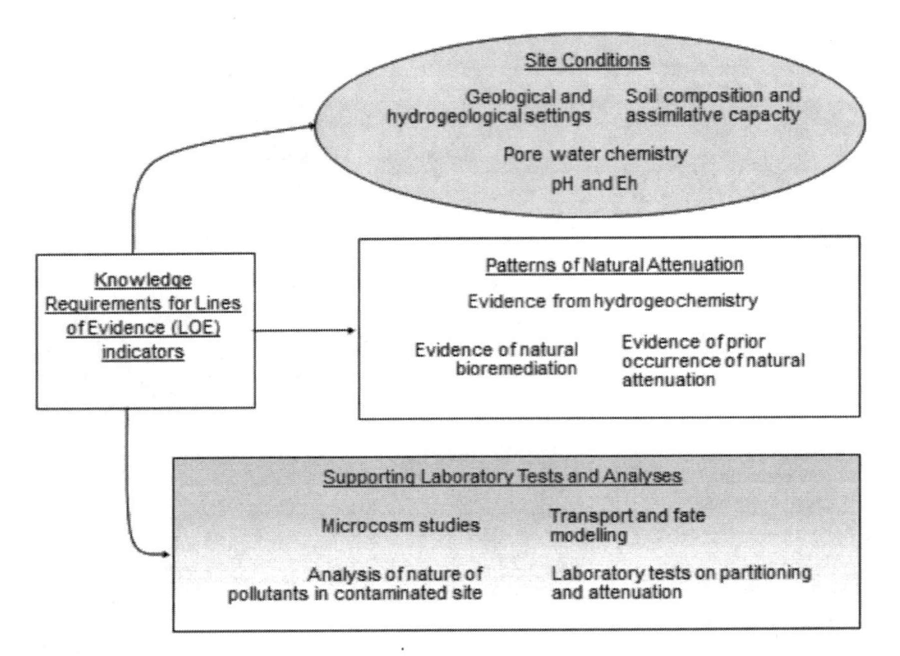

FIGURE 10.17 Required information and analyses for *lines of evidence* (LOE) indicators.

- Hydrocarbons – including the PHCs (petroleum hydrocarbons), the various alkanes and alkenes, and aromatic hydrocarbons such as benzene, MAHs (multicyclic aromatic hydrocarbons), e.g., naphthalene; and PAHs (polycyclic aromatic hydrocarbons), e.g., benzo-pyrene;
- Organohalide compounds – of which the chlorinated hydrocarbons are perhaps the best known. These include TCE (trichloroethylene), carbon tetrachloride, vinyl chloride, hexachloro-butadiene, PCBs (polychlorinated biphenyls) and PBBs (polybrominated biphenyls).
- Miscellaneous compounds – including oxygen-containing organic compounds, such as phenol and methanol, and nitrogen-containing organic compounds such as TNT (trinitrotoluene).

As we have seen in Chapter 2, the density of these compounds in comparison to that of water has direct control on their transport in the subsoil. We classify non-aqueous phase liquids (NAPL) into the light NAPLs identified as LNAPLs, and the dense ones called the DNAPLS. LNAPLs include gasoline, heating oil, kerosene, and aviation gas. DNAPLs include the organohalide and oxygen-containing organic compounds such as 1,1,1-trichloroethane, chlorinated solvents such as tetrachloroethylene (PCE), trichloroethylene (TCE) and carbon tetrachloride (CT), PCBs, PCPs (pentachlorophenols) and TCPs (tetrachlorophenols). As shown in Figure 2.7 in Chapter 2, since LNAPLs are lighter than water and the DNAPLs are heavier than water, NAPLs will likely stay above the water table, and DNAPLs tend to sink through the water table and come to rest at an impermeable bottom (bedrock).

The various results of transformations and biodegradation of organic chemicals have been discussed in various forms in the earlier part of this chapter. The significant outcome of NA as a tool for mitigation of impact is the evidence of occurrence of biodegradation and transformation of the target organic chemicals in the NA process. The *indicators* that need to be prescribed in the LOE relate to specific decreases in concentration of the contaminants and transformations (conversions and biodegradation) of organic chemical contaminants. Determination of the nature and composition of the transformed products of the original organic chemical contaminants is required. Knowledge of the products obtained via abiotic and biotic processes is essential. A good example of this is, for example, recognizing that abiotic transformation products are generally other kinds of organic chemical compounds, whereas transformation products resulting from biotic processes are mostly seen as stages (intermediate products) towards mineralization of organic chemical compounds. Procedures for the identification and characterization of abiotic processes for chlorinated hydrocarbons have been reviewed (USEPA, 2009). Biologically mediated transformation processes are the only types of processes, which can lead to mineralization of the subject organic chemical compound. Complete conversion to CO_2 and H_2O (i.e., mineralization) does not always occur. However, intermediate products can be formed during the mineralization. New techniques are being developed to study the sources of the contaminants and the biodegradation process such as Compound Specific Isotope Analysis (CSIA) on dissolved organic contaminants, such as chlorinated solvents, aromatic petroleum hydrocarbons, and fuel oxygenates isotope (USEPA, 2008b).

Two important mechanisms for PFAS retention include chemical and geochemical retention (Newell et al. 2021a). Contrary to many organic contaminants, biodegradation is not viewed as a major mechanism. Intermediate products PFAS and/or the final products, perfluoroalkyl acids (PFAAs) may occur.

Munitions constituents (MC) are "any materials originating from unexploded ordnance (UXO), discarded military munitions (DMM), or other military munitions, including explosive and non-explosive materials and emission, degradation, or breakdown elements of such ordnance or munitions (10 U.S.C. 2710(e)(3))." While MNA has been applied at nitroaromatic explosive (Pennington et al., 1999) and perchlorate sites (ITRC, 2005) a report by Rectanus et al. (2015) summarized the state-of-technology of MNA for all MC categories. Biological, physical, and chemical mechanisms for MNA were reviewed, in addition to eight case studies. Three lines of evidence (LOE) are used to evaluate the potential of MNA for MC. They include (1) plume stability and geometry (i.e, nature and level of contamination over time and space), (2) geochemical conditions (i.e., pH, redox potential, leaching potential), and (3) microbial indicators (e.g., microcosms, CSIA). Enhanced MNA can be achieved via addition of chemicals (e.g., reductants), nutrients and/or microbes. Overall, sorption of MC in soils and organic matter controls the transport in soil. Although RDX is less soluble and has a lower sorption potential compared to TNT, it is more mobile. The redox potential has a significant influence on the biotransformation of MC. Sorption and precipitation are major mechanisms on the natural attenuation of metal components. The LOE approach with site-specific data is recommended for evaluating the potential of NA at MC sites for Naval Facilities.

10.8.2 Metals

At the very least, prescription of the indicators for the lines of evidence in respect to heavy metals requires determination of (a) the nature and concentration of sorbed metal ions, (b) porewater chemistry including pH and *Eh*, and (c) the environmental mobility of heavy metals. The environmental mobility of heavy metals is dependent to a very large extent upon whether they are in the pore water as free ions, complexed ions or sorbed onto the soil particles. Prescription of *indicators* for LOE should take into account the assimilative capacity of the subsoil and the nature and fate of the heavy metals in the subsoil. So long as the full assimilative potential of the soil for HM is not reached, attenuation of the HMs will continue. Metals that are sorbed onto the soil particles are held by different sets of forces – determined to a large extent by the soil fractions and the pH of the soil-water system. The various types of soils and their different soil fractions have different sorption capacities, dependent on the nature and distribution of the HMs and pH of the system.

A four-tiered approach has been developed by the USEPA (2007a) for inorganic contaminants as follows:

I. It must be demonstrated that the ground-water plume is not enlarging and that if immobilization is a dominant attenuation process, then sorption of the contaminant onto aquifer solids is occurring.

II. Both the *rate and mechanism* of the attenuation process must be determined.
III. The *capacity* of the aquifer to attenuate the contaminant within the plume and the irreversibility of the immobilization must be determined.
IV. The monitoring program and a contingency plan must be designed based on the determined mechanisms for the attenuation process and site characteristics.

A second volume in the series looked at specific MNA information for various non-nuclides including arsenic, cadmium, chromium, copper, lead, nickel, nitrate, perchlorate, and selenium (USEPA, 2007b). The third volume addresses radio-nuclides (USEPA, 2010) including tritium, radon, strontium, technetium, uranium, iodine, radium, thorium, cesium, and plutonium-americium. In addition to the previously mentioned mechanisms, radioactive decay processes are emphasized for these contaminants.

Precipitation of HMs as hydroxides, sulphides, and carbonates generally classify as part of the assimilative mechanism of soils because the precipitates form distinct solid material species, and are considered as part of the attenuation process. Either as attached to soil particles or as void pluggers, precipitates of HMs can contribute significantly to attenuation of HMs in contaminant plumes. Ion exchange can also play a role. Determination of the chemistry including pH and *Eh* for assessment of lines of evidence should not neglect examination of possibilities of precipitation and solubilization of metals as part of the evidence phase. Hydroxide precipitation is favoured in alkaline conditions as for example when $Ca(OH)_2$ is in the groundwater in abundance. With available sulphur and in reducing conditions, sulphide precipitates can be obtained. Sulphide precipitates can also be obtained as a result of microbial activity – except that this will not be a direct route. Sulphate reduction by anaerobic bacteria will produce H_2S and HCO_3^- thus producing the conditions for formation of metal sulphides.

Significant knowledge has been gained, particularly with regards to site characterization (USEPA, 2008a). For example, at the US Department of Energy Handford Site, liquid waste with uranium entered an unconfined aquifer. It was estimated that natural attenuation could reduce the uranium levels to less than 20 μg/L in the next 3 to 10 years. Characterization was revised to reduce the uncertainty in the natural attenuation processes. The newer components include laboratory tests for sorption-desorption testing measuring mass fluxes in the smear zone, and evaluating reversibility of sorption, uranium speciation measurement with x-ray spectroscopy/diffraction and electron microscopy with chemical extraction tests, identification of clay minerals in the aquifer and determining uranium distribution according to particle size. By controlling the flux of uranium into the aquifer and by better characterizing the mass and speciation of uranium, it is estimated that MNA can play an important role in the remediation.

The role of natural attenuation was evaluated in a reservoir in Brazil contaminated with radionuclides and metals as the result of acid mine drainage from a uranium mine (Chaves et al., 2023). High contents of acid volatile sulfide (AVS) and changes in isotopes demonstrated the chemical stabilization of ^{238}U, ^{226} Ra, and zinc in the sediments of the reservoir. However, sequential extraction tests indicated that the radionuclides are associated with the labile fraction of the sediments and thus monitoring is needed to ensure migration of the contaminants is not occurring.

10.9 EVIDENCE OF SUCCESS

Evidence of success (EOS) is a requirement specified by Yong and Mulligan (2019) as testimony to the success of utilization of NA as a tool for remediation of contaminated sites. Whilst monitoring is a necessity in application of MNA, there is a need for one to have knowledge of whether the "signs and signals" registered in the monitoring programme testifies to a successful MNA treatment programme. In essence, EOS takes the role of *indicators* of success or steps towards success by MNA in remediation of contaminated sites.

With the same rationale, EOS can be used in impact mitigation and management programmes. Figure 10.18 takes the MNA protocol shown in Figure 10.13 as the basis for determination whether NA can be successfully used as an impact mitigation and management tool. The first two levels of protocol are similar to the MNA steps. At the third step or level, a clear knowledge of the kinds of impact, and mitigation and management requirements need to be articulated. These are combined with information from laboratory tests that are designed to provide the kinds of information necessary to determine material parameters and interaction processes. Supporting predictions on fate and transport are necessary pieces of information. All of these combine to provide one with the tools to determine whether NA can be successfully used to meet the requirements for impact mitigation and management. Negative responses will require that NA be rejected as a tool, or used in conjunction with other technological tools to provide the necessary impact mitigation and management solution.

A positive response will require structuring implementation procedures and strategies in combination with a competent monitoring scheme to track progress. A very necessary part of the implementation scheme is the specification or prescription of *success indicators* shown in the ellipse at the bottom right-hand corner of Figure 10.18. These indicators also serve as markers for performance assessment of the impact mitigation scheme. We have shown the requirements in terms of intermediate stage indicators and final success indicators. Since the time required for processes contributing to the natural attenuation to fully complete their functions, it is necessary to prescribe intermediate indicators as tracking indicators and as performance assessment markers. Note that we have used *italicized* notation for *indicators* when we mean them as markers, and have left them without italicization when we discuss them as general items.

Monitoring and sampling of porewater and soils are needed in the contaminant attenuation zone. The choice of type of monitoring wells and sampling devices and their spatial distribution and/or

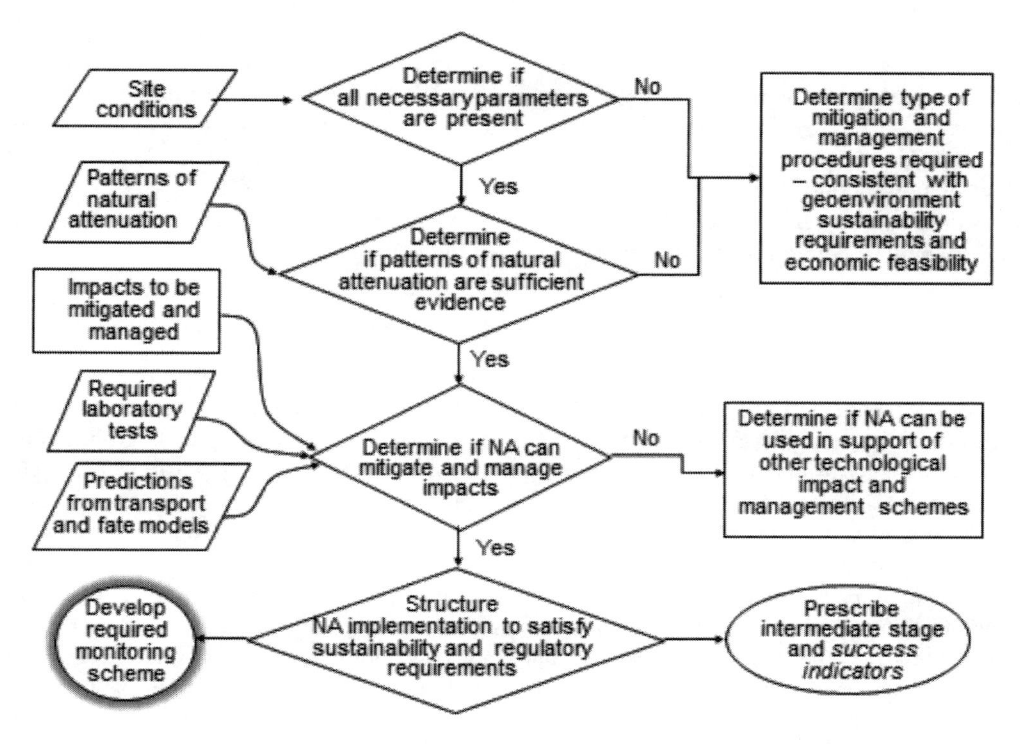

FIGURE 10.18 General protocol to determine feasibility and application of NA as a tool for impact mitigation and management.

location will depend on the purpose for the wells and devices. At least three separate and distinct monitoring-sampling schemes need to be considered:

- Initial site characterization studies. Site characterization monitoring and sampling provide information on site subsoil properties and hydrogeology. Subsurface flow delineation provides one with the information necessary to anticipate transport direction and extent of contaminant plume propagation. With a proper knowledge of the requirements of the verification and long-term monitoring-sampling schemes, a judicious distribution of monitoring wells and sampling devices upgradient and downgradient can be made such that the information obtained can be used to service the requirements for all the three monitoring-sampling schemes.
- Verification monitoring – this requires placement of monitoring wells and soil sampling devices within the heart of the contaminant plume and also at positions beyond the plume. Figure 10.14 gives an example of the distribution of the wells and devices. Obviously, assuming that the wells and devices are properly located, the more monitoring and sampling devices there are, the better one is able to properly characterize the nature of the contaminant plume. Monitoring wells and sampling devices placed outside the anticipated contaminant plume will also serve as monitoring wells and sampling devices for long term conformance assessment.
- Long-term conformance monitoring. This is essential to verify success of mitigation scheme and for long term management of the potential impact.

Analyses of samples retrieved from monitoring wells will inform one about the concentration, composition, and toxicity of the target contaminant. A knowledge of the partition coefficients and solubilities of the various contaminants, together with the monitoring well information will provide one with the opportunity to check the accuracy of predictions from transport-fate models. For organic chemicals detected in the monitoring-sampling program, laboratory research may be required to determine the long-term fate of the transformed or intermediate products. This is not a necessary requirement if modelling predictions and especially if the *indicators* for the intermediate show good accord with the sampling values of contaminant concentrations. Tests on recovered soil samples from the sampling program should determine the environmental mobility of the contaminants and also the nature and concentration of contaminants sorbed onto the soil particles (soil solids). Detailed discussion of many of the bonding mechanisms and their reactions to changes in the immediate environment has been developed in Chapters 2 and 9.

10.10 ENGINEERED MITIGATION-CONTROL SYSTEMS

As we have indicated at the outset, the use of technological schemes for mitigation and management of contaminants in the ground is probably best utilized for limited and well defined source locations of contaminants. Good examples of these are waste landfills, leaking underground storage tanks, spills, and discharges, and containment ponds. Many of these are shown in the diagram in Figure 2.6 in Chapter 2. Technological solutions for management and control range from construction of impervious barriers that would intercept the plume to removal of the pollution source and the entire affected region. To a very large extent, the methods chosen or designed to manage and control contaminant advance in the subsoil are necessarily site and situation specific. Also, to a large extent, the nature of the threats posed by the contaminants and the pathways to the various receptors are considerations that will dictate the type and kind of technological solution sought. Finally, the control-management technological solution sought will always be analyzed within the framework of risks-reward and cost-effectiveness.

The record shows that there are some very difficult-to-treat contaminated sites. By and large, these are sites contaminated with organic chemical compounds that are severe threats to human health. For these kinds of contaminated sites, containment with confining structures have been constructed. These allow these sites to be isolated whilst awaiting effective and economic remediation solutions. In the case of impervious barriers, these are generally constructed from sheet piles lined in the interior with membranes to deny lateral advance of the pollution plume. Difficulties arise in controlling the downward advance of the pollution plume when the plume arrives at the lateral impervious barrier. Suggestions range from driving the sheet piles down into an impervious layer as shown for example in Figure 10.19, to injection grouting to develop an impervious base at some depth in the ground, to inclined-to-horizontal placement of sheet piles using techniques similar to the oil industry for inclined drilling.

If an impervious clay layer can be found directly below the contaminated region, the methodology shown in Figure 10.19 is probably the most expeditious means for controlling the escape of fugitive contaminants. By and large, for situations such as the one depicted in the figure, the contaminants resident in the contaminated site would likely be various kinds of organic chemicals. Heavy metals associated with these chemicals will likely be sorbed by the soil solids and will not be very mobile. Hence, the nest of treatment wells sunk into the contaminated region will be geared towards bioremediation of the organic chemical contaminants. Monitoring wells placed outside the confining sheet-pile wall, particularly downstream, will provide continuous information on the efficiency of the containment system. Note that the schematic representation of the nest of wells (treatment and monitoring) is relatively crude. Wells should be placed with varying vertical and horizontal sampling points and locations. If extra precaution is sought, a treatment zone using enhanced NA capability as described in Section 10.6.1 outside the confining sheet-pile wall can be introduced.

For control of contaminant plumes during treatment as part of impact management, several options are available. The contaminant source is well delineated and defined, and if site hydrogeology is well understood, the solution shown in Figure 10.20 is one that utilizes the capabilities of enhanced NA in combination of the permeable reactive barrier (PRB) previously described.

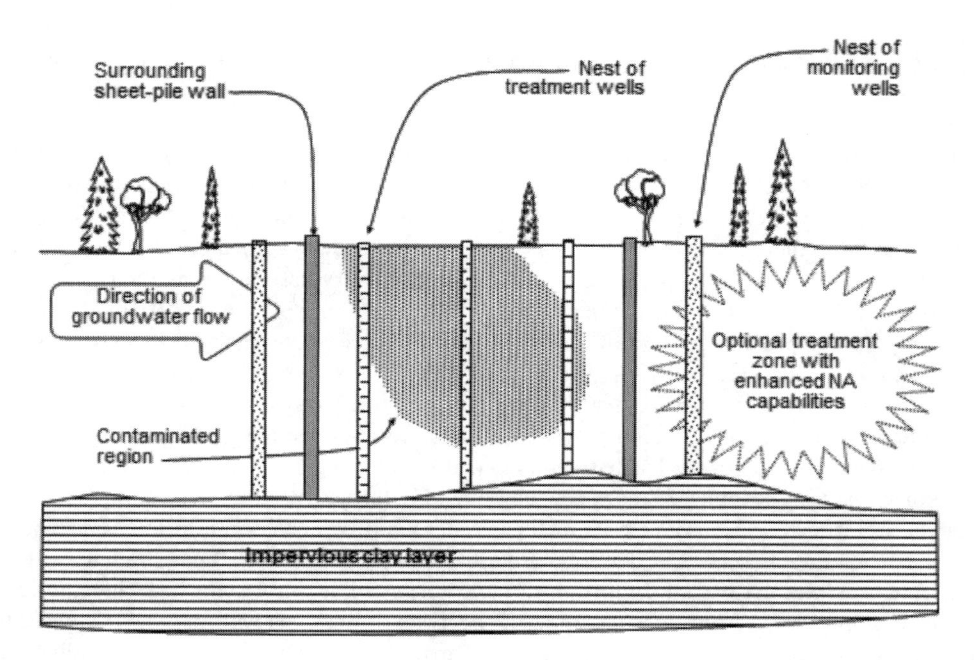

FIGURE 10.19 Containment of contaminants in a contaminated site using a confining sheet-pile wall that surrounds the contaminated region. The sheet-pile wall is sunk into the impervious clay layer to prevent bottom escape of contaminants. Treatment wells are sunk into the contaminated regions, and monitoring wells are placed downstream with some upstream also. Optional treatment zones using enhanced NA capabilities can be used, if needed.

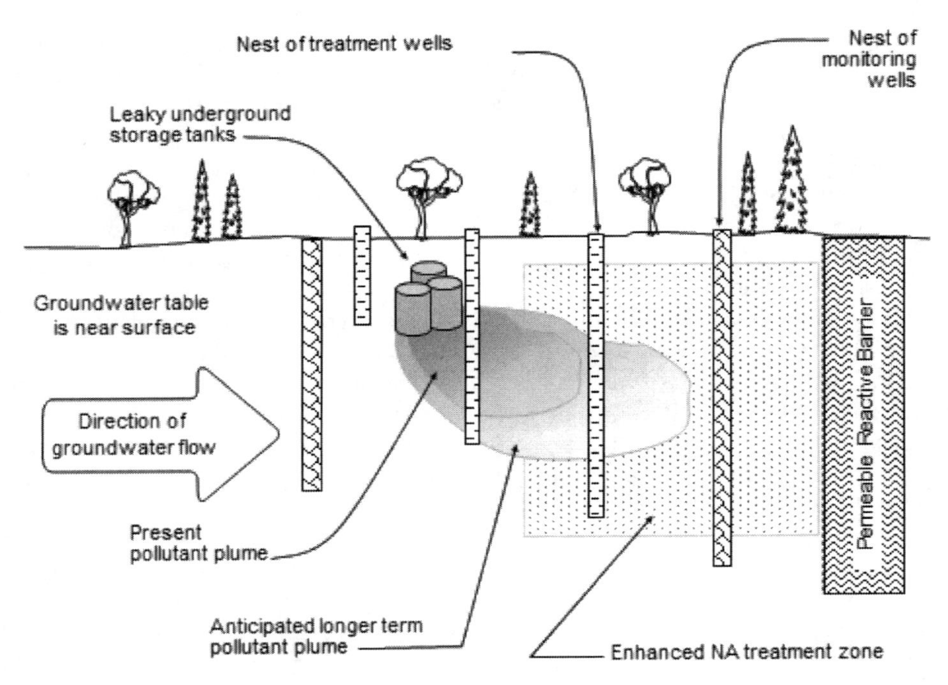

FIGURE 10.20 Use of treatment wells and enhanced natural attenuation treatment zone in combination with permeable reactive barrier to mitigate and manage impact from leaky underground storage tanks. Treatments for enhancement can be any or all of the following: geochemical intervention, biostimulation, and bioaugmentation. Treatment occurs in the pollutant plume and downgradient from the plume.

This procedure is both a mitigation and management tactic for management of contaminant impact. Note again that the monitoring and treatment wells are somewhat simplistic in illustrative portrayal.

The engineered barrier systems for a municipal and a hazardous waste landfill shown in Figure 1.10 are a good demonstration of the extent to which composite barrier systems can be designed and engineered to meet the requirements for management and control of contaminants. The details of the filter, membrane, and leachate collection system are specified by regulatory *command and control* requirements or by performance requirements. In the case of the municipal solid waste (MSW) landfill liner system shown in the bottom right-hand corner of Figure 1.10, the soil material comprising the engineered clay barrier underlying the synthetic membrane must possess hydraulic conductivity values that are below the maximum permissible values. The basic idea in the design details of the engineered barrier for the MSW landfill is that if leachates inadvertently leak through the high-density polyethylene membrane (HDPE) and are not captured by the leachate collection system, the contaminants in the leachate plumes will be attenuated by the engineered clay barrier. The engineered clay barrier serves as the second line of defense or containment.

For the hazardous waste (HW) bottom liner system shown in the bottom left-hand corner of Figure 1.10, there are two lines of defense before the soil sub-base. The HDPE (high density polyethylene) acts as the first barrier. Before this, the leachate collection system is designed to collect leachates draining down from the waste pile. If the system functions well, i.e., as designed, there should be very little leachate reaching the HDPE barrier. If, however, leachate does collect at the HDPE barrier, and if this barrier is somehow breached, the underlying synthetic membrane (most likely another HDPE) is designed to prevent the leachate from escaping. Above this synthetic membrane there is a leak detection system that will alert the managers of the landfill that the first HDPE has been breached and that the leachate collection system is most likely malfunctioning. If the second membrane fails, the underlying soil subbase can be designed as an attenuation barrier using NA principles.

10.10.1 Remediation as Control-Management

Technically speaking, remediation of sites and regions contaminated with noxious substances and contaminants belongs to a category separate from contaminant impact mitigation. We have included this here because the *treatment wells* shown in Figures 10.19 and 10.20 are in fact wells or devices that introduce remediation aids. In addition, it can be legitimately argued that remediation of a contaminated site in effect removes the contaminant source – assuming of course that the remediation-treatment process is successful.

The priority requirement in remediation-treatment of a contaminated site is to eliminate the health and environmental threat posed by the presence of contaminants in the contaminated site. Traditionally, this objective is met with the *dig and dump* technique. Replacement with clean fill material will now ensure that all the contaminants have been removed from the affected site or region. If total removal of all contaminants is not an option, minimization of the risk posed by the presence of the contaminants is the next priority. This latter course can take several forms. The basic factors to be considered include the following:

- Contaminants: Type, concentration, and distribution in the ground.
- Site: Site specificities, i.e., location, site constraints, substrate soil material, lithography, stratigraphy, geology, hydrogeology, fluid transmission properties, etc.
- Economics and Risks: Cost-effectiveness, timing, and risk management.

The techniques that can be considered fall into five groups. These include (a) physico-chemical, (b) biological, (c) thermal, (d) electrical-acoustic-magnetic, and (e) combination. Physico-chemical techniques rely on physical and/or chemical procedures for removal of the contaminants. These include precipitation, desorption, soil washing, ion exchange, flotation, air stripping, vapour and vacuum extraction, demulsification, solidification, stabilization, reverse osmosis, etc.

Biological techniques are generally used to treat organic chemicals, but as we have pointed out previously, these can also be used for remediation of heavy metal contaminated sites. The techniques used include bacterial degradation and/or transformation, biological detoxification, aeration, fermentation, and biorestauration. Thermal procedures include vitrification, closed-loop detoxification, thermal fixation, pyrolysis, super critical oxidation, etc. Electrical-acoustic-magnetic methods include electrochemical oxidation, electrokinetics, electrocoagulation, ultrasonic, and electroacoustics. Finally, the last group that specifies *combination* implies that any of the four previous groups may be combined in a series-type technical solution to provide the necessary remediation treatment. This is sometimes called a *treatment train.*

Remediation of contaminated sites is a very large challenge that offers innumerable opportunities for technological innovation. The basic means for treatment given in the preceding paragraph have been used in many different technologically clever ways to effect remediation of contaminated sites. The reader is advised to consult specialized manuals and textbooks devoted exclusively to remediation and treatment of contaminated sites. Bioremediation occupies perhaps one of the aspects of greatest attention by researchers and practitioners attending to remediation. Much research is being conducted and reported in the various specialized journals. Chapter 11 will discuss remediation techniques in more detail.

10.11 PROTOCOLS DEVELOPED FOR NATURAL ATTENUATION

There are various technical guidelines and protocols that have been established for MNA. A summary of some are as follows. More detail can be found in Yong and Mulligan (2019). *Designing*

Monitoring Programmes to Effectively Evaluate the Performance of Natural Attenuation was developed by the Air Force Centre for Environmental Excellence (Weidemeier and Haas, 1999). The EPA 1998 *Technical Protocol for Evaluation of Natural Attenuation of Chlorinated Solvents in Ground Water* (Weidemeier et al., 1998) was established to demonstrate mechanisms of chlorinated solvent natural attenuation. The protocol by the ASTM applies to petroleum contamination from underground storage tank releases into the groundwater (ASTM, 2015).

To evaluate if natural attenuation is occurring, *lines of evidence* have been established (NRC, 2000). They include the following:

- Decreases in contaminant concentration or amount over time.
- Chemical indicators of microbiological activity in the groundwater.
- Laboratory microcosm studies to determine if bacteria present at the site are biodegrading the contaminants.

It is suggested by the NRC (2000), that secondary or tertiary lines are only necessary if primary lines are not conclusive.

The EPA guidance document for RCRA, and Superfund sites and for underground storage tanks (UST) covers all contaminants in soil and groundwater (EPA, 1999). A subsequent guidance was issued that focuses on inorganic contaminants (USEPA, 2015). In general, the NRC (2000) noted that only the DOE guidance (Brady et al., 1998) gives some guidance regarding the sorption and sequestration of inorganics. The U.S. Air Force *Technical Protocol For Implementing Intrinsic Remediation with Long-Term Monitoring for Natural Attenuation of Fuel Contaminants* (Weidemeier et al., 1995) was evaluated by the NRC (2000) as one of the most scientifically sound, as it describes ways to estimate biological activity, dilution, sorption, and dispersion but does not address MTBE. The EPA (2001) report has made various recommendations concerning MTBE. A technical protocol by the API (API, 2007) builds on previous reports of the USEPA and ASTM, using a tiered approach for evaluating natural attenuation.

Three reports were prepared by the US EPA (2007a, 2007b, 2010) to address inorganics including MNA of metals, nitrate, perchlorate, and radionuclides, which was followed by another to specifically address inorganic contaminants (USEPA, 2015). NJDEP (2012) and Truex et al. (2011) and have also has issued guidance for inorganic and radionuclide contaminants, and organic compounds, such as petroleum hydrocarbons and chlorinated solvents.

Few protocols exist for soil with the exception of the USEPA (1999) and those by the DOE. More recently some efforts have been put forward. A framework for inorganic non-volatile contaminant was prepared for evaluation of MNA in the vadose zone (Truex and Carroll, 2013), Magar et al. (2009) developed a technical guidance for natural recovery of sediments. This was followed by USEPA (2014) describing methodologies for assessment and prediction of natural processes in sediments. The majority of the available protocols for organic contaminants address only fuel hydrocarbons or chlorinated solvents and infrequently PAHs, PCBs, explosives and pesticides. More recently however, Rectanus et al. (2015) has summarized the mechanisms of MNA in soil and groundwater for munitions.

One of the more recent guidelines of MNA was developed by Danko et al. (2021) for several organics including 1,4 dioxane, and chlorinated solvents (1,1,1-Trichloroethane 1,1,1-TCA, 1,1-dichloroethane 1,1-DCA, and 1,1-dichloroethene 1,1-DCE). A fate and transport model and decision flowcharts were developed to guide the lines of evidence for MNA for these compounds. The model, the MNA Rate Constant Estimator, was developed to obtain 1,4-dioxane and chlorinated ethane degradation rate constants for predicting the rates of source and plume degradation. The tools were then integrated into the BioPIC software. The model and frameworks were then validated. It was determined that the ^{14}C assay was successful at nine of the ten field sampling sites, and the model could determine 1,4-dioxane biodegradation rate constants for eight of nine sites. The cost to gather the data and perform the MNA evaluated was estimated at $30,000 US, which could be substantially less than other remediation alternatives.

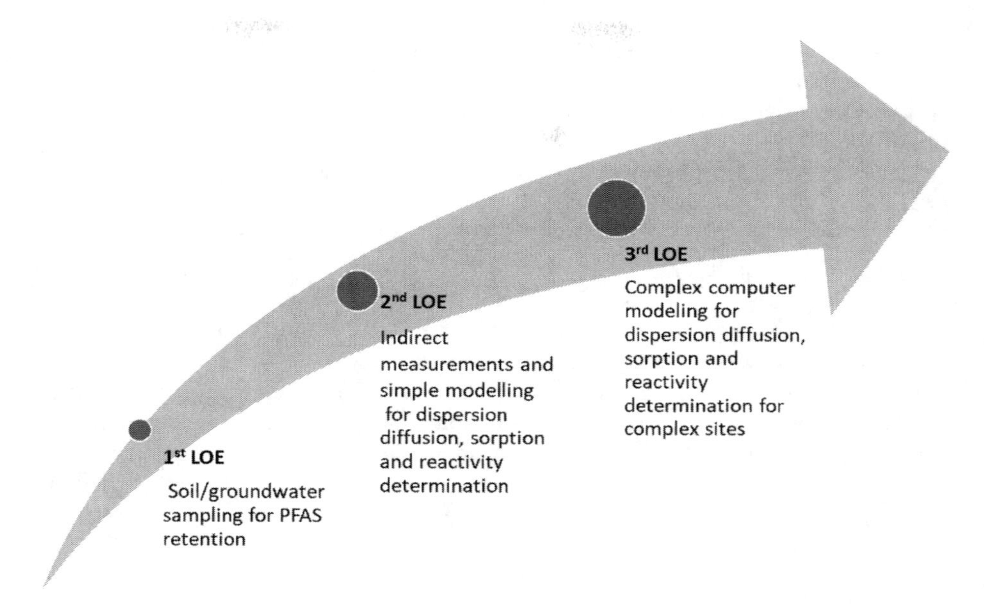

FIGURE 10.21 Three-tiered approach for evaluating MNA of PFAS. (Newell et al. 2021b.)

TABLE 10.1
Required Data for MNA Suitable for PFAS in the Vadose Zone

Data Required	Methodology to Obtain the Data
Geology and PFAS quantity	Soil sampling
Sorption of PFAS and other parameters (CEC, organic content, bulk density)	Lab analyses after soil sampling by various methods
Water table determination	Soil and groundwater sampling with direct push technology
Mass discharge and recharge rates	Soil type, precipitation data, leachate determination, lysimeter data

Source: Newell et al. (2021b).

Some potential MNA management guidelines for PFAS in groundwater have been presented by Newell et al. (2021b). The three-tiered approach potential LOE are shown in Figure 10.21. Data requirements for the vadose zone characterization are listed in Table 10.1. Ten MNA questions and tools for addressing the questions are identified. They address retention of PFAS in the vadose and saturated zones and plume migration. In addition, if the MNA is not sufficient, enhancement methods have been suggested such as methods to reduce PFAS leaching from the vadose zone (capping, reducing soil moisture, sorbents, phytoremediation, stabilization of the water table fluctuations). Although there is potential for the MNA of PFAS, future research is needed for implementation at contaminated sites for MNA and Enhanced MNA.

10.12 CONCLUDING REMARKS

Geoenvironment impact mitigation requires one to have information not only on the kinds of stressors responsible for the impacts, but also on the nature of their effects on the land compartment of the geoenvironment (which is the subject of interest of this book). Whilst mitigation of stressor impacts is a necessary requirement and goal in protection of the health of biotic receptors and the geoenvironment, impacted site restoration is also another pressing issue that needs to be addressed. To do so, it is first necessary to determine the goals of site restoration – i.e., the type of site functionality that restoration should achieve. Determination of site functionality requires attention to the kinds of attributes and indicators required.

We have shown, from the discussions in the first part of this chapter and in previous chapters that our knowledge of the nature of impacts from mechanical-type and most hydraulic-type stressors have resulted in the development of standard practice guidelines and codes in civil and geotechnical engineering. We have also learnt that perhaps the biggest source of threats to the geoenvironment and its land compartment is the presence of chemical contaminants in the ground. The previous chapters have shown that the various discards, spills and loss of materials (chemicals, etc.) and discharge of wastes, either in liquid form or as solids, are common to all types of human activities – associated with such entities as (a) households, (b) cities, (c) industries, (d) farms, and (e) mineral and hydrocarbon exploitation. These pose significant threats to the land environment and the receiving waters that are well-perceived. Not as well perceived are the threats presented by atmospheric-based non-point-sources, such as those shown in illustrative form in Figure 10.22. Under rainfall conditions, pesticides, herbicides, and other pest control chemical aids, have the potential not only to combine with the rainfall run-off to contaminate the receiving waters, but they also have the potential to infiltrate into the ground and threaten groundwater supplies. In addition to the non-point-sources that originate on the land surface (land-based), there are the non-point-sources that originate from precipitation through the atmosphere containing noxious gas emissions and airborne contaminants (as particulates) from offending smokestacks and other types of smoke discharges. These can be

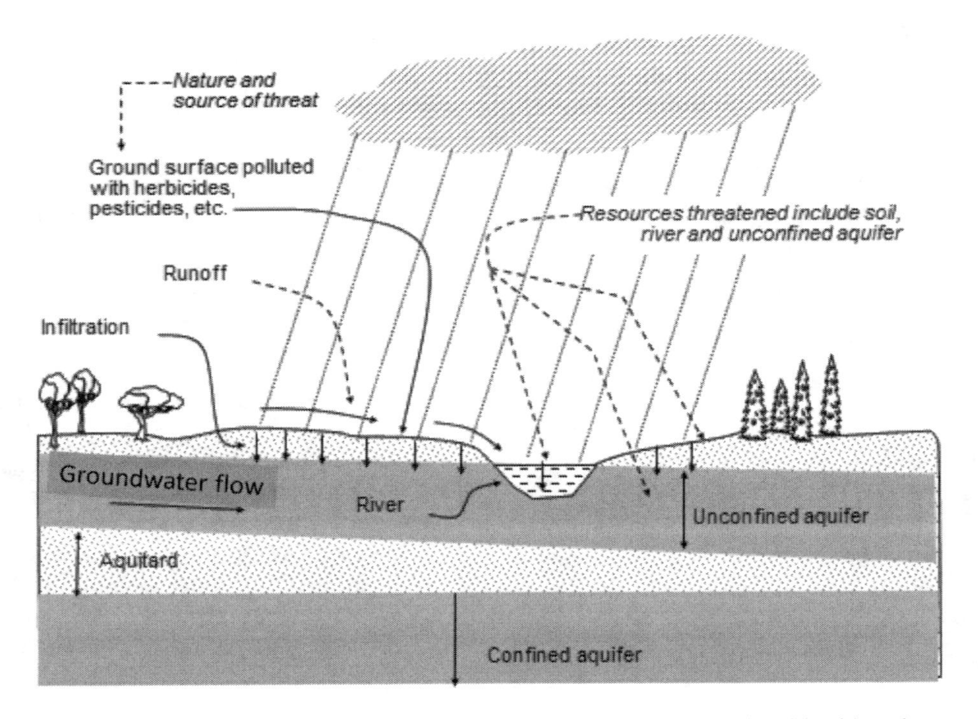

FIGURE 10.22 Demonstration of threats and impacts from atmospheric-based and land-based non-point-sources. Resources threatened include the river, soil, and unconfined aquifer. We presume that the aquitard is sufficient to protect contamination of the confined aquifer. If this is incorrect, the confined aquifer will eventually be contaminated by the water in the unconfined aquifer.

called atmospheric-based non-point pollution or contaminant sources. Included in this list are NO_x (nitrogen oxides), SO_2 (sulphur dioxide), CO (carbon monoxide), Pb and other metals such as Al, As, Cu, Fe, La, Mg, Mn, Na, Sb, V, and Zn (Lin et al., 1997) as airborne particulates, and VOCs (volatile organic chemicals) and PAHs (polycyclic aromatic hydrocarbons), such as benzene, toluene, and xylene, and particulate matter PM_{10} and $PM_{2.5}$. Particulate matter $PM_{2.5,}$ i.e., particulate matter less than 2.5 μm in size will in all likelihood remain suspended in the ambient air whilst PM_{10} (particulate matter less than 10 μm but greater than 25 μm) will be deposited eventually or with the aid or precipitation. Whilst acid rain is one of the outcomes of precipitation through this type of atmosphere, deposition of the airborne contaminants has not received the attention that is deserved. Airborne particulate matter is a great concern to public health in the ambient air because of their effect of these particulates on the hearts and lungs of those who are exposed to these.

The impacts to the health and quality of biotic receptors and the land environment presented by atmospheric-based and land-based non-point-sources of pollution cannot be readily mitigated by limited point-directed technological solutions. We recognize that the broad-based nature of the affected regions, i.e., land surfaces and water bodies, make containment, and management of the spread of contaminants prohibitively difficult and costly. In consequence, the use of the natural in-place soil as a tool for mitigation of the impact of such contaminant sources is a solution that needs to be exploited. To do so, we need to develop a better appreciation of the assimilative properties of soils, and also of the various geochemical and biological aids that will increase or enhance the assimilative capability of the soils. This has been the focus of this chapter. Contaminated land detracts considerably not only from one's ability to provide the necessary food supply, but also compromises the receiving waters and sources of water supply for humans. In summary, the main issues addressed include the following

- Impacts from contaminants in the ground need to be mitigated and managed as a beginning step towards protection of the resources in the environment, and also as a first step towards achievement of a sustainable geoenvironment.
- Using the properties of the natural soil-water system as the primary agent for such purposes allows one to address pollution sources that encompass the range from point-source to both land-based and atmospheric-based non-point-sources.
- Enhancing the natural attenuation properties of the subsoil to make it more effective as a control tool allows the subsoil to remain in place as a mitigation-management tool. This is the essence of a semi-passive remediation-treatment of contaminated sites.
- The physical, mechanical, chemical and biological properties of soil made it a good resource material for management of contaminants and waste products. These properties are responsible for the assimilative capacity of soils and the natural attenuation capability of the soil.
- Mitigation and management of contaminants in the subsoil should seek to reduce and eliminate the presence of contaminants in the soil. Engineering the natural attenuation capability of soils, through enhancements of the attenuation capability with geochemical, biological and nutrient aids will provide greater management options.
- Reduction in the concentration and toxicity of contaminants is the ultimate goal. Achievement of this goal can be obtained using various strategies involving technology and the properties of the soil. The various options discussed are by no means the complete spectrum of capabilities. More innovative schemes are being developed to address the problems of contamination of the land environment.

Until the threats and problems of land environment pollution are successfully managed and essentially rendered harmless, the road to sustainability will not be clear. Whilst the treatment in this book does not address the monumental problem of depletion of non-renewable resources, it nevertheless argues that the impacts from contamination of the land environment need to be mitigated and managed – as a very necessary step towards the road to sustainability of the geoenvironment.

REFERENCES

Alvarez-Cohen, L., and McCarty, P.L. (1991), Effects of toxicity, aeration and reductant supply on trichloroethene transformation by a mixed methanotropic culture, *Applied and Environmental Microbiology*, 57(1): 228–235.

Amin, M.M., Khanahmad, H., Teimouri, F., et al. (2017), Improvement of biodegradability of explosives using anaerobic- intrinsic bioaugmentation approach, *Bulgarian Chemical Communications,* 49: 735–741.

Apajalahti, J.H.A., and Salkinoja-Salonen, M.S. (1987), Complete dechlorination of tetrahydroquinone by cell extracts of pentachloro-induced *Rhodococcus chlorophenolicus*, *Journal of Bacteriology*, 169: 5125–5130.

Ara, I., and Mulligan, C.N. (2015), Reduction of chromium in water and soil using a rhamnolipid biosurfactant, *Geotechnical Engineering Journal of the SEAGS & AGSSEA*, 46(4): 25–31.

API. (2007), Technical Protocol for Evaluating the Natural Attenuation of MTBE, API Publication, Washington DC, 4761.

ASTM D5084-16a. (2016), *Standard Test Methods for Measurement of Hydraulic Conductivity of Saturated Porous Materials using a Flexible Wall Permeameter*, ASTM International, West Conshohocken, PA.

ASTM. (2015), *Standard Guide for Remediation of Ground Water by Natural Attenuation at Petroleum Release Sites,* ASTM International, West Conshohocken, PA. www.astm.org.

Bader, J.L., Gonzales, G., Goodell, P.C., Pilliand, S.D., and Ali, A.S. (1996), Bioreduction of Hexavalent Chromium in Batch Cultures Using Indigenous Soil Microorganisms. In: *HSRC/WERC. Joint Conference on the Environment*, Albuquerque, NM, 22–24 April.

Bartha, R. (1986), Biotechnology of petroleum contaminant biodegradation, *Microbial Ecology*, 12: 155–172.

Borden, R.C., Daniel, R.A., LeBrun, L.E., and Davis, C.W. (1997), Intrinsic biodegradation of MTBE and BTEX in a gasoline-contaminated aquifer. *Water Resources Research,* 33: 1105–1115.

Bradley, P.M., Chapelle, F.H., and Landmeyer, J.E. (2001), Effect of redox conditions on MTBE biodegradation in surface water sediments, *Environmental Science and Technology*, 35(23): 4643–4647.

Brady, P.V., Spalding, B.P., Krupka, K.M., Waters, R.D., Zhang, P., Borns, D.J., and Brady, W.D. (1998), *Site Screening and Technical Guidance for Monitored Natural Attenuation at DOE Sites*, Sandia National Laboratory, Dept. of Energy, Washington, D.C., August 30.

Caschetto, M., Robertson, W., Petitta, M., and Aravena, R. (2018), Partial nitrification enhances natural attenuation of nitrogen in a septic system plume, *Science of the Total Environment*, 625: 801–808.

Chaves, R.D.A., Rodrigues, P.C.H., de Souza, L.R., et al. (2023), How natural attenuation can benefit the environment: A case study of a water reservoir in Brazil, *Journal of Radioanalytical and Nuclear Chemistry,* 332(2): 301–2316. https://doi.org/10.1007/s10967-023-08868-7.

Coleman, N.V., Nelson, D.R., and Duxbury, T. (1998), Aerobic biodegradation of hexahydro-1,3,5-trinitro-1,3,5-triazine (RDX) as a nitrogen source by a Rhodococcus sp., strain DN22. *Soil Biology and Biochemistry*, 30: 1159–1167.

Coles, C.A., and Yong, R.N. (2004), Use of Equilibrium and Initial Metal Concentrations in Determining Freundlich Isotherms for Soils and Sediments. In: R.N. Yong and H. R.Thomas, (eds.), *Geoenvironmental Engineering: Integrated Management of Groundwater and Contaminated Land*, Thomas Telford, London, pp. 20–28.

Danko, A., Adamson, D., Newell, C., Wilson, J., Wilson, B., Freedman, D., and Lebron, C. (2021), Development of a Quantitative Framework for Evaluating Natural Attenuation of 1,1,1-TCA, 1,1-DCA, 1,1- DCE, and 1,4-Dioxane in Groundwater, ESTCP Project ER-201730, 360 pp,

Dasu, K., and Lee, L.S. (2016), Aerobic biodegradation of toluene-2,4-di(8:2 fluorotelomer urethane) and hexamethylene-1,6-di(8:2 fluorotelomer urethane) monomers in soils, *Chemosphere,* 144: 2482–2488.

Ehrlich, H.L. (1996), *Geomicrobiology*, 3rd ed., Marcel Dekker, New York, 719 p.

Esteve-Núñez, A, and Ramos, J.L. (1998), Metabolism of 2,4,6-trinitrotoluene by Pseudomonas sp. JLR11, *Environmental Science and Technology*, 32: 3802–3808.

Eweis, J.B., Ergas, S.J., Chang, D.P.Y., and Schroeder, E.D. (1998), *Bioremediation Principles*, WCB McGraw-Hill, Boston.

Frankenberger, W.T. Jr., and Losi, M.E. (1995), Applications of Bioremediation in the Cleanup of Heavy Metals and Metalloids. In: H.D. Skipper and R.F. Turco, (eds.), *Bioremediation Science and Applications*, SSSA, Special Publication Number 43, SSSA, Madison, WI, pp. 173–210.

Harkness, M.R., McDermott, J.B., Abramawicz, D.A., Salvo, J.J., Flanagan, W.P., Stephens, M.L., Mondella, F.J., May, R.J., Lobos, J.H., Carrol, K.M., Brennan, M.J., Bracco, A.A., Fish, K.M., Warmer, G.L., Wilson, P.R., Dietrich, D.K., Lin, D.T., Morgan, C.B., and Gately, W.L. (1993), In situ stimulation of aerobic PCB biodegradation in Hudson River sediments, *Science*, 259: 503–507.

Higgins, I.J., and Gilbert, P.D. (1978), *The Biodegradation of Hydrocarbons. The Oil Industry and Microbial Ecosystems*, Heyden and Son, Ltd., London, pp. 80–115.

Interstate Technology & Regulatory Council (ITRC). (2005), *Perchlorate: Overview of Issues, Status, and Remedial Options,* PERCHLORATE-1, Washington, DC. www.itrcweb.org.

Landmeyer, J.E., Chapelle, F.H., Bradley, P.M. Pankow, J.F., Church, C.D., and Tratnyek, P.G. (1998), Fate of MTBE relative to benzene in a gasoline-contaminated aquifer (1993-1995*), Ground Water Monitoring Review*, 18(4): 93–102.

Lin, Z.Q., Schemenauer, R.S., Schuepp, P.H., Barthakur N.N., and Kennedy, G.G. (1997), Airborne metal contaminants in high elevation forests of southern Quebec, Canada, and their likely source regions, *Journal of Agricultural and Forest Meteorology,* 87: 41–54.

Liu, J. and Avendaño, S.M. (2013), Microbial degradation of polyfluoroalkyl chemicals in the environment: A review. *Environment International*, 61:98–114.

MacDonald, E. (1994), Aspects of competitive adsorption and precipitation of heavy metals by a clay soil, M.Eng Thesis, McGill University, Montreal, Canada.

Magar, V.S., Chadwick, D.B. Bridges T.S., Fuchsman P.C., Conder J.M., Dekker, T.J. Steevens, J.A., Gustavson, K.E., and Mills, M.A.. (2009), Technical Guide: Monitored Natural Recovery At Contaminated Sediment Sites, Environmental Security Technology Certification Program (ESTCP), Project ER-0622, 276 pp. www.clu-in.org/products/tins/tinsone.cfm?id=42952962

Manahan, S.E. (1990), *Fundamentals of Environmental Chemistry*, 4th ed., Lewis Publishers, Boca Raton, 612 pp.

Martin, J.P., and Focht, D.D. (1977), Biological Properties of Soils. In: L.F. Elliot and F.J., Stevenson, (eds.), *Soils for Management of Organic Wastes and Waste Waters,* American Society of Agronomy, Madison, WI, pp. 115–172.

Martínez-Hernández, V. Meffe, R., Kohfahl, C., and de Bustamante, I. (2017), Investigating natural attenuation of pharmaceuticals through unsaturated column tests, *Chemosphere*, 177: 292–302, https://doi.org/10.1016/j.chemosphere.2017.03.021

Nakano, M., and Yong, R.N. (2013), Overview of rehabilitation schemes for farmlands contaminated with radioactive cesium released from Fukushima power plant, *Engineering Geology*, 155: 87–93.

Newell, C.J., Adamson, D.T., Kulkarni, P.R., Nzeribe, B.N., Connor, J.A., Popovic, J., and Stroo, H.F. (2021a), Monitored natural attenuation to manage PFAS impacts to groundwater: Scientific basis, *Groundwater Monitoring and Remediation,* 41: 76–89. https://doi.org/10.1111/gwmr.12486.

Newell, C.J., Adamson, D.T., Kulkarni, P.R., Nzeribe, B.N., Connor, J.A., Popovic, J., and Stroo, H.F. (2021b), Monitored natural attenuation to manage PFAS impacts to groundwater: Potential guidelines, *Remediation,* 31: 7–18. www.onlinelibrary.wiley.com/doi/full/10.1002/rem.21697.

New Jersey Department of Environmental Protection (NJDEP). (2012), *Monitored Natural Attenuation Technical Guidance*, Site Remediation Program, 175 pp.

NRC (National Research Council). (2000), Natural Attenuation for Groundwater Remediation. In: *Committee on Intrinsic Remediation, Water Science and Technology Board and Board on Radioactive Waste Management, Commission on Geosciences, Environment and Resources*, National Academy Press, Washington, DC, 274 pp.

Nies, L., and Vogel, T.M. (1990), Effects of organic substrates on dechlorination of Arochlor 1242 in anaerobic sediments, *Applied and Environmental Microbiology*, 56: 2612–2617.

Nyffeler, U.P., Li, Y.H., and Santschi, P.H. (1984), A kinetic approach to describe trace-element distribution between particles and solution in natural aquatic systems, *Geochimica Cosmochimica Acta*, 48: 1513–1522.

Pennington, J.C., Bowen, R., Brannon, J.M., Zakikhani, M., Harrelson, D.W., Gunnison, D., Mahannah, J., Clarke, J., Jenkins, T.F., and Gnewuch, S. (1999), *Draft Protocol for Evaluating, Selecting, and Implementing Monitored Natural Attenuation at Explosives-Contaminated Sites*, U.S. Army Engineer Research and Development Center, Vicksburg, MS, 156 p.

Pirnik, M.P. (1977), Microbial oxidation of methyl branched alkanes, *CRC Critical Reviews in Microbiology*, 5: 413–422.

Pitter, P., and Chudoba, J. (1990), *Biodegradability of Organic Substances in the Aquatic Environment,* CRC Press, Boca Raton, FL, 306 pp.

Preuss, A., Fimpel, J., and Dickert, G. (1993), Anaerobic transformation of 2,4,6-trinitrotoluene (TNT), *Archives of Microbiology*, 159: 345–353.

Reineke, W., and Knackmuss, H.J. (1988), Microbial degradation of haloaromatics. *Annual Review of Microbiology*, 42: 263–287.

Rectanus, H., Darlington, R., Kucharzyk, K., and Moore, S. (2015), Attenuation Pathways for Munitions Constituents in Soils and Groundwater, TR-NAVFAC EXWC-EV-1503, 81 pp.

Schwarzenbach, R.P., Gschwend, P.M., and Imboden, D.M. (1993), *Environmental Organic Chemistry,* Wiley and Sons, Inc. New York, 681pp.

Soesilo, J.A., and Wilson, S.R. (1997), *Site Remediation Planning and Management*, Lewis Publishers, New York, 432 pp.

Spain, J.C., and Gibson, D.T. (1991), Pathway for biodegradation of p-nitrophenol in a Moraxella sp. *Applied and Environmental Microbiology*, 57(3): 812–819.

Steiert, J.G., and Crawford, R.L. (1986), Catabolism of pentachlorophenol by a *Flavobacterium sp., Biochemical and Biophysical Research Communications*, 141: 825–830.

Strauss, H. (1991), *Final Report: An Overview of Potential Health Concerns of Bioremediation*, Env. Health Directorate, Health Canada, Ottawa, 54 pp.

Suflita, J.M., Robinson, J.A., and Tiedje, J.M., (1983), Kinetics of microbial dehalogenation of haloaromatic substrates in methanogenic environments, *Applied and Environmental Microbiology*, 45: 1466–1473.

Thayer, A.M. (1991), Bioremediation: Innovative technology for cleaning up hazardous waste, *Chemical and Engineering News*, 69 (34): 23–44.

Truex, M.J., and Carroll. K.C. (2013), Remedy Evaluation Framework for Inorganic, Non-Volatile Contaminants in the Vadose Zone, Pacific Northwest National Laboratory Richland, Washington, PNNL-21815; RPT -DVZ-AFRI-004, 67 pp.

Truex, M., Brady, P., Newell, C., Rysz, M., Denham, M., and Vangelas, K. (2011), The Scenarios Approach to Attenuation-Based Remedies for Inorganic and Radionuclide Contaminants, Savannah River National Laboratory Savannah River Nuclear Solutions, LLC Savannah River Site Aiken, SC, SRNL-STI-2011-00459, 111 pp,

USEPA Office of Solid Waste and Emergency Response (OSWER). (1999), Use of Monitored Natural Attenuation at Superfund, RCRA Corrective Action and Underground Storage Tank Sites, Directive Number 9200.4-17P, April 21, 32 p.

USEPA. (1999), Information on OSWER directive 9200.4-17p – Use of monitored natural attenuation at Superfund, RCRA corrective action, and underground storage tank sites, USEPA-540-R-99-009.

USEPA. (2004), Cleaning Up the Nation's Waste Sites: Markets and Technology Trends (2004 Edition), Office of Solid Waste and Emergency Response, EPA 542-R-04-015, September 2004. Available at www.epa.gov/tio clu-in.org/marketstudy

USEPA. (2007a), *Monitored Natural Attenuation of Inorganic Contaminants in Ground Water, Volume 1, Technical Basis for Assessment,* R.G. Ford, R.T. Wilkin, and R.W. Puls (eds.), EPA/600/R-07/139, Office of Research and Development. Cincinati, OH.

USEPA. (2007b), *Monitored Natural Attenuation of Inorganic Contaminants in Ground Water, Volume 2, Assessment for Non-Radionuclides Including Arsenic, Cadmium, Chromium, Copper, Lead, Nickel, Nitrate, Perchlorate, and Selenium. R.G. Ford,* R.T. Wilkin and R.W. Puls, (eds.), EPA/600/R-07/140 , October, Office of Research and Development. Cincinatti, OH.

USEPA. (2008b), A Guide for Assessing Biodegradation and Source Identification of Organic Ground Water Contaminants using Compound Specific Isotope Analysis (CSIA), EPA 600/R-08/148, Office of Research and Development, National Risk Management Research Laboratory, Ada, Oklahoma.

USEPA. (2009), *Identification and Characterization Methods for Reactive Minerals Responsible for Natural Attenuation of Chlorinated Organic Compounds in Ground Water*, Office of Research and Development National Risk Management Research Laboratory, Ada, Oklahoma, 74820 pp.

USEPA. (2010), *Monitored Natural Attenuation of Inorganic Contaminants in Ground Water Volume 3Assessment for Radionuclides Including Tritium, Radon, Strontium, Technetium, Uranium, Iodine, Radium, Thorium, Cesium, and Plutonium-Americium,* Ford, R. G, and Wilkin, R. T. (eds.), EPA/600/R-10/093, September, Office of Research and Development, Cincinnati, OH.

USEPA. (2014), Technical Resource Document on Monitored Natural Recovery U.S. EPA, EPA 600-R-14-083, National Risk Management Research Laboratory, Cincinnati, OH, 251 pp.

US EPA. (2015), Use *of Monitored Natural Attenuation for Inorganic Contaminants in Groundwater at Superfund Site, U.S.* Environmental Protection Agency, Office of Solid Waste and Emergency Response, Directive 9283.1-36, August 2015.

Weidemeier, T., Swanson, M.A., Moutoux, D.E., Gordon, E.K., Wilson, J.T., Wilson, B.H.,,Kampbell, D.H., Haas, P.E., Miller, R.N., Hansen, J.E., and Chapelle, F.H. (1998), *Technical Protocol for Evaluating Natural Attenuation of Chlorinated Solvents in Ground Water,* EPA/600//R-98/128, September, EPA Office of Research and Development, Washington, DC.

Weidemeier, T.H. and Haas, P.E. (1999), *Designing Monitoring Programmes to Effectively Evaluate the Performance of Natural Attenuation,* U.S. Air Force Center for Environmental Excellence, San Antonia, TX.

Weidemeier, T.H., Wilson, J.T., Kampbell, D.H., Miller, R.N., and Hansen, J.E., (1995), Technical Protocol for Implementing Intrinsic Remediation With Long-Term Monitoring for Natural Attenuation of Fuel Contamination Dissolved in Groundwater (Revision 0), Air Force Center for Environmental Excellence. April, 1995.

Widdel, F. (1988), Microbiology and Ecology of Sulphate- and Sulphur-Reducing Bacteria. In: J.B. Zehner, (ed.), *Biology of Anaerobic Microorganisms*, John Wiley and Sons, New York, pp. 469–585.

Wilson, J.T., Kaiser, P.M. and Adair, C. (2005), Monitored Natural Attenuation of MTBE as a Risk Management Option Leaking Underground Storage Tank Sites. Office of Research and Development, Cincinnatti, OJ EPA 600-R-04-179.

Wujcik, W.J., Lowe, W.L., Marks, P.J. and Sisk, W.E. (1992), Granular activated carbon pilot treatment studies for explosives removal from contaminated groundwater, *Environmental Progress,* 11(3): 178–189.

Yong, R.N., Boonsinsuk, P., and Tucker, A.E. (1984), A study of frost-heave mechanics of high clay content soils, *Transactions of ASME*, 106: 502–508.

Yong, R.N., and Mulligan, C.N. (2004), Natural Attenuation of Contaminants in Soils, Lewis, Boca Raton, FL, 319 pp.

Yong, R.N., and Mulligan, C.N., (2019), *Natural and Enhanced Attenuation of Contaminants in Soils*, Lewis Publishers, Boca Raton.

Yong R.N., and Warkentin, B.P. (1975), *Soil Properties and Behaviour*, Elsevier Scientific Publishing Co., Amsterdam, 449 pp.

Yu, Y., Liu, Y., & Wu, L. (2013), Sorption and degradation of pharmaceuticals and personal care products (PPCPs) in soils. *Environmental Science and Pollution Research*, 20, 4261–4267.

Zeeb, P., and Wiedemeier, T. H. (2007), *Technical Protocol for Evaluating the Natural Attenuation of MtBE API PUBLICATION 4761*, Regulatory and Scientific Affairs Department, May.

11 Remediation and Management of Contaminated Soil

11.1 INTRODUCTION

The dumping of materials, bankrupt and abandoned manufacturing plants, insufficient methods for waste storage, treatment and disposal facilities have contributed to the contamination of many sites as indicated in previous chapters. Chemical waste categories include organic liquids, such as solvents from dry cleaning, oils including lubricating oils, automotive oils, hydraulic oils, fuel oils and organic sludges/solids and organic aqueous wastes and wastewaters. Most soil contamination is the result of accidental spills and leaks, generation of chemical waste leachates and sludges from cleaning of equipment, residues left in used containers and outdated materials and indiscriminant dumping. Smaller generators of chemical contaminants include improperly managed landfills, automobile service establishments, maintenance shops, and photographic film processors. Household wastes including pesticides, paints, cleaning and automotive products may also contribute significantly as sources of organic chemicals (LaGrega et al., 2004). The more common heavy metals include lead (Pb), cadmium (Cd), copper (Cu), chromium, (Cr), nickel (Ni), iron (Fe), mercury (Hg), and zinc (Zn).

A variety of *in situ* and *ex situ* remediation techniques exists to manage the contaminated sites. For evaluation of the most appropriate technique, the procedure in Figure 11.1 should be followed. *Ex situ* techniques include excavation, contaminant fixation or isolation, incineration or vitrification, washing and biological treatment processes. *In situ* processes include (a) bioremediation, air or steam stripping or thermal treatment for volatile compounds, (b) extraction methods for soluble components, (c) chemical treatments for oxidation or detoxification, and (d) stabilization/solidification with cements, limes, and resins for heavy metal contaminants. Phytoremediation although less developed has also been used. The most suitable types of plants must be selected based on pollutant type and recovery techniques for disposal of the contaminated plants. Other technologies related to nanotechnologies are also being developed.

Most *in situ* remediation techniques are potentially less expensive and disruptive than *ex situ* ones, particularly for large, contaminated areas. Natural or synthetic additives can be utilized to enhance precipitation, ion exchange, sorption, and redox reactions (Mench et al., 2000). The sustainability of reducing and maintaining reduced solubility conditions is key to the long-term success of the treatment. *Ex situ* techniques are expensive and can disrupt the ecosystem and the landscape. For shallow contamination, remediation costs, worker exposure, and environmental disruption can be reduced by using *in situ* remediation techniques.

In this chapter, various soil remediation technologies will be described. In Chapter 3, groundwater remediation techniques are described and in Chapter 10, monitored natural attenuation is addressed and thus they will not be repeated here. Therefore, the focus will be on *in situ* and *ex situ* techniques for soil and sediment remediation.

DOI: 10.1201/9781003407393-11

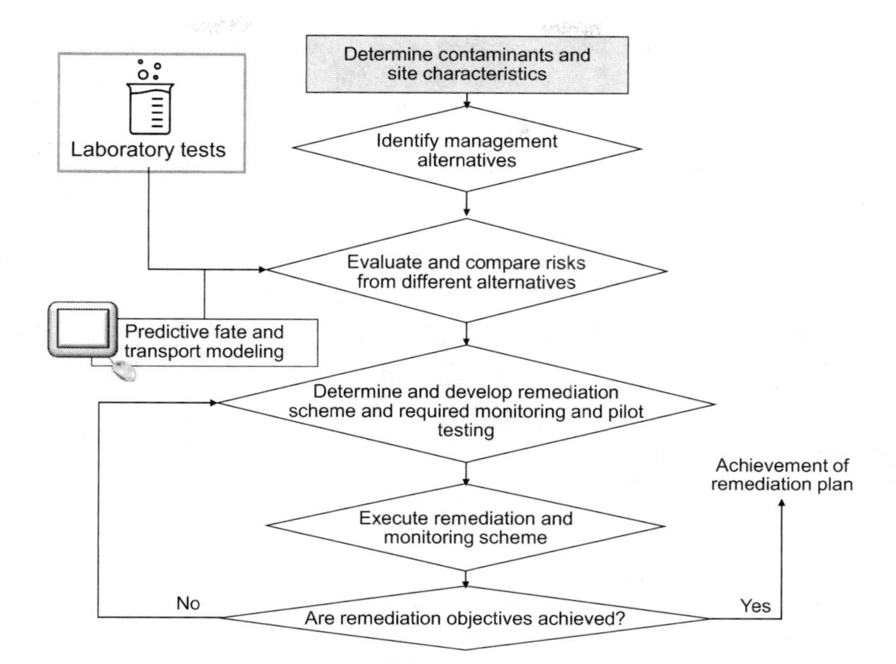

FIGURE 11.1 Flow chart demonstrating development of a programme to ensure achievement of remediation objectives.

11.2 PHYSICAL/CHEMICAL REMEDIATION TECHNOLOGIES

Two options are available for disposal of contaminated soil: (a) disposal in a secure landfill or disposal facility, and (b) treatment of the contaminated soil and reuse of the treated soil. Option (a) is not a preferred option as it is not sustainable. Treatment of contaminated soil can be an expensive procedure, especially when the quantities are large.

11.2.1 Isolation

Contaminated soil can be isolated to prevent further movement of the contaminants. This management scheme is usually of low to medium cost. Steel, slurry, cement, or bentonite barriers and grout walls and synthetic membranes can be used singly or in combination to reduce the permeability of the soil to less than 10^{-7} cm/s. They are considered as passive containment systems and can be temporary to complement remediation systems or be permanent. Vertical or horizontal barriers or capping are modes of reducing the movement. Vertical barriers should extend (or be keyed in) to an impermeable clay or bedrock layer so that the contaminants will not bypass the barrier. The materials of the barrier must be compatible with the contaminants (US EPA, 1989). Slurry materials can include Portland cement, soil-bentonite, and cement-bentonite mixtures that are pumped into an excavated trench by a backhoe. Grout curtains are made of materials such as Portland cement, sodium silicate, or consist of polymers or bitumen. The materials are injected through a series of drilled holes. The spacing of the holes is dependent on the soil permeability and type of grout used. Sheet piling involves driving steel or concrete sheets. The sheets are connected through sealing or interlocking. This method is often used during removal of underground storage tanks (UST) or for prevention of erosion.

Capping can be used to reduce the infiltration of surface water to provide stability over the contamination, prevent mobilization of contaminants or to improve the appearance of the site. The capping can be simple or consist of multiple layers and can be temporary or permanent. Ditches and berms can be employed to manage surface water run-off. Synthetic membranes can be used by installing sheets that are overlapped and seamed together. They can be made of polymers, fabrics, rubber, and other materials. The installation must be done properly without puncturing or tearing. Weathering and root penetration over time can be problematic. Other materials for capping include low permeability (10^{-6} to 10^{-7} cm/s) soils and clays upon compaction. A clean soil can then be added over the cap as a topsoil to allow for vegetation of the site. Liming, fertilization, and seeding of plants, such as grasses without the potential for deep roots will prevent erosion and not penetrate the cap. For landfills, multi-layer caps as previously discussed can be employed. A groundwater extraction system may be used to reduce the level of the water table. The barriers can also be placed either upstream or downstream or totally around the contaminated area. Horizontal barriers through horizontal drilling and grout injection can be used to restrict downward movement.

11.2.2 Confined Disposal

In the case that soil, or sediment, is contaminated, *in situ/ex situ* remediation can be performed. If remediation is not possible, the excavated or dredged material can be disposed in a confined disposal facility in diked near the shore, island, or on the land facilities. The facilities must be designed to contain the contaminants. In the case of sediments, they must be previously dewatered such as in a contained disposal facility since landfill facilities cannot handle slurries. Large volumes cannot usually be accommodated since landfills do not have the capacity. Potential mechanisms for contaminant release are due to leachates, run-off, effluents, volatilization, uptake by plants, and ingestion by animals. Therefore, pre-treatment by stabilization/ solidification may be necessary. Containment facilities can be used for storage, dewatering, and pre-treatment for other processes. These costs are usually less than those for landfill. Contained aquatic disposal, the placement of material in a confined aquatic area called a confined disposal facility (CDF). These areas can be strategically placed in depressions and confined by dikes. This technique can be used for disposal of contaminated soils or sediments. Clean material can be placed above and at the edges. The USACE and USEPA (2003) have reviewed the use of CDFs for dredging projects in the Great Lakes.

Confined aquatic disposal (CAD) is used for placement of contaminated materials in a natural or excavated depression. It has been used mainly for navigational purposes, such as in Boston Harbour, not disposal of contaminated material. It may be appropriate if landfill disposal or *in situ* capping is not possible. Maintenance costs are low and there can be an increased resistance to erosion. Depths can be a few to more than 10 m and widths are in the range of 500 to 1500 m. As they are filled, capping is used.

Another approach is to place the material in woven or non-woven permeable synthetic fabric bags, geotubes, or containers (NRC, 1997). The contaminants must not seep through the fabric into the water and these uncertainties must be further investigated.

11.2.3 Physical Separation

Physical separation processes are generally technically simple methods for separation of solids on the basis of size and density and are often used as pre-treatments. These processes have been applied for the separation of contaminated fractions from the clean coarser particles. As coarser particles, such as sand and gravel fractions have less contamination on their surfaces, washing is often enough to clean for beneficial use. This is important to reduce the amount of material to be disposed of. The most contaminated fractions may require further treatment or restricted disposal. The volume of the fine residuals may be minimized using mechanical dewatering techniques (Olin-Estes and Palermo, 2001).

Physical separation processes include centrifugation, flocculation, hydrocyclones, screening, and sedimentation. Hydrocyclones separate the larger particles greater than 10 to 20 micrometrer by centrifugal force from the smaller particles, fluidized bed separation removes smaller particles at the top (less than 50 micrometres) in the countercurrent overflow in a vertical column by gravimetric settling, and flotation is based on the different surface characteristics of contaminated particles. Addition of special chemicals (flotation agents) and aeration causes these contaminated particles to float. Screening is most applicable for particles larger than 1 mm. Magnetic extraction also may be used. If the solids' content is high, mechanical screening can be used. Gravity separation or sedimentation is applicable if the contaminated fraction has a higher specific gravity that the rest of the soil fraction. The expense is only justified if the soil contains more than 25% sand, which is rare (NRC, 1997). Physical techniques only concentrate the contaminants in smaller volumes and are thus useful before thermal, chemical, or other processes.

In Japan, similar techniques and processes are used to obtain aggregates from soils for concrete. The soils taken from mountainous areas are washed. The fine and light fractions are separated from coarse particles (concrete aggregates) in centrifugal tank and dewatered using the filter presser or belt presser. The water content is usually controlled as about 40% from the energy cost and treatability of the materials. The technology can also be used for dredged materials.

11.2.4 Soil Vapour Extraction (SVE)

Soil vapour extraction (SVE) (Figure 11.2) may be utilized for the vaporization of volatile and semi-volatile components in the unsaturated zone (Yong, 1998; Rathfelder et al., 1991). Soil can be decontaminated by applying a vacuum to pull the volatile emissions through the soil pore spaces. The air may be then treated at the surface with activated carbon filters, chemical oxidation, or biofilters. This technique is applicable to highly permeable soils and volatile contaminants such as gasoline or solvents. Other parameters such as the octanol/water coefficient, Henry's law constant and solubility of the contaminant, and moisture and organic contents of the soil also affect the removal efficiencies.

Soil vapour extraction (SVE) (Figure 11.2) involves the removal of VOCs and some fuels with Henry's law constant greater than 0.01 or a vapour pressure greater than 0.5 mm Hg through either air injection or vacuum vapor extraction. SVE is an *in situ* unsaturated (vadose) zone soil remediation technology in which a vacuum is applied to the soil to induce the controlled flow of air and remove volatile and some semi-volatile contaminants from the soil (US EPA, 1997). The extracted vapour may then be treated to recover or destroy the contaminants, depending on applicable regulations. The area of the extraction is called the zone of influence. Vertical extraction vents are typically implemented to depths of greater than 1.5 metres up to as much as 91 metres. Horizontal extraction vents (trenches or horizontal borings) can be used as warranted by contaminant zone geometry, drill rig access, or other site-specific factors.

The treatment is usually *in situ* for highly permeable soils. Groundwater levels may require lowering to decrease the moisture content. The contaminants pass through the void space in the soil by vapourization and are captured for further treatment on the soil surface by condensation, combustion, oxidation, incineration, activated carbon absorption, or biofiltration. Field and pilot studies are usually necessary to determine the feasibility and subsequently the design of the method as well as to obtain information necessary to design and configure the system. The process may be used in combination with other methods such as a bioremediation. A surface seal consisting of a geomembrane, concrete or asphalt caps or natural materials such as clay or bentonite can be employed to control vapour flow. Typically, *in situ* SVE processes can require 1 to 3 years. Costs vary significantly between sites.

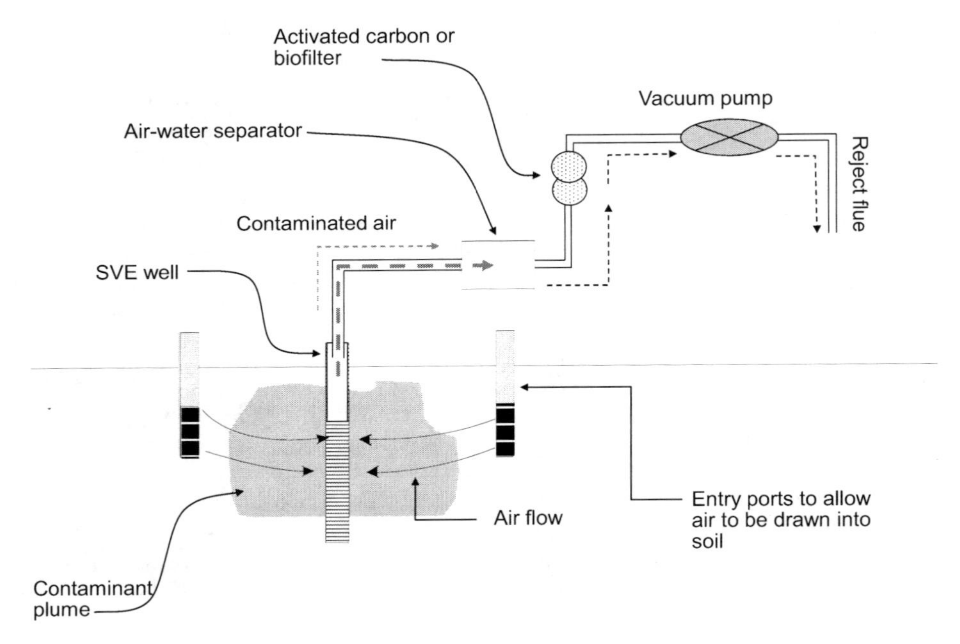

FIGURE 11.2 Schematic of a SVE process. A series of SVE wells can be introduced into the ground – connected in series or in parallel. The number of SVE wells that can be introduced will depend on the capacity of the vacuum pump system. (Adapted from Yong, 1998.)

11.2.5 Fracturing

Fracturing differs from the method for extraction of oil or gas and is used to enhance the efficiency of other *in situ* technologies in difficult conditions, such as silts, clays, shale, and bedrock (US EPA, 2012a). Most fracturing for remediation purposes does not exceed depths of 30 m. Existing fissures are enlarged or new fractures are introduced particularly horizontally After fracturing, vapour extraction, or forced air injection is performed. Technologies commonly used in soil fracturing include pneumatic fracturing (PF), blast-enhanced fracturing and Lasagna™ process (US EPA, 1996). Blast-enhanced fracturing is used at sites with fractured bedrock formations. In the Lasagna™ process, *in situ* electroosmosis is combined with hydraulic fracturing is used to enhance sorption/degradation zones horizontally in the sub-surface soil. For the PF process, fracture wells in the contaminated vadose zone are drilled and short bursts (~20 seconds) of compressed air are injected to form fractures and repeated at various intervals.

11.2.6 Soil Washing and Soil Flushing

Soil remediation of the soil can be performed with or without excavation via soil washing or *in situ* flushing (Mulligan et al., 2001). Solubilization of the contaminants can be performed with water alone or with additives. The solubility of the contaminant is thus a key factor. Contaminants, such as trichloroethylene (TCE), polycyclic aromatic hydrocarbons (PAHs), and polychlorinated biphenyls (PCBs), are of very low solubility.

To remove non-aqueous phase liquids (NAPLs) from the groundwater, extraction of the groundwater can be performed by pumping to remove the contaminants in the dissolved and/or free phase NAPL zone in a pump and treat system. However, substantial periods of time can be required, and effectiveness can be limited. Drinking water standards of the extracted water can be achieved after treatment with water treatment systems, such as activated carbon, ion exchange, membranes, and

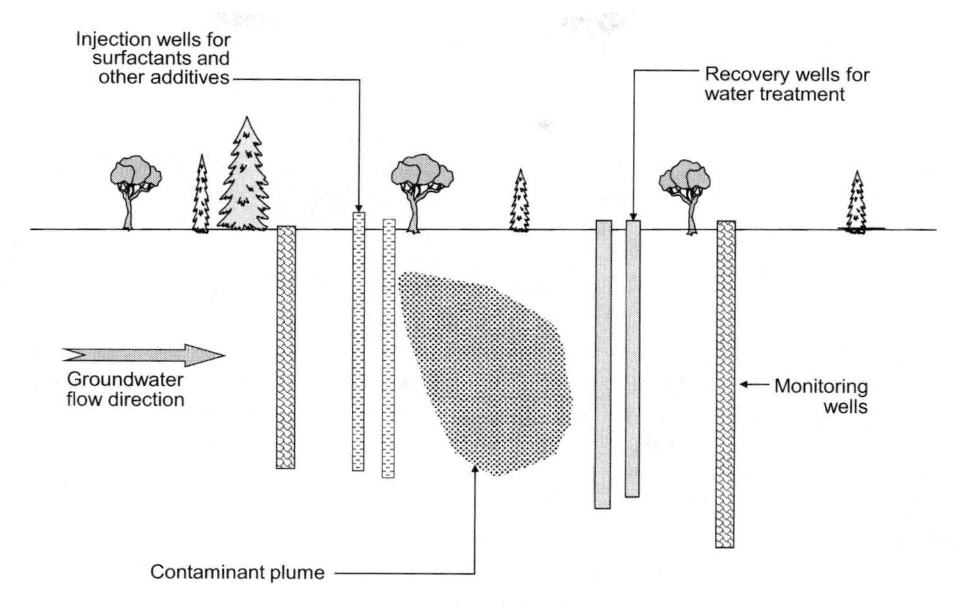

FIGURE 11.3 Schematic of a soil flushing process for removal of contaminants.

other methods. To treat the contaminated soil, extraction solutions can be introduced into the soil using surface flooding, sprinklers, leach fields, horizontal or vertical drains to enhance the removal rates of the contaminants. Water alone can be utilized for water soluble salts, and anions, such as arsenate, arsenite, cyanide, nitrate, and selenate. Surfactant or solvent solutions are utilized to solubilize and extract the less water-soluble contaminants as shown in Figure 11.3 in soil flushing. Additives can include organic or inorganic acids or bases, water soluble solvents, complexing or chelating agents, such as ethylenediaminetetraacetic acid (EDTA) or nitrilotriacetic acid (NTA) and surfactants. As EDTA has been found to be not very degradable and can add toxicity to the soil, other alternatives are being investigated (Pinto et al., 2014). For example, biosurfactants which are very biodegradable and of low toxicity have been investigated as reviewed by Mulligan (2021).

To reduce further environmental problems due to the sorption of residuals during flushing, the additives must be of low toxicity and biodegradable. Various factors including soil pH, type, porosity and moisture content, cation exchange capacity, particle size distribution, organic matter content, permeability and the type of contaminants can influence the effectiveness of the treatment.

Highly permeable soils (greater than 1×10^{-3} cm/s) are more amenable for treatment as the washing solution must be pumped through the soil by injection wells or surface sprinklers or other means of infiltration. Depth to groundwater can increase costs. The washing solution should be treated to remove and/or recover the contaminants and reuse the water through recovery wells or drains. However, the spreading of contaminants and the fluids must be contained and recaptured. Control of these infiltrating agents may be difficult, particularly if the site hydraulic characteristics are not well understood. Emissions of volatile organic compounds (VOCs) should be monitored and treated if required. Recycling of additives is desirable to improve process economics and reduce material use. Metals, VOCs, polychlorinated biphenyls (PCBs), fuels, and pesticides can be removed through soil flushing.

In choosing the most appropriate remediation technology, factors to be considered must include exposure routes, future land use, acceptable risks, regulatory guidelines, level and type of contaminants, site characteristics and resultant emissions. Laboratory and field treatability tests should be performed to obtain site-specific information. A schematic illustration of the criteria and tools for evaluating technologies and protocols for environmental management of contaminated soils and sediments is shown in Figure 11.2. *In situ* flushing was selected at 19 Superfund sites between 1982 and 2008 (EPA's 2010 Superfund Remedy Report (Thirteenth Edition)). The Seventh Edition (US EPA, 2023) showed that flushing was selected at two sites from 2018 to 2020, which was less than previous years.

A variation of soil flushing is foam injection (Wang and Mulligan, 2004). Foam consists of tiny bubbles, making an emulsion-like two-phase system where the mass of gas or air cells is dispersed in a liquid. Surfactants assist in creating and stabilizing the foams. A number of interesting applications have been investigated as it is applicable to remove various soil contaminants and is compatible with pump-and-treat systems, bioremediation and even nanoremediation (Vu and Mulligan, 2022). As an innovative technology, there are various requirements for future development including the effect of soil matrix characteristics, contaminant speciation, pulsed operation, and surfactant partitioning on the efectiveness of *in situ* foam flushing to the sub-surface conditions. Site geologic conditions must be investigated and proper selection of the foaming surfactant and its concentration must be determined. The mechanism of the surfactant on the remediation of contaminated soils is still not clear. Development of predictive mathematical models will be helpful for optimal surfactant selection for the subsurface.

Soil washing is applicable for soils contaminated with metals, and/or organic contaminants (El-Shafey and Canepa, 2003). Soil washing is an *ex situ* process that uses water to remove contaminants from soil and sediments by physical and/or chemical techniques. Soil washing involves the addition of a solution with the contaminated soil to transfer the contaminants to the wash solution. It is most appropriate for weaker bound metals in the form of hydroxides, oxides, and carbonates. Mercury, lead, cadmium, copper, nickel, zinc, and chromium can be recovered by electro-chemical processes if the levels of organic compounds are not significant. Metals can also be removed from precipitation or ion exchange. Precipitation is not applicable for metal sulfides. Pre-treatment to remove uncontaminated coarser fractions can be used. Various additives can be employed such as bases, surfactants, acids, or chelating agents. Nitric, hydrochloric, and sulphuric acids can be used. However, if sulphuric acid is used, 50% of the amount is required compared to hydrochloric acid (Papadopoulos et al., 1997).

Figure 11.4 illustrates a typical soil washing process where the separation consists of size separation, washing, rinsing, and other technologies similar to those used in the mineral processing industry. Larger particles are separated from the smaller ones as they have lower contamination levels. The smaller volumes of soil can be treated less expensively. Surfactants may be added in the washing water. The more contaminated size range is 0.24 to 2 mm due to the surface charges of the soil clay particles that attract anionic metal contaminant and the organic fraction that binds organic contaminants. Wash water and additives should be recycled or treated prior to disposal. The mechanical dewatering of particles is performed via a filter press, conveyer filtration, centrifugal separation, etc., froth flotation by the introduction of air bubbles into a slurry may also be used. The disposal of the treated fine particles varies depending on the type and levels of the contaminants.

Mixtures of metals and organic contaminants may require sequential washing with different additives to target the various contaminants. Soil washing processes generally use hot water to reduce the viscosity of hydrocarbons. The increased temperature also increases the solubilities of metal salts. The treated soil can then be washed to remove any residual wash solution prior to disposal. Ideally the wash solution should be recycled. Although extensively used in Europe, full-scale processes are less common in the US laboratory, feasibility tests should be conducted to determine optimal conditions (chemical type and dosage, contact time, agitation, temperature, and extraction steps to meet regulatory requirements). As spent wash water can be a mixture of soluble contaminants and fine particles, treatment is thus required to meet reuse or disposal requirements. Full-scale demonstrations may be required to demonstrate the feasibility of newly developed treatment processes. Presently, wastewater management systems act as a foundation for modern public health

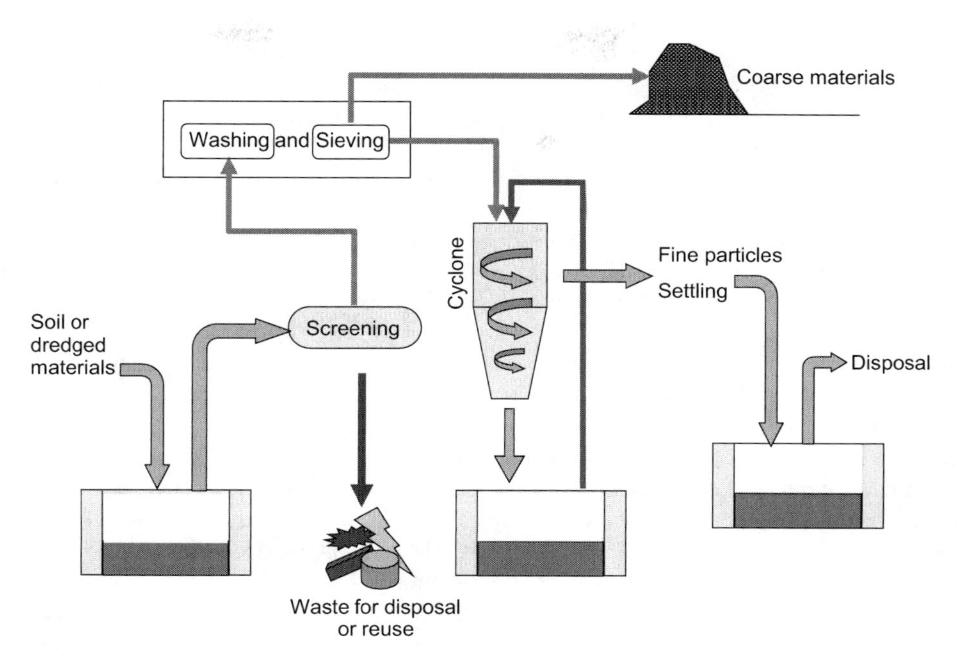

FIGURE 11.4 Schematic diagram of a soil flushing process for removal of contaminants.

and environment protection. The idea of most suitable wastewater management systems is to use less energy, allow for elimination, or beneficial reuse of biosolids, restore natural nutrient cycles, have much smaller footprints, be more energy efficient, and be designed to eliminate exposed wastewater surfaces, odours, and hazardous by-products (Daigger and Crawford, 2005). Some of the incentives for the industry to incorporate sustainability into their wastewater solutions are as follows (Mosley, 2006). In addition to the technical aspects of a wastewater treatment technology, selection of a particular technology should be based on all aspects that determine its sustainability.

11.2.7 Oxidation

Oxidation (US EPA, 2006) has been used *in situ* or *ex situ* for the complete chemical destruction of many toxic organic chemicals and partial oxidation prior to bioremediation or MNA. The compounds become less hazardous or toxic. The oxidants can include ozone, permanganate, hydrogen peroxide, sodium persulfate, and Fenton's reagent. High treatment efficiencies of greater than 90% can be achieved for unsaturated aliphatics, such as TCE and aromatic compounds, such as benzene. The efficiency of degradation depends on the concentration of the oxidant, the contaminant properties and concentration, and site characteristics, such as pH, temperature, and the concentration of other oxidant-consuming substances, including natural organic matter, minerals, carbonate, and other scavengers of free radicals.

The oxidant, known as Fenton's reagent, destroys a variety of wastes and generates no harmful by-products. Fenton's reagent was invented by Fenton in 1894. Today there are several methods known as "modified" Fenton's reaction where different additives increase the oxidizing efficiency by increasing the pH tolerance, increasing the reaction time and producing more and more stable radicals.

$$H_2O_2 + OH^- \Rightarrow HO_2^- + H_2O \text{ (perhydroxyl radical)} \quad (11.1)$$

$$HO_2^- => H^+ + O_2^{2-} \text{ (superoxide radical anion)} \tag{11.2}$$

$$HO_2^\bullet + O_2^{2-} => HO_2^- + O_2 \text{ (hydroperoxide anion)} \tag{11.3}$$

The co-existing oxidation-reduction reactions associated with a modified Fenton's process promote enhanced desorption and degradation of recalcitrant compounds (Fenton, 1893, 1895, 1899). These include compounds such as carbon tetrachloride and chloroform, which were previously considered untreatable by Fenton's chemistry. There is a complete mineralization of organic matter. The breakdown is fast – within days, typically minutes – hours, depending on the concentration of H_2O_2. The process has some effects on the residual free phase.

Modified Fenton's reagent, hydrogen peroxide, and potassium permanganate were applied polycyclic aromatic hydrocarbons (PAHs) contaminated sediments (Ferrarese et al., 2008). They concluded that the optimal oxidant dosages determined were quite high, as sorbed PAH mineralization requires very vigorous oxidation conditions, especially for soils and sediments with high organic matter content. Their results indicated that the optimal oxidant dose must be carefully determined under site-specific conditions. Kellar et al. (2009) have used a sodium-based Fenton reagent in the US in *in situ* and *ex situ* applications. The method has been proposed for the remediation of sediments near a chlorinated solvent site in Pennsylvania.

Wet air oxidation requires high temperatures and pressures but is capable of destroying PCBs and PAHs. Large quantities of water are not detrimental to the process. Costs are high at large scale, however.

Oxidation/reduction of heavy metals is another method for remediating in and *ex situ* soils, sediments sludges, and ash. A detoxification technology called TR-DETOX involves the percolation of inorganic and organic reagents to reduce heavy metals to their lowest valence state. The stabilized solids achieve the TCLP requirements and are considered as no longer leachable. The TR-DETOX technology can also be applied along with biological treatment (bioremediation) to ensure degradation of organic contaminants. One of the main chemicals is sodium polythiocarbonate that forms a precipitate that becomes less soluble over time. Lime, silicates and Portland cement are not added and costs are usually about one-quarter of stabilization/solidification processes. A unique characteristic is electronic addition of reagent. Pilot tests are required to determine the most appropriate formulation (Mulligan et al., 2001).

11.2.8 Nanoremediation

Nanoremediation involves the application of nanomaterials to transform or reduce the toxicity of the contaminants. The advantage of the addition of the nanoparticles to soil is that they can be transported further due to their small size. Zero-valent iron (nZVI) is the most common with particles ranging from 10 to 100 nm in size. Bimetallic (iron combine with gold, nickel, palladium, or platinum) nanoparticles or emulsified (EZVI) particles have also been used for the treatment of chlorinated compounds such as TCE, PCE, 2,2-dichloroethane (DCA) and vinyl chloride 1-1-1-tetrachloroethane (TCA), polychlorinated biphenyls (PCBs), halogenated aromatics, nitroaromatics, metals, such as arsenic and chromium (US EPA, 2008a), and nitrate, perchlorate, sulfate, and cyanide. Other types of nanoparticles include zinc oxide (ZnO), titanium dioxide (TiO2) and carbon-based. For soil remediation, nanoparticles can reduce, immobilize, and convert heavy metals and degrade organic contaminants (Araújo et al., 2015). Carbon-based nanomaterials and their applications were reviewed by Vu and Mulligan (2023b).

The treatment efficiency organic contaminants in soil by surfactant-stabilized nanoparticles is higher than only surfactant (less than 90%) or nanoparticles (less than 80%) due to the synergistic effects between surfactants and nanoparticles. The materials can be injected directly via a direct push technology or wells. The materials could also be received from the groundwater after injection and then they could be reinjected. Make-up nanoscale materials would likely be needed to compensate for losses. Nanomaterials could also be used in permeable reactive barriers for groundwater treatment.

Information from 45 sites was reviewed as a representative of the total projects under way using nanomaterials for site remediation (Karn et al., 2009). They concluded that nanoremediation could reduce costs and the time of cleaning up large-scale contaminated sites but eliminating the need disposal of contaminated soil and reducing some contaminant concentrations to negligible levels. However, potential adverse environmental impacts must be avoided due to the unknowns regarding nanoparticle transport and loss of activity.

An example of a nanotreatment remediation site (U.S. EPA OSWER Selected Sites Using or Testing Nanoparticles for Remediation) is at the BP Prudhoe Bay Unit, North Slope, AK, abandoned oil field RCRA Pilot Site. The soil is organic-rich over an alluvial gravel and was contaminated with TCA and diesel fuel. The maximum initial TCA concentration was 58,444 μg/kg, TCA reduction was 96% for a shallow test (up to 1.2 m depth) after 1 year of application (mean concentration) whereas for the deep test (up to 2.3 m depth), TCA reduction was 40%. Treatment and control plots for each delivery method were subjected to identical treatment processes to validate the results. Two different mixing methods were used: For the shallow test, physical mixing with lake water and for the deep test, pressurized injection via 20 injection points. The work was performed by PARS Environmental, Inc. and Lehigh University.

Rhamnolipid and sophorolipid biosurfactants were evaluated with nanoparticles to remove petroleum hydrocarbons from soil (Vu and Mulligan, 2023a). Nanoparticles have gained attention as promising materials for soil remediation, in many studies, both at bench and field scales. The results showed that high oil removal rates could be obtained with batch experiments (up to 84% removal). A rhamnolipid biosurfactant: Nanoparticle ratio of 10:1 (wt%: wt%) was optimal. The rhamnolipid biosurfactant performed better than the sophorolipid biosurfactant and ultraplex surfactant. In addition, nanoparticles could be reused for three cycles with rhamnolipid biosurfactant. The results indicated that a biosurfactant/Fe-Cu nanoparticle suspension had potential for oil-contaminated soil treatment and could be environmentally friendly. A biosurfactant foam/nanoparticle mixture was evaluated by Vu and Mulligan (2022) to remediate an oil-contaminated soil. A rhamnolipid biosurfactant foam/nanoparticle combination, a rhamnolipid biosurfactant solution/nanoparticle, and only rhamnolipid biosurfactant were compared, and resulted in 67%, 59%, and 52% removal of the oil, respectively. Increasing the flow rate decreased removal rates.

11.2.9 Electrokinetic Remediation

Electrokinetics involves the use of electrodes and electrical current to mobilize inorganic contaminants. It is more effective for treatment of silty soils than for clay soils where energy requirements can be substantial. Energy levels must be higher than the energy that binds the contaminants to the soil. Electro-osmosis and electrophoretic phenomena are the principal mechanisms in the treatment process. Conditioning fluids are required to enhance contaminant ion movement, and electrode dissolution or fouling is a substantial problem.

The electrokinetic remediation process can remove metals and organic contaminants from low permeability soil, sludge, dredged sediment, and waste materials. The technique uses electrochemical and electrokinetic processes to desorb, and then remove the contaminants. Electrochemical remediation uses a low DC current or a low potential gradient to electrodes that are inserted into the contaminated soil or sediment (Virkutyte, 2002). When DC electric fields are applied to the contaminated soil, ions migrate toward the corresponding electrodes (Figure 11.5). Cations are attracted to the cathode, and anions to the anode. An electric gradient initiates movement by electromigration (charged chemicals movement), electro-osmosis (movement of fluid), electrophoresis (charged particle movement), and electrolysis (chemical reactions due to electric field) (Rodsand and Acar, 1995). For example, under an induced electric potential, the anionic form of Cr(VI) migrates towards the anode, while the cationic

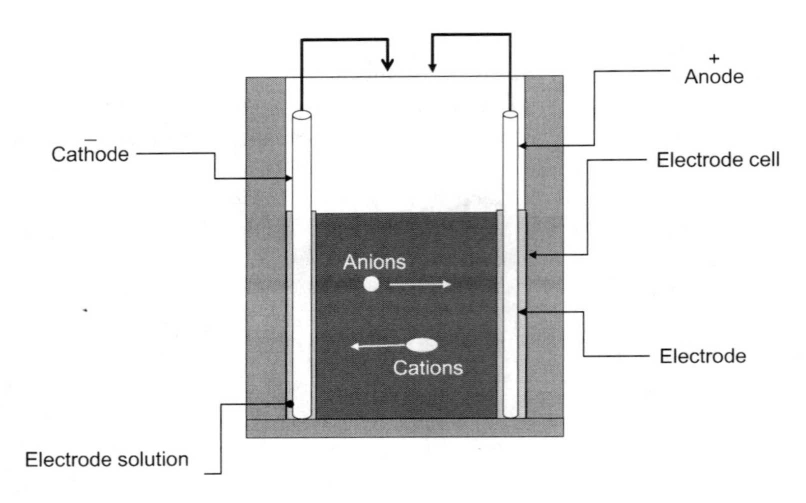

FIGURE 11.5 Electrokinetic treatment of contaminated soil.

forms of Cr(III), Ni(II), and Cd(II) migrate toward the cathode. Contaminants that accumulate at the electrodes are extracted by methods such as pumping water near the electrodes, precipitation/co-precipitation, electroplating, or complexation with ion-exchange resins (Reddy et al., 2001). This method is well suited for fine-grained soil and dredged sediment.

Control of the pH and electrolyte conditions within the electrode casings is essential in the optimization of the process efficiency. The process can be used to recover ions from soils, muds, dredging, and other materials (Acar et al., 1993). Metals as soluble ions and bound to soils as oxides, hydroxides, and carbonates are removed by this method. Other non-ionic components can also be transported with the flow. Unlike soil washing, this process is effective with clay soils.

Demonstrations of this technology have been performed but are limited in North America (Mulligan et al., 2001). A review by Lima et al. (2017) showed that the technology can be combined with other techniques, such as bioremediation, *in situ* chemical oxidation or reduction, and phytoremediation. At pilot and full-scale, the electrokinetic technology has been tested for demonstration purposes at the following sites: (1) Louisiana State University, (2) Electrokinetics, Inc., (3) Geokinetics International, Inc., and (4) Battelle Memorial Institute. Geokinetics International, Inc. (GII) has successfully demonstrated the *in situ* electrokinetic remediation process in five field sites in Europe for copper, zinc, lead, arsenic, cadmium, chromium, and nickel. In the UK, it was evaluated for treatment of highly contaminated mercury in canal sediments. Other ions such as cyanide, nitrate, and radionuclides such as uranium and strontium can also be treated by electrokinetics. New applications are also being developed for pesticides and PFAS.

Interferences include large metal objects, moisture content, temperature and some other contaminants. Metal recovery from highly contaminated soils could improve the process economics.

Soil and sediment particles have buffer capacity and release adsorbed substances from the surfaces when the value of pH decreases. Therefore, acidification may be a very effective method to solubilize the metal hydroxides and carbonates, other species adsorbed onto sediment particles, as well as protonate organic functional groups (Yong et al., 2006). Generally, in an electrochemical remediation process, the development of an acidic front is often coupled with a successful remediation (Nystroem et al., 2006). However, because of the higher buffering capacity of sediments, acidification of dredged materials may not be an acceptable method. Surfactants can increase the solubility and mobility of heavy metals during electrochemical remediation, depending on its function on decreasing the ζ potential of sediment and then reducing the van der Waals interactions (Nystroem et al., 2006). Therefore, using surfactants improves metal removal (Abidin and Yeliz, 2005). Direct costs have been estimated at \$15/m^3 with energy expenditure of \$0.03 per kilowatt hours, together with the cost of enhancement, could result in direct costs of \$50/m^3 or more. Another study has estimated full-scale costs at \$117/m^3. For remediation of metal-contaminated fine-grained and heterogeneous soils, this technique could potentially be competitive (FRTR, 2007). Energy, waste generation, and water use are low.

11.2.10 Solidification/Stabilization

The purpose of solidification/stabilization (S/S) processes is to reduce the mobility of the contaminants by addition of an agent that solidifies and then immobilizes the metals or hydrocarbons. Binders include cement, fly ash, sodium silicates, lime, sulphur-based binders, and organic-based binders and pozzolans are added *in situ* or *ex situ* (US EPA, 2000). Other processes or groups include bituminization, emulsified asphalt, polyethylene extrusion, pozzolan/Portland cement, and soluble phosphates. *Ex situ* S/S requires disposal of the stabilized residue. Solidification/stabilization is often utilized for metal contamination as there are few destructive techniques available for metals. Some metals such as As, Cr(VI), Pb, Cu, Ni, Zn, and Hg are suitable for this type of treatment. The metals are hydrolyzed to form hydroxides, oxides, carbonates, and sulphates, etc., that are of limited solubility. Liquid monomers that polymerize and cement are injected to encapsulate the soils. Leaching of the contaminants must, however, be carefully monitored as is the case for vitrification, the formation of a glassy solid. Cement- or silicate-based (5 to 10% by weight additives) processes are useful for soils and sediments and are economical as the mixing equipment and materials are readily available. Other materials containing iron (red mud, sludge from a water treatment plants, bog iron ore, unused steel shot, and steel shot waste) have been evaluated (Mulligan and Kamali, 2003) for immobilizing cadmium and arsenic contaminants in sediments. All were effective in reducing the bioavailability of the metals to plants, but the safest was sludge from a drinking water plant with low levels of As. However, if there are different types of metals present, the treatment may not be as effective.

Water contents greater than 20% or chlorinated hydrocarbon contents greater than 5% increases the amount of agents required. Variability in the water content, grain size, and the presence of debris can make handling of the materials difficult and decrease the efficiency of the solidification process. In addition, since immobilization leads to an increase in volume, larger areas of land are required for disposal. Thus, smaller volumes for treatment are more appropriate (Hazardous Waste Consultant, 1996).

For organic contaminants such as oil or gasoline, thermoplastic binder and organic polymerization has utilized. The most commonly used thermoplastic material is polyethylene or asphalt. The organic materials can include polyethylene, polypropylene, urea formaldehyde, or paraffin. The contaminated soil is dried and mixed into the polymer that is cooled to form a solid. The mix can be extruded into a metal drum for easy transport and disposal. Full-scale projects have been performed in the US, Canada, Japan, and Belgium. If the process is performed *ex situ*, the soil is usually screened to remove large materials, mixed with the binder and water (e.g., in a rotary drum) and then transferred to a disposal area. Off gases would need to be treated for dust or volatile organic contaminants. The process can be performed in a mobile unit or at a fixed site.

For *in situ* applications, the reagents must be prepared, an auger can be used to mix the binder directly into the soil and off gases should be treated, particularly for organic contaminants. Dust generation is minimal, and costs related to excavation and transport are eliminated. Boulders, bedrock, and clay can cause mixing problems. Contamination of the groundwater must be avoided. Therefore, mobile contaminants such as Cr(VI) should be converted first to the less mobile Cr(III) form. After solidification, reuse of the land for buildings may be possible. Backfill may be required as a cover for revegetation. Durability testing will be required. In all cases, compatibility testing should be performed to determine the most appropriate binding material and TCLP tests after the binding will indicate the leachability of the contaminants.

Halogenated semi-volatiles, non-halogenated semi-volatiles and non-volatiles, volatile and non-volatile metals, low level radioactive materials, corrosives and cyanides have been treated effectively. Newer applications have concerned PFAS contaminated soils (Sörengård et al., 2021). Activated carbon addition is needed and showed a removal rate greater than 98% for PFASs with perfluorocarbon chain lengths of more than 6. In the Netherlands, a rotating drum was used in a full-scale experiment (Rienks, 1998). In total, 680 tonnes of dewatered sediment were treated at 600 °C for 38.5 h for mineral oil, PAHs, and mercury. Mercury levels decreased by 80% from 1.5 to 0.3 mg/kg while mineral oil and PAHs decreased by greater than 99.8%. Leaching of arsenic, molybdenum, and fluoride increased after thermal treatment, which can have implications in the reuse of the treated sediments as road or construction materials. At a former wood-preserving plant in Fresno, California, contaminated groundwater with As, Cr, Cu, and PCP was found (US EPA, 2005). Various hydrocarbons, PAHs were found in the first 2 m of the soil. The Silicate Technology Corporation (STC) immobilization with proprietary organophilic materials was used. The soil was removed from the unlined disposal pond and transported to the processing area for mixing, addition of reagents and discharge into concrete forms. Neutral conditions were required for As stabilization as mobility increased at lower pH values. Arsenite could not be converted to arsenate by the process. PCP and other organic contaminants were below TCLP threshold levels before and after treatment. Mixing had to be thorough and dust was a problem that had to be minimized.

11.3 THERMAL REMEDIATION

11.3.1 Vitrification

Another immobilization technique is vitrification, which involves the insertion of electrodes into the soil, which must be able to carry a current to heat the soil up to 1400 to 2000 °C and then to solidify as it cools. Toxic gases from the organic contaminants can also be produced during vitrification and must be collected and treated. Volume reduction of 20 to 50% occurs leading to subsidence above the melt and thus backfill is required to fill the volume (Weston, 1988). The heavy metal and radionuclide contaminants remain in the glass-like substance (US EPA, 2006). Costs can be high since fuel values are low and moisture contents are high (above 20%). Vitrification operation costs are according to water and electricity consumption and treatment depth and area.

A technology was developed for the remediation of organic contaminants and immobilization of metals in a glassy matrix and evaluated on the dredged sediments from New York/New Jersey harbour (IGT, 1996). A plasma torch is used to heat the sediments. Feeding of the wet sediments into the plasma reactor and adjustment of residence times can be difficult, however. Cadmium, mercury, and lead levels were reduced efficiently (97, 95, and 82%). Glass tiles and fibre glass materials were produced and could be used as valuable end products.

Temperatures higher than 1200 °C possibly degrade organic compounds and volatilize heavy metals. As the solids like minerals will melt at this temperature, the technique which utilizes this temperature range is called vitrification technique or GeoMelt process (www.nuclearsolutions.veolia.com/en/our-expertise/technologies/our-geomelt-vitrification-technologies-stabilize-waste, accessed Apr. 2024). The materials can be burned, electrically melted, or other means. In the US, Japan, and Australia, the GeoMelt process has been in commercial use since the 1990s and has treated more than 26,000 metric tons of waste, including remediating sites contaminated with heavy metals, radioisotopes, pesticides, herbicides, solvents, PCBs, dioxins, and furans.

The GeoMelt processes are designed to be an *in situ* or mobile container thermal treatment process that involves the electric melting of contaminated soils, sludges, or other earthen materials and debris either *in situ* or *ex situ* for the purpose of permanently destroying, removing, and/or immobilizing hazardous and radioactive contaminants. *Ex situ* technology for vitrification is illustrated as shown in Figure 11.6.

Excavated materials are first dried and transported into a forge. The materials are melted at a temperature higher than 1200 °C. The produced gas is cooled down and treated with activated carbon.

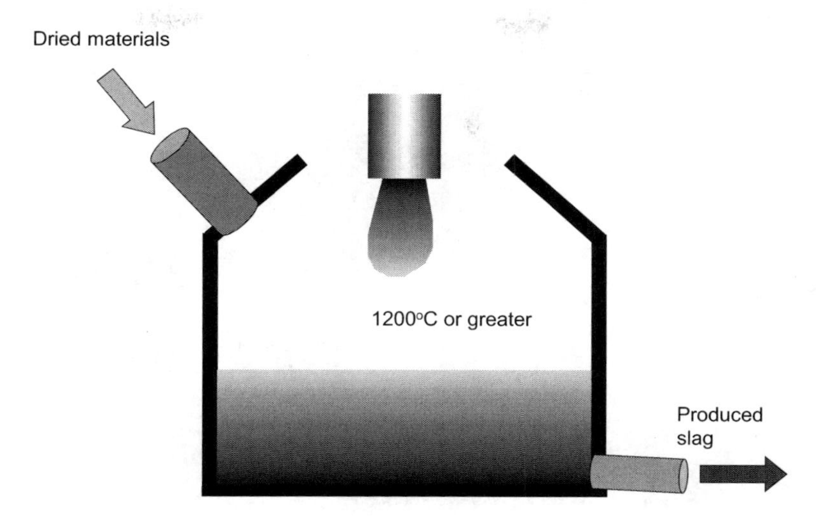

FIGURE 11.6 Vitrification of soil or sediment.

After contaminants are removed by the activated carbon, the gas is released into the air. Since hazardous materials, such as organic compounds and heavy metals, in the materials are vapourized, the solids after vitrification are usually clean. This technique is recommended as one of the treatment techniques of sediments contaminated with dioxin, in the Japanese technical guideline of sediments contaminated with dioxin. The process flow is shown in Figure11.7.

11.3.2 Incineration

Incineration by rotary kiln has been used to treat hazardous materials including waste, contaminated soils, and sediments. Other types of incineration of fluidized bed are circulating bed combustors and infrared combustion. The heat is supplied with a burner in a kiln, as shown in Figure 11.8, and the materials can be carbonized, when the contaminants can be released by vapourization. Flue gas treatment is required in this case. Temperatures for organic contaminants are in the range of 800 to 1200 °C (US EPA 1998). On site mobile units or off-site treatment facilities are employed. In Japan, this rotary kiln technique is also recommended as one of the treatment techniques for sediments contaminated with dioxin. PCBs can also be treated.

There are soils contaminated with hydrocarbons in many industrial sites and oil refineries. Thermal treatments are the most popular and versatile because they can be effectively applied to a wide range of organic contaminants (Merino and Bucalá, 2007).

At different temperatures (150–800 °C), results showed that at about 300 °C the hexadecane can be removed almost completely from the soil matrix (99.9% destruction removal efficiency, DRE), and that temperatures above this value do not improve the removal efficiency noticeably (Merino and Bucalá, 2007). In Japan, treatment of excavated and dredged materials is required for materials with dioxin levels higher than 3000 pg-TEQ/g. The recommended techniques are as follows:

(1) Melting (Geo-melting technique) (>1200 °C).
(2) Incineration (rotary kiln incinerator) (1100 °C).
(3) Low temperature thermal degradation (400–600 °C).
(4) Chemical decomposition (350 °C).

These techniques have different advantages and disadvantages.

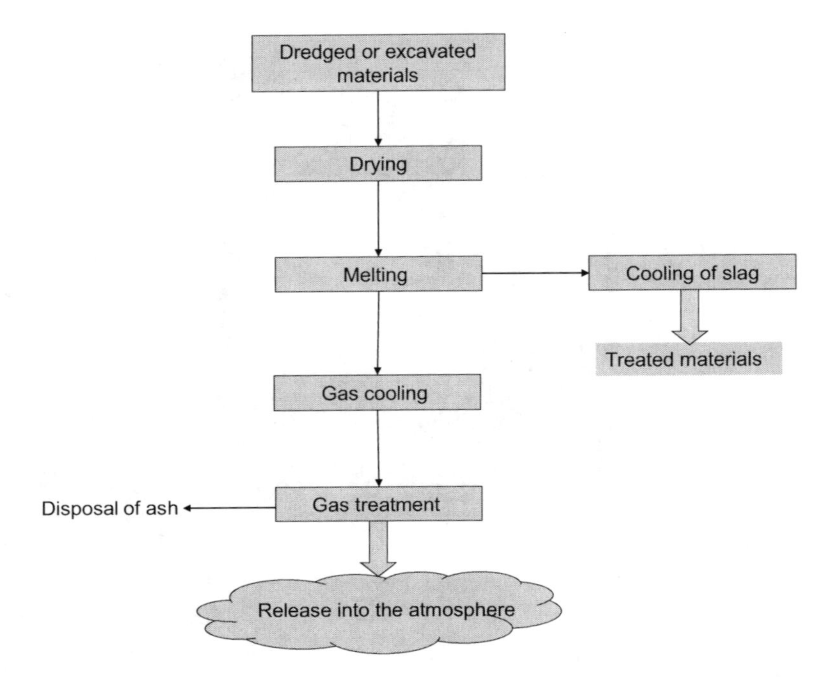

FIGURE 11.7 Japanese process for dioxin-contaminated sediments. (Adapted from Mulligan et al., 2010.)

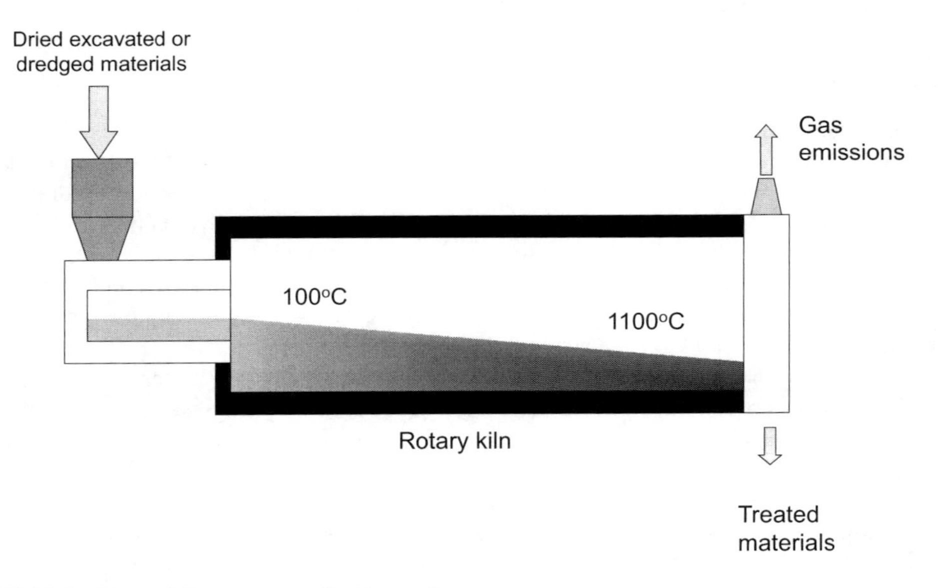

FIGURE 11.8 Rotary kiln treatment of soil or sediments.

11.3.3 Thermal Extraction

In situ thermal treatment or thermal enhancement of SVE is used to enhance volatilization or decompose contaminants and can be performed using a variety of techniques including electrical resistance, conductive heating, electromagnetic/fibre optic/radiofrequency heating or hot-air, water or steam injection to increase the volatilization rate of semi-volatiles, DNAPLs, LNAPL, and facilitate extraction. These methods can be completed in the short to medium term such as less than 40 days. Examples are three- and six-phase soil heating (US EPA, 2012b). Large-scale *in situ* projects employ three-phase soil heating while six phases is for the demonstration phase. Electrical resistance heating uses electrodes to produce a current that heats the soil above 100 °C for steam generation, causes soil drying and fracturing. Vacuum extraction then removes the contaminants.

The radiofrequency technique can heat soils to over 300 °C and enhance SVE by (1) increasing the contaminant vapour pressure and diffusivity, (2) drying the soil which increases permeability, (3) increasing the volatility of the contaminant by stripping with the water vapour; and (4) increasing the mobility by decreasing the viscosity. The current stops once the soil is dried. Incineration or granular activated carbon is used to treat the extracted vapour.

Steam injection or steam-enhanced extraction (SEE) heats the soil and groundwater and can destroy contaminants. The steam drives the contaminants that can be removed by groundwater and vapour extraction. Thermal conduction heating (TCH) destroys contaminants by electrical conductance or evaporates them for subsequent removal by a carrier gas or a vacuum system.

Another technique is an *ex situ* process, thermal desorption by heating the soil to 90 to 320 °C for VOCs or 320 to 540 °C for SVOCs removal. The low-temperature process works well for oil contaminated soils. A major advantage is that the decontaminated soil retains its physical properties and organic matter in the soil is not damaged. Therefore, there is the potential for reinstating the past ability of the soil to support future biological activity. The vapours subsequently must be treated by thermal oxidation for complete destruction (USEPA, 2012b).

With regards to heavy metals mercury, arsenic, and cadmium and its compounds can be evaporated at 800 °C with the appropriate air pollution control system. Some of the metals remain in the solid residues and will have to be properly disposed of. Thermal extraction is applicable mainly for mercury since this metal is highly volatile.

Mercury Recovery Services (MRS) developed and commercialized a process that mixes a proprietary material and the mercury contaminated material at temperatures of 150–650 °C (Weyand et al., 1994). The process can be mobile or fixed, batch, continuous or semicontinuous. Unit capacities ranged from 0.5 to 10 t/h. The mercury can be as an oxide, chloride, and sulphide. No liquid or solid secondary products were generated. The treated material contained less than 1 ppm of mercury. The process consisted of two stages, feed drying and mercury desorption, which was then condensed as a 99% pure metallic form from the vapour phase. Air emissions did not contain mercury. Costs were high, in the range of US$ 650–1000/t. Soil from approximately 6000 metring sites along the natural gas pipeline system in the western US was treated using the MRS process. Over 18 months, a 12-tonne per day mobile unit processed a total of 6000 tonnes of soil with 100–2000 mg/kg of mercury. The treated soil contained less than 2 mg/kg of Hg, The results from the TCLP tests indicated that leaching was minimal at less than 0.0025 mg/kg, which is substantially below the 0.2 mg/kg EPA limit. In addition, more than 3500 pounds of metallic mercury were recovered for sale and recycling.

Costs of various thermal treatments were reviewed by Vidonish et al. (2016). Hot air and steam injection and thermal desorption are in general lower cost than incineration, vitrification, and microwave heating. Opportunities for alternative energy sources were also discussed, in addition to consideration of water needs, ecological, and soil quality.

11.4 BIOLOGICAL REMEDIATION

For heavily contaminated soils, various approaches can be used to enhance the rate of bioremediation. Substances must be biodegradable and not toxic for treatment. *Ex situ* bioremediation has been more successful than *in situ* processes due to easier control of environmental parameters such as mixing that allows uniform nutrient and oxygen contents. Proprietary biological mixtures for bioaugmentation are also available. *Ex situ* biotreatment systems include the use of slurry bioreactors, biopiles, landfarming, and composting. In general, the more sophisticated the process, the more expensive the treatment. Treatability studies are usually performed to determine the efficiency of the bioremediation for the type of contaminants and sediments at the site. The microbial population, nutrient levels, pH, moisture content, contaminant type and concentration, and sediment characteristics must be determined and followed. Bench, pilot, and demonstration scale tests are needed to properly design the remediation technology.

Microorganisms have been effective in treating organic contaminated sediments such as PAHs. Zhao and Hawari (2008) have demonstrated that anaerobic degradation of RDX was possible in a Halifax sediment. Degradation rates of TNT>RDX>HMX were found. *Shewanella* and *Halomonas* bacterial isolates were found (Zhao and Hawari, 2008). Khodadoust et al. (2009) showed that PCBs could be degraded anaerobically with the periodic addition of iron (0.01 to 0.1g/g).

11.4.1 Slurry Reactors

Slurry bioreactors use 5 to 30% solid content in a highly agitated treatment. Mass transfer, aeration, and environmental conditions such as pH can be optimized more easily than for *in situ* remediation. This type of treatment is particularly applicable for compounds of low biodegradability such as SVOCs, VOCs, ordnance, pesticides, PCBs and PAHs of soils and sediments. Slurry methods can be used since dewatering is not required (Figure 11.9). Bioremediation is a low-cost technology and therefore has the potential for wide use. However, metal remediation technologies are not as developed as organic treatments. After treatment, water must be separated from the solids and may require subsequent treatment.

Surfactants can be added to enhance contaminant solubility, or the natural bacteria could be stimulated to produce natural biosurfactants. The latter approach was investigated for an oil and heavy metal contaminated harbor soil (Jalali and Mulligan, 2008) and has shown potential. Results showed that over the 50-day experiment, nutrient amendments enhanced biosurfactant production to up to three times their critical micelle concentration (CMC). Limiting the inorganic source of nitrogen showed an enhancement of biosurfactant production by 40%. The biosurfactants produced

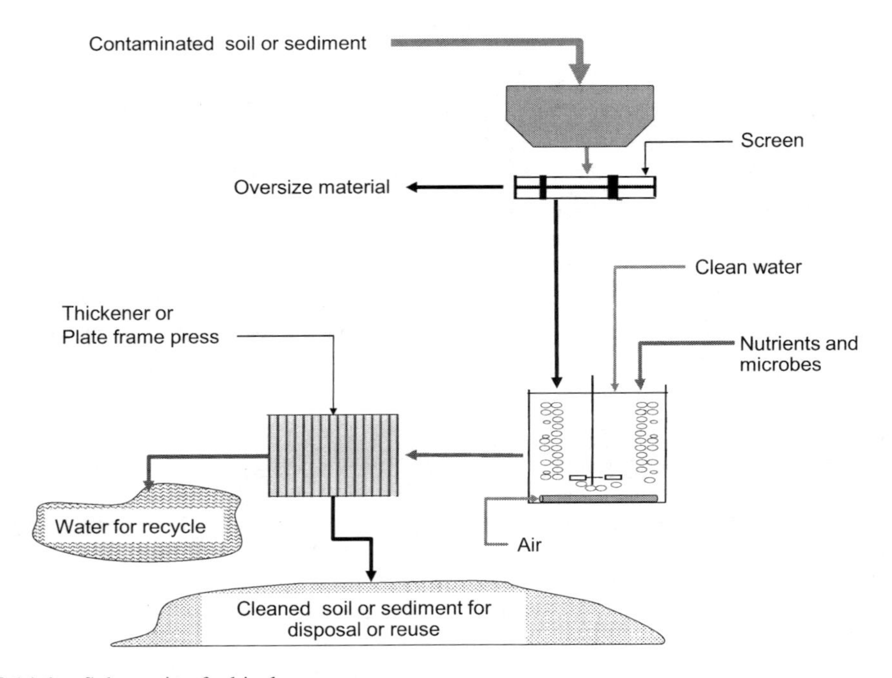

FIGURE 11.9 Schematic of a bioslurry process.

by the indigenous soil microorganisms were also able to solubilize 10% of TPH and 6% of the metal content of the soil and enhance biodegradation of petroleum hydrocarbons.

11.4.2 Landfarming

Landfarming includes mixing the surface layer of soil with the contaminated sediment (Rittmann and McCarty, 2001). Soil microorganisms are utilized for biodegradation of the contaminants including oil sludge, wood preserving wastes, coke wastes, JP-5, fuel oils, diesel fuel, and some pesticides (FRTR, 2007). The resulting product is compost. Moisture must be monitored, and nutrients can be added to enhance biodegradation. Occasional turning of the soil increases the oxygen content and permeability of the sediment/surface soil mixture. Bulking materials and nutrients may also be added. The process is simple but could lead to contaminant volatilization and leaching. Therefore, monitoring is required. Land requirements can be extensive. In the US and Belgium, bioremediated dredged materials has been mixed with compost and/or municipal sewage sludge to produce soil for landscaping projects and in Germany it has been used in orchards.

An additive that has been used with landfarming is DARAMEND™ (https://active-oxygens.evonik.com/en/products-and-services/soil-and-groundwater-remediation/daramend-reagent). It is a solid phase amendment (Figure 11.10) to promote anoxic conditions to enhance the bioremediation of pesticides and explosives such as toxaphene, DDT, dieldrin, TNT, RDX, PCE, TCE, DDD, DDE, and chlorinated VOCs (CVOCs). The reduction in the redox potential enhances the dechlorination of organochlorine compounds. With tilling equipment, the material can be mixed into a depth of 0.6 m. Hydrated lime is used to maintain the pH between 6.6 and 8.5. Redox potential and moisture were monitored at an evaluated at a Superfund Site (Montgomery, AL) of a soil/sediment contaminated with pesticides (US EPA, 2004a). Approximately 4500 tonnes were treated and all contaminated reached specified levels. Santiago et al. (2003) evaluated DARAMEND for PAH contaminated sediment. However, PAH concentrations were higher than expected (average of 900 ppm) and thus could be reduced by bioremediation to CCME criteria (260 ppm) in bench-scale experiments. Thermal treatment was successful, however.

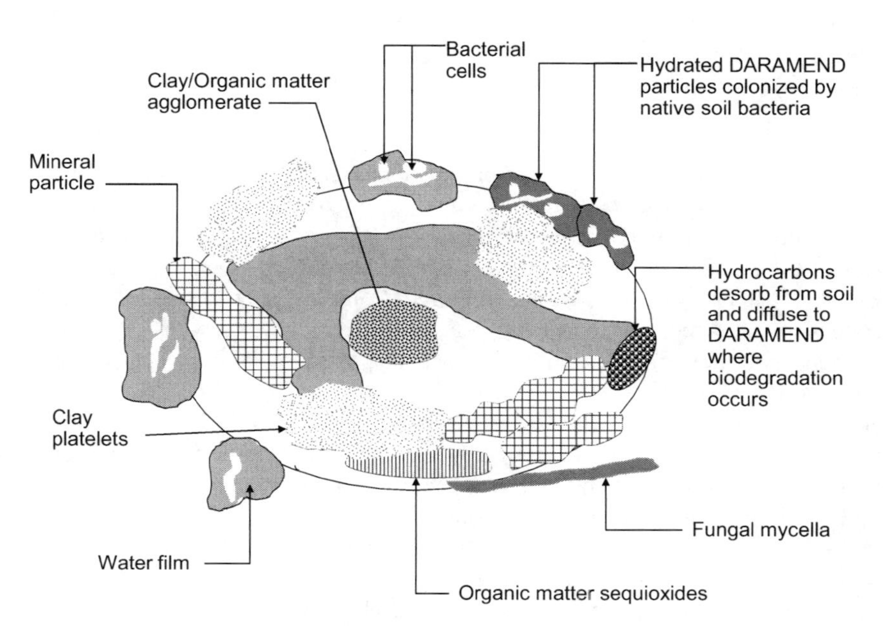

FIGURE 11.10 Schematic of the DARAMEND technology. (Adapted from Mulligan, 2002.)

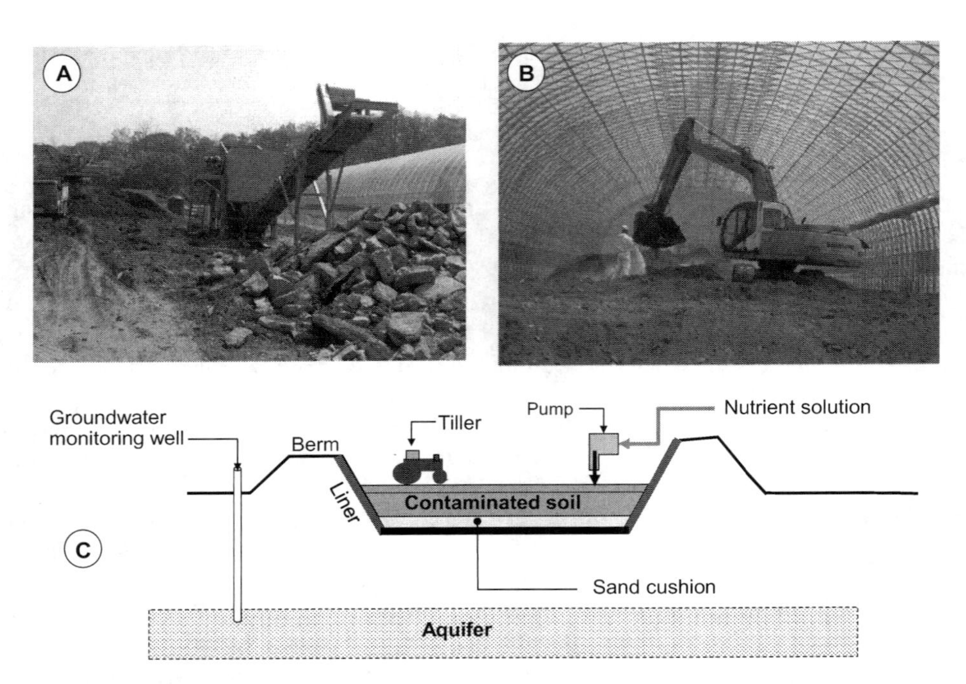

FIGURE 11.11 Photos of (a) removal of large debris, (b) soil mixing equipment, and (c) schematic of a landfarming process.

In Asia, a landfarming process was performed on hydrocarbon-contaminated soil (Figure 11.11). The TPH content of 3700 mg/kg was reduced to less than 450 mg/kg. The BTEX was reduced from greater than 80 mg/kg to undetectable levels in 90 days. Commercial microbes and nutrients were added, and the soil was tilled three times a week. The soil moisture was maintained at 40 to 60% soil capacity. Weekly monitoring for TPH, BTEX, nitrates, phosphorus, ammonia, aerobic respiration, pH, microbial counts, and moisture contents. Volumes of greater than 4000 m^3 were treated in each batch.

11.4.3 Composting

Composting involves the biodegradation of organic materials to produce carbon dioxide and water for soils and sediments. Typical temperatures are in the range of 55 to 65 °C due to the heat from the biodegradation process. Animal or vegetable wastes such as sewage sludge are often used as organic amendments. Bulking agents to increase the porosity of the material such as woods chips are added. Moisture content and temperature must be monitored. Composting processes include windrows and biopiles, and in vessel composting. Gaseous emissions and leachates may be produced and thus will need to be managed. Thermophilic composting can be applied to the treatment of explosive or PAH contaminated soils.

Composting of contaminated sediment was evaluated (Khan and Anjaneyulu, 2006). A mixture of 10 kg of sediment with 0.5% fertilizer and 50% compost was used. The sediment contaminants included phenols (16–24 mg/kg) and benzene (3.4 mg/kg). Fertilizer was added as a nutrient and compost was used to inoculate with microorganisms. Wood chips were added for support and aeration in the pile for composting. The parameters, pH, total volatile solids, microbial count, temperature and contaminant concentration, were monitored over the period of five weeks. Whereas benzene was almost completely biodegraded, lower levels of phenol degradation were obtained (80 to 85%) Therefore, composting was shown to be technically feasible at lab scale.

TABLE 11.1
Comparison of Bioremediation Technologies

Parameter	Windrow Composting	Landfarming	Biopile Composting
Applicability	PAHs, explosives,	Fuel oil, diesel fuel, PCBs, pesticides	Fuels, solvents
Site requirements	Excavation and special mixing equipment	Excavation and earthmoving equipment, liners	Excavation and earthmoving equipment, aeration, liners
Limitations	Bulking agents that increase volume and may need to be removed	Permanent structures required	Static process without mixing
Cost	Moderate	Low	Low to moderate

Source: Adapted from Myers and Williford (2000) and FRTR (2007).

Myers and Williford (2000) examined the bioremediation of contaminated sediments in a CDF. Composting (windrows and biopiles), landfarming and land treatment were examined for PAHs, PCBs, and PCDDs/F (Table 11.1). Land treatment is similar to landfarming except that the contaminated soil or sediments interact with the surrounding soil. Monitoring for potential leaching and volatilization of contaminants is essential. Composting and land treatment have the potential to be cost-effective but require pilot and demonstration studies. Composting tests were not successful for remediating PAHs, but PCB degradation may be more promising (Myers et al., 2003).

11.4.4 Bioconversion Processes

Microorganisms are also known to oxidize and reduce metal contaminants. Mercury and cadmium can be oxidized while arsenic and iron can be reduced by microorganisms. This process (called mercrobes) has been developed and tested in Germany for removal of concentrations greater than 100 ppm from 95% to 99%. Since the mobility is influenced by its oxidation state, these reactions can affect the contaminant mobility. Organic contaminants such as benzene can also be degraded.

Chromium conversion is also affected by the presence of biosurfactants. A study was conducted by Massara et al. (2007) on the removal of Cr(III) to eliminate the hazard imposed by its presence of kaolinite. The effect of the addition of negatively-charged biosurfactants (rhamnolipids) on chromium contaminated soil was studied. The sequential extraction results showed that rhamnolipids remove Cr(III) mainly from the carbonate, and oxide/hydroxide portions of the soil, stable forms from the soil. The rhamnolipids have also the capability of reducing close to 100% of the extracted Cr(VI) to Cr(III) over a period of 24 days.

11.4.5 Phytoremediation

Phytoremediation is the use of plants to remove, contain, or render harmless environmental contaminants. Constructed wetlands use aquatic plants such as water hyacinths to remove nutrients and contaminants from water. The various mechanisms involved in phytoremediation include (a) phytoextraction, (b) uptake of contaminants through the roots and subsequent accumulation in the plants, (c) phytodegradation, (d) metabolism of contaminants in the leaves, shoots and roots, (e) release of enzymes and other components for stimulation of bacterial activity or biochemical conversion and rhizodegradation, and (f) mineralization of contaminants in the soil by microbial activity in the rhizosphere. Phytoremediation is a low-cost *in situ* technology that causes minimal disturbance and is aesthetically pleasing. It is acceptable to the public and generates low amounts of waste. This technology is presently being developed for a treatment of a wide variety of organic and

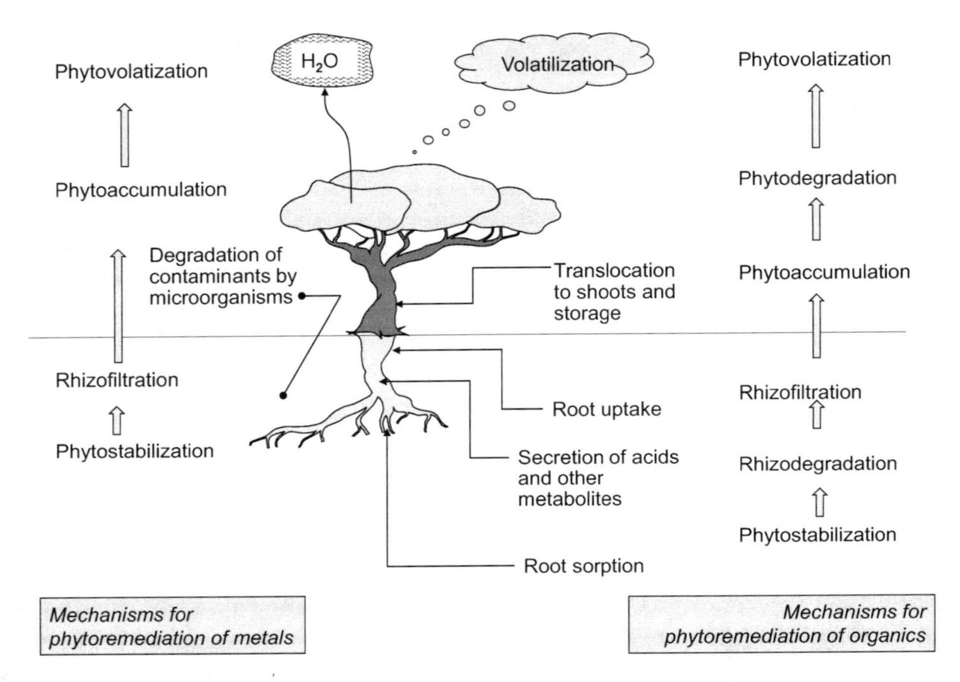

FIGURE 11.12 Mechanisms of phytoremediation. Adapted from Mulligan, 2002.)

inorganic contaminants. Better understanding of contaminant uptake by the plants, development of hyperaccumulators, and increased field testing are needed.

Some plants have been shown to retain contaminants in their roots, stems, and leaves via phytoextraction, phytodegradation, phytostabilization, and biodegradation in the rhizosphere (Hazardous Waste Consultant, 1996) (Figure 11.12). Phytoaccumulation is the transport of contaminants from the roots to the shoots and leaves. Contaminants are metabolized in the plant via enzymes for phytodegradation. Phytostabilization immobilizes contaminants by excretion of various chemicals from the roots. Around the plants root, microorganism growth is stimulated by the nutrients in the soil. These microorganisms can then biodegrade the contaminants in the soil. Various examples are shown in Table 11.2.

Vegetative caps consisting of grasses, trees, and shrubs can be established in shallow fresh water. The resulting vegetative mat can hold sediments in place. The construction of wetlands is growing for wastewater treatment and thus the knowledge on wetland configurations is growing. However, vegetative caps have not yet been applied to the remediation of sediments (Mulligan et al., 1999). However, phytoremediation could be implemented where dredged sediments have been placed in contained areas and a wetland is then constructed to remediate and contain the sediments. Lee and Price (2003) indicated that although phytoextraction of Pb with chelates may be troublesome due to potential leaching into groundwater, immobilization, and phytostabilization can be appropriate in CDFs. The site could potentially be restored for beneficial use as a wildlife habitat.

11.4.6 In Situ Bioremediation

Bioremediation involves the use of microorganisms to degrade organic contaminants completely or to less toxic components (US EPA, 2013). Usually, an electron donor or energy sources, an electron acceptor and nutrients are required for a successful process. Naturally occurring microorganisms are often used as these are adaptable to changing conditions. Biostimulation is the seeding of known organisms if the site is deficient in required organisms. Nutrients can be pumped in or percolated.

TABLE 11.2
Phytotechnology Mechanisms

Mechanism	Description
Phytodegradation	Uptake and degradation of contaminants within plant tissues by enzymatic activity for remediation
Phytoextraction	Uptake of contaminants by plants and sequestration into the plant tissue
Phytohydraulics	Ability of plants to take up and transpire water to control hydrology
Phytosequestration	Ability of plants to sequester certain contaminants into the rhizosphere through release of phytochemicals, and sequester contaminants on/ into the plant roots and stems through transport proteins and cellular processes
Phytovolatilization	Ability of plants to take up, translocate, and subsequently volatilize contaminants in the transpiration stream for remediation
Rhizodegradation	Ability of released phytochemicals to enhance microbial biodegradation of contaminants in the rhizosphere for remediation

Source: Adapted from Interstate Technical Regulatory Council (ITRC) (2009).

For aerobic processes, dissolved oxygen is added either by aerating water for saturating and adding ozone or hydrogen peroxide. For shallow soils, nutrients and oxygenated water can be added via infiltration galleries or spray irrigation. For deeper soils, injection wells are required. High temperatures enhance bioremediation rates. However, at lower temperatures bioremediation still occurs but more slowly. Heat blankets to cover the soil surface can increase the soil temperature and subsequently the biodegradation rate. Bioremediation can take several years to clean the site.

Under anaerobic conditions, methane, carbon dioxide, and trace amounts of hydrogen gas will be produced. Various electron acceptors are necessary in place of oxygen including, sulphate, iron (III), or nitrate. Under sulphate-reduction conditions, sulphide or elemental sulphur is produced, and under nitrate-reduction conditions, nitrogen gas is the final product. A disadvantage of anaerobic procedures is that contaminants can be degraded to products that are as or more hazardous than the original contaminant. For example, of this is the biodegradation of TCE to the more toxic vinyl chloride. To avoid this problem, aerobic conditions can be created to biodegrade the vinyl chloride.

Bioremediation techniques have been successfully used to remediate soils, sludges, and groundwater contaminated with petroleum hydrocarbons, solvents, pesticides, wood preservatives, and other organic chemicals. Bench- and pilot-scale studies have demonstrated the effectiveness of anaerobic microbial degradation of nitrotoluene in soils contaminated with munitions wastes. Bioremediation is especially effective for remediating low-level residual contamination in conjunction with source removal.

The most common contaminants treated by bioremediation include PAHs, non-halogenated SVOCs and BTEX. Superfund sites are commonly bioremediated if they contain wastes associated with wood preserving (creosote), and petroleum refining and reuse (BTEX).

Excavation of contaminated soil is not required. Bioremediation is often less costly compared with other technologies including thermally enhanced recovery with heating, chemical treatment with expensive chemical reagents, and *in situ* flushing (which may require further treatment of the flushing water) and thermal desorption and incineration, which require excavation and heating.

Although bioremediation cannot degrade inorganic contaminants, bioremediation can be used to change the valence state of inorganics and cause adsorption, immobilization onto soil particulates, precipitation, uptake, accumulation, and concentration of inorganics. These techniques are promising for immobilizing or removing inorganics from soil and other wastes. Bacteria (Bader et al., 1996) and biosurfactants (Massara et al., 2007) can reduce Cr(VI) to Cr(III), which is less toxic and mobile. Sulfate-reducing bacteria can form insoluble metal sulfides. Heap leaching and *in situ* leaching have been used by the metal industry for copper recovery (Rawlings, 1997). *Thiobacillus* bacteria produces sulfuric acid that can be used to solubilize metal sulfides to metal sulfates. Fungus such as *Aspergillus niger* can produce citric and gluconic acids (Mulligan et al., 1999a).

11.4.7 Bioventing

Bioventing (Figure 11.13) is an *in situ* process that involves forced aeration in the vadose zone to enhance the biological degradation of SVOCs and non-volatile contaminants with Henry's law coefficients less than 0.1. These contaminants can include petroleum hydrocarbons, nonchlorinated solvents, some pesticides, wood preservatives, and other organic chemicals. A high soil permeability (greater than 10^{-6} cm/s) is needed and sufficient nutrients must be supplied. Feasibility testing should evaluate air permeability of soil and *in situ* respiration rates. Aeration levels are substantially lower than in SVE processes to avoid emissions and enhance biodegradation. However, as some volatilization will occur, then the emissions must be monitored and captured and treated if required. Air treatment can include biofiltration, activated carbon or catalytic oxidation, or thermal treatment. High soil moisture contents can decrease process efficiency. Cleanup times can range from a few months to several years. Bioventing is becoming more common, and most of the hardware components are readily available due to experience with soil vapour extraction (SVE). The Air Force is demonstrating bioventing at 135 sites (FRTR, 2007).

11.4.8 Biosparging

For the biosparging process, air or oxygen are injected with nutrients into the saturated zone to allow microorganisms to degrade contaminants. The process is more effective in highly permeable

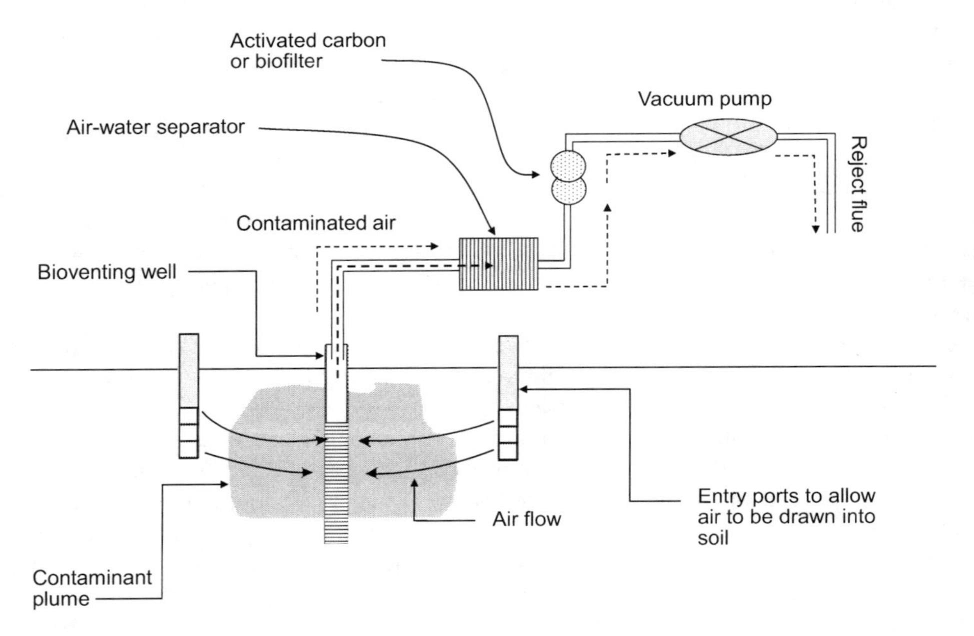

FIGURE 11.13 Schematic of a bioventing process. Note that although the bioventing process is similar to SVE, the air flow is lower to allow bioremediation.

soil (US EPA 1994). Dissolved petroleum in groundwater and at the capillary fringe can be treated by this method. Biosparging can be combined with soil vapour extraction (SVE) or bioventing to capture volatilized components by the vapour extraction system. This method was described in Chapter 3.

11.4.9 Microbial-Induced Mineral Precipitation

The term "biomineralization" appeared as terminology in a book "*On Biomineralization*" written by Lowenstam and Weiner in 1989. At that time, the biomineralization was distinguished into biological controlled mineral precipitation (BCMP) and biological-induced mineral precipitation (BIMP).

From a technical point of view, the role of those innovative microbial-induced mineral precipitation (MIMP) in BIMP technologies can be separated into two fields, i.e., mineralization of pollutants, such as heavy metals and immobilization of CO_2 in the ground as a component of carbonate minerals. The former is dealt with in this chapter, and Chapter 13 is provided in the latter. In Chapter 13, the mineralization with $CaCO_3$ is utilized for the solidification of ground for soil improvement. What these two groups have in common is that both require appropriate species of microorganisms and practitioners need to be familiar with the cultivation methods and properties of the species.

In environmental remediation techniques, one of the newest technologies may be those of microbial immobilization of pollutants. This remediation technique started as decomposition and/or fixation of pollutants nearly at the same time when microbial solidification technology started to be developed around 2000. Because the fundamentals of these techniques are similar, there are researchers who have studied both the immobilization of pollutants and solidification technologies for soils. From an academic point of view, both technologies belong to a new field of biotechnology in geoenvironmental engineering, which is an interdisciplinary field.

11.4.9.1 Microbial Immobilization of Cations

Microbial-induced carbonate precipitation (MICP) is an engineering tool used for soil stabilization. Note that MICP geomechanics is covered in Chapter 13 in this book. It has promise, however, as a bioremediation method that aims to immobilize or remove metal(loid)s (Wilcox et al., 2024). To immobilize cations in treated soil or to solidify soils, carbonation technology, i.e., microbial induced carbonate precipitation (MICP) technology can be used alone.

Han et al. (2020) reviewed research for soil solidification and heavy metal stabilization, using bacterial-induced mineralization (BIM). The term BIM and MICP are not very different. The latter term seems to concentrate on carbonate or calcite to distinguish from other biominerals. In this book, common terminology is used like MICP, which means "microbial induced carbonate precipitation", where the term "induced" is against "controlled" or "templated". The letter C in MICP is often used as an initial of calcite. In this book C is used as vaterite and aragonite and also other types of carbonates such as carbonates with other types of metals like siderite. Carbonate includes calcium carbonates (calcite, vaterite, and aragonite) and other carbonate containing metals and inorganic matter (with Fe, Mg, Sr, Cd, Cu, Ni, Pb, Mn, Zn, Co, Ba, NH_4, etc.). Similarly, the abbreviations for bio-mineralization of phosphate and sulfate minerals are expressed as MIPP and MISP, respectively. Accordingly, the different abbreviations indicate that the species of microorganisms is different. For example, phosphate-mineralization bacteria or sulphate reduction bacteria are required, respectively.

Lin et al. (2023) reviewed that heavy metals and some other toxicity in the environment can be removed by different microbial mineralizations using the appropriate species of microorganisms and applicable chemical reactions. In this review, the ultimate products are separated into carbonate (MICP), phosphate (MIPP), and sulphate (MISP).

11.4.9.2 Microbial-Induced Mineralization

Zhu and Dittrich (2016) reviewed MICP in terms of six types of metabolism, such as (a) photosynthesis, (b) ureolysis, (c) denitrification, (d) ammonification, (e) sulphate reduction, and (f) methane oxidation. These metabolic reactions induce $CaCO_3$, as indicated below.

(a) Photosynthesis by *Cyanobacteria*

$$2HCO_3^- + Ca^{2+} \rightarrow CH_2O + CaCO_3 + O_2 \quad (11.4)$$

(b) Ureolysis by urease-producing bacteria

$$CO(NH_2)_2 + 2H_2O + Cell + Ca^{2+} \rightarrow 2NH_4^+ + Cell\text{-}CaCO_3 \quad (11.5)$$

(c) Denitrification by nitrate-reducing bacteria

$$CH_2COO^- + 2.6H^+ + 1.6\ NO^- \rightarrow 2CO_2 + 0.8N_2 + 2.8H_2O \quad (11.6)$$

$$Ca^{2+} + CO_2 + 2OH^- \rightarrow CaCO_3 + H_2O \quad (11.7)$$

(d) Ammonification by *Myxobacteria*
(e) Sulphate reduction by sulphate-reducing bacteria

$$SO_4^{2-} + 2(CH_2O) + OH^- + Ca^{2+} \rightarrow CaCO_3 + CO_2 + 2H_2O + HS^- \quad (11.8)$$

(f) Methane oxidation by methanogens

$$\text{anaerobic reaction: } CH_4 + SO_4^{2-} + Ca^{2+} \rightarrow CaCO_3 + H_2S + H_2O \quad (11.9)$$

$$\text{aerobic reaction: } CH_4 + 2O_2 \rightarrow CO_2 + 2H_2O \quad (11.10)$$

11.4.9.3 MICP and Biomineralization of Heavy Metals

Carbonates are by far the most common biominerals. At present, MICP using urease-producing bacteria has been the most popular. MICP by urease-producing bacteria has been studied by many researchers in the field of soil improvement. In their studies, microbially precipitated carbonates are utilized as binding materials for soil particles. The merits in the use of ureolysis by urease-producing bacteria is that firstly, a large amount of calcium carbonate can be produced at once. Urea is an organic substance but is artificially produced. The CO_2 generated by hydrolysis immediately changes to CO_3^{2-} and $2NH_3$ change into $2NH_4^+$ by the existing water (Mobley and Hausinger, 1989). The initial by-product, ammonium carbonate, raises the pH, and if the hydrolysis rate is moderate, $CaCO_3$ with binding power as a binder is generated. Note that $CaCO_3$ cannot be produced if the pH is lower than 7, but the hydrolysis of urea solves the problem. In addition, microorganisms play a role as the nuclei of $CaCO_3$ crystals (Zhu and Dittrich, 2016).

The carbonate ions precipitate around each bacterial cell. Metal cations may selectively react with carbonate ions. The simplest reaction is ideally expressed by

$$M^{2+} + CO_3^{2-} \rightarrow M\ CO_3 \quad (11.11)$$

and possibly

$$Ca_m^{\ 2+} + M_{(1-m)}^{\ 2+} + CO_3^{2-} \rightarrow Ca_m M_{(1-m)}\ CO_3 \quad (11.12)$$

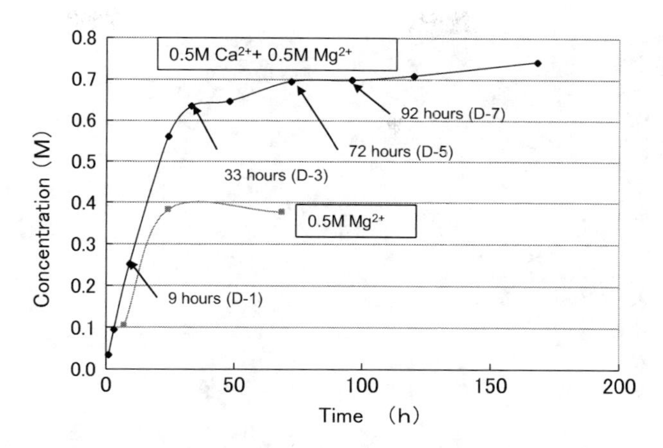

FIGURE 11.14 Carbonate mineralization with elapsed time in MICP, using Ca^{2+} and Mg^{2+}, and Mg^{2+} alone. D-1, D-3, D-5, D-7 are elapsed times when the photos were taken (see Figure 11.15).

where M denotes metals other than Ca^{2+}. It is noted that the effects of adsorption selectivity, inhibition, and retardation due to metal cations can affect the quantity of products (Fukue et al., 2023) as also demonstrated in Figures 11.14–11.16. According to Fukue et al. (2023), inhibition and retardation in ureolytic MICP may occur by relatively low urease activity, which is dependent on viability of strains and their concentration, against Ca^{2+} or other metal concentrations.

If a simple reaction is considered like the reaction (11.11), the products are simpler as shown in Figure 11.17. The photo shows the simplest calcite crystals in the case that Ca^{2+} is used as M in the reaction (11.11). However, a slight change in pH, temperature, bacteria concentration, some other conditions, etc., may produce amorphous calcite, vaterite, or polycrystalline, as shown in Figure 11.18. In addition, the crystal size produced varies with Ca^{2+} concentration and bacteria concentration, depends on the aggregation size of bacteria with Ca^{2+}ions. Since the urease-producing bacterial size is around 1 μm by 5 μm, and the bacterium could be the nuclei of a carbonate crystal, the crystal size is at least that of clay particles or larger. In MICP, carbonate crystals grow larger than 100 μm.

In application of reaction (11.11), many studies showed high precipitation rates of metal carbonate products (Chen et al., 2021; Kumar et al., 2023; Li et al., 2013; Lin et al., 2023; Qian et al., 2018; Song et al., 2022; Zeng et al., 2021; Zeng et al., 2022). For example, Li et al. (2013) investigated biomineralization of Ni, Cu, Pb, Co, Zn, and Cd, using six species of urease-producing bacteria. In the tests, different metal carbonate precipitations were confirmed using 1 mL of inoculum, 0.5 M urea and 2 g/L $NiCl_2$, $CuCl_2$, $PbCl_2$, $CoCl_2$, $ZnCl_2$, or $CdCl_2$ at a room temperature for 48 h, respectively. As a result, the precipitation rates of all metal carbonates were higher than 89.5%, and 100% at a maximum. It was remarkable that the precipitation rate of Pb at 48 h was 100% by all bacteria.

Zeng et al. (2021) also obtained very high removal rates of Pb and Cd, i.e., almost 100 %, using MICP. Furthermore, Zeng et al. (2022) applied MICP for the removal of Cd, and concluded that Cd^{2+} was immobilized by the forms of $Cd(OH)_2$, $CdCO_3$, and $Ca_xM_{(1-x)}CO_3$ with high immobilized rate. The processes due to those metabolisms are so-called microbial-induced carbonate precipitation (MICP), as was denoted earlier. Exceptionally, two studies on biomineralization of Cu were reported (Li et al., 2013; Duarte-Nass et al., 2020; Seplveda et al., 2021).

On the other hand, Duarte-Nass et al. (2020) obtained a Cu precipitation rate of only 10%. In their discussion, it was described that the urea used was at a too high concentration at a ratio of Cu^{2+}/urea in M/M of 666. There may be an optimal blend for Cu^{2+}, urea and bacteria (viable OD).

For this problem, it was explained that copper (II) has a high affinity for ammonia, forming coordination complexes keeping the metal soluble (Liu et al., 2019), and thereby preventing copper

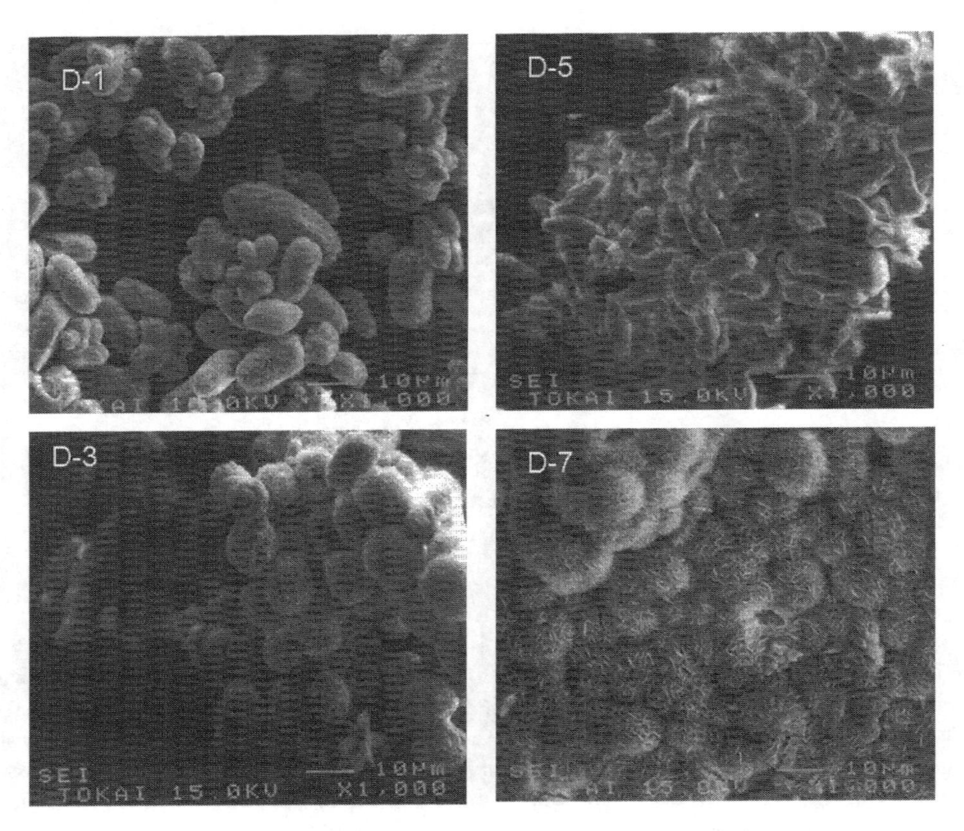

FIGURE 11.15 Morphology on the MICP process using 0.5 M Ca^{2+} and 0.5 M Mg^{2+}, showing from amorphous particles to the precursor of dolomite with time.

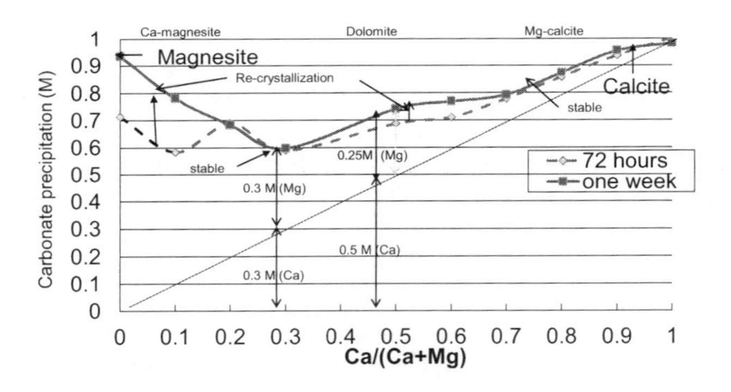

FIGURE 11.16 Precipitation ratios in the MICP process using various ratio of Ca^{2+} to (Ca^{2+} + Mg^{2+}).

carbonate from forming. It is considered that if excess urea is applied in a ureolytic system, excess NH_3 and CO_2 are produced. As a result, the pH increases rapidly. It was pointed out that under this condition, the Cu (II) complex tends to be more dominant than $CuCO_3$. In MICP, the pH varies with the process progress, which cannot be usually controlled. Therefore, another mechanism of MIPP was proposed for the biomineralization of Cu^{2+}, which is described later.

As a special example, remediation of cyanide is given. Cyanide tailings contain large amounts of heavy metals and highly toxic cyanides. Wang et al. (2021) proposed a "two-step" process to remediate cyanide tailings (CT), first using microorganisms to break down cyanide, and then using MICP to

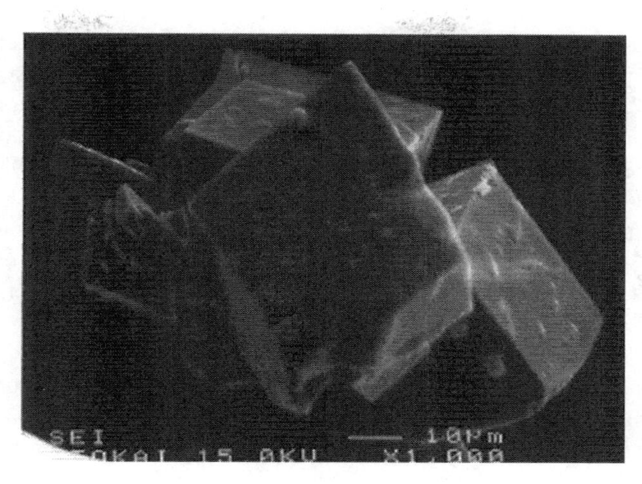

FIGURE 11.17 Single calcite crystals.

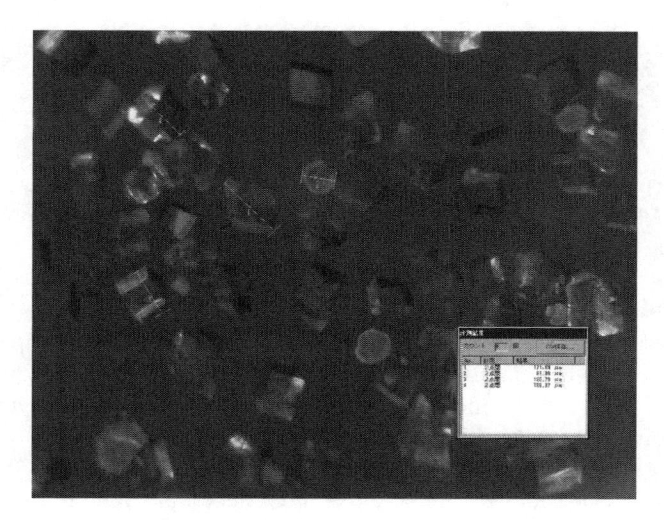

FIGURE 11.18 Various morphological aspects produced by MICP in the same glass tube using the same agents.

solidify CT. They isolated bifunctional bacteria with cyanide-degrading and high urease activity from CT and identified the *Anairinibacillus tyrosinisorbens* strain (named JK-1). Furthermore, as a result of treating CT in a "two-step" process with JK-1 bacteria, the degradation of free cyanide (F–CN) and total cyanide (T-CN) in CT reached 94.54% and 88.13%, respectively. After MICP treatment, $CaCO_3$-induced solidified CT into blocks of calcite and sphalerite crystals, and the unconfined compressive strength (UCS) reached 0.74 MPa. Thus, compared with chemical treatment, the treatment of CT by this new process is very efficient and green, which can result in CT coagulation, cyanide decomposition, and heavy metal immobilization at the same time (Wang et al., 2021).

11.4.10 Microbial-Induced Phosphate Precipitation (MIPP)

Except for carbonates, phosphate is a biomineral that is abundant in nature. Accordingly, it is no wonder that the microbial-induced phosphate precipitation (MIPP) can also be used as the technology for the remediation of heavy metal contamination in soils and leachate.

The MIPP process produces PO_4^{3-} by the decomposition of organic or inorganic phosphate reduction. Next, the phosphate ions are reacted with heavy metals including radionuclides, and mineralized to remediate the contaminated soil and/or water. Therefore, it is a potential technology that can simultaneously solidify and remediate the soil, similar to MICP. Note that MIPP can also be applied to some metal pollutants, which cannot be easily treated with MICP. For instance, Cu^{2+} reacts with PO_4^{3-} and precipitates like $Cu(OH)_3PO_4$ (Zhao et al., 2019) and $Cu(PO_4)_2(OH)_4$ (Do et al., 2020). The precipitation rates of Cu were higher than 75% in both studies. Such metals produced as phosphates are Cu, Ni, Pb, U, Zn, Cd, etc., and are presented in the forms of $M_i(PO_4)_j$, M_iHPO_4, $(PO_4)_j$ $M_i(PO_4)_j(OH)_k$, $M_i(PO_4)_j$ nH_2O. In addition, Ca, Cl, H, or MH_4 sometimes are present in the given forms (Lin et al., 2023).

It was described earlier that the removal of Cu cannot be easy using MICP process, while MIPP led to a higher rate of removal and remediation of Cu. Zhao et al. (2019) could produce phytase and alkaline phosphatase to degrade phytic acid by using *Rahnella sp.* (strain LRP3). By that, soluble phosphate can be released to the bacterial culture and the pH increased. Finally, the phosphate crystal of $Cu_3(OH)_3PO_4$ with the mean diameter of 10 μm was induced through the process of biomineralization.

It is known that some carbonates like siderite ($FeCO_3$) and dolomite ($CaMgCO_3$) are more stable than calcite ($CaCO_3$). In general, phosphates are more stable than carbonates chemically and physically. Therefore, MIPP technology has been increasingly studied recently (George and Wan, 2023; Jiang et al., 2020; Teng et al., 2019; Xia et al., 2023; Zhang et al., 2019; Zhao et al., 2019). Notably, a great deal of research has been done in China and it seems that research is progressing rapidly. However, Xia et al. (2023) pointed out that most successful approaches in the remediation techniques are performed under the controlled conditions in the laboratories. Actual application requires the optimization of many parameters, such as microbial competition, physical and chemical parameters of soils, pollutant concentrations, existence of co-pollutants, etc.

11.4.10.1 Microbial-Induced Sulfide Precipitation (MISP)

Sulphate-reducing bacteria is a general term for microorganisms that decomposes organic matter in an anaerobic environment and reduces sulphates using the generated electrons. Sulphates are reduced by such bacteria to sulphide ions, which can be generated as hydrogen sulphide or react with metal ions to form sulphides. Since sulphur is not incorporated into biological substances in this reaction, it is called a dissimilatory sulphate reduction reaction. By MISP, a technology that utilizes such bacteria, sulphides can mineralize under anaerobic conditions. This type of metal mineralization is being mainly studied for an environmental remediation (water and soils) and also metal recovery technology (Kimber et al., 2020; Li et al., 2023; Park et al., 2019; Xu and Cheng, 2020).

Su et al. (2022) pointed out the existence and importance of different bioreactions that can ionize heavy metals by promoting the oxidation of metal sulphide minerals, and immobilize heavy metal ions as metal sulphides, which were further converted to metal sulphide nanoparticles (NPs) for contaminant treatment. Note that metal sulphide oxidation bacteria are required for oxidation (Hu et al., 2023; Linssen et. al., 2023), and sulphate-reducing bacteria (SRB) are needed for reduction (Kimber et al., 2020; Li et al., 2023). Accordingly, by the sulphate oxidation or the sulphide reduction, the recovery of metal from sulphide or metal immobilization from contaminated water can be achieved. In addition, the conversion to sulphide nanoparticles (NPs) can be applied for preparation of the semiconductor ZnS NPs. In particular, low-cost binary metal sulphides including Sb_2S_3, SnS, PbS, Cu_2S, Ag_2S, Bi_2S_3, and FeS_2 have been of interest as light absorbing materials in thin film solar cells (Moon et al., 2019). Thus, metal sulphides have semi-conductive properties, which are economically valuable.

On the other hand, zinc oxide nanoparticles (ZnO NPs) are attracting much interest due to their potential toxicity and ubiquity in consumer products. Thus, it is important to know that metals are circulating via microbial actions, and microbes play important roles in natural environments (Chen et al., 2021).

Su et al. (2022) explained the mechanism of sulphate reduction in their review, as follows. In an anaerobic environment, sulphate reduction generates H_2S through dissimilatory reduction of sulphate with organic compounds (such as lactic acid, acetate, etc.) as electron donors, which combines with heavy metals to form metal sulphide nuclei, as follows:

$$2CH_2O + SO_4^{2-} \rightarrow 2HCO_3^- + H_2S \quad (11.13)$$

$$H_2S + M^{2+} \rightarrow MS(S) + 2H^+ \quad (11.14)$$

where CH_2O, M, and MS represent organic matter, heavy metal, and metal sulphides, respectively.

According to the literature on sulphide reduction bacteria (SRB), the potential and perspectives in the future seem to be bright. However, Xu and Cheng (2020) stated that the limitation of SRB in the treatment of heavy metals is that high concentrations of heavy metals possibly inhibit SRB. In the treatment of industrial wastewater, which was polluted at low pH with heavy metals, the activity of sulphate reduction bacteria can be greatly decreased during a long-term exposure of the heavy metals, and thus the sulphate reduction will be reduced. Furthermore, they added that SRB are closely related to the production of methylmercury in the aquatic environment, which is the most toxic species of mercury and is prone to bioaccumulation and biomagnification in aquatic biota. There are several recent articles concerned with mercury (Li et al., 2023; van Rooyen et al., 2023; Singh et al., 2023).

Zhang et al. (2021) demonstrated that the element sulphur (S^0) with both reducing and oxidizing properties is a suitable material for water treatment technology. It is because biochemical oxidation and reduction is a principle in biological water and wastewater treatment, where electron donors and/or acceptors must be provided. They also demonstrated such S^0-based biotechnologies as providing a cost-effective and attractive alternative to traditional biological methods for water and wastewater treatment.

Similarly, Li et al. (2023) pointed out the imperfection in SRB remediating heavy metal pollution in the mining areas and suggested potential merits to use sulphur-reducing bacteria instead of sulphate-reducing bacteria. These include improvement of technical efficiency and cost-effectiveness of processing against the inhibition associated with the pH level and metal ion concentration, the formation of toxic by-products, and the consumption of organic electron donors. Some examples can be drawn from the following reactions.

For sulphate reduction due to sulphate-reducing bacteria:

$$0.5\ SO_4^{2-} + C_{org} + 0.5\ H^+ \rightarrow 0.5\ HS^- + CO_2 \quad (11.15)$$

$$n\ HS + 2\ M^{n+} \rightarrow M_2S_n \downarrow + nH^+ \quad (11.16)$$

where C_{org} is organic matter, n is the electronic valence. For sulphur-reducing bacteria,

$$2\ S^0 + C_{org} + 2\ H_2O \rightarrow 2HS^- + CO_2 + 2\ H^+ \quad (11.17)$$

The difference between sulphate reduction and sulphur reduction can be made by comparing reactions (11.15) and (11.17) (Li et al., 2023). As a result, using the same amount of organic matter, the sulphur reduction can produce four-fold more HS^-, in comparison to that by the sulphate reduction. The sulfidogenic bacteria (34 genera) are widespread within the lineage of *Deltaproteobacteria*. The rest were affiliated with the lineages of *Euryarchaeota*, *Aquificae*, *Nitrospirae*, *Thermodesulfobacteria*, *Clostridia*, *Epsilonproteobacteria*, and *Gammaproteobacteria*. Some genera have been reclassified recently, based on a widely used classic NCBI taxonomy (https://ncbi.nlm.nih.gov/taxonomy) (Li et al., 2023).

11.4.10.2 Radionuclides

Radionuclides have been released into environment due to human activities related to the nuclear industry, nuclear accidents, or nuclear weapon tests. Radionuclides are probably the most important threat to human health among heavy metals. Previous studies showed promise in remediation by living organisms, especially microorganisms (Lin et al., 2023).

The biomineralization and immobilization of radionuclides has been studied more extensively than expected. In particular, research on Sr has been conducted from an early stage, and research has been conducted on fixation and removal using MICP (Achal et al., 2012; Fujita et al., 2004; Zhao et al., 2022). Some readers may wonder why Cs is not presented here. The authors think because Cs can be strongly adsorbed on the soil particles, it can be removed easily without using a complicated microbiological technique in comparison to other radionuclides. Another reason may be due to high emissions if an accident happens. Therefore, radioactive Cs is dealt with in Section 8.8.

The biomineralization of uranium cannot be induced by MICP, but it is possible by MIPP and MISP. However, uranium can reach the environment through mining activities, weathering of uranium-containing minerals, or accidental release, is biologically and chemically severely toxic, is highly radioactive in soil, usually as UO_2^{2+}, easily dissolves as migrtile, and poses a serious threat to human health (Lin et al., 2023).

Biomineralization of uranium was studied using various microbes (Beazley et al., 2007; Liang et al., 2015; Tu et al., 2019). Beazley et al. (2007), in relation to uranium contamination at the Department of Energy's Field Research Center in Oak Ridge, Tennessee, studied phosphate biomineralization under less-studied aerobic conditions, or aerobic precipitation of the U(VI)-phosphate phase promoted by microbial enzymatic activity. The fungi used isolated three heterotrophic bacteria from the field soil. As a result, the two bacterial strains hydrolyzed enough organophosphate to precipitate 73–95% of the total uranium after 120 hours of incubation in simulated groundwater. It should be noted that the highest rates of uranium precipitation and phosphatase activity were observed between pH 5.0 and 7.0. From the EXAFS spectra, the uranyl phosphate precipitate was identified as a mineral in the autunite/metaauchunite group.

11.5 COMPARISON OF TREATMENT TECHNOLOGIES

11.5.1 Treatment Technologies Overview

The overall trend in remediation technologies has been overviewed by US EPA (2004b) in the US. Between 1982 and 2002, 863 treatment technologies were applied at 638 Superfund sites. *In situ* technologies were used at 42% of the sites. The majority of the sites were treated using SVE, bioremediation, and solidification/stabilization in descending order. The majority of *ex situ* remediation technologies were by S/S, incineration, thermal desorption, and bioremediation in descending order. Over the years, *in situ* remediation has been increased from 31% (in 1985 to 1989) to 49% (1998 to 2002). The reasons included cost-effectiveness, decreased exposure to contaminants as no excavation is required and willingness by professionals to use this approach. In fiscal year (FY) 2009–2011, 300 decisions were made for Superfund source materials. About half were *in situ* treatment, SVE, chemical treatment, bioremediation, S/S, multi-phase extraction, bioremediation, and *in situ* thermal treatment were common. Off-site methods included physical separation, S/S, off-site treatment, and recycling. Physical separation includes sieving, sifting, removal of solid media, dewatering (for sediments), and decontamination. A variety of technologies are utilized as summarized in Table 11.3.

11.5.2 Design of a Remediation Process

In choosing the most appropriate remediation, the following approach shown in Figure 11.19 should be used. The first step of any remediation programme is to develop a conceptual site

TABLE 11.3
Soil Treatment Technologies Utilized for Superfund Sites (Superfund Remedy Report 17th Edition, 2023)

Technology	% Decision Documents (FY* 2005–2008)	% Decision Documents (FY* 2009–2011)	% Decision Documents (FY* 2018–2020)
In situ			
Chemical treatment	21	21	4
SVE	7	14	10
S/S	9	9	6
Thermal treatment	9	6	11
Multiple phase extraction	7	3	2
	4	3	
Constructed wetland	0	2	
Reactive cap	0	2	
Flushing	1	1	2
Fracturing	1	1	
Phytoremediation	1	0	
Bioremediation			3
Soil amendments			2
MNA			1
Containment			35
Total number	48	80	38
Ex situ			
Physical separation	21	28	5
S/S	19	13	5
Pump and treat	12	11	
Off-site treatment	7	9	
Recycling	10	8	
On-site treatment	1	5	
Phytoremediation	0	4	
Chemical treatment	3	3	
Bioremediation	3	3	
NAPL recovery	1	1	
Thermal desorption	2	1	2
Disposal			53
Other	9	0	16
Institutional control			75
Total number of sites	65	67	131

* FY Fiscal year

model (CSM) to evaluate the potential for applying bioremediation at a site. The nature and extent of contamination and site characteristics; site hydrogeology, geochemistry and oxidation-reduction conditions; biodegradation potential; contaminant fate and transport; and receptor and exposure pathways must be determined. Once the CSM is established and refined, activities undertaken prior to the implementation of a remediation programme often involve treatability studies, examination of soil to ensure that undesirable reactions with the contaminants or their degradation products are prevented. The success of a remediation application highly depends on characterization and monitoring, and an effective model for lowering natural resource consumption and waste generation, completed before and during its implementation. Monitoring and operation should also be designed to reduce emissions and energy requirements. For example,

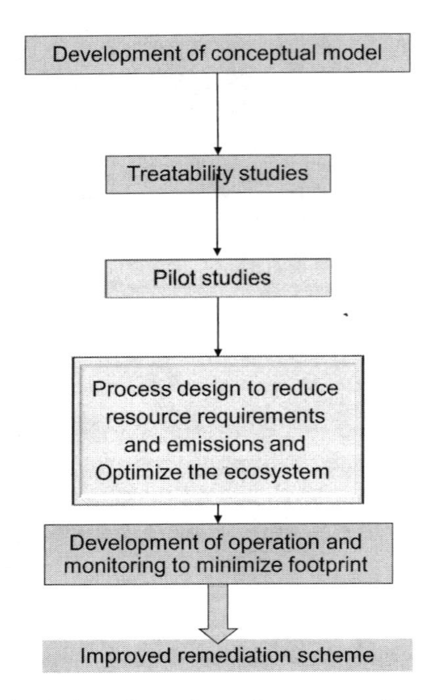

FIGURE 11.19 Flow sheet for the design of an enhanced green remediation.

low-temperature processes require less energy. The advantages of the various options must be weighed. In addition to past practices to obtain a more sustainable remediation approach, other aspects must also be evaluated to

- enhance land management practices and maintain biodiversity and the ecosystem population,
- reduce water consumption and pollution,
- reduce total energy use and increase the use of renewable energies,
- reduce air and GHG emissions.

Demonstration of the designed process will enable larger-scale evaluation of various factors before final design. Bioremediation has been suggested as a green remediation scheme (US EPA, 2010). An example was shown for bioaugmentation at the MAG-1 Site, Fort Dix, NJ. Laboratory tests were used to evaluate the potential of a bacterial culture for treating chlorinated VOCs. A new well system was then devised to disperse the inoculums in the groundwater recirculation systems. The system was optimized to reduce material consumption and maintenance of the equipment. The contaminants were reduced by 99% with negative impacts.

Pilot-scale field tests of composting (Former Joliet Army Ammunition Plan, Will County, IL) were performed to optimize amendments for composting and the turning frequency (US EPA, 2010). Additives included manure, wood chips, spent digester waste, and stable bedding. A 8.1 hectare windrow composting facility was constructed to treat 280,000 tonnes of explosives contaminated soil. Soil cleanup was three years ahead of schedule. The land will be used for a business park and an engineer training centre. Another site in Upper Arkansas River used municipal biosolids and limes for composting mine waste, which led to complete vegetative cover.

A case study by Aecom involved a sediment cleanup alternative analysis at Lower Duwamish Waterway, Seattle, WA (Fitzpatrick and Woodward, 2013). The site was a 8.3 km waterway contaminated by industrial inputs situated in an economic corridor. Issues involved tribal fishing rights and proximity to residential neighbourhoods. Recontamination was occurring due to urban sources. EPA preferred dredging but agreed to a green sustainable remediation (GSR) evaluation. Alternatives cost between $200 million to 1.2 billion. Carbon dioxide emissions were compared for the various alternatives. Dredging/landfilling/backfilling exhibited the highest level (5500 tonnes) compared to minimal amounts from monitored natural recovery (MNR).

11.6 GREEN REMEDIATION

According to the New York State Department of Environmental Conservation (NYSDEC, 2011) DER-31 green remediation is defined as "the practice of considering all environmental effects of remedy implementation and incorporating options to minimize the environmental footprint of cleanup actions". Some practices are easily implementable. Renewable energy can be used as much as possible. Idling of vehicles must be limited. Materials should be reused as base or fill to reduce waste as much as possible. Ideally the alternatives that are the most sustainable will have the following characteristics:

- Little impact on the environment in the short and long term.
- Low GHG emissions.
- Smaller environmental footprint.
- Complete and permanent cleanup.
- Optimal reuse.
- Reduction of toxicity, mobility, and volume of contamination.
- Achievement of remedial objectives.

According to the EPA Green Remediation Best Management Practices Mining Sites (US EPA, 2008b), potential practices include a better understanding of the contamination behaviour, and implementation of passive treatment systems for acid mine drainage (slippery rock watershed, agricultural waste (mushroom compost and limestone)) was used for neutralizing 82 kg of acidity per day. About 2 tonnes of MnO_2 was recovered over 8 years. Other waste materials such as industrial by-products like chitin, and food waste as biosorbents could also be used. Other approaches can also be implemented. These include the following:

- employment of renewable energy for cleanup and land reuse,
- soil amendments such as compost soil covers consisting of non-invasive grasses or shrubs, draught resistant plants to reduce exposure to contaminated soil and waste,
- metal (copper, zinc, gold, nickel) recovery from waste materials,
- phytoremediation to treat soil and water.

The USEPA has developed 14 metrics for the environmental footprint of remediation (www.clu-in.org/greenremediation/) for greener cleanup. Water (groundwater, wastewater, public water, or other), energy (electricity, diesel gasoline, or other), waste (hazardous and non-hazardous) and materials (refined and unrefined) are the categories with metrics in brackets.

Profiles of green remediation are found at clu-in.org/greenremediation/profiles. An example is the Continental Steel Corp site at Kokomo, Indiana (Table 11.4) (https://clu-in.org/greenremediation/profiles/continentalsteel, accessed Apr. 2024). Soil and groundwater were contaminated with VOCs, PCBs, PAHs, and metals including lead. The contamination was in the creeks, lagoons, quarry, plant, slag processing, and disposal areas. Source control, land report, and treatment of the groundwater were required. The main green remediation strategy included onsite renewable energy production, stormwater management optimization, reuse or recycling of construction and demolition wastes, and minimization of raw material imports and offsite disposal.

TABLE 11.4
Summary of Green Remediation Results (https://clu-in.org/greenremediation/profiles/continentalsteel)

Category	Selected Results
Renewable energy generation	60% grid offset by wind turbine installation for groundwater extraction Solar farm of 21,000 PV panels to produce 9.1 million kWh solar energy
Water quality	Minimization of runoff with chipped wood from onsite trees and field stone for creek bank stabilization Stormwater storage for remediation of quarry pond
Material recycling and reuse	45,873 m^3 of slag used as backfill Steel scrap salvaged and sold to generate ($1.6 million) to defray remediation costs and other salvaged materials to fill in basements and voids

The ASTM has developed guidelines for green remediation (ASTM, 2016). Aspects for integration into the project include the following:

- Reducing use of energy and maximization of renewable energy.
- Reducing air pollutants and greenhouse gas emissions.
- Reducing water use and impact on water quality.
- Reducing waste and materials and increasing material reuse and recycling.
- Land and ecosystems' protection.

The ASTM guidelines are mainly focused on the US due to regulatory aspects, but the process can be applied in other locations. A suggested process is indicated in Figure 11.20. Best management practices should be adopted to reduce the environmental footprint of the remediation. The steps include the site assessment, evaluation of the technologies for remediation, carrying out the remediation and monitoring and optimization of the process. Some elements to reduce the footprint are summarized in Table 11.5. A more extensive greener cleanup BMP table for various technologies is included in the guideline.

Twenty-two of the BMPs can be used for PCB site remediation (US EPA, 2022). As an example, a cleanup was initiated at the Boeing Everett Plant, Everett, Washington. The site was needed for construction of the 777× composite wing plant. At this site various BMPs were employed such as the following:

- Green requirements were included in requests for proposals, and contracts.
- Covers were used to control dust instead of water spraying.
- Disposable material use was minimized.
- Reusable and recycling containers were favoured for product selection.
- Local laboratories were used to reduce transportation.

Pump and treatment, biobased products, can be used such as biological surfactants. Mulligan (2023) has shown that biodegradable, non-toxic products called biosurfactants (e.g., rhamnolipids and sophorolipids) can be produced from waste materials and can be employed for soil flushing or washing for metal and organic contaminants or for enhanced biodegradation of organic pollutants. Biosurfactant applications for remediation of contaminated soil and water are promising due to their biodegradability, low toxicity, and critical micelle concentration (CMC) and high effectiveness in enhancing biodegradation and affinity for metals. Biosurfactants can

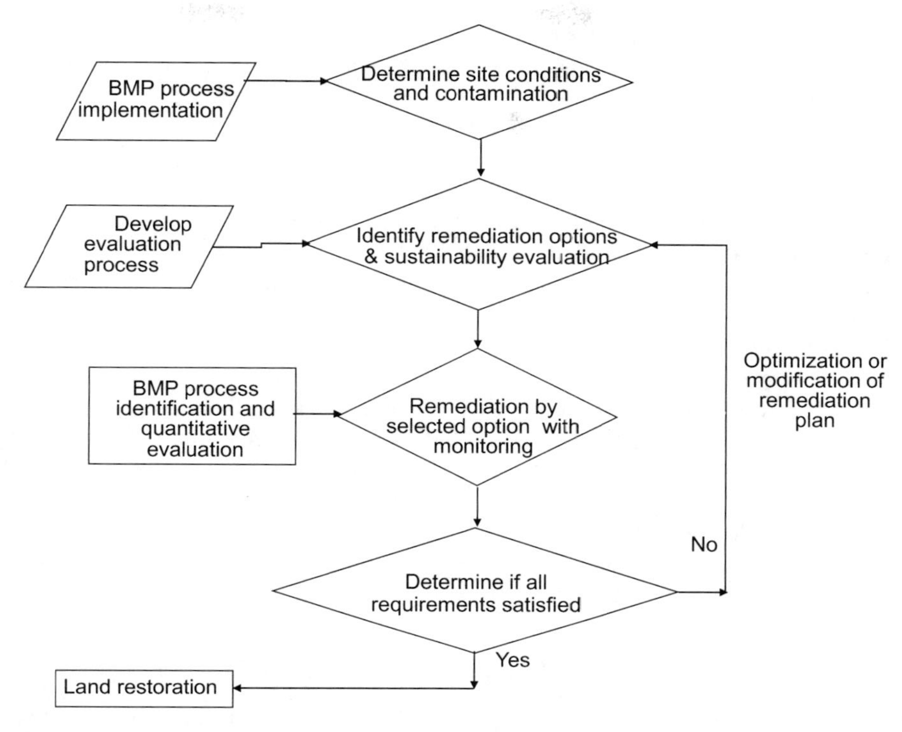

FIGURE 11.20 Greener remediation scheme by ASTM (adapted from ASTM, 2013). The BMP process includes BMP opportunity assessment, prioritization, selection, implementation, and documentation.

TABLE 11.5
Means of Reducing the Remediation Environmental Footprint

Objective	Means of Achieving Objective
Reduction of energy use and increasing alternative energies	Enhancement of energy-efficient equipment, and wind or solar energy
Reduction of emissions into the air	Minimization of dust, increasing emission controls, use of EVs or hybrids
Reduction of water and resource	Use of water-efficient processes, water reclamation, green infrastructure for erosion, and run-off control
Increased reuse and recycling and reducing material use	Use of renewable, recycled, certified products, or locally generated materials
Land and ecosystem protection	Minimization of ecosystem disturbance or destruction, biodiversity restoration

Source: ASTM (2016).

remove heavy metals through the mechanisms of solubilization, complexation, and ion exchange. Most research has involved rhamnolipids. Other biosurfactants and process scale-up need further investigation.

Other components to be incorporated in green remediation can include

- use of recycled concrete for pipe bedding, landscaping,
- use of natural gas, low emission and noise or clean diesel generators or renewable energy for equipment,
- employment of gravity flow where possible,
- recharging groundwater with uncontaminated groundwater,
- reduction of waste materials with reusable equipment,
- use of on-site analysis to avoid the need for off-site shipping,
- use of native plants to restore biodiversity,
- minimization of tree and vegetation removal and traffic routes to reduce site disturbance,
- use of pervious materials for pavements,
- capture of rainwater for dust control and other uses.

At the North Ridge Estates Superfund Site, asbestos was the main contaminant, along with arsenic at an old military barracks (US EPA, 2015). For the cleanup, the USEPA employed the ASTM Guidance for Greener Cleanups. Thirty BMPs were selected and applied at the site. Twelve were employed for the design of the remediation. Retrofitting the diesel technology for many pieces of equipment in particular was expected to reduce particulate matter (PM), hydrocarbon, CO and NO_x emissions substantially. Monitoring of fuel and water consumptions was also to be initiated.

11.7 SUSTAINABLE REMEDIATION FRAMEWORKS AND TOOLS

There is a growing need to incorporate sustainability into projects due to pressure from all stakeholders. To integrate this into projects, decision support tools are needed. According to the EPA (2008b), minimization of energy use, air emissions, water impacts, material and waste use, land and ecosystems is required. Only a few of the frameworks, however, consider social aspects. Since the CLARINET project on sustainable remediation (Bardos et al., 2002), the concept has gained popularity. The Sustainable Remediation Forum (SURF) was subsequently initiated in the US in 2002. Worldwide SURF developed in many countries. Pressure from all stakeholders is increasing to incorporate sustainability into projects. The Network of Industrially Contaminated Sites in Europe (NICOLE) (www.nicole.org/, accessed Apr. 2024) has established a framework for sustainable remediation for Europe with SURF-UK. SURF US was the first SURF initiative. Guidelines are shown on their web site (SURF 2009) (www.sustainableremediation.org/library/, accessed Aug. 2024). According to the SURF framework (Holland et al., 2011), sustainable remediation is defined as "sustainable approaches to the investigation, assessment and management (including institutional controls) of potentially contaminated land and groundwater". SURF initiatives are in Australia, Brazil, Canada, China, Colombia, Italy, Japan, Mexico, the Netherlands, New Zealand, and Taiwan. The key principles of sustainable remediation are related to the below:

- Human health and environment protection.
- Safe working practices for workers and the local communities.
- Decision-making in a clear, consistent, and reproducible manner.
- Clear, understandable transparent record keeping and reporting.
- Stakeholder involvement in a clear practice.
- Decision making based on scientific relevant and accurate data.

The overall framework is summarized as sustainability integration in the following steps over the project life: Site investigation, remediation selection, design and implementation, operation and maintenance and closure. To evaluate the environmental, economic, and social aspects, the boundaries, objectives, type, and indicators of the assessment, stakeholder involvement, and the sensitivity of the analysis all must be determined. The SURF UK framework was used in the tender phase to evaluate alternatives for a site 7 km northwest of Bicester, Oxfordshire (Bardos et al., 2013). As it was a former airbase, it was designated as a heritage site. The 505-hectare site was targeted for redevelopment for 1000 homes and related infrastructure while maintaining the heritage interest and included a petroleum, oil, and lubrication (POL) system with 13 km of pipework and 71 tanks. After fuel removal, the system was filled with water and alkaline solution to inhibit corrosion. Some oily water though still existed in the system. The risk analysis indicated that there was potential groundwater contamination from oily water and sludge and thus remediation was required. For the site remediation selection, preservation of the ecological and heritage characteristics, minimization of site disruption, and consideration of potential unexploded ordinance (UXOs) on site were required. The sustainability assessment at the tender stage was employed to show that the selection retained was the most sustainable.

Initially remediation options were selected to achieve the remediation goals (breaking contaminant pathways, effectiveness, practicality and ease of operation. The scoring system was distributed equally. After an initial screening process, a more detailed assessment was performed. Qualitative indicators were identified and quantitative indicators such as embodied carbon data from the ICE database, Environmental Agency carbon calculator and data for suppliers were calculated. The selected option included filling of the pipelines with foam (not the most cost-efficient) and the tanks with a PFA grout, water treatment was on site with discharge to the land. The approach allowed the involvement of stakeholders and allowed the choice of the most balanced approach, not necessarily the most economic.

More recently, SURF has examined how the effects of climate change and extreme events could influence contaminated sites and how mitigation strategies could be developed (Maco et al., 2018). They determined that climate change and extreme events could have significant impacts on site remediation and contaminant management. Social vulnerability was another aspect that needed to be considered. Harclerode et al. (2015) has discussed techniques for the integration of the social dimension into remediation decision-making. Overall, it was recommended that sustainable remediation and climate change adaptation be co-practiced to ensure resilient remediation.

Reddy and Adams (2010) have also indicated that efforts are being made towards standardization of sustainability frameworks. Therefore, specific indicators could include energy and water consumption, greenhouse gas emissions, waste generation, cost of remediation. Various assessment tools have been developed. Some include screening matrices and others are life-cycle assessments. Various public agencies including the Minnesota Pollution Control Agency, Illinois EPA and the California DTSC have developed preliminary assessment tools. For the Air Force Centre for Engineering and the Environment (AFCEE), GSI Environment Inc., developed the Sustainable Remediation Tool (SRT). The EPA has a GHG calculator tool. Life-cycle assessment is another approach. However, it can be complicated and expensive to determine the appropriate data. Comparison of various remediation options at a baseline is a simplified and logical approach.

The ITRC has published a *Technical and Regulatory Guidance: Green and Sustainable Remediation: A Practical Framework, GSR-2, 2011.* (www.itrcweb.org/GuidanceDocuments/GSR-2.pdf, accessed Apr. 2024). It has defined green and sustainable remediation as "the site-specific use of products, processes, technologies, and procedures that mitigate contaminant risk to receptors while balancing community goals, economic impacts, and net environmental effects". A more recent document (Sustainable Resilient Remediation (SRR-1)) has now been issued following the previous guidance. It incorporates aspects of resiliency, particularly in light of climate change (ITRC, 2021). It can be found at https://itrcweb.org/teams/training/sustainable-and-resilient-remediation (accessed Apr. 2024).

Some of the requirements for successful implementation of sustainability assessment tools are as follows:

- Involvement and training of stakeholders.
- Clear guidance documents regarding definitions.
- Standardized metrics and validation.
- Well-defined frameworks.
- Documentation of sustainable remediation practices.
- Regulatory and/or financial incentives.

Implementation of the sustainability assessment tools will allow evaluation and selection of the remediation technologies that will minimize environmental impact and reduce resource requirements. In addition, it will enable and promote implementation of more sustainable practices.

An ASTM guide for integration of sustainable objectives into Cleanup (E2876-13) has been developed, which includes the environmental, social, and economic aspects. The various aspects of a project are shown in Figure 11.21. The first phase involves planning and scoping of the project. This is followed by information gathering including stabling the sustainability objectives. Data needs and activities are identified for the project. The sustainable core elements include air emissions, community involvement, economic impacts, to the local community and government, efficient and economic cleanup, and minimization of energy use. Other elements include enhancing the human environment, reduction of land and ecosystem impacts, vitality of the local community, minimization of materials and wastes and water impact and inclusion of stakeholders.

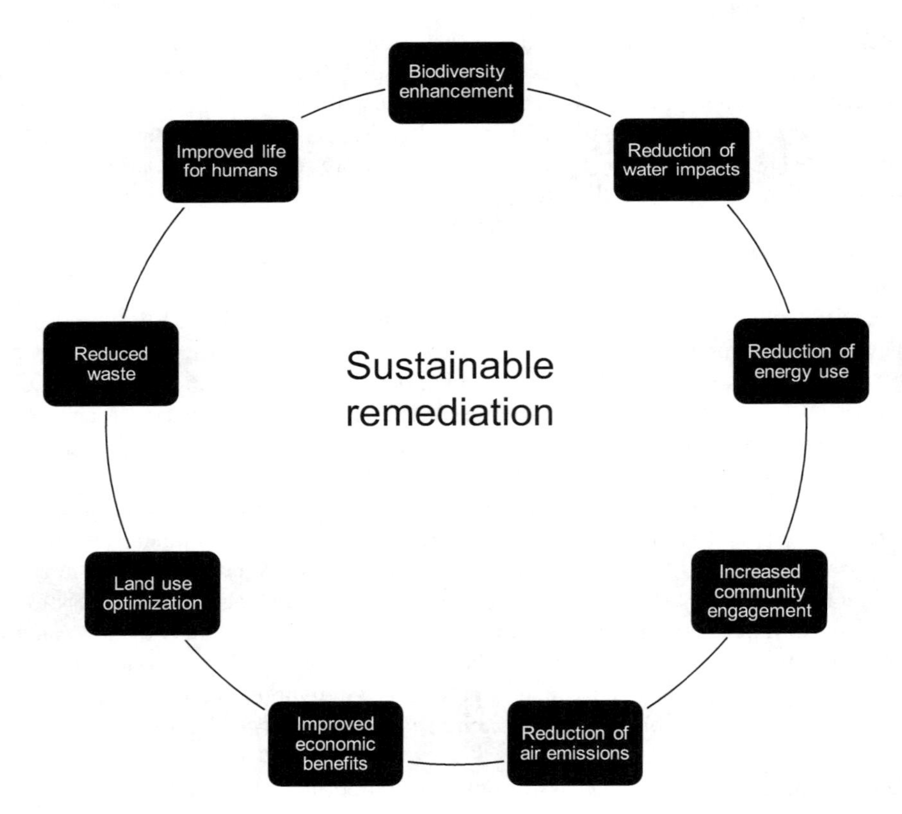

FIGURE 11.21 Incorporation of various elements to enable sustainable remediation. (Adapted from ASTM, 2013.)

In Canada, some sustainable remediation initiatives are led by the Federation of Canadian Municipalities (Green Municipal Fund, https://greenmunicipalfund.ca/resources/tool-sustainable-remediation-and-risk-management-options).

Various tools can be used including to assist in sustainable remediation such as LCA (USEPA Life Cycle Assessment, Principles and Practice, EPA/600/R-06/060 (May 2006)) or footprint analysis (USEPA Methodology for Understanding and Reducing a Projects Environmental Footprint, EPA 540-R-12-002 (Feb. 2012b)).

An ISO standard has been developed on sustainable remediation practices (ISO 18504:2017), providing procedures on sustainable remediation based on world-wide interest. Information on assessment of sustainable remediation, addition to standard methodologies and terminologies are provided but not which indicators or weights should be chosen in an environmental, legal, policy, and socio-economic context.

Hou et al. (2023) reviewed sustainable remediation technologies for contaminated sites. Remediation technologies need to be beneficial not detrimental to the environment. Some of the concerns from existing technologies include secondary pollution from leaching of contaminants and waste generation, high energy consumption and carbon footprint, dust and odor issues, and disturbance to soil biota. Reduction of the environmental footprint is particularly important. Through various efforts, greenhouse gas emissions can be reduced by up to 80%. Immobilization, passive barriers, bioremediation, and *in situ* chemical treatments can be designed to be more sustainable. LCA analyses can help in this process. Nature-based solutions are particularly beneficial for carbon sequestration. These can include construction of large spaces and infrastructure, such as wetlands and green landscaping. Plants, however, can accumulate contaminants and thus monitoring is crucial. Generation of sustainable energies can also be incorporated into the remediation (e.g., biomass, solar, wind energies, or heat pumps) to reduce carbon footprints. Efforts to understand long-term resilience and social-ecological impacts must be undertaken. Research and collaboration are needed to ensure future success.

11.8 CASE STUDIES USING A SUSTAINABILITY APPROACH

GoldSET© (Golder Sustainability Evaluation Tool), a sustainability decision support tool for project planning and design was created by Golder Associates (www.wsp.com/en-gl/services/goldset (Mulligan et al., 2013). It is a robust and transparent framework to embed sustainable development practices into design, construction, and operational decision-making phases of any project. The sustainability tool has been applied for various applications, such as site remediation and mining tailings around the world. The tool operates through the assessment of different project options against a number of quantitative and qualitative indicators for each of the three dimensions of sustainability: environment, society, and economy. Indicators provide a way of describing the situation surrounding the project with a weighting scheme allowing the relative importance of each indicator to be reflected.

The first step in an evaluation involves the description of the site, the definition of the project objectives and the identification of key issues of concern to all stakeholders. The second step is to identify and elaborate on various remediation options and/or alternatives that are thought to be suitable for the site specificities and project restrictions. Those options will then be assessed from an economic, social, and environmental viewpoint, and in some cases from a technical viewpoint. The scoring scheme attached to each indicator provides a mechanism to assess the performance of each option with respect to the indicator. Together, the entire set of indicators should be representative of a project's performance, impacts, and cost.

All qualitative indicators have scoring schemes minimally consisting of three levels. Quantitative indicators have both relative and absolute scoring schemes. For the specific quantitative indicators like greenhouse gas emissions and net present value (NPV), the framework is adopted to a level of detailed calculators. Relative scoring schemes assign a score of zero to the lowest performing option, while assigning 100 to the best performing option. Absolute scoring schemes have a fixed scoring scale independent of the options and score the options relative to this fixed scale. These fixed values were adopted from accredited organizations (UNEP, WHO, etc.) as benchmarking values for consumption of natural resources or concentration of pollutants in the media.

11.8.1 Case Study for a Benzene-Contaminated Site

A benzene spill occurred along the highway near a small town of 1600 residents following a train derailment. A certain amount of benzene then leached into the ground water aquifer used by the residents for their domestic well water. The area of the affected site was 6 hectares. The concentration of benzene in ground water was 55 μg/L which needed to be reduced to its maximum contaminant level (MCL) of 5 μg/L for drinking water. The software was used for comparison of remediation options for benzene contaminated groundwater at the given site.

Identified options included the following:

- Pump and treat followed by activated carbon treatment of the groundwater.
- Pump and treat and followed by air stripping of the groundwater.
- Biosparging and soil vapour extraction (SVE).
- Monitored natural attenuation.

Table 11.6 provides a list of the indicators used for the analysis. Weights (1 to 3) were assigned to each indicator of Table 11.6. Although a life-cycle assessment is not performed by the software, the indicators should be designed to consider the changing technological and environmental conditions and at different stages in the process. Results showed that, with respect to the environment, natural attenuation, and biosparging and SVE appear to be more appropriate for this site, since they have minimum impact on soil quality, on fauna and flora resulting from the project,

TABLE 11.6
List of Selected Sustainability Indicators Considered for the Analysis

	Type of Indicator	
Environmental Aspects	**Economic Aspects**	**Social Aspects**
Soil quality	NPV of option costs	Community health and safety
Sediment quality	Potential litigation	Worker's health and safety
Contaminated soil erosion	Financial recoveries	Drinking water supply
Groundwater quality	Environmental reserve	Direct local employment
Free product	Standards, laws and regulations	Opportunities for local business generation
Surface water quality	Service reliability and performance	Public disruption
Waterborne contaminant migration	Reuse of the property	Quality of life
Water usage	Corporate image	Public use
Impact on fauna during project	Reliability	Cultural heritage
Impact on fauna after project	Technological uncertainty	Impact on the landscape
Soil vapour intrusion	Logistics	Management practices
Greenhouse gas emissions		
Energy consumption		
Quantity of wastes		
Hazardous wastes		
Residual impact of technology		

TABLE 11.7
Comparison of the Sustainability of Remediation Technologies for a Benzene-Contaminated Site

Alternative	Environment	Society	Economics	Lifetime	Cost ($ in Millions)
Biosparging /SVE	74%	67%	63%	2 years	4.5
MNA	87%	67%	76%	20 years	3.2
Pump and treat with air stripping	48%	64%	74%	5 years	3.5
Pump and treat with activated carbon	53%	74%	68%	2 years	4.5

soil vapour intrusion, greenhouse gas emissions, energy consumption, and quantity of wastes. On the other hand, activated carbon and air stripping seem to be less appropriate, because of the disturbance to the natural condition of the site due to excavation and groundwater extraction, higher energy consumption due to pumping of groundwater, and higher production of greenhouse gases due to the heavy truck usage.

Activated carbon and air stripping have the highest initial costs while natural attenuation and biosparging have higher monitoring costs. However, the selection of natural attenuation may require additional institutional controls to adhere to local regulations and continuous monitoring. Natural attenuation and biosparging would have less impact on community, while activated carbon and air stripping have more impact on community due to potential accidents, truck traffic, and excavation of wells. Furthermore, there is no limitation in terms of remediation time, natural attenuation and biosparging appears to be the best options among the selected options for remediation of the contaminated aquifer.

On the other hand, if the contaminated aquifer is the only drinking water supply for the society, other remediation options such as activated carbon or air stripping should be considered. The required costs for site remediation are higher than bioremediation options due to the urgency for an acceptable source of drinking water.

The overall comparison can be shown in Table 11.7 and Figure 11.22. It is not surprising that natural attenuation is the most sustainable and balanced approach in terms of % for environment, society, and economics due to the low impact on the site and costs. However, the duration of the remediation, which is approximately 20 years, would be a major issue if the site is needed in the near future. In this case, then biosparging would be the next most sustainable option that can be accomplished in a reasonable time frame.

11.8.2 European Case Studies

Sweeney et al. (2024) reviewed ten sustainable remediation case studies. Each of the case studies is downloadable at www.claire.co.uk/concawe (accessed Apr., 2024). Trends in the implementation of sustainable remediation were identified. All case studies were done at the option analysis phase and thus risk management aspects were already determined. All included conceptual site models and emphasized engagement of multiple parties. Recommendation for sustainable remediation implementation included the following:

- The life of the site should be considered.
- Sustainable management practices should be incorporated.
- Regulator interest should be encouraged.
- SURF-UK guidance or ISO 18504:2017 should be followed for the basics of sustainability assessments.

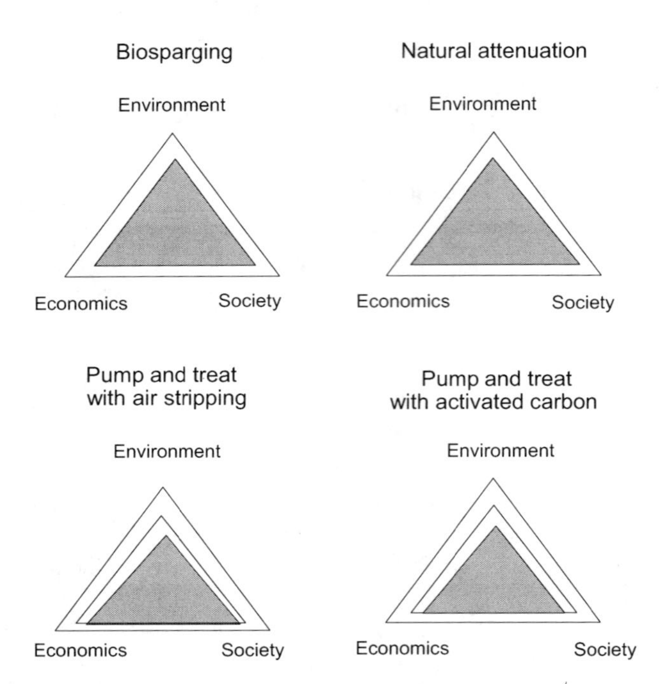

FIGURE 11.22 Comparison of various options using the GoldSET sustainability tool. (Adapted from Mulligan et al., 2013.)

- Boundary conditions must be consistent for option comparisons.
- Individual sustainability criteria should be closely considered for improved sustainability analyses.

11.8.3 SURF Case Study

Pfizer initiated sustainable remediation at a commercial/industrial site using the ASTM Greener Cleanup Standard Guide (E 2893-13), in addition to social and economic aspects (ASTM, 2013). Local stakeholders such as citizen groups and associations were engaged in the remediation and future redevelopment plans for the site. The site contained VOCs, SVOCs, PCBs, and metals in the DNAPL area, soil and groundwater and contaminants such as PCBs in the sediment. The following strategy was initiated. *In situ* thermal treatment was used for the DNAPL area, groundwater was treated biologically, by filtration and chemically with coagulants and UV. Soil covers were used to isolate contaminated soils. An ecological habitat was enhanced to treat and manage stormwater. Some BMPs included use of on-site practices to minimize transportation, carbon footprint analyses, geotextile bags for sediment dewatering, groundwater extraction with a pulsed mode to match river tidal level, job creation and workshops for the local community, local sourcing of supplies, and services and ecological restoration.

11.9 CONCLUDING REMARKS

The oldest method in dealing with contaminated soils is referred to as excavation and disposal or dig and dump, an unsustainable method of managing contaminated soils. Remediation is essential for land reuse. The remediation techniques include the processes of washing, solidification/stabilization, decomposition or/and biodegradation. The techniques for soils and sediments are similar.

TABLE 11.8
Summary of Selected Technologies for Site Remediation

	Technology						
Factor	**Vitrification**	**Thermal Desorption**	**Soil Washing/Flushing**	**Stabilization/ Solidification**	**Bioremediation**	**SVE**	**Phytoremediation**
Effect on contaminants	Thermal oxidation and metal immobilization	Heating of the soil to degrade or volatile contaminants	Metal and chemical removal and possible organic oxidation	Immobilization within matrix	Biodegradation of organic contaminants or conversion of metals	Vapourization of contaminants	Biodegradation and integration in to the plants
Commercial viability	Commercially available	Commercial scale	Some custom design needed	Commercially available	Commercially available	Commercially available	Commercially available
Beneficial use	Glass aggregate	Soil can potentially be reused	Manufactured soil by addition of bulking agents	Compacted fill, capping material	Soil can potentially be reused	Potential recovery of solvents	Potential for biomass energy
Waste streams generated	Debris from screening and wastewater	Off-gas treatment for emissions	Debris from screening, wastewater, sludge from wastewater treatment	Debris from screening,	Wastewater, air emissions, debris from screening	Gaseous emissions that require treatment	Potential waste generated and emissions through biomass handling
Requirements	Plasma arc facility/ dewatering/ Desalination, water, fuel, power	*In* or *ex situ*, power requirements	Water, power	Land area, mixing equipment, additives, power	Water, additives power for mixing equipment if slurries or aeration systems, land if landfarming	Power for generating air flow, wells, gaseous treatment	Harvesting of plants and some maintenance with water
Costs	High	High	Low	Moderate to high	Low to high	Low to moderate	Low

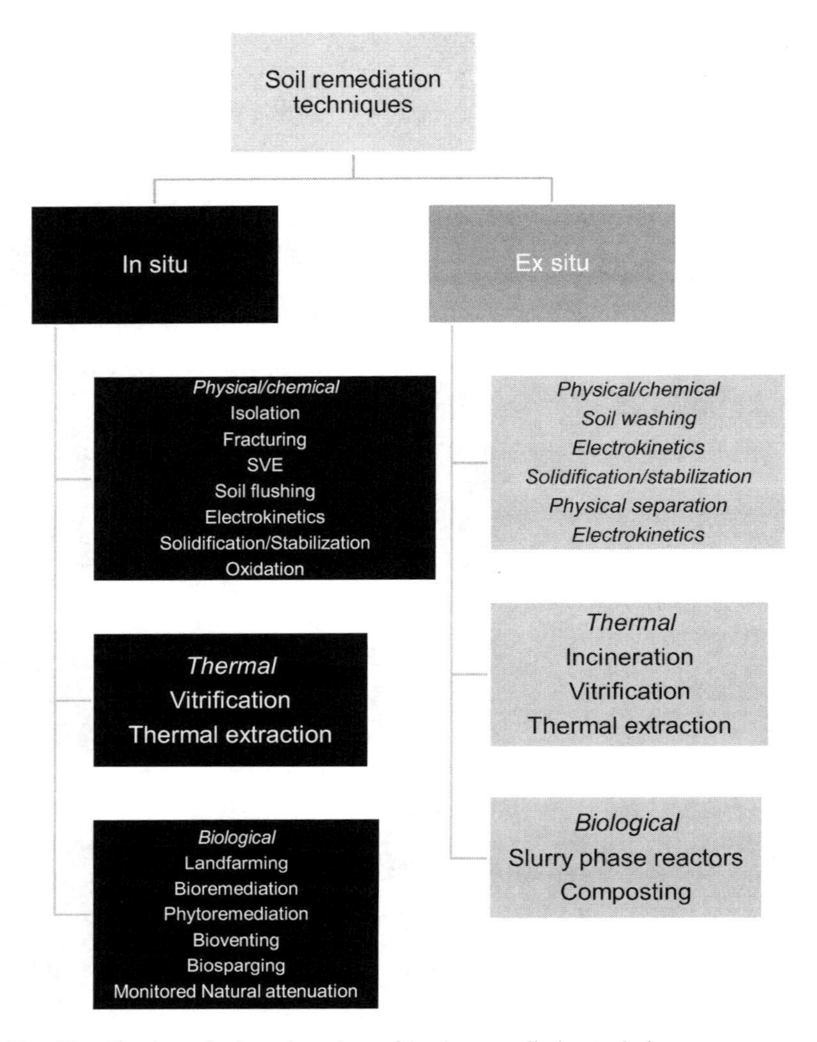

FIGURE 11.23 Classification of selected *ex situ* and *in situ* remediation techniques.

When dredged materials are obtained from the sea, the existence of salt may prevent use of some of the techniques established for soils.

The cost performance of the remediation processes is complex in terms of availability of disposal site, price of construction materials, remediation operation cost and the international and domestic constraints in the environmental situation and time required for remediation and site characteristics. A summary is shown in Table 11.8 and Figure 11.23. *Ex situ* bioremediation allows superior control of environmental parameters such as pH, oxygen, and mixing than *in situ* bioremediation. However, there are still many challenges related to bioavailability of the contaminants. This issue can be remedied through the use of biological surfactants. *In situ* techniques offer decreased costs and increased worker safety. Site characterization and conceptual model development are essential for consideration of alternative remediation approaches in addition to laboratory and pilot testing. To reduce the impacts of the remediation technologies and enhance sustainability, new efforts are required and thus frameworks are being developed to ensure reduction in resource requirements and reduction in emissions and waste generation to protect the public and environment. The evaluation for sustainable remediation should be transparent and allow consultation with the various stakeholders to enable incorporation of the environmental, social, and economic aspects. Innovative techniques need to be developed and demonstrated to optimize land reuse. More tools and guidelines are available for the implementation of sustainable remediation. Incorporation of resilience due to climate change is another important factor that is increasingly being considered.

REFERENCES

Abidin, K., and Yeliz, Y. (2005), Zeta potential of soils with surfactants and its relevance to electrokinetic remediation, *Journal of Hazardous Materials*, 120: 119–126.

Acar, Y.B., Alshawabkeh, A.N., and Gale, R.J. (1993), Fundamentals aspects of extracting species from soils by electrokinetics, *Waste Management,* 12: 141–151.

Achal, V., Pan, X., and Zhang, D. (2012), Bioremediation of strontium (Sr) contaminated aquifer quartz sand based on carbonate precipitation induced by Sr resistant, Halomonas sp., *Chemosphere,* 89: 764–768. http://dx.doi.org/10.1016/j.chemosphere.2012.06.064.

Araújo, R., Castro, A.C.M., and Fiúza, A. (2015), The use of nanoparticles in soil and water remediation processes, *Materials Today: Proceedings*, 2: 315–320. http://dx.doi.org/10.1016/j.matpr.2015.04.055.

ASTM. (2016), Standard Guide for Greener Cleanups E2893-16, West Conshohocken, PA, May 2016.

ASTM. (2013), Standard Guide for Integrating Sustainable Objectives into Cleanup E2876-13, West Conshohocken, PA, June 2013.

Bader, J.L., Gonzales, G., Goodell, P.C., Pillaiand, S.D., and Ali, A.S. (1996), Bioreduction of Hexavalent Chromium in Batch Cultures Using Indigenous Soil Microorganisms. In: *HSRC/WERC Joint Conference on the Environment*, Albuquerque, NM, April 22–24.

Bardos, R.P., Lewis, A., Nortcliff, S., Mariotti, C., Marot, F., and Sullivan, T. (2002), Review of Decision Support Tools for Contaminated Land Management, and their use in Europe, Final Report, Austrian Federal Environment Agency, 2002 on behalf of CLARINET, Spittelauer Lände 5, A-1090, Wien, Austria.

Bardos, P., Bakker, L., Cundy, A., Edgar, S., Harries, N., Holland, K., Muller, D., Nathanail, P., Pijls, C., and Smith, J. (2013), SPS 9. Common Themes and Practice in Achieving Sustainable Remediation Worldwide, with Case Studies and Debate. In: *AquaConSoil 2013*, 16–19 April 2013, Barcelona.

Beazley, M.J., Martinez, R.J., Sobecky, P.A., Webb, S.M., and Taillefert, M. (2007), Uranium biomineralization as a result of bacterial phosphatase activity: Insights from bacterial isolates from a contaminated subsurface, *Environmental Science & Technology*, 41: 5701–5707.

Chen, C., Unrine, J.M., Hu, Y., Guo, L., Tsyusko, O.V., Fan, Z., Liu, S., and Wei, G.W. (2021), Responses of soil bacteria and fungal communities to pristine and sulfidized zinc oxide nanoparticles relative to Zn ions, *Journal of Hazardous Materials,* 405: 124258. https://doi.org/10.1016/j.jhazmat.2020.124258.

Daigger, G.T., and Crawford, G.V. (2005), *Wastewater Treatment Plant of the Future—Decision Analysis Approach for Increased Sustainability*, IWA Publishing, London, U.K., Unique ID No.200504039.

Do, H., Che, C., Zhao, Z., Wang, Y., Li, M., Zhang, X., and Zhao, X., (2020) Extracellular polymeric substance from *Rahnella sp.* LRP3 converts available Cu into $Cu(PO_4)_2(OH)_4$ in soil through biomineralization process. *Environmental Pollution,* 260: 114051. https://doi.org/10.1016/j.envpol.2020.114051.

Duarte-Nass, C., Rebolledo, K., Valenzuela, T., Kopp, M., Jeison, D., Rivasf, M., Azocar, L., Torres-Aravena, A., and G Ciudad, G. (2020), Application of microbe-induced carbonate precipitation for copper removal from copper-enriched waters: Challenges to future industrial application, *Journal of Environmental Management,* 256: 15, 109938. https://doi.org/10.1016/j.jenvman.2019.109938.

El-Shafey, E.I., and Canepa, P. (2003), Remediation of a Cr (VI) contaminated soil: Soil washing followed by Cr (VI) reduction using a sorbent prepared from rice husk, *Journal de physique. IV, International Conference on Heavy Metals in the Environment N°12, Grenoble, FRANCE (26/05/2003)*, vol. 107 (1), pp. 415–418.

Federal Remediation Technologies Roundtable (FRTR). (2007), *Remediation Technologies Screening Matrix and Reference Guide*, 4th ed. www.frtr.gov/matrix2/section3/3_2.html.

Fenton, H.J.H. (1893), The oxidation of tartaric acid in presence of iron, *Proceedings Chemical Society,* 9(I): 113.

Fenton, H.J.H. (1895), New formation of glycolic aldehyde, *Chemical Society Journal* (*London*), 67, 774: 5.

Fenton, H.J.H., and Jackson, H. (1899), Oxidation of polyhydric alcohols in the presence of iron, *Journal of the Chemical Society*, 75: 1–11.

Ferrarese, E., Andreottola, G., and Oprea, I.A. (2008), Remediation of PAH-contaminated sediments by chemical oxidation, *Journal of Hazardous Materials,*152: 128–139.

Fitzpatrick, A., and Woodward, D. (2013), SURFs Up Canada and Going Global: Integrating Sustainable Principles into Site Remediation SITE REMEDIATION IN B.C., FROM POLICY TO PRACTICE Vancouver, BC, March 7-8, 2013.

Fujita, Y., Redden, G.D., Ingram, J.C., Cortez, F., Ferris, G., and Smith, R.W. (2004), Strontium incorporation into calcite generated by bacterial ureolysis, *Geochimica et Cosmochimica Acta,* 68(15): 3261–3270. https://doi:10.1016/j.gca.2003.12.018.

Fukue, M. Lechowicz, Z., Fujimori, Y., Emori, K., and Mulligan, C.N. (2023), Inhibited and retarded behavior by Ca^{2+} and Ca^{2+}/OD loading rate on ureolytic bacteria in MICP process, *Materials*, 16: 3357. https://doi.org/10.3390/ma16093357.

George, S.E., and Wan, Y. (2023), Microbial functionalities and immobilization of environmental lead: Biogeochemical and molecular mechanisms and implications for bioremediation, *Journal of Hazardous Materials,* 457: 131738. https://doi.org/10.1016/j.jhazmat.2023.131738.

Han, L., Li, J., Xue, Q., Chen, Z., Zhou, Y., and Poon, C.S. (2020), Bacterial-induced mineralization (BIM) for soil solidification and heavy metal stabilization: A critical review, *Science of the Total Environment,* 746: 140967. https://doi.org/10.1016/j.scitotenv.2020.140967.

Harclerode, M., Ridsdale, D.R., Darmendrail, D., Bardos, P., Alexandrescu, F., Nathanail, P., Pizzol, L., and Rizzo, E. (2015), Integrating the social dimension in remediation decision-making: State of the practice and way forward, *Remediation*, 26(1): 11–42. https://doi.org/10.1002/rem.21447.

Hazardous Waste Consultant. (1996), Remediating Soil and Sediment Contaminated with Heavy Metals. In: *Hazardous Waste Consultant,* Elsevier Science, Vol. 1, p. 4, November–December.

Holland, K.S., Lewis, R.E., Tipton, K., Karnis, S., Dona, C., Petrovskis, E., Bull, L.P., Taege, D., and Hook, C. (2011), Framework for integrating sustainability into remediation projects, *Remediation*, 21(3): 7–38. https://doi.org/10.1002/rem.20288.

Hou, D., Al-Tabbaa, A., O'Connor, D., et al. (2023), Sustainable remediation and redevelopment of brownfield sites, *Nature Review Earth Environmental,* 4: 271–286. https://doi.org/10.1038/s43017-023-00404-1.

Hu, X., Yang, H., Fang, X., Shi, T., and Tan, K. (2023), Recovery of bio-sulfur and metal resources from mine wastewater by sulfide biological oxidation-alkali flocculation: A pilot-scale study, *Science of the Total Environment,* 876: 162546. http://dx.doi.org/10.1016/j.scitotenv.2023.162546.

Institute of Gas Technology (IGT). (1996), process literature, June.

Interstate Technical Regulatory Council (ITRC). (2009), Phytotechnology Technical and Regulatory Guidance and Decision Trees, Revised. PHYTO-3. Washington, D.C.: Interstate Technology & Regulatory Council, Phytotechnologies Team, Tech Reg Update. www.itrcweb.org

Jalali, F., and Mulligan, C.N. (2008), Enhanced Bioremediation of a Petroleum Hydrocarbon and Heavy Metal Contaminated—Soil by Stimulation of Biosurfactant Production. In: *Geo-Environmental Engineering 2008*, Kyoto, June 10–12.

Jiang, L., Liu, X., Yin, H., Liang, Y., Liu, H., Miao, B. Pengc, Q., Meng, D., Wang, S., Yang, J., and Guo, Z. (2020), The utilization of biomineralization technique based on microbial induced phosphate precipitation in remediation of potentially toxic ions contaminated soil: A mini review, *Ecotoxicology and Environmental Safety* 191: 110009. https://doi.org/10.1016/j.ecoenv.2019.

Karn, B., Kuiken, T., and Otto, M. (2009), Nanotechnology and in situ remediation: A review of the benefits and potential risks, *Environmental Health Perspectives*, 117: 1832–1831.

Kellar, E.M., Arenberg, E.D., Mickler, J.C., Robinson, L.I., and Smith, S.D. (2009), EN Rx Oxidation Process for Slow-Release Sediment, Soil and Groundwater Remediation. In: *5th International Conference on Remediation of Contaminated Sediments, Feb. 2-5*, Jacksonville, FL.

Khan, Z., and Anjaneyulu, Y. (2006), Bioremediation of contaminated soil and sediment by composting, *Bioremediation Journal,* 16(2): 109–122.

Khodadoust, A.P., Varadhan, A.S., and Bogdan, D. (2009), Enhanced Anaerobic Biodegradation of PCBs in Contaminated Sediments Using Periodic Amendments of Iron. In: *5th International Conference on Remediation of Contaminated Sediments*, Feb. 2-5, Jacksonville, FL.

Kimber, R.L., Bagshaw, H., Smith, K., Buchanan, D.M., Coker, V.S., Cavet, J.S., and Lioid, J.R. (2020), Biomineralization of Cu_2S Nanoparticles by *Geobacter sulfurreducens*, *Applied and Environmental Microbiology,* 86: 18. https://doi.org/10.1128/AEM.00967-20.

Kumar, A., Song, H.-W., Mishra, S., Zhang, W., Zhang, Y.-L., Zhang, Q.-R., and Yu, Z-G. (2023), Application of microbial-induced carbonate precipitation (MICP) techniques to remove heavy metal in the natural environment: A critical review, *Chemosphere,* 318: 137894. https://doi.org/10.1016/j.chemosphere.2023.

LaGrega, M.D., Buckingham, P.K., and Evans, J.C. (2004), *Hazardous Waste Management*, 2nd ed., McGraw Hill, New York.

Lee, C.R., and Price, R.A. (2003), Review of Phytoreclamation and Management Approaches for Dredged Material Contaminated with Lead, DOER Technical Notes Collection (EDRD TN-DOER-C29), U.S. Army Engineer Research and Development Center, Vicksburg, M.S. www.wes.army.mil/el/dots/doer.

Li, Y., Zhao, Q., Liu, M., Guo, J. Xia, J., Wang, J., Qiu, Y., Zou, J., He, W., and Jiang, F. (2023), Treatment and remediation of metal-contaminated water and groundwater in mining areas by biological sulfidogenic processes: A review, *Journal of Hazardous Materials* 443: 130377. https://doi.org/10.1016/j.jhazmat.2022.130377.

Liang, X., Hillier, S., Pendlowski, H., Gray, N., Ceci, A., and Gadd, G.M. (2015), Uranium phosphate biomineralization by fungi, *Environmental Microbiology,* 17(6): 2064–2075.

Lima, A.T., Hofmann, A., Reynolds, D., Ptacek, C.J., Van Cappellen, P., Ottosen, L.M., and Pamukcu, S. (2017), Environmental electrokinetics for a sustainable subsurface, *Chemosphere,* 181: 122–133.

Linssen, R., Slinkert, T., Buisman, C.J.N., and Klok, J.B.M. (2023), Annemiek ter Heijne, Anaerobic sulphide removal by haloalkaline sulphide oxidising bacteria, *Bioresource Technology,* 369: 128435. https://doi.org/10.1016/j.biortech.2022.128435.

Li, M., Cheng, X. and Guo, H. (2013), Heavy metal removal by biomineralization of urease producing bacteria isolated from soil, *International Biodeterioration & Biodegradation* 76, 81e85, http://dx.doi.org/10.1016/j.ibiod.2012.06.016

Liu, F., Zhou, K., Chen, Q., Wang, A., and Chen, W. (2019), Application of magnetic ferrite nanoparticles for removal of Cu(II) from copper-ammonia wastewater, *Journal of Alloys and Compounds*, 773: 140–149. https://doi.org/10.1016/j.jallcom.2018.09.240.

Maco, B., Bardos, P., Coulon, F., Erickson-Mulanax, E., Hansen, L.J., Harclerode, M., Hou, D., Mielbrecht, E., Wainwright, H.M., Yasutaka, T., and Wick, W.D. (2018), Resilient remediation: Addressing extreme weather and climate change, creating community value, *Remediation*, 29(1): 7–18. https://doi.org/10.1002/rem.21585.

Massara, H., Mulligan, C.N., and Hadjinicolaou, J. (2007), Effect of rhamnolipids on chromium contaminated soil. *Soil Sediment Contamination: International Journal.* 16: 1–14.

Mench, M., Manceau, A., Vangronsveld, J., Clijsters, H., and Mocquot, B. (2000), Capacity of soil amendments in lowering the phytoavailability of sludge-borne zinc, *Agronomie*, 20: 383–397.

Merino, J., and Bucalá, V. (2007), Effect of temperature on the release of hexadecane from soil by thermal treatment, *Journal of Hazardous. Materials,* 143: 455–461.

Mobley, H.L.T., and Hausinger, R.P. (1989), Microbial urease: Significance, regulation, and molecular characterization, *Microbiology Review*, 53: 85–108.

Moon, D.G., Rehan, S., Yeon, D.H., Lee, S.M., Park, S. ., Ahn, S., and Cho, Y.S. (2019), A review on binary metal sulfide heterojunction solar cells, *Solar Energy Materials and Solar Cells*, 200: 109963, https://doi.org/10.1016/j.solmat.2019.109963.

Mosley, E. (2006), *Developing a Sustainability Rating Tool for Wastewater Systems*, WEFTEC, Washington, DC, October 21–25.

Mulligan, C.N. (2002), *Environmental Biotreatment,* Government Institutes, Rockville, MD.

Mulligan, C.N. (2021), Sustainable remediation of contaminated soil using biosurfactants, *Frontiers in Bioengineering and Biotechnology*, 9: 635196. https://doi.org/10.3389/fbioe.2021.635196.

Mulligan, C.N., and Kamali, M. (2003) Bioleaching of copper and other metals from low-grade oxidized mining ores by *A. niger, Journal of Chemical Technology and Biotechnology,* 78: 497–503.

Mulligan, C.N., Yong, R.N., and Gibbs, B.F. (2001), Heavy metal removal from sediments by biosurfactants, *Journal of Hazardous Materials*, 85: 111–125.

Mulligan, C.N. Yong R.N., and Gibbs B.F. (1999), On the use of biosurfactants for the removal of heavy metals from oil-contaminated soil,. *Environmental Progress,* 18: 50–54.

Mulligan, C.N., Fukue, M., and Sato, Y. (2010), *Sediments: Contamination and Sustainable Remediation,* Taylor and Francis, CRC Press, Boca Raton, 320 pp.

Mulligan, C.N. Dumais, S., and Noel-de-Tilly, R. (2013), Sustainable Remediation of Contaminated Sites. In: *Coupled Phenomena in Environmental Geotechnics CPEG*, Torino, Italy, July 1-3.

Mulligan, C.N. (2023), Sustainable Production of Biosurfactants Using Waste Substrates, Chapter 3. In: R. Aslam et al. (eds.), *Advancements in Biosurfactants Research,* 3 Springer Nature, Switzerland AG . https://doi.org/10.1007/978-3-031-21682-4.

Myers, T.E., and Williford, C.W. (2000), Concepts and Technologies for Bioremediation in Confined Disposal Facilities, DOER Technical Notes Collection (ERDS TN-DOER-C11)), U.S. Army Engineer Research and Development Center, Vicksburg, MS. www.wes.mil/el/dots/doer.

Myers, T.E., Bowman, D.W., and Myers, K.F. (2003), Dredged material composting at Milwaukee and Green Bay, WI, confined disposal facilities, DOER Technical Notes Collection (ERDC TN DOER-C33), U.S. Army Engineer Research and Development Center, Vicksburg, MS. www.erdc.army.mil/el/dots/doer.

New York State Department of Environmental Conservation (NYDEC). (2011), DER-31 / Green Remediation, DEC Program Policy. 9 pp, www.dec.ny.gov/docs/remediation_hudson_pdf/der31.pdf, accessed Apr. 2024.

NRC, National Research Council. (1997), *Contaminated Sediments in Ports and Waterways, Cleanup Strategies and Technologies,* National Academic Press, Washington.

Nystroem, G.M., Pedersen, A. J, Ottosen, L.M., and Villumsen, A. (2006), The use of desorbing agents in electrodialytic remediation of harbour sediment, *Science of the Total Environment,* 357(1): 25–47.

Olin-Estes, T.J., and Palermo, M.R. (2001), Recovery of dredged material for beneficial use: the future role of physical separation processes, *Journal of Hazardous Materials,* 85: 39–51.

Papadopoulos, D., Pantazi, C. Savvides C., Harlambous, K.J. Papadopoulos A., and Loizidou, M. (1997), A Study on heavy metal pollution in marine sediments and their removal from Dredged Material, *Journal of Environmental Health,* A32(2): 347–360.

Park, I., Tabelin, C.B., Jeon, S., Li, X., Seno, K., Ito, M., and Hiroyoshi, N. (2019), A review of recent strategies for acid mine drainage prevention and mine tailings recycling, *Chemosphere*, 219: 588e606. https://doi.org/10.1016/j.chemosphere.2018.11.053.

Pinto, I.S.S., Neto, I.F.F., and Soares, H.M.V.M. (2014), Biodegradable chelating agents for industrial, domestic, and agricultural applications—A review, *Environ Science and Pollution Research,* 21**:** 11893–11906. https://doi.org/10.1007/s11356-014-2592-6.

Qian, C., Yu, X., and Wang, X. (2018) Potential uses and cementing mechanism of bio-carbonate cement and bio-phosphate cement, *AIP Advances,* 8: 095224. https://doi.org/10.1063/1.504073.

Rathfelder, K., Yeh, W.W.G., and Mackay, D. (1991), Mathematical simulation of soil vapour extraction systems: Model development and numerical examples, *Journal of Contaminant Hydrology,* 8: 263–297.

Rawlings, E., ed. (1997), *Biomining: Theory, Microbes and Industrial Processes.* Georgetown TX, Springer-Verlag/Landes Bioscience.

Reddy, K.R., Xu, C.Y., and Chinthamreddy, S., (2001), Assessment of electrokinetic removal of heavy metals from soils by sequential extraction analysis, *Journal of Hazardous Materials,* 84(2–3): 85–109.

Reddy, K.R., and Adams, J.A. (2010), Towards Green and Sustainable Remediation of Contaminated Site. In: *6th International Congress on Environmental Geotechnics*, New Delhi, India, pp. 1222–1227.

Rittmann, B.E., and McCarty, P.L. (2001), *Environmental Biotechnology: Principles and Applications,* McGraw Hill, New York.

Rienks, J. (1998), Comparison of results for chemical and thermal treatment of contaminated dredged sediments, *Water Science and Technology,* 37: 355–342.

Rodsand, T., and Acar, Y.B. (1995) Electrokinetic extraction of lead from spiked Norwegian marine clay, *Geoenvironment,* 2: 1518–1534.

Santiago, R., Inch, R., Jaagumagi, R., and Pelletier, J.-P. (2003), Northern Wood Preservers Sediment Case Study. In: *2nd International Symposium on Contaminated Sediments*, pp. 297–303.

Singh, A.D., Khanna, K., Kour, J., Dhiman, S., Bhardwaj, T., Devi, K., Sharma, N., Kumar, P., Nitika Kapoor, N., Sharma, P., Arora, P., Sharma, A and Bhardwaj, R. (2023), Critical review on biogeochemical dynamics of mercury (Hg) and its abatement strategies, *Chemosphere*, 319: 137917. https://doi.org/10.1016/j.chemosphere.2023.137917.

Song, H., Kumar, A., and Zhang, Y. (2022), Microbial-induced carbonate precipitation prevents Cd2+ migration through the soil profile, *Science of the Total Environment,* 844: 157167. http://dx.doi.org/10.1016/j.scitotenv.2022.157167.

Sörengård, M., Gago-Ferrero, P., Kleja, D.B., and Ahrens, L. (2021), Laboratory-scale and pilot-scale stabilization and solidification (S/S) remediation of soil contaminated with per- and polyfluoroalkyl substances (PFASs), *Journal of Hazardous Materials*, 402: Article 123453. http://dx.doi.org/10.1016/j.jhazmat.2020.123453.

Sepúlveda, S., Duarte-Nass, C., Rivas, R., Azócar, L. Ramírez, A., Toledo-Alarcón, J., Gutiérrez, L., Jeison, D. and Torres-Aravena, Á. (2021), Testing the Capacity of *Staphylococcus equorum* for Calcium and Copper Removal through MICP Process, *Minerals,* 11, 905. https://doi.org/10.3390/min11080905

Su, Z., Li, X., Xi, Y., Xie, T., Liu, Y., Liu, B., Liu, H., Xu, W., and Zhang, C. (2022), Microbe-mediated transformation of metal sulfides: Mechanisms and environmental significance, *Science of the Total Environment*, 825: 153767. http://dx.doi.org/10.1016/j.scitotenv.2022.153767.

Sweeney, R., Harries, N., Bardos, P., and Vaiopoulou, E. (2024), Analysis of sustainable remediation techniques and technologies based on 10 European case studies, *Remediation*, 34(2): e21773. https://doi.org/10.1002/rem.21773.

Teng, Z., Shao, W., Zhang, K., Huo, Y., Zhu, J., and Li, M. (2019), Pb biosorption by *Leclercia adecarboxylata*: Protective and immobilized mechanisms of extracellular polymeric substances, *Chemical Engineering Journal,* 375: 122113. https://doi.org/10.1016/j.cej.2019.122113.

Tu, H., Yuan, G., Zhao, C., Liu, J., Li, F., Yang, J., Liao, J., Yang, Y., and Liu, N. (2019), U-phosphate biomineralization induced by *Bacillussp*.dw-2 in the presence of organic acids, *Nuclear Engineering and Technology,* 51: 1322e1332. https://doi.org/10.1016/j.net.2019.03.002.

US ACE and USEPA. (2003), Great Lakes Confined Disposal Report to Congress, US Army Corps of Engineers-Great Lakes National Program Office.

US EPA. (1989), Corrective Action: Technologies and applications, Seminar publication, Center for Environmental Research Information, Cincinnati, OH, EPA/625/4-89/020.

US EPA. (1994), Biosparging. From Chapter 8 of How to Evaluate Alternative Cleanup Technologies for Underground Storage Tank Sites: A Guide for Corrective Action Plan Reviewers, EPA 510-R-04-002.

US EPA. (1996), Lasagna™ Public-Private Partnership, Office of Research and Development, Office of Solid Waste and Emergency Response, EPA Report EPA/542/F-96/010A.

US EPA. (1997), Analysis of Selected Enhancement for Soil Vapor Extraction, OSWER, Office of Solid Waste and Emergency Response, September EPA542-R-97-007.

US EPA. (1998), On-Site Incineration, Overview Superfund Operating Experience OSWER, Office of Solid Waste and Emergency Response (OSWER) March EPA-542-R-97-012.

US EPA. (2000), Solidification/stabilization use at Superfund Sites, OSWER September EPA 542-R-00-010.

US EPA. (2004a), TH agricultural & Nutritional Company Site, RODS Abstract Information, Superfund Information Systems. www.epa.gov/superfund.

US EPA. (2004b), Treatment technologies for Site Cleanup (Eleventh Edition) EPA-542-R-03-009, February, Office of Solid Waste and Emergency Response.

US EPA. (2005), Region 9 Superfund, Selma treating Co. www.yosemite.epa.gov/r9/sfund/r9sfdocw.nsf/vwsoalphabetic/Selma+Treating+Co.?

US EPA. (2006), In situ treatment technologies for contaminated soil, Engineering Forum Issue Paper, OSWER, November, EPS 542-F-06-013.

US EPA. (2008a), Nanotechnology for Site Remediation Fact Sheet, US EPA Office of Solid Waste and Emergency Response, EPA 542-F-08-009, October 2008, www.semspub.epa.gov/work/HQ/139541.pdf.

US EPA. (2008b), Green Remediation: Incorporating Sustainable Environmental Practices into Remediation of Contaminated Sites USEPA, April. U.S. Environmental Protection Agency Office of Solid Waste and Emergency Response April 2008, EPA 542-R-08-002

US EPA. (2010), Green Remediation Best Management Practices: Bioremediation, US EPA Office of Solid Waste and Emergency Response EPA 542-F-10-006, March.

US EPA. (2012a), A Citizen's Guide to Fracturing for Site Cleanup, OSWER September, EPA 542-F-12-008.

US EPA. (2012b), A Citizen's Guide to Thermal Desorption, OSWER, September, EPA 542-F-12-020.

US EPA. (2013), CLU-IN Bioremediation Technology. www.clu-in.org.

US EPA. (2015), Greener Cleanups Bulletin: Application of the ASTM Standard Guide for Greener Cleanups at the North Ridge Restates Superfund Site, Office of Solid Waste and Emergency Response, Office of Superfund Remediation and Technology Innovation, EPA 542-F-15-011, Sept. 2015.

US EPA. (2022), Greener Cleanups Best Management Practices: PCB Cleanups, PCB Information and Reference Series Fact Sheet, EPA 530-F-22-005, December 2022.

US EPA. (2023), Superfund Remedy Report (SRR) EPA-542-R-23-001 Office of Land and Emergency Management, 17th Edition, January 2023. www.epa.gov/remedytech/superfund-remedy-report, accessed May 2024.

van Rooyen, D., Erasmus, J.H., Gerber, R. Nachev, M. Sures, B. Wepener, V., and Smit, N.J. (2023), Bioaccumulation and trophic transfer of total mercury through the aquatic food webs of an African sub-tropical wetland system, *Science of the Total Environment,* 889: 164210. http://dx.doi.org/10.1016/j.scitotenv.2023.164210.

Vidonish, J.E., Zygourakis, K., Masiello, C.A., Sabadell, G., and Alvarez, P.J.J. (2016), Thermal Treatment of Hydrocarbon-Impacted Soils: A Review of Technology Innovation for Sustainable Remediation, *Engineering*, 2(4): 426–437. https://doi.org/10.1016/J.ENG.2016.04.005.

.Virkutyte, J., Sillanpää, M., and Latostenmaa, P. (2002), Electrokinetic soil remediation: Critical overview, *Science of the Total Environment,* 289: 97–121.

Vu, K.A., and Mulligan, C.N. (2022), Utilization of a biosurfactant foam/nanoparticle mixture for treatment of oil pollutants in soil, *Environmental Science and Pollution Research,* 29: 88618–88629.

Vu, K.A., and Mulligan, C.N. (2023a), Remediation of organic contaminated soil by Fe-based nanoparticles and surfactants: A review, *Environmental Technology Reviews,* 12(1): 60–82. https://doi.org 10.1080/21622515.2023.2177200.

Vu, K.A., and Mulligan, C.N. (2023b), Synthesis of Carbon-Based Nanomaterials and Their Use in Bio or Nanoremediation. In: F. Fernández-Luqueño, F. López-Valdez, and G. Medina-Pérez, (eds.), *Bio and Nanoremediation of Hazardous Environmental Pollutants*, Taylor & Francis Group Publisher, p. 33. https://doi.org/10.1201/9781003052982.

Wang, S., and Mulligan, C.N. (2004), Surfactant foam technology in remediation of contaminated soil, *Chemosphere,* 57: 1079–1089.

Wang, W., Duan, Y., Wu, Y., Huang, Y., Gao, F., Wang, Z, and Zheng, C. (2021), Harmless treatment of cyanide tailings by a bifunctional strain JK-1 based on biodegradation and biomineralization, *Journal of Cleaner Production,* 313: 127757. https://doi.org/10.1016/j.jclepro.2021.127757.

Weston, R.F. (1988), *Remedial Technologies for Leaking Underground Storage Tanks*, Chelsea, MI, Lewis.

Weyand, T.E., Rose, M.V., and Koshinski, C.J. (1994), Demonstration of Thermal Treatment Technology for Mercury Contaminated Waste, June 1994.

Wilcox, S.M., Mulligan, C.N., and Neculita, C.M. (2024), Microbially induced calcium carbonate precipitation as a bioremediation technique for mining waste. *Toxics*, 12: 107. https://doi.org/10.3390/toxics12020107.

Xia, L., Tan, J., Huang, R., Zhang, Z., Zhou, K., Hu, Y., Song, S., Xu, L., Farías, M.E., and Sánchez, R.M.T. (2023), Enhanced Cd(II) biomineralization induced by microalgae after cultivating modification in high-phosphorus culture, *Journal of Hazardous Materials,* 443: 130243, https://doi.org/10.1016/j.jhazmat.2022.130243.

Xu, Y.-N., and Chen, Y. (2020), Advances in heavy metal removal by sulfate-reducing bacteria, *Water Science and Technology*, 81(9): 1797–1827.

Yong, R.N. (1998), Compatible Technology for Treatment and Rehabilitation of Contaminated Sites, NNGI Report No. 5, Nikken Sekkei Geotechnical Institute, Japan, pp. 1–33.

Yong, R.N., Mulligan, C.N., and Fukue, M. (2006), Geoenvironmental Sustainability, CRC Press, Taylor & Francis, Boca Raton, FL.

Zeng, Y., Chen, Z., Du, Y., Zeng, Lyu, Q., Yang, Z., and Yan, Z. (2021), Microbiologically induced calcite precipitation technology for mineralizing lead and cadmium in landfill leachate, *Journal of Environmental Management,* 296: 113199. https://doi.org/10.1016/j.jenvman.2021.113199.

Zeng, Y., Chen, Z., Lyu, Q., Wang, X., Du, Y., Huan, C., Liu, Y., and Yan, Z. (2022), Mechanism of microbiologically induced calcite precipitation for cadmium mineralization, *Science of the Total Environment*, 852: 158465. http://dx.doi.org/10.1016/j.scitotenv.2022.158465.

Zhang, L., Qiu, Y-Y., Zhou, Y., Chen, G.-H. van Loosdrecht, M.C.M., and Jiang, F. (2021), Elemental sulfur as electron donor and/or acceptor: Mechanisms, applications and perspectives for biological water and wastewater treatment, *Water Research,* 202: 117373. https://doi.org/10.1016/j.watres.2021.117373.

Zhang, K., Xue, Y., Xu, H., and Yao, Y. (2019), Lead removal by phosphate solubilizing bacteria isolated from soil through biomineralization, *Chemosphere,* 224: 272–279. https://doi.org/10.1016/j.chemosphere.2019.

Zhao, J.-S., and Hawari, J. (2008) Biodegradation of Nitramines in Marine Sediment. In: *In situ Contaminated Sediments Workshop*, Concordia University, Montreal, Quebec, September 10.

Zhao, J., Csetenyi, L., and Gadd, G.M. (2022), Fungal-induced $CaCO_3$ and $SrCO_3$ precipitation: A potential strategy for bioprotection of concrete, *Science of the Total Environment* 816: 151501. https://doi.org/10.1016/j.scitotenv.2021.151501.

Zhao, X., Do, H.T., Zhou, Y., Li, Z., Zhang, X., Zhao, S., Li, M., and Wu, D. (2019), *Rahnella sp.* LRP3 induces phosphate precipitation of Cu (II) and its role in copper-contaminated soil remediation, *Journal of Hazardous Materials,* 368: 133–140. https://doi.org/10.1016/j.jhazmat.2019.

Zhu, T., and Dittrich, M. (2016), Carbonate precipitation through microbial activities in natural environment, and their potential in biotechnology: A review, *Frontiers in Bioengineering and Biotechnology*, 20, Sec. Microbiotechnology 4. https://doi.org/10.3389/fbioe.2016.00004.

12 Sustainable Nitrogen and Carbon Cycles

12.1 INTRODUCTION

12.1.1 Organic Matter Elimination and Decomposition

One of the biggest environmental problems on a global scale today is the excess release of CO_2, which has been attributed to the use of fossil fuels formed from the degradation of living organisms over a long period of time. Accordingly, the diversity of living organisms and a large amount of organic matter were lost from the conventional mass cycle. In other words, the biodiversity has been changed and some food chains are impacted.

As a result of the rapid use of fossil fuels over the last few decades and the development of industry, excessive amounts of CO_2 are emitted compared to the amount absorbed by plants. CO_2 and NH_3 are released by the decomposition of organic matter. For example, urea ($(NH_2)_2$ CO) in urine decomposes into CO_2 and $2NH_3$. Thus, the conventional cycles of carbon and nitrogen are based on the decomposition of organic matter in nature.

Originally, biodiversity is maintained only when the organic elements of living organisms are cycled in the order of excretion, decomposition, and inoculation. In other words, biodiversity refers to a situation in which the excrement of living organisms is circulated in a well-balanced manner. When the cycle is out of balance, certain organisms become extinct, and the cycle becomes unstable, gradually reducing the number of species. In other words, a sustainable cycle of nutrients is of importance for the sustainability of the life of living organisms. Therefore, it is necessary to understand the state of the current environment and to consider what should be done to normalize the balance.

12.1.1.1 Nutrients

Three major components necessary for life support are carbohydrates, lipids, and proteins. Carbohydrates are represented by the chemical formula $C_n(H_2O)_m$. Energy is produced by oxidation of carbohydrates *in vivo*, which is broken down into CO_2 and H_2O. Lipids *in vivo* are classified into simple lipids, compound lipids, and derived lipids. A protein has about more than 40 amino acids connected by peptide bonds. Smaller compounds with less peptide bonds are often called polypeptides. It is noted that the amino acid contains nitrogen. Therefore, nitrogen and carbon are the most important elements for living organisms. From this point of view, the nitrogen and carbon cycles are considered in terms of marine environment sustainability.

12.1.1.2 Waste Discharge

As ammonia at high concentrations is toxic to living organisms. Therefore, they have a mechanism for release ammonia from their body. For example, teleost fish release ammonia directly from their gills, but in water the ammonia is converted into ammonium ions and reducing its toxicity. On the other hand, cartilaginous fishes excrete nitrogen waste as less toxic urea, not ammonia, but as in post-metamorphic amphibians and mammals. This urea is used by cartilaginous fish to regulate body fluids.

In mammals, to avoid ammonia the liver is mainly responsible for excrement formation. Some of the waste products metabolized in the liver are returned to the vascular system and urea is excreted

DOI: 10.1201/9781003407393-12

through the renal urine excretion process, while others are secreted into the bile ducts to become bile and excreted into the duodenum. Complete aquatic species and most amphibian larvae require ammonia excretion and die easily when the water content is restricted. In adults, urea is produced and excreted. Insects and reptiles, including birds, excrete in the form of "uric acid", which is more complex than ammonia. The same "cloaca" pores are used for faces, urine, sperm, and eggs. However, reptiles excrete ammonia and urea in addition to uric acid. In addition, the rate of nitrogen urine excretion varies depending on the species.

12.1.2 Decomposition of Organic Matter

Organic matter consists of organic compounds of biological origin derived from the degradation of plants and animals or waste materials. Urea is the primary organic matter in excrement and also the first artificially synthesized organic compound. Organic matter is simply defined as any of the carbon-based compounds found in nature, and consists of mainly carbon C, nitrogen N, oxygen O, and hydrogen H. The decomposition of organic matter produces NH_4^+ and CO_2.

Figure 12.1 shows the nitrogen and carbon cycles. The main processes of nitrogen and carbon are the food chain, decomposition of the wastes from animals and humans, and natural formation of limestone, carbonates, such as calcite and aragonite. The excess CO_2 which is a risk for the earth and living organisms is shown in the red circle, which denotes the atmosphere. At present, eliminating the red circle is a pivotal mission, but it is extremely difficult.

12.2 NITROGEN CYCLE

Nitrogen absorbed in living organisms is an element of amino acids (proteins) and nucleotides (nucleic acids). Urea is not simply an end product for excretion but is also an important factor for environmental adaptation. Figure 12.2 is a simplified diagram of the nitrogen cycle in agriculture (Sieczka and Koda, 2016). Non-agricultural examples of the nitrogen cycle are shown in Figure 12.3. It illustrates the following items: (1) gasification into ammonia, (2) mineralization/immobilization, (3)nitrification-denitrification, (4) leaching of NH_4^+, (5) eutrophication, and (6) acid rain.

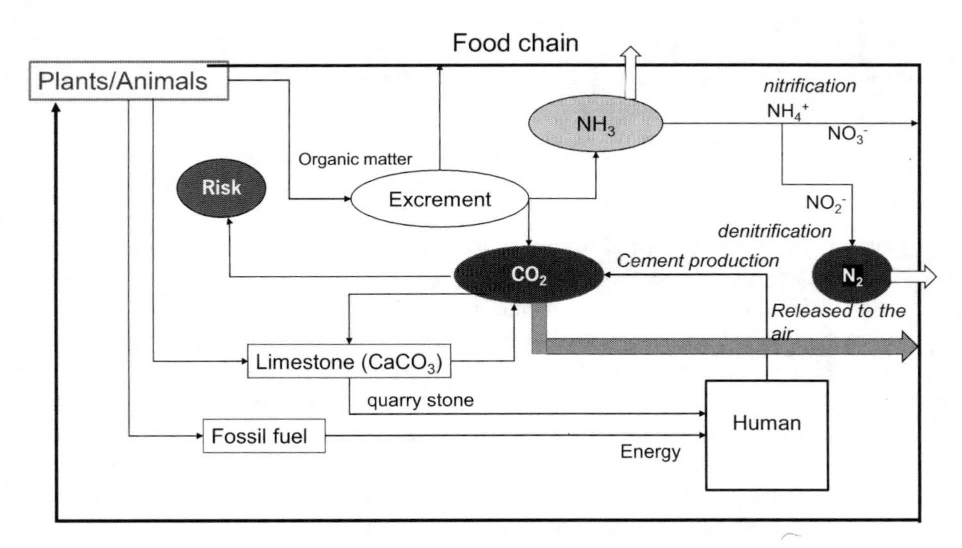

FIGURE 12.1 Illustration of nitrogen and carbon cycles.

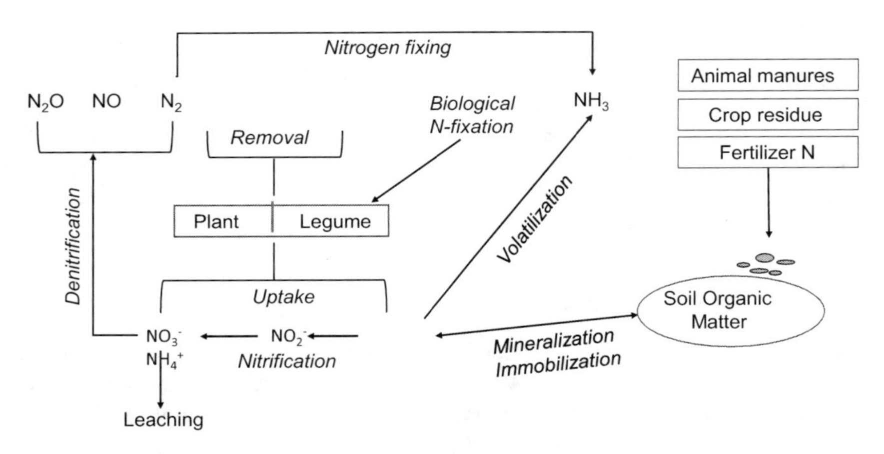

FIGURE 12.2 A simplified diagram of the nitrogen cycle in agriculture. (Sieczka and Koda, 2016.)

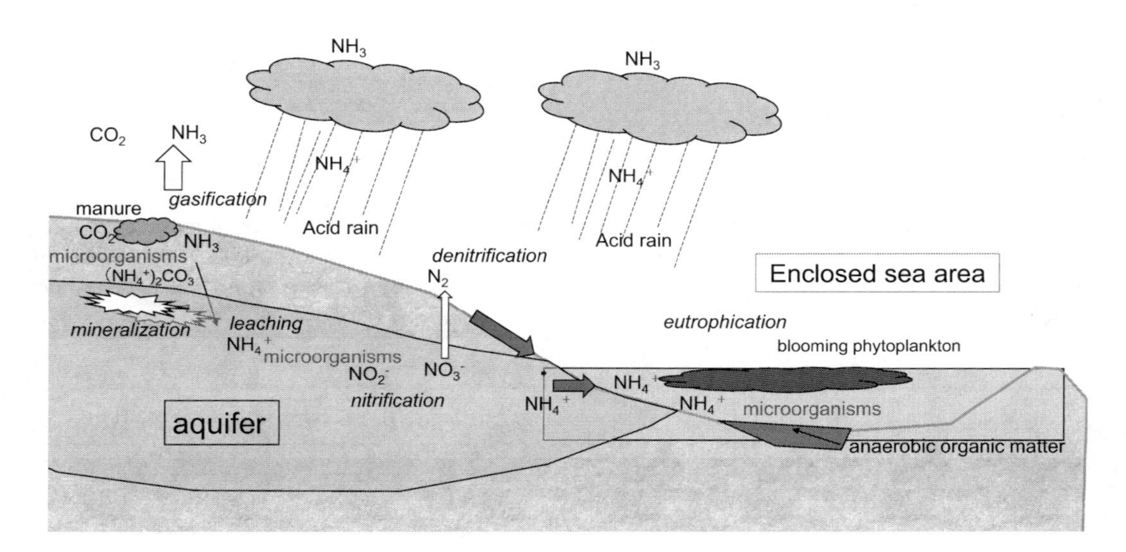

FIGURE 12.3 Non-agricultural examples of the nitrogen cycle.

Since manure consists of proteins, amino acids, urea, nucleic acids, etc., which are not ingested by plants, it is necessary to be decomposed into ammonia nitrogen and nitrate nitrogen. In the figure, the gasification into ammonia means the change from manure to ammonia and its emission into the air, will lead to acid rain. This is described later. Ammonium ions can be adsorbed on soil particles and can also react with carbonate ions to produce mineral ammonium carbonate. Ammonium ions are converted into nitrite or nitrate with oxidation via nitrification by microorganisms. After nitrification, under anaerobic conditions the reduction by denitrifying bacteria can lead to denitrification, which means the release of nitrogen gas, N_2O or N_2, into the air. This is discussed later. The ingestion of nitrogen by plants including phytoplankton is the primary process in the nitrogen cycle. In general, the ingestion is possible when the nitrogen is in the form of NH_4^+ and NO_3^-.

When excess nitrogen flows into groundwater or aquifers, negative effects on human health and the environment may occur. According to the World Health Organization (WHO), the water quality standards are provided for drinking water with nitrite ions and nitrate ions. Furthermore, as the environmental negative effects, nitrogen and phosphorus have caused eutrophication in enclosed sea areas. The eutrophication causes phytoplankton blooms, which may consume oxygen.

12.2.1 Guidelines for Nitrogen Compounds in Drinking Water

The WHO guidelines (WHO, 2017) give the following values:

Nitrate: 50 mg/L as nitrate ion, to prevent methaemoglobinaemia and thyroid effects in the most sensitive subpopulation, bottle-fed infants, and, consequently, other population subgroups.

Nitrite: 3 mg/L as nitrite ion, to prevent methaemoglobinaemia induced by nitrite from both endogenous and exogenous sources in bottle-fed infants, the most sensitive subpopulation, and, consequently, the general population.

Combined nitrate plus nitrite: The sum of the ratios of the concentrations of nitrate and nitrite so its guideline value should not be exceeded.

Occurrence

> Nitrate levels vary significantly, but levels in well water are often higher than those in surface water and, unless heavily influenced by surface water, are less likely to fluctuate. Concentrations often approach or exceed 50 mg/l where there are significant sources of contamination. Nitrite levels are normally lower, less than a few milligrams per litre.

Ammonia nitrogen is converted to nitrite nitrogen or nitrate nitrogen by an action called nitrification. As long as these concentrations are low, they will eventually be released into the atmosphere as N_2 and NO_2 by denitrification. However, in Japan, as a result of excessive use of nitrogen fertilizers and production activities, the concentration of nitrite nitrogen and nitrate nitrogen in groundwater increased, and it was found that they were a health problem depending on the amount contained in drinking water. As a result, the total nitrogen concentration of nitrate nitrogen and nitrite nitrogen was set at 10 mg-N/L as the standard for the quality of Japan drinking water, and 0.04 mg-N/L was set as the standard for nitrite nitrogen. Under these circumstances, it is necessary to consider measures to prevent nitrous acid and nitrate problems from occurring by human activities.

It is clear that nitrogen is one of the nutrients and is indispensable for living organisms. However, regarding the harmfulness of inorganic substances in general, it has been shown that excessive intake can lead to toxicosis, and if it is too low, there is a risk of deficiency. In other words, whether a substance is safe depends on the appropriate intake value, and not necessarily the type of substance. From this point of view, Chapter 8 shows the depth distribution of heavy metals in seafloor sediments, which shows that data of past environments are embedded in sediments, and that there are background values in the environment. Historically, pre-industrial sediments show concentrations at almost constant background values, but newer sediments show higher concentrations due to human activities.

12.2.2 Nitrogen and Nitrogen Compounds

Nitrogen is one of the most important nutrients for all living things, and its supply is made up of the cycle of nitrogen compounds that extend from the local to the global environment. Proteins and amino acids are nitrogen-based compounds, and for their sustainability, it is necessary to have a complete cycle.

In other words, since the circulation system has been constructed, it cannot be said that the species is conserved. Compounds such as ATP, enzymes, etc., in the cells of living organisms are made of proteins.

Urea decomposes even at high temperatures, but at low temperatures it is hydrolyzed by enzymes (urease) to produce ammonia and carbon dioxide. It is well known that urease is contained in many species of urease-producing microorganisms, legumes, such as Jack beans and soybeans, and most fresh plant leaves. Watanabe et al. (1983) examined urease activity and nickel contents in fresh leaves on arbitrarily selected 134 plants and found that most of them showed urease activity.

Catabolism is the breakdown of high-molecular weight organic matter or inorganic matter into simple small molecules, such as water and ammonia, and energy is obtained in the process to synthesize adenosine triphosphate (ATP) (Mobley and Hausinger, 1989). ATP is in the cells of all plants, animals, and microorganisms to provide energy for metabolic processes, such as cell proliferation, muscle contraction, plant photosynthesis, fungal respiration, and yeast fermentation. Thus, all organic matter (living organisms or remains of living organisms), including food, bacteria, mould, and other microorganisms, contain ATP.

In adults, about 30 g of urea are excreted as urine per day. On the other hand, in bony fish, ammonia accumulated in the blood enters the gill cells through proteins of the erythrocyte cell membrane by osmotic pressure and is again excreted through the proteins of the gill cell membrane. CO_2 can freely enter and exit the cell membrane. In water, ammonia becomes ammonia ions (Wright and Wood, 2009). However, when ammonium ions are converted to ammonia due to rising water temperature and pH, they become toxic to fish. Except for birds, many reptiles excrete ammonium in the form of uric acid.

In agriculture, urea, an organic substance, was produced for the first time after World War I, and now is often used as a nitrogen fertilizer. However, in hydrolysis, ammonia is released into the atmosphere as a gas at a large rate, which can be problematic. To prevent this, calcium chloride was used to produce ammonium chloride, to keep it in the soil even under dry conditions (Fenn and Miyamoto, 1981).

On the other hand, the loss of fertilizer is thought to cause eutrophication. In other words, increasing the amount of NH_4^+ contributes to the proliferation of phytoplankton, which is one of the causes of red tides in enclosed waters. The dead bodies of phytoplankton are deposited on the seafloor, whose degradation causes the lack of oxygen in the seawater and sediments. The lack of dissolved oxygen in the seawater yields sulfide, which sometimes leads to the occurrence of blue tides. However, the increase in phytoplankton is not a bad thing in itself, and it is necessary to consider it holistically, including the decrease in zooplankton that ingest it and the lack of control of nutrient inflow into confined waters.

Non-agricultural urea applications include ice melting agents (mainly for aircraft), skin moisturizers, and diesel engines. Methods for using urea as a biocement technology are now being studied around the world. However, only a few studies have been conducted on the by-product of the carbonate formation process (ammoniacal nitrogen).

12.2.3 Ammonia and Ammonium Ion

After hydrolysis of urea, NH_3 gas released from the cell membrane changes to NH_4^+ in water. However, when the ambient temperature or pH are high, some of the NH_4^+ changes to NH_3 in the water. The effect of temperature and pH on the ratio of NH_3 and NH_4^+ was evaluated by Emerson et al. (1975). From the values obtained by Emerson et al. (1975) the following information is attained.

(1) If the temperature is below 20 °C and the pH is below 7, NH_3 is almost non-existent.
(2) If the temperature is 40 °C and the pH is above 10, most of the NH_4^+ changes to NH_3. Under these conditions, ammonia gas can be removed from the ammonium solution.

Note that Emerson et al. (1975) also described that the ionic strength affected their evaluation.

Rain containing ammonia is basic, but after permeating the soil as rainfall, it acidifies the soil by nitrification, and so it is considered a type of acid rain, which acts as a toxin. However, when NH_3 is discharged into water, ammonia changes from NH_3 to NH_4^+ under the above conditions. Total nitrogen concentration can be measured, and the ratio of NH_3 and NH_4^+ using the temperature and pH values can be estimated (Emerson et al., 1975).

12.2.4 Mineralization and Immobilization

NH_4Cl is a colorless crystal and a water-soluble mineral that exists in nature as ammonium chloride. Therefore, a change in crystallization from ammonium chloride solution occurs due to drying.

Immobilization

NH_4^+ and K^+ (potassium) are difficult to hydrate in pore solutions in soil, unlike Ca and Mg ions. The cation exchange ability on the surface of the soil particles is low, and it is not easily desorbed from soil particles.

Materials with high adsorption capacity

When using natural zeolite by Lebedynets et al. (2004), the equilibrium concentration increased sharply to 40 mg/L and the adsorption capacity increased sharply to about 7.0 mg/g. The equilibrium concentration gradually increased to about 500 mg/L to 11 mg/g. At an equilibrium concentration of 500 mg/L or more, the capacity was almost constant (about 11 mg/g). Thus, the adsorption of NH_4^+ depends on the type of adsorbent.

Fixation of NH_4^+ to soil particles

Nieder et al. (2011) used five key words, such as fixed $NH_4^+(NH_4^+{}_f)$, native $NH_4^+{}_f$, recently fixed NH_4^+, N dynamics, and $NH_4^+{}_f$ available to plants and microflora, to explain the sorption and desorption of ammonium to soil particles. According to the authors, NH_4^+ which is adsorbed to the soil as a conventional fertilizer, is non-desorption ($NH_4^+{}_f$). The desorption of NH_4^+ ($NH_4^+{}_f$) fixation and NH_4^+ in soil is a challenge in many studies and includes conflicting results. They point out that the conventional findings vary depending on methodology, soil type, mineralogical composition, and agro-climatic regulations. They reviewed the following:

(a) Methods for the determination of $NH_4^+{}_f$.
(b) Mechanism of NH_4^+ fixation.
(c) Contents of $NH_4^+{}_f$ in soils and influencing factors.
(d) Availability of fixed NH_4^+ to plants and microflora.
(e) Seasonal dynamics of $NH_4^+{}_f$.

Several methods have been proposed for determining $NH_4^+{}_f$ in soil. The principle of most of the methods is to remove organic matter and exchangeable NH_4^+. Then, dissolution/extraction of the residual sample is conducted for release of $NH_4^+{}_f$.

NH_4^+ adsorption properties are similar to metals. In general, the cation exchangeable capacity (CEC) and specific surface area (SSA) are factors. In the review by Nieder et al. (2011), crystal structures of clay minerals are discussed as an influencing factor on the adsorption properties. The contents of $NH_4^+{}_f$ -N in the plough layer of arable soils vary substantially. The data show that the parent material has a major influence on the $NH_4^+{}_f$ -N pool in soils developed from different parent materials, which increases in the order sand (diluvial sand and red sandstone) $<$ basalt $\approx$ granite $<$ loess $\approx$ ground moraine $<$ alluvial sediment $<$ limestone $<$ marsh sediment.

Soil can retain NH_4^+ in the pore water. Nommik and Vahtras (1982) express the balance of exchangeable and fixed ammonium ions, as:

$$NH_4^+{}_s = NH_4^+{}_e + NH_4^+{}_f \quad (12.1)$$

where subscripts, *s* and *e* denote ions in soil solution and exchangeable ions, respectively. Then, Equation (12.1) means total ammonium ions are separated into exchangeable and fixed ions. It is not easy to determine the equilibrium condition in Equation (12.1), because of methodological limitations (Nieder et al., 2011). The seasonal dynamics of $NH_4^+{}_f$ -N is expressed by roughly 200 ± 30 mg/kg from Mengel and Scherer (1981). Therefore, ±15 % of $NH_4^+{}_f$ -N varies according to season and soil depth.

12.2.5 Anammox Nitrification/Denitrification

Nitrification is a phenomenon in which ammonia is generated by the decomposition of nitrogen compounds in the soil is oxidized by microorganisms, such as nitrifying bacteria and changes to nitric acid or nitrous acid. Winogradsky theorizes that nitrification involves two types of aerobic bacteria: Ammonia-oxidizing bacteria that produce NO_2^- and nitrite-oxidizing bacteria that produce NO_3^-. It was revealed in the 1890s. After about more than 100 years, the Anaerobic Ammonium Oxidation (ANAMMOX) process was discovered by Mulder et al. (1995), which is an anaerobic ammonium oxidation process, and a shortcut to the nitrogen cycle. As the anaerobic oxidation process was unexpected, it was a significant discovery for the scientific community. The difference between the conventional nitrification/denitrification and anammox denitrification is illustrated in Figure 12.4. Processes commonly found in wastewater treatment processes (Hosseinpour et al., 2021; Saborimanesh et al., 2019) and in soils, such as ammonification, sorption, volatilization, nitrification, denitrification, ANAMMOX, and nitrate reduction, may all occur in landfills (Shalini and Joseph, 2012).

The Anammox bacterium was later named *Candidatus "Brocadia Anammoxidans"* (Brochier and Philippe, 2002). This bacterium was recognized as of the first anammox species. After that almost the entire length of the genome was decoded, and metabolic pathways were identified (Strous et al., 2006).

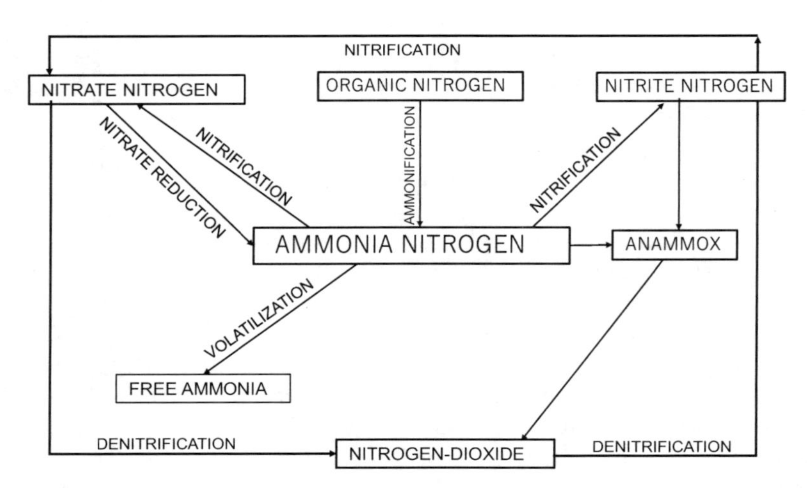

FIGURE 12.4 Potential nitrogen transformation pathways/nitrogen cycle that may occur in bioreactor landfills. (Shalini and Joseph, 2012.)

Anammox is a coccus with a diameter of 1 μm. Anammox bacteria grows at a growing temperature of 6–42 °C and a pH of 6.7–8.3 is optimal. In 100 mM ammonia and nitric acid the anammox reaction is not inhibited, but in the presence of methanol, it is. If the nitrite concentration is 7 mM or more, the anammox reaction is inhibited (Strous et al., 1998). Therefore, the sewage treatment plants are changing from the conventional nitrification-denitrification to the Anammox. As a result, the cost is reduced and the operating efficiency is dramatically improved (Gong et al., 2021: Ji et al., 2022; Johnston et al., 2023; Shalini and Joseph, 2012; Weralupitiya et al., 2021; Xing et al., 2016; Yuan et al., 2022). These studies involve relatively new technological developments. The main issues are related to sludge or leachate, anammox performance, reactor equipment, and operation.

Strous et al. (1998) proposed the following anammox reaction in a reactor,

$$NH_4^+ + 1.32\ NO_2^- + 0.066\ HCO_3^- + 0.13\ H^+ \rightarrow 1.02\ N_2 + 0.26\ NO_3^- + 2.03\ H_2O + 0.066\ CH_2O_{0.5}N_{0.15} \quad (12.2)$$

On the other hand, to explain the anammox reaction, the following reaction formula are used, for the partial nitrification,

$$2NH_4^+ + 1.5O_2 \rightarrow NH_4^+ + NO_2^- + H_2O + 2H^+ \quad (12.3)$$

and for the anammox reaction,

$$NH_4^+ + NO_2^- \rightarrow N_2 + 2H_2O \quad (12.4)$$

Adding Equations (12.3) and (12.4),

$$2NH_4^+ + 1.5O_2 \rightarrow N_2 + 3H_2O + 2H^+ \quad (12.5)$$

After the study by Brochier and Philippe (2002), the basic sequence of *Candidatus "Brocadia Anammoxidans"* was determined to detect Anammox bacteria. Since then, Anammox bacteria in natural ecosystems have also been found in various places. Dalsgaard et al. (2003) confirmed that the Anammox reaction occurred in the deep sea of Golfo Duluce, Costa Rica. In addition, they found that 50% of the denitrification process in nitrogen-contaminated bays was due to the Anammox reaction.

12.2.6 Uptake of NH_4^+ and NO_3^- by Plants

The three major nutrients for living organisms are proteins, lipids, and carbohydrates. Among these, the nitrogen source is organic nitrogen (for example, urea), inorganic ammonia nitrogen, inorganic nitrate nitrogen, etc., originally decomposed by proteins. On the other hand, nitrogen gas N_2, which accounts for 4/5 of the atmosphere, is said to be available to very few plants.

In terms of the flow of the nitrogen cycle, if animal excrement is urea or NH_3, it can be initially thought of as ammoniacal nitrogen. Under aqueous conditions, NH_3 becomes NH_4^+ and is absorbed from plant roots. As mentioned in the previous section, nitrifying microorganisms take up NH_4^+ to produce NO_2^- or NO_3^-. This nitrification process plays a role in reducing the toxicity of ammonia and makes it possible to incorporate nitrate nitrogen (NO_3^-) by a pathway different from ammoniacal nitrogen. Ammonia nitrogen and nitrate nitrogen are important factors for plant crop production (biomass production). On the other hand, the loss of fertilizer is thought to cause eutrophication.

12.3 CARBON CYCLE

12.3.1 Carbon Neutral

Carbon is an important element for all living organisms as a forming element of proteins and carbohydrates. Furthermore, carbon plays a role as a major element of carbonate minerals and fossil fuels derived from living organisms.

The limestone that forms part of the earth's surface are carbonate minerals, which are of biological origin. On the other hand, fossil fuels such as coal, oil, natural gas, and carbon hydrate are also derived from living organisms. Therefore, it can be seen that much of the carbon from living organisms is accumulated in carbonate minerals and fossil fuels. By heating limestone to remove CO_2, quicklime CaO can be produced, which is the main raw material for cement. CO_2 must be released to obtain energy from fossil fuels.

It would be ideal if all CO_2 emitted was incorporated into plants (organic matter) as part of the global carbon cycle but given the current demand for concrete and energy and the rate of decline of plants (forests) that consume CO_2, the current situation is unbalanced since the Industrial Revolution. For this reason, carbon neutrality has become the slogan of environmental issues around the world.

Currently, the most feared global environmental problem is global warming, which is caused by excess carbon dioxide released into the atmosphere. The negative effects are manifested in abnormal weather and associated disasters caused by rising air and seawater temperatures. Therefore, it is necessary to avoid carbon dioxide emissions as much as possible, and also to reduce carbon emissions to zero through individual efforts, such as by using or fixing the released carbon dioxide. By increasing biomass, this promotes of carbon neutrality. The formation of calcareous substances produced by biomass is considered to cause a carbon neutral effect that nature contributes in a major way. These range from gigantic coral reefs and limestone mountain ranges to micro-biomineralization by the action of microorganisms (Castanier et al., 1999). Thus, to consider the global environmental problem, it is important to understand the cycle of carbon. On the other hand, it is of importance to know the adverse effects of biomass reduction and acid rain. It has been reported that an increase in atmospheric CO_2 concentration reduces the pH in seawater (Pelejero et al., 2010). This will affect the carbon cycle.

12.3.2 Carbon Cycle

Carbon is found in all organic matter, such as most foods, and living and dead organisms. In this sense, fossil fuel is considered to be also organic. Petroleum consists of C and H, and coal is composed of C, H, and O, similar to carbohydrates.

The carbon cycle diagrams show different steps as presented in websites, respectively. As the basic steps are similar, in this book, the main steps are introduced. The largest amount of carbon transfer would be absorption from the atmosphere by plants. Plant carbon absorption is the most desirable method for absorbing and fixing excess CO_2. However, due to the vicious cycle caused by global warming, deforestation due to wildfires and weather disasters is increasing in frequency and size.

Carbon transfer from plants to animals is a part of the food chain. Examples include insects eating grass and a person eating vegetable salads. Birds eat nuts and carry seeds far away. The work of collecting nectar by bees is necessary for pollination and is necessary for fruit and vegetable production.

Oceans absorb CO_2 from the atmosphere. The transfer of CO_2 into the oceans needs the waves of seawater due to wind, rains, waves from boats, and the movement of waterfowl and whales. The use of fossil fuels releases CO_2 directly into the atmosphere. Fossil fuels support most of the world's energy, so it's hard to curb it. Future development of renewable energy is desired. When living organisms breathe, carbon dioxide is emitted. In addition, when the so-called excrement (organic matter) of living organisms is decomposed, CO_2 and ammonia are released. The transfer of carbon from the land to the ocean can be made in the form of organic matter via water flows. The organic matter becomes food for organisms, and some are decomposed into ammonia and carbon dioxide by microorganisms.

There is a process of transfer from ocean water to seafloor sediments via living organisms. Shellfish, corals, and foraminifera use calcium carbonate shells to protect themselves from foreign enemies. At the time of the generational change, the shell becomes seabed sediments. The sediments are consolidated by their self-weight, and become limestone (Wetzel, 1989). Thus, carbon has been fixed at the sea bottom.

Non-organic carbon matter is composed only of carbon, such as diamond and black coal, as well as carbon monoxide, carbon dioxide, and limestone (carbonate minerals such as calcite, aragonite, and dolomite, etc.). The carbonates can be formed by inorganic reactions or organic reactions. On the earth, most carbonates formed in nature are of biological origin.

Carbonate minerals are produced by the largest number of organisms as biominerals. They are salts in which carbonate ions and metals are ionically bonded and are amorphous or crystalline. The crystals can be single crystals and polycrystals. In addition, depending on the conditions, hydrated crystals and amorphous minerals are generated. On the other hand, the production of carbonates by inorganic reactions is not very significant.

12.3.3 Types of Biominerals

There are two types of carbonate biomineralization, i.e., biologically controlled carbonate precipitation (BCCP) and biologically induced carbonate precipitation (BICP). In BCCP, minerals such as internal organs and shells for the protection of living organisms are produced. On the other hand, in MICP, it refers to a process generated by an induced reaction caused by environmental conditions rather than intentionally by the organism. This type of precipitation concerns microorganisms. For example, in the case of urease-producing bacteria, which will be discussed in Chapter 13, it is thought that the purpose is to hydrolyze urea to release ammonia and make the surrounding area alkaline and improve the habitat. Thus, for urease-producing bacteria, it is not the original purpose to produce carbonates. In this case, the process is called microbially induced carbonate (or calcite) precipitation (MICP).

Some carbonate crystals have the same chemical composition, which is called isomorphism but show different crystal structures and shapes. Specifically, calcium carbonate, which can be expressed by the chemical formula of $CaCO_3$, includes three minerals, such as calcite, aragonite, and vaterite. Since their crystals are different, the density and hardness are also different. The process of this carbonate precipitation is recognized as carbon fixation (Guida et al., 2017).

Mg ions have a specific role in the formation of aragonite. Since the ionic diameter of Mg is smaller than that of the Ca ion diameter, the Mg ion is sandwiched between CO_3 ions as the portion narrows. Therefore, the crystal growth in the crystal-axis direction stops, and the crystal structure is different from calcite (Folk, 1974). Furthermore, Mg ions can be incorporated into calcite crystals, and the crystal structure changes to become aragonite (Kawano et al., 2015). Amorphous carbonate is a structure in which calcium carbonate microcrystals are irregularly gathered. This is called amorphous calcium carbonate (ACC).When ACC contains water, they are called amorphous hydrated carbonates.

12.3.4 Fate of Carbon in BCCP

12.3.4.1 Foraminifers and Coccolithophores

Foraminifera are a group of protists with predominantly calcareous shells and reticulated pseudopods. The size of foraminifera is usually smaller than 1 mm, while the largest ones may be

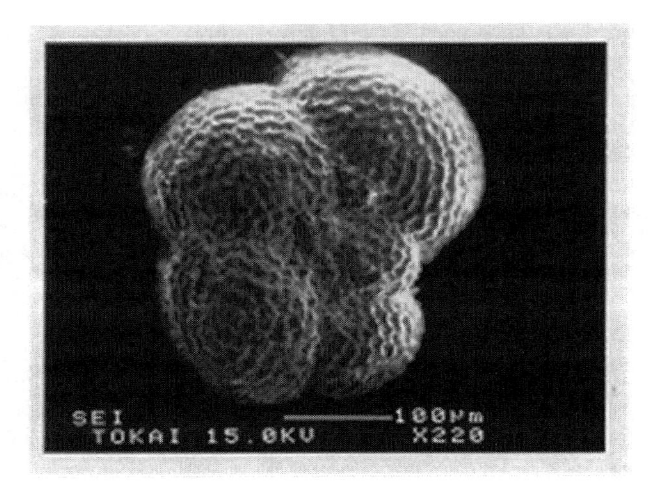

FIGURE 12.5 An example of a foraminifer, protists with predominantly calcareous shells, and reticulated pseudopods.

20 cm (Figure 12.5). The classification of foraminifera is carried out by the shape and composition of the shell. In general, many have a calcium carbonate shell. The shapes are diverse, and it is said that there are more than 200,000 species including fossil species.

Coccolithophores are spherical cells about 5 to 100 micrometres in diameter, surrounded by calcareous plates called coccolith, which are about 2 to 25 micrometres in diameter. Coccolithophores are an important group of about 200 marine phytoplankton species covered with calcium carbonate shells. They are of ecological and biogeochemical importance, but the reasons for their calcification remain elusive.

The classification of foraminifera is carried out by the shape and composition of the shell. In general, many have a calcium carbonate shell. The shapes are diverse, and it is said that there are more than 200,000 species including fossil species. Coccolith is a structure of calcium carbonate that covers the cell surface of coccolithophores (Figure 12.6). The most common shapes are circular and elliptical.

The formation of calcareous shells of foraminifera and coccolith is by the biologically controlled carbonate precipitation (BCCP), as well as MICP. Research on MICP has increased rapidly since the beginning of 2000, and the number of microorganisms known to be involved has increased.

12.3.4.2 Accumulation Rate of Soil Particles and Carbonates on the Sea Bottom

The larger the granular soil particles flowing into the sea from rivers, the faster they settle and accumulate around the estuary. The finer the particles, the farther they are carried from the estuary. Fine particles such as silt and clay are well dispersed in river water, but in seawater, sodium and calcium ions are adsorbed by the surface charge of soil particles, causing a reduction in repulsive forces between particles. As a result, the deposition of aggregates is accelerated. For example, the accumulation rate of solids varies from more than 10 cm/y at large river mouths (Fisk and McClelland, 1959) to 0.25 cm/1000 y on deep ocean bottoms (Arrhenius, 1952) or to a non-deposition location such as a channel. The accurate observation or measurement of the accumulation rate of marine sediments cannot, however, be easily achieved. Nevertheless, extensive effort has been made to study deep sea sediments. According to Broecker (1974), deep sea sediments have accumulated at remarkably similar rates over the last few million years (indicated by the potassium-argon calibrated magnetic reversal chronology), as compared with the past few hundred thousand years (indicated by the ^{230}Th and ^{231}Pa dating methods), and as compared with the past few tens of thousands of years (indicated

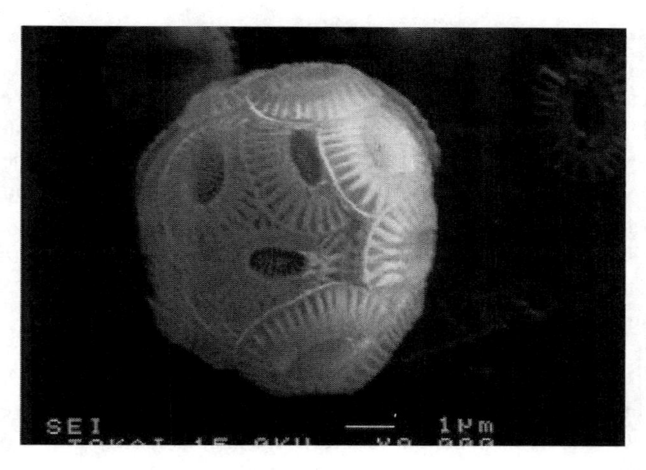

FIGURE 12.6 An example of a coccolith structure of calcium carbonate that covers the cell surface of coccolithophores.

by ^{14}C dating). This indicates that the accumulation rate of deep-sea sediments is almost the same for at least 1,000,000 years. Since some deep-sea sediments mainly consist of calcareous materials, it can be roughly determined that the accumulation rate of calcareous materials is also constant for the same period. Other than the origin and precipitation rate of carbonate, the content of calcium carbonate varies from almost 0% in river mouth sediments to more than 80% in deep sea sediments. This is mainly because of the burying effect of the noncalcareous materials, as stated above.

Fukue et al. (1996) investigated the carbonate contents of sea sediments on the coasts around Japan. The results showed that calcareous materials content in marine sediments increases with the distance from the coast. This trend was the same in bays, inland seas, and the Pacific coast. It is well known that the calcareous materials in marine deposits is predominantly composed of the skeletal structure of plants and animals, such as coccoliths and foraminifers. These calcareous structures contain variable amounts of other substances, such as magnesium carbonate. However, the percentage of materials other than calcium carbonate is generally small and for this reason it is commonly neglected (Sverdrup et al., 1972).

The precipitation rate of calcium carbonate has been estimated by geochemists. Broecker (1974) obtained values ranging between 0.8 and 1.2 g/1000 y/m^2 for precipitation rates of calcium carbonate at the ocean bottom, based on the age determination as mentioned above. This range is often referred to as an average rate of calcium carbonate deposition.

Fukue et al. (1996) concluded that one of the factors governing the sedimentary environment is the accumulation rate of solid particles, which can be estimated approximately from the following relation, $m_{tv} = (100/C)\, m_{cv}$ where m_{tv} is the accumulation rate of calcareous and noncalcareous materials, C is the calcium carbonate content and m_{cv} is the precipitation rate of calcium carbonate alone. Then, the average sedimentation rate of carbonate based on rate measurements on sediments in various seas was estimated as 0.95 g/1000 y/cm^2. This value agreed with Broeker's results. Considering the total area of the sea, the carbonate fixed on the sea bottom is estimated as 34.2×10^8 t/y. The conversion from $CaCO_3$ to CO_2 indicates 15.0×10^8 t/y. This value is comparable to the fourth largest amount of carbon dioxide emissions in 2018.

12.3.4.3 Carbonate Diagenesis in Marine Sediments

Unfortunately, geotechnical engineers have not paid attention to the carbonate content in soil. It may be because it has been thought that carbonates are not binding materials. In fact, carbonate fractions themselves act as soil particles. The dead bodies of protozoa such as foraminifera which were formed by MCCP are not able to be bind materials in soils unless no carbonate diagenesis occurred, i.e., a post-depositional chemical, physical, and/or mechanical process.

Soil constituents usually consist of various types of materials, such as minerals, organic matter, amorphous materials, sulphates, carbonates, etc. These materials have interactions with each together and also with other substances, such as ions and compounds, etc. The type and content of the constituents depend on both the origin and history of the soil. It is no doubt that the cementation of soils results from the varied nature of interactions and chemical bonds between constituents, and with other substances. These interactions and chemical bonds will determine the mechanical strength characteristics of the soils. Therefore, many studies have been performed to understand the soil interactions and cementation mechanisms in soils.

Scandinavian quick clays which show a very high sensitivity were believed to be formed by the leaching of salts (Rosenqvist, 1953; Mourn et al., 1971), while it was thought that amorphous materials and organic matter played an important role in the cementation of sensitive clays (Pusch and Arnold, 1969). On the other hand, carbonates were also considered to be one of the cementing agents in soils (Sergeyev et al., 1973). This is logical given the fact that limestone consists of calcium carbonate. In fact, at the ocean bottom, carbonate-rich sediments (ooze) petrify and form limestone via chalk by self-weight consolidation (Wetzel, 1989). However, the marine limestone formation process from coral sand-gravel is unlikely to be due to the physical consolidation alone. Sergeyev et al. (1973) described how a ring reinforced skeleton of carbonate formed in loess soils and showed that a similar ring skeleton of hydroxides of iron in loams caused an increase in strength of about 40%.

The evidence that the carbonates originated to organisms are necessarily contained in marine sediments, and are reflected for the profile of cementation process was demonstrated by Fukue et al. (1999). The sediment cores were taken from various sea areas including estuary, bays, inland sea and ocean bottoms, and the effects of carbonate contents are investigated on the cementation strength in terms of sediment depth. As a result, it was shown that the effects of carbonate on the strength were much stronger than the effects of consolidation process. One example of the results is shown in Figure 12.7, which indicates the profiles of carbonate content and vane shear strength of surface sediments in Tokyo Bay (Fukue et al., 1999).

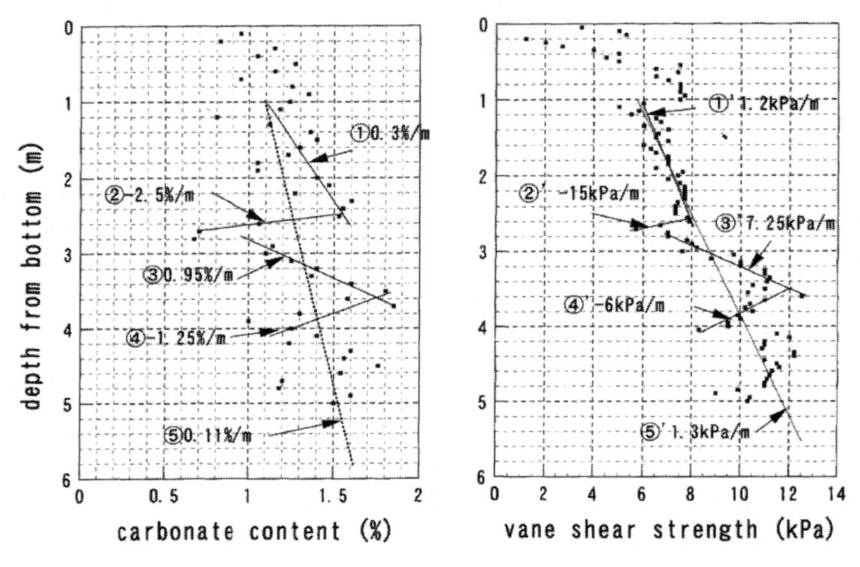

FIGURE 12.7 The profiles of carbonate content and vane shear strength of surface silty sediments in Tokyo Bay.

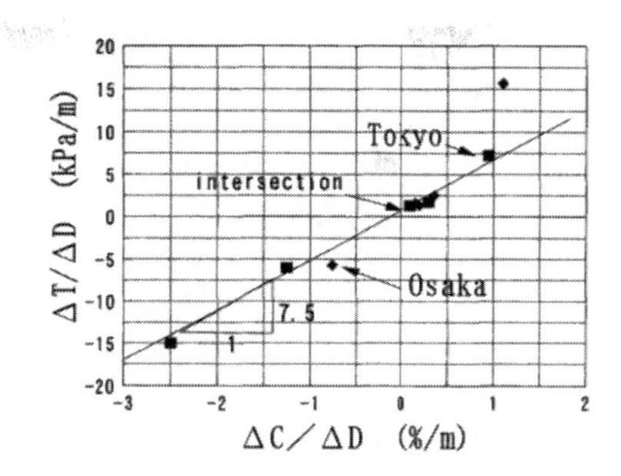

FIGURE 12.8 Vane shear strength vs carbonate content for the sediments of Tokyo and Osaka Bays. (Fukue et al., 1999.)

If the data on the carbonate profile was not obtained and only the vane shear strength profile is shown, the consolidation process can be demonstrated, i.e., increase in density. However, the data on carbonate content may contradict the explanation by the consolidation process. Note that the situation is the same for the case of Osaka Bay sediments (Fukue et al., 1999). The profile of carbonate content indicates the influence of marine regression in the Alluvial period. Therefore, the marine regression increased the deposition materials from the land in the bay and coastal areas, where the carbonate (e.g., foraminifera and coccolith) content will be decreased. This process appeared as the decrease in carbonate content from deep to shallow, which can be seen as the total trend of content profile.

Approximate lines for the plots of Figure 12.8 may be considered as the trend in a short time, because it was found that the carbonate content and strength showed similar slopes (Fukue et al., 1999). Since both the slopes of carbonate content and strength relationships can be compared at the same soil depth, the carbonate content and strength relationships can be obtained using both slopes, as indicated in Figure 12.8.

It is remarkable that the vane shear strength profile $\Delta T/\Delta D$ is a linear function of carbonate content profile $\Delta T/\Delta D$. Therefore, the slope of the relationship can be expressed by the ratio $\Delta T/\Delta C$ that is $(\Delta T/\Delta D)/(\Delta C/\Delta D)$, which is the increasing rate of vane shear strength against the increasing rate of carbonate content. The slope of the relationship was 7.5 kPa/% with an intersection of approximately 1 kPa/m. This means that if there is no carbonate effect, the increasing rate of vane shear strength per 1 m of depth was about 1 kPa/m, which is the intersection given by $\Delta C/\Delta D = 0$. However, another interpretation can be possible if there is no development of skeletons due to carbonates, then the consolidation of sediments could be promoted.

The investigation showed that carbonates originating from organisms are contained in sea sediments. It was shown that the carbonates played an important role in the cementation process in sediments. Since the surface sediments are relatively loose and soft, the example here is situated at an initial stage of carbonate diagenesis under a marine condition. Therefore, it is interesting that the relationship is almost the same for Tokyo and Osaka Bay surface sediments. Furthermore, continuous carbonate diagenesis was obtained for the deeper sediments near the Haneda International Airport, as shown in Figures 12.9 and 12.10.

Compared to a vane shear strength of less than 12 kPa for the Tokyo Bay surface sediments, the unconfined compressive strength of sediments deeper than 42.5 m from the bottom varied from approximately 100 to 180 kPa. The $\Delta qu/\Delta C$ ratio was approximately 64 kPa/%. It is interesting that this value accords with the $\Delta qu/\Delta C$ ratio (= a value) obtained on loose sand using the MICP test (Chapter 14).

Since the unconfined compressive strength of soil is theoretically converted to double that of the undrained vane shear strength, from Figures 12.8 and 12.10, it is considered that the $\Delta qu/\Delta C$ ratio

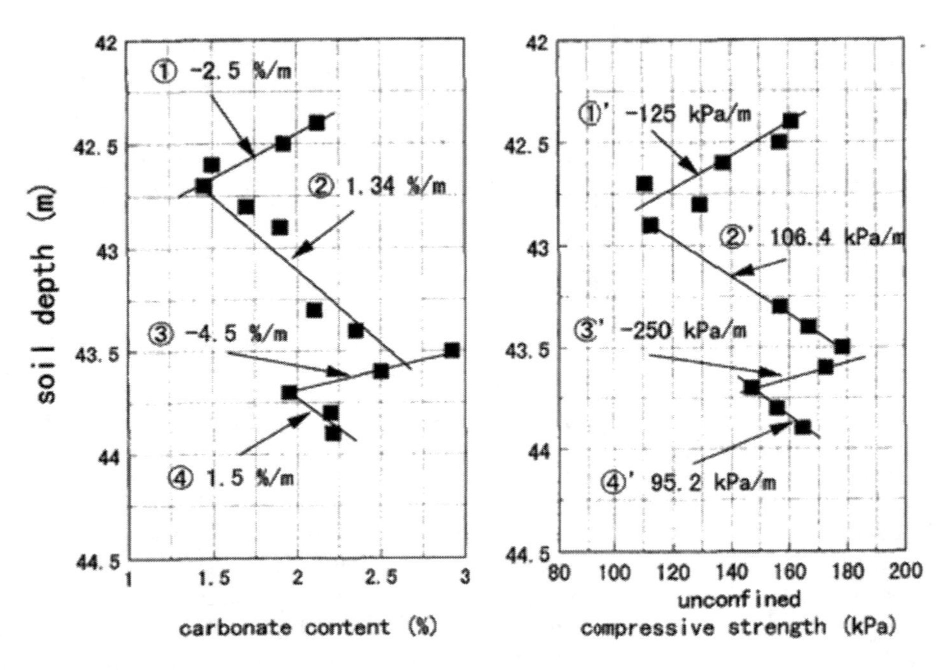

FIGURE 12.9 Unconfined compressive strength and carbonate content for silty sediments samples obtained from a soil depth of 42.5 to 44 m, near Haneda International Airport. (Fukue et al., 1999.)

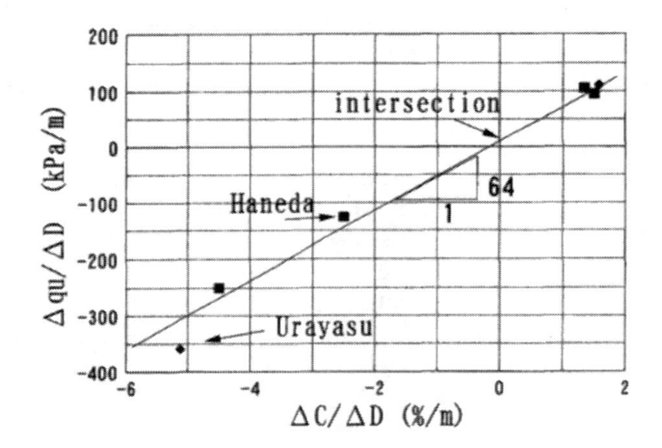

FIGURE 12.10 $\Delta qu/\Delta C$ ratio obtained based on Figure 12.9. (Fukue et al., 1999.)

has increased 4.2 times. As the initial conditions of both surface and deeper sediments are similar because of the same sea area, the different depths indicate an elapsed time. Therefore, physically the deeper sediments have been more consolidated with a heavy self-weight and a longer time. In addition, it shows that the cementitious effect of carbonate has a strong effect (Chapter 14), Then, the chemical mechanism for carbonate diagenesis becomes important.

12.3.4.4 Coral Reef and Limestone Formation

It is known that limestones are formed from coral gravels, shells, and coral reefs (Wang et al., 2020), as well as the foraminifera and coccolith at the ocean bottom (Wetzel, 1989). Calcium carbonate that is primarily composed of limestone can be transformed by the changes in pH, redox potential, and CO_2 partial pressure (Garrels and Christ, 1965; Friedman, 1993; Ahm et al., 2018; Bhanwar et al., 2023). In addition, it is possible that microorganisms may be active (Castanier et al., 1999; De Boever et al., 2017). Therefore, the change in the local environment on the surfaces of limestones can possibly lead to the dissolving and precipitating of calcium carbonates. This slow process possibly relieves the damage due to the self-consolidation of the cementation for the skeleton of the sediment matrix.

De Boever et al. (2017) presented an invited review paper about early diagenesis of non-marine carbonates. They concluded as follows: Diagenesis starts immediately after the formation of the first crystals. The onset of diagenesis often takes places in niches or 'micro-scale environments', in relation to metastable phases, like ACC, vaterite, Ca-xalates, hydrous Mg-carbonates and aragonite. That is, in MICP, the first products within a few days, are diverse and are ACC, vaterite, hydrated carbonate, aragonite, Mg-calcite, etc. (see Chapter 13). However, it may be important that the products were under the existence of Mg^{2+}. Their other conclusion was about the activities of micro-organisms and organic matter. Another message reads as follows: Early diagenesis seems to be mostly a constructive (and not destructive) process. It often results in a coarsening of the fabrics and even seems to homogenize initially very different microfabrics. Thus, their message seems to indicate that the carbon will continue to be fixed as carbonate in the future.

Coral bleaching means the death of corals at a high rate. The bleaching phenomenon is said to have a strong effect of global warming (Fukunaga et al., 2022, Glynn, 1991, Majumdar et al., 2018). The bleaching of the Shikisei Lagoon in Okinawa is expanding, and NHK accompanied the investigation by the Japanese Ministry of the Environment on the first day of the survey. Currently appealing for the coral reef crisis on the Web (NHK).

Abe et al. (2021) developed a high-resolution coral bleaching simulation under a warming climate. As a result, they were able to simulate current temperatures and coral bleaching and mortality at +1.5 °C and +2.0 °C. They described specific examples of how the simulation can contribute to future planning. In addition to bleaching, other reasons for coral loss include predation by crown-of-thorns starfish, soil pollution, and deterioration of water quality.

Nine researchers from five organizations made a literature review to suggest how to better conserve Bali's coral reef ecosystem (Boakes et al., 2022). The literature review provided an in-depth analysis of the tools used to conserve Bali's coral reefs and compared them to those used in other countries. The suggestions are as follows:

(1) increasing its designation of official Marine Protected Areas (MPAS) and strengthening management of existing ones,
(2) creating an MPA network,
(3) substantially reducing marine plastic pollution,
(4) continuing artificial reef construction in degraded habitats,
(5) continuing to develop Bali as an ecotourism destination,
(6) increasing engagement in global science to inform marine conservation decision making, and
(7) developing more marine monitoring programs.

12.3.4.5 Oceanic Crust Decarbonation

There is a well-known theory called "plate tectonics" that indicates that the surface of the earth is covered with bedrock several tens of km thick called plates (lithosphere). Accordingly, more than 10 pieces of the plates come together to envelop the entire surface of the Earth. Moreover, each plate is moving in a different direction at a rate of several cm per year. These plates are powered by convection in the solid mantle of the Earth's interior. Earthquakes are caused by the destructive movement caused by the enormous force generated by the movement of the plates.

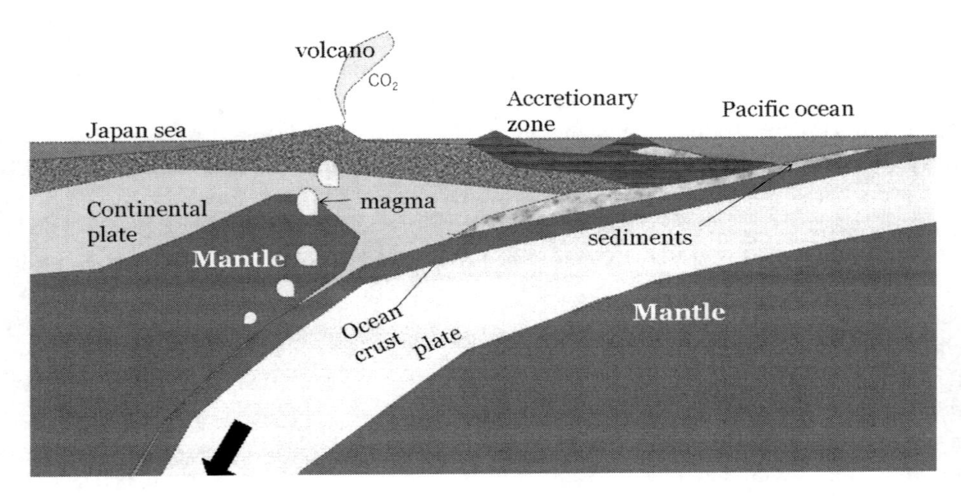

FIGURE 12.11 A general subduction style and morphology of arc basins. (Modified from Gordon, 2010; Dickinson and Seely, 1979.)

The upper mantle layer constitutes the lower layer of the plate, which is about 2900 km thick. Mantle convection is produced by the heat of the Earth's centre, which slowly rises, cools near the surface, and descends. The rate is in several cm per year. Some of the components of the elevated mantle melt into magma, which erupts and solidifies from the submarine mountains called ridges or ocean rises. It is the formation of oceanic plates.

Mantle convection causes the formation of ocean ridges and trenches. The ridge is the place of birth for a plate, and the trench becomes the inlet of the subduction plate into the Earth's interior (Figure 12.11). The plate moves slowly from the ridge to the trench, while increasing the volumes of oceanic sediments.

Subducted plates involving rich or poor-carbonate sediments transport CO_2 from the lithosphere, hydrosphere, and atmosphere to the Earth, where it can be released to the surface through processes such as arc volcanism or stored at depths over geological time (Arzilli et al., 2023).

12.3.5 Fate of Carbon in MICP

Microbially induced carbonate precipitation (MICP) is distinguished from BICP as described earlier. Zhu and Dittrich (2016) reviewed six carbonate precipitation processes in nature. They reviewed MICP in terms of six types of metabolism, such as (a) photosynthesis, (b) ureolysis, (c) denitrification, (d) ammonification, (e) sulphate reduction, and (f) methane oxidation.

The ureolysis concerns the decomposition of animal excrement, so that the product of CO_2 can be significant. Photosynthesis leads to the formation of beachrocks. Therefore, these two mechanisms cannot be neglected in describing the carbon cycle.

12.3.5.1 Urease-Producing Bacteria

As far as urea excretion is concerned, urease-producing bacteria are the most important decomposers. Because urea can be quickly lost from the system via ammonia volatilization or nitrate leaching following hydrolysis of urea by urease producing soil microorganisms, Alizadeh et al. (2017) investigated lowering the urease activity of the urease-producing microorganisms in New Zealand. For the investigation, 31 species in fungi, actinomycetes and yeast species, and 30 bacteria were subjected to the investigation. The hydrolysis of urea in the existence of Ca^{2+} is expressed as:

$$(NH_2)_2\,CO + (e) + Ca^{2+} + 2H_2O \rightarrow 2NH_4^{\;+} + CaCO_3 \downarrow \tag{12.6}$$

where (e) denotes the enzyme (urease). Note that the urease concentration (e) affects inhibition and retardation of reaction (12.6) (Fukue et al., 2023). The carbon in ureolysis is immobilized in the $CaCO_3$ precipitates in the reaction (12.6). Thus, microbially induced carbonate precipitation (MICP) occurs, as a natural fixation phenomenon of carbon in nature. Note that the reaction (12.6) has been used for the artificial MICP (see section 11.4.9 and Chapter 13).

Reaction (12.6) shows urea is decomposed into $2NH_4$ and CO_2 in the water. In this reaction, if no water remains in some soil pores, $2NH_3$ and CO_2 are released into the air. The CO_2 can be intaken in plants, but ammonia cannot be used in living organisms because of its toxicity. This has been an agricultural problem when urea is used. The existence of water changes NH_3 into NH_4^+. Then, plants can intake NH_4^+, as described in the earlier section. The nitrogen cycle starts from NH_4^+ after leaching into rivers and seas (as described earlier).

12.3.5.2 Photosynthesis by Cyanobacteria

It was explained that beachrocks were coastal sedimentary formations resulting from a relative rapid cementation of beach sediments by the precipitation of carbonate cements (Arrieta et al., 2011). At that time, there was no information on the formation mechanism of beachrocks. McCutcheon et al. (2016) suggested that based on the other researcher's studies, extracellular polymeric substances (EPS) of *cyanobacteria* played a major role in nucleation in the formation of beachrocks. They concluded that *cyanobacteria* dissolved $CaCO_3$ to supply cations. On the other hand, the metabolism of *cyanobacteria* in plants, which causes the uptake of CO_2 and release of O_2 (Castanier et al., 1999), is

$$2HCO_3^- + Ca^{2+} \rightarrow CH_2O + CaCO_3 + O_2 \quad (12.7)$$

where CH_2O is formalin (Neumeier, 1999, Pomar and Hallock, 2008).

As the metabolism of *cyanobacteria* is oxygenogenic photosynthesis, visible light is required. That is why there are beachrocks only near the seacoast and in shallow water near the coast (Johnston et al., 2023; Neumeier, 1999; Vousdoukas et al., 2007).

12.3.6 Carbon Cycle Via Methane

In the carbon cycle a small percentage of the biomass generated via photosynthesis each year is remineralized via methane. Methane is generated in anaerobic environments mainly from cellulose by a syntrophic association of anaerobic bacteria, protozoa and fungi, and methanogenic archaea. Methane is further oxidized to CO_2 either by anaerobic microorganisms with sulphate, Fe^{3+}, Mn^{4+}, and nitrite or by aerobic microorganisms with O_2 as electron acceptors. Of the approximately 1 billion tons (1 Gt) of methane generated by methanogenic archaea each year, almost half escapes into the atmosphere where most of it is photochemically oxidized to CO_2 (Thauer, 2011).

In the first two decades after methane is emitted, it is approximately 80 times more powerful than CO_2 as a GHG, but it is removed from the atmosphere much more quickly – after about a decade, whereas CO_2 remains in the atmosphere for centuries. Methane is also a precursor to tropospheric ozone (O_3), and thus contributes to air pollution worldwide (Mar et al., 2022).

Recently the thawing of permafrost has raised serious concerns due to the potential release of CH_4 under the permafrost. There are serious concerns associated with greenhouse gases (GHG) fluxes in high latitude ecosystems and how the permafrost thawing may potentially affect the global climate, through the alteration of carbon dioxide (CO_2) and methane (CH_4) emissions (Masyagina and Menyailo, 2020).

Ueyama et al. (2023) evaluated season CH_4 and CO_2 emissions from a forest floor of a lowland black spruce forest on permafrost in the interior of Alaska using automated-closed chambers, anaerobic incubation of peat soils and next-generation sequencing. As a result, they concluded that the future trajectory of CH_4 emissions could substantially change with moisture conditions, deep soil temperatures, and sedge compositions.

Xie et al. (2023) warned that the anthropogenic carbon dioxide emissions have been pushing the accelerated thawing of permafrost, which promotes methane emissions that further intensifies permafrost thawing and global warming. In addition, they concluded that such a positive-feedback at an inflection point of Earth climate may break the glacial-interglacial cycle and initiate a long-lasting greenhouse period.

The risk of CH_4 emissions from Alaska and Siberia is strongly dependent on future climate change. If CH_4 emissions from these areas becomes apparent, it cannot be turned back. When CH_4 is reduced by sulphate-reducing bacteria under anaerobic conditions, HCO_3^- is produced (Castanier et al., 1999). In addition, in the case of oxidation by combustion, CO_2 is generated, as described above.

12.4 CONCLUDING REMARKS

Nitrogen and carbon cycles are extremely important factors for evaluating the current global environmental problem. Comparing the global environment before and after the Industrial Revolution, the situation is clearly very different. As far as recent several years are concerned, weather conditions and associated disasters have intensified in type and magnitude, and the frequency seems to be accelerating.

In this chapter, it was shown that nitrogen and carbon are cycling for living organisms. Accordingly, it was explained that the excess CO_2 released is attributed to the burning of fossil fuel. Because ecosystems maintain a balance as a whole while constantly changing their components through material cycles in the air, water, soil, etc., and food-chains between organisms. Therefore, even if a specific substance is newly added to the ecosystem from the outside, the added amount is no longer necessary.

On the other hand, carbon is biologically fixed and immobilized as biologically produced carbonate precipitates. This may be a naturally prepared countermeasure against the CO_2 released due to accidental wildfires and volcanic eruptions. In other words, the biological carbonate precipitation may be for living organisms to balance the carbon cycle. Otherwise, the biological carbonate precipitation process is unsustainable. In any case, sustainability is considered to have existed in nature, in which biodiversity is essential.

A major event in Earth's history removed nitrogen and carbon in past geological periods, and buried them deep underground, resulting in fossil fuels. This means that historically there has been a gap in the C/N cycle. Although it is necessary to fill the gap, forests continue to be developed, and many lives, property, and forests are being burned by wildfires.

REFERENCES

Abe, H., Kumagai, N. H., Yamano, H., and Kuramoto, Y. (2021), Coupling high-resolution coral bleaching modeling with management practices to identify areas for conservation in a warming climate: Keramashoto National Park (Okinawa Prefecture, Japan), *Science of the Total Environment*, 790: 148094, https://doi.org/10.1016/j.scitotenv.2021.148094.

Ahm, A.C., Bjerrum, C.J., Bättler, C.L., Swart, P.K., and Higgins, J.A. (2018), Quantifying early marine diagenesis in shallow-water carbonate sediments, *Geochimica et Cosmochimica Acta*, 236: 140–159.

Alizadeh, H., Kandula, D.R.W., Hampton, J., Stewart, A., Leung, D.W.M., Edwards, Y., and Smith, C. (2017), Urease producing microorganisms under dairy pasture management in soils across New Zealand, *Geoderma Regional*, 11: 78–85, http://dx.doi.org/10.1016/j.geodrs.2017.10.003.

Arrhenius, G. (1952), *Sediment Cores from the East Pacific: Part I. – Properties of the Sediment and Their Distribution: Report of the Swedish Deep-Sea Expedition*, 1947-1948, vol. 1, No. 1, Goteburg, Sweden, Elanders Boktryckeri Aktiebolag, 91 pp.

Arrieta, N., Goienaga, N., Martínez-Arkarazo, I., Murelaga, X., Baceta, J.I., Sarmiento, A., and Madariaga, J.M. (2011), Beachrock formation in temperate coastlines: Examples in sand-gravel beaches adjacent to the Nerbioi-Ibaizabal Estuary (Bilbao, Bay of Biscay, North of Spain), *Spectrochimica Acta, Part A*, 80: 55–65.

Arzilli, F., Burton, M., La Spina, G., Macpherson, C.G., van Keken, P.E., and McCann, J. (2023), *Earth and Planetary Science Letters*, 602: 117945, https://doi.org/10.1016/j.epsl.2022.117945 001.

Bhanwar, P., Reddy, A.S., and Dave, T.N. (2023), A spreadsheet-based decision support system for selection of optimal soil liquefaction mitigation technique, *Decision Analytics Journal*, 6: 100154.

Boakes, Z., Hall, A. E., Ampou, E.E., Jones, G.C.A., Suryaputra, I.G.N.A., Mahyuni, L.P., Prasetijo, R., and Stafford, R. (2022), Coral reef conservation in Bali in light of international best practice, a literature review, *Journal for Nature Conservation*, 67: 126190, https://doi.org/10.1016/j.jnc.2022.126190.

Brochier, C., and Philippe, A. (2002), A non-hyperthermophilic ancestor for bacteria, *Nature*, 417: 244.

Broecker, W.S. (1974), *Chemical Oceanography*, Harcourt Brace Jovanovich, Inc., New York/Chicago/San Francisco/Atlanta, 85 pp.

Castanier, S., Métayer-Levrel, G.L., and Perthuisot, J.-P. (1999), Ca-carbonates precipitation and limestone genesis — The microbiogeologist point of view, *Sedimentary Geology*, 126: 9–23.

Dalsgaard, T., Canfield, D. E., Petersen, J., Thamdrup, B., and Acuña-González, J. (2003), N_2 production by the anammox reaction in the anoxic water column of Golfo Dulce, Costa Rica, *Nature*, 422(6932): 606–608, April 10..

De Boever, E., Brasier, A.T., Foubert, A., and Kele, S. (2017), What do we really know about early diagenesis of non-marine carbonates? *Sedimentary Geology*, 361: 25–51, https://doi.org/10.1016/j.sedgeo.2017.09.011 00.

Dickinson, W.R., and Seely, D.R. (1979), Structural and stratigraphy of forearc regions, *American Association of Petroleum Geologists Bulletin*, 63: 2–31.

Emerson, K., Russo, R.C., Lund, R.E., and Thurston, R.V. (1975), Aqueous ammonia equilibrium calculations: Effects of pH and temperature, *Journal of the Fisheries Research Board of Canada*, 32: 2379–2383.

Fenn, L.B., and Miyamoto, S. (1981), Ammonia loss and associated reactions of urea in calcareous soils, *Soil Science Society of America Journal*, 45: 537–540.

Fisk, H.N., and McClelland, B. (1959), Geology of continental shelf off Louisiana: Its influence on off-shore foundation design, *Bulletin of the Geological Society of America*, 70: 1369–1394.

Folk, R.L. (1974), The natural history of crystalline calcium carbonate: Effects of magnesium contents and salinity, *Journal of Sedimentary Petrology*, 44: 40–53.

Friedman, G.M. (1993), A revised classification of limestones—Discussion, *Sedimentary Geology*, 84, 1–4: 241–242.

Fukue, M., Lechowicz, Z., Fujimori, Y., Emori, K., and Mulligan, C.N. (2023), Inhibited and retarded behavior by Ca^{2+} and Ca^{2+}/OD loading rate on ureolytic bacteria in MICP process, *Materials*, 16: 3357, https://doi.org/10.3390/ma16093357.

Fukue, M., Nakamura, T., and Kato, Y. (1999), Cementation of soils due to calcium carbonate, *Soils and Foundation*, 39: 55–64.

Fukue, M., Nakamura, T., Kato, Y., and Naoe, K. (1996), Correlation among carbonate content, accumulation rate and topography of seabed, *Soils and Foundations*, 36(1): 51–60.

Fukunaga, A., Burns, J.H.R., Pascoe, K.H., and Kosaki, R.K. (2022), A remote coral reef shows croalgal succession following a mass bleaching event, *Ecological Indicators*, 142: 109175, https://doi.org/10.1016/j.ecolind.2022.109175.

Garrels, R.M., and Christ, C.L. (1965), *Solution, Minerals, and Equilibria*, Harper & Row, New York, 450 pp.

Glynn, P.W. (1991), Coral reef bleaching in the 1980s and possible connections with global warming, *TREE*, 6(6): 175–179.

Gong, Q., Wang, B., Gong, X., Liu, X., and Peng, Y. (2021), Anammox bacteria enrich naturally in suspended sludge system during partial nitrification of domestic sewage and contribute to nitrogen removal, *Science of the Total Environment*, 787: 147658, https://doi.org/10.1016/j.scitotenv.2021.147658.

Gordon, G. S. (2010), *Stratigraphic Evolution and Reservoir Quality in a Neogene Accretionary Forearc Setting: Eel River Basin of Coastal Northwestern California*, Search and Discovery Article #10249, AAPG Annual Convention and Exhibition, New Orleans, Louisiana, April 11–14, 2010.

Guida, B.S., Bose, M., and Garcia-Pichel, F. (2017), Carbon fixation from mineral carbonates, *Nature Communications*, 8: 1025, DOI: 10.1038/s41467-017-00703-4.

Hosseinpour, B., Saborimanesh, N, Yerushalmi, L., Walsh, D., and Mulligan, C.N. (2021). Start-up of oxygen-limited autotrophic partial nitrification-anammox process for treatment of nitrite-free wastewater in a single-stage hybrid bioreactor, *Environmental Technology*, 42(6): 932–940, https://doi.org/10.1080/09593330.2019.1649467

Ji, S., Gu N., Li, Y.-Y., and Liu, J. (2022), Rapid proliferation of anaerobic ammonium oxidizing bacteria using anammox-hydroxyapatite technology in a pilot-scale expanded granular sludge bed reactor, *Bioresource Technology*, 362: 127845, https://doi.org/10.1016/j.biortech.2022.127845.

Johnston, W.G., Cooper J.A.G., and Olynik J. (2023), Shoreline change on a tropical island beach, Seven Mile Beach, Grand Cayman: The influence of beachrock and shoreprotection structures, *Marine Geology*, 457: 107006, https://doi.org/10.1016/j.margeo.2023.107006.

Kawano, J., Sakuma, H., and Nagai, T. (2015), Incorporation of Mg^{2+} in surface Ca^{2+} sites of aragonite: An ab initio study, *Progress in Earth and Planetary Science*, 2: 7, DOI 10.1186/s40645-015-0039-4.

Lebedynets, M., Sprynskyy, M., Sakhnyuk, I., Zbytniewski, R., Golembiewski, R., and Buszewski, B. (2004), Adsorption of ammonium ions onto a natural zeolite: Transcarpathian clinoptilolite, *Adsorption Science & Technology*, 22(9): 731–741.

Majumdar, S.D., Hazra, S., Giri, S., Chanda, A., Gupta, K., Mukhopadhyay, A., and Roy, S.D. (2018), Threats to coral reef diversity of Andaman Islands, India: A review, *Regional Studies in Marine Science*, 24: 237–250, https://doi.org/10.1016/j.rsma.2018.08.011.

Mar, K.A., Unger, C., Walderdorff, L., and Butler, T. (2022), Beyond CO_2 equivalence: The impacts of methane on climate, ecosystems, and health, *Environmental Science and Policy*, 134: 127–136, https://doi.org/10.1016/j.envsci.2022.03.027.

Masyagina, O.V., and Menyailo, O.V. (2020) The impact of permafrost on carbon dioxide and methane fluxes in Siberia: A meta-analysis, *Environmental Research*, 182: 1090, https://doi.org/10.1016/j.envres.2019.109096.

McCutcheon, J., Nothdurft, L.D., Webb, G.E., Paterson, D., and Southam, G. (2016), Beachrock formation via microbial dissolution and re-precipitation of carbonate minerals. *Marine Geology*, 382: 122–135, http://dx.doi.org/10.1016/j.margeo.2016.10.010 0025-3227.

Mengel, K., and Scherer, H.W. (1981), Release of nonexchangeable (fixed) soil ammonium under field conditions during the growing season, *Soil Science*, 131: 226–232.

Mobley, H.L.T., and Hausinger, R. P. (1989), Microbial urease: Significance, regulation, and molecular characterization, *Microbiologicaly Reviews*, 53(1): 85–108.

Mourn, J., Loken, T., and Torrance, J.K. (1971), A geotechnical investigation of the sensitivity of a normally consolidated clay from Drammen, Norway, *Géotequenique*, 21(4): 329–340.

Mulder, A., van de Graaf, A.A., Robertson, L.A., and Kuenen, J.G. (1995), Anaerobic ammonium oxidation discovered in a denitrifying fluidized bed reactor, *FEMS Microbiology Ecology*, 16: 177–184.

Neumeier, U. (1999), Experimental modelling of beachrock cementation under microbial influence, *Sedimentary Geology*, 126: 35–46.

Nieder, R., Benbi, D.K., and Scherer, H.W. (2011), Fixation and defixation of ammonium in soils: A review, *Biology and Fertility of Soils*, 47: 1–14, DOI 10.1007/s00374-010-0506-4.

Nommik, H., and Vahtras, K. (1982), Retention and Fixation of Ammonium and Ammonia in Soils. In: F.J. Stevenson (ed.), *Agronomy Monographs*, John and Wiley & Sons, Inc. https://doi.org/10.2134/agronmonogr22.c4.

Pelejero, C., Calvo, E., and Hoegh-Guldberg, O. (2010), Paleo-perspectives on ocean acidification, *Trends in Ecology and Evolution*, 25(6): 332–344.

Pomar, L., and Hallock, P., (2008), Carbonate factories: A conundrum in sedimentary geology, *Earth-Science Reviews*, 87: 134–169.

Pusch, R., and Arnold, M. (1969), The sensitivity of artificially sedimented organic-free illitic clay, *Engineering Geology*, 3(2): 135–148.

Rosenqvist, I. Th. (1953), Considerations on the sensitivity of Norwegian quick clays, *Géotechnique*, 3(5): 195–200.

Saborimanesh, N., Yerushalmi, L., Walsh, D., and Mulligan, C.N. (2019), Pilot-scale application of a single-stage hybrid airlift BioCAST bioreactor for treatment of ammonium from nitrite-limited wastewater by a partial nitrification/anammox process, *Environmental Science and Pollution Research*, 26(25): 25573–25582, DOI: 10.1007/s11356-019-05754-2.

Sergeyev, Y.M., Budin, D.Y., Osipov, V.I., and Shibakova, V.S. (1973), The Importance of the Fabric of Clays in estimating Their Engineering Geological Properties. In: *Proceedings of the International Symposium on Soil Structure, Gothenburg 3*, pp. 243–251.

Shalini, S.S., and Joseph, K. (2012), Nitrogen management in landfill leachate: Application of SHARON, ANAMMOX and combined SHARON–ANAMMOX process, *Waste Management*, 32: 2385–2400.

Sieczka, A., and Koda, E. (2016), Kinetic and equilibrium studies of sorption of ammonium in the soil water environment in agricultural areas of central Poland, *Applied Sciences*, 6: 269, doi:10.3390/app610026.

Strous, M., Heijnen, J.J., Kuenen, J.G., and Jetten, M.S.M. (1998), The sequencing batch reactor as a powerful tool for the study of slowly growing anaerobic ammonium oxidizing microorganisms, *Applied Microbiology and Biotechnology*, 50: 589–596.

Strous, M., Pelletier, E., Mangenot, S., Rattei, T., Lehner, A., et. al. (2006), Deciphering the evolution and metabolism of an anammox bacterium from a community genome, *Nature*, 440: 790–794, doi: 10.1038/nature04647.

Sverdrup, H.U., Johnson, M.A., and Fleming, R.H. (1972), *The Oceans, Modern Asia Editions*, Prentice-Hall, Inc. New York, 998 pp.

Thauer, R.K. (2011), Anaerobic oxidation of methane with sulfate: on the reversibility of the reactions that are catalyzed by enzymes also involved in methanogenesis from CO_2, *Current Opinion in Microbiology*, 14: 292–299, DOI 10.1016/j.mib.2011.03.003.

Ueyama, M., Iwata, H., Endo, R., and Harazono, Y. (2023), Methane and carbon dioxide emissions from the forest floor of a black spruce forest on permafrost in interior Alaska, *Polar Science*, 35: 100921, https://doi.org/10.1016/j.polar.2022.100921.

Vousdoukas, M.I., Velegrakis, A.F., and Plomaritis, T.A. (2007), Beachrock occurrence, characteristics, formation mechanisms and impacts, *Earth-Science Reviews*, 85: 23–46.

Wang, X., Shan, S., Wang, X., and Zhu, C. (2020), Strength characteristics of reef limestone for different cementation types, *Geotechnical and Geological Engineering*, 38:79–89, https://doi.org/10.1007/s10706-019-01000-1.

Watanabe, Y., Sarumaru, H., and Shimada, N. (1983), Distribution of urease in the higher plants, *Technical Bulletin of Faculty of Horticulture, Chiba University*, 32: 37–43 (text in Japanese with English Abstract).

Weralupitiya, C., Wanigatunge, R., Joseph, S., Athapattu, B.C.L., Lee, T.-H., Biswas, J.K., Ginige, M.P., Lam, S.S., Kumar, P.S., and Vithanage, M. (2021), Anammox bacteria in treating ammonium rich wastewater: Recent perspective and appraisal, *Bioresource Technology*, 334: 125240, https://doi.org/10.1016/j.biortech.2021.125240.

Wetzel, A. (1989), Influence of heat flow on ooze/chalk concentration; Quantification for consolidation parameters in DSDP sites 504 and 505, sediments, *Journal of Sedimentary Petrology*, 59: 539–547.

World Health Organization (WHO). (2017), *Guidelines for Drinking-Water Quality (GDWQ)*, 4th Ed., Geneva, 631 pp. www.who.int/publications/i/item/9789241549950, accessed May 2024.

Wright, P.A. and Wood, C.M. (2009), Review, A new paradigm for ammonia excretion in aquatic animals: Role of Rhesus (Rh) glycoproteins, *Journal of Experimental Biology*, 212: 2303–2312, Published by The Company of Biologists, doi:10.1242/jeb.023085.

Xie, G.-Z., Zhang, L.-P., Li, C.-Y., and Sun, W.-D. (2023) Accelerated methane emission from permafrost regions since the 20th century, *Deep-Sea Research Part I*, 195: 103981, https://doi.org/10.1016/j.dsr.2023.103981.

Xing, B.-S., Guo, Q., Jiang, X.-Y., Chen, Q.-Q., He, M.-M., Wu, L.-M., and Jin, R.-C. (2016), Long-term starvation and subsequent reactivation of anaerobic ammonium oxidation (anammox) granules, *Chemical Engineering Journal*, 287: 575–584, https://doi.org/10.1016/j.cej.2015.11.090.

Yuan, Q., Zhang, Y., Xue, X., Wang, C., Ding, N., Xu, H., and Sun, Y. (2022), Morphological, kinetic, and microbial community characterization of anammox bacteria with different inoculations and biofilm types for low-ammonium wastewater treatment, *Journal of Water Process Engineering*, 47: 102748, https://doi.org/10.1016/j.jwpe.2022.102748.

Zhu, T., and Dittrich, M. (2016), Carbonate precipitation through microbial activities in natural environment, and their potential in biotechnology: A review, Frontiers in Bioengineering and Biotechnology, 4: 1–21. https://doi.org/10.3389/fbioe.2016.00004.

13 Ureolytic Microbial Carbonate Precipitation

13.1 INTRODUCTION

As mentioned in Chapter 12, the problem of global warming is expanding accompanied by the unprecedented frequency of catastrophes. On the other hand, as mentioned in Chapter 11, global pollution, which is attributed to various industrial and municipal wastes is becoming increasingly serious. A major challenge is how to rely on nature to solve these problems in a more natural, sustainable, and environmentally friendly way. Originally, the Earth's environment is protected by the interaction of organisms, due to biodiversity. By understanding this, we are sure to find more environmentally friendly solutions. In addition to the engineering issues that have been addressed, it is necessary to understand that environmental measures against global warming, such as carbon neutrality, must be incorporated at the same time.

One of the major sets of challenges in the constructed environment is the problem of the inability of the natural ground to provide proper supporting capability for overlying structures, and vulnerability of slopes and ground to catastrophic ground movement due to metastability of the soil mantle. Failure to recognize the challenges or failure to properly account for the potential hazardous and catastrophic natural or anthropogenic events can lead to geo-disasters. Chapter 10 has defined *geo-disasters* as disasters that occur in the land compartment of geoenvironment as a result of natural and human-associated catastrophic events, such as earthquakes, floods, hurricanes, landslides, embankment, and dam failures, etc.

To overcome deficiencies in ground support capability, or to strengthen the capability of the soil mantle to withstand the physical forces resulting from hazardous and catastrophic events, ground improvement techniques have been devised – the intent of which are to strengthen the ground (soil mantle) so as to provide the proper resistance to undesirable and unplanned ground movement. The discussion in this chapter is the use of a new innovative and sustainable technique for ground improvement to withstand the stressors resulting from the geo-disasters, and hence to prevent or mitigate geo-disasters. The basis for the new innovative technique discussed in this chapter requires a brief review of the natural processes involved in producing the type of ground conditions that are resistant or vulnerable to geo-disaster stressors. What most people do not know is that many of the world's largest cities are located on coastal planes, where their ground was once the ocean floor. Even experts may not be aware of the bearing capacity of the former seabed ground that supports such large cities. That is because the carbonates produced by sea organisms form the skeletal structure of the particles.

Carbonate minerals produced by living organisms are divided into two types. One is called biological-controlled carbonate minerals (BCCM) such as shells and organs that are needed by living organisms, and the other is biological-induced carbonate minerals (BICM) that are not directly related to biological reactions but are produced by chemical reactions that depend on environmental conditions. These minerals are called biominerals, and their production mechanisms and processes are called biomineralization. Specific examples of induced minerals include microbial-induced carbonate minerals in this chapter.

Biominerals range from carbonates, sulphates, phosphates, etc. They are diverse, including metal species, amorphous, or hydrates. Environmentally, it is broadly divided into aerobic and anaerobic

DOI: 10.1201/9781003407393-13

processes. In terms of quantity, it is known that carbonates are very large, particularly, due to biomineralization, which contributes to the sequestration of CO_2 (Chapter 12).

In fact, marine deposited soils have cementation due to carbonates (Fukue et al., 1999). The intensity of the cementation due to carbonates is relatively strong. The analysis showed that for silty clays obtained from a depth of 42 to 44 m under the sea bottom off-shore from Haneda International Airport, Japan, 1% of carbonate increased the unconfined compressive strength by about 60 kPa.

Most of the carbonates are biological products (foraminifera and coccolith) in the seawater as mentioned in Chapter 12. The limestone and marble have been used for building materials. On the other hand, limestone has been used for the raw materials of cement, which can be used for concrete and mortar. Furthermore, the cement is used as the grout for soft soil. Note that in the processes of cement production, much of the CO_2 by chemical decomposition of the limestone is released. The CO_2 released is due to the following calcination (IUPAC Compendium of Chemical Terminology, 3rd ed. International Union of Pure and Applied Chemistry; 2006. Online version 3.0.1, 2019. https://doi.org/10.1351/goldbook.C00773 accessedAug. 2024)

$$CaCO_3 \rightarrow CaO + CO_2 \quad (13.1)$$

where CaO is calcium oxide or quick lime, which is used as a concrete matrix. Note that the heat needed for the calcination is 848 °C. In order to bring down the temperature to this level, it was necessary to improve the technology. In comparison, the biocement causes carbonates themselves to be formed in the ground, thereby binding soil particles. The result is the same as that of carbonate diagenesis. Therefore, carbon neutrality can be achieved.

It is also noted that heavy metals can form the microbial minerals in the form of MCO_3, where M denotes a heavy metal (Chapter 11). Accordingly, biological carbonate precipitation can be used to remove and immobilize the heavy metals from water.

13.1.1 Carbonate Diagenesis and Microbially Induced Carbonate Precipitation (MICP)

Diagenesis is the physical and chemical changes in sediments after weathering process caused by water-soil particle interactions, microbial activity, and consolidation after their deposition. In general, the actions of high temperature and high pressure are not included in diagenesis, but in metamorphism. The term "carbonate diagenesis" can be used for the hardening process of sediments by the actions of carbonate cementation, in which there are various pathways.

From a microbiogeologist's point of view, the natural calcium carbonate precipitation and limestone genesis were discussed by Castanier et al. (1999). They described that the environmental conditions of heterotrophic bacterial metabolic pathways are diverse (aerobiosis, anaerobiosis, microaerophilicity). They also summarize that in aqueous environments, apart from the deep ocean, the potential efficiency of heterotrophic bacterial carbonatogenesis in Ca-carbonate sedimentation is much higher than autotrophic or abiotic processes. This suggests that heterotrophic bacteria are more useful for applying artificial carbonate precipitation. The intention of this chapter is artificial application of carbonate diagenesis by MICP, as an improvement of soil for geo-disaster prevention and construction.

13.1.2 Development of Biocement Technology

Since about 2000, the dominance of aerobic heterotrophic bacteria has been revealed, and interest in microbially induced carbonate precipitation, MICP has increased. MICP and BICM are almost synonymous, but BICM is commonly used in the field of biomineralization and MICP in applications using microorganisms.

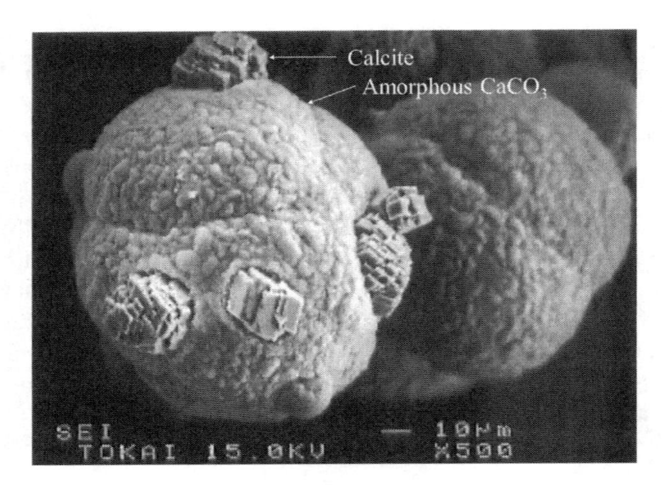

FIGURE 13.1 Aggregated mass with a few thousand microbes covered with amorphous calcite and growing calcite crystals from the inside of the mass (SEM image).

Unlike inorganic reactions, mineral formation by MICP is relatively slow and crystal growth takes time. Under this condition, it is not uncommon that the first product is amorphous or hydrated minerals. If high concentrations of Ca^{2+} with some Mg^{2+} and enough bacteria are used as a solution, a large number of amorphous particles often gather and gradually form a crystal in solution, as shown in Figure 13.1. If such a phenomenon occurs between soil particles, carbonates can bind the soil particles during crystallization.

Ever since 2000, precipitation of calcite in soils has been studied for soil improvement. First, the clogging of sand to stop groundwater flow and then the solidification of soil became the object of study in civil engineering (DeJong et al., 2006; Ferris et al., 2003; Stocks-Fischer et al., 1999; Whiffin et al., 2007).

13.1.3 Urease Activity and Induced Carbonates

In MICP, there are six types of metabolism, such as (a) photosynthesis, (b) ureolysis, (c) denitrification, (d) ammonification, (e) sulphate reduction, (f) methane oxidation (Section 11.4.9). These metabolic types are achieved by different microorganisms, respectively. Among these, ureolysis is the most familiar and convenient to deal with. Furthermore, it is realistic to use ureolysis in the application of carbonate precipitation in terms of materials and microorganisms.

As described in Chapter 12, in nature, ureolytic MICP is based on the hydrolysis of urea. The catalyzing action of ureases plays an important role in the chemical reaction. Urease was the first enzyme to crystallize from jack bean (in 1926), and its substrate, urea, was also the first organic molecule synthesized in the laboratory (in 1828) (Sirko and Brodzik, 2000). Urease is found in most plants, fungi, bacteria, and some invertebrates (Watanabe et al., 1983; Polacco and Holland, 1993; Sirko and Brozik, 2000). Urease plays a major role in the nitrogen and carbon cycles because it allows organisms to use urea as a source of nitrogen and carbon (Chapter 12).

Since the rate of hydrolysis depends on the concentration of bacteria, urease activity is generally presented by mM urea/min/OD_{600nm}, where OD_{600nm} is defined as the optical density of biomass measured with 600 nm wavelength. However, urease activity varies over time in terms of viability of bacteria. Therefore, it is doubtful that the urease activity can be used as a fundamental property of bacteria for engineering purposes. This is discussed in this chapter.

13.1.4 Present Problems in MICP Technology

Hydrolysis considerably increases pH. Under this condition, homogeneity of metal precipitation as a result of biomineralization can be obtained. Calcite precipitation induced by ureolytic bacteria was studied by many researchers, such as Ciurli et al. (1996), Le Métayer-Levrel et al. (1999), Tiano et al. (1999), Ferris et al. (2003), Whiffin et al. (2007), De Muynck et al. (2008), Lian et al. (2006) and Jimenez-Lopez et al. (2008). These studies concern urease activity, calcite precipitation, and their application to engineering purposes. They are biocementation for soil improvement and the coating and protection of stones and concrete. Hammes et al. (2003) studied a variety of urease activities with different species and strains. They showed that the rate of urea hydrolysis depends on the strains.

The enzyme activity, i.e., ureolysis due to the urease is well known as:

$$(NH_2)_2CO + [e] + H_2O \rightarrow 2NH_3 + CO_2 \tag{13.2}$$

Reaction (13.2) is known as hydrolysis with the catalysis of urease [e]. A microbe's system is shown in Figure 13.2, in which the first products are $2NH_3$ and CO_2 (Mobley and Housinger, 1989). However, the properties and activity in terms of [e] is usually uncertain because of the unknown viability of bacteria. This means that the production of $2NH_3$ and CO_2 is not predictable. Note that this can cause discrepancies and inconsistencies in research results among researchers. This is discussed throughout this chapter.

Reaction (13.2) is a symbolic expression that if organic matter is decomposed, NH_3 and CO_2 are always produced. If reaction (13.2) is achieved, then the $2NH_3$ and CO_2 can be transported outside of the cells, and react with H_2O, as follows:

$$2NH_3 + CO_2 + H_2O \rightarrow 2(NH_4)^+ + CO_3^{2-} \tag{13.3}$$

This reaction may not be MICP, but MCCP, because the hydrolysis can be intentional for the urease-producing bacteria. The intention of the bacteria may be to increase the pH, like *H. pylori*. However, the reaction after this is undoubtedly induced.

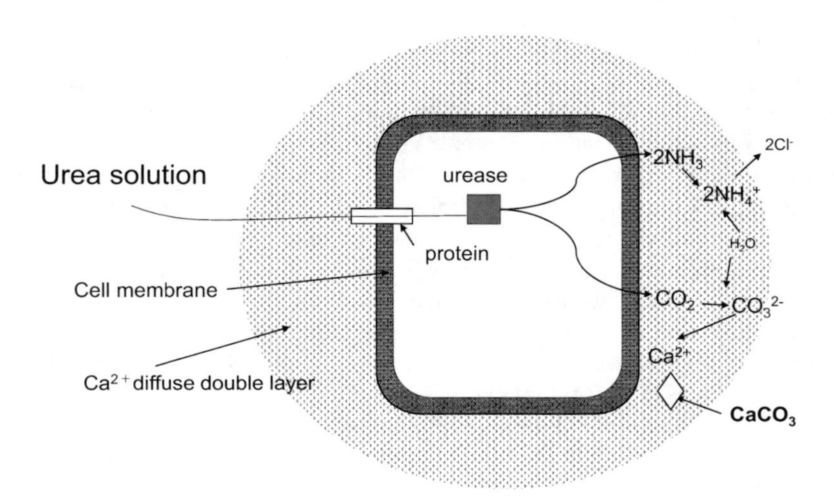

FIGURE 13.2 Microbial hydrolysis of urea, showing that the first products are CO_2 and NH_3 inside the microbe's cell, and that carbonates are produced first in the diffuse double layer outside of the cell.

It should be remembered that the enzyme [e] in reaction (13.2) is not given appropriately in terms of quantity and quality. The species and strains of ureolytic microbes, their concentrations, and micro-environments play an important role for the urease activity (Mobley and Housinger, 1989; Benini et al., 1996; Sun et al., 2019; Omoregie et al., 2018; Kahani et al., 2020). For example, the urease activity of biomass increases proportionally to the number of cells (Benini et al., 1996). This implies that the concentration of microbes plays an important role in the MICP process.

The product, $NH_{3.}$ of the hydrolysis increases the pH outside of the cell (Fenn and Miyamoto, 1981)and helps the reaction (13.4) in the presence of Ca^{2+}. Reaction (13.4) indicates that 1M $CaCO_3$ can be produced when 1M urea and 1M $CaCl_2$ are used, as far as no inhibition of urease activity occurred in reaction (13.2). Accordingly, the concentration of Ca^{2+} becomes a dominant factor for the $CaCO_3$ precipitation rate.

$$Ca^{2+} + CO_3^{2-} \rightarrow CaCO_3 \tag{13.4}$$

The production of $CaCO_3$ does not change the pH of the system. The Ca^{2+} is usually supplied in the form of calcium chloride, $CaCl_2$. However, NH_4 Cl^-, which are the product by the adverse reaction decreases the pH of the solution. It is noted that if Ca^{2+} is not present or calcium carbonate production has not yet been completed, ammonium carbonate $(NH_4)_2 CO_3$ is generated. During that time, ammonium carbonate is considered to maintain a high pH of the system.

For the improvement of soils, rock and concrete using microbial carbonate precipitation, there are basically three problems such as reliability, repeatability, and cost performance. This may result from the drawback in using microbes whose properties and characteristics are influenced by the change in cell viability. The problem becomes serious directly or indirectly when scale-up and commercial applications are considered (Omoregie et al., 2018). In many studies, bacteria and media have been injected into soils for *in situ* cultivation. In those cases, bacteria cultivation has not always been well controlled. Accordingly, the viability and the bacteria concentration (OD) cannot be known or the design of the blending for biocement solution (BCS) cannot be easily achieved. Under this condition, the reliability and repeatability are difficult to obtain. Therefore, it is important to solve the problem concerning the unknown quality of the biocement solution, BCS. Unless *in situ* cultivation can be achieved under control, the urease-producing bacteria have to be cultivated in an incubator.

Considering every factor from a realistic point of view, using a lightly equipped incubator (plastic bag method), avoiding a large amount of culture solution, a long-term storage (at ultra-low temperature –80 °C at least 4 years), and reducing the volume (centrifugation), the solution was properly obtained (Fukue et al., 2022, 2023). To do this, it was also necessary to establish methods that enable the definition and evaluation of microbial urease activity in MICP.

The quality of stored microbes is usually unknown, because of many governing factors, such as the varied nature in the original culture solution, i.e., inherent properties, centrifugal effects, amounts of microbes in the supernatant during centrifugation, effects of added glycerin, ageing, thawing effects, etc. Most of these factors reduce the number of viable cells. Accordingly, the urease activity in the BCS will decrease. To characterize the properties of BCS for engineering purposes, it is required to establish a technique to determine the quality of microbes including the carbonate precipitation rate (CPR). Herein, CPR is defined as the mass of precipitated carbonates, or the mass of precipitated carbonates at a given time.

In microbiology, optical density (OD) is used as the most popular method to account for the growth of microorganisms, as mentioned earlier. The number of cells is measured by absorbance of light called optical density, OD. However, the disadvantage is that the OD cannot distinguish viable and dead cells. Therefore, OD, which was measured on aged bacteria cannot represent the number of viable cells.

13.1.5 Perspectives for MICP Technology

It is true that many researchers were praised for the MICP technology. There are many advantages for environmental geotechnical engineers, and the future prospects are bright. On the other hand, in engineering methodology, there are still problems to be solved. For example, two urgent problems are to establish an engineering evaluation method on bacteria in relation to reaction (13.2), and to understand conditions of inhibition and retardation by Ca^{2+}. Presently, due to these two problems, it is not possible to evaluate test results in a unified manner, which can also become a difficulty of MICP. This opinion may be similar to the intension of discussion by Lai et al. (2021) and Fu et al. (2023).

As far as ground improvement by MICP is concerned, it has not yet been technically systematized. For example, there is little established about how agreement on how to design to obtain desired strength, whether it is necessary to test, and if so, whether the results can be evaluated correctly, and how much it will cost. The most important aspects for engineering must be safe, reliable, and superior results in comparison to the conventional technologies. In order to demonstrate these three requirements, many engineers must work on them. The following must be prepared.

(1) An evaluation method for bacterial enzymes required for reaction (13.2).
(2) Confirmation of Ca^{2+} inhibition and retardation conditions related to MICP and its evaluation method.
(3) A test method and analysis method for (1) and (2).

These subjects were studied and published by Fukue et al. (2022, 2023), which will be introduced in this chapter.

13.2 MICROBIAL-INDUCED CARBONATE PRECIPITATION (MICP)

There are many species of microorganisms which can be used for MICP technology. However, considering metabolism, safety, treatability and efficiency, urease-producing bacteria may be better than other microbes. In fact, *Sporosarchina pasteurii* seems to be the most common microbial species used in research papers on MICP. In this chapter, a new species of bacteria is mainly utilized to investigate the basic properties and behaviour of the bacteria in MICP, in order to industrialize this eco-friendly, economical, and sustainable environmental technology.

13.2.1 Isolation and Cultivation of Bacteria

Most of the new findings presented in this section will be based on experimental results obtained using the NO-A10 strain. Therefore, detailed information on this strain will be provided here.

To find urease-producing bacteria, bored core samples were collected from 11 sites in different locations in Japan. Different concentrations of $CaCl_2$ or $CaOH_2$ were added to the soil samples and kept for a few days to eliminate some species of bacteria, which cannot live in the presence of $CaCl_2$ or $CaOH_2$. In this procedure, about 150 strains were extracted and their capacity for urease activity was examined. As a result, NO-A10 and NO-N10 were selected. Finally, the strongest strain of *Sprosarcina* sp., a new species, alkalophilic NO-A10, was chosen. The strains used in this study were cultivated with an electron donor compound (EDC) at a pH of 9. It is noted that the NO-A10 strains can be cultivated under a pH higher than 7, while NO-N10 can be cultivated under a lower range in pH. Figure 13.3 shows the basic sequence of NO-A10, which is an unknown species.

The example of growth curves of the strains using a culture medium at pH 9 are shown in Figure 13.4. NO-N10 shows urease activity but the rate is relatively slow. NO-A10 shows a relatively slow growth, in comparison to NO-N10. However, the growth rate is not necessarily correlated with

```
gcgaattgaa gggagcttgc tccctgatat tagcggcgga cgggtgagta acacgtgggc   60
aacctgccct gcagatgggg ataactccgg gaaaccgggg ctaataccga ataatcagtt  120
cttccgcatg gaagaactct gaaagacggt ttcggctgtc actgcaggat gggcccgcgg  180
cgcattagct agttggtggg ataacggcct accaaggcga cgatgcgtag ccgacctgag  240
agggtgatcg gccacactgg gactgagaca cggcccagac tcctacggga ggcagcagta  300
gggaatcttc cacaatggac gaaagtctga tggagcaacg ccgcgtgagc gaagaaggtt  360
ttcggatcgt aaagctctgt tgcgagggaa gaacaagtac gggagtaact gcccgtacct  420
tgacggtacc tcgtcagaaa gccacggcta actacgtgcc agcagccgcg gtaatacgta  480
ggtggcaagc gttgtccgga attattgggc gtaaagcgcg cgcaggcggt cctttaagtc  540
tgatgtgaaa gcccacggct caaccgtgga gggtcattgg aaactggagg acttgagtac  600
agaagaggaa agcggaattc cacgtgtagc ggtgaaatgc gtagagatgt ggaggaacac  660
cagtggcgaa ggcggctttc tggtctgtaa ctgacg                             696
```

FIGURE 13.3 The basic sequence of NO-A10.

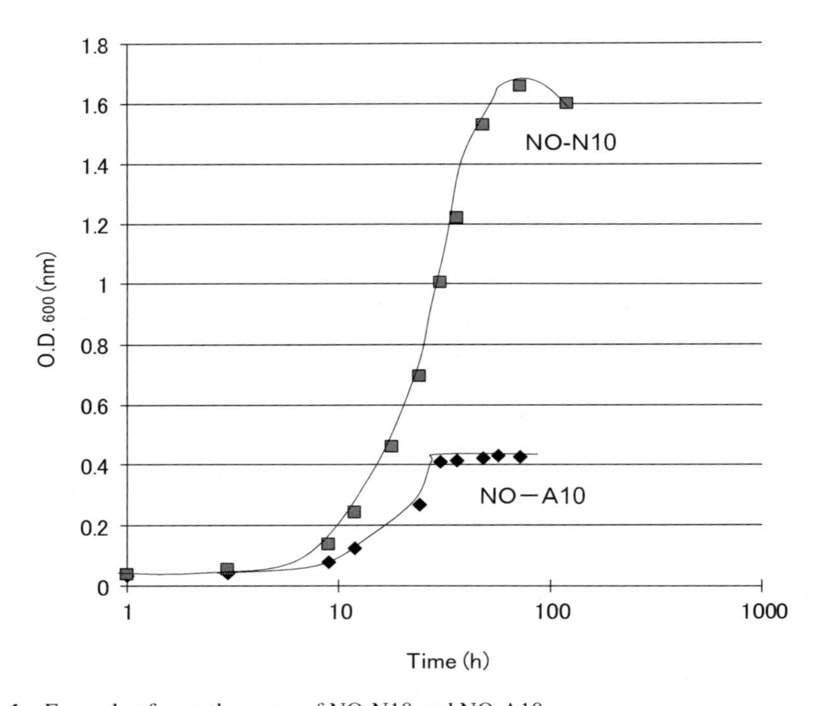

FIGURE 13.4 Example of growth curves of NO-N10 and NO-A10.

higher urease activity, as described later. Note that NO-A10 can grow faster by an ordinary cultivation using peptone or polypeptone, which is a mixture of polypeptides and amino acids formed by the partial hydrolysis of protein.

13.2.2 Urease Activity in the Presence of Calcium Ions

Both strains, A10 and N10, were cultivated with a culture medium under the same conditions, except for pH. The cultivated bacteria were centrifuged to separate them from the medium. After the supernatant was removed, a certain volume of 5% NaCl solution was added to the bacteria. This is called the bacterial solution (B solution) in the study. Urea and calcium chloride solution with a similar molar concentration was prepared using the buffer solution 10 mM (NH_4OH + NH_4Cl). This is called the reaction solution (R solution). Note that for the R solution, Mg ions can also be added if it is required. The concentrations of urea and $CaCl_2$ ranged from 0.8 to 3.0 M. At the final stage, the same volumes of the B and R solutions were mixed. The concentrations of urea and $CaCl_2$ ranged from 0.4 to 1.5 M. A 5 ml aliquot of B and R solution was put into 5 mL glass test tubes and the amount of carbonate (calcite) precipitated with time was measured by the following procedures.

After a certain time, the liquid in the test tubes was filtered with a 1 μm filter. The filter with solid residues was dried in an oven (at 110 °C). The amount of carbonate (calcite in this case) in the liquid was determined by subtracting the mass of the filter from the total weight of the filter with carbonate. Most of the precipitated calcite adhered to the wall of the tube. After the liquid was removed, the tube was dried in an oven (at 110 °C) until the mass becomes constant. The precipitated calcite on the wall was determined by subtracting the mass of the tube from the total dry mass of the tube with calcite.

The fundamental MICP was examined using NO-N10 and -A10, which shows relatively high urease activity (Fukue et al., 2011). In this section, MICP with elapsed time, temperature, pH, and OD are discussed. Note that the OD values used here are relative ones and different from the viable OD value defined later.

13.2.3 Urease Activity and Carbonate Precipitation Rate

The carbonate precipitation rate in inorganic reaction is extremely rapid. However, the rapid precipitates have no binding capacity. On the other hand, since the microbially induced reactions occurs via hydrolysis due to the enzyme reaction, it is slow in comparison to the inorganic reaction. Therefore, it is important to know that a relatively slow precipitation is required to have a good binding capacity of carbonates. This is one of the main reasons to use microbial precipitation of carbonates as binding materials.

Calcite precipitation was examined using different concentrations of urea and $CaCl_2$ in the test tubes. Because 1 M-urea produces 1 M $CaCO_3$ (usually calcite), as indicated by reactions (13.2) to (13.4), the initial concentrations of urea and $CaCl_2$ were prepared at the same molar concentration. The temperature was maintained at 20 °C. The concentration of bacteria (OD) was sufficiently high to complete the reaction. The microbial precipitation of carbonate induced by NO-A10 with elapsed time is shown in Figure 13.5.

The legend shows the type of strain and the concentration of the $CaCl_2$ solution used. For example, "A10-0.4M" means that the type of strain is NO-A10 and the $CaCl_2$ concentration is 0.4 M. For comparison, the cases of NO-N10 are also shown. Although some bacteria lose urease activity under the presence of Ca ions, NO-A10 showed an ideal reaction, i.e., 100% production of calcite. This means that the cell viability per unit volume was sufficient in A10.

Figure 13.5 shows that NO-A10 had a high (faster) urease activity even for 1.5 M $CaCl_2$. On the other hand, the carbonate precipitation by NO-N10 is relatively slow and urease activity by N10 was relatively low. Figure 13.5 also shows that the reaction rate by NO-A10 was almost independent of the concentrations of $CaCl_2$ and the reaction was completed within 25 hours for any concentration of $CaCl_2$ up to 1.5 M. This type of CPR can be explained by the Ca^{2+}- control precipitation. In other words, the maximum carbonate precipitation is controlled by Ca^{2+}. On the other hand, the carbonate precipitation by NO-N10 was not complete after 100 hours. However, the carbonate precipitation rate in MICP will vary with many factors, such as quality of the urease-producing bacteria, temperature and pH, which affect the enzyme reaction, viability and cell number, blending of agents or materials, etc., as demonstrated later. This will be discussed later.

13.2.4 Temperature

Urease activity is affected by temperature. Ciurli et al. (1996) investigated temperature stabilities of the free and the adsorbed urease. The results showed that a strong stabilization of the adsorbed

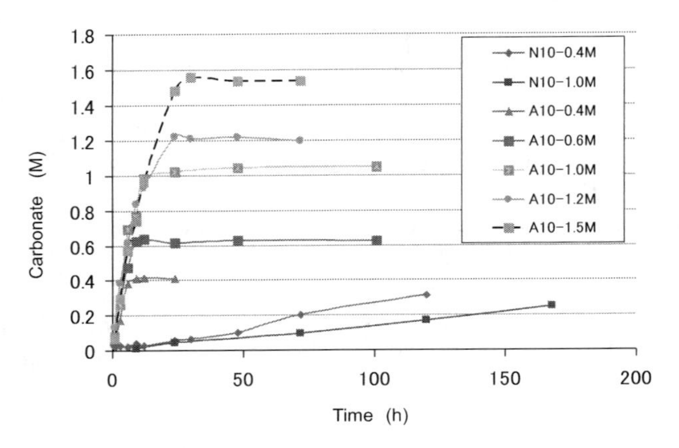

FIGURE 13.5 Carbonate precipitation rate with elapsed time, in terms of Ca^{2+} concentrations, using N10 and A10 strains.

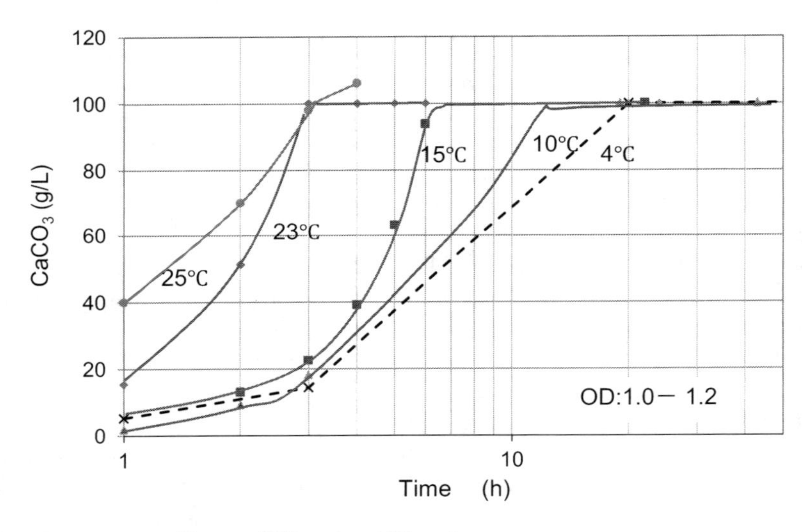

FIGURE 13.6 Temperature effect on CPR, using A10 strains.

form of the enzyme was observed. At 50 °C the activity of both the free and adsorbed enzymes started to decrease, and at 70 °C both forms of the enzyme were completely inactivated. This may be because high temperature damages the urease (enzymes), which is a protein. No data was presented below 30 °C.

The effect of temperature on CPR is shown in Figure 13.6. At a temperature lower than 30 °C, the results showed that the higher the temperature, the higher the degree of CPR production. It is unclear whether this is due to urease activity or the effect of temperature on the carbonate formation process.

Sun et al. (2019) investigated temperature effects on urease activity, CPR and unconfined compressive strength with time. The results showed that the measurements of CPR and unconfined compressive strength were dependent on the urease activities with elapsed time. However, as the inhibition and retardation effects were not known, the ultimate values on CPR and unconfined compressive strength were not clear.

13.2.5 pH

Both the change in pH and high temperatures can damage urease. However, it is possible that the proper range of suitable pH and temperature for urease activity depends on the species of bacteria. It is noted that the effect of pH is different for urease activity and MICP process, because urease activity acts even below pH lower than 7, but not MICP. Hydrolysis of urea by microorganisms increases pH by ammonia production. Therefore, as long as pH is not too high, urea hydrolysis by enzymes occurs. A good example is the hydrolysis of urea by *H. pylori*, which increases pH in stomach acid. Therefore, MICP is not necessarily related to urease activity.

In acidic solutions, $CaCO_3$ is dissolved. In general, the optimum pH for MICP ranges from 7.5 to 9.0 (Armad et al., 2023). However, it is common that the pH value changes by the NH_3 or $(NH_4)_2 CO_3$ concentrations, which also result from the hydrolysis of urea. NH_3 will convert immediately to NH_4^+. Under the existence of CO_3^{2-}, the product $(NH_4)_2 CO_3$, which is not usually stable, can be alkaline (Mobley and Hausinger, 1989). However, by eliminating CO_3^{2-} the byproduct NH_4^+ may decrease the pH in the solution. In MICP, the ultimate pH can significantly impact if $CaCO_3$ concentration is remained or not. To make sure that the pH value of pore solution is sufficiently high, it is suggested that a buffer solution is utilized and finally the pore solution is removed. On the other hand, Wang et al. (2023) investigated activity by soybean urease. They have obtained urease activity at pH from 3 to 11. Figure 13.7 shows an example of the effect of initial pH in MICP for NO-A10 strains. The results show that carbonate can precipitate at a pH from 7 to 10.5, It is natural that the effective pH range in MICP is narrow in comparison to urease activity.

13.2.6 Optical Density (OD)

The OD value in BCS provides an optical density, degree of bacteria density in solution, which provides on information of an intensity of urease activity. In this section, the OD value used was calculated depending on the dilution, based on the OD (optical density) measured on the cultivated solution. Therefore, the comparison of the data was only effective for the series of the experiments using a fresh culture solution. In other words, the culture solution produced another day cannot be compared with one obtained on different day, unless the initial OD values are in accordance. In this sense it is very difficult to compare the bacterial data. Accordingly, the data obtained by different researchers should not be compared quantitatively but only qualitatively. Thus, there is no unified scale of OD, because of different initial conditions and many factors for ageing over time and various events.

The trend of OD-CPR relation can be obtained easily, as follows (Lai et al., 2021; Wen et al., 2020). However, the results obtained by different resources of OD values cannot be used to compare OD effects on CPR, as mentioned above. This is a very important point for the future of MICP technology. In other words, CPR is possibly a significant quantifiable factor that can be unified in MICP technology. For that, the OD, which can provide bacteria capacity, has to be unified and defined.

The maximum CPRs, which can be produced by various unified ODs as defined by Fukue et al. (2022), are shown in Figure 13.8. The various OD values were prepared by dilution of the 1.0 OD solution. The initial pH value used was approximately 9.5 due to the 20 mM ammonium buffer. The ambient temperature was 25 °C.

The results showed that the CPR increases with increasing of OD, and when OD is higher than approximately 0.25, the carbonate precipitation reaches 1.0 M. Note that the CPR was obtained at a reaction time of 24 hours, and that if the time limit is removed, the precipitation rate will be linearly increased from the zero point of CPR (Fukue et al., 2022). Detailed discussion will also be made later. It is noted that the CPRs greater than 1.0 M can be obtained with greater OD values. This is because hydrated carbonates tend to be produced when the reaction rate is high (Rodriguez-Blanco et al., 2017).

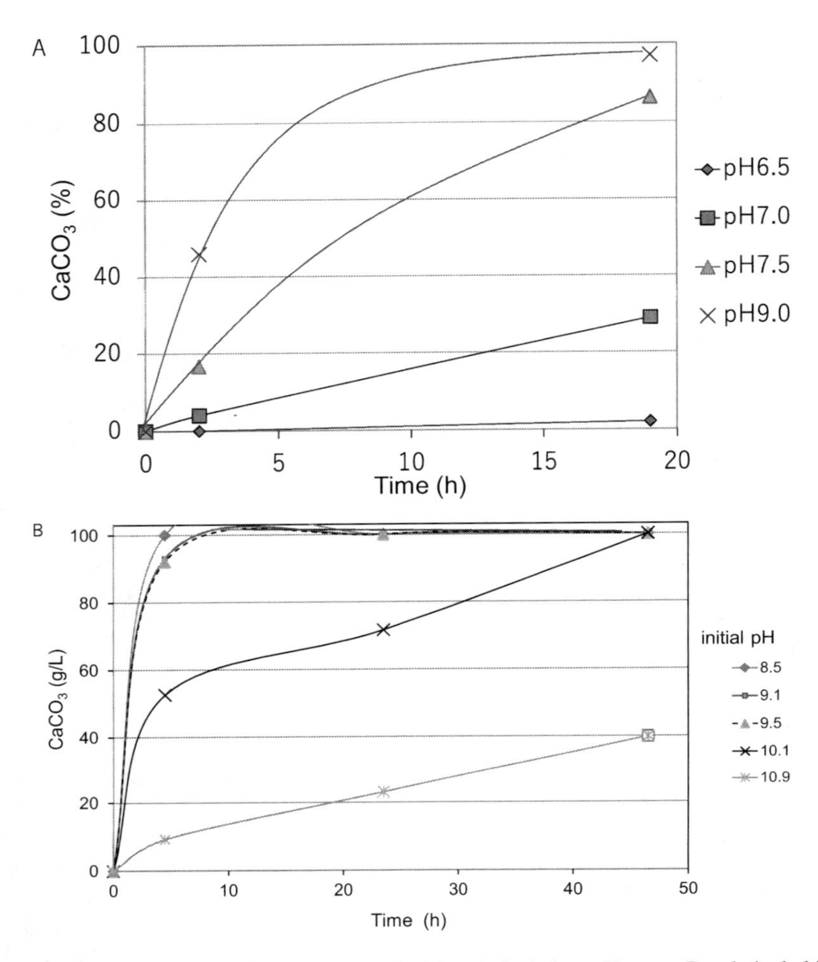

FIGURE 13.7 Effect on pH in MICP for NO-A10 strains. A: relatively low pH range, B: relatively high pH range.

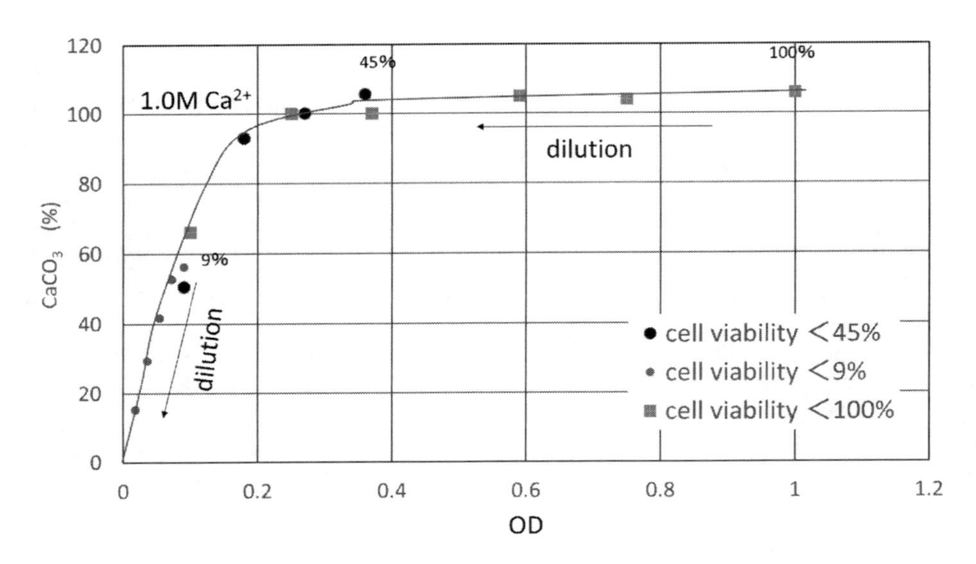

FIGURE 13.8 The maximum CPR by various unified OD, showing the limited capacity of bacteria in MICP process.

13.3 CONCEPT OF SUSTAINABLE MICP

About 20 years have passed since microbiologically induced carbonate precipitation technology was proposed. The number of researchers in the field of MICP has rapidly increased. However, commercialization has not progressed, and an academic system has not been established. There are a few possible reasons for this as follows:

(1) The underlying specialties are divided into several fields, and few researchers are familiar with all of them. For example, microbiology, geochemistry, material science, geology, geotechnical, and geoenvironmental engineering.
(2) There is difficulty in handling living microorganisms for engineers. The problems such as production, storage, and transportation of bacteria must be solved.
(3) Commercialization requires good cost performance, technical establishment, public relations, and a wide range of human resources.
(4) The countermeasures against undesirable by-products, such as ammonium have to be established.

Furthermore, it is a main problem that at present, there is no consensus in the basic experimental results (Lai et al., 2021; Fu et al., 2023).

13.3.1 Fundamental Approach to the MICP Mechanism

In fact, there is no difference between soils solidified by MICP technology and soils naturally cemented, because most natural soils have been more or less solidified with carbonates. In Chapter 12, it was demonstrated that marine sediments have been solidified by carbonates (Fukue et al., 1999). On the other hand, there are many examples of sandstone, which have been solidified with carbonates. Their solidification processes may be varied, which may be possibly called carbonate diagenesis (Burley and Worden, 2003). To date, carbonate content has rarely been measured as a strength factor in geotechnical engineering, but carbonates have come to be highlighted with respect to MICP. However, apart from the field of MICP, there is still no sign of carbonate measurement in natural soil. In this chapter, the amount of carbonate produced in soils, i.e., carbonate content, is assumed as one of the most important factors governing the strength of soils.

13.3.2 Optical Density and MICP

In microbiology, the optical density $OD_{600\,nm}$ of culture media is important as a representative value of the number of cells, and many studies have been conducted on its reliability (Francois et al., 2005). Although the OD value has come to be used because of the time-consuming method of measuring cell count both methods are still viewed as problematic from the point of view of engineers.

The OD value is measured in the same way as the principle of turbidity measurement, and the definitions are similar. However, since this method cannot distinguish between live bacteria and dead bacteria, it is a disadvantage that aged bacteria cannot be evaluated by the OD method for the urease activity, as mentioned earlier. Then, the first problem was how to check the viable bacteria rate. It is more difficult to measure the viable bacteria directly. In this book, MICP mechanics are discussed based on bold hypotheses and assumptions, at least to the extent that they do not contradict the experimental results. For this purpose, the first and most basic hypothesis is proposed as follows.

13.3.3 Standard OD-CPR Relationship

An OD-CPR relationship was determined in terms of $OD_{600\,nm}$ values by dilution using a fresh culture solution adjusted to OD of 1.0, where the standard viability corresponding OD of 1.0 was

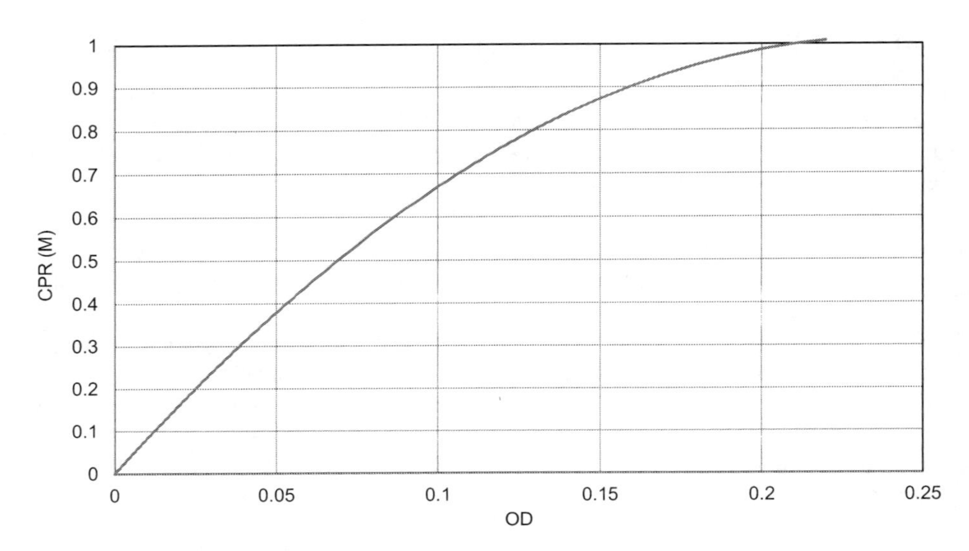

FIGURE 13.9 24 h standard OD-CPR relationship in MICP, based on Figure 13.8 at 25 ℃ and $0 < Ca^{2+} \leqq 1.0$ M. (Fukue et al., 2022.)

defined as 100 % (Fukue et al., 2022). As a result, the 24 h OD-CPR relationship is presented by the quadratic function of OD (Figure 13.9), which is based on Figure 13.8, and given by

$$CPR = \lambda_1 \, OD - \lambda_2 \, (OD)^2 \tag{13.5}$$

where λ_1 (M) and λ_2 (M) are the constants, 8.46 M and 17.633 M, respectively (Fukue et al., 2022).

Equation (13.5) is then defined as the 24 h standard OD-CPR relationship. When the OD value is given, then CPR at a 24 h reaction time can be determined, and the reverse is also true. Figure 13.10 illustrates the respective relationship by infinite time and limited elapsed times. It is noted that the retardation is a function of OD^2, as indicated in Equation (13.5). The 24 h standard relationship becomes a calibration for conversion from an apparent optical density OD* of unknown bacteria to the OD value, when an apparent viability Rcv is known. The terms OD* and Rcv are discussed with experimental results in later sections.

13.3.4 Micromechanisms in MICP

The hypothesis and assumption made must be proved by theoretically and experimentally. The proof may require micro- and macro-points of view. On the other hand, visualization of $CaCO_3$ formation is important to understand the micro-mechanism in MICP. Microbial $CaCO_3$ formation has been observed with microscopes, i.e., using SEM and digital microscope, for many years. The results were not often consistent and cannot be sometimes summarized. Some phenomena are similar to the behaviour of clay minerals dispersed in water. This may be because both bacteria and clay particles have similar sizes and negative charges on their surfaces. In an electrolyte solution, both bacteria and clays possibly have the diffuse double layer of cations around their surfaces (Cruz et al., 2017; Poortinga et al., 2002).

When dispersed bacteria solution is mixed with a reactive solution, bacteria repulse each other. With time, bacteria form electrical double layers with Ca^{2+}. The bacteria surrounded with Ca^{2+} become aggregates and can move by advection in the biocement solution (BCS). Some of the aggregates adsorb or adhere on to the solid surface. Then, carbonates formed in suspended aggregates will increase their density, and they start settling and some aggregates coexist in BCS.

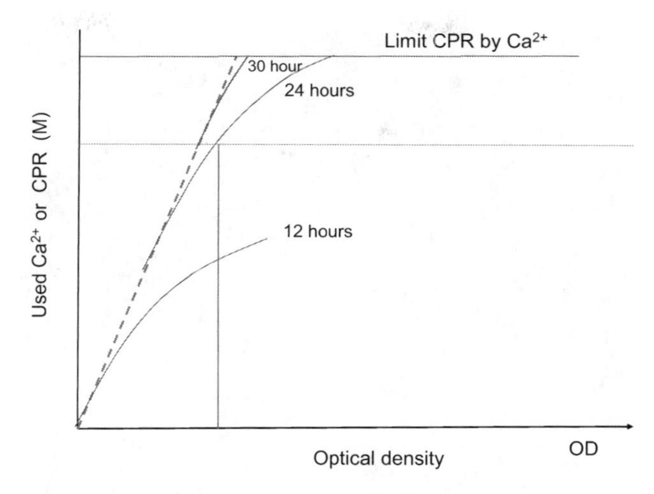

FIGURE 13.10 Interpretation of OD-CPR relationships in MICP.

FIGURE 13.11 Early chemical reaction, and aggregation of bacteria with Ca^{2+} in MICP. A photo taken with a digital microscope.

Figure 13.11 shows a photo in a video of BCS in a glass tube, taken with a digital microscope, when many aggregates are still invisible because no crystal has formed yet. In the photo, some of suspended aggregates with crystals are adsorbed onto the glass surfaces. After two and a half hours, suspended aggregates are not visible but the adsorbed aggregates (crystals) become greater, as shown in Figure 13.12. Figure 13.13 shows the growth of crystals after 24 h.

Bacteria play the role of nuclei for crystalline formation. Crystalline calcites grow with time, depending on the remaining Ca^{2+} and urea in the solution. These processes can be observed with a digital microscope and the crystal size can be measured also with the digital microscope.

13.3.5 Viable OD and Apparent OD*

It is assumed that even if the quality of bacteria is high or not, the OD-CPR relationship for a given bacteria species is unique, as defined earlier. The quality of the bacteria is theoretically reflected by

FIGURE 13.12 Crystalline growth on the glass surface without suspended aggregates, after two and a half hours (digital microscope).

FIGURE 13.13 Example of growth of crystalline minerals after 24 h reaction (digital microscope).

OD* and Rcv. The key point is that the theoretical relationship is expressed by CPR = *f*(OD) = *g*(Rcv OD*). Since OD is equal to Rcv OD*, the conversion from OD* to OD is simple. In order to obtain the OD from Rcv OD*, the OD* to be used has to be able to feedback from OD to OD*. To achieve this, it is required to reconciliate the state of bacteria sample and OD*. In other words, using the concentrated bacteria, the preparation method of BCS with a given OD* value is required. Since the OD value of cultivation solution is obtained as routine work, the feedback point was set at the OD value of cultivation solution, which is denoted as OD_i. The OD_i was not always constant but varied for each cultivation. Therefore, OD_i is defined as 4.0, which is the approximate average and target value for cultivation. The following is a definition of OD* and the required bacteria, and its calculation formula.

First, consider the state in which the OD is lowered to the initial value of 4.0 based on the volume V_{bac} of the frozen bacteria. The initial optical density at that time is OD_i (= 4.0). Since OD_i and OD* are the density of bacteria in terms of optical density, the mass balance of bacteria can be considered from the conservation law of bacterial mass. Considering the balance between the biomass for the initial cultivation solution and BCS, the balance can be given by

initial volume of cultivation solution × initial OD = the volume of BCS × OD* (to be used)

$$(V_{bac} \times C_f) \times OD_i = BCS \times OD^* \tag{13.6}$$

where V_{bac} is required bacteria for BCS, C_f is the centrifugal concentration factor, and BCS is the volume of the biosolution. Then, the required volume of bacteria is expressed by:

$$V_{bac} = BCS \times OD^*/(OD_i \times C_f) \tag{13.7}$$

where the concentration factor C_f is approximately 250, and OD_i is about 4.0.

Next, the CPR* when the apparent OD* of the microorganism is used is measured. Here, assuming that CPR* = CPR, the OD value can be obtained from the standard OD-CPR relationship, so that the rate of cell viability, Rcv can be obtained (Fukue et al., 2022).

$$Rcv = OD/OD^* \tag{13.8}$$

The biggest advantage of this concept is that OD* can be converted to OD to handle microbial activity in a unified manner, so that the results can be correctly analyzed and compared. It is not important if Equation (13.7) is correct or not, but it is a conversion rule. The most important thing is that the CPR is the key, because it provides both OD and Rcv OD*. Relation OD = Rcv OD* has to obey a rule, which is Equation (13.7) in our case.

13.4 R_{CV} TEST AND OD CONVERSION

As described earlier, it has been extremely difficult to determine the viability of bacteria as a material in MICP technology. However, without overcoming this problem, the MICP technology cannot be fully applied. The following is an easy-to-understand summary of what has been understood about MICP until very recently, rather than in chronological order.

The most important aspect of the engineering application of MICP technology is to establish a method to obtain the desired CPR by using an ideal blend of materials. Since CPR is defined by the induced $CaCO_3$ (M), the maximum CPR is limited by the Ca^{2+} concentration used. On the other hand, the CO_3^{2-} value that is the counter ion of Ca^{2+} is induced by CO_2 from the hydrolysis of urea, as given by reaction (13.2). However, there was no established method about how to predict the concentration of bacteria (enzymes) in BCS. The existence and formulation of the standard OD-CPR relationship have given us further idea and knowledge relating OD and Ca^{2+}/OD (Fukue et al., 2023). The apparent viability Rcv and apparent optical density OD* were defined to use the standard OD-CPR relationship as a calibration to determine the Rcv value.

13.4.1 One-Point Rcv Test and OD Conversion Method

The CPR experimental results can be analyzed using OD, which is converted from OD*. The OD* values used are tentative, to obtain OD*-CPR relationship. When the OD* value varies from 0 to about 0.22, the maximum precipitation of $CaCO_3$ increases from 0 to 1.0 M (100 %), as shown in Figure 13.9. For that, a Rcv value to be required can be mathematically obtained using an OD* value. The OD* values used are tentative, which neglect the effects of unknown factors on the Rcv value, such as loss of bacteria in the supernatant during centrifuging, the effects on Rcv by adding glycerin to concentrated bacteria, and unexpected temperature changes. In addition, the OD_i and C_f

values are not constant, but vary. In this study, both are fixed as 4.0 and 250, respectively, which are approximate averages of the actual data. The OD conversion enables solving of the problem previously mentioned because OD is expressed by "Rcv OD*", which can be converted from the one measured CPR for a given OD* at an elapsed time of 24 hours.

13.4.2 Evaluation of Rcv by a Single Point Method

The 24 h OD-CPR relationship Equation (13.5) is expressed with experimental results by the following quadratic equation:

$$17.633\,(\text{Rcv OD*})^2 - 8.46\,(\text{Rcv OD*}) + \text{CPR} = 0 \qquad (13.9)$$

The quadratic formula is

$$X = \frac{8.46 - \sqrt[2]{(8.46)\wedge 2 - 4(17.633)\text{CPR}}}{2(17.633)} \qquad (13.10)$$

Thus, X = OD can be obtained using only CPR or directly from Equation (13.8).

$$X = \text{Rcv OD*} = \text{OD} \qquad (13.11)$$

Taking the maximum CPR for three evaluations, each Rcv was determined by

$$\text{Rcv} = X/\text{OD*} \qquad (13.12)$$

where X and OD* are known. If Rcv is determined, any OD* can be converted to OD, i.e., Rcv times OD*. Equation (13.11) was already defined by Equation (13.8).

$$\text{OD} = \text{Rcv OD*} \qquad (13.13)$$

The example of Rcv determination using a single point method is indicated in Figure 13.14. The experimental data are presented in Table 13.1. Refrigerated NO-A10 strains were subjected to CPR tests at 35 days after thawing from a frozen state. That was why the Rcv values were low (0.09–0.13), in contrast to about 0.3 for the strains younger than 7 days after thawing.

In Figure 13.14, the square points indicate the measured CPR corresponding to the respective OD* value. The three estimated lines are 24h OD*-CPR relationships corresponding to the respective Rcv values. In other words, it shows that a specific Rcv value can be determined for a single OD* and CPR relation, and that the OD*-CPR relationship can be predicted when the Rcv is assumed to be constant for various OD*.

It is noted in Table 13.1 that test No. 8 took 8 days. Therefore, for the analysis to calculate Rcv, Equations (13.10)–(13.12) could not be used, but the following simple linear relationship was used. Neglecting the first term in Equation (13.9), we obtain,

$$8.46\,(\text{Rcv OD*}) - \text{CPR} = 0 \qquad (13.14)$$

Then,

$$\text{Rcv} = \text{CPR}/(8.46\ \text{OD*}) \qquad (13.15)$$

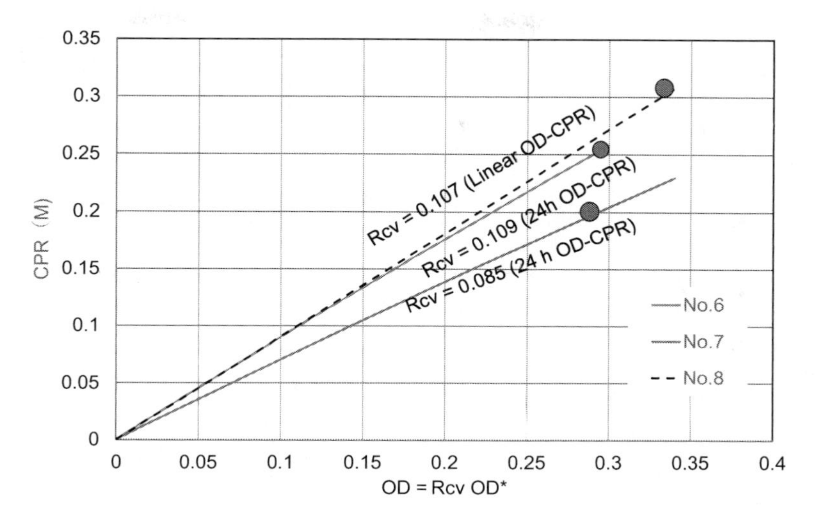

FIGURE 13.14 Example of CPR test for determining Rcv by a single point method.

TABLE 13.1
Experimental Results Obtained by the Single-Point Method for Determining Rcv

No	Time (h)	Tube (g)	Filter (g)	$CaCO_3$ (g/10 ml)	OD*	CPR (M)	Ca^{2+} (M)	X	Rcv (X/OD*)
6	24 h	0.239	0.012	0.251	0.292	0.251	0.3	0.0318	0.109
7	24 h	0.186	0.009	0.195	0.286	0.195	0.3	0.0243	0.085
8	8 days	0.297	0.001	0.298	0.328	0.298	0.3	0.0352	0.107

Tube: dry weight of $CaCO_3$ in glass tube, filter: dry weight of filtered $CaCO_3$.

Thus, Rcv was determined using CPR, OD* and a constant of 8.46, for No. 8 in Table 13.1.

13.4.3 Multi-Point Rcv Test and OD Conversion Method

Detailed methodology on the determination of Rcv is shown in the Appendix. Figure 13.15a) shows the OD*-CPR relationship. However, the relationships by fitting were obtained using the two Rcv values, i.e., 0.65 and 0.6. As seen in the figure, a Rcv of 0.6 is more likely for fitting the data points than 0.65. The converted OD-CPR relationship from OD*-CPR relationship in Figure 13.15a is presented in Figure 13.15b. The non-linear relationships have been generally obtained by other researchers (Lai et al., 2021).

The initial Ca^{2+}/OD after OD conversion was approximately 8.3 M, as presented in Figure 13.15b, which is close to the linear standard line (LSL), given by that CPR = 8.46 OD. This value satisfies the condition of non-inhibition, i.e., initial $Ca^{2+}/OD < 8.46$ (Fukue et al., 2023). It is noted that this condition provided for non-inhibition and only common retardation induced as a function of OD^2 (Fukue et al., 2023). Figure 13.15b shows the fitted OD-CPR relationship, which is in good agreement with the 24 h standard OD-CPR relationship. Note that the 24h OD-CPR relationship is provided for the short-time CPR or Rcv test for engineering purposes. However, in the design phase in engineering, it is necessary to consider linear relationships which can be in steady state.

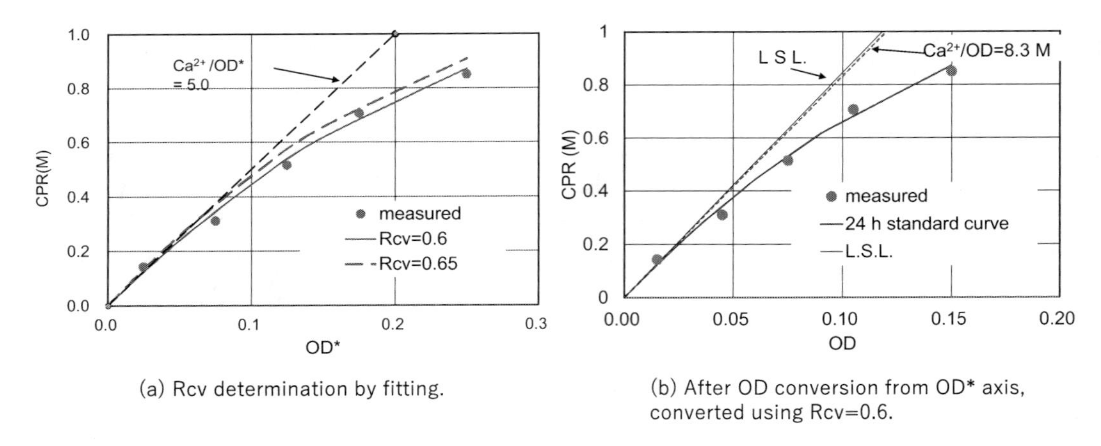

(a) Rcv determination by fitting.

(b) After OD conversion from OD* axis, converted using Rcv=0.6.

FIGURE 13.15 Conversion from 24 h OD*-CPR relationship (a) to 24 h OD-CPR relationship (b).

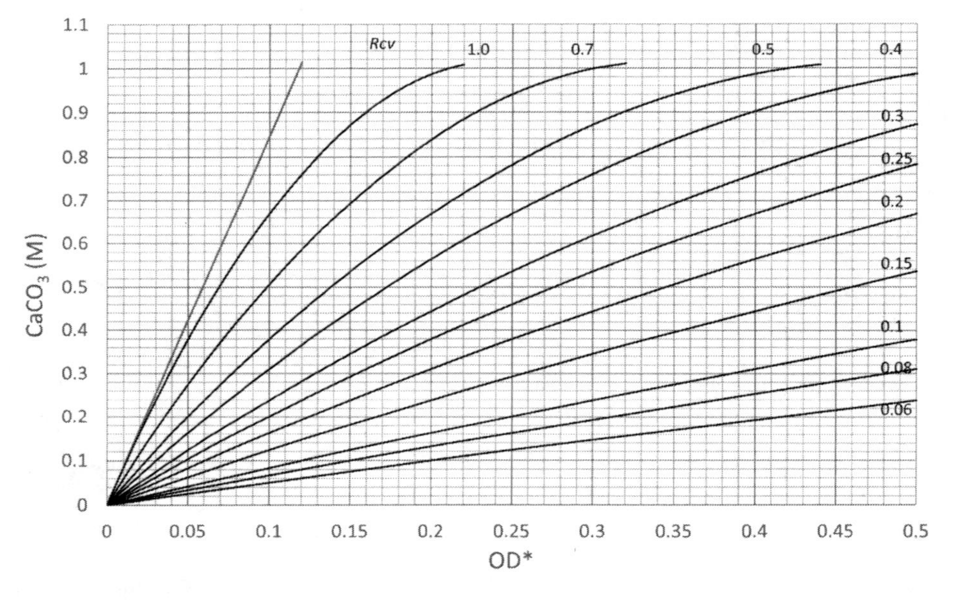

FIGURE 13.16 Calibration curves for Rcv determination by graphical solution using experimental 24 h OD*-CPR relationship.

When OD*-CPR relationship is obtained, the Rcv value will be obtained by comparison with the 24 h standard OD-CPR calibration curves theoretically prepared (Figure 13.16).

For the NO-A10 strains, the inhibition of MICP occurs when initial Ca^{2+} is higher than 1.5 M for any OD value. On the other hand, under an initial Ca^{2+} of 1.0 M, the inhibition occurs after the precipitation of approximately 0.1 M $CaCO_3$. However, when the initial Ca^{2+}/viable OD ratio becomes lower than a certain value, the inhibition will change into the retardation of MICP (Fukue et al., 2023).

Research on MICP technology has developed rapidly after 2000. However, there seems to be major obstacles to its development. Contradictions and a lack of understanding in the results of each study have been revealed (Lai et al., 2021). Fukue et al. (2022) thought that the main reason for this was due to the difficulty of evaluating the ageing of bacteria. Therefore, it was attempted to develop a concept and test method to quantify the capability of bacteria to Ca^{2+} concentration in a short time.

13.4.4 Rcv Change Due to Ageing

The bacteria viability in the bio-solution reduces due to ageing. In this section, the ageing of bacteria includes various factors, which cause the change in viability, such as the changes in pH and temperature, the effects of centrifugation, even unknown factors, etc., but the planned dilution and concentration are not included.

As demonstrated in the earlier sections, the viability of bacteria for engineering purposes was defined by Rcv, and the test method was established. As the initial Rcv test specimen, the thawed sample was used. In addition, to investigate the effect of elapsed time, chilled samples after thawing were used. Note that frozen samples involved 15% glycerol. On the other hand, two chilled samples without freezing were used to see the effect of glycerol. A 15 % of glycerol was added for one and no glycerol was used for the other.

Figure 13.17 shows the Rcv changes for the concentrated bacteria samples with various qualities. The Rcv values decrease by ageing, but little by freezing at –80 °C. The poor quality of bacteria showed low Rcv values, i.e., low viability. The poor quality may be due to contamination during cultivation, but sometimes the cause was unknown. The adverse effect of glycerol was clearly seen for the un-frozen sample. This may be because glycerol has similar properties of ethanol. However, this adverse effect was not realized in the results for frozen samples.

Figure 13.17 shows that the Rcv values varies with different culture solutions. This is the main reason why it is required to have a test method for the viability of bacteria, i.e., Rcv value. Under this situation, proper evaluation cannot be achieved in MICP, because the same quality of bacteria cannot be used to do the tests. In many cases, it is said that differences in the results were attributed to the bacterial species used. However, this is not the solution. The role of the urease-producing bacteria is hydrolysis of urea. Accordingly, the species of bacteria may affect the rate of hydrolysis. However, the rate depends on the bacterial concentration. However, considering that the MICP process starts

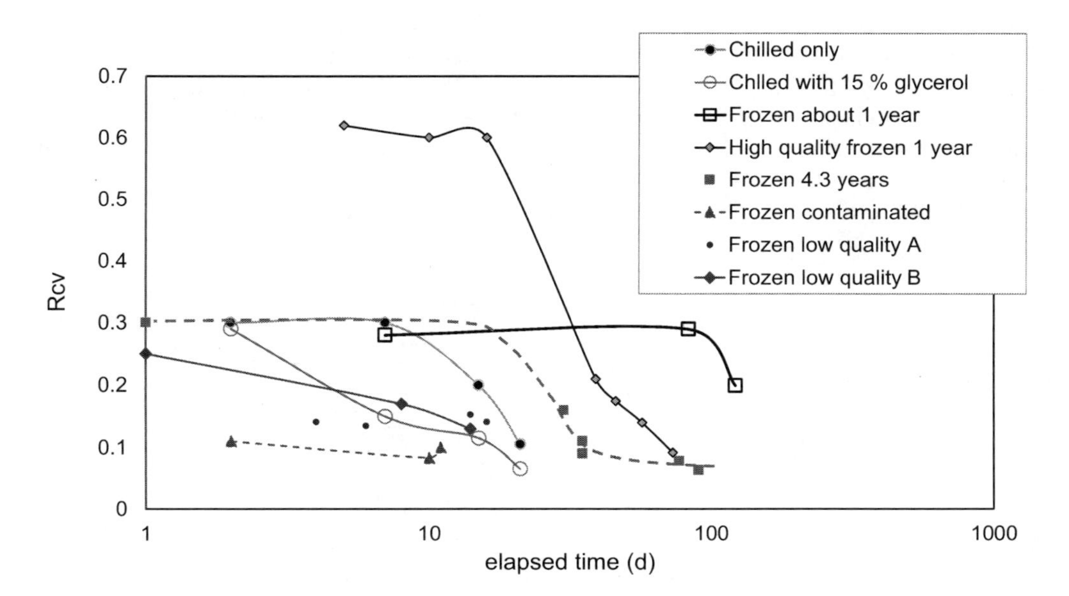

FIGURE 13.17 Rcv changes due to aging and other effects.

from the $CaCO_3$ formation, the difference in the results in the MICP process may not be due to the species of bacteria. This is because there is no clear standard for bacteria or strains capacity to precipitate carbonate.

13.5 INHIBITION AND RETARDATION IN MICP

13.5.1 Inhibition and Retardation

The retardation effects due to Ca^{2+} in MICP was discussed by Lai et al. (2021). They described that it is important to understand retardation in the MICP process, for getting a consensus on the experimental results in MICP. In fact, there is no consensus even on simple matters related to MICP (Lai et al., 2021). However, it is not sure whether the retardation is necessarily related to the consensus or not, because of the many factors affecting the experimental results in MICP.

The inhibition and retardation in the MICP process must be comprehensively interpreted. For that, it may be important to examine the experimental results with viable OD, which should be related to CPR. This is because the retardation occurs not only by Ca^{2+} but also OD, and possibly Ca^{2+}/OD. If the OD-CPR relationship is ambiguous, the experimental result in MICP becomes uncertain.

In the studies by Fukue et al. (2022, 2023), it is an empirical hypothesis that the 24 h standard OD-CPR relationship consists of two terms, which are a linear increase in CPR with OD and the retardation effect as a function of OD^2. The first term in the equation concerns the capacity of bacteria on carbonate precipitation, which means how much Ca^{2+} the bacteria can process (Fukue et al., 2023). The critical condition was given by Ca^{2+}/ OD = 8.46 M (Fukue et al., 2023). However, it is important to note that the concept of the existence of the critical condition due to Ca^{2+}/OD can be applied for other species of urease-producing bacteria. However, a constant number of 8.46 presented may vary for different species of bacteria, because the definition of OD-CPR relationship may not be unified. In addition, the range in OD values used are varied for bacteria species and researchers.

13.5.2 Inhibition and Retardation Due to 1.0 M Ca2+

Figure 13.18 shows the result of the 24 h CPR tests using an initial Ca^{2+} of 1.0 M and various OD values converted from the OD* (Fukue et al., 2023). If Ca^{2+}/OD is greater than 8.46 M, the precipitation CPR is about 0.1 M, which does not increase with elapsed time. This phenomenon is defined as the inhibition of MICP in this book, while at a Ca^{2+}/OD smaller than 8.46 M, the MICP increases with time. The boundary between inhibition and relatively slow rate precipitation seems to be given by an OD of 0.118, which results from the equation that Ca^{2+}/OD = 8.46 M. The slow precipitation is defined as the retardation by Ca^{2+}/OD. A further decrease in Ca^{2+}/OD value, i.e., increase in OD increases CPR. It should be remembered that the maximum CPR is equal to 8.46 OD. Thus, the

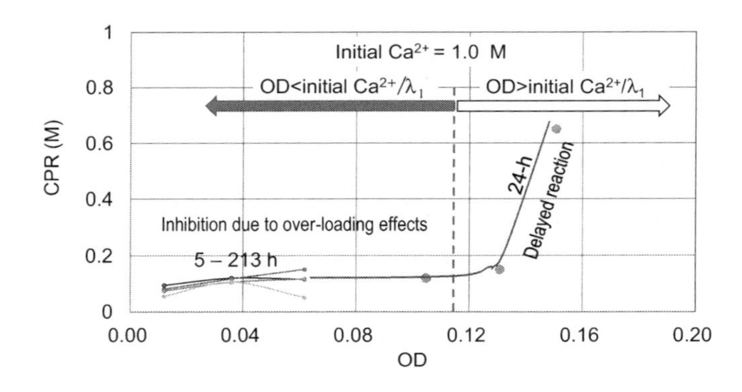

FIGURE 13.18 Inhibition and retardation in MICP using 1.0 M Ca^{2+} and a relatively low OD.

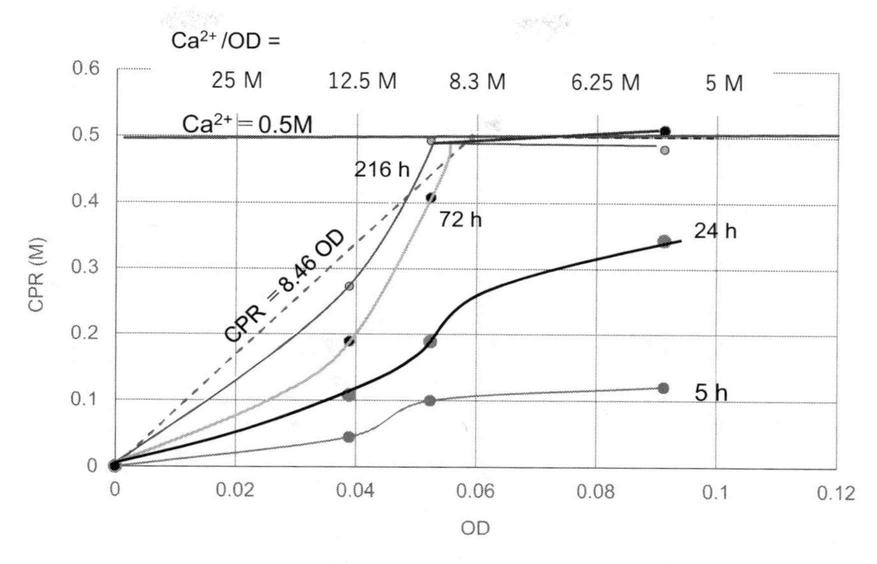

FIGURE 13.19 OD-CPR relationships using 0.5 M Ca^{2+}.

inhibition and retardation due to Ca^{2+} were affected by the Ca^{2+}/OD value. It should be noted that the results obtained here are due to the availability of OD conversion.

Such a phenomenon was gradually clarified by using the OD values of viable bacteria, and the results of the experimental studies conducted without OD evaluation would not only hinder understanding, but would also possibly undermine the credibility of MICP technology.

13.5.3 0.5 M Ca^{2+}

The lower initial Ca^{2+} results in lower Ca^{2+}/OD ratios. Figure 13.19 shows the OD-CPR relations with time using an initial 0.5 M Ca^{2+}. In comparison to the case of 1.0 M Ca^{2+}, it is understood that it is not inhibition due to Ca^{2+}, but retardation. As far as the low range in OD is concerned, the inhibition in Figure 13.18 seems to change into retardation. Thus, the reduction of the initial Ca^{2+} or Ca^{2+} /OD value prevents the inhibition. The retardation makes the reaction rate slower, but the induced CPR possibly reaches the value provided by the standard OD-CPR relationship (Fukue et al., 2023).

13.5.4 0.3 M Ca^{2+}

The OD-CPR relationships for further reduction of initial Ca^{2+} is shown in Figure 13.20, which was obtained using an initial 0.3M Ca^{2+}. The deviation in the OD-CPR relationships seems to be the effect of temperature, because a higher temperature leads to a higher reaction rate. Note that the retardation of Ca^{2+}/OD will be induced by the condition $Ca^{2+}/OD > 8.46$, i.e., at a relatively low OD to Ca^{2+} ratio. It is noted that the original experimental data are always obtained using OD*, and the analysis has to be on the converted values to those for OD values (Fukue et al., 2022, 2023). Thus, in MICP application, the OD conversion, which enables a unified evaluation of precipitation activity is quite important.

13.5.5 Retardation due to Constant Ca^{2+}/OD Ratio

The condition of constant Ca^{2+}/OD ratio means the increasing loading condition of Ca^{2+} with increasing OD values. Therefore, the OD-CPR relationship is influenced by both initial Ca^{2+}/OD ratio and the initial Ca^{2+} value itself. In addition, in order to understand the MICP process under a

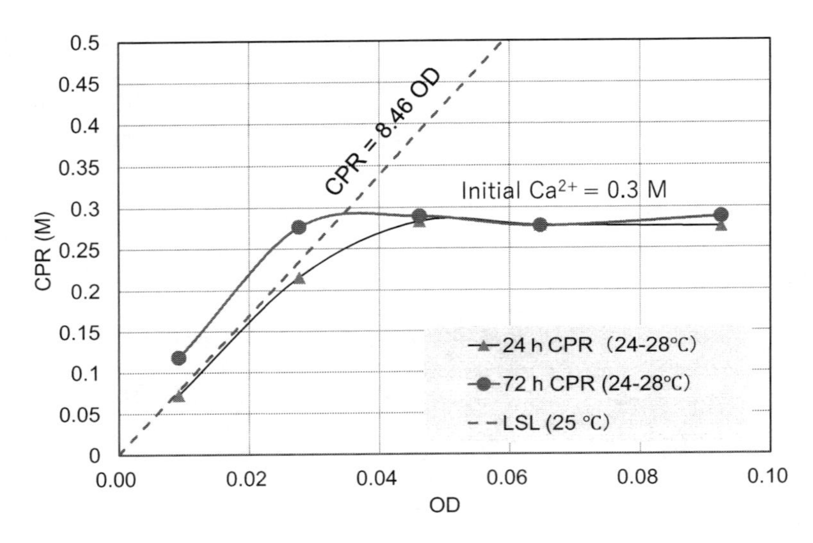

FIGURE 13.20 OD-CPR relationships under 0.3 M Ca^{2+}.

constant Ca^{2+}/OD, the linear and non-linear OD-CPR relationships should be taken into account. In the linear OD-CPR relationship, the time is assumed as infinite (first term of Equation (13.5)), while the non-linear relationship involves the second term of Equation (13.5). However, the maximum CPR can be obtained within a few days.

In the analysis for a non-linear relationship, the elapsed time should be limited to 24 h. The second term of Equation (13.5) is the retardation effect concerning OD^2, which may be dissimilar to the other types of retardation. This type of retardation is negligible when OD is relatively low. It is possible that the retardation of this type may be due to competition among a large number of bacteria to take in Ca^{2+} ions, which are decreasing.

Even if the initial Ca^{2+}/OD ratio is constant, the retardation is induced if the ratio is higher than 8.46 M, as shown in Figure 13.21. In addition, the figure shows four types of OD-CPR relationships, which were converted from the originally OD*-CPR relationships to the OD-CPR relationships, respectively. The LSL is the unique linear standard line for NO-A10, and the 24 h OD-CPR is also a unique relationship, experimentally obtained at 24 hours of reaction using NO-A10 (Fukue et al., 2022). The fourth OD-CPR relationship is recognized at 24 h measured values, indicated by the circle points. The relationship shows retardation behaviour at relatively high OD values.

In general, the measured OD-CPR relationship has to agree with the 24 h OD-CPR relationship. However, the measured values shown in the figure are out of the 24h OD-CPR relationship. It is because that the Ca^{2+}/OD is greater than that on the LSL (Fukue et al., 2023). If the Ca^{2+}/OD is equal or less than that of the LSL, the measured values of OD-CPR have to agree with the 24 h OD-CPR relationship. Thus, retardation can occur at a relatively high range of Ca^{2+}/OD ratio. However, it is noted that this type of retardation can be recovered within one more day. In other words, the CPR will be 1.0 M within 2 days in this case.

The inhibition and/or retardation can be avoided at an OD higher than 1/8.46. Thus, the inhibition and retardation in MICP is comprehensively explained case by case, based on the critical Ca^{2+}/OD value. However, CPR is influenced by other factors such as temperature, pH, etc. Therefore, these factors should be taken into account when the range of factors varies widely. It is important to note that the OD here is not a general optical density with bacteria, but a particular optical density defined in this chapter.

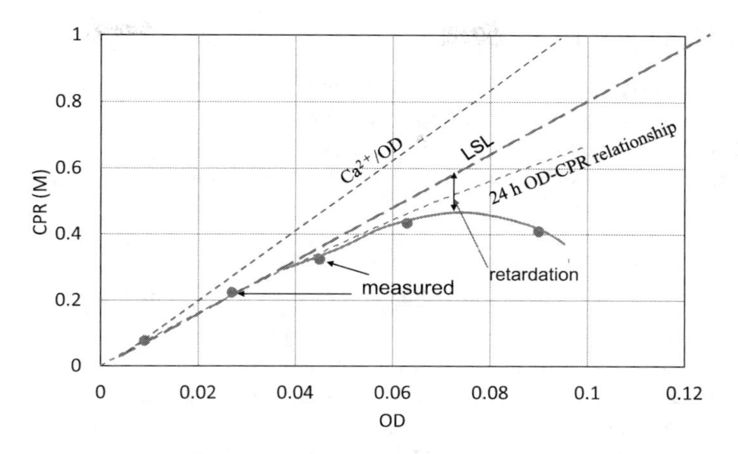

FIGURE 13.21 Various OD-CPR relations, and retardation which occurred at relatively high Ca^{2+}/OD values.

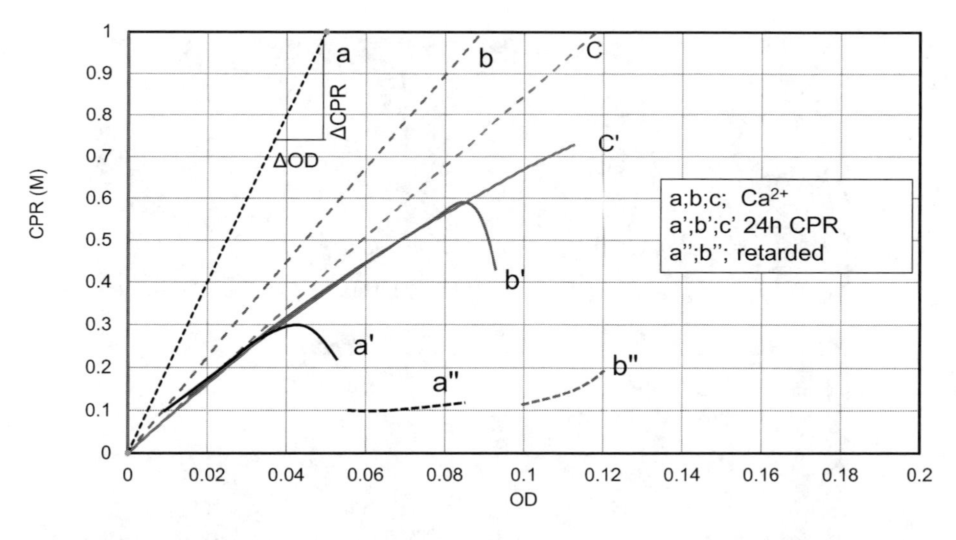

FIGURE 13.22 Effects of Ca^{2+} and Ca^{2+}/OD in MICP. (1) lines a, b, and c: Ca^{2+} used for BCS, (2) line c: L.S.L, (3) a', b', and c': measurement CPR, (3) 24 h Standard OD-CPR line, (4) a', b': include retardation, (5) a'', b'': retardation.

Figure 13.22 illustrates a summary of retardation behaviour in MICP in terms of initial Ca^{2+}/OD. In the figure, lines a, b and c indicate constant Ca^{2+}/OD ratios with various Ca^{2+} and OD values, respectively. Lines a', b', and c' are generally experimentally measured corresponding to lines a, b, and c, respectively. As far as line a is concerned, the constant Ca^{2+}/OD ratio is very high. When the Ca^{2+} is in a low range, OD-CPR measurement follows the 24 h OD-CPR relationship as shown in line a', but the retardation behaviour occurs when the Ca^{2+}/OD approaches the CPR limit for bacteria OD of 8.46. The retardation appears as the reduction of CPR from the designated value. If the Ca^{2+}/OD clearly exceeds 8.46 M, the CPR drops to approximately 0.1 M, which is due to inhibition induced by a very high Ca^{2+}/OD ratio. If the measurement is continued using Ca^{2+} = 1.0 M, the result is shown in line a'' in Figure 13.22 . Note that experimentally, the Figure 13.22 can be obtained using OD values converted from OD*. For that, Rcv values have to be used.

13.6 MORPHOLOGY IN MICP

The morphology in MICP has not been well understood, because various crystals form under the same condition. For example, the morphology in the bio-solution varied even in the slightly leaning test glass tubes (Fukue et al., 2023). Therefore, considering *in situ* soil conditions, it is considered as meaningless to evaluate the *in situ* carbonate morphology based on the laboratory experiments. In particular, under the presence of other metal cations, the morphology of products may be complicated. That is why further research is needed. Figures 13.23 to 13.25 show morphologically interesting SEM images.

Thus, calcite in MICP can be initially induced by amorphous particles, and then changes to calcite crystal(s), depending on the initial states of aggregation of bacteria with Ca^{2+}. Since the MICP process proceeds with the hydrolysis of urea, the calcite precipitation cannot instantaneously occur. This is of importance when MICP is used as a binding material for soil improvement, because no binding effects can act for instantaneous precipitation of $CaCO_3$. The morphological study in the MICP process can be assisted by both SEM and digital microscope images. The identification of the images may require the use of both methods (Fukue et al., 2011, 2022, 2023). The MICP may

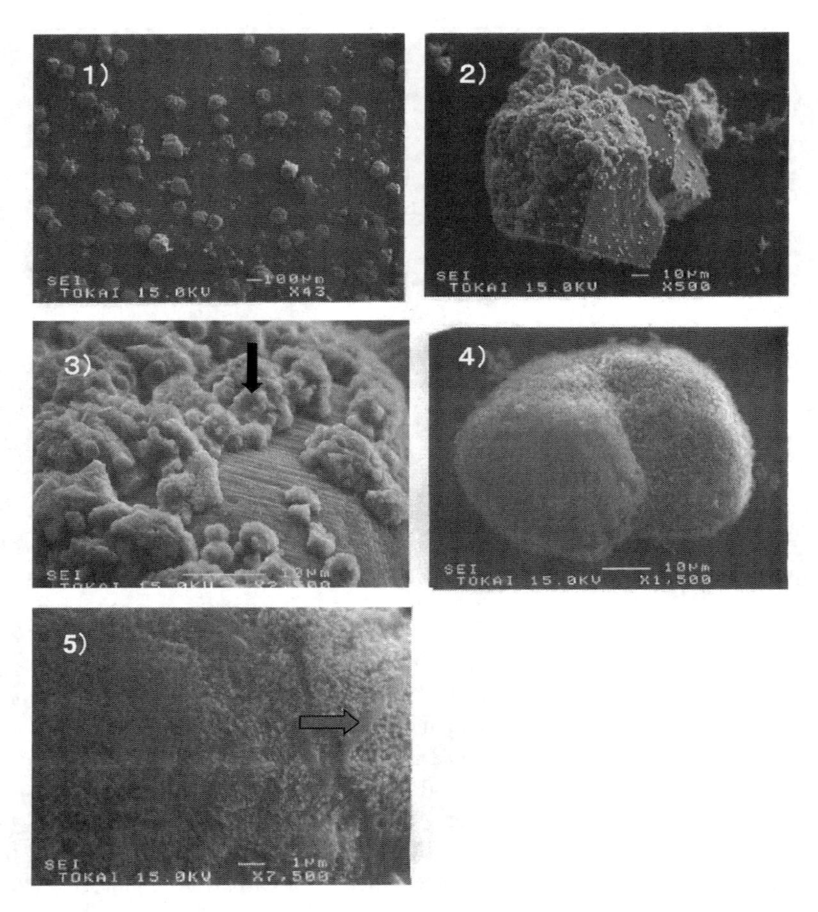

FIGURE 13.23 SEM images of calcite precipitated on a glass plate in the BC solution, with various magnifications. (1): many scattered particles greater than 100 μm, (2): Single particles greater than 100 μm, (3): amorphous calcite on the calcite crystal, (4): a large number of very small amorphous calcite particles, with a high magnification, bacterium size: approximately ϕ1 μm × 5 μm.

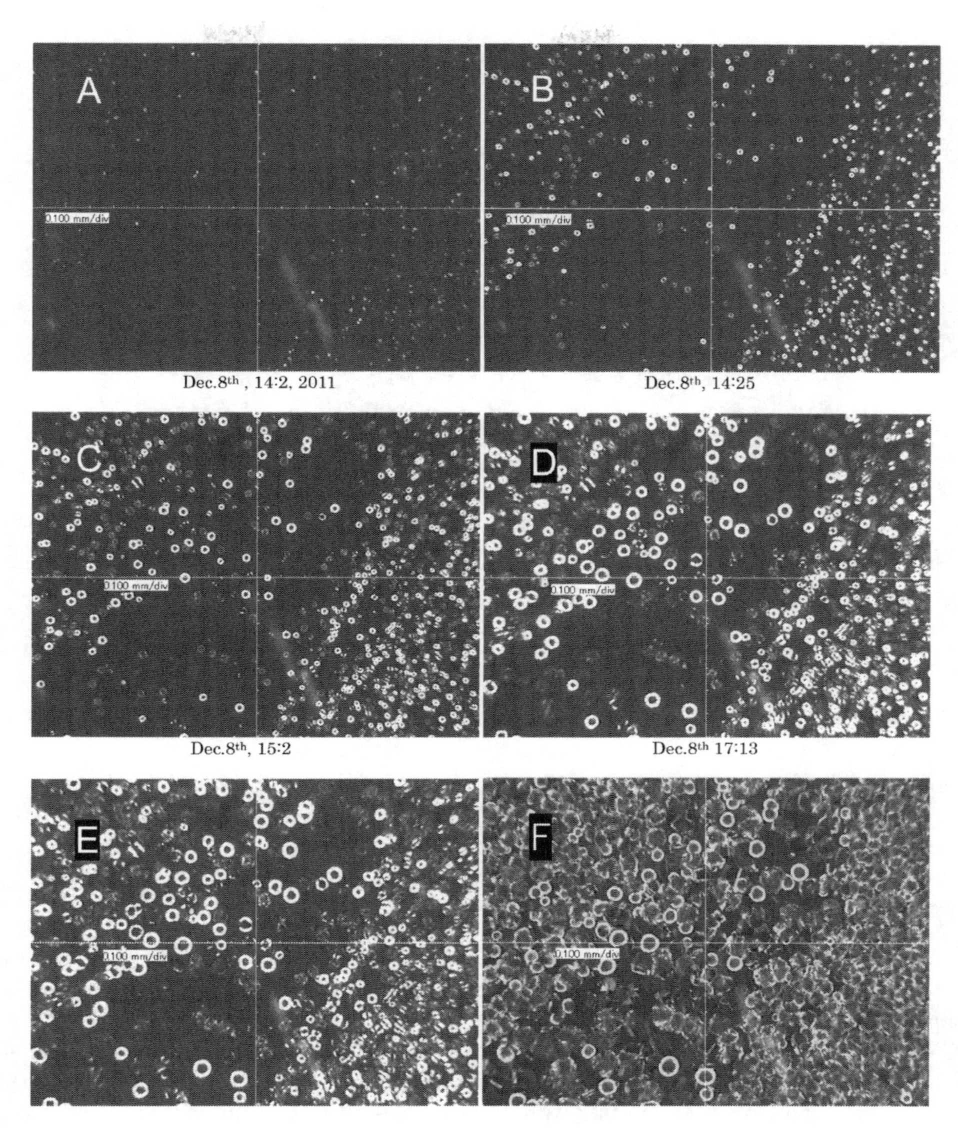

FIGURE 13.24 Digital microscope SEM images of spherical aggregates of bacteria with growing amorphous calcites and growing calcite crystals from aggregates, (A): Aggregates of bacteria with growing amorphous calcites, and many aggregates can be invisible, (B–E): Both growing aggregates and crystalline particles, (F): Crystallization from aggregates.

be affected if ions other than Ca^{2+} are involved, the style and morphological situation may change completely. For example, incorporation of Mg^{2+} is demonstrated in Figure 13.25.

13.7 CONCLUDING REMARKS

Since about 2000, MICP has been increasingly studied in the field of geotechnology. However, a rapid increase in researchers and published articles do not necessarily achieve significant advances from a practical point of view. Borrowing or imitating microbiology practices does not succeed in applied technology development. MICP may be totally the theme of biomineralization or

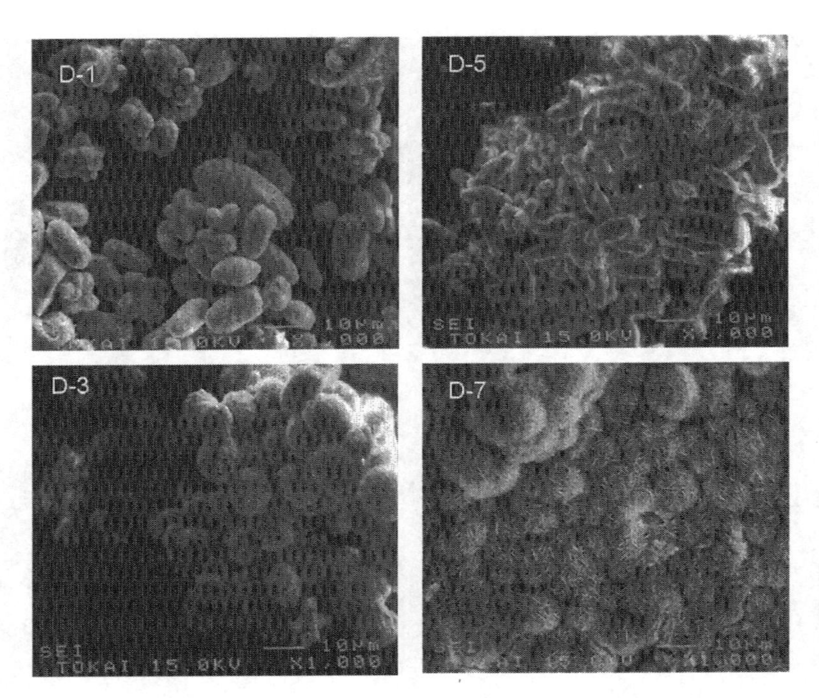

FIGURE 13.25 SEM images of the change with time in morphological features in MICP using 0.5 M Ca^{2+} and 0.5 M Mg^{2+}. (D-1) (Fukue et al., 2011): Reaction time of 1 h, amorphous calcite particles, (D-3): Reaction time of 9 h, disturbance of the surfaces of particles due to incorporation of Mg^{2+}, (D-5): Reaction time of 33 h, change to aragonite particles, (D-7): Reaction time of 72 h, change to proto dolomite by continuous effects of Mg^{2+}, incorporation of Mg^{2+} more than one month.

microbiology. However, it is difficult to establish some new field technology, using only knowledge and methods from biomineralization or microbiology. To move forward, it should be determined whether a traditional method can be used in a new field, and if that is not possible, then a new method can be devised.

In this chapter, the first topic was how to evaluate bacteria capacity for precipitation of carbonate, but not the urease activity. It is because urease activity is the capacity for hydrolysis of urea but not carbonate precipitation. In other words, it becomes possible to measure bacteria precipitation capacity of carbonates, the measurement of urease activity is not needed. In fact, this subject was solved by developing the test method of Rcv value, which is a viability of bacteria used in the test. Accordingly, the quantitative capacity of bacteria can be denoted on [e] in Equation (13.2). However, it is noted that the concept of Rcv can be used for any urease-producing bacteria, but the new standard calibration equation or curve is required for other bacteria.

The measurement of viability in microbiology took a long time and could not be applied for engineering purposes, and the conversion of the measurement items from urease activity to carbonate production amount was a big benefit. To date, the test results have been compared and evaluated by each researcher's OD value without being able to evaluate aged bacteria. In this situation, it is difficult to reach a consensus and to proceed to the next step (Lai et al., 2021; Fu et al., 2023).

In the second half of this chapter, it was shown how to convert the experimental results into unified OD values using Rcv. As a result, the conditions of inhibition and retardation in the MICP reaction were clarified. In this chapter, the most basic properties and behaviour of biocement necessary for MICP were dealt with. The results are summarized as follows.

(1) There is inhibition and retardation as the behaviour of biocement consisting of ureolytic bacteria, urea, Ca^{2+}, etc.
(2) The bacteria are sensitive to time, so that the rapid evaluation of bacteria is urgently required. The conventional method using an OD value is not applicable for engineering purposes.
(3) To estimate an optimum number of bacteria for the carbonate precipitation, a new method was established by using a standard OD – carbonate precipitation relationship.
(4) To estimate the number of bacteria, viability of bacteria (Rcv) was defined as OD = Rcv OD*, where OD* is the apparent OD value of aged (unknown) bacteria and also used as the optimum number of bacteria.
(5) The Rcv test was established to obtain the Rcv value for aged bacteria. Then, simply test, Rcv OD* → OD design → Rcv OD*→OD* (engineering application).
(6) Converting from (OD*, CPR) to (OD, CPR), a unified analysis became possible based on old data in MICP.
(7) Inhibition and retardation were conditioned by the Ca^{2+}/OD value.

REFERENCES

Ahrmad, J., Khan, M.A., and Ahmad, S. (2023), State of the art on factors affecting the performance of MICP treated fine aggregates, *Materials Today: Proceedings*, 24, 4130–4142. https://doi.org/10.1016/j.matpr.2023.04.087.

Benini, S., Gessa, C., and Ciurli, S. (1996), *Bacillus pasteurii* urease: A heteropolymeric enzyme with a binuclear nickel active site, *Soil Biology & Biochemistry*, 28: 819–821.

Burley, S.D., and Worden, R.H. (eds.), (2003), *Sandstone Diagenesis-Recent and Ancient*, Reprint Series Volume 4 of the International Association of Sedimentologists, Black Well Publishing Ltd, Oxford, UK, 649 pp.

Castanier, S., Métayer-Levrel, G.L., and Perthuisot, J.-P. (1999), Ca-carbonates precipitation and limestone genesis-the microbiogeologist point of view, *Sedimentary Geology*, 126: 9–23.

Ciurli, S., Marzadori, C. Benenel, S. Delana, S., and Gessa, C. (1996), Urease from soil bacterium *Bacillus pasteurii*: Immobilization on Ca-polygalacturonate, *Soil Biology* & *Biochemistry*, 28(6): 811–817.

Cruz, R.C.D., Segadães, A.M., Oberacker, R., and Hoffmann, M.J. (2017), Double layer electrical conductivity as a stability criterion for concentrated colloidal suspensions, *Colloids and Surfaces A: Physicochemical and Engineering Aspects*, 520: 9–16.

De Muynck, W., Cox, K., De Belie, N., and Verstraete, W. (2008), Bacterial carbonate precipitation as an alternative surface treatment for concrete, *Construction and Building Materials*, 22: 875–885.

DeJong, J.T., Fritzges, M.B., and Nüsslein, K. (2006), Microbially induced cementation to control sand response to undrained shear, *Journal of Geotechnical and Geoenvironmental Engineering*, 132: 1381–1392.

Fenn, L.B., and Miyamoto, S. (1981), Ammonia loss and associated reactions of urea in calcareous soils, *Soil Science Society of America Journal*, 45: 537–540.

Ferris, F.G., Phoenix, V. Fujita, Y., and Smith, R.W. (2003), Kinetics of calcite precipitation induced by ureolytic bacteria at 10 to 20°C in artificial groundwater, *Geochimica et Cosmochimica Acta*, 67(8): 1701–1722.

Francois, F., Devlieghere, F., Standaert, A.R., Geeraerd, A.H., Cools, I., Van Impe, J.F., and Debevere, J. (2005), Environmental factors influencing the relationship between optical density and cell count for Listeria monocytogenes, *Journal of Applied Microbiology*, 99: 1503–1515.

Fu, T., Sarachob, A.C., and Haigha, S.K. (2023), Microbially induced carbonate precipitation (MICP) for soil strengthening: A comprehensive review, *Biogeotechnics*, 1: 1–23, https://doi.org/10.1016/j.bgtech.2023.100002.

Fukue, M., Lechowicz, Z., Fujimori, Y., Emori, K., and Mulligan, C.N. (2022), Incorporation of optical density into the blending design for a biocement solution, *Materials*, 15, https://doi.org/10.3390/ma15051951.

Fukue, M. Lechowicz, Z. Fujimori, Y. Emori, K. Mulligan, C.N. (2023), Inhibited and retarded behavior by Ca^{2+} and Ca^{2+}/OD loading rate on ureolytic bacteria in MICP process, *Materials* 16: 3357. https://doi.org/10.3390/ma16093357

Fukue, M., Nakamura, T., and Kato, Y. (1999), Cementation of soils due to calcium carbonate. *Soils and Foundations*, 39: 55–64.

Fukue, M., Ono, S., and Sato, Y. (2011), Cementation of sands due to microbiologically induced carbonate, *Soils and Foundations*, 51: 83–93.

Hammes, F.N., Boon, G., Clement, G., de Villiers, S.G., Verstraete, W., and Siciliano, S.D. (2003), Strain-specific ureolytic microbial calcium carbonate precipitation, *Applied and Environmental Microbiology*, 69(8): 4901–4909.

Jimenez-Lopez, C., Jroundi, F., Pascolini, C., Rodriguez-Navarro, C., Piñar-Larrubia, G., Rodriguez-Gallego, M., and González-Muñoz, M.T. (2008), Consolidation of quarry calcarenite by calcium carbonate precipitation induced by bacteria activated among the microbiota inhabiting the stone, *International Biodeterioration & Biodegradation*, 62: 352–363.

Kahani, M., Kalantary, F., Soudi, M.R. and Pakdel, L., and Aghaalizadeh, S. (2020), Optimization of cost-effective culture medium for *Sporosarcina pasteurii* as biocementing agent using response surface methodology: Up cycling dairy waste and seawater, *Journal of Cleaner Production*, 253: 120022, https://doi.org/10.1016/j.jclepro.2020.120022.

Lai, H.-J., Cui, M.-J., Wu, S.-F., Yang, Y., and Chu, J. (2021), Retarding effect of concentration of cementation solution on biocementation of soil, *Acta Geotechnica*, 16: 1457–1472, https://doi.org/10.1007/s11440-021-01149-1.

Le Métayer-Levrel, G., Castanier, Ł.S., Orial, G., Loubiére, J.-F., and Perthuisot, J.P. (1999), Applications of bacterial carbonatogenesis to the protection and regeneration of limestones in buildings and historic patrimony, *Sedimentary Geology*, 126: 25–34.

Lian, B., Qiaona Hu, Q., Chen, J., Ji, J., and Teng, T. (2006), Carbonate biomineralization induced by soil bacterium *Bacillus megaterium*, *Geochimica et Cosmochimica Acta*, 70: 5522–5535.

Mobley, H.L.T., and Hausinger, R.P. (1989), Microbial urease: Significance, regulation, and molecular characterization, *Microbiological Reviews*, 53: 85–108.

Omoregie, A.I., Ginjom, R.H., and Nissom, P.M. (2018), Microbially induced carbonate precipitation via ureolysis process: A mini- review, *Transaction on Science and Technology*, 5: 245–256.

Polacco, J.C., and Holland, M.A. (1993), Roles of Urease in Plant Cells. In: K.W. Jeon and J. Jarvik (eds.), *International Review of Cytology*, Academic Press, Inc., San Diego, vol. 145, pp. 65–103.

Poortinga, A.T., Bos, R., Norde, W., and Busscher, H.J. (2002), Electric double layer interactions in bacterial adhesion to surfaces, *Surface Science Reports*, 47: 1–32.

Rodriguez-Blanco, J.D., Sand, K.K., and Benning, L.G. (2017), ACC and Vaterite as Intermediates in the Solution-Based Crystallization of $CaCO_3$. Chapter 5. In: A. VanDriessche, M. Kellermeier, L. Benning, and D. Gebauer (eds.), *New Perspectives on Mineral Nucleation and Growth*, Springer International Publishing, Cham, Switzerland, pp. 93–111.

Sirko, A., and Brodzik, R. (2000), Plant ureases: Roles and regulation, *Acta Biochimica Polonica*, 47(4): 1189–1195.

Stocks-Fischer, S., Galinat, J.K., and Bang, S.S. (1999), Microbiological precipitation of $CaCO_3$, *Soil Biology and Biochemistry*, 31: 1563–1571.

Sun, X., Tong, T., and Wang, C. (2019), Study of the effect of temperature on microbially induced carbonate precipitation, *Acta Geotechnica*, 14: 627–638, https://doi.org/10.1007/s11440-018-0758-y.

Tiano, P., Biagiotti, L., and Mastromei, G. (1999), Bacterial bio-mediated calcite precipitation for monumental stones conservation: Methods of evaluation, *Journal of Microbiological Methods*, 36(1999): 139–145.

Wang, Y., Wang, Z., Chen, Y., Cao, T., Yu, X., and Rui, P. (2023), Experimental study on bio-treatment effect of the dredged Yellow River silt based on soybean urease induced calcium carbonate precipitation, *Journal of Building Engineering*, 75: 106943, https://doi.org/10.1016/j.jobe.2023.106943.

Watanabe, Y., Sarumaru, H., and Shimada, N. (1983), Distribution of urease in the higher plants, *Technical Bulletin of Faculty of Horticulture, Chiba University*, 32: 37–43 (text in Japanese with English Abstract).

Wen, K., Yang, L., Farshad, A., and Lin, L. (2020), Impact of bacteria and urease concentration on precipitation kinetics and crystal morphology of calcium carbonate. *Acta Geotechnology*, 15: 17–27.

Whiffin, V.S., van Paassen, L.A., and Harkes, M.P. (2007), Microbial carbonate precipitation as a soil improvement technique, *Geomicrobiology Journal*, 24: 417–423.

14 MICP Soil Improvement

14.1 INTRODUCTION

Since around 2000, the MICP technology for geotechnical engineering have been developed. Therefore, the MICP mechanisms and application has been increasingly studied in many countries. Accordingly, the number of the articles concerning MICP has been rapidly increased and they are published in various journals and books (Omoregie et al., 2021). Similarly, as other applications of MICP, the remediation and immobilization technologies of heavy metals have also been developed in many countries (Lin et al., 2023a). The mechanisms of MICP were dealt with in Chapter 13, and the remediation and immobilization of heavy metals were dealt with in Chapter 11.The basic properties and behaviour of biocement solution were demonstrated in Chapter 13, which include the optimum blending of biocement solution, the inhibition and retardation of MICP, based on new concepts. These were the important subjects that are lacking in the application of MICP so far. Chapter 14 deals with sustainable soil improvement technology based on the MICP-soil system. This is a coupling between MICP technology and geotechnical engineering.

Simply, the technology can be described as the solidification of granular materials. However, it is environmentally, methodologically, and effectively different from the conventional technologies. For example, the new technology can form almost arbitrary strength of rock. Moreover, it is very effective because it increases both the adhesive force, c and the internal friction angle, ϕ. If both c and ϕ increases together, the bearing capacity of soils is remarkably increased. This is easily confirmed by the theory of bearing capacity.

From the viewpoint of being environmentally friendly, the comparison between ordinary cement and biocement shows a clear answer, as described later. Methodologically, it is easy to inject biocement solution into granular soil, because of the low viscosity of the solution. This does not need heavy equipment. Construction (injection) can be stopped and resumed. The cultivation of bacteria can be achieved with a 1.0 m^3 plastic bag. The centrifuged bacteria were stored safely in freezers at least more than 4 years at –80 °C (Fukue et al., 2022, 2023).

The production of carbonates was discussed in Chapter 13. In other words, it was shown that the carbonate precipitation rate (CPR) can be optimally blended with the ratio of Ca^{2+} and bacterial viability. Therefore, Chapter 14 derives the relationship between the carbonate content of the soil and CPR and uses it in the design.

MICP technology has not been commercially used because of the lack of academic and technical preparation. In this chapter, the aim is to proceed solving these subjects for commercialization. The purpose of geotechnical engineering and environmental geotechnical engineering to which MICP is applied are to stabilize brittle ground by generating carbonates in the ground. It is also the purpose to create a safe environment by fixing harmful substances in the ground (Chapter 11).

14.2 SOIL-MICP SYSTEM

14.2.1 Definition of Elements of Soil-MICP System

The carbonate precipitation rate, CPR, is defined as the MICP rate (M) by a BC solution, where M is $CaCO_3$ in mole/L of BCS. On the other hand, the carbonate content, C, is defined by mass of carbonates precipitated (m_c)/ mass of soil particles in soil volume (m_s) × 100 (%), as shown in Figure 14.1.

DOI: 10.1201/9781003407393-14

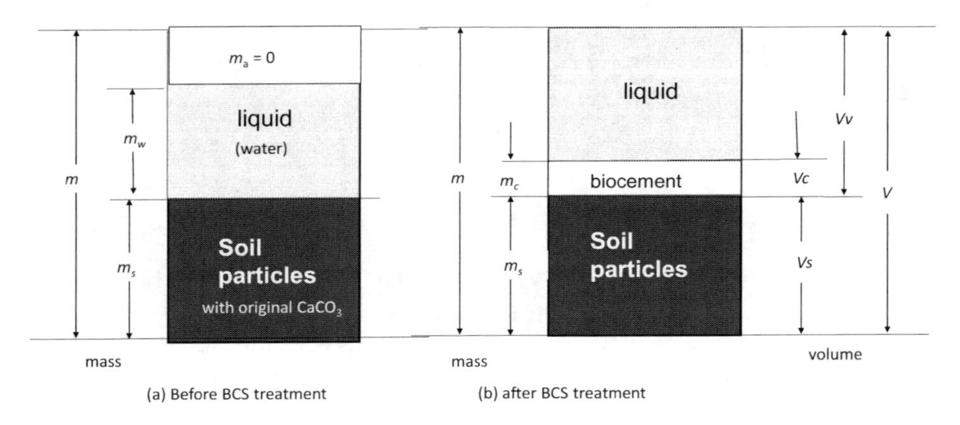

FIGURE 14.1 Definition of elements used in soil-MICP solution system.

TABLE 14.1
Definition of Various Elements of the Soil-MICP System

V_v/V: Initial porosity	m_s/V: Dry density of soil
V_s/V: Solid volume ratio	$C = m_c/m_s \times 100$ (%): Carbonate content of soil (%)
V_c/V_s: Volumetric carbonate content	$m_c = C\, m_s/100$
V_c/V: Carbonate volume ratio	$(m_s + m_c)/V = m_s/V + m_c/V = m_s/V + (C/100)\, m_s/V$
$(V_v\text{-}V_c)/V$: Final porosity	$=(m_s/V)(1+C/100)$: Dry density of BC soil
m_s/V_s: Density of soil particles	m_c/V: Carbonate density of BC soil

Before the BCS treatment, soil consists of three elements, such as solid, liquid, and gas. The solids are usually minerals and organic matter, the liquids are usually water including seawater and other types of solutions such as the MICP solution, and the gases are mostly air and other types of gases, such as NH_3. After the BCS treatment, biocement, i.e., $CaCO_3$ is added to the solid, and the pore water is replaced by ammonium chloride solution as a by-product of the reaction of MICP. Note that general soil contains some $CaCO_3$ as a part of solid, and limestones are mostly composed of $CaCO_3$. If the effect of the $CaCO_3$ by the BCS treatment is concerned, the original $CaCO_3$ content of the sample should be distinguished from the $CaCO_3$ induced by MICP. This is apparent by the definition of carbonate content. In Figure 14.1, the volume of soil particles as solid, V_s includes the originally contained carbonates. The V_c is the volume of biocement, and the volume of solids, $V_s + V_c$, is the total solid volume. Using the definitions of the elements shown in Figure 14.1, the following terms in Table 14.1 are defined with the volumetric characteristics of each element.

14.2.2 Measurement Method for Carbonate Content

The carbonate content can be measured using the dissolution reactions of carbonates with acid. The carbonate content was determined by measuring the CO_2 gaseous pressure produced from the following reaction;

$$CaCO_3(s) + 2H^+ \rightarrow Ca^{2+} + H_2O + CO_2(aq) \tag{14.1}$$

Subsequently, according to Henry's law, the dissolved carbon dioxide will be in equilibrium with the gaseous CO_2.

$$CO_2(aq) \rightarrow CO_2(g) \tag{14.2}$$

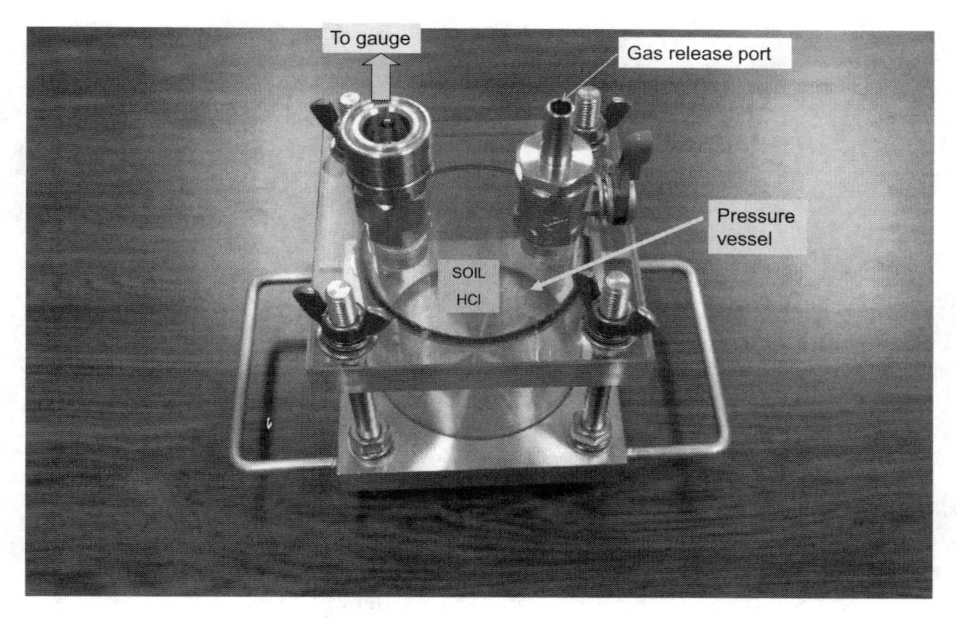

FIGURE 14.2 Device for carbonate content of soils using CO_2 gaseous pressure.

To investigate the reaction, a calcite–acid reactor was used (Fukue et al., 1999). The device is shown in Figure 14.2. The device for carbonate content test consists of a reactor chamber, pressure metre, and valve for the exhaust gas. The calibration curve obtained using calcium carbonate agent is shown in Figure 14.3. The carbonate content C is then defined as

$$C = \frac{carbonate}{dry\ mass\ of\ soil} \times 100(\%) \tag{14.3}$$

The hydrochloric acid concentration used was usually 3.0 M. Calcium carbonate sample of about 0.1 g were used for obtaining the calibration curve and the CO_2 gaseous pressure produced from the reaction was measured over time.

A few grams of dry soil specimen were used to measure the CO_2 gaseous pressure produced. The mass of calcium carbonate was obtained using the calibration curve (Fukue et al., 2011).

14.3 MICROBIOLOGICAL SOIL MECHANICS

14.3.1 Definition of Strength for Ordinary Soil and BCS Treated Soil

In the practice of geotechnical engineering, soil strength is necessary for design, so the strength constant of the *in situ* soil is determined, or the strength is obtained by appropriate testing of the sampled soil. In biocement technology, the aim is to stabilize the ground and structures on the ground by increasing the bearing capacity of the ground. The degree of intensity to be increased depends on how much CPR or carbonate content C is used. The carbonate content C is directly calculated from the CPR. Therefore, it can be adjusted depending on the concentration of the biocement liquid (BCS) used or the number of injections into the ground. However, the effect of biocement also depends on the type of soil. If it is needed, the pre-injection test can be used to examine the effects of biocement.

The concept of material strength in construction engineering varies depending on the material type. For geotechnical materials, the Coulomb or Mohr-Coulomb failure criterion is usually used. In this criterion, the concept used is that the soil resists external forces by means of adhesive and

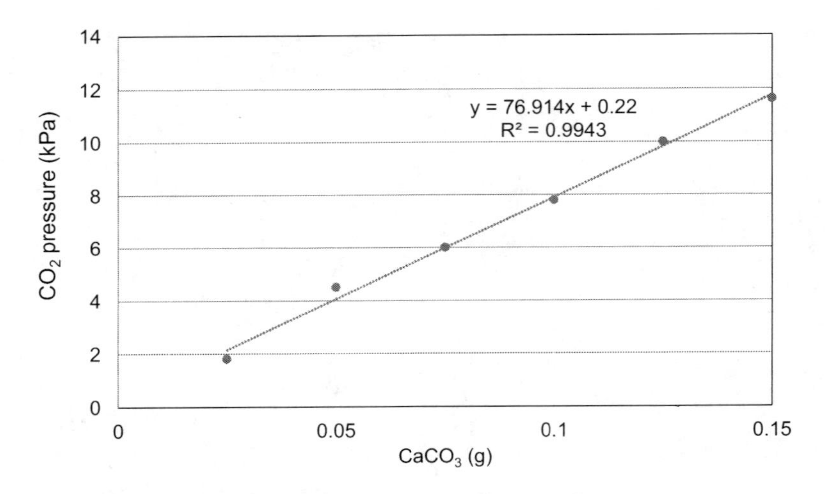

FIGURE 14.3 Calibration for carbonate content determination.

frictional forces. So, the following Coulomb failure criterion should be checked to see if it is available.

$$\tau = c + \sigma \tan \varphi \tag{14.4}$$

where τ is the shear strength (kPa), c is the cohesion (kPa), σ is the normal stress (kPa) acting on the failure plane, and ϕ is the internal frictional angle (°) of soil.

Soil can be generally divided into three types, depending on the types of interaction between the particles. Clay and silt have almost no frictional resistance but their cohesion force resists external force. On the other hand, granular soils like sand and gravel do not have cohesion but the sliding friction between their particles can resist the external force. The third type of soil has both the cohesive and frictional resistant forces, which consists of mixtures of the first and second types of soils. However, it does not mean that the third type is stronger than the first or second type of soil. These three types of soils are always weaker than any type of rock, because of the limited cohesion. In addition, the cohesive soil particles like clay and silt may reduce the frictional resistance as part of the third type of soil.

The shear strength of typical cohesive soils, such as clay and silt, is expressed by Equation (14.5), while for granular soil, it is shown in Equation (14.6).

$$\tau = c \tag{14.5}$$

$$\tau = \sigma \tan \varphi \tag{14.6}$$

For the third type, the shear strength is expressed by

$$\tau = c + \tan \varphi \tag{14.7}$$

Note that Equations (14.5)–(14.7) are independent.

When granular soil is treated in MICP, biocement bonding occurs at the interparticle contact area, and the binding mechanism is not dissimilar qualitatively to the cohesion of soils. However, the biocement bonding force and the cohesion of soils are quantitatively different, because the former is the same as limestone, while the latter is due to the adsorbed water molecules on soil particle surfaces. This is discussed later.

This biocement bond forms the skeletal structure of the granular soil and makes the arch effect working on the granular soil stronger. In addition, the binding particles to the soil particle surface can increase the frictional effect of the particle surface. As a result, the shear strength of the BC-treated soil can be expressed as follows, unlike the original granular soil.

$$\tau^* = c^* + \sigma \tan \varphi^* \tag{14.8}$$

where "*" means the BC treatment effects on respective strength components.

The failure criteria are written as a function of principal stresses, as follows:

$$\tau^* = \frac{\sigma_1 - \sigma_3}{2} \sin 2\alpha \tag{14.9}$$

$$\sigma = \frac{\sigma_1 + \sigma_3}{2} + \frac{\sigma_1 - \sigma_3}{2} \cos 2\alpha \tag{14.10}$$

These relationships are illustrated in terms of Mohr's circle as in Figure 14.4.

Coulomb's failure criterion is a part of Mohr-Coulomb's failure criteria. The former expresses a linear criterion, while the latter assumes non-linear criteria. Hard rocks and strong BC-treated soils need non-linear failure criteria (van Paassen, 2009).

14.3.2 Failure Criteria and Strength Constants of BCS-Treated Soils

In general, BC-treated soil solidified with calcium carbonate crystals exhibit brittle properties. This is because carbonate crystals between soil particle bonds are fragile even with a slight strain. The properties due to BC treatment are quite different from that of the cohesive soils, which shows plasticity.

In general, soil fails by shear on the shear plane, as shown in Figure 14.4. However, in the case of BC treated loose granular soil, as shown in Figure 14.5, the failure pattern varies as shown in

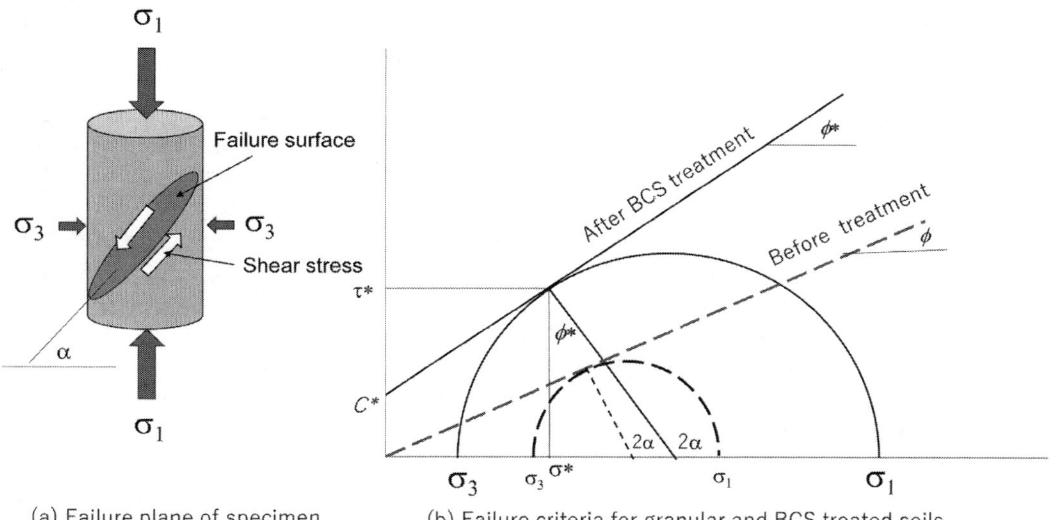

FIGURE 14.4 Relationships among principal stresses, shear strength, and normal stresses with Mohr's stress circle.

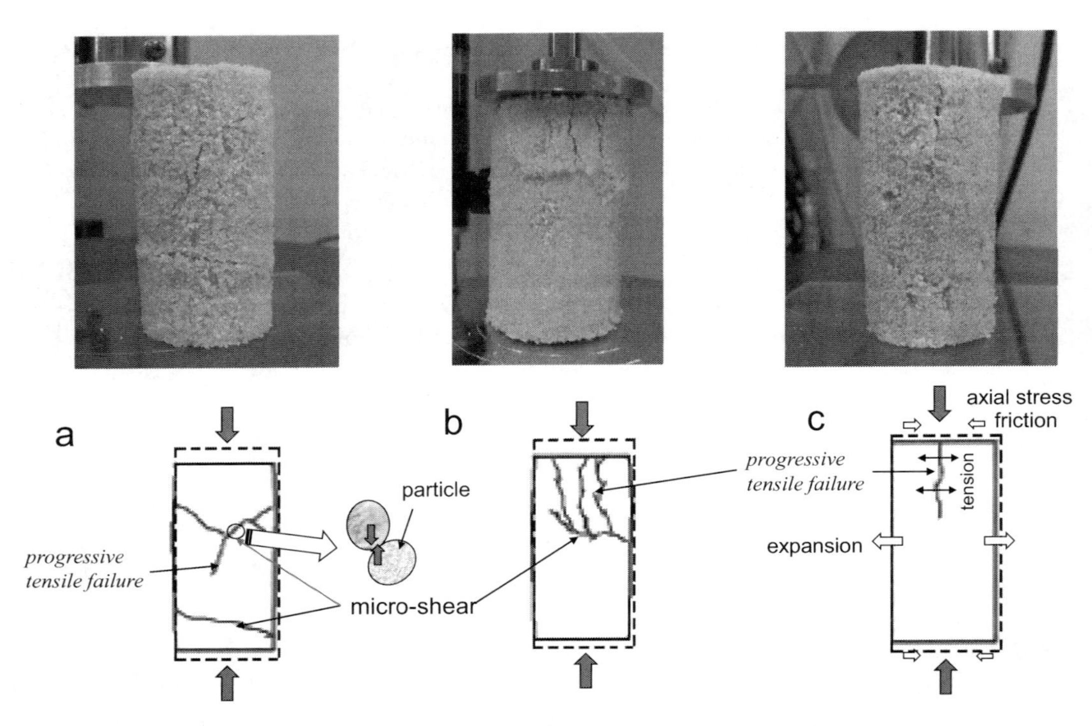

FIGURE 14.5 Micro-tensile fracture at the cementation part between sand particles.

Figure 14.5. The failure types shown in the figure can be defined as progressive micro-failure and micro-tensile failure. This behaviour may be due to compressive and ductile properties of the materials under unconfined stress condition. In fact, a slight confined stress can recover the three-dimensional skeleton structure, which covers compressive and ductile properties. This is discussed later.

In the case of a low dry density soil, even if a uniaxial compression test can be performed, the phenomenon of tensile fracture described above occurs. The fractures are micro-scopically caused by tension, and the macro-scopic phenomenon often appears as a vertically divided or horizontal fracture. Therefore, the measurement in the uniaxial compression test cannot be evaluated as uniaxial compressive strength. For example, when the actual measurement results are compared with the uniaxial compressive strength obtained from the triaxial compression test, it is less than 1/10 as shown in Figure 14.6. Figure 14.6 also shows important information for applying MICP technology in soil improvement.

(a) Mohr-Coulomb's failure criterion can be used for BCS-treated loose fine sand.
(b) Unconfined compressive strength (q_u) of BCS treated loose fine sand obtained from uniaxial compression test is too low, because the specimen was not fractured in compression but in tensile force.
(c) Unconfined compressive strength (UCS) of BCS-treated loose fine sand measured with pocket penetrometer (ASTM WK27337) agreed well with Mohr-Coulomb's failure criterion.
(d) Mohr's circle obtained using a low confined stress σ_3 of 10 kPa agreed with the failure criterion.
(e) Unconfined compressive strength (q_u) of BCS-treated loose sand by the conventional test method can not be utilized as design strength, because the value is too conservative.

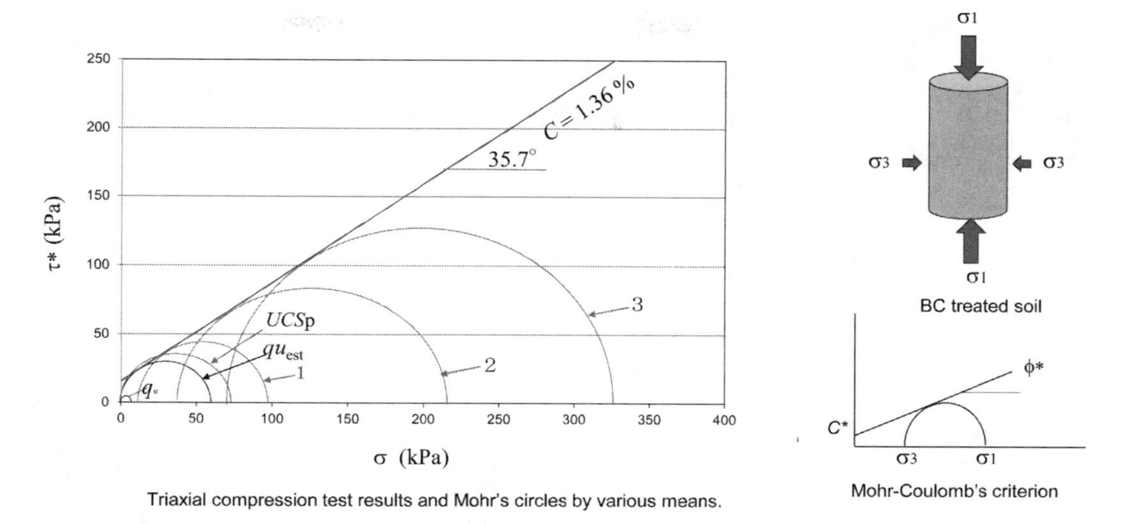

FIGURE 14.6 Failure criterion in triaxial compression showing various Mohr's circles in triaxial and unconfined compression test, and unconfined compressive strength estimated by pocket penetration.

14.3.3 Carbonate Content and Dry Density for MICP-Treated Soils

Carbonate content in marine sedimentary soil is the key factor for the strength characteristics (Fukue et al., 1999) and BCS-treated soils also show clear carbonate dependency effects. However, it has been pointed out by many studies that the relationships between carbonate content and strength vary widely (Fu et al., 2023).

Fu et al. (2023) comprehensively reviewed the progress of more than a decade of research on the application of MICP in soil strengthening. They conducted a literature survey of more than 280 papers and pointed out the challenges and problems of future MICP technology and concluded that it is inappropriate to understand the mechanical behavior of bio-cemented soils only by their carbonate content. On the other hand, van Paassen (2009) obtained both carbonate content C–UCS and initial dry density ρ_d–UCS relations in their large-scale MICP experiments for fine sand and concluded that the degree of correlation of ρ_d–UCS relation is higher than C–UCS. In addition, Lai et al. (2021) described that there was no consensus for the results of MICP study and suggested that the retardation in the MICP reaction might affect the result.

Their conclusion and opinion may be right. The carbonate content should be a good and new factor to understand soil strength in geotechnical engineering. If UCS is expressed only by carbonate content, it is natural that large deviations appeared by the difference in initial density of the specimen, as shown in the C–UCS relation diagram shown by Fu et al. (2023). If the initial dry density of the specimen is known, UCS can be expressed as a function of both the initial dry density and the carbonate content, and its behavior can be seen with completely different eyes. In that case, the dry density and the carbonate content can be expressed by the following relationship.

If induced carbonate content alone increases the dry density of soil, the dry density and carbonate content relationship is expressed by

$$\rho_d{}^* = \rho_d\left(1+\frac{C}{100}\right) \quad (14.11)$$

where ρ_d is the initial dry density of soil, ρ_d * is the dry density of the BC-treated soil and C is the carbonate content due to MICP.

Fukue et al. (1999) investigated the profiles of C–UCS relations of natural marine sediments. The results showed that the strength of sediments showed clear effects of carbonate content. In marine sediments, carbonate has been necessarily contained and the strength has been attributed to density and carbonate content. Next, typical two examples are considered.

14.3.4 BCS-Treated Soil, Limestone, and Sediments

The MICP technique can be learnt from the sequential carbonate diagenesis in nature. The process of formation of sandstone, marine sediments, and limestone may be good examples. This is because these rocks are formed through carbonation processes, which may be concerned with MICP (Castanier et al., 1999) and or similar mechanisms (Fukue et al., 1999). The formation of beach rock can also be in the similar category in rock formation.

The strength of silt ground in coastal plains has been developed with a little carbonate content, but it is not well known (Fukue et al., 1999). Therefore, it can be said that most geotechnical engineers and researchers have already dealt with soils containing carbonates produced by MICP, without knowing it. Similarly, natural carbonate sandstones have been formed by the MICP process (Burley and Worden, 2003). These natural solidification processes of soils may be called carbonate diagenesis of soils. Therefore, artificial soil improvement may be in the category of carbonate diagenesis. In this chapter, from a similar point of view, natural limestone formation and artificial soil improvement are identified.

The relationship between ρ_d and UCS as the limestone formation process from coral sand to limestone was obtained in Okinawa, Japan (Kogure et al., 2005), while the relationship between ρ_d and UCS for fine sand was obtained by MICP (van Paassen, 2009). Both the relationships were fitted by respective curves with relatively high R^2 values, as follows.

The ρ_d values of the specimens were approximately in the range of 1.7 and 2.4 t/m³. The fitted curve for limestone obtained by Kogure et al. (2005) was given as

$$\text{UCS} = 2.2 \times 10^{-2} \rho_d^{8.8} \tag{14.12}$$

with a R^2 of 0.78. On the other hand, UCS for fine sand is presented by van Paassen (2009), as

$$\text{UCS} = 4 \times 10^{-5} \exp(6 \rho_d) \tag{14.13}$$

where R^2 was 0.84.

The fitted relations using Equations (14.12) and (14.13) are shown in Figure 14.7.

These typical examples suggest that there is a respective upper limit for UCS with ρ_d. It is considered because the full solidification of a given skeleton structure is constrained by (or a function of) ρ_d under the carbonate diagenesis.

The dry density for soil improving is arbitrary from the minimum density to maximum density by means of compaction of soils. Minimum density and initial density of BCS-treated soil is defined in Figure 14.8. The arbitrary selected dry density becomes the initial dry density for soil improvement. Accordingly, the UCS–ρ_d relationships in Figure 14.7 do not include the incremental UCS process from the initial dry density.

If the UCS incremental processes from the different initial densities are added, the number of the relationships are infinite. On the other hand, if the axis of carbonate content is also added in Figure 14.7, the projected UCS–C relations for various initial dry densities will deviate widely as shown in Figure 14.8, which indicates UCS–C and UCS–ρ_d relationships projected UCS–C and UCS–ρ_d planes, respectively. Thus, UCS–C relationships show a strong ρ_d dependency, so there is no point to discuss the uniqueness of the UCS–C relationship.

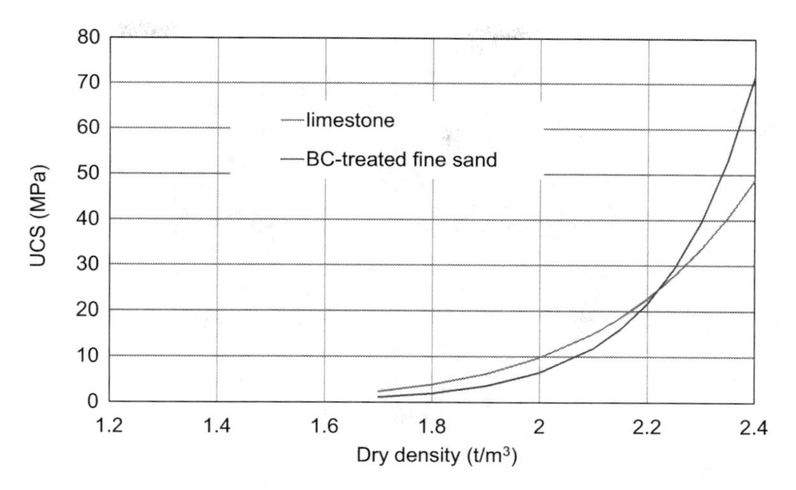

FIGURE 14.7 UCS–ρ_d relationships for natural limestone (Kogure et al., 2005) and BCS-treated fine sand (van Paassen, 2009).

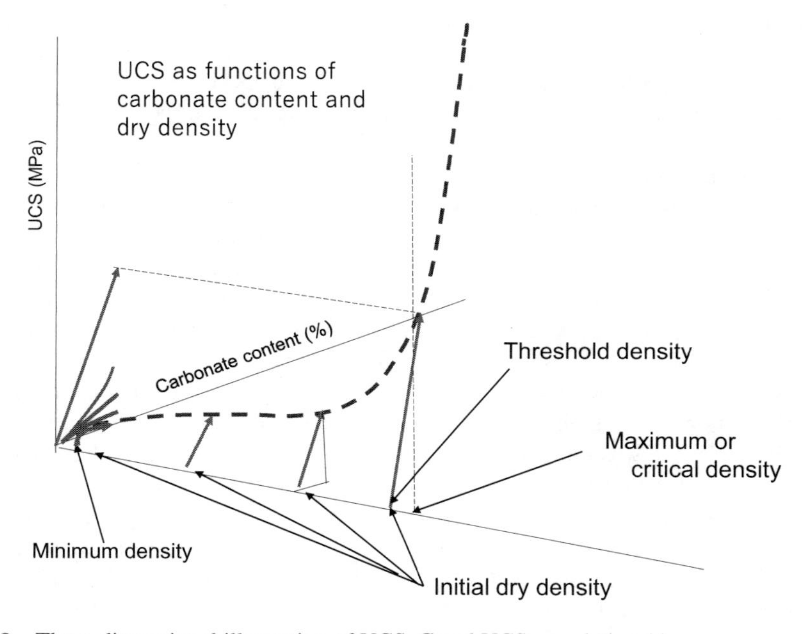

FIGURE 14.8 Three-dimensional illustration of UCS–C and UCS–ρ_d relationship.

14.3.5 UCS, Carbonate Content, and Dry Density

The advantage of using biocement is that it provides bonding force to sand and gravel, and that increases frictional resistance force. Thereby, the bearing capacity of these materials is dramatically increased mechanically. Unconfined compressive strength can be used to evaluate the strength of granular soil with bonding. However, it is often sometimes difficult to prepare the specimen of such brittle substances, as indicated in Figures 14.5 and 14.6. Even if the specimen is available, the micro-failures of specimen due to tension gives a very low strength, as shown in Figure 14.9. Therefore, the unconfined compression test on loose MICP-treated soils is often useless.

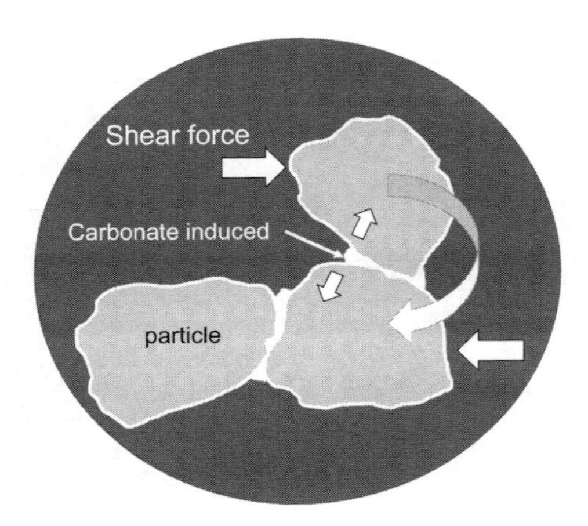

FIGURE 14.9 Illustration of local tensile failure for weakly bonded sand specimens with biocement.

The target UCS for the reduction of soil liquefaction potential even at most may be a few hundred kPa. Therefore, it seems that as can be seen in Figure 14.6, the pocket penetrometer can be used conveniently to measure the UCS of BCS-treated soils. The pocket penetrometer does not need the preparation of specimen but can also accumulate data easily. The test method for pocket penetrometers is provided by ASTM International to obtain UCS of viscous soils. Note that the pocket penetrometer provided by ASTM is not applicable for soils stronger than 460 kPa.

(1) UCS of Various Sands in Terms of Carbonate Content and Initial Dry Density

The properties of soils used for measuring UCS by pocket penetrometers are presented in Table 14.2. The "initial C_0 (%)" is the carbonate content naturally contained, which is not attributed to cementation. "Vietnam sand" was crushed rock and had electrostatic charges so strong that the particles could bounce in the firing polystyrene box. In the table, the fine sand **c** means the compacted fine sand **b**. Fine sand **c** was used to investigate the effects of densification on MICP in sands. That is why the dry density of fine sand **c** was relatively high.

Six of the seven types of dry sand samples were loosely packed in the polystyrene box. The initial density of the samples was obtained as an average value of dry sample weight divided by the total volume. Fine sand **c** was compacted to investigate the effect of initial density as mentioned. The Ca^{2+} BC solution was injected one, two, or three times, into the sample packed in the respective polystyrene box. After an appropriate time, UCS was measured with the pocket penetrometer and several grams of destructed soil sample were used to measure the carbonate content, by means of CO_2 gaseous pressure method (section 14.2.2). The UCS is presented in terms of ρ_s and C, using Equation (14.9), as shown in Figure 14.10. Thus, the effect of both the dry density and carbonate content on UCS can be expressed at the combined relationship.

The UCS presented in Table 14.2 was measured using the pocket penetrometer (ASTM WK27337), which is shown in Figure 14.11. The measurements with the pocket penetrometer agreed well with the unconfined compressive strength defined by Mohr-Coulomb's failure criterion. Note that for weak BCS-treated soils, the measurements with the penetrometer are more appropriate than that obtained from the unconfined compression test, as described earlier.

Figure 14.10 also demonstrates that MICP technology can be applied at any initial dry density of granular soils and that the C–UCS relationship is only unique at a given initial dry density. Generally, the C–UCS relationships at different initial dry densities do not agree, as can be seen by the comparison between fine sand **b** and fine sand **c** (densified fine sand **b**) in Table 14.2. In other words, the densification of sand makes the $\boldsymbol{a}$-value greater, where $\boldsymbol{a}$ is defined as,

$$\boldsymbol{a} = \mathrm{UCS}/C \tag{14.14}$$

TABLE 14.2
Physical Properties of Sandy Soils Used for Measuring UCS and the *a*-value

Type of Sand	Particle's Density (t/m³)	D_{max} (mm)	D_{60} (mm)	D_{30} (mm)	D_{10} (mm)	Uniformity U_c	ρ_{dmax} (t/m³)	ρ_{dmin} (t/m³)	Initial C (%)	Initial ρ_d (t/m³)	a (kPa/%)
Vietnam sand	2.662	2.00	0.610	0.48	0.34	1.79	1.695	1.416	0.22	1.4	186
coarse sand	2.650	2.00	1.35	1.05	0.87	1.55	1.568	1.336	0.2	1.28	34.5
fine sand a	2.693	2.00	0.303	0.205	0.135	2.24	1.719	1.410	0.3	1.4	64.4
medium sand	2.664	4.75	0.33	0.28	0.2	1.65	1.619	1.308	0.47	1.26	44.0
fine sand b	2.679	4.75	0.295	0.195	0.13	2.27	1.683	1.377	0.2	1.38	53.9
fine sand c	2.679	4.75	0.295	0.195	0.13	2.27	1.683	1.377	0.2	1.58	120
river sand	_	4.75	0.50	0.301	0.155	3.23	1.752	1.404	0.38	1.56	68.7

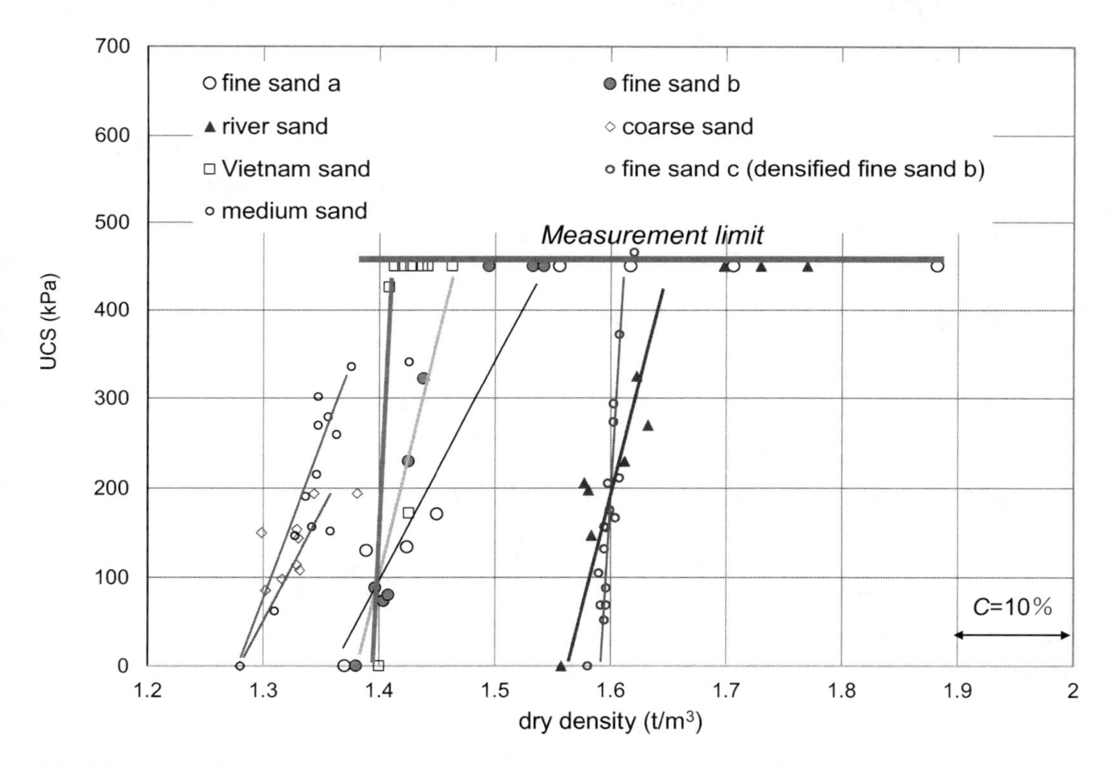

FIGURE 14.10 The effects of initial dry density and carbonate content on UCS.

An average ***a***-value is seen as the slope of fitting curves in Figure 14.10. The average ***a***-value is dependent on the initial dry density alone as seen in Figure 14.10, while the ***a***-values vary with soil type as shown in Table 14.2. This may be because that for some soils, the slope tends to decrease at a high UCS. This means that there is a limit of the linearity for the UCS–C relationship, though some did not reach the limit because of the measuring limit for the pocket penetrometer.

The grain size distribution can be represented by the uniformity of soil, the relationship between ***a***-value and uniformity is shown in Figure 14.12. The uniformity of soil is defined as D_{60}/D_{10}, where D_{60} and D_{10} denote the 60% grain size and 10% grain size, respectively. It is interesting that the relationship between uniformity and ***a***-value is linear for the loosest dry density, which may be controlled by particles geometry. On the other hand, soils which have special characteristics like strong interfacial effects (Vietnam sand) and a large number of particle's contacts (densified sand) seem to

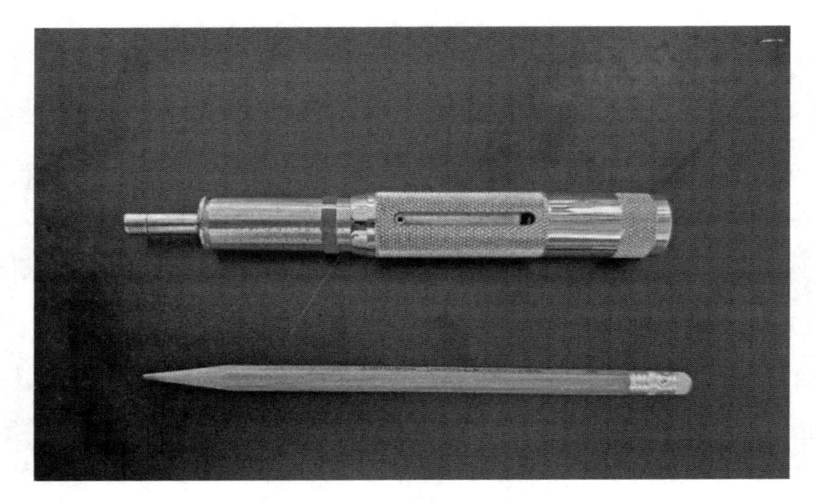

FIGURE 14.11 Pocket penetrometer provided by ASTM WK27337.

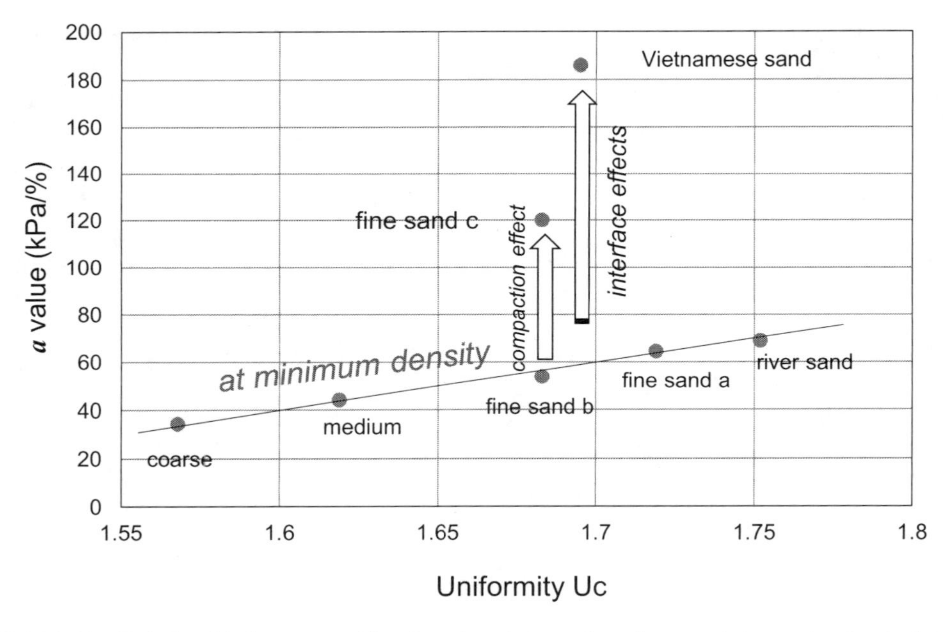

FIGURE 14.12 Relationship between uniformity of particles and ***a***-value, at minimum dry density level.

be depart from the relationships under the condition which provide the lowest ***a***-value, as shown in Figure 14.12. It is interesting that the lowest ***a***-values may provide the geometrical and interfacial potentials of particles, which were obtained at the largest volumes as respective materials.

Note that the lowest ***a***-value is provided by the lowest dry density, because of the lowest number of contacts between particles. In Table 14.2, there is contradiction that for coarse sand and medium sand, the initial density is lower than the minimum dry density. This is considered to be because large pore spaces were formed in a rectangular box during the sample preparation.

The relationship between initial density and ***a***-value is presented in Figure 14.13. The two exceptional data indicated are for Vietnamese sand and river sand. As mentioned earlier, Vietnamese sand

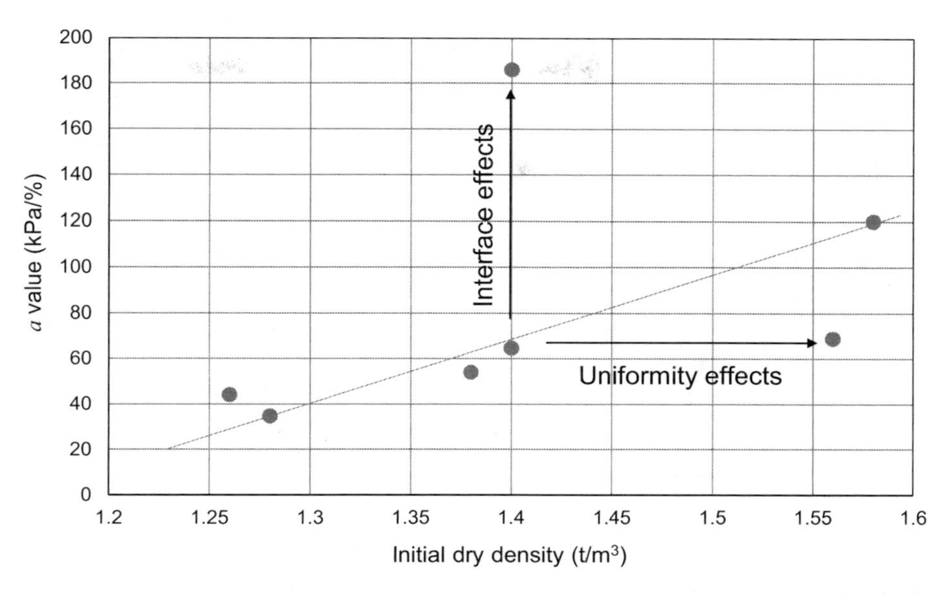

FIGURE 14.13 Relationship between initial dry density and ***a***-value, and two exceptional data points due to relatively high electric charges and uniformity.

had a relatively very high electric charge on the surface of particles, while river sand contained small gravel particles, caused a relatively high minimum density by a relatively high uniformity (Table 14.2). Note that if all the particles have the same size, the uniformity is 1.0, a wide distribution of particle size makes uniformity high.

(2) Transformation from First to Second Stages

In this section, the principles of the carbonate formation in soils is presented as a basis of practical application.

Two UCS–ρ_d relationships in limestone diagenesis and the MICP process in fine sands are compared (Kogure et al., 2005; van Paassen, 2009). However, these two relationships are limited by a higher density level, i.e., approximately greater than 1.7 t/m^3. To compare these two fitted curves and actual data shown in Table 14.2, the fitted curves are elongated by the two fitted Equations (14.12) and (1413). The result is shown in Figure 14.14.

Figure 14.15 shows the test results of Table 14.2 added to Figure 14.14. The horizontal axis is ρ_d, but it can be presented as C also, as shown by the scale in the figure. In Figure 14.15, the linear fittings of added data become the data of the first stage in MICP, which are rapid incremental processes of the carbonate and UCS on a changing dry density. The second stage in the MICP process starts after the linear increase in the UCS–ρ_d relationship. The beginning of the second stage can be recognized as a rapid change from the linear slope of UCS–ρ_d curve to along one of the dotted curves. The data on coarse sand and fine sand **a** in Figure 14.15 are for examples. Note that the data of UCS of 460 kPa is due to the measuring limit of the pocket penetrometer.

14.3.6 Determination of *C* for Design Strength UCS

Because sand and gravel do not inherently have cementation between particles, their UCS becomes a barometer of the bonding force generated by biocement. Therefore, UCS can be used as a key factor for soil improvement in terms of carbonate content. For example, the design strength to avoid liquefaction of sand has often been given by around a UCS of 100 kPa, based on the empirical

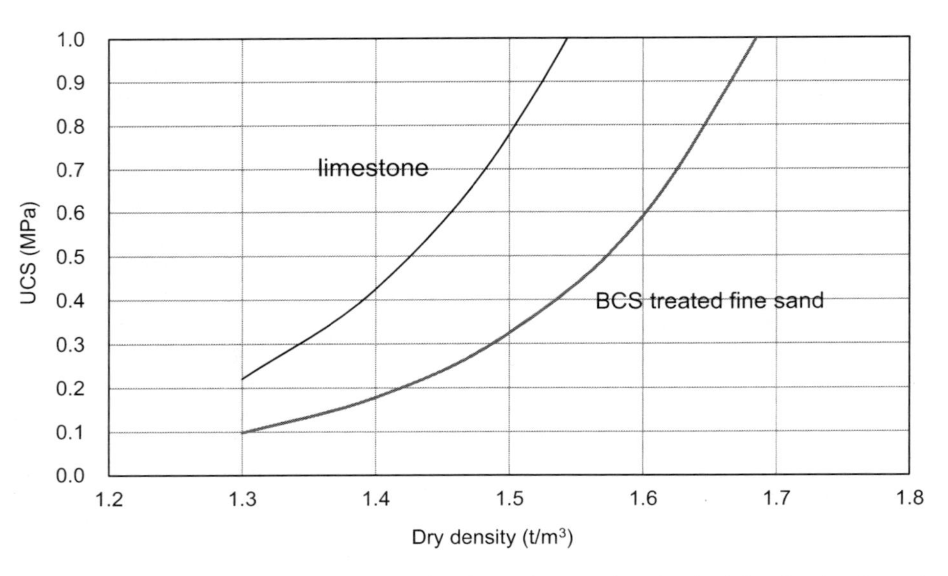

FIGURE 14.14 Density-UCS curves estimated by elongation to a low level of dry density, from the fitted curves on the Ryukyu limestone (Kogure et al., 2005) and the BCS-treated fine sand (van Paassen, 2009).

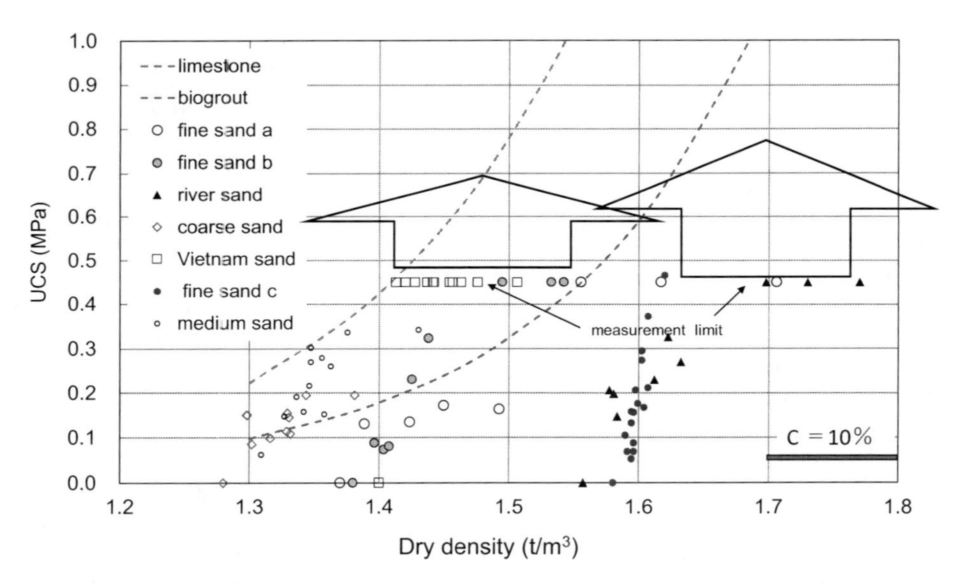

FIGURE 14.15 UCS on the first and second stages of various sandy soils, and comparison to the fitted curves for limestone and fine sand.

investigation. Thus, typical design strength is generally much lower than that targeted in the MICP study. In general, the target strength for soil improvement for an ordinary house may be less than a few hundred kPa. Furthermore, in geotechnical engineering too strong soil is usually avoided, and BCS-treated soil, i.e., rock that can be reverted into soil is one of the environmental objectives in MICP technology. From an economical point of view, to create strong rock by MICP technology will not be cost-effective.

For geotechnical engineering purposes, the design may start by so-called design strength. If the soil improvement is concerned, UCS is often selected as design strength, because it is obtained by a simple and easy test, i.e., the unconfined compression test. The situation may be similar in MICP technology. In MICP technology, there are some difficulties to prepare specimens and to conduct the test due to ductile materials, as explained earlier. However, handy type tests are conveniently used, which provide many merits in terms of cost and time savings.

To use design strength, the concept of relationships among UCS, ρ_d, ρ_d*, C and the ***a***-value, which include first and second stages are demonstrated in Figure 14.16.

The first stage in the MICP process is the solidification from the initial state (initial dry density), which starts from the granular state. In the first stage, the UCS increases linearly with C. If the target strength is on the first stage, then the target C is given by

$$C(\%) = \frac{\text{UCS}}{a} \tag{14.15}$$

where UCS is a target (design) strength.

The ***a***-values of ordinary sandy soils are possibly given by the uniformity and initial density of soil (Figures 14.12 and 14.13). Further study is required to accumulate data.

In the second stage, UCS (the design strength) is given as Figure 14.16. If UCS is given as a function of dry density, ρ_d* for the target UCS is obtained by Equation (14.16).

$$\rho_d{}^* = \rho_d\left(1 + \frac{C_f}{100}\right) \tag{14.16}$$

where C_f is the carbonate content to obtain the target UCS.

Then, the carbonate content required to get the target UCS is

$$C_f(\%) = 100\left(\frac{\rho_d{}^*}{\rho_d} - 1\right) \tag{14.17}$$

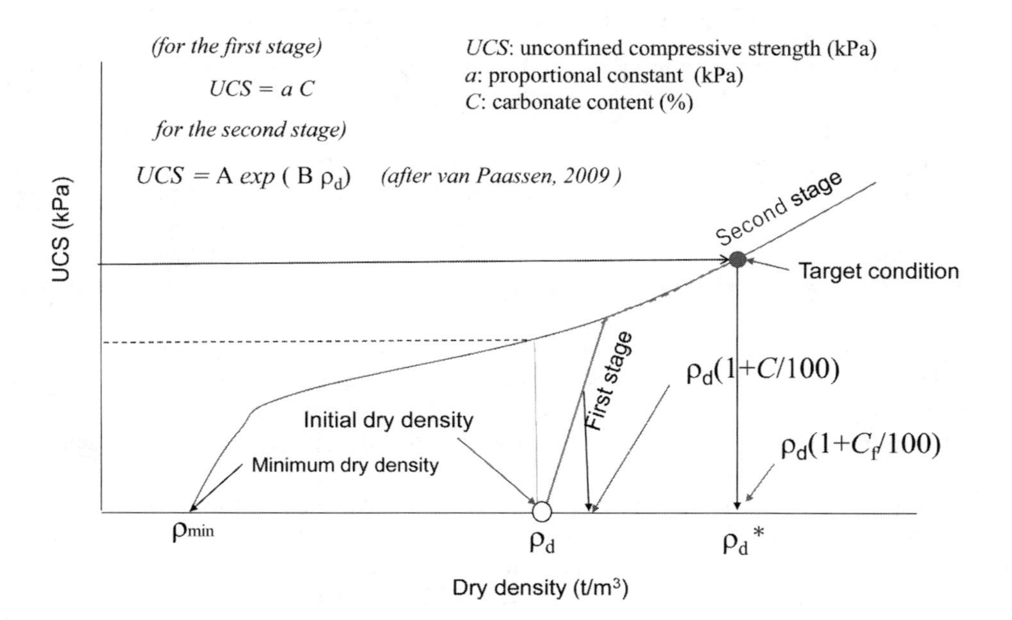

FIGURE 14.16 Relationships among UCS, ρ_d, ρ_d*, C and the ***a***-value.

where ρ_d* is the target dry density. For example (van Paassen, 2009),

$$\text{target UCS} = A \exp(B\,\rho_d^*) \quad (14.18)$$

where A and B are constants. The constants can be obtained experimentally, if it is needed. Note that the UCS can easily be obtained using the pocket penetrometer.

The estimation of design strength can vary case by case. When the design strength in UCS is decided, the target UCS can be converted into C, by testing or assuming the ***a***-value of the soil. The direct conversion from UCS into C can be obtained by UCS = a C.

14.3.7 Conversion between CPR and C

It is noted that CPR is different from C. The CPR (M or g/L) is the carbonate precipitation rate in BCS, while C (%) is the carbonate content in soil. Assuming saturated flow of BCS, the carbonate content C (%) is defined in terms of CPR by

$$C = \frac{m_c}{m_s} \times 100 = \frac{n}{(1-n)} \frac{(0.1)\text{CPR}}{\rho_s} \times 100 \qquad (\%) \quad (14.19)$$

where n is the porosity (in decimal), ρ_s is the density of soil particles (t/m^3). The units of CPR are mole/L. Equation (14.19) gives approximately $C = 2.5$ %, when $n = 0.4$, CPR = 1.0 M, and ρ_s is 2.65 t/m^3. However, CPR is unknown, depending on the adsorption of aggregates consisting of bacteria with Ca^{2+} in BCS. Thus, it is convenient to remember that in ordinary granular sandy soils, 1.0 M Ca^{2+} can produce at most 2.5% of carbonate (calcite). The conversion between C and CPR can be made using Equation (14.19). For the optimum blending of BCS, the CPR should be known by the conversion from the target C. Then, the volume of concentrated bacteria will be determined for use in BCS. The CPR and optimum blending were dealt with in Chapter 13. The process from the design of UCS to construction is demonstrated in Figure 14.16.

14.4 INJECTION AND SEEPAGE TECHNIQUES – SATURATED FLOW

The viscosity of BCS is low in comparison to other agents, such as water, or sodium silicate solution. Injection of the agent with a high viscosity produces swelling of soil and/or cracks in the soil. A low viscous agent like biocement solution can be used with low injection pressure or free injection.

In general, the permeability of soils is considered from the particle size or void ratio. However, strictly speaking, soil permeability depends on the specific surface area of particles, which is attributed to the grain size (surface area) and its distribution. The higher the specific surface area, the greater the friction loss is. Therefore, when the grain size is greater, the permeability of soil increases Thus, most of the surface phenomena of soil is related to the specific surface area of soil particles.

The properties and behaviour of bacteria adsorption require understanding adsorption theory. The NO-A10 can adsorb on solid surfaces like glass, soil, and rock minerals, some metals etc. This is possibly because of gram-positive bacteria (Kagawa, 1973). Theoretical and experimental studies on bacteria adsorption have been conducted by Kagawa (1973), Hermansson (1999), Fowle and Fein (1999), and van Paassen (2009).

The properties and behaviour of bacteria adsorption during injection of BCS in terms of aggregation of bacteria are considered to be complicated in the presence of Ca^{2+} ions and crystallization of $CaCO_3$ during infiltration. Nevertheless, the distribution of $CaCO_3$ (bacteria) is roughly expressed with an exponential function of flow distance, as shown in Figure 14.17.

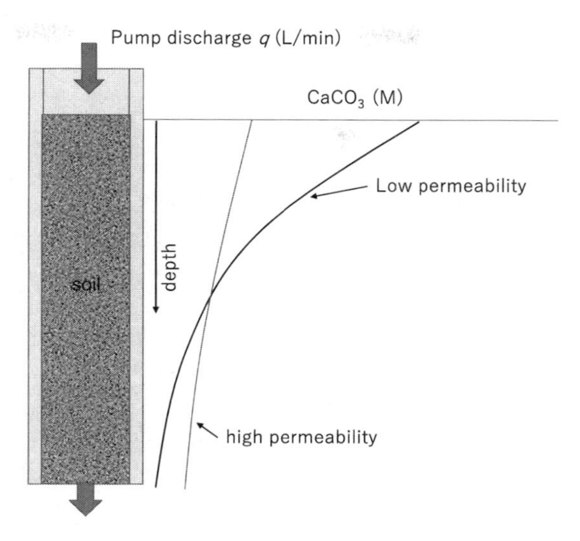

FIGURE 14.17 Theoretically ideal $CaCO_3$ distribution in confined one-dimensional BCS infiltration.

FIGURE 14.18 Solidified sand and gravel in laterally unconfined and confined conditions. (a) Laterally confined, fine and coarse sands, square sectional area: $W220 \times D220$ (mm), and (c) laterally confined, gravel, $W900 \times D900 \times H700$ (mm), where W, D, and H are width, depth, and height.

This type of injection method has been used to examine the vertical $CaCO_3$ concentration and/or strength due to MICP (van Paassen, 2009). Basically, the depth profiles of $CaCO_3$ concentration and UCS are approximately expressed as an exponential function. However, a thickness of a layer of a few cm from the surface of the sand specimen shows without exception lower strength and $CaCO_3$ content than the layer below. The reason why is that the surface layer has almost no overburdened pressure and injected water can always move soil particles. Accordingly, less cementation occurs near the surface. Solidified sand and gravel are demonstrated in Figure 14.18.

14.4.1 Vertical Injection-Saturated Plug Flow

Vertical injection of BCS under gravity flow may be realistic. It can be easier if the injection depth is shallow. When the injection depth is deep, the injection pipe has to be penetrated deeply. Then, the

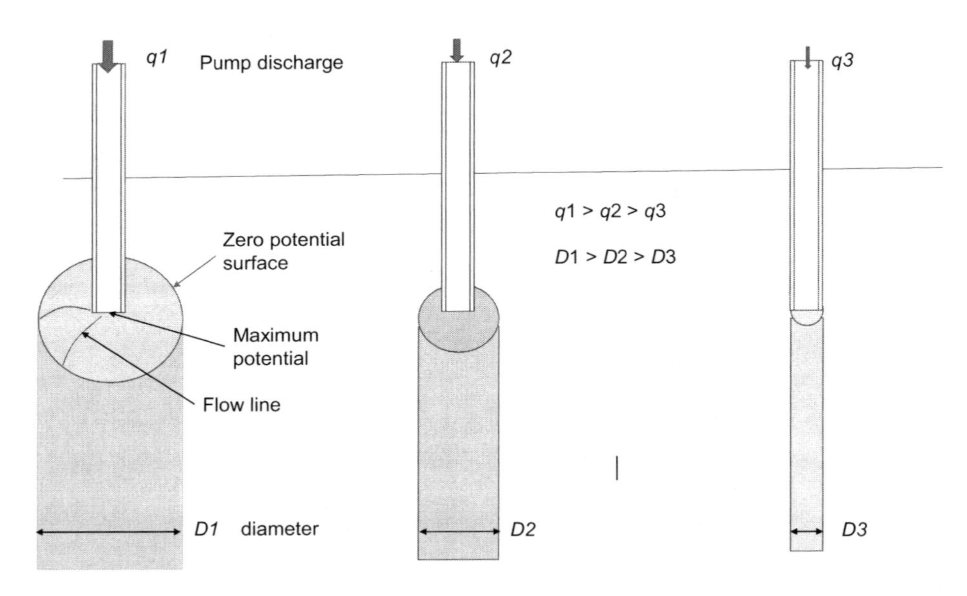

FIGURE 14.19 Flow pattern for vertical infiltration of BCS into granular soil.

flow pattern of BCS out from the injection mouth has to be predicted. It is convenient to use boring rods for the transport of BCS with a pump from the ground. In general, the flow pattern depends on the pump discharge and permeability of the soil. A simple image of the flow pattern in terms of pump discharge and soil permeability is illustrated in Figure 14.19.

From Figure 14.19, a simple assumption is as follows: At steady state, the injected BCS q (m^3/s) is equal to q that flow in the sphere of BCS, and the flow rate of the unit time at any depth (cross-section) of the cylinder in the figure. Under the continuous infiltration, the injection of BCS keeps the spherical liquid body constant at the outlet of the injection pipe. At the cross section of the cylindrical liquid body, Darcy's law is applicable.

$$q = kAi \tag{14.20}$$

where k (m/s) is the coefficient of permeability of the soil, A (m^2) is the cross-sectional area of the cylindrical liquid body, and i is the hydraulic gradient, which can be assumed as 1.0 under the free infiltration. Then, Equation (14.20) is given by

$$q = k\,\pi l^2/4 \tag{14.21}$$

The diameter of the cylindrical liquid body is presented by

$$l = 2\sqrt{\frac{q}{k\pi}} \tag{14.22}$$

where $l(\mathrm{m})$ is the sectional diameter. Thus, if k is constant, the greater the q, the greater the diameter of flow section, as shown in Figure 14.19.

The permeability of soil is a key factor for the injection of BCS. If a lower permeable layer exists at some depth, the BCS flow will be reduced by the layer. In such a case, it is suggested to penetrate the injection pipe into the lower layer, if a deeper layer has to be improved (Figure 14.20). The injection technology of liquid into soil has been established in geotechnical engineering. The technical

FIGURE 14.20 Solidified sand column in an embankment using a pipe and pump injection of BCS.

problems, which can be faced in BCS injection, can be solved by using technologies developed in some other fields.

14.4.2 Spraying on a Soil Surface

When solidifying the ground surface, the spray method is suitable. In such a case, drying of BCS and the movement of soil particles by sprayed solution should be controlled. The examples of application have been performed for the measures of wind erosion (Wang et al., 2018; Meng et al., 2021) and soil surface erosion (Chen et al., 2021; Liu et al., 2021; Sun et al., 2022).

Wind erosion is a major problem in deserts. The measure is thought to help for solving dust and sand storm (DSS) problems. However, it may be difficult to implement unless it is a project that spans a large international area.

It seems possible to use the BCS spray method to solidify the surface ground to prevent erosion of the Earth's surface by rainwater and waves. However, considering that the ground is vulnerable to rainwater, it is necessary to do so under conditions where surface flow does not occur.

14.4.3 Infiltration Below the Groundwater Table

Since the density of BCS is usually greater than 1.05 t/m^3, it flows downward in groundwater and seawater, unless the density of groundwater is greater than 1.05 t/m^3. This was examined experimentally using coloured BCS. Groundwater seepage, however, may cause the advection of solutes and bacteria. Therefore, the advection and diffusion of the components of BCS should be taken into account for the design of construction.

14.4.4 Lateral Infiltration

Below the groundwater table, lateral infiltration can be used to transport BCS using an injection pipe and well pump. BCS, injected from a tank into the pressurizing pipes, is penetrated into the soil to be solidified and the remaining liquid is pumped up from the pumping well as illustrated in Figure 14.21. Van Paassen (2009) used lateral infiltration for large-scale experiment.

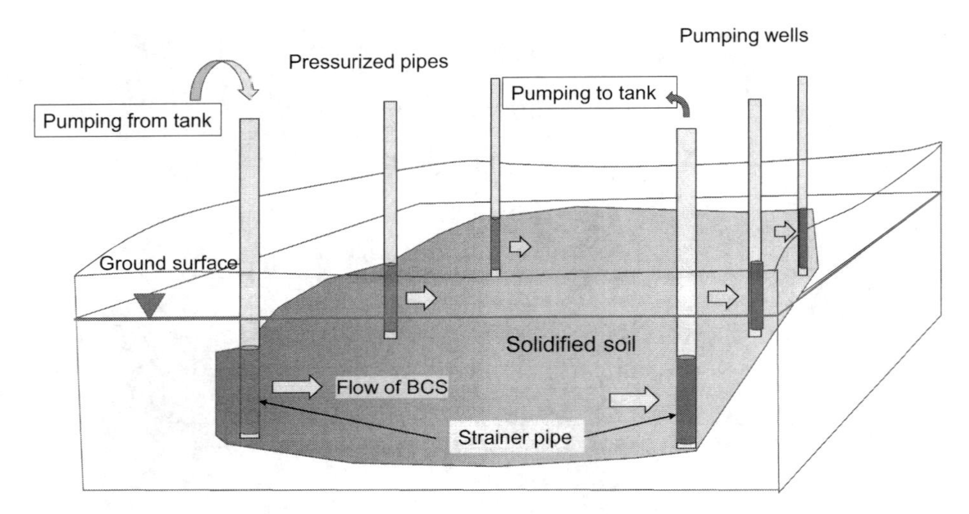

FIGURE 14.21 Model of lateral transport of BCS using injection and pumping wells to solidify sand and/or gravel below the groundwater level.

The technologies of lateral transport of liquid in soils can be seen in the literature concerning the remediation of polluted soils. At present, the problem of heterogeneity has not been solved. However, there is always a heterogeneity problem in natural soils. The solution has been solved in other ways, such as the concept of safety factor.

As for the problem of heterogeneity of BCS, it is naturally inhomogeneous because it is composed of particles. Therefore, the carbonate produced in the soil are also heterogeneous. In sand, there are so-called idle particles in which stress is not transmitted that does not contribute at all to the formation of the skeleton. Originally, soils have heterogeneous particle size distribution, and there is also a heterogeneity problem between micro-pores and macro-pores. Considering the addition of geological stratigraphy, it seems that our predecessors have solved most of the heterogeneity problems and thus it seems essential to use their knowledge. Accordingly, how to consider the microstructural heterogeneity against external forces transmitted by the homogeneous macroscopic structures is our challenge.

14.5 APPLICATION OF MICP TO SOILS

The MICP technique in ground improvement can be divided into two methods: Cultivating bacteria in the ground and injecting bacteria cultivated indoors. This section deals with a method belonging to the latter. A feature of this method is that the MICP process can be controlled using optimum blending of BCS as described in Chapter 13 to enable the prediction of the carbonate content to be induced, the strength, and bearing capacity of soils.

The significance in MICP application consists of the increment of soil strength. Therefore, soil improvement is basically an increase in soil strength. The most significant in MICP technology is to bring the soil strength to the desired value. The strength here is shear strength (c, ϕ), UCS, N value or q_C, etc., in geotechnical engineering. These values have been used for the design of all kinds of structures. Accordingly, the goal in MICP is very simple and clear.

14.5.1 Bearing Capacity

The bearing capacity of soil is the capacity to bear the loads applied to the ground. The first theory for the evaluation of the ultimate bearing capacity was proposed by Terzaghi (1943). Since then, revised opinions on the basic formula have been proposed by some researchers (Meyerhof, 1951). The formula developed for determining bearing capacity of soils has been used in practice and education in the world, not only by geotechnical engineers but also students. Therefore, in this section, it is restricted to dealing with the formulation, except for key points.

Similar to ordinary soils, the ultimate bearing capacity of BCS-treated soil is presented as a function of cohesion c^*, frictional angle ϕ^*, embedment D_f, surcharge, soil self-weight γ_1, γ_2, etc., is given as,

$$R_u = i_c \alpha c N_c + i_\gamma \beta \gamma_1 B \eta N_\gamma + i_q \gamma_2 D_f N_q \tag{14.23}$$

where

N_c, N_γ, N_q: Factors of bearing capacity.
α, β: Shape factor.
η: Correction factor by foundation size.
B: Width of foundation.
D_f: Depth of footing.
i_c, i_γ, i_q: Correction factors by loading inclination. These factors are negligible in the case of no loading inclination.

In Equation (14.23), the factors of bearing capacities, N_c, N_γ, and N_q are a function of only internal frictional angle ϕ^*. The factors are empirically derived that usually correlates with the angle of internal friction of the soil. Therefore, there are several tables of bearing capacity factors under various conditions (GIW, 2024. http://geotechnicalinfo.com/bearing_capacity_factors.html, accessed May 2024).

The three terms on the right-hand side of Equation (14.23) contain a factor of N_c, N_r, and N_q, respectively. Since there is no cohesion for granular sandor gravel, the first term is removed in the case without MICP treatment. Furthermore, in the case of no footing depth the third term is also removed in the case without MICP treatment. Accordingly, the comparison of the ultimate bearing capacity before and after the MICP process is expressed by

$$\frac{R_u^*}{R_u} = \frac{(c^* \alpha N_c + \beta \gamma_1 B \eta N_{\gamma^*})}{(\beta \gamma_1 B \eta N_\gamma)} = \frac{c^* \alpha N_c}{\beta \gamma_1 B \eta N_\gamma} + \frac{\Delta N_{\gamma^*}}{N_\gamma} + 1 \tag{14.24}$$

The first and second terms on the right-hand side of Equation (14.24) are always greater than 1.0, which means that the incremental bearing capacity occurs due to MICP treatment. In particular, the first term contributes to the increase in bearing capacity by the synergy of c^* and ϕ^*.

(1) Estimation of Ultimate Bearing Capacity for Fine Sand

The bearing capacity after MICP treatment for fine sand was evaluated using the formula of bearing capacity. The evaluation was made theoretically assuming the plate loading test, using Equation (14.23) and the data from the triaxial compression test. The data used are presented in Table 14.3. The measured strength constants were of c^* and ϕ^*. The γ_1 value is arbitrary here. The bearing capacity factors are obtained using ϕ^*, and others are default values *(Bearing Capacity Factors on the Geotechnical Information Website*). (http://geotechnicalinfo.com/bearing_capacity_factors.html, accessed May 2024).

The results showed that the ultimate bearing capacity R_u of fine sand is estimated as

$$R_u = \beta \gamma 1 B \eta N_{\gamma^*} = 0.3 \times 13.7 \times 0.3 \times 1.5 \times 13.1 = 24.2 \text{ kN/m}^2 \tag{14.25}$$

TABLE 14.3
Data Used for Evaluation of Bearing Capacity for Fine Sand, Assuming Plate Load Test

	c^* (kN/m^2)	ϕ(°)	α	β	η	γ_1 (kN/m^3)	B (m)	N_c	N_γ	N_{γ^*}
before	–	26.2	–	0.3	1.5	13.7	0.3	–	13.1	–
after	12.7	40	1.2	–	–	–	–	75.3	–	109.4

TABLE 14.4
Physical Properties of Ground Soil

Particle Density	Grain size (mm)			
(t/m^3)	D_{10}	D_{30}	D_{60}	D_{max}
2.77	0.6	2.7	9	18

TABLE 14.5
Ca^{2+} Concentration and Volume of BCS

Ca^{2+}	First	Second	Volume/single injection/area
(M)	injection (m^3)		(m^3/m^2)
0.5	1.2	1.2	0.3

On the other hand, R_u^* after MICP process is estimated as

$$R_u^* = c^* \alpha N_c + \beta \gamma_1 B \eta N_{\gamma^*} = 12.7 \times 1.2 \times 75.3 + 0.3 \times 13.7 \times 0.3 \times 1.5 \times 109.4 = 1349 \text{ kN/m}^2 \quad (14.26)$$

Then, $R_u^*/R_u = 55.7$.

Thus, it was found that the bearing capacity of fine sand for plate load test numerically increased by 55.7 times by the MICP treatment in comparison to that of fine sand without the MICP process. Note that the effect of MICP treatment depends on how much c^* will be increased.

(2) Example of a Plate Load Test on Gravel Ground

To examine the effects of the MICP technique experimentally, a plate load test was performed on the gravel ground. The soil ground consisted of gravel with a small amount of coarse sand, as presented in Table 14.4. The gravel depth was 0.5 m and the area was 4 m^2. The biocement solution, BCS was injected from the surface of the ground. The injected BCS is described in Table 14.5.

The change in soil properties during the test was investigated by the measurement of surface wave velocity. The result is shown in Figure 14.22. The size of disk plate used was 0.3 m in diameter. The BCS was injected from the surface of the ground using a pump at approximately 20 L/min.

The evaluation of the MICP treatment was also made by the change of surface wave velocity. Comparing the surface wave velocity V_r before and after BC processing, the average values were 71 m/s and 182 m/s, respectively, in the two geodesies. The surface wave velocity V_r can be used as non-destructive test to evaluate strength of MICP treated soils. The UCS cannot be defined for granular soils, but it is possible for MICP treated soils which have cohesion c^* and internal friction angle ϕ^*. At present, there is no direct conversion method from V_r to UCS. However, it is found that the S-wave velocity V_s has been correlated to shear strength and UCS.

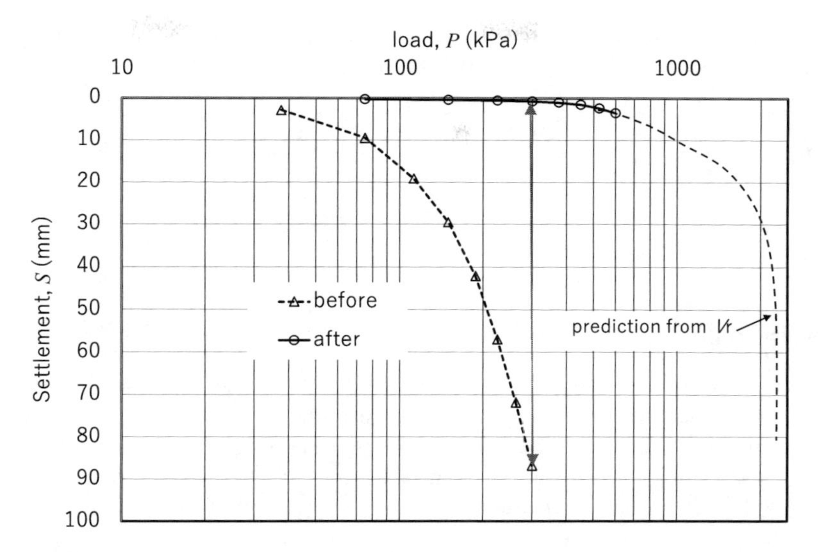

FIGURE 14.22 Comparison between the results in plate loading tests for non-treated and BC-treated gravel conditions.

The correlation between UCS (kPa) and V_s (m/s) was obtained for MICP treated fine sand. The result was demonstrated without considering units by

$$\text{UCS} = 0.9\ V_s \tag{14.27}$$

From the theoretical correlation between V_s and V_r, the following empirical relationship was obtained:

$$\text{UCS} \approx V_r = 0.9\ V_s \tag{14.28}$$

At present, Equation (14.28) is tentative, but it may be worth evaluating, because of the non-destructive method.

Figure 14.22 shows the results of the disk plate load test performed on the gravel before and after BC treatment. It is clear that the amount of settlement at the load of 300 kPa of the arrow is very different for before and after the treatment. The yield point after the BC treatment is estimated to be about 450 kPa. However, it is considered that before the BC treatment, progressive destruction continues, and no clear yield point can be obtained.

The experimental results were examined by theoretical estimation as follows. The relationship among the strength constants c^*, ϕ^* and UCS is geometrically obtained using Mohr's circle, as shown in Equations (14.29) and (14.30).

$$c^* = \frac{\text{UCS}}{2}\frac{1-\sin\phi^*}{\cos\phi^*} \tag{14.29}$$

$$\phi^* = 36.4 \pm 2.7^\circ \tag{14.30}$$

Therefore, if two of these three values are known, the remaining one can be determined. For ϕ^*, Equation (14.30) is under study when the UCS is high, because the failure criterion possibly becomes non-linear (van Paassen, 2009). Now, assuming $\phi^* = 40°$, Equation (14.29) gives $c^* = 24.0$ kPa. Then, taking $N_c = 75$, $N_\gamma = 100$, the following R_u^* is obtained.

$$R_u^* = c^* \alpha N_c + \beta \gamma_1 B \eta N_{\gamma^*} = 24 \times 1.2 \times 75 + 0.3 \times 13.7 \times 0.3 \times 1.5 \times 100 = 2335 \text{ kN/m}^2 \quad (14.31)$$

Thus, there is no doubt that the biocement can drastically improve the bearing capacity of granular soils, which can be proven theoretically and practically.

14.5.2 Soil Structure and Infrastructure

Biocement technology is a method of injecting BCS into the ground and makes the soil skeleton harder by coating the skeleton of soil with carbonate precipitates. However, it cannot be applicable for low permeable soils because BCS cannot be injected into the soils. In soft clayey and silty soils, sand piles are installed in the clay and silt, so that BCS can be injected into the sand pile. Then, the solidified pile groups can support various loadings, which are soils, vehicles, buildings, and any other structure. Although there are still a few examples of application to earth structures and infrastructure, considering the characteristics of the technology, it is considered that the range of applications for stabilization of earth structures such as river embankments is extremely large, including applications in the field of conventional ground improvement. In the current situation where it is common to use surrounding soil as a material for earth structures, it is necessary to improve the quality of the material (Lin et al., 2023b). Biocement can be used for this purpose. Furthermore, when the bearing capacity of a retaining wall or a foundation pile is decreased, it can be applied as a bearing capacity problem in the preceding paragraph.

The main features of biocement technologies promoted in this book are as follows.

(a) Ground with the required strength can be formed *in situ*.
(b) Since cement substances are carbonates, they are not foreign to the ground.
(c) Accordingly, it is environmentally friendly and can be used as a permanent material. Low total life-cycle cost, because process and products are similar to carbonate diagenesis as described in Chapter 12.

The features of biocement technologies from the viewpoint of ground improvement technology are as follows.

14.5.3 Reduction of Earth Pressure

In civil engineering, earth retaining work is very important. Slope stabilization and ground stabilization associated with excavation are prerequisites not only for safe works, but also protecting people's lives livelihoods from ground disasters through the safe infrastructures. There are few examples that MICP technology has been used for earth retaining works. In this section, potential use of MICP for earth retaining works is proposed.

The stability of soils and ground can be achieved by solidification of backfill soils. The earth pressure is attributed to the sliding force of the wedge of backfill soil along the sliding surface (Figure 14.23). Therefore, the MICP treatment is suggested along the predicted failure plane of soil. The treatment may be effective not only for the entire sliding surface, but also partially. The effect is due to the increase in strength along the failure plane. If the sliding surface avoids the treated zone, the stability of the failure plane must be improved according to the failure analysis. In

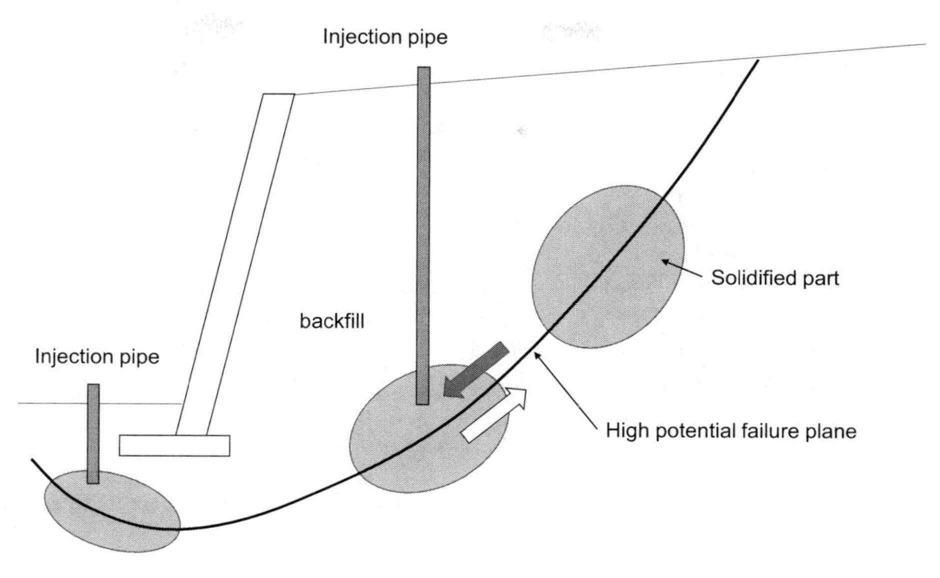

FIGURE 14.23 Retaining wall stabilized with back fill solidified by MICP.

actual performance, the stability of the backfill soil has to be quantitatively analyzed using the shear strength which is a function of c^* and ϕ^* of the treated zones. The final analyses for slope stability will be basically similar to those used in geotechnical engineering.

The MICP technique can strengthen the soils supporting the structures. Accordingly, the MICP treatment can reduce the earth pressure acting on the structures. Thus, the MICP technology can be used for various structures in contact with soil. The merit using MICP for earth retaining work may be cost-effective. In fact, the reduction of earth pressure can provide low cost for construction cost of retaining wall including the cost of retaining wall.

14.5.4 Surface Erosion

Research on the use of MICP as a ground erosion control has only recently begun. Since solidification of sand-gravel ground by MICP is clearly possible, more quantitative research is required to prevent erosion and stabilize slopes (Cheng et al., 2021; Liu et al., 2021; Sun et al., 2022; Zúniga-Barra et al., 2022).

Gowthaman et al. (2020) investigated the effect of MICP on the freeze–thaw durability. They concluded that the MICP treated slope soils are durable to the freeze–thaw induced erosion, using average high carbonate precipitates of 11–13% ($CaCO_3$ by mass) and 20–23%. They also described that these two ranges of carbonate contents caused an erodibility of 50 and 2% mass losses, respectively, under certain conditions. Liu et al. (2021) studied a countermeasure against erosion due to rainfall on the viscous soil slope. They added sands to solidify viscous soil slopes by MICP. They described in conclusion that instead of the intense gully erosion that occurred in the control sample treated with deionized water, only minor gully erosion and sheet erosion were observed for the MICP-treated sample. They concluded that the higher sand content in the MICP-treated soil samples resulted in a higher erosion resistance.

The feasibility, mitigation mechanism, and the effects of MICP treatment cycle and cementation solution concentration were investigated (Cheng et al., 2021), through the rainfall erosion test and penetration test. They concluded that taking into account the overall effectiveness, efficiency and cost, five cycles of MICP treatment with 1.0 M cementation solution is optimal for mitigating the rainfall erosion of the tested loess. It is found that the loess treated by 1.0 M cementation solution could have the best erosion resistance, followed by loess treated by 1.5 M cementation solution, and the lowest one is the loess treated by 0.5 M cementation solution.

Surface rainfall erosion resistance and freeze–thaw durability of bio-cemented and polymer-modified loess slopes were studied by Sun et al. (2022). The freeze–thaw (FT) durability of MICP-polyacrylamide (PAM) treated loess slopes was also studied. The obtained results showed that MICP-PAM treatment improved erosion resistance and addition of 1.5 g/L PAM achieved the best erosion control and highest surface strength. It was summarized that the high erosion resistance of MICP-PAM treated slopes could be attributed to the stable spatial structure of precipitation, and PAM addition conveyed stronger resistance to tension or shear force.

14.5.5 Coastal Erosion

There are many coasts which have been eroded by waves. If the sea levels continue to rise, coastal erosion is expected to become even more severe. There are a few articles concerning the study on coastal erosion (Salifu et al., 2016; Kou et al., 2020; Li et al., 2022). Salifu et al. (2016) as described as follows.

MICP produced 120 kg calcite per m^3 of soil, filling 9.9% of pore space. Cemented sand withstood up to 470 kPa unconfined compressive stress and showed significantly improved slope stability; both slopes showed negligible sediment erosion. With efforts towards optimization for scaling up and further environmental considerations (including effect of slope saturation on MICP treatment, saline water and estuarine/coastal ecology amongst others), the MICP process demonstrates promise to protect shore slope sites.

14.5.6 Wind Erosion

Study on wind erosion using MICP has been targeted for deserts (Chae et al., 2021; Meng et al., 2021; Wang et al., 2018). These studies used unconfined compressive strength or bearing capacity for evaluating the resistance to wind erosion. Wang et al. (2018) described that an unconfined compressive strength of 4 MPa was required. However, it seems that compressive strength of 4 MPa is comparable to rock, which may be too high to avoid wind erosion.

Chae et al. (2021) applied MICP as a countermeasure against wind erosion of four soil types (i.e., medium sand, fine sand, loamy fine sand and loam). They used MICP treatment by pouring and mixing methods, endowing wind erosion resistance to all soil samples.

Meng et al. (2021) described in their study that the optimal cementation solution (with equimolar urea and $CaCl_2$) concentration and spraying volume were 0.2 M and 4 L/m^2, respectively. Under the conditions, the thickness of soil crusts was 12.5 mm, the $CaCO_3$ content of surficial soil was 0.57%, and the bearing capacity of desert soil exceeded 300 kPa (as measured with a 6 mm-diameter handheld penetrometer). However, the value measured by the penetrometer may mean unconfined compressive strength, but not the bearing capacity. Generally, the bearing capacity of MICP treatment soil is about 10 times of UCS (section 14.5.1), depending on the different failure types. It is noted that 6 mm-diameter handheld penetrometer may be a similar instrument which has been provided by ASTM International to measure unconfined compressive strength of cohesive soils. If the penetrometer used is the same type of that provided by ASTM International, the term "bearing capacity" used can be identified as UCS, as demonstrated in this chapter.

It seems that there is no consensus in the studies introduced above. The mechanism of wind erosion needs clarification. It is considered that the water-flow erosion of sediments can be evaluated in terms of shear strength. In wind erosion, it should be examined whether or not unconfined compressive strength is appropriate.

14.5.7 Reduction of Soil Liquefaction Potential

Soil liquefaction may occur due to a combination of conditions of soil and external cyclic loading. The liquefaction occurs when soil skeleton is destroyed by cyclic loading and the soil loses the effective stress, and only when the soil surcharge is supported by the developed excess porewater pressure (Figure 14.24).

Liquefied soil is the same as that in which the shear strength does not exist, and mechanically it is the same as the liquid on which excess water pressure acted. In other words, liquefaction does not occur if even one of the necessary conditions for occurrence is missing. In this sense, soil liquefaction can be regarded as a rare physical phenomenon. In other words, soil liquefaction can be prevented if one of the necessary conditions will be eliminated. For this reason, various technical countermeasures have been developed. The main technologies are roughly divided into the following:

(a) Replacement of soil.
(b) Soil improvement.
(c) Drainage improvement.
(d) Decrease in groundwater level.

The respective methods are also divided into many technologies. For example, for soil improvement there are methodologically densification, solidification, and dissipation of excess pore water, they are further divided into different technologies. The research report of the Construction Technology Research Committee of the Japan Society of Civil Engineers introduces a total of 164 construction technologies, in 2012. The details can be found in other literature. The MICP technology was not suggested in the report in 2012 and there was no MICP technology, which has been established for reduction of soil liquefaction potential in Japan.

MICP technology is considered to fall into the category of an infiltrating solidification method which can be distinguished from the high-pressure injection methods using chemical agents. In the report in 2012, 15 methods are introduced in the osmotic category.

At present, the predominance of MICP technology for reduction of liquefaction potential of soils was supported by many studies (DeJong et al., 2022; Riveros and Sadrekarimi, 2020; Sharma and

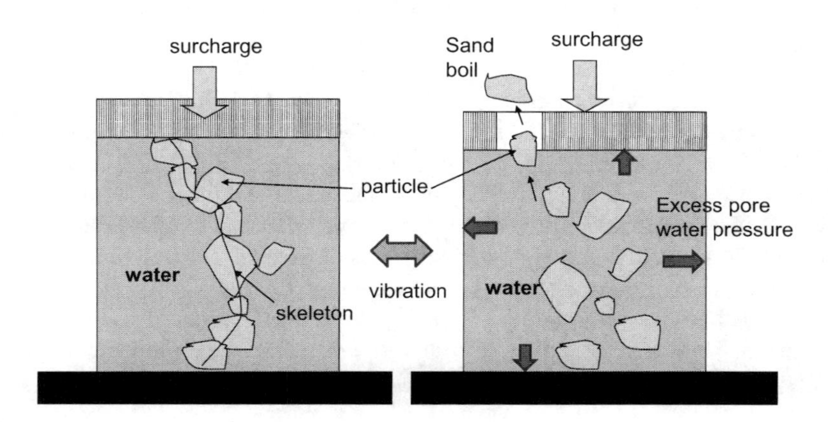

FIGURE 14.24 Mechanism of soil liquefaction.

Satyam, 2021; Sharma et al., 2022; Shan et al., 2022; Xiao et al., 2018; Zamani and Montoya, 2019; Zhang et. al., 2020; Zúniga-Barra et al., 2022; Wang et al., 2023a, 2023b).

In particular, DeJong et al. (2022) provided a state-of-the-art lecture in 20th International Conference on Soil Mechanics and Foundation Engineering. Their lecture was concerned with soil improvement and the reduction of soil liquefaction potential, and their paper exceeded more than one hundred pages. This means that MICP has made an impact as an applied technology for ground improvement and secured its potential.

In fact, anyone who understands MICP will believe that the MICP technology can help prevent liquefaction. The main reasons are as follows:

(1) It is an infiltration method, and in-situ solidification is possible.
(2) Suitable for sandy and gravel soils, it can be used below the water table.
(3) The technology can be used effectively for repetition and rework.

The liquefaction is triggered by the collapse of particles skeleton in immersed soils. The term liquefaction is defined as the state that the effective stress on the skeleton under gravity is instantaneously vanished by collapse and instead porewater pressure takes over. Such soil collapse is caused by shock and vibration induced by mostly earthquakes. Therefore, the aim of the MICP application is to make particle skeleton stronger. This can be confirmed by measuring the strength of soil, such as shear strength, unconfined compressive strength, *S* waves or surface waves, *N* value by the standard penetration test, qc value by the cone penetration test, and measurements of resistance by handy penetrometers such as pocket penetrometer, etc.

The infiltration solidification method due to MICP can be used to reduce liquefaction potential. The approach is simply to increase cohesion of sand and gravel which has potential of liquefaction. Therefore, in the MICP method, it is most important to decide what is the design strength and how much carbonate content is required for that. The products in MICP treatment are the same as a natural soil mineral, carbonates. Other than that, it can be treated and controlled using established knowledge and methods in geotechnical engineering. As is well-known, carbonate is a prime component of limestone.

14.6 CONCLUDING REMARKS

Research to incorporate MICP technology into ground improvement began around 2000. Since this is still less than 25 years old, it is difficult to say that it is an established technology. In particular, there are many difficulties for geotechnical engineers to proceed with research on microorganisms. At the same time, there are also many difficulties for microbiology specialists to deal with ground improvement.

Since around 2010, the number of researchers interested in MICP has started to increase rapidly, and although the range of research content that can be seen from the published papers seems to have expanded, it cannot be said that the essential problems have been solved from the viewpoint of technological progress (Lai et al., 2021; Fu et al., 2023).

In Chapter 13 prior to proceeding with Chapter 14, the essential problems in MICP were discussed and some solutions were provided. In particular, the standardization of OD-CPR and the inhibition and retardation by Ca^{2+} were comprehensively explained. Chapter 14 dealt with the application of MICP, in particular for soil improvement. The main intent in this chapter was how to connect MICP technology to geoenvironmental engineering and geotechnical engineering. To that end, we took note of two things. That is, what to take over and to where, and what the benefits of using MICP are. Finally, a graphical conclusion is presented in Figure 14.25, because there are too many items to be explained.

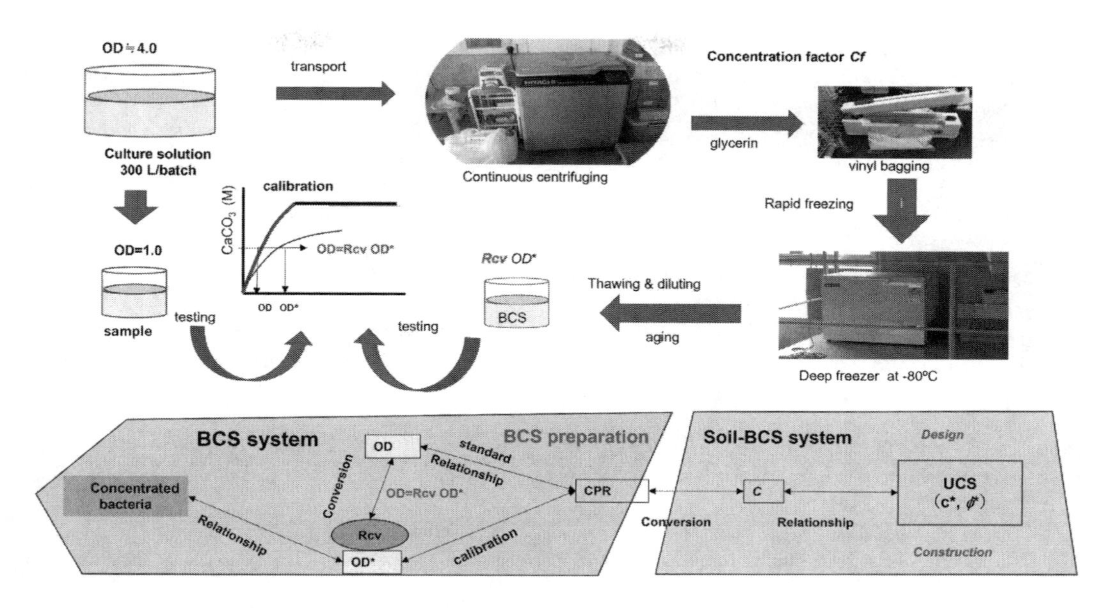

FIGURE 14.25 Graphical conclusions of Chapters 13 and 14, from bacteria cultivation to soil improvement.

REFERENCES

Burley, S.D., and Worden, R.H. (eds.). (2003), *Sandstone Diagenesis-Recent and Ancient*, Reprint Series Volume 4 of the International Association of sedimentologists, Black Well Publishing Ltd., Berlin, Germany.

Castanier, S., Métayer-Levrel, G.L., and Perthuisot, J.-P. (1999), Ca-carbonates precipitation and limestone genesis-the microbiogeologist point of view, *Sedimentary Geology*, 126: 9–23.

Chae, S.H., Chung, H., and Nam, K. (2021), Evaluation of Microbially Induced Calcite Precipitation (MICP) methods on different soil types for wind erosion control, *Environmental Engineering Research*, 26(1): 190507, https://doi.org/10.4491/eer.2019.507.

Chen, M., Gowthaman, S., Nakashima, K., Komatsu, S., and Kawasaki, S. (2021), Experimental study on sand stabilization using bio cementation with wastepaper fiber integration, *Materials*, 14: 5164.

Cheng, Y., Tang, C., Pan, X., Liu, B., Xie, Y., Cheng, Q., and Shi, B. (2021), Application of microbial induced carbonate precipitation for loess surface erosion control, *Engineering Geology*, 294: 106387.

DeJong, J.T., Gomez, M.G., San Pablo, A.C.M., Graddy, C.M.R., Nelson, D.C., Lee, M., Ziotopoulou, K. Kortbawi, M. E., Montoya, B., and Kwon, T.-H. (2022), State of the Art: MICP Soil Improvement and Its Application to Liquefaction Hazard Mitigation. In: *Proceedings of the 20th International Conference on Soil Mechanics and Geotechnical Engineering, Sydney, Australia, 1–5 May 2022*, vol. 1, pp. 405–508, ISBN 978-0-9946261-6-5.

Fowle, D.A., and Fein, J.B. (1999), Competitive adsorption of metal cations onto two-gram positive bacteria: Testing the chemical equilibrium model, *Geochimica et Cosmochimica Acta*, 63(19/20): 3059–3067.

Fukue, M. Lechowicz, Z. Fujimori, Y. Emori, K., and Mulligan, C.N. (2022), Incorporation of optical density into the blending design for a biocement solution, *Materials*, 15: 1951, https://doi.org/10.3390/ma15051951.

Fukue, M. Lechowicz, Z. Fujimori, Y. Emori, K., and Mulligan, C.N. (2023), Inhibited and retarded behavior by Ca^{2+} and Ca^{2+}/OD loading rate on ureolytic bacteria in MICP process, *Materials*, 16: 3357, https://doi.org/10.3390/ma16093357.

Fukue, M. Nakamura, T. and Kato, Y. (1999), Cementation of soils due to calcium carbonate, *Soils and Foundations*, 39: 55–64.

Fukue, M. Ono, S. and Sato, Y. (2011), Cementation of sands due to Microbiologically induced carbonate, *Soils and Foundations*, 51: 83–93.

Fu, T., Saracho, A.C., and Haigh, S.K. (2023), Microbially induced carbonate precipitation (MICP) for soil strengthening: A comprehensive review, *Biogeotechnics*, 1: 1–23. https://doi.org/10.1016/j.bgtech.2023.100002.

GIW. (2024), Bearing Capacity, Geotechnical info com. Creer Development and Resources for Geotechnical Engineers. http://geotechnical info.com/bearing_capacity_factors.html, accessed May 2024.

Gowthaman, S., Nakashima, K., and Kawasaki, S. (2020), Freeze-thaw durability and shear responses of cemented slope soil treated by microbial induced carbonate precipitation, *Soils and Foundations*, 60: 840–855.

Hermansson, M. (1999), The DLVO theory in microbial adhesion, *Colloids and Surfaces B: Biointerfaces*, 14: 105–119.

Kagawa, H. (1973), Adsorption of bacteria by soil, *Soil Science and Nutrition*, 17(4): 150–153.

Kogure,T., Aoki, H., Maekado, A., and Matsukura, Y. (2005), Effect of specimen size and rock properties on the uniaxial compressive strength of Ryukyu limestone, *Journal of the Japan Society of Engineering Geology*, 46(1): 2–8.

Kou, H., Wu, C, Ni, P., and Jang, B. (2020), Assessment of erosion resistance of biocemented sandy slope subjected to wave actions, *Applied Ocean Research*, 105: 102401, https://doi.org/10.1016/j.apor.2020.102401.

Lai, H.-J., Cui, M.-J., Wu, S.-F., Yang, Y., and Chu, J. (2021), Retarding effect of concentration of cementation solution on biocementation of soil, *Acta Geotechnica*, 16: 1457–1472, https://doi.org/10.1007/s11440-021-01149-1.

Li, Y., Guo, Z., Wang, L., Yang, H., Li, Y., and Zhu, J. (2022), An innovative eco-friendly method for scour protection around monopile foundation, *Applied Ocean Research*, 123: 103177.

Lin, H., Suleiman, M., Jabbour, H.M., and Brown, D. (2023b), Bio-grouting to enhance axial pull-out response of pervious concrete ground improvement piles, *Canadian Geotechnical Journal*, 55: 119–130.

Lin, H., Zhou, M., Li, B., and Dong, Y. (2023a), Mechanisms, application advances and future perspectives of microbial-induced heavy metal precipitation: A review, *International Biodeterioration & Biodegradation*, 178: 105544, https://doi.org/10.1016/j.ibiod.2022.105544.

Liu, B., Xie, Y., Tang, C., Pan, X., Jiang, N., Singh, D., Cheng, Y., and Shi, B. (2021), Bio-mediated method for improving surface erosion resistance of clayey soils, *Engineering Geology*, 293: 106295.

Meng, H., Gao, Y., He, J., Qi, Y., and Hang, L. (2021), Microbially induced carbonate precipitation for wind erosion control of desert soil: Field-scale tests, *Geoderma*, 383(1): 114723, https://doi.org/10.1016/j.geoderma.2020.114723.

Meyerhof, G.G. (1951), The ultimate bearing capacity of foundations, *Géotechnique*, 2: 301–332.

Omoregie, A.I, Palombo, E.A., and Nissom, P.M. (2021), Bioprecipitation of calcium carbonate mediated by ureolysis: A review, *Environmental Engineering Research*, 26(6): 200379, https://doi.org/10.4491/eer.2020.379.

Riveros, G.A., and Sadrekarimi, A. (2020), Liquefaction resistance of Fraser River sand improved by a microbially-induced cementation, *Soil Dynamics and Earthquake Engineering*, 131: 106034.

Salifu, E., MacLachlan, E., Lyer, K.R., Knapp, C.W., and Tarantino, A. (2016), Application of microbially induced calcite precipitation in erosion mitigation and stabilization of sandy soil foreshore slopes: A preliminary investigation, *Engineering Geology*, 201: 96–105.

Shan, Y., Zhao, J., Tong, H., Yuan, J., Lei, D., and Li, Y. (2022), Effects of activated carbon on liquefaction resistance of calcareous sand treated with microbially induced calcium carbonate precipitation, *Soil Dynamics and Earthquake Engineering*, 161:107419.

Sharma, M., and Satyam, N. (2021) Strength and durability of biocemented sands: Wetting-drying cycles, ageing effects, and liquefaction resistance, *Geoderma*, 402: 115359, https://doi.org/10.1016/j.geoderma.2021.115359.

Sharma, M., Satyam, N., and Reddy, K.R. (2022) Large-scale spatial characterization and liquefaction resistance of sand by hybrid bacteria induced biocementation, *Engineering Geology*, 302: 106635 https://doi.org/10.1016/j.enggeo.2022.106635.

Terzaghi, K. (1943), *Theoretical Soil Mechanics*, John Wiley & Sons, New York.

Sun, X., Miao, L., Chen, R., Wang, H., and Xia, J. (2022), Surface rainfall erosion resistance and freeze-thaw durability of bio-cemented and polymer-modified loess slopes, *Journal of Environmental Management*, 301: 113883.

van Paassen, L. (2009), *Biogrout, Ground Improvement by Microbially Induced Carbonate Precipitation*, Delft University Technology, 195. ISBN 978-90-8147181-7.

Wang, K., Wu, S. and Chu, J. (2023a), Mitigation of soil liquefaction using microbial technology: An overview, *Biogeotechnics*, 1, (1): 100005, https://doi.org/10.1016/j.bgtech.2023.100005.

Wang, Y., Konstantinou, K., Tang, S. and Chen, H. (2023b), Applications of microbial-induced carbonate precipitation: A state-of-the-art review, *Biogeotechnics*, 1(1), https://doi.org/10.1016/j.bgtech.2023.100008.

Wang, Z. Zhang, N., Ding, J., Lu, C., and Jin, Y. (2018), Experimental study on wind erosion resistance and strength of sands treated with microbial-induced calcium carbonate precipitation, *Advances in Materials Science and Engineering*, 1(1), 1–19, Article ID 3463298, https://doi.org/10.1155/2018/3463298.

Xiao, P., Liu, H., Xiao, Y., Stuedlein, A.W., and Evans, T.M. (2018), Liquefaction resistance of bio-cemented calcareous sand, *Soil Dynamics and Earthquake Engineering*, 107: 9–19, https://doi.org/10.1016/j.soildyn.2018.01.008.

Zamani, A., and Montoya, B.M. (2019). Undrained cyclic response of silty sands improved by microbial induced calcium carbonate precipitation, *Soil Dynamics and Earthquake Engineering*, 120: 436–448, https://doi.org/10.1016/j.soildyn.2019.01.010.

Zhang, X., Chen, Y., Liu, H., Zhang, Z., and Ding, X. (2020), Performance evaluation of a MICP-treated calcareous sandy foundation using shake table tests, *Soil Dynamics and Earthquake Engineering*, 129: 105959, https://doi.org/10.1016/j.soildyn.2019.105959.

Zúniga-Barra, H., Toledo-Alarcon, J., Torres-Aravena, A., Jorquera, L., Rivas, M., Gutiérrez, L., and Jeison, D. (2022), Improving the sustainable management of mining tailings through microbially induced calcite precipitation: A review, *Minerals Engineering*, 189: 107855, https://doi.org/10.1016/j.mineng.2022.107855.

15 Towards Geoenvironmental Sustainability

15.1 INTRODUCTION

Land and water resources are the principal physical capital items that constitute the geoenvironment. Their quality and health are vital issues because they provide the habitat and also the basis for life-support systems for plants, animals, and humans. These are natural resources that are basically renewable resources. In the absence of human-created stressors generating negative impacts, the various resource elements have the ability to maintain their natural quality through natural processes and/or through replenishment. These resource elements have the ability to flourish and grow and are integral parts of the overall environment. We consider a significant part of *sustainability of the geoenvironment* to include sustainability of land and water resources. Sustainability is obtained when all the resource elements of the various land and water ecosystems are renewed, replenished, re-charged, and re-stocked – to a level that will continue to meet the needs of those that depend on these resources. This will only occur when the health and quality of the land and water resources are protected, maintained, and allowed to flourish. Failure to do so will lead to a degradation of the quality of these two major capital resources, and in turn will imperil and diminish the capability of these resources to allow the elements and habitants of the multitude of land and water ecosystems to renew and replenish themselves.

The focus of this book has been directed towards the major impacts or deterrents to sustainability objectives arising from human activities. The primary sustainability concerns relating to industry, agricultural and urban interactions with the land environment and its receiving waters are impacts from

- exploitation and extraction of renewable and non-renewable resources,
- depletion of non-renewable resources and mis-use and mis-management of renewable resources,
- noxious discharges from industrial and agricultural operations and urban activities.

A variety of implications and impacts arising from these concerns have been discussed in the previous chapters. As noted in the concluding section of the previous chapter, the point-source pollution impacts from urban and industrial activities account for one major component of the large ground contamination problem. The other major components are atmospheric-based and land-based non-point source pollution. Figure 15.1 shows some of the elements of these two major components and the natural resources threatened by both kinds of contaminant sources.

15.1.1 Unsustainable Actions and Events

Unsustainable actions refer to those actions and circumstances that generate negative impacts to, or in, or on the geoenvironment. Discussions on many of these impacts or activities and events contributing to such impacts have been given in the previous chapters. The following brief discussion in this section provides a set of examples that demonstrate the problem at hand and also why the quality and health of the land and water resources are central issues.

DOI: 10.1201/9781003407393-15

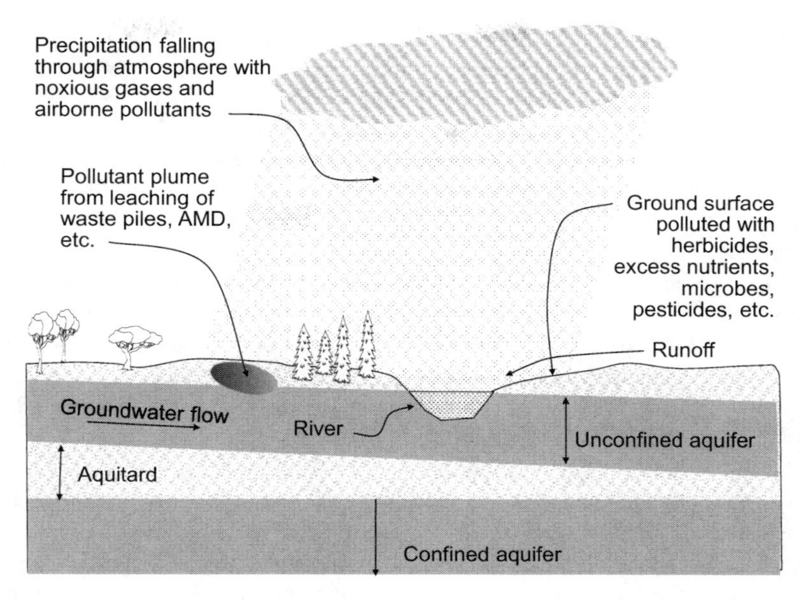

FIGURE 15.1 Some of the more prominent causes of pollution of recharge water for rivers, other receiving waters, and groundwater (aquifers). Contamination of the confined aquifer depends on whether communication is established with the unconfined aquifer.

15.1.1.1 Iron and Coal Mining

A recounting of the 1966 major tip failure (slippage) at Aberfan that was responsible for the deaths of over a hundred individuals can be found in McLean and Johnes (2000). The height of the piles together with the slope of the piles and the weather conditions contributed to the disastrous slippage problems. Even without the disaster of tip failure, these tips, heaps and piles constitute a blight to the land environment and leaching of these piles will produce conditions that will (a) render the landscape sterile (Haigh, 1978) because of the acidity and deficiencies in nitrogen and phosphate, and (b) threaten the groundwater and water resources if and when the leachates reach these resources.

15.1.1.2 Oil and Petroleum

Exploration, drilling, extraction, and production of petroleum can impact the environment at numerous points. Contamination of shores after oil spills can lead to oil that can remain for decades at beaches or marshes. Pressurized hoses with cold or hot water are used for cleanup of spills on beaches, etc., may create more problems – as for example in the case of the Exxon Valdez spill where the more than 200,000 tonnes of disposable diapers, pads, clothing, and other waste materials used in washing individual stones on the cobble beaches required landfilling or burning (Graham, 1989). In addition, the hot spray used in the cleaning activities is reported to have caused damage to the benthic fauna.

Oil spills and pipeline leaks may also contaminate the soil and groundwater. Drilling mud stored in improperly lined pits may also leak into the soil. Well blowouts from over-pressurized zones can lead to spreading of the petroleum on the surface of the land or water if at sea. On March 29, 1980, 468,000 tonnes of crude oil (the largest one in history) rushed into the Gulf of Mexico after a blowout, thereby affecting Mexican and Texas beaches. Approximately 0.8% of the oil reached the Texas shoreline – with about 5% of the original amount remaining after one year (Payne and Phillips, 1985).

Another more significant event in 2010, one the largest oil spills, known as the Deepwater Horizon oil spill, occurred in the Gulf of Mexico when 780,000 m^3 of oil were released from an oil well blowout and explosion until the oil rig was capped after over 87 days. Despite the beach and wetland protection via floating booms, skimmer ships, controlled burners, and Corexit dispersant addition, which seemed to increase oil toxicity substantially, marine wildlife, fishing, tourism, the economy due to shutdown of the oil rig and human health were adversely affected. More than 2.1 million kg (of oily mass were collected in 2013, and half of that amount was collected in 2012 from the Louisiana beaches. It has been estimated that more than 1728 km of shoreline from Louisiana, Mississippi, Alabama, and Florida received oil from the spill. Sand sifting, manual, and mechanical removal were used for sand beaches whereas cutting of the vegetation, vacuum and pumping and bioremediation were used for marsh remediation. Sediments at the bottom of the seafloor were also affected by the non-degrading oil. Overall, as of Aug. 4, 2010, 26% of the oil was residual, 25% was evaporated or dissolved, 17% recovered at the well head, 16% naturally dispersed, 8% chemically dispersed, 5% burned and 3% skimmed (Schrope, 2010).

At Lac Megantic in the Province of Quebec in Canada on Jul. 6, 2013, a runaway freight train loaded with crude oil derailed and exploded, killing 47 people and destroying half of the downtown region. More than 30 buildings were destroyed from the fire including many irreplaceable historic buildings. The train had been left unattended on the main track with insufficient brakes on an incline. Oil contamination of the land was extensive affecting more than 31 hectares (www.thestar.com/news/canada/lac-megantic-environmental-report-details-extent-of-contamination/article_ecc4d6a0-06f6-52fa-91bd-0a674ce0da0c.html). Many businesses had to be relocated. More than 100,000 L of oil reached the Chaudière River and traveled more than 120 km. The amount of oil involved in the accident totaling more than 5.6 million liters. The total cost is estimated at $500 million for the remediation of water and soil in the area.

15.1.1.3 Medical Wastes

In the US, the total amount of municipal solid waste produced was 264 million tonnes in 2018 (US EPA, 2022). Approximately 5.35 million tonnes of medical waste were produced in the US (www.medicalwastepros.com/blog/how-much-medical-waste-is-produced/, accessed Jun. 2024), which is approximately 2% of the municipal waste. COVID-19 in particular had a significant effect on medical and plastic waste generation as discussed in Chapter 7. Pathogen transport and survival is possible if these types of wastes are disposed as MSW. However, there is little experimental data available for the determination of pathogens in leachates or in runoff from a landfill.

15.1.1.4 Pulp and Paper

After significant improvements in the production processes of the pulp and paper industry, the water consumption was reduced significantly over the past century from 500–1000 m^3 of water per tonne of paper to 13 m^3/tonne of paper produced in countries such as Germany. Water usage is even lower in general in recycled fibre-based mills. However, still more than 15% of paper mills use more than 100 m^3/tonne of water (Esmaeeli and Sarrafzadeh, 2023). In addition, there are 700 organic and inorganic compounds pollutants in the wastewater, ranging from lignin, VOCs, various chlorinated compounds, nutrients (N and P), sulfate, methyl tertiary butyl ether (MTBE), and metals. Sludge generation is in the range of 40 to 50 kg per tonne of paper. The generation of dry sludge is approximately 70% primary and 30% secondary (Bajpai, 2015). Total sludge generation is about 200 million tonnes per year (FAO, 2022), www.fao.org/faostat/en/#data/FO, accessed May 2024).

Chlorine and compounds of chlorine are used in the bleaching of wood pulp particularly by the kraft or sulfite processes. Processes using elemental chlorine were producing significant quantities of dioxins by replacement of elemental chlorine by chlorine dioxide has reduced dioxin production substantially. From 1990 elemental chlorine free (ECF) and totally chlorine free (TCF) processes have replaced element chlorine bleaching. In 2005, elemental chlorine decreased by over 90% since 1990.

Energy consumption trends have shifted in recent years. The United States Environmental Protection Agency has found that paper recycling instead of producing virgin paper has reduced water pollution by 35% and air pollution by 74% (US EPA, 2007). Water and energy consumption are reduced by 9%–25% and 28%–70%, respectively, for recycled paper compared to virgin paper (Ramezani et al., 2011). Recycling one tonne of paper can save up to 17 trees and 26,500 L of water (https://paperexcellence.com/making-a-difference-the-environmental-benefits-of-recycling-paper/ , accessed Jun. 2024). Chemical use is also reduced during paper recycling since the process is mainly water based.

15.1.1.5 Cement, Stone, and Concrete

Globally more than 30 billion tonnes of concrete are produced each year, an increase of 3 times that of 40 years ago (Monteiro et al., 2017). In addition, its production accounts for at least 8% of global emissions. Cement is the main ingredient of concrete. A typical mix is 12% Portland cement, 34% sand, 48% crushed stone, and 6% water, all of which are abundant. However, transportation can be a main issue, and water requirements for washing and reduction of the impact of dust. Fly ash, a waste from coal-fired power plants, is now present at a proportion of about 15 to 30% in cement. Energy consumption is considerable in cement production. Addition of fly ash with concrete can make the process more energy efficient. Since coal is often used for energy emissions of carbon dioxide, sulphur and nitrous oxide are high, carbon dioxide emissions are estimated at 1200 kg CO_2 per tonne of cement, 60% from energy use, and the other 40% from calcining. According to Fayomi et al. (2019) for every 1 kg of cement produced, 0.500–900 kg of CO_2 is emitted for tonne of cement produced. Dust emissions are also significant (180 kg/tonne of cement produced) but should be controlled as much as possible by water sprays, hoods, etc. Other pollutants include sulphur dioxide and nitrous oxide from both fuels and raw materials. Water pollution from washwater of high pH is generated. Settling ponds are used to remove the solids. Reduction of the pH below 12 renders the wastewater not hazardous. Water usage has been reduced through recycling in the plant in a closed loop. Construction and demolition waste according to the US EPA (2015) consisted mainly of Portland cement concrete (67%) and asphalt concrete (18%). More and more of this waste will be used in road aggregate. It has been indicated that recycled concrete (77%) is being used in "backfill/road base" whereas only 23% specified utilization in producing new concrete (Jin and Chen, 2015). Pre-casting concrete at a central facility can reduce materials used and wastewater generated. Other stone wastes are also generated such as marble, granite, stone, and limestone, which could potentially be used as a replacement for sand in mortar and for hazardous waste containment (Gehlot and Shrivastava, 2023).

15.1.1.6 Various Stressors and Impacts

Waste generation and disposal demonstrate the problems of stressors and impacts. For example, of the 40.4 million tonnes of commercial and industrial (C&I) waste produced in the UK in 2020, 33.8 million tonnes (84%) were generated in England (www.gov.uk/government/statistics/uk-waste-data/uk-statistics-on-waste). These amounts can be compared to the total was generated in the UK of 222.2 million tonnes of total waste in 2018, with England responsible for 84% of the UK total (187.3 million tonnes).

In Chapter 7, we have discussed the impact of urbanization and urban sprawl. Urban development is a major consumer of land. Natural landscape areas around the cities are converted into housing estates, industrial parks, and other kinds of facilities designed to serve the community. Land is typically used for housing, businesses, industry, surface and subsurface infrastructures such as roads, wastewater supply, sewers, and power lines, and for recreational purposes such as parks and playgrounds.

According to Appunn (2018), Germany's agricultural land has decreased from 2000 to 2016, by 6970 km^2 as the result of urban sprawl and new infrastructure. In addition, infrastructure construction of houses or roads in Germany consumes 66 hectares every day. Over the period of 20 years, (1992–2012) about 12.6 million hectares of farmland has been lost due to expanding urban areas (59%) and low-level residential development (41%). In addition, 62% of all development occurs using agricultural land (www.agweek.com/business/31-million-acres-lost-development-cuts-into-u-s-farmland). In the US, 70 hectares of farm and ranchland are lost every hour for housing and other industries (https://farmlandinfo.org/publications/farms-under-threat-the-state-of-the-states/).

Site contamination is also a major issue worldwide. For example, in the European Union (EU), approximately 2.8 million sites are contaminated. Of these, it has been estimated by the European Environment Agency (EEA) that 300,000 contaminated sites require clean-up (www.who.int/europe/news-room/fact-sheets/item/industrially-contaminated-sites, accessed Jun. 2024). Although site contamination is common to all countries, developing countries will be impacted in the near future to a greater extent. For example, the number of contaminated sites in China is estimated at 200,000 (Li et al., 2015). Soil biodiversity is another aspect that must be protected as it provides a release of nutrients for use by plants and other organisms, assists in the removal of contaminants and pathogens, and participates in the carbon cycle, and it is a source of genetic and chemical resources (http://eusoils.jrc.ec.europa.eu/). Soil biodiversity is highly threatened in areas of high population density and/or areas of intense agricultural activity as discussed in section 6.4.

More than 360,000 tonnes of waste were generated in 1993 by the fabricated metals industry (US EPA, 1995), which now decreased to less than 290,000 tonnes in 2019. New sources reduction strategies have been introduced. However, from 2009 to 2019, quantities have decreased by 9%. According to the Toxic Release Inventory (TRI), approximately 73.6% of the waste was recycled, 17.1 % was treated, energy was recovered on-site for 1.6% and another 7.8 % was either released to the environment or transferred off site (US EPA, 2021).

The impacts from contaminants together with impacts from utilization of energy and other natural resources on the geoenvironment are the main geoenvironmental issues addressed as barriers to sustainability goals. In this chapter, we will (a) discuss the use or exploitation of non-renewable non-living renewable natural resources, (b) look at some typical case histories and examples of sustainability actions, and (c) present the geoenvironmental perspective of the present status of *"where we are in the geoenvironmental sustainability framework"*, with a view that points towards *"where we need to go"*.

15.2 EXPLOITATION AND STATE OF RENEWABLE NATURAL RESOURCES

By all accounts, metal and mineral resources together with fossil fuels are non-renewable natural resources. There is also a case to be made for classifying water from deep-seated aquifers as non-renewable resources. Their continued extraction will not only result in their depletion, but will ultimately lead to their exhaustion. Fossil fuels are of particular concern since they are used not only for fuels, but are the main source of raw material for countless numbers of products ranging from plastics and synthetic fibres to pharmaceutical and other consumer products. This eventuality has been recognized and efforts have been and are being made by industry and consumers to (a) reduce and/or find more efficient fuel consumption engines, and (b) find renewable substitutes. It is not within the purview of this book to discuss alternative fuels for powering engines and motors. There are numerous textbooks and research literature dealing with advances and innovative ideas in this subject.

In respect to generation of electricity outside of hydroelectric facilities and oil and coal generating plants, various alternative and renewable sources of electricity-generating are available. These include tidal-wave, solar thermal, solar photoelectric, geothermal, winds onshore and offshore, and biomass. According to the Renewables 2023 Global Status Report (www.ren21.net/gsr), renewable energy production continued to grow to over 12.6% of total energy consumption from 8.8% in 2011.

There are two specific classes of renewable natural resources, namely living and non-living. Living renewable natural resources include land and aquatic animal, forests, native plants, etc., whilst non-living renewable natural resources include water and soil. By definition, renewable natural resources refer to those resources that have the capability to regenerate, replenish, and renew themselves, either naturally or with human intervention within a reasonable time period. *Sustainability* as an objective requires that full regeneration-replenishment of renewable resources must be obtained. It is recognized that when consumption (use, exploitation, etc.) exceeds regeneration-replenishment rate, sustainability of the renewable resource is not obtained. This does not mean that the renewable resource will not or cannot renew itself. It simply means that the amount or rate of the resource that is renewed is insufficient to meet the demands placed on it. In recognition of this, we need to distinguish between (a) unsustainable renewable natural resources, i.e., renewable resources that by virtue of circumstances cannot be fully renewed or replenished, and (b) sustainable renewable natural resources, i.e., renewable resources that can be totally regenerated and replenished. When consumption rate is greater than the rate of regeneration and/or replenishment, etc., the amount or nature of the particular renewable natural resource will be depleted and may eventually become extinct. Striking examples of this are over-fishing and over-use of groundwater (water from aquifers).

15.2.1 Sustainability of Renewable Non-Living Natural Resources

There is a further distinction or differentiation needed in discussing renewable natural resources. One needs to distinguish between *natural* and *developed* resources. Differentiation between renewable *natural* resources and renewable *developed* resources is necessary to distinguish between the renewable natural capital items (water, soil, land and aquatic animals, native plants) and restocking and regeneration of man-made capital such as fish farms and agricultural output from land farming. As has been noted previously, just because a resource is renewable does not make it sustainable. Two necessary, but not sufficient, conditions for sustainability of renewable natural resources are: (a) replenishment and regeneration of the natural capital items in a reasonable time frame either through natural processes or through sound management practise, and (b) renewed natural resources are sufficient and will continue to be sufficient to meet the demands placed on these resources. Impediments to sustainability are due to (1) rate of recharge or regeneration or replenishment being outpaced by over-exploitation of the natural resource, and (2) corruption, degradation and/or pollution of the natural resource.

The renewable non-living natural resources of prime importance are water and soil. They are in essence renewable dynamic resources – characterized by recharge and replenishment of these resources. However, when recharge and replenishment cannot overcome the deficits in the non-living renewable resources, these resources are no longer sustainable. A good example of the preceding is the excessive use, over-exploitation and pollution of water resources. A full treatment of these and other issues relating to sustainable water use will be found in the textbooks dealing with this particular problem. The geoenvironmental concerns for water and soil are in respect to degradation in water and soil quality due to their misuse and also due to pollution of these capital items. Discussions in the previous chapters have shown that water and groundwater pollution, together with soil pollution and loss of soil quality are the major downfall of sustainability of water and soil resources – other than over-use, abuse and misuse of these non-living renewable resources by humans.

The "beetle-type" diagram in Figure 15.2 provides a very simple illustration of some of the major stressors on water, groundwater, and soil resources responsible for the unsustainable outcome of the non-living renewable natural resources. Other than the use, misuse, abuse, etc., by industry and humans shown in the top left-hand side of the illustration, most of the stressors and

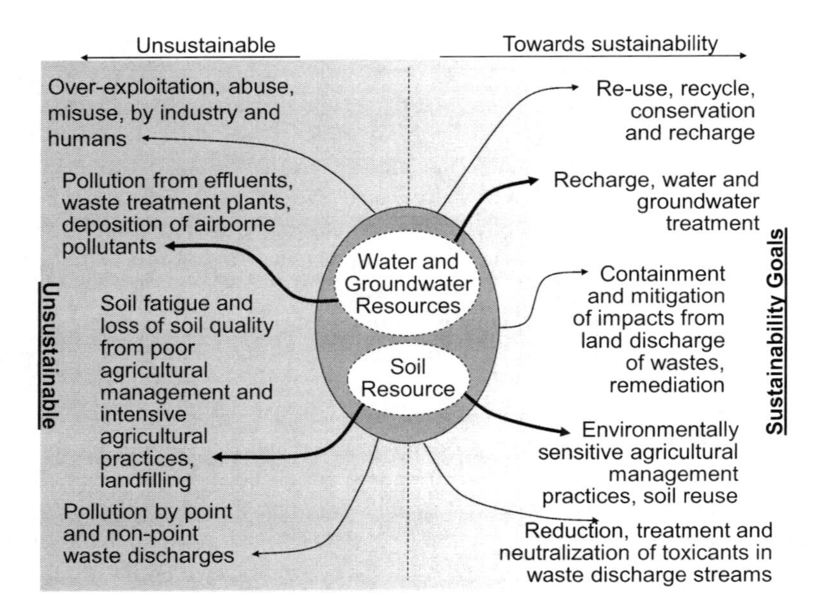

FIGURE 15.2 Some of the major stressors of groundwater, water, and soil resources responsible for the unsustainable state of these non-living renewable natural resources (shown on left). A sample of the types of actions needed to permit these renewable natural resources to meet sustainability goals and objectives.

their impacts have been identified and discussed in the previous chapters. Some of the key actions needed to drive the renewable natural resources of water, groundwater, and soil towards sustainability goals and objectives are shown on the right-hand side of the illustration. The major impact of pollutants on both water and soil resources is degradation of the quality of these resources. In the case of soils, for example, degradation of soil quality will lead directly to loss of productivity and lower agricultural yields. For water, degradation of water quality leads directly to loss of drinking water status at the upper end of water usage, to relegation for agricultural and industrial use, and finally to non-usable.

Pollutants are chemical stressors that can severely affect water, groundwater, and soil quality. These include land-based non-point source pollutants such as herbicides, pesticides, fungicides, etc., and point-source pollutants from effluents, waste treatment plants, and liquid discharges as wastes and spills from industrial plants (e.g., heavy metals and organic chemicals). Some of the methods and procedures for mitigating of the major impacts from pollutant stressors, together with treatment and remediation options have been discussed in the previous chapters. Undoubtedly, there will be more advanced and sophisticated methods developed in the very near future to meet the challenges posed by these stressors. Unsustainable renewable natural resources become non-renewable, renewable natural resources – a condition that should not be allowed to happen.

15.2.2 Geoenvironment and Management of Renewable Resources

From a geoenvironmental perspective, the actions shown on both sides of Figure 15.2 are some of the more significant actions that detract from or lead to conservation and recovery of non-living renewable resources. The arrows that emanate from the total elliptical shell indicate actions affecting the total land environment, i.e., both soil and water-groundwater resources. Those arrows emanating specifically from the top or bottom resources (water-groundwater resources and soil resources) indicate actions affecting those specific resources. Not all the pertinent or necessary actions for all of these are shown. Only the more significant ones are depicted in the diagram.

15.2.2.1 Unsustainable Actions

Many of the impacts due to the unsustainable actions shown on the left-hand side of Figure 15.2 have been discussed in the previous chapters and have been referred to in Section 15.1.1. The points of note include (a) over-exploitation, abuse, and misuse of water and soil resources by humans are actions requiring corrective measures that extend beyond the geoenvironmental sphere, (b) pollution from effluents, waste treatment plants, deposition of airborne pollutants are both point and non-point sources of contaminants that degrade water quality and create contaminated ground. The extent and seriousness of health threats from the pollutants and the degree of degradation of both water and land resources are functions of both the quantity/concentration and toxicity of the discharges and atmospheric depositions on land. Run-offs over land surface and industrial chemical spills can severely pollute receiving waters and groundwaters.

The bacterial contamination in the water in the town of Walkerton (Ontario, Canada) that led to the death of seven residents in the period from May to the end of Jul., 2000, and referred to in Chapter 3, is testimony to the effects of run-off and infiltration of polluted recharge rainwater. Reported evidence indicated that heavy rains on May 12 transported bacteria from cattle manure into Walkerton's shallow town well, and that town residents were exposed to *E. coli* over the next few days. As reported, more than 500 residents reported *E. coli* symptoms and a further 150 residents sought hospital treatment (http://camillasenior.homestead.com/Walkerton_Chronology-2004.pdf, accessed May 2024).

Another example of note is the Love Canal problem that surfaced in the 1970s. As is well known, this is probably the first well-publicized hazardous waste dumping site in North America. (www.epa.gov/archive/epa/aboutepa/love-canal-tragedy.html, accessed May 2025). Prior to this period, awareness of the problems and seriousness of indiscriminate dumping of toxic materials were not appreciated by the general public, and it was claimed that the Love Canal site was the recipient of such hazardous materials for a period of at least 20 years. Tests conducted by the New York Department of Environmental Concern showed severe pollution of ground and waters in the area, resulting in the declaration of a state of emergency by the Governor of the State of New York, and the closing of schools and relocation of several families.

Many other examples of water and ground contamination from inadvertent and deliberate dumping can be cited as indicated by the Teshima case described in Chapter 7. There are other cases of pollution however that are indirectly caused by man-made activities. A good example of this is leaching of exposed sulphide ores and rocks (Chapter 5) and waste piles in landfills will also produce polluted recharge water. Other causes of pollution of water resources have been discussed in Chapter 3.

Soil fatigue and loss of soil quality can occur from natural causes such as those leading to aridification and desertification. Long periods of rainfall deprivation leading to aridification and finally desertification are conditions of nature. Desertification can also result from a prolonged process of degradation of a once-productive soil. The root causes for desertification are deemed to be a complex mix of various degradative actions. The present concern is soil fatigue and loss of soil quality from human activities leading to poor forest and agricultural management and intensive forestry and agricultural practise are conditions that will eventually render the soil useless for production of plants and crops. In respect to agriculture, as we have noted in Chapter 2, *soil quality* is a determinant of the capability of a soil to sustain plant and animal life and their productivity, and any diminution of soil quality will impact on its capability to provide the various functions such as plant and animal life support, forestry and woodland productivity, and will undoubtedly result in the loss of biological activity and biodiversity, and depletion of nutrients in the soil.

We should note that soil quality as a measure of the functionality or capability of a soil is not confined exclusively to the agricultural usage. Soil can also serve other kinds of functions. These include (a) containment and management of wastes and waste streams, (b) resource material for production of building blocks, and (c) sub-base support for structures and facilities. The determinants for soil quality for these types of functions will differ from the classic definition which was developed for agricultural use. This is discussed further in Section 15.3.

15.2.2.2 Towards Sustainability

For sustainability management of the land environment and particularly of the water and soil resources, a necessary requirement is for recharge materials and processes to be devoid of pollutants and other detrimental and degrading agents. This is particularly acute for reuse and recycle of process water. At all times, the quality of water and soil needs to be maintained and even improved. For this to occur, we need to establish water and soil quality indices and to further establish baseline values for these indices. These indices will require analyses involving indicators – both status and material indicators. A more detailed discussion of these and the quality indices will be given in a later section in this chapter.

Recharge of Water Resources: The sources of natural recharge of receiving waters and groundwater are direct precipitation (rain, snow, hail, and sleet) and snowmelt delivered as percolation and infiltration. The chemistry of precipitation that defines the quality of the precipitation is a function of the nature, chemistry and concentration of airborne particulates through which precipitation occurs. As noted in Chapter 10, the toxic substances in the atmosphere derived from man-made activities include: NO_x (nitrogen oxides), SO_2 (sulphur dioxide), CO (carbon monoxide), Pb and other metals such as Al, As, Cu, Fe, La, Mg, Mn, Na, Sb, V, and Zn, as airborne particulates, VOCs (volatile organic chemicals), aromatic hydrocarbons such as benzene, toluene, and xylene and PAHs such as anthracene and naphthalene. The same holds true for the chemistry of the snowpack that serves as the storage for recharge as snowmelt. Thus, for example, Nanus et al. (2003) reports that high-elevation areas in the Rocky Mountains annually receive large amounts of precipitation, most of which accumulates in a seasonal snowpack. They maintain that all accumulated atmospheric deposition is delivered in a very short period of time to the ground and receiving waters during spring snowmelt. The presence of the noxious gases together with other airborne particulates, ensures that the pH of rainfall onto the ground surface will be acidic. Spatial variations in atmospheric deposition of acid solutes are the result of precipitation amount in combination with concentration, and that deposition does not necessarily reflect variations in concentration alone (Nanus et al., 2003).

Deposition of airborne particulates with rainfall will also ensure that these will be carried with the surface run-off and also with infiltrating water. The other causes of pollution of precipitation recharge water include (a) run-offs and from polluted land surfaces, as might be found on agricultural lands, and (b) infiltration into subsurface through land surface polluted with pesticides, fungicides, pesticides, other surface wastes, organic debris, heaps, leach piles, sulphide rock piles, etc., as illustrated in Figure 15.2. The evidence shows that in regions where urbanization, industrialization, and exploitation are present, it is difficult to find precipitation recharge devoid of pollutants and airborne pollutants. Furthermore, in these regions, it is also difficult to rule out pollution of the receiving waters and groundwaters from contaminated run-offs and infiltration. For regions remote from the effects of industrialization and urbanization, and also far remote from airborne pollutants, one would have better chances of finding uncontaminated recharge precipitation. Treatment of polluted or contaminated recharge precipitation is not generally practical since it is more than likely that pollution already exists in the water and land receptors in urbanized and industrialized regions. Instead, passive treatment using natural processes together with aggressive treatment of extracted water are used to provide safe drinking water (see Chapter 3).

Improvement of Soil Quality for a Sustainable Soil Resource: Soil is an important resource material. It contains most of the nutrients required for plant growth and is rich with microorganisms. Besides being the most critical medium for agricultural food production and also production of other kinds of crops and trees such as cotton and palm trees, it is also a very important tool for management of wastes and waste discharges in the ground – as seen in the previous chapter. It serves as a dynamic resource not only for production of food and raw materials, but also for the soil microorganisms contained in the soil. These microorganisms not only play an important role in the natural bioremediation of harmful organic chemicals in the ground, they participate intimately in the recycling of carbon, nitrogen, phosphorus, and other elements in the soil. In essence, they are significant contributors to the control or management of greenhouse gases, water flow in soils, soil quality and through all of these the life-support systems for humankind.

It is recognized that loss of nutrients, loss of biodiversity, loss of soil organics, salinization, acidification, and degradation of physical, chemical, and biological properties of soil occur with time – through leaching processes, intensive agricultural practice, erosion, and through over-use and poor land management. Natural and/or man-assisted recharge of soil as a resource material, i.e., recharge of soil quality, is required if sustainability of soil quality is to be achieved. We consider *recharge of soil quality* to consist of any or all of the types of physical, chemical, and biological amendments, methods, and processes that serve to increase soil quality. Figure 15.3 shows a schematic description of soil quality changes with time. The ordinate shown in the diagram represents the soil quality index (SQI). The SQI is a composite index that incorporates analyses that include consideration of the physical, chemical, and biological indicators relating to the soil resource application under review. Thus, for example, the specific component indicators for the physical, chemical, and biological indicators leading to calculations of the SQI for waste management would differ from those obtained for agricultural production, or for forestry. A more detailed discussion of the SQI and the various indicators will be found in a later section in this chapter.

Artificial recharge, i.e., recharge through human intervention, involves the use of soil conditions, fertilizers, added nutrients, biological agents, and good agricultural land management practice. The term *soil amendments* is used as a general "catch-all" term to include all the preceding items and processes. These soil amendments, when introduced into a tired soil, are designed to improve the soil qualities through improvement of the physical, chemical, and biological properties of the soil – to allow for better use of the soil as a resource material. Improvements in soil permeability,

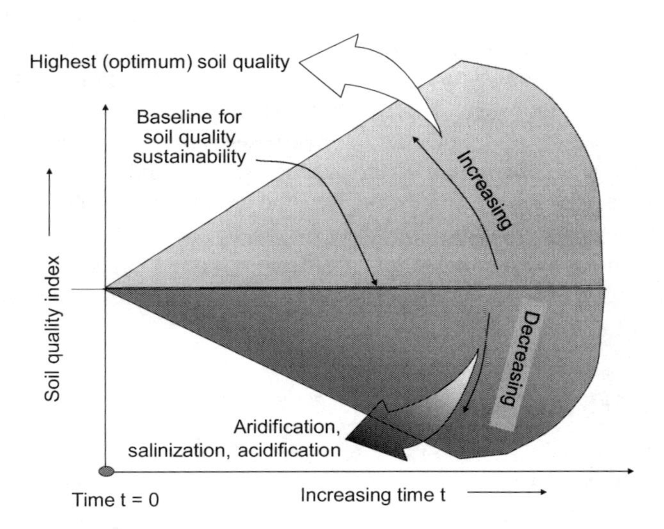

FIGURE 15.3 Illustration of increasing (enhanced) and decreasing (diminished) soil quality in relation to time. The soil quality index is a composite index determined on the basis of measures of achievement of levels required by prescribed physical, chemical and biological indicators. Note that different soil quality indices are needed, depending on whether one is concerned with agricultural production or for example the use of soil for waste management.

infiltration, water retention, nutrient-holding capacity, and soil structure – on top of the addition of nutrients and fertilizers – constitute the artificial soil recharge process.

15.2.2.3 Protection of Soil and Water Resources

The impacts of significant consequence on the soil and water resources in the geoenvironment are due to physical disturbances of the land environment, and direct and/or contamination of the soil and water resources – as shown in Figure 15.2. The various factors, conditions, and circumstances wherein impacts are generated have been discussed in the previous chapters. The three different categories for protection of the soil and water resources in the land environment are as follows:

- Category 1 – Direct and specific protective measures to ensure no degradation of soil and water qualities. This category of actions and measures assumes that the soil and water qualities are at acceptable levels, and that with proper management, they will be sustainable.
- Category 2 – Measures and actions to mitigate and minimize detrimental impact to both soil and water qualities. We assume that the impacts are managed to the extent that their effects do not degrade both soil and water qualities, and also do not pose health threats to humans and other biotic receptors.
- Category 3 – Application of treatment and remediation technologies to return soil and water qualities to levels acceptable for use. This category applies to situations where soil quality and/or water quality have degraded to the state that treatment and remediation are required to return them to the levels of quality required for use.

The actions included in the various measures undertaken for the three categories of protection and management of soil and water resources range from passive to aggressive. The physical and chemical buffering properties of the soil are central to the effectiveness of the passive protection technique. At the other end of the scale are physical protection barriers such as liner-barrier systems that prevent the migration of leachates and pollutants, and treatment and remediation techniques that require aggressive physical, chemical, and even biotechnological intervention. Application of any of these techniques depends on (a) the nature and scope of the perceived threat, (b) the resource being threatened and its functions or use, (c) the predicted intensity and type of damage done to the resource by the impact, (d) the extent of resource protection needed, and (e) the economic impact. In Figure 15.1, for example, assuming the absence of noxious gases and airborne particles, the perceived source of threat is represented by the pesticides, insecticides, and other surface pollutants that will move toward the river and also infiltrate into the ground during periods of precipitation. The resources being threatened are the river, surface layer soil, and the unconfined aquifer. Assuming that the river and the unconfined aquifer serve as drinking water sources, and further assuming that the soil is an agricultural soil resource, the need and extent of protection required for these resources will be evident.

15.3 WATER AND SOIL QUALITY INDICATORS

The discussion in this section extends the discussion on indicators in Section 9.2 of Chapter 9. Figure 9.1 in Chapter 9 depicts the role of indicators in the situation created by precipitation falling through airborne noxious gases and particulates. The water and soil quality indicators identified as monitoring targets include both system status and material performance- or material property-status types. Water quality indicators and soil quality indicators are essentially material property-status indicators. They are meant to indicate the quality of the material (water or soil). The quality of the material under consideration or analysis is established with specific reference to its intended function. In regards to soil quality, for example, we have seen from the previous section that the classic definition developed for agricultural use needs to be broadened to encompass the use of soils for various other purposes – from waste management to building supplies and construction. This is also true for water quality indicators. The range of usage starts from the top with drinking water standards setting the height of the water quality bar. At the low end of water usage would be water for agricultural purposes and other similar functions. Indicators for all the various functions of both water and soil would vary both in form (type) and detail.

There are several levels of specificity (i.e., levels of detail) in prescription of water quality and soil quality indicators. These depend on (a) the intended function and management goals; as for example drinking water usage or irrigation purposes, (b) ability to obtain all the necessary data sets, (c) available and applicable remedial and/or corrective technological capabilities, (d) scale and risk tolerance, and (e) economic factors. Perhaps the overriding factors in all of these are *management goals* and *risk tolerance*.

15.3.1 Soil Quality Index

In Figure 15.3 and in the previous section, soil quality index (SQI) is shown as a measure of soil quality. Similar to the different intended functions for water and soil, determination of soil quality index (SQI) and water quality index (WQI) will also depend on many of the same factors described in the preceding paragraphs. Development of indices requires full consideration of the many different properties and influences that ultimately combine to produce the material status. Since this is a dynamic process dependent on applications or processes being applied to the soil, internal soil reaction rates and elapsed time, the indices will also vary in accord with circumstances and time. Quantification of SQI and WQI permits one to arrive at determinations that show whether the material and finally, the system itself, will be sustaining. Taking the SQI as an example and referring to Figure 15.3, when calculations show that the SQI at any one particular time is greater than the baseline value, we will have increasing soil quality, and we can be assured of sustainability of the function served by the soil. Evaluation and quantification of SQI is application or function-specific, i.e., they depend on management goals for the material (water and soil).

15.3.1.1 Example of SQI development

To illustrate the procedures that one would follow to evaluate and determine the appropriate SQI, we will use the role of soil as a resource material for management of the impacts from contaminant discharge into the ground. We recall from the previous chapter that the basic properties contributing to the development of the assimilative capability of soils are physical, chemical, and biological. From this starting point, determination of what pertinent attributes are significant and measurable is required. Furthermore, it needs to be determined whether or how these attributes vary with circumstances specific to the problem at hand – i.e., functions or use of the soil. Figure 15.4 is a schematic illustration of the physical, chemical, and biological properties that are considered to be significant in the development of the assimilative capability of the soil. If one were to compare the kinds of attribute data sets with the information given in Figure 9.19 in Chapter 9, it would be immediately evident that many of the basic interactions developed between contaminants-pollutants have been incorporated in the measured attributes. We use the term *attribute* in the discussion to mean the property or characteristic being measured.

The data obtained from tests and other kinds of measurements (field and laboratory) of the physical, chemical, and biological attributes can be used as (a) input to compare with individual attribute indicators, thus leading to immediate comparison of the sustaining capability of each individual attribute, and/or (b) input to statistical and analytical models developed to produce a *lumped* (combined or total) index parameter. Prescription of individual attribute indicators is performed once again on the basis of intended function of the soil. Take for example the set of attributes in the physical properties listed in Figure 15.4. Density, porosity, calculated exposed surface area, and hydraulic conductivity have been chosen as the set of pertinent attributes. Consider two specific

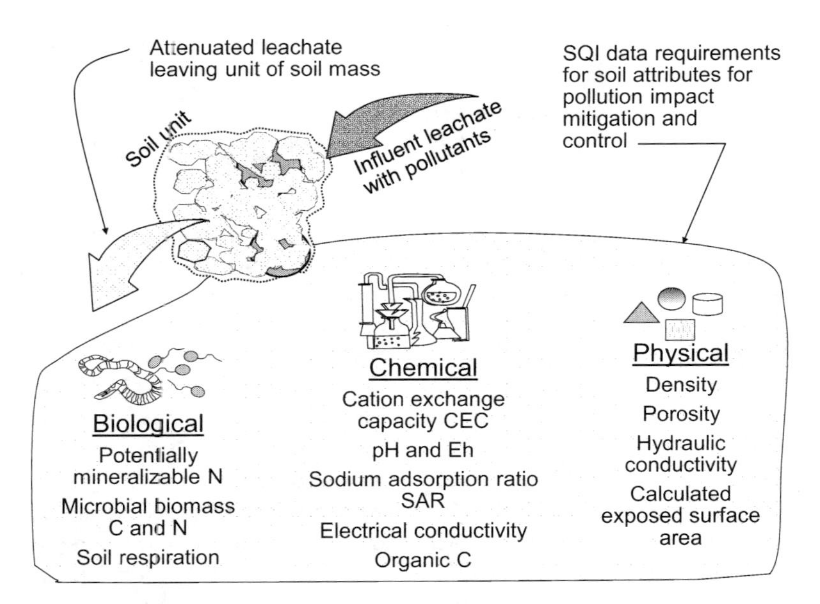

FIGURE 15.4 Soil properties pertinent for use of soil as a management tool for mitigation of impact from contaminants discharged in the ground. Data from these attributes serve as input to determination of contamination mitigation soil quality index (SQI).

applications for the soil: (a) use as permeable reactive barrier (PRB) material – as in Figure 15.5, and (b) use as an engineered clay barrier (ECB) in the liner system shown in Figure 1.11 in Chapter 1. The primary controlling property in both the PRB and ECB applications is the hydraulic conductivity. For the PRB, one permits the transporting fluid to penetrate the PRB at a rate, which allows for partitioning and transformation processes to occur. Residence time in the PRB is paramount. This is controlled by an appropriate soil permeability and thickness of the PRB. In general, one might want to design a wall thickness in conjunction with a Darcy coefficient of permeability k in the range of 10^{-5} to 10^{-7} cm/s – depending on the partitioning processes envisaged. In the case of the ECB application, a k value of considerably less than 10^{-7} cm/s is generally sought. The prescription of k indicators for desired objectives can now be obtained. The prescription of attribute indicators and their application can be seen in Figure 15.5 for the example of hydraulic conductivity. Given that hydraulic conductivity (as characterized by the Darcy k value) is a direct function of density and porosity, i.e., $k = f(\gamma, n)$ where γ refers to soil density and n refers to porosity, the *weighting* of data for γ and n becomes important. Application of weighting factors in such situations is to a large extent based on knowledge of previous behaviour.

Determination of weighting factors to be used for all the data sets relating to the physical, chemical, and biological attributes can be a challenging task. Much depends on the experience and knowledge of the analyst. The results of the weighted data are used in a deterministic model that is designed to produce a lumped index known as the quality index. As stressed previously, the quality index will have a prefix that denotes the function of the material, as for example, *drinking water quality index* and *pollution mitigation soil quality index*.

15.3.2 Water Quality Index (WQI)

Research and development of water quality indices have not received the same level of attention as for SQI in soil science. Instead, attention has been focused more on the establishment of national standards. Because of the very direct relationship between the quality of drinking water and human

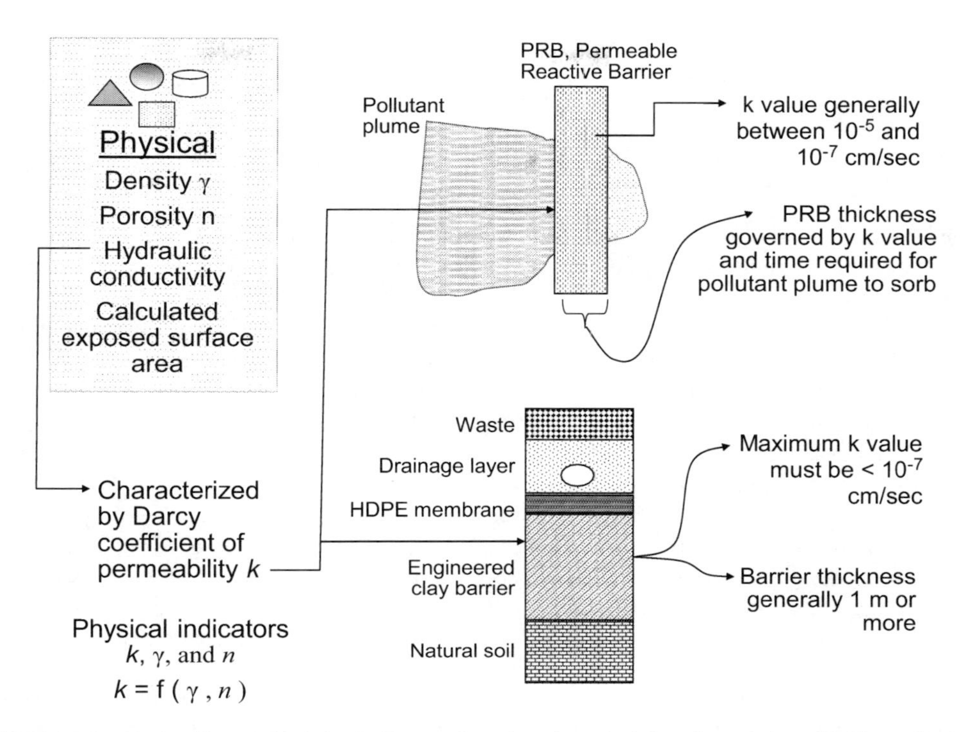

FIGURE 15.5 Hydraulic conductivity indicators based on intended function of the soil. Since the k value determined is a function of density and porosity, the weighting factor for k is considerably larger than those for density and porosity. The calculated exposed surface area is a parameter of interest and may be neglected. The thicknesses of the PRB and ECB are determined by situation-specific conditions. The 1 m or more thickness for the ECB is quite common for containment of municipal solid waste landfills.

health, drinking water quality standards are set by the regulatory bodies for most, if not all, of the countries in the world. As discussed in Section 3.3.1 in Chapter 3, the main parameters usually monitored for drinking water are BOD, colour, turbidity, N, P, suspended solids, odour, heavy metals, VOCs, pesticides, emerging chemicals of concern, bacterial level (such as coliform forming units, CFU), and perhaps other microorganisms. Infectious diseases caused by pathogenic bacteria, viruses and protozoa in drinking water are by far the greatest health threat.

Whilst these national standards set the basis for the water quality for the various countries, production of a national drinking water quality index for each and every country will require considerable effort in developing the relationships and weighting functions. To a large extent, this is because of the risk-benefit approach adopted by many responsible authorities in articulating quantitative values for the parameters chosen for monitoring. The risk-benefit consideration is not economically driven – at least not directly. Rather, it is driven by the need to provide acceptable drinking water to the most people without endangering public health. Setting standards that may be considered "too stringent" may on the one hand be prudent and safe, but on the other hand may make drinking water unavailable to a large percentage of the population – especially in regions of water-deprivation. Because of these kinds of factors, development of drinking water quality indices becomes more than a challenge.

15.4 SUSTAINABILITY CASE STUDIES

We will look at some case studies in this section. These are either cases demonstrating sustainable practices and the implementation of methods to evaluate the sustainability of a project, or process with regards to the geoenvironment. At least one case study is presented for each of the sectors, urbanization, resource exploitation, food production, industrial development and the marine environment.

15.4.1 Rehabilitation of Airport Land

Between 150,000 and 350,000 m^3 of soil in the contaminated site needed to be remediated at a former Norwegian airport (Ellefsen et al., 2001). A method was used to incorporate environmental effects into an evaluation of different remedial options. The environmental costs and benefits were determined which became part of the decision assessment. One of the main environmental targets was the reuse of the treated soil for landscaping. Asphalt and concrete would also be reused. To perform the assessment, a model developed by the Danish National Railway Agency and the Danish State Railways was utilized. A life-cycle approach for the remediation was used. Consumption of materials, fuel and energy, effects of noise, odour and other annoyances to humans and emissions to air, soil and water were calculated.

The site (including soil and groundwater) was contaminated with diesel and heating oil between 3.5 to 5 m below the surface due to leaking storage tanks and runoff water. Free phase oil was also found. Two remedial options were chosen and compared. They were (a) excavation followed by biological treatment, and (b) *in situ* treatment by biosparging and removal of six tanks.

Energy consumption for the excavation option was found to be five times higher than the *in situ* procedure. Additionally, emissions of greenhouse gases were estimated to be three times more than the *in situ* option. For the *in situ* option, electricity consumption was considered to be the main environmental cost of the biosparging process. However, as hydroelectricity is the major source of electrical power, negligible amounts of CO_2 were emitted. Material consumption was equal in both cases but differed in origin. Iron and manganese were required for excavation machines while nickel and copper were used for the air injection pipes and electrical materials. Biosparging was thus selected as the remedial option. The area is used for parking and storage facilities and will be used in the future for housing and parks. Other environmental assessments in the future will be performed using this approach.

A follow-up report by Ellefsen et al. (2005) indicated that the cleanup was completed in 2003. The area will become a green area with reserves for nature. Another challenge of the area was the management of the asphalt and sub-base contamination from PAHs from an old runway. Instead of transporting the waste to a hazardous waste facility, bitumen at a level of 3% was added to 20,000 tonnes of the soil via a cold mix process. The stabilized mixture was then used as a road foundation in the area. Leaching tests with water and road salt indicated that the material was appropriate for reuse. Another 80,000 tonnes of the contaminated soil were used without stabilization in the same road, while 60,000 m^3 were used for other road construction and 80,000 tonnes for new terrain construction. In total 200,000 m^3 of PAH contaminated soil were used with only 15,000 m^3 requiring hazardous waste disposal.

Materials from the demolition were also used for recycling at onsite recycling plant. The onsite plant enables a reduction of 80,000 truckloads of materials during construction. A test road built with recycled asphalt and concrete showed better behaviour than natural aggregates. Overall, 50,000 m^3 of C&D waste, 120,000 m^3 of old runways, 150,000 m^3 of excavated material and 300,000 m^3 of blasted rock was used for 450,000 m^3 of roads, buildings and ditches. Another 60,000 m^3 of composted sludge, 120,000 m^3 of excavated materials, and 150,000 m^3 of sand was for 400,000 m^3 of new soil and green areas while 888,000 m^3 of the 1,050,000 m^3 excavated material went to landfill. For energy, energy in seawater was exploited by heat pumps. The target was to supply 50% of the energy by renewable energy. Wetlands were conserved for migrant birds and other landscaping for aesthetics and public recreation. Low impact transport such as public transportation, footpaths and cycle paths was prioritized (Statsbygg, Norwegian Directorate of Public Construction and

Property). Several buildings on the former airport site were renovated. Presently, several companies have their offices at the site. There are also housing projects.

15.4.1.1 Sustainability Indicators – Observations and Comments

There are several indicators that can be used to determine whether the remediation project meets the aims or principles of *sustainability*. These include the following:

- Land use – the results show that if one uses the contaminated land as a starting point, the original plan for remediation and rehabilitation of the land to permit usage as parking and storage, office and housing facilities is a step upward, i.e., better land use. The subsequent report indicating use of the rehabilitated land as green space is a positive step towards sustainability goals. We need to note that the land use indicators here are not in reference to the initial airport land use. Because of the new intended green space land use, the sustainability indicators can now be cast in terms of "return to nature" indicators.
- Energy utilization – conservative energy use as a target for remediation procedures does not always produce results that will support complete site remediation. Comparing two specific remediation procedures for energy use is a good procedure in minimizing depletion of energy resources – especially non-renewable energy resources. Since the energy resource to be used for both remediation and rehabilitation procedures is hydroelectric based, and assuming that this is fully renewable, the sustainability feature here can be viewed more as a conservation measure. The energy indicators are referenced specifically to the remediation-rehabilitation processes, and not to the production of hydroelectricity.
- Noxious emissions – the use of hydroelectricity as the source of power has essentially limited noxious emissions. Since the impact of CO_2 discharges has been minimized with the type of energy used, one will need to accept that emissions indicators for full sustainability cannot be realistically set. A set of realistic parameters and values for emissions indicators needs to be prescribed.
- Non-renewable and renewable resource materials – the metals used are non-renewable resources. Whilst reuse of the asphalt pavement material and the underlying and contiguous contaminated soil for road construction shows a positive approach to the principles of sustainability, there is a requirement for monitoring to ensure that these materials do not present future contamination problems to the immediate environment. As with the situation of emissions, a set of appropriate parameters and values for material and system status indicators is required.

The use of recycled products in construction of roads has also been recently demonstrated in a project in Finland designed to show that sensitivity to sustainability objectives in construction of roads (Lahtinen et al., 2005). New types of road construction materials based on the industrial by-products, fly ash, and fibre ash were evaluated in new roads. Fibre ash was evaluated in light traffic paths and for the widening of safety lanes. The pilot construction took place in 2002 and 2003. Monitoring of the road performance is underway until the end of 2005. The fly ash was obtained from the incineration of bark, peat and/or sludge and the fibre ash is a fibre sludge from the paper industry with fly ash and cement binder. To date, the results of using the recycled materials in the road construction is positive both technically and economically and provides a potential way of saving virgin, non-renewable resources.

15.4.2 Sustainable Mining Land Conversion

Mount Cenis was established as a coal mine near Herne in the Ruhr District of Germany in 1871 (Genske, 2003). Subsequent coal washing and coking facilities were then built. The mine was one

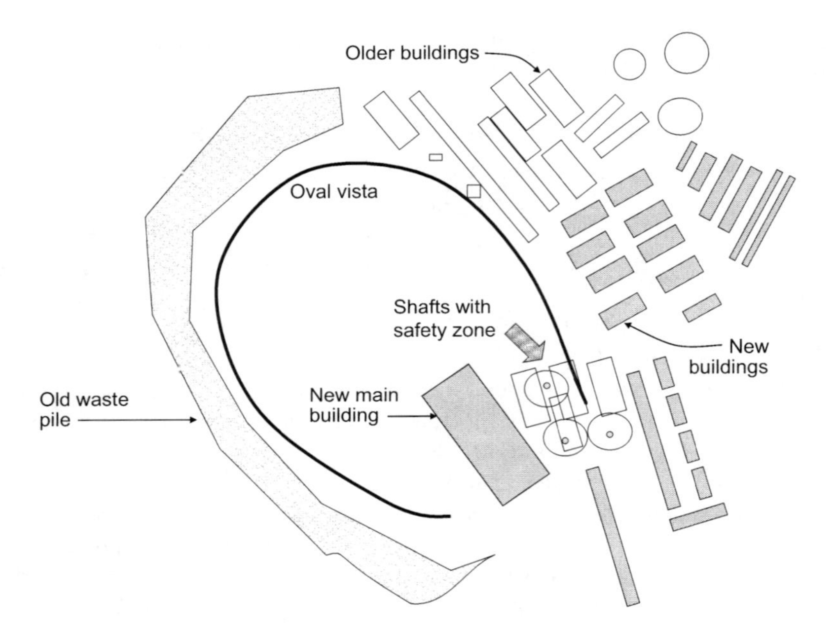

FIGURE 15.6 Redeveloped Mont Centis site. New buildings are in gray and older buildings are in white. The three abandoned shafts (encircled with a safety zone) are used to recover energy. (Adapted from EMC, Mont-Cenis, Report of the Entwicklungsgesellshaft, Mont-Cenis, Herne D, 44 pp., 1998.)

of the largest in the area but it was closed in 1978 due to the coal and steel crisis in Europe. At the site, 26 hectares of land were contaminated. There was subsidence, acid mine drainage, mine gas leakage and many underground structures. However, in 1990, a large project was conceptualized to remediate and reuse the land for companies and enterprises. The main features of the project shown in Figure 15.6 (EMC, 1998), included the following:

- An academy for the Ministry of Interior (the largest building).
- Various public service buildings such as a meeting hall, civic administration buildings, and a library.
- Shops and services for a shopping mall that existed already.
- 250 housing units.
- Conference space, hotel, restaurant, and recreation park.

Wood was chosen for the structures due to resource efficiency as detailed in a report for the Club of Rome of 1995 (Weizäker et al., 1997). Forests close to the construction site were chosen as the source of the wood to reduce transportation requirements. Concrete buildings were chosen to reduce climate control requirements. Energy savings of 23% for heating enabled an 18% reduction of the CO_2 emissions. Approximately 10,000 m^2 (3800 PV cells) of solar cells covered the roof of the wood and glass structure for the academy, hotel, living quarters and sports facilities. Natural ventilation was incorporated into the building. Energy consumption was reduced to 32 kWh per year. The power plant on the roof provides 1 MW of power, more than twice needed by the centre (EMC, 1998).

Approximately, 120 million m^3 of methane are generated from the abandoned mines. The general practice of burning the methane releases approximately 8 million tonnes of CO_2. Therefore, it was decided to capture the gas containing 60% methane. This was converted to 2 million kWh of electricity and 3 million kWh of heat for the nearby buildings (Backhaus, 2017). The mine gas and solar panels generate energy for 130,000 homes. A power station is used for energy storage.

A supplementary natural gas plant (1800 kWh) and a hot water storage tank were constructed to ensure adequate energy and heat due to the fluctuating nature of the methane production. As a backup, connection was made to the municipal energy system. This also was used for discharge of excess energy. Rainwater was collected in an underground cistern from buildings for use in toilets, showers, washing solar panels, and for watering gardens.

Infiltration of the water was allowed only where the soil was low in contamination. Excavation of the contaminated soil was not performed since this fills landfills and transfers the problem to another place. The soil was instead placed on clay liners or membranes to prevent leaching of the contaminants to the groundwater. Herb gardens are grown on top of the contaminated land. Gravel and sand filters are placed above the liners to collect the precipitation. The entire project cost 110 million Euros and was shared between the community and private investors.

The project currently includes an award-winning building that is one of the largest building-integrated solar facilities of the world. The area includes a high-density housing development, a park, and a kindergarten. One of the last parts established was a residential space and neighbourhood shopping centre. A public market is also present. On top of the supermarket, there is a roof garden with a children's playground. The residential area houses 40 flats suitable for the elderly and seven penthouse flats. The former Mont Cenis coal mine supplies methane for the heating of this complex and other neighbouring buildings (Backhaus, 2017).

15.4.2.1 Sustainability Indicators – Observations and Comments

As with Section 13.4.1, there are several indicators that can be examined to evaluate whether sustainability or the path towards the goals of sustainability has been taken.

- Land use – increased land use capability has been achieved with the remediation-rehabilitation scheme. As with the land use case in Section 13.4.1, the land use indicators chosen are in specific reference to the initial condition prior to remediation and rehabilitation, and not before mining.
- Energy sources – the multi-source energy input, from solar to methane gas capture and reuse, coupled with more efficient climate control in the buildings show good attention to energy conservation and greenhouse gas emission reduction. This agrees with the requirement for reduction in depletion rate of non-renewable energy resources as a goal towards *conservation for sustainability.*
- Water use – as with the conservation strategy for energy, utilization, and re-use of water show good accord with water sustainability indicators.
- Remediation of contaminated land – the concern for not dumping contaminated soil into another landfill as a reason for placing the contaminated soil on secure membranes and left on-site is a responsible attitude. One assumes that the appropriate requirements for (a) monitoring of the contaminated soil facility, (b) prescribing indicators for safety-recovery status of the contaminated soil facility, on the assumption that the contaminated soil will be remediated through intrinsic remediation processes, and (c) development of appropriate risk-management procedures.

15.4.3 Agriculture Sustainability Study

de la Rosa et al. (2003) have described soil quality decision support tools for the protection of agricultural zones, particularly in the Mediterranean region. Soil quality indicators are not sufficient for land management as soil, biological components, climate, water and land use must all be considered. MicroLEIS is an agro-ecological decision support system available at the web site of FAO (2024). Databases, statistics, expert systems, neural networks, Web and GIS, and other technologies are integrated into the scheme (Figure 15.7). Information on the impact of soil use such as tillage on soil

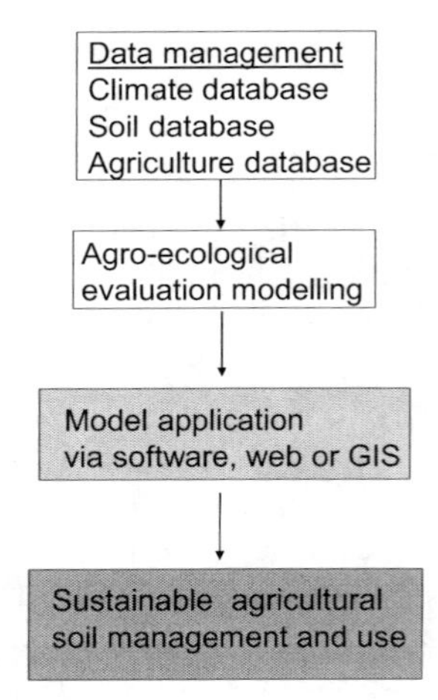

FIGURE 15.7 Components of the MicroLEIS Decision Support System. (Adapted from de la Rosa et al., 2005.)

TABLE 15.1
Soil Quality Indicators

Group	Soil Quality Indicator
Physical	Aggregate strength and stability
	Bulk density
	Soil compaction
	Soil texture
	Soil structure
	Topsoil depth
Chemical	Plant nutrients
	pH
	Salinity
Biological	Organism population
	Respiration rate

Source: USDA (2006).

properties, soil quality, and crop production is still needed. The concept of soil quality according to the Soil Quality Institute (USDA, 2006) is related to sustainable soil use and management. Some indicators are shown in Table 15.1. Physical, chemical, and biological factors are included in soil quality. Soil functions can be defined for crop growth, soil erosion, or soil contamination and thus other indicators must be added to address soil contamination in particular. For example, heavy metal contents as the result of sewage soil application or for contaminated site remediation are well defined by legislation. Biological indicators are less advanced. New DNA extraction techniques or ribosomal RNA are developing (Thies, 2006). The soil assessment approach can be seen as Figure 15.7 as developed by de la Rosa et al. (2005). de la Rosa and Sobral (2007) have indicated that the general principles of sustainable agriculture practices should aim to

- Increase organic matter content.
- Decrease erosion.

- Improve water infiltration.
- Decrease sub-soil compaction.
- Increase water holding capacity.
- Decrease the leaching of agrochemicals in to the groundwater.

A variety of innovated agro-ecological innovations could be employed (Uphoff et al., 2006) for sustainable agriculture. Some biological techniques include biorehabilitation of the soil, composting and vermi-composting and bioremediation of organic contaminants such as pesticides.

For an area in Azarbaijan, six agro-ecological land evaluation models of the MicroLEIS DSS software were used to evaluate and compare strategies for land use (Shahbazi et al., 2008). Results from soil morphology and analytical data were studied. It was determined that 45% of the area was classified as suitable for agricultural use, whereas 12% must be reforested and not used for agriculture, to minimize the land degradation. The crop rotation of wheat-alfalfa-soybean was selected as the best. Overall, the tool was appropriate for evaluation of sustainable agro-ecological systems in semi-arid regions.

15.4.3.1 Sustainability Indicators – Observations and Comments

The use of *soil quality* as an indicator for sustainability in agricultural production has a history of almost 30 years. The use of soil quality as a tool for assessing the health of a soil with particular reference to agricultural purposes was first discussed in the late 1970s (Warkentin and Fletcher, 1977). The term *soil health* which was (and may still be) used by farmers refers to the functional capability of the soil to support crops and other plants. With the quantification procedures indicated in this section, it is seen that a structured effort is being made in agriculture to meet the goals and requirements for sustainable agriculture. The production of quantification techniques for soil quality provides the means for comparing the dynamic state of the agricultural soil and offers the opportunity to develop the methodology for determination of soil quality indices. As with the other soil quality indices developed for the various soil function discussed in Section 15.3.1, these agricultural soil quality indices are the necessary constituents of the indicators for agricultural soil sustainability.

15.4.4 Petroleum Oil Well Redevelopment

The Damson Oil Site in California near Venice Beach is an abandoned oil well that stopped production in 1989 (CCLR, 2000). Damson deconstructed the facilities in 1991. However, after the oil wells were capped, the company filed for bankruptcy leaving soils contaminated with hydrocarbons, sumps with oil and sludge from the extraction process, vaults with oil and several miles of pipeline. Further contamination occurred as a result of deliberate dumping of debris by passers by. Since it was deemed too expensive to restore the site to a sandy beach, it was decided to establish an in-line skating facility. The oil site was to be capped with concrete and the remediation costs, and the improvements made would enable the facility to be economically viable. The plan consisted of (a) an environmental site assessment, (b) waste removal for all surface soils, liquid wastes and sludges, (c) demolition plan for the pipeline and other structures and tanks, (d) construction plan for a skating facility, (e) possible restoration of the beach, (f) establishment of other facilities, and (g) negotiation of a risk management plan with the Regional Board for remediation objectives and standards. By 2000, environmental assessments and cost estimates were completed and the construction of the boardwalks was initiated. The brownfield blocked the view of the ocean. The Department of Recreation and Parks acquired the site for public use and then the City's Brownfields Program provided the funds for the site remediation of $100,000 that demolishing the structures, removing soil contaminated with metals and crude oil, completing lead and asbestos abatement, and installing groundwater monitoring wells. After remediation, the California Regional Water Quality Control Board performed a quarterly groundwater monitoring. A skate park was built on some of the reclaimed land. The Brownfields Program provided funding for oil storage and drilling infrastructure cleanup, including underground storage tank removal. Besides the skate park, the multi-purpose recreation center includes picnic areas, play areas, sports fields, etc. (City of Los Angeles, 2024).

Another former oil refinery was converted to a business and recreational opportunity in Casper, Wyoming (Applegate et al., 2005; US EPA, 2007). The refinery had operated since the early 1900s but closed in 1991 due to environmental liabilities related to oil spills, sludges, and underground pipelines. The cost of the site remediation was estimated at $350 million US. Various risk assessments were undertaken. To protect the river and remediate the groundwater, a horizontal wall for air sparging and venting was designed and installed, in addition to a sheet-pile barrier wall. Pipes were also removed to eliminate the pollution source. Final remediation strategies included (a) removal of sediment from the lake, (b) removal of tanks, pipes, concrete and other material from the refinery area. Cleaning of the groundwater involved oil recovery, sparging, venting, phytoremediation, and monitored natural attenuation in the refinery and tank farm areas. Engineered wetlands and ponds were employed for removal of iron and dissolved hydrocarbon contaminants after oil/water separation. All concrete (nearly 272,000 tonnes) that was removed was crushed for reuse at the site and most of the pipes were sent to recyclers. Some examples of concrete reuse include (a) for use as drainage in the water treatment system, (b) construction of a barrier that prevented animals from entering the waste depository near the lake, and (c) for construction of roads. It is estimated that oil recovery at the site will take approximately 25 years – on the basis of analysis of the mobility of the oil strongly adsorbed to the alluvium. As a golf course was also constructed at the site, oil recovery wells had to be designed so that they would not be placed in the fairways and greens (https://archive.epa.gov/epawaste/hazard/web/pdf/casper11-07.pdf, accessed Jun. 2024).

Water at the site was also to be reused. Therefore, a system (one of the largest engineered wetlands) was set up that included management of the stormwater, irrigation of the golf course, and pumping of the water into the lake for the migratory birds. The schematic of the water management system is shown in Figure 15.8 and can handle between 1890 to 5670 L/min. It is mainly hidden and integrated in the golf course. Nearby Soda Lake was also contaminated and thus 150,000 m^3 of contaminated sediment had to be removed and efforts were made to promote the bird and

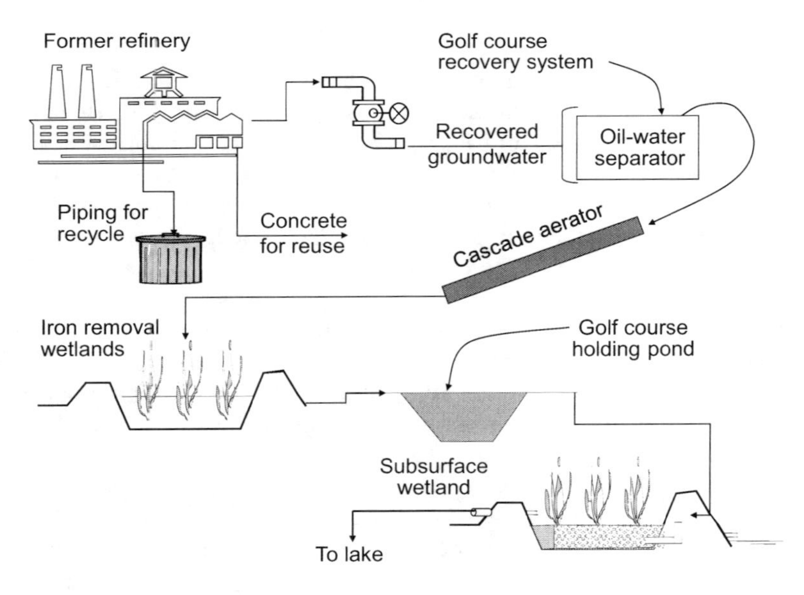

FIGURE 15.8 Golf course water treatment system for recovered groundwater from former refinery and tank farm. (Adapted from Applegate et al., 2005.)

wildlife population. Institutional controls were necessary to manage the remaining contamination. A kayaking course was also placed in the river. To implement phytoremediation, 2000 trees were planted. There was extensive public participation in the project. There have been numerous awards for this project.

15.4.4.1 Sustainability Indicators – Observations and Comments

From a land use sustainability perspective, the ongoing project in the first example together with remediation of the contaminated site in the second example and use of recycled materials, it appears that improved land use has been obtained. The remediation-rehabilitation plan in the first example provides opportunities for prescription of indicators for sustainability. It appears that site restoration will be performed in a fashion that will return the site to conditions and usage beyond initial sandy beach conditions. From a land use standpoint, the remediation-rehabilitation scheme is a positive step. One presumes, however, with the initial environmental site assessment that a proper accounting has been given to avoidance of negative impacts from the development and operation of all new facilities. Prescription of *facilities-operation indicators* would be useful.

15.4.5 Mining and Sustainability

The Sullivan Mine in Kimberley, British Columbia, Canada was discovered in 1892 (Teckcominco, 2001). Whilst the community was previously dependent on a single industry (mining), since the closure in 2001, the community became more diversified and the area is now a resort destination. To determine the viability of the community, a set of indicators was developed based on the following guidelines: (a) Canadian Mining Association Guidelines for Sustainable Development, (b) Australian Minerals Industry, (c) the Global Reporting Initiative, and (d) the World Business Council for Sustainable Development. These indicators were divided into economic, environment and social categories for several components. The environment indicators for sustainability were (a) ecosystem health based on soil erosion and species diversity, (b) the availability of natural resources such as minerals, (c) the effect of the company on the ecological amenity, and (d) sustainability. The features included for the environment sector were a tailings pond reclamation, and a drainage water treatment plant. Reuse, recycling and safe handling of all products are practiced. A solar power facility has been constructed on reclaimed mine land. Forest areas are treated to minimize the wildfire risk. The Mark Creek is now been carefully landscaped with a golf course. Hiking and mountain bike trails have been established. A Sullivan Liaison Committee was formed to ensure public participation in the process. More information can be found at www.teck.com/operations/canada/legacy/sullivan-mine/ (accessed May 2024).

Four fatalities occurred at this site in May, 2006 in a seepage monitoring station that was hydraulically connected to an acid generating waste rock. Various studies were completed to determine the cause of the asphyxiation deaths in the shed (Hockley et al., 2009). Reactions occur within the waste piles that lead to hazardous gas production. Gas flows are induced due to temperature differences and depletion of oxygen and moisture in the pore spaces. Therefore, physical and chemical factors must be considered for storage of waste rock.

Another example of the practice of sustainability (Stanton-Hicks, 2001) can be found in an iron industry at another site in Australia (Pilbara). Mining was initiated in the region in the 1960s with large-scale technology and inland mining operations. A number of programmes have been established since 1990 to train and work with the Indigenous employees and contractors in the area. However, there is still a lack of engagement. The life-cycle cost approach is being used with eventual rehabilitation of the site to be considered at an early stage. Fugitive dust is a particular issue for mine dumps. In addition to using wind block and conveyor coverings, the appropriate amount of water must be added to ensure that dust does not become airborne. New opportunities are being developed at the mining area due to the growing interest in the critical minerals such as lithium and

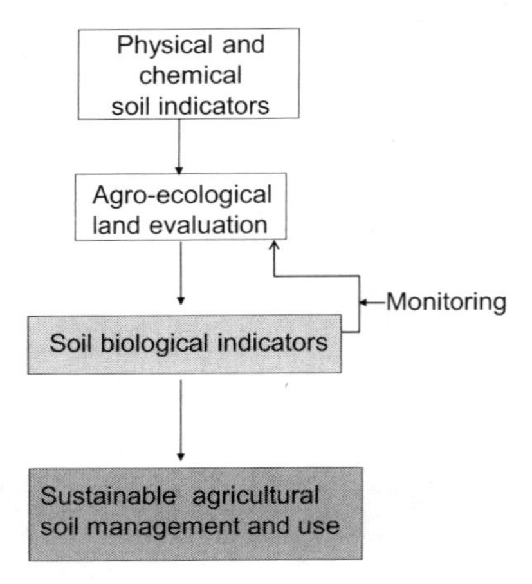

FIGURE 15.9 Schematic of agroecological approach for sustainable agricultural soil use and management. (Adapted from Moller et al., 2006.)

manganese necessary to the green energy transition. Recommendations for environmental sustainability in the area by an OECD report include reduction of greenhouse gas emissions by energy transition, promotion of partnerships with First Nation communities and universities to monitor environmental impacts and provide accessible and easily readable reports (OECD, 2023).

Moller et al. (2006) presented an approach for sustainable development during mine closure for the Western Australia to minimize impact on humans and the environment. The four stages included understanding the aspects related to closure, evaluation of land use options, development of the strategy and finally development of the management plan. The principles are shown in Figure 15.9.

15.4.5.1 Sustainability Indicators – Observations and Comments

The aspects of sustainability include the following:

- Maximization of economic, ecological, and socio-cultural efficiencies through integration.
- Extension of mine life through new techniques and development of new markets such as critical minerals.
- Land-use agreements to benefit economic and socio-economic sustainability.
- Engagement of the local communities to ensure project redevelopment viability.
- Pastoral management practices to improve environmental management, impact assessment and cooperation with land management agencies.

15.4.6 Organic Urban Waste Management in Europe

In Europe more than 40% of all waste is biologically treated via composting or anaerobic digestion (ECN, 2019). This is currently short of the target of 65% by 2035. In the EU, this amounts to around 47.5 million tonnes of food and garden residues. Composting is the main method of treatment (30.5%), followed by anaerobic digestion (12.4%), and a combination of both (4.6%). The composts can serve as an organic fertilizer or soil improver. Open windrows are the main technology for European countries. Another technology, anaerobic digestion, on the other hand, is divided into dry digestion and wet digestion, which treats more than 12,400 million tonnes per year of waste. Treatment in Slovenia, the Netherlands, Belgium, and Sweden are the highest in Europe .

Various policies with the EU Commission also will promote biological treatment directly or indirectly (Figure 15.10). The Landfill Directive was initiated in 1999 and consolidated in 2018

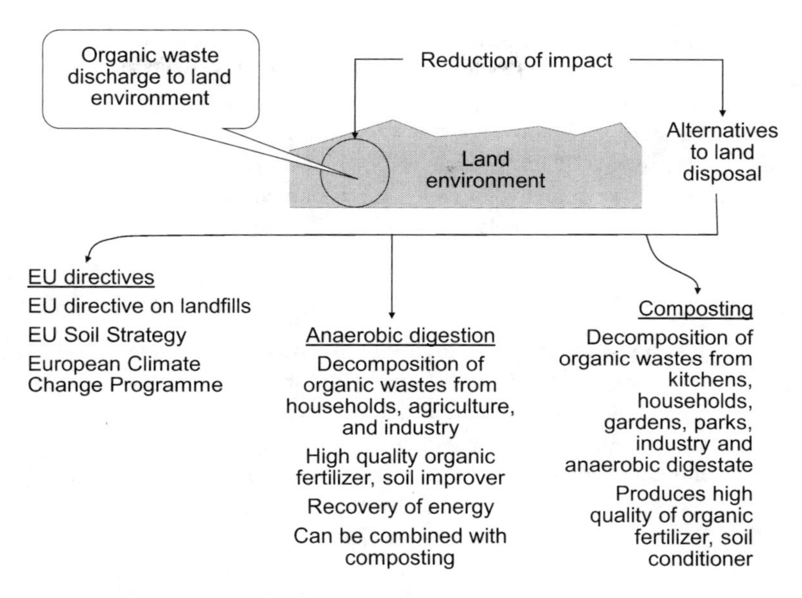

FIGURE 15.10 European initiatives for managing organic waste. (Barth, 2005.)

(http://data.europa.eu/eli/dir/1999/31/2018-07-04). It indicates that landfilling must be restricted for materials that can be recycled or recovered for energy after 2030, and for all municipal waste materials to 10% by 2035. In addition, EU countries are required to develop and implement strategies to reduce the amount of biodegradable waste sent to landfills. Therefore, composting and digestion rates will need to be increased.

The European Soil Strategy (European Commission, 2021) promotes the sustainable protection of soil by mitigating the three threats to soil including decreased organic matter, soil contamination, and erosion due to desertification and includes prevention soil degradation. The addition of composts and residues of organic matter to soil can clearly enhance organic matter and restore the agronomical and microbiological properties of soil. In addition, the addition of these types of organic matter to the soil can mitigate climate change by providing a carbon sink in the soil. Since the adoption of the strategy in 2006, a report was issued in 2012 from the Commission to the European parliament, the Council, the European Economic and Social Committee and the Committee of the Regions to implement the Soil Thematic Strategy and ongoing activities. This report also presents trends in soil degradation both in Europe and worldwide, and challenges for the future protection of the soil. The aim of the new EU Soil Strategy under development and under consultation in 2021 is to work towards land degradation neutrality by 2030, a target of the Sustainable Development Goals (SDGs, 2015). In particular, it is focused on soil fertility improvement, reduction of erosion, increasing soil organic matter, restoring contaminated soils, and evaluating its ecological status and improving soil quality monitoring.

In addition, the strategy will support the EU's international commitments, feeding into the EU position at the upcoming global biodiversity negotiations in COP 15 of the UN Convention for Biological Diversity (CBD), under the UN Framework Convention on Climate Change (UNFCCC), the UN Convention to Combat Desertification (UNCCD), the FAO' Global Soil Partnership and more broadly for EU external action and development cooperation.

The European Climate Change Programme (ECCP) promotes the buildup of carbon in the soil by organic fertilizers which can thus serve as a carbon sink (via the process of carbon sequestration) of up to 2 gigatonnes of carbon per year. This is compared to 8 gigatonnes of carbon per year that are emitted into the atmosphere. Therefore, the practice of biological waste treatment of separated wastes that is developing in Europe is clearly more sustainable than landfilling solid waste. The first ECCP ran from 2002 to 2004. ECCP (II) was initiated in 2005. It focuses on implementing aspects highlighted in ECCP(I). Some of the aspects being investigated are particularly relevant to the geoenvironment such as carbon capture and geological storage, agriculture, and the sinks sub-group on agricultural soils and forest-related sinks. Current aspects regarding adaption and mitigation to climate change for the EU are found at https://commission.europa.eu/energy-climate-change-environment/topics/climate-change_en. The EU has a goal of climate resiliency by 2050.

15.4.7 Sediment Reuse: Orion Project, Port of New York, and New Jersey

Sediments must often be removed by dredging to maintain waterways and ports. Approximately 5 to 10% of these sediments are contaminated (Urban Harbor Institute, 2000). Management of these materials must be planned carefully with environmental protection and economic viability. An initial draft report was of a Dredged Material Management Plan (DMMP) was prepared in 1999 and finalized with Environmental Impact Statement in 2008 The Port of New York and New Jersey has to dredge approximately 0.8 to 1.6 million cubic metres of sediment every year. An interim plan is in preparation in 2024 with the objective to develop a plan to meet all the dredged material placement capacity within the Port of New York and New Jersey to the end of 2029 and identify new beneficial use options www.nan.usace.army.mil/Portals/37/Interim%20DMMP%20Update%20Fact%20Sheet.pdf, accessed Jun. 2024). Since one-third of the sediments do not meet standards, alternate uses are required. More than 1.2 million cubic metres have been used as foundation fill for a parking lot. The sediments were dredged, transported to a screening facility and then pumped on to the shore for mixing with cement kiln dust to improve compressive strength. The mixture was placed on a 24 hectare lot and asphalt was used to cover the fill. No virgin material was required for the lot foundation. In the DMMP, numerous options are considered for dredged sediments. They include the following:

- Contaminant remediation, removal, treatment of containment by capping.
- Use of dredged material to remediate the ocean site of Historic Area Remediation Site (HARS).
- Use of dredged materials for restoration and creation of wetlands, benthic and bird habitats, reefs for marine life.
- Remediation of abandoned mines, quarries and landfills with dredged materials.
- Reuse of treated dredged materials, for construction grade cement, light weight aggregate and manufactured soil.
- Confined aquatic disposal.

15.4.8 Example of the Use of a Multi-Geosynthetic Approach for a Pathway

This is a case study (Furey et al., 2023) on technical, sustainability, and practical benefits of a multi-geosynthetics approach in ʔapsčiik t̓ašii (Ups-cheek Ta-shee) of a 25-km-long, multi-use pathway on Vancouver Island, BC, Canada. The geosynthetic-based solution facilitated a reduced net CO_2 footprint primarily through protection of the sensitive rainforest ecosystem. The Wayii Segment of the pathway meanders between Veteran Class trees down a 23-m-high foreshore slope with poor foundation soils, equipment access limitations, historical landslide activity and high seismicity. The design was flexible to navigate tree root zones and ground conditions. Multiple types of geosynthetics were applied: Knitted and polymeric coated, polyester uniaxial geogrids, high stiffness, polypropylene biaxial geogrids, biaxial geogrid composites, non-woven geotextiles, and HDPE geocells.

The conceptual design for the trail project began in the spring of 2016 with a team of technical consultants and local First Nations communities. In addition to a carefully selected trail alignment, micro-routing during construction in response to local environmental, archeological, culturally sensitive and/or geotechnical conditions, was facilitated through use of geosynthetics. The "wandering

FIGURE 15.11 Photo illustrating the finished switchback trail from above. (Photo courtesy of Parsons Inc.) Note the mature trees in the photo that were able to be protected while simultaneously providing both vertical support to the trail and a slope stabilizing mass (buttress) using geotextiles.

path" style (Figure 15.11) that was incorporated into the constructed trail resulted in a significant reduction in the number of trees impacted by the project from an estimated 25,000 trees in early path alignment iterations to approximately 1200 trees in the final version.

The use of geosynthetics was only a part of the extensive sustainable aspects of the project. Protection of the adjacent sensitive rainforest ecosystem and cultural resources was introduced as a key project objective at the outset and then emphasized throughout design and construction. During planning and design thorough environmental site characterization was completed and facilitated strategic route selection. The Environmental Protection Plan (EPP) was reviewed by registered professional biologists who also provided quality assurance reviews and were present on site throughout construction to make sure the EPP was implemented and adjusted according to the site conditions.

In addition to the multiple advantages of utilizing geosynthetics, the broader trail project included the following environmentally beneficial elements: Three bridges over fish bearing streams, 370 lineal meters of elevated boardwalks over sensitive wetlands, three amphibian highway box culvert crossings, 60 amphibian crossing culverts along the trail, and 11 fishery enhancement sites that resulted in a net increase in spawning habitat. The highway amphibian tunnels reduced the annual mortality rate of northern red-legged tree frogs, a species at risk.

This project demonstrates that while geosynthetics are an important contributor, they should be used in combination with, and in support of, other methodologies and techniques to achieve maximum sustainability benefits. Geosynthetics solutions addressed technical challenges associated with low bearing capacity, differential settlement, earth retention, slope stability, global stability, and sub-surface drainage while simultaneously reducing environmental impacts. An interesting feature of this project was the use of multiple types of geosynthetics including knitted and polymeric coated polyester uniaxial geogrids, polypropylene biaxial geogrids, biaxial geogrid composites, nonwoven geotextiles and geocells. Geosynthetics helped to minimize excavations, reduce the footprint of slopes and enabled the construction of retaining structures and slope stabilization by working around existing trees and thereby helped to complete the project with minimum disturbance to the ecologically sensitive landscape. The project also demonstrated that simple, innovative, and sustainable solutions using geosynthetics are possible with detailed planning and close and continuous interaction between geosynthetic specialists, geotechnical designers, biologists and the contractors.

15.5 SUSTAINABILITY FRAMEWORKS AND TOOLS

15.5.1 Sustainability Frameworks and Guidelines

Various frameworks have been developed to facilitate sustainability. Some include the Global Reporting Initiative (GRI), and the triple bottom line. The triple bottom line includes the three aspects of environmental, social, and economic. Frameworks for green and sustainable remediation have been discussed in Section 11.7.

It has been used for measuring sustainability performance for a region or business. Without measurements, sustainability is only a vague concept to many that cannot be realized. Indicators are often used to measure progress. They are data that can be measured to describe a condition or trend. Some indicators include waste recycling rates, water and energy consumed, employment rates, greenhouse gas emissions, and water quality. For sustainability purposes, social and economic indicators are also required. The Global Reporting Initiative has been used widely for organizations of any size for measuring and reporting on the environmental, economic, and social dimensions via various indicators. Performance indicators cover environmental, social, and economic aspects. Elements of the reports include the organization description, management approach, indicators employed, the impacts and boundaries and management of the impact.

The Global Reporting Initiative has been used widely for organizations of any size for measuring and reporting on the environmental, economic, and social dimensions via various indicators. Performance indicators cover environmental, social and economic aspects. Elements of the reports include the organization description, management approach, indicators employed, the impacts and boundaries and management of the impact.

The World Business Council for Sustainable Development, formerly the Business Council for Sustainable Development (WBCSD, 2005) was initiated in 1992. It developed principles of eco-efficiency. They include the following:

- Reduction of material and energy intensity for goods and services.
- Reduction of dispersion of toxic materials.
- Increased ability to recycle materials.
- Maximization of resource use.
- Extending the durability of products.
- Increasing the service intensity of goods and services.

The WBCSD, GRI, and UN Global Compact, published SDG Compass (2015) to indicate tools and indicators for businesses to work towards the SDGs. The World Federation of Engineering Organizations represents more than 20 million engineers with extensive expertise to contribute to the achievement of the UN Sustainable Development 2030 Goals (WFEO, 2015). The WFEO Model Code of Practice for Sustainable Development and Environmental Stewardship (WFEO, 2013) was developed and adopted at the September 2013 General Assembly.
The Ten Principles of the Code of Practice include the following:

1. Maintaining and continuously improving awareness and understanding of environmental stewardship, sustainability principles and issues related to the field of practice.
2. If knowledge is not adequate to address environmental and sustainability issues, consult with others with the required expertise.
3. Global, regional, and local societal values should be incorporated to include local and community concerns, quality of life and other social concerns related to environmental impact. Traditional and cultural values must be included.
4. Sustainability aspects should be incorporated at the earliest possible stage and employ applicable standards and criteria.

5. Costs and benefits of environmental protection, eco-system components, climate change and extreme events and sustainability should be incorporated in the economic viability of the work.
6. Environmental stewardship and sustainability planning should be over the life cycle for planning and management activities and efficient, sustainable solutions should be employed.
7. A balance between environmental, social, and economic factors should be achieved to contribute to healthy surroundings in the built and natural environment.
8. An open, timely, and transparent engagement process for both external and internal stakeholders should solicit input in an, and respond to economic, social and environmental concerns.
9. Regulatory and legal requirements must be met and exceeded by applying best available, economically viable technologies and procedures.
10. In the cases where there are threats of serious or irreversible damage and scientific uncertainty, risk mitigation measures should be implemented in a timely fashion to minimize environmental degradation.

15.5.2 Sustainability Tools

To calculate the emissions for projects, there are various tools available online. One of them is the free carbon calculator spreadsheet specific for the geotechnical industry. It was developed by the European Federation of Foundation Contractors (EFFC) and the Deep Foundations Institute (DFI) and can be downloaded from geotechnicalcarboncalculator.com. This tool is simple to use, and certified to ISO, GHG Protocol, and PAS 2050. For certification, a third party would be needed for the carbon calculations.

An example of the calculator use for emission reduction was for ground improvement design. Two designs were compared, including a base bid and a value-engineered design for installation below a school in Massachusetts (Yamamoto and Martin, 2022). A rigid inclusion ground improvement solution comprised the base bid, whereas for the value-engineered design, an aggregate pier ground improvement replaced a portion of the rigid inclusion. Emissions by source were determined for both cases and showed that there was a reduction of about 50% of the emissions for the value-engineered design compared to the base bid design. The base bid design totaled 1300 tonnes of carbon emissions, which is the equivalent of emissions from 260 US cars running for one year and requiring a forest the size of 835 football fields to absorb it all.

Another tool is a rating system such as Envision or BREEAM Infrastructure (formerly CEEQUAL) that can assist in the assessment of infrastructure project through goals and indicators. These are under continual development. Envision (ISI, 2023) is used to evaluate the sustainability of infrastructure projects by designers, constructors, community groups, owners, and policy makers. Bronze, silver, gold, and platinum levels can be obtained according to the points reached. Institute for Sustainable Infrastructure (ISI) was founded by the American Public Works Association (APWA), ASCE and the American Council of Engineering Companies (ACEC). Envision can be used for planning, design, construction, and operation of various types of infrastructure projects related to airports, bridges, dams, roads, landfills and water treatment systems among others. A quick assessment can be done with a checklist for the early stages of a project. A certified evaluator must do the full assessment. In Canada, the tool is promoted by Envision Canada in partnership with the Canadian Society for Civil Engineering (CSCE).

Another example of a tool is GoldSET© (Golder Sustainability Evaluation Tool), a sustainability decision support tool for project planning and design was created by Golder Associates (now WSP Golder). It is a framework to embed sustainable development practices at various phases of any project. It has been applied for various applications such as site remediation and mining tailings around the world and includes a number of quantitative and qualitative indicators for the three sustainability dimensions, environment, society, and economy. An example was shown in Section 11.5.3.

15.5.2.1 Lower Manhattan Coastal Resiliency Enhancement-Battery Project, New York, New York

Folllowing Hurricane Sandy in 2012, efforts to reduce the risk and impact of coastal storms, heat waves and a rise in sea level from climate change, a project were initiated in the lower Manhattan area (https://envisioncanada.comproject-awards/lower-manhattan-coastal-resiliency-the-battery/, accessed Aug. 2024). The Envision framework was employed throughout the project for measuring and improving sustainability and received the Platinum award. The main objectives of the project were to reduce waste sent to landfills, enhance stormwater quality and quantity, and to reuse and use local and low carbon materials. Materials such as wood, granite, and metals, in particular were targeted. A carbon-based procurement method for concrete was developed to enable GWP of the mixes. An estimated 54% reduction in embodied carbon emissions was achieved compared to the initial design. A recycled fill material was selected which enable a $3 million savings and a reduction of 75% in construction waste going to the landfill. Consultation with various groups including the New York State Historic Preservation Office (SHPO), the Stockhouse-Munsee Indigenous Group, the Maritime Association of New York, the National Park Service and original artists developed a plan for revitalization and preservation of the park features. Art works, pedestrian access, public seating, trees and gardens were improved.

15.6 CONCLUDING REMARKS – TOWARDS SDGS RELEVANT TO THE GEOENVIRONMENT

It is clear that so long as depletion of the non-renewable natural resources contained within the geoenvironment occurs, sustainability of the geoenvironment cannot be attained. When one adds the burden of natural catastrophic disasters and their consequences together with physical, chemical, and biological impacts to the geoenvironment from the various stressors described in the previous chapters, it becomes all the more evident that geoenvironmental sustainability is an impossible dream. One has two simple choices: (a) to concede that the sustainability of the geoenvironmental resources that provide society with its life-support systems cannot be realized and prepare to face the inevitable, or (b) to correct those detrimental elements that can be corrected and to find substitutes and alternatives to replace the depleting geoenvironmental resources. The material in this book is a first step in a long series of required steps in adoption of the second choice.

It has been argued that if the global population were to be reduced to some small limiting size, and if replacements or substitutes for the non-renewable resources can be found, sustainability can be achieved. Working on the assumption that this will not likely happen, we have chosen to address the problem of protection of the geoenvironmental base that provides society with its life-support systems. The need to protect the environment and especially the natural resources that provide the basis for the sustenance and well-being of society is eminently clear. The subject addressed in this book is a difficult one, not only from the viewpoint of the basic science-engineering relationships involved in ameliorating adverse impacts on the geoenvironment, but as much or more so from the fact that many crucial elements contributing to the generation of these same impacts could not be properly addressed. This is a fact and a realization that many of these elements were either not within the purview of this book (especially the critical subject of biological diversity), or were elements that were dictated by forces influenced by business, public awareness, and political will. Prominent amongst these are (a) social-economic factors and business-industrial attitudes and relationships, (b) public attitudes, awareness, sensitivity and commitment, and (c) political awareness and will.

In adopting the second choice, we have focussed on the importance of the geoenvironment as a resource base for (a) provision of the required sustenance of the human population, and (b) production of energy and goods. We have attempted to develop a better understanding of the stressors on the geoenvironment and to lay emphasis on the need to better manage the geoenvironmental natural renewable and non-renewable resources. Again, the absence of discussion relating to the direct primary sources of stressors such as the decision-makers responsible for the upstream and downstream industries, means that this book can only provide the geoenvironmental perspective on results of the main impacts resulting from these stressors. The oceans and the coastal marine environments are also significant resource bases, and are essential components of the life-support for the human population and must not be neglected.

Outside of the calamitous natural events in the very recent years, in the form of earthquakes, hurricanes, landslides, floods, etc. that have caused death and severe distress to countless numbers of unfortunate humans, it is seen that pollution of air, land and water resources is the greatest anthropogenically-derived threat to the human population and the geoenvironment. The various discards, spills and loss of materials (chemicals, etc.) and discharge of wastes, either in liquid form or as solids, are common to all types of human activities within (a) the built urban environment, (b) mineral and hydrocarbon exploitation, (c) agricultural ecosystem and (d) industries. Some of these activities include wastewater discharges, use of non-renewable resources as energy input and also as raw materials for the industries, injection wells, leachates from landfills and surface stockpiles, open dumps, illegal dumping, improper waste disposal, underground storage tanks, pipelines, train accidents, irrigation practices, gaseous and noxious particulate airborne emissions, production wells, hydraulic fracking, use of plastics, pesticides and herbicides, urban run-off, mining activities, etc. These pose significant threats to the land environment and the receiving waters, and to the inhabitants of these environments.

The degree of environmental impact due to pollutants in a contaminated ground site is dependent on (a) the nature and distribution of the pollutants, (b) the various physical, geological and environmental features of the site, and (c) existent land use. Through management and education, the sources of pollution must be controlled to maintain water quality and supply for future generations. Environmental management including various remediation and impact assessment and avoidance tools have been developed that technology can develop such as the development of renewable resources replacements for the non-renewable resources that are being depleted. Mitigation and management of pollutants in the subsoil should seek to reduce and eliminate the presence of pollutants in the soil. Engineering the natural attenuation capability of soils, through enhancements of the attenuation capability with geochemical, biological and nutrient aids will provide greater management options. Considerable attention needs to be paid to many of these issues by researchers, policy makers, and other professionals to alleviate the stresses to the geosphere and seek sustainability and innovative and non-conventional ways are needed for society to live in harmony with the environment now and in the future.

Figure 15.12 shows how geoenvironmental engineers can work towards the SDGs. Engineers must work with organizations, other experts, governmental organizations, and the communities to develop technologies, policies, and frameworks. Engineers in many disciplines must also work together toward sustainable engineering. Sharing of case studies can assist in promoting and understanding sustainable practices and implementation.

To address climate change and sustainability challenges, engineers need to work together and be more involved in decision-making at all stages of the project. They should become more involved in local or regional activities to assist in the decision making. They need to consult with stakeholders for input regarding concerns and to adapt to local conditions. Even during the construction and/or operation phases, engineers should be able to address concerns and provide advice on the sustainability of a project. Research is needed to develop innovative solutions to this changing world under the influence of climate change and increasing uncertainty, deteriorating infrastructure, introduction of new contaminants into the environment, and growing population to name a few. Engineers have an ethical requirement to rise to this challenge and thus they must be involved in sustainable engineering practices. Some tools such as carbon calculators, Envision and GOLDSET are available and are constantly being improved. Material selection is also an important and appropriate material selection can substantially reduce embodied carbon and GHG emissions. Various partnerships are

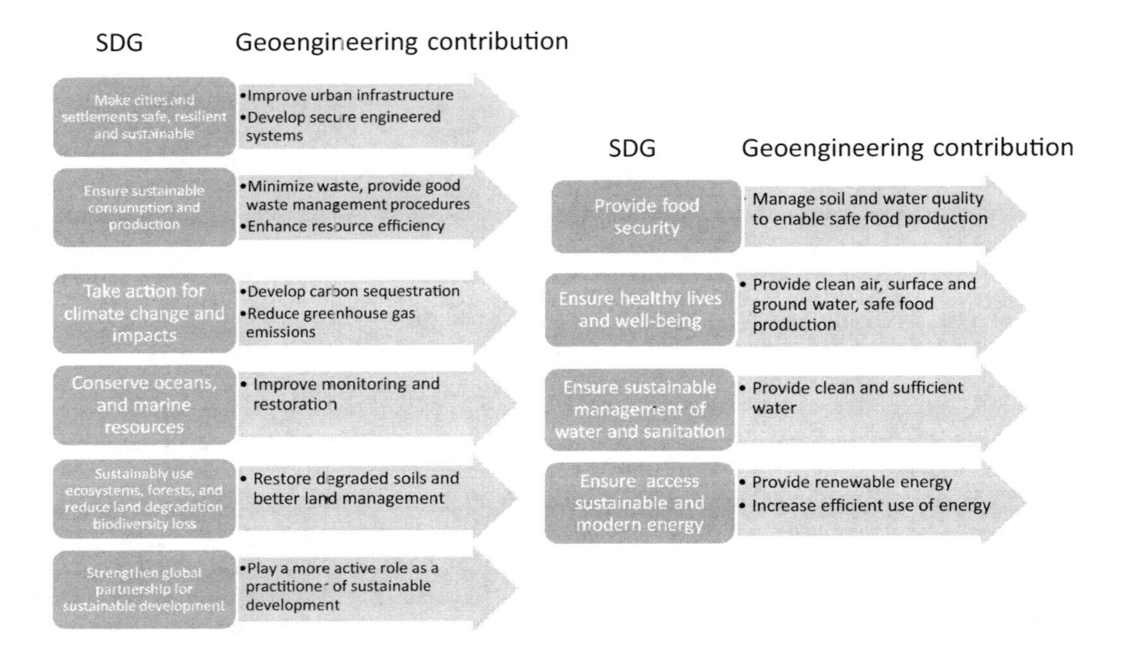

FIGURE 15.12 Contributions of geoenvironmental engineers to SDGs. (Adapted from Mulligan, 2019.)

needed. Engineers of various expertise needed to work together. Engagement of local communities, in particular Indigenous communities, is essential. The role of engineers in sustainable development has been undervalued but it is critical. However, action is needed now.

REFERENCES

Applegate, D., Degner, M., Deschamp, J., and Haverl, S. (2005), Highly refined, *Civil Engineering*, 75(6): 44–49.

Appunn, K. (2018), Climate Impact of Farming, Land Use (Change) and Forestry in Germany, Factsheet, 30 Oct 2018. www.cleanenergywire.org/factsheets/climate-impact-farming-land-use-change-and-forestry-germany, accessed Jun. 2024.

Backhaus, C. (2017), Experience with the Utilization of Coal Mine Gas from Abandoned Mines in the Region of North-Rhine-Westphalia, Germany. https://unece.org/fileadmin/DAM/energy/se/pdfs/cmm/cmm12/Workshop_2017/7.Mr._Backhaus.pdf, accessed Jun. 2024.

Bajpai, P. (2015), Generation of Waste in Pulp and Paper Mills Waste. Springer, Cham. https://doi.org/10.1007/978-3-319-11788-1_2..

California Center for Land Recycling (CCLR). (2000), *Brownfield Redevelopment Case Studies*, CCLR, San Francisco, CA, 47 pp.

City of Los Angeles (2024), Brownfields Success Stories. www.lacitysan.org/san/faces/wcnav_externalId/s-lsh-es-si-b-bss?_adf.ctrl-state=1a8cxin4i8_5&_afrLoop=22279294578414305#!, accessed Jun. 2024.

De la Rosa, D. (2005), Soil quality evaluation and monitoring based on land evaluation. *Land Degradation & Development*, 16: 551–559.

de la Rosa, D., Diaz-Pereira, E., and Mayol, F. (2003), Restrictions of agricultural zones and identification of soil protection practices by using soil quality decision support tools. With special reference to the Mediterranean region. Alicante SCAPE Workshop, Sevilla, 14–16 June, 2003.

de la Rosa, D., and Sobral, R. (2007). Soil Quality and Methods for Its Assessment. In: A.K. Braimoh and P.L.G. Vlek (eds.), *Land Use and Soil Resources*, Springer Press, Dordrecht, Germany.

European Compost Network. (2019) ECN Status Report 2019, European Biowaste Management, Overviews of Bio-Waste Collection, Treatment & Markets Across Europe. www.compostnetwork.info/download/ecn-status-report-2019/.

Ellefsen, V., Westby, T., and Andersen, L. (2001), Sustainability: The Environmental Element-Case Study 1. In: *Clarinet Final Conference, Sustainable Management of Contaminated Land, Proceedings, Vienna, Austria, June 21 to 22, 2001*.

Ellefsen, V., Westby, T. and Systad, R.A., (2005), Remediation of Contaminated Soil at Fornebu Airport-Norway, Stabilization and Re-Use of PAH-Contaminated Soil. In: *Proceedings of the 16th ICSMGE, Osaka, Japan*, pp. 2365–2370.

EMC. (1998), *Mont-Cenis, Report of the Entwicklungsgesellschaft*, Mont-Cenis, Herne D, 44 pp.

Esmaeeli, A., and Sarrafzadeh, M.-H. (2023), Reducing freshwater consumption in pulp and paper industries using pinch analysis and mathematical optimization, *Journal of Water Process Engineering*, 53: 103646, https://doi.org/10.1016/j.jwpe.2023.103646.

European Commission. (2021) Commission Consults on New EU Soil Strategy - European Commission (europa.eu). https://environment.ec.europa.eu/news/commission-consults-new-eu-soil-strategy-2021-02-02_en#:~:text=The%20aim%20of%20the%20new%20EU%20Soil%20Strategy,fertility%2C%20reduce%20erosion%20and%20increase%20soil%20organic%20matter, accessed Jun. 2024.

FAO. (2024) Land Evaluation Decision Support System (MicroLEIS-DSS) for Agricultural Soil Protection. www.fao.org/land-water/land/land-governance/land-resources-planning-toolbox/category/details/en/c/1109820/#:~:text=The%20Mediterranean%20Land%20Evaluation%20Information%20System%20%28Micro-LEIS%29%20is,decision-making%20in%20a%20wide%20range%20of%20agro-ecological%20schemes, accessed Jun. 2024.

Fayomi, G.U., Mini, S.E., Fayomi, O.S.I., and Ayoola, A.A. (2019), Perspectives on environmental CO_2 emission and energy factor in cement industry, *IOP Conf. Ser. Earth Environ. Sci.*, 331(1), 10.1088/1755-1315/331/1/012035.

Furey, D., Bhat, S., and Thomas. J (2023) A multi-geosynthetics sustainable solution case study- ʔapsčiik t̓ašii (Upscheek Tashee), *GeoSaskatoon*, 2023(Oct.): 1–4.

Genske, D.D. (2003), *Urban Land: Degradation, Investigation, Remediation*, Springer-Verlag, Berlin, 331pp.

Gehlot, A.R., and Shrivastava, S. (2023), Utilization of stone waste in the development of sustainable mortar: A state of the art review, *Materials Today: Proceedings*, https://doi.org/10.1016/j.matpr.2023.03.149.

Graham, E. (1989), Oilspeak, common sense and soft science. Audubon (Sept. 1989) 102–11.

GRI, UN Global Compact, and WBCSD. (2015), SDG Compass 2016. SDG Compass: The Guide for Business Action on the SDGs. https://sdgcompass.org/wp-content/uploads/2015/12/019104_SDG_Compass_Guide_2015.pdf, accessed May 30, 2023.

Haigh, M.J. (1978), Evaluation of Slopes on Artificial Landforms Blaenavon, UK, Research Papers 183, Department of Geography, University of Chicago, Chicago, Ill,.

Hockley, D., Kuit, W., and Philip, M. (2009), Sullivan Mine Fatalities Incident: Key Conclusions and Implications for Other Sites, *Securing the Future and 8th ICARD*, June 22–26, 2008. Skelleteå, Sweden.

ISI. (2023), Envision Overview. https://sustainableinfrastructure.org/about-isi/, accessed Jun. 2024.

Jin, R., and Chen, Q. (2015), Investigation of concrete recycling in the U.S., *Construction Industry Procedia Engineering*, 118: 894–901, https://doi.org/10.1016/j.proeng.2015.08.528.

Lahtinen, P.O., Maijala, A., and Kolkka, S. (2005), Environmentally Friendly Systems to Renovate Secondary Roads. Life Environment Project: Kukkia Circlet. LIFE02 ENV/FIN/000329. In: *Proceedings of the 16th ICSMGE, Osaka, Japan*, pp. 2407–2410.

Li, X.N., Jiao, W.T., Xiao, R.B., Chen, W.P., and Chang, A.C. (2015). Soil pollution and site remediation policies in China: A review. *Environmental Reviews*, 23(3): 263–274, https://doi.org/10.1139/er-2014-0073.

McLean, I., and Johnes, M. (2000), *Aberfan: Disasters and Government*, Welsh Academic Press, Cardiff, 274 pp.

Moller, M., Flugge, R., and Murphy, D. (2006), Pilbara Iron's approach to sustainable development during mine close-The case study of Greater Tom Price and Pannawonica Operation. Unpublished report, Sinclair Knight Merz, Australia, pp. 1–6.

Monteiro, P., Miller, S., and Horvath, A. (2017), Towards sustainable concrete. *Nature Materials*, 16: 698–699, https://doi.org/10.1038/nmat4930.

Mulligan, C.N. (2019). *Sustainable Engineering Principles and Implementation*, CRC Press, Boca Raton, FL, USA, 216 pp.

Nanus, L., Campbell, D.H., Ingersoll, G.P., Clow, D.W., and Mast, M.A. (2003), Atmospheric deposition maps for the Rocky Mountains. *Journal of Atmospheric Environment*, 37: 4881–4892.

OECD. (2023) *Mining Regions and Cities Case of the Pilbara*, Australia. OECD Rural Studies, OECD Publishing, Paris, 167 pp., https://doi.org/10.1787/a1d2d486-en

Payne, J.R., and Phillips, C.R. (1985), *Petroleum Spills in the Marine Environment*, Chelsea, Michigan, Lewis, 148 pp.

Ramezani, O., Kermanian, H., Razmpour, Z., and Rahmaninia, M. (2011), Water Consumption Reduction Strategies in Recycled Paper Production Companies in Iran. http://ceur-ws.org/Vol-1152/paper77.pdf, accessed May 2024.

Schrope, M. (2010), Upbeat oil report questioned. *Nature*, 466: 802.

Shahbazi, F., De la Rosa, D., Anaya-Romero, M., Jafarzadeh, A.A, Sarmadian, F., Neyshabouri, M., and Oustan, S. (2008), Land use planning in Ahar area (Iran), using MicroLEIS DSS. *International Journal of Agrophysics*, 22: 277–286.

Stanton-Hicks, E. (2001), *Sustainability and the Iron Ore Industry in the Pilbara: A Regional Perspective. Sustainability and Mining in Western Australia*, Institute for Sustainability and Technology Policy, Murdoch University.

Teckcominco. (2001), *The Sullivan Mine: A Case Study on Mining and Sustainability*, Mineral Councils of Australia Environmental Workshop, October.

Thies, J.E. (2006), Measuring and Assessing Soil Biological Properties. In: N. Uphoff, A. Ball, E. Fernandes, H. Herren, O. Husson, M. Laing, Ch. Palm, J. Pretty, P. Sanchez, N. Sanginga, and J. Thies (eds.), *Biological Approaches to Sustainable Soil Systems*, \Taylor & Francis/CRC Press, Boca Raton, FL, pp. 655–670.

UN (United Nations). (2015), Sustainable Development Goals. https://sdgs.un.org/goals, accessed May 10, 2024.

Uphoff, N., Ball, A., Fernandes, E., Herren, H., Husson, O., Laing, M., Palm, Ch., Pretty, J., Sanchez, P., Sanginga, N., and Thies, J. (eds.). (2006), *Biological Approaches to Sustainable Soil Systems*, Taylor & Francis/CRC Press, Boca Raton, FL.

Urban Harbors Institute. (2000). Green ports: Environmental Management and Technology at U.S. Ports, USEPA 825706-01-0, March.

US EPA (1995), *Profile of the Fabricated Metal Products Industry*, Office of Compliance Sector Notebook Project, Office of Compliance, U.S. EPA, Washington, DC, EPA 310-R-95-007, September.

US EPA. (2007), BP Former Refinery, Casper, Wyoming. https://archive.epa.gov/epawaste/hazard/web/pdf/casper11-07.pdf, accessed Jun. 2024.

US EPA. (2015), Advancing Sustainable Materials Management: Facts and Figures 2013, 186. https://digital.library.unt.edu/ark:/67531/metadc949098/, accessed Jun. 2024.

US EPA. (2021), TRI National Analysis, Fabricated Metals Manufacturing Waste Management Trend. https://19january2021snapshot.epa.gov/trinationalanalysis/fabricated-metals-manufacturing-waste-man agement trend_.html#:~:text=From%202018%20to%202019%3A%201%20Production-related%20 waste%20managed,was%20managed%20through%20treatment%2C%20energy%20recovery%2C%20 and%20recycling, accessed Jun. 2024.

US EPA. (2022) National Overview: Facts and Figures on Materials, Wastes and Recycling, US Environmental Protection Agency. www.epa.gov/facts-and-figures-about-materials-waste-and-recycling/national-overv iew-facts-and-figures-materials, updated July 31, 2022, accessed May 2024.

USDA. (2006), Soil Quality Institute, Natural Resources Conservation Services. www.nrcs.usda.gov/conservat ion-basics/natural-resource-concerns/soil/soil-science.

Warkentin, B.P., and Fletcher, H.F. (1977), Soil Quality for Intensive Agriculture. In: *Proc. Int. Seminar on Soil Environment and Fertility Management in Intensive Agriculture, Society. Science, Soil and Manure*, National Institute of Agricultural Science, Tokyo, pp. 594–598.

WBCSD (World Business Council for Sustainable Development) (2024), What we do. www.wbcsd.org/what-we-do/accessed Oct. 2024.

WBCSD, GRI, and UN Global Compact (2015), SDG Compass, The guide for business actions on the SDGs. https://sdgcompass.org/wp-content/uploads/2015/12/019104_SDG_Compass_Guide_2015.pdf, accessed Oct. 2024.

Weizäker, E.U., von, Lovins, A.B., and Hunter Lovins, L. (1997), *Factor Four: Doubling Wealth Halving Resource Use*, Earthscan, London, 322 pp.

WFEO. (2013), WFEO Model Code of Practice for Sustainable Development and Environmental Stewardship Think Global and Act Local. www.wfeo.org/the-code-of-practice-for-sustainable-development-and-environmental-stewardship/, accessed May 30, 2023.
WFEO. (2015), WFEO Engineers for a Sustainable Post 2015. World Federation of Engineering Organization, UN Scientific and Technological Communities Major Group, July 22, 2015, Version 1.6.
Yamamoto, K., and Martin, K. (2022), Implementing Sustainability in the Geotechnical Industry, ASCE Boston Section. www.bsces.org/news/org/implementing-sustainability-in-the-geotechnical-industry-4509, accessed Jun. 2024.

Appendix

DETAILED METHODOLOGY FOR R_{CV} DETERMINATION

By using OD as a representative value of the number of viable bacteria, the following relationship is established. That is, it can be expressed as follows in the same way as the dilution rate and the concentration rate. OD is related to CPR, while OD* is also related to CPR.

Preparation of Bacteria

Frozen bacteria in a closed bag or a tube is thawed by immersion in 30 °C water.

The mass of bacteria is weighed by the following ways.

$$\text{Mass of bacteria } m_b \text{ (g)} = \rho_{BC}\, V_{BC}\, \text{OD}^* / (\text{OD}_i\, C_f) \quad (A.1)$$

where ρ_{BC} is the density of frozen bacteria, V_{BC} is the volume of the biocement solution, OD* is the target apparent optical density, OD_i is the initial optical density of culture, and C_f is the concentration factor in the centrifugation.

Example: $\rho_{BC} = 1.07$ g/cm^3, $BCS = 100$ mL, OD* = 0.5, $OD_i = 4.0$ and $C_f = 250$.
Calculation: $m_b = 1.07 \times 100 \times 0.5/(4 \times 250) \fallingdotseq 0.0535$ g
$V_{bac} = m_b/1.07 = 0.050$ mL

Methodology

The BC stock solution is prepared according to the above calculation, and 10 mL diluted test solution in five steps is taken in a glass test tube. Note that the initial Ca^{2+}/OD* remains constant by the dilution method. This is a way to avoid inhibition and retardation due to Ca^{2+}. To avoid inhibition, the initial Ca^{2+}/OD should be lower than 8.46 M, but it is an unknown OD at this stage.

In this book, CPR obtained by direct tests corresponds to the OD*, which is apparent optical density. Then, the optical density "OD" without an asterisk expresses optical density after OD conversion from OD*.

(2) Cover the glass test tube with a lid to prevent evaporation and let it stand for 24 hours at a predetermined temperature while it is placed in a test tube stand.
(3) After 24 hours, the solution of the glass test tubes is filtered with filter paper, and the filter paper and the test tube are furnace dried at 110 °C for 2 hours. The filter papers to be used are dried in the drying furnace in advance to reduce the mass of water. The test tubes are also dried before the mass of carbonate is measured.
(4) Measure the mass of dry test tubes and filter paper and record them on the data sheet.
(5) Enter the calculated numerical values, OD*, and CPR on the data sheet.
(6) Calculate each X value from the CPR calculated using Equation (13.10).
(7) Rcv is calculated from the corresponding OD* with the average of X values as OD.

Index

C

D

G

H

N

T

U

V

W

X

Y

Z